BMA

ibr

D1389704

Fenner and White's Medical Virology

Fenner and White's Medical Virology

Fifth Edition

Christopher J. Burrell
School of Molecular and Biomedical Science,
University of Adelaide, Adelaide, South Australia, Australia

Colin R. Howard
College of Medicine and Dentistry, University of Birmingham, United Kingdom

Frederick A. Murphy
University of Texas Medical Branch, Galveston, Texas, USA

AMSTERDAM • BOSTON • HEIDELBERG • LONDON
NEW YORK • OXFORD • PARIS • SAN DIEGO
SAN FRANCISCO • SINGAPORE • SYDNEY • TOKYO
Academic Press is an imprint of Elsevier

Academic Press is an imprint of Elsevier
125 London Wall, London EC2Y 5AS, United Kingdom
525 B Street, Suite 1800, San Diego, CA 92101-4495, United States
50 Hampshire Street, 5th Floor, Cambridge, MA 02139, United States
The Boulevard, Langford Lane, Kidlington, Oxford OX5 1GB, United Kingdom

Notices
Knowledge and best practice in this field are constantly changing. As new research and experience broaden our
understanding, changes in research methods, professional practices, or medical treatment may become necessary.

Practitioners and researchers must always rely on their own experience and knowledge in evaluating and using any
information, methods, compounds, or experiments described herein. In using such information or methods they
should be mindful of their own safety and the safety of others, including parties for whom they have a professional
responsibility.

To the fullest extent of the law, neither the Publisher nor the authors, contributors, or editors, assume any liability
for any injury and/or damage to persons or property as a matter of products liability, negligence or otherwise, or
from any use or operation of any methods, products, instructions, or ideas contained in the material herein.

British Library Cataloguing-in-Publication Data
A catalogue record for this book is available from the British Library

Library of Congress Cataloging-in-Publication Data
A catalog record for this book is available from the Library of Congress

ISBN: 978-0-12-375156-0

For Information on all Academic Press publications
visit our website at https://www.elsevier.com

Working together
to grow libraries in
developing countries

www.elsevier.com • www.bookaid.org

Publisher: Sara Tenney
Acquisition Editor: Jill Leonard
Editorial Project Manager: Pat Gonzalez
Production Project Manager: Julia Haynes
Designer: Matt Limbert

Typeset by MPS Limited, Chennai, India

Dedication

Frank J. Fenner

David O. White

This book is dedicated to our friends Frank J. Fenner (1914–2010) and David O. White (1931–2004), the founders of the series of books now spanning five editions of *MEDICAL VIROLOGY* and five editions of *VETERINARY VIROLOGY*. They set a standard of scholarship that is impossible to match and a *joie de vivre* that made the writing and editing almost fun.

They taught us that the subject of virology must be seen within the context of society as a whole as well as within the context of science. They envisioned virology as extending broadly, from its roots in the science of microbiology and the practice of infectious disease, through its key roles in the development of molecular and cell biology, to become an independent scientific and medical discipline and a significant contributor to human, animal, and environmental well-being. We hope our students will come to understand the "big picture" of medical virology as well as Frank and David achieved throughout their outstanding careers.

We would also like to dedicate this book to our wives and families, our teachers and mentors, together with our students, all of whom have shaped our thinking and provided us inspiration over the years in so many different ways.

Christopher J. Burrell
Colin R. Howard
Frederick A. Murphy

Contents

Part II
Specific Virus Diseases of Humans

16. Poxviruses

17. Herpesviruses

18. Adenoviruses

25. Orthomyxoviruses

26. Paramyxoviruses

27. Rhabdoviruses

28. Filoviruses

29. Bunyaviruses

Foreword

To say that this new version of the classic *Medical Virology* text by Frank Fenner (1914–2010) and David White (1931–2004) is timely is, to say the least, an understatement. It's been 21 years since the fourth edition, with authors Chris Burrell, Colin Howard, and Fred Murphy building on that much respected foundation to produce a greatly refreshed fifth edition. The basic philosophy and structure that made this such a useful volume for many generations of advanced medical students, clinicians, and pathologists remain but, as all who have been even peripherally involved in molecular science over the past two decades will realize, there was an enormous amount of new information that had to be read, sifted, and selected for relevance to the information needs of this target audience.

Basic virology has, of course, been well served by the many revisions of Bernie Fields' (1938–1995) exhaustive text and, following the example set by *Field's Virology*, the fifth edition will appear as *Fenner and White's Medical Virology*. It is a fitting tribute. Taken together, the original authors, and those responsible for this latest version, have variously been active and publishing on one or the other aspect of virus-induced disease and/or pathology since 1948: and the virology lineage goes back even further!

Frank Fenner was the student of MacFarlane Burnet (1899–1985) who, applying technical and conceptual approaches learned from his early studies with bacteriophage, pioneered quantitative mammalian virology, and genetics. David White began his career as a virologist in the university microbiology department headed by Frank Fenner. Both Fred Murphy and I worked for a time and Chris Burrell completed his PhD in that same department, where Rolf Zinkernagel and I did the experiments in viral immunity that led to a Nobel Prize. Reflecting on lineages and the solid foundations that enable the work of subsequent generations, I realize that I've met all the people named above in one or other professional or personal context and have benefitted from either the institutional structures that they established, or the insights they developed. My current office is in the Department of Microbiology and Immunology at the University of Melbourne, which was headed for many years by David White. Perhaps, as Isaac Newton (1642–1726) said "we stand on the shoulders of giants" though, when I've met the "giants" of science, they've pretty much turned out to be hard working, smart, and dedicated people who quietly and systematically built a body of knowledge and a tradition of scholarship. That very much describes the discipline of virology.

Thus, while reflecting on the up-to-date understanding of the authors, and of other researchers who have provided key insights concerning each particular pathogen, *Fenner and White's Medical Virology* reflects a cumulated wisdom that goes back to the early days of 20th century virology. Back then, the thinking and technical range of mammalian virologists was very much focused on pathogenesis and disease process. Now, after decades of illuminating the molecular characteristics of viruses and infected cells, it is gratifying to see these precise investigative tools being applied to the understanding of what actually happens in an infected individual. This is a great time for any medical researcher who is fascinated by the linked issues of disease process, pathology, and treatment.

This fifth edition provides an accessible and informative account of medical virology, based on both contemporary science and what went before. It constitutes an excellent resource for both the physician and the research investigator. New discoveries are constantly being made and, while much remains unclear, the rate of advance in understanding is extraordinary. Hopefully, we will not need to wait another 20 years for the next edition of this seminal text.

Peter C. Doherty
University of Melbourne
December 10, 2015

Preface

Enormous strides have taken place since the fourth edition of this book in regard to our understanding of virus properties and how viral diseases can be controlled, treated, and prevented. Considerable advances have been made across the whole spectrum of virology, particularly in relation to virus structure and replication, "point-of-care" diagnostics and the wide use of antiviral therapies. This progress has done much to contribute to a substantial improvement in public health across the globe.

At the same time, we have seen the emergence of new viral agents, and older agents in new guises, sometimes creating a need for radically new concepts to guide our understanding and clinical and public health actions. Notable examples include HIV/AIDS; SARS and its recently appearing relative MERS; Hendra and Nipah viral diseases; the more recent epidemic manifestations of Ebola; the appearance in humans of novel influenza A viruses; and the ever-increasing range of manifestations of "older" viruses in immunosuppressed individuals. Paradoxically, other aspects of virology, for example, the clinical descriptions of common infections and the principles of management, remain as valid as when they were first made. Our challenge is to transmit hard-learned experiences, within the context of new advances, to an ever-widening audience of newcomers entering the medical, scientific, and related professions.

In keeping with the advice of many colleagues, we have renamed this edition *Fenner and White's Medical Virology*. We have retained the layout of chapters as first set out in 1970 by Frank Fenner and David White, as we believe this structure still provides an excellent framework for the discipline.

Part I deals with the principles of animal virology, and includes chapters on basic virology and viral replication, host immunity, pathogenesis of infection, viral oncogenesis, viral diagnostics, vaccinology, chemotherapy, epidemiology, surveillance, and emergence.

Part II systematically examines in turn each of the virus families containing human pathogens. Chapters in Part II are for the most part set out in a standardized format to allow the reader more rapid access to the information being sought. Of course, virus infections in clinical practice present as syndromes and do not identify themselves by their virus family, so a key final chapter associates common viral syndromes with the etiological agents usually responsible. Much of the approach adopted has been shaped by the authors' experience in teaching virology to science and medical students and graduates—which has itself been shaped by earlier editions of this book.

This edition is designed to meet the needs of advanced medical students, clinicians and pathologists, university teachers, researchers, and public health workers who are seeking a single accessible source of key information about the range of major human viruses and the options for treatment and control. We have included historical perspectives, so that readers gain some insight into the paths and the personalities involved in our reaching this current state of knowledge; we have also highlighted gaps in knowledge, and unmet challenges in the management and control of virus diseases. However, in a book of this size it is not feasible to include all details, and thus readers seeking a more complete discussion of the diagnosis and management of a particular condition should refer to the appropriate source. Similarly, although not intended as a comprehensive summary of current research into a particular viral agent, we hope to prime the reader's enthusiasm to explore further aspects in the scientific literature. We are acutely aware of the need to foster the next generation of virologists, and if reading this book plays a small part in this process, then we will be gratified.

We wish to acknowledge a number of friends and colleagues who kindly agreed to review certain chapters: Peter Balfe (replication), Michael Beard (innate immunity, flaviviruses), Shaun McColl (adaptive immunity), Geoff Higgins (laboratory diagnosis), Yu-Mei Wen (oncogenic viruses, adenoviruses, paramyxoviruses, coronaviruses), Tony Cunningham (herpesviruses), Wendy Howard (influenza), Stephen Locarnini (hepadnaviruses), David Shaw (retroviruses), Sharon Lewin (retroviruses), Barbara Coulson (reoviruses), Lorena Brown (orthomyxoviruses), Peter McMinn (picornaviruses), Bill Rawlinson (caliciviruses, astroviruses), and Mike Catton (viral syndromes). We also thank our graphic artist Richard Tibbitts for his work on the diagrams, and Steven Polyak for the chemical structures of antiviral drugs in the chapter "Antiviral Chemotherapy."

Christopher J. Burrell
Colin R. Howard
Frederick A. Murphy

Part I

Principles of Virology

Chapter 1

History and Impact of Virology

Infectious disease is one of the few genuine adventures left in the world. The dragons are all dead and the lance grows rusty in the chimney corner.... About the only sporting proposition that remains unimpaired by the relentless domestication of a once free-living human species is the war against those ferocious little fellow creatures, which lurk in the dark corners and stalk us in the bodies of rats, mice, and all kinds of domestic animals; which fly and crawl with the insects, and waylay us in our food and drink and even in our love.

So wrote the great microbiologist Hans Zinsser in his book *Rats, Lice and History*, written in 1935, as he reflected on his life in infectious disease research. Zinsser's thoughts have stimulated generations of students and professionals ever since. Infectious diseases of today present challenges that are different but just as demanding as those facing Zinsser over 80 years ago.

This book presents the subject of medical virology from the perspective of its traditional base as a life science and its application to clinical practice and public health. It is the perspective established by Frank Fenner and David White, who in 1970 conceived the rationale for this book, and maintained it through the previous four editions. It is the perspective that many others have used to teach and learn medical virology.

The foundations of the science of medical virology are intertwined with the other life sciences, particularly microbiology and infectious diseases. Medical virology has a relatively brief history, spanning just over a century, but it is crowded with intriguing discoveries, stories of immense personal courage and numerous practical applications, many of which have had an overwhelmingly positive benefit on humankind. Its origins involved the replacement of centuries-old beliefs and theories with discoveries borne out of rigorous scientific investigation. Targeted prevention and control strategies could only be developed and implemented once the concept of *the specificity of disease causation* had been accepted, namely that infectious diseases are caused not by some common *miasma* (a mysteriously poisonous substance), but rather by specific agents. In a wider sense, the microbial sciences have played a pivotal role in the development of medical thought overall, particularly in applying scientific rigor in understanding pathological processes. Advances in understanding infectious agents have led to improvements in human health and well-being that arguably have exceeded the contribution of any other branch of science. Indeed, no less than 35 workers in this and closely related fields have been awarded the Nobel Prize in Physiology or Medicine in recognition of their achievements.

Infectious disease discoveries have had a profound effect on life expectancy and well-being across the world. For example, epidemics of smallpox, yellow fever, and poliomyelitis, commonplace until well into the 20th century, have been virtually eliminated by the application of various prevention and control strategies. However, hitherto unrecognized diseases have emerged over the past half-century at the rate of at least one per year. Many of the viruses dealt with in this edition were unknown when the first edition was published over 45 years ago. The epidemiology of other viruses has radically changed as humans continue to alter the environment in so many ways. Meeting the challenges posed by emerging diseases requires the medical virologist to acquire ever more increasing expertise and access to ever more complex technologies. Today diseases such as HIV/AIDS, hepatitis C, influenza, and diarrheal diseases represent significant threats to public health. Tomorrow it will be other diseases, the nature and means of control for which are largely unpredictable. One positive note is that all emerging viral diseases of recent years have been found to be caused by members of previously recognized families of viruses. Thus a thorough knowledge of representative members of each family is likely to facilitate and inform the rapid development of knowledge about any new pathogen.

WHY STUDY VIROLOGY?

As many bacterial infections have succumbed to treatment with antibiotics, viral infections now pose proportionally a much greater threat to global public health than was the case, say, a half-century ago. Viral diseases exact a particularly heavy toll among young children and infants in

Fenner and White's Medical Virology. DOI: http://dx.doi.org/10.1016/B978-0-12-375156-0.00001-1

the economically less developed nations where healthcare resources are limited. Ironically, there is a resurgence of interest in viruses that target bacteria (bacteriophages) as an alternative strategy for the control of some increasingly drug-resistant bacterial infections (e.g., cholera).

Although this book focuses on viral infections of medical significance, the reader needs to be aware that viruses are a major threat to livestock and plant species, and thereby of great importance in human nutrition and food supply. Human adaptation to diseases of livestock and crops has played a major role in the development of all civilizations.

Virology is much broader than linking a particular disease to a specific pathogen: there are literally hundreds of new viruses being discovered that do not apparently relate to any known pathological condition of either animals or humans. Many of these may in the future be linked to human illnesses and thus the reader needs to be aware of the wider scope of the virological landscape, if not in detail at least to the point of "expecting the unexpected." Conversely, the tantalizing goal remains to clarify what role, if any, viruses may play in well-known diseases of uncertain etiology, for example, multiple sclerosis.

The vast majority of new viral threats emerging annually either originate from an animal host (zoonosis) or are the result of host range extension (that is, "host species jumping"), or other changes in the epidemiology, ecology, and/or pathogenicity of the etiological agent. Since the last edition of this book, virus emergence has become a major focus of virological research.

The discovery of a new human pathogen often stimulates the discovery of related, but hitherto unidentified agents that may, or may not, present threats to human health at some point in the future. A prime example is the emergence of SARS virus, a coronavirus, and the subsequent explosion in our knowledge of coronaviruses of animals. This helped in the later rapid recognition of another human respiratory coronavirus—MERS coronavirus (Middle East Respiratory Syndrome coronavirus).

A BRIEF HISTORY OF VIROLOGY

The history of virology can be divided into a number of eras: these span (1) the discovery of viruses as entities distinct from other disease-causing pathogens, (2) the association of many major human diseases with causative viruses, (3) the development of methods for virus isolation and characterization, (4) the defining of the chemical properties of viruses, and (5) the design and application of vaccines and therapeutics. A summary of the major milestones in the development of virology is given in Table 1.1.

Virology has its foundations in the initial discoveries of bacteria and related diseases. Up to the 19th century the prevailing view was that diseases of humans and animals were the result of miasmas and other environmental influences.

This was despite the thesis of Girolamo Fracastoro who suggested as early as 1546 that epidemic diseases were disseminated by minute particles carried over long distances. Anton van Leeuwenhoek first saw bacteria through his microscope in 1676 and Lazarro Spallanzani first grew bacteria in culture in 1775. Remarkably, Edward Jenner developed vaccination against smallpox in 1796 against a backdrop of prevailing opinion that such diseases were caused by environmental factors rather than specific microscopic agents.

The establishment of microbiology as a scientific discipline owes much to the work of Louis Pasteur, who in 1857 discovered the specificity of microbial fermentation, who then went on in 1865 to elaborate the nature of diseases of silkworms. But it was his work on rabies that signaled the start of the virus discovery era. In 1885, Pasteur looked on as his first rabies vaccine was given to a boy, Joseph Meister, bitten severely by a rabid dog, thus opening up the strategy of vaccine development through a process of virus attenuation (Fig. 1.1).

The early pioneering work of the 19th century linking disease to specific bacteria was greatly assisted by the earlier development of the unglazed porcelain ultrafilter by Charles Chamberland who worked in Pasteur's laboratory. These filters originally were used to sterilize water and other fluids by preventing the passage of bacteria. Dimitri Ivanovsky (1892) and Martinus Beijerinck (1898) showed that the agent causing mosaic disease in tobacco plants (now known to be tobacco mosaic virus [TMV]) passed through ultrafilters retaining bacteria. Beijerinck realized he was dealing with something other than a microbe but erroneously thought that the entity that passed through the ultrafilter was an infectious liquid and not a particle—he called it a "contagium vivum fluidum." Friedrich Loeffler and Paul Frosch were the first to correctly conclude that an ultrafilterable infectious agent was indeed a submicroscopic particle. Studying the cause of foot-and-mouth disease of cattle, Loeffler and Frosch found that the causative agent passed through a Chamberland ultrafilter but not the finer Kitasato ultrafilter. Thus these first virologists saw ultrafiltration in a new way—they focused attention on what passed through the ultrafilter rather than what was retained, and thereby established an experimental methodology widely adopted in the early 20th century. In quick succession, further diseases were shown to be caused by ultrafilterable agents: in 1900 the first human virus, yellow fever virus, and its mosquito transmission cycle was discovered by Walter Reed, James Carroll and the US Army Yellow Fever Commission in Havana, Cuba, a discovery that was guided by the earlier work of the Cuban physician Carlos Findlay (Fig. 1.2).

The concept of ultrafilterable infectious agents became more widely acceptable when Karl Landsteiner and Erwin Popper showed conclusively in 1909 that poliomyelitis was caused by an ultrafilterable agent. Importantly, as early as

TABLE 1.1 Some Milestones in the History of Virology

Date	Discoverer(s)	Discovery(ies)
1796	E. Jenner	Application of cowpox virus for vaccination against smallpox
1885	L. Pasteur	Development of rabies vaccine
1892	D. Ivanovsky, M. Beijerinck	Ultrafiltration of tobacco mosaic virus
1898	F. Loeffler, P. Frosch	Ultrafiltration of foot-and-mouth disease virus—clear proof of virus etiology of disease—discovery of the first virus
1898	G. Sanarelli	Discovery of myxoma virus
1900	W. Reed, J. Carroll, A. Agramonte, J. Lazear, C. Finlay	Discovery of yellow fever virus and its transmission by mosquitoes
1903	M. Remlinger, Riffat-Bay, A. di Vestea	Discovery of rabies virus
1907	P. Ashburn, C. Craig	Discovery of dengue viruses
1909	K. Landsteiner, E. Popper	Discovery of polioviruses
1911	P. Rous[a]	Discovery of the first tumor virus: Rous sarcoma virus
1911	J. Goldberger, J. Anderson	Discovery of measles virus
1915	F. Twort, F. d'Herelle	Discovery of bacterial viruses (bacteriophages)
1918		Beginning of global pandemic of influenza
1919	A. Löwenstein	Discovery of herpes simplex virus
1930	K. Meyer, C. Haring, B. Howitt	Discovery of Western equine encephalitis virus
1931	M. Theiler[a]	Attenuation of yellow fever virus—vaccine development
1933	C. Andrews, P. Laidlaw, W. Smith	Isolation of human influenza viruses in ferrets
1933	R. Muckenfuss, C. Armstrong, H. McCordock, L. Webster, G. Fite	Discovery of St. Louis encephalitis virus
1934	C. Johnson, E. Goodpasture	Discovery of mumps virus
1934	M. Hayashi, S. Kasahara, R. Kawamura, T. Taniguchi	Discovery of Japanese encephalitis virus
1935	W. Stanley[a]	Purification/crystallization of tobacco mosaic virus
1936	C. Armstrong, T. Rivers, E. Traub	Discovery of lymphocytic choriomeningitis virus
1937	L. Zilber, M. Chumakov, N. Seitlenok, E. Levkovich	Discovery of tick-borne encephalitis virus (Russian spring summer encephalitis virus)
1938	B. von Borries, H. Ruska, E. Ruska	First electron micrograph of viruses (ectromelia, vaccinia viruses)
1939	E. Ellis, M. Delbrück	Development of one-step growth curve—bacteriophage
1940	K. Smithburn, T. Hughes, A. Burke, J. Paul	Discovery of West Nile virus
1941	G. Hirst	Discovery of agglutination of red blood cells by influenza virus
1945	M. Chumakov, G. Courtois, colleagues	Discovery of Crimean-Congo hemorrhagic fever virus
1948	G. Dalldorf, G. Sickles	Discovery of Coxsackieviruses
1949	J. Enders[a], T. Weller[a], F. Robbins[a]	Development of cell culture methodology for polio, measles, and other vaccines
1950	L. Florio, M. Miller, E. Mugrage	Discovery of Colorado tick fever virus
1952	R. Dulbecco, M. Vogt	Development of plaque assay for animal viruses—polioviruses, Western equine encephalitis virus
1953	W. Rowe	Discovery of human adenoviruses

(Continued)

TABLE 1.1 Some Milestones in the History of Virology (Continued)

Date	Discoverer(s)	Discovery(ies)
1954	J. Salk, J. Youngner, T. Francis	Development of inactivated polio vaccine
1958	J. Lederberg[a]	Discovery of genetic recombination and the organization of the genetic material of bacteria
1959	A. Sabin, H. Cox, H. Koprowski	Development of attenuated live-virus polio vaccine
1962	A. Lwoff, R. Horne, P. Tournier	Classification of the viruses based on virion characteristics
1964	M. Epstein, B. Achong, Y. Barr	Discovery of Epstein–Barr virus and its association with Burkitt's lymphoma
1965	D. Tyrrell, M. Bynoe, J. Almeida	Discovery of human coronaviruses (B814 and 229E)
1965	F. Jacob[a], A. Lwoff[a], J. Monod[a]	Discoveries of genetic control of enzymes and virus synthesis: the operon
1967	B. Blumberg[a], H. Alter, A. Prince	Discovery of Australia antigen and its link to hepatitis B
1969	M. Delbrück[a], A. Hershey[a], S. Luria[a]	Discoveries related to the replication mechanism and the genetic structure of viruses
1970	H. Temin[a], D. Baltimore[a], R. Dulbecco[a]	Discoveries related to the interaction between tumor viruses and the genetic material of the cell—reverse transcriptase
1972	A. Kapikian, colleagues	Discovery of Norwalk virus (norovirus)
1973	R. Bishop, G. Davidson, I. Holmes, T. Flewett, A. Kapikian	Discovery of human rotaviruses
1973	S. Feinstone, A. Kapikian, R. Purcell	Discovery of hepatitis A virus
1975	Y. Cossart, A. Field, A. Cant, D. Widdows	Discovery of parvovirus B-19 and its association with aplastic crisis in hemolytic anemia
1975	P. Sharp[a], L. Chow, R. Roberts[a], T. Broker	Discovery of RNA splicing and split genes (adenovirus)
1976	D. C. Gajdusek[a]	Discovery of transmissible spongiform encephalopathies
1976	K. Johnson, P. Webb, J. Lange, F. Murphy, S. Pattyn, W. Jacob, G. Van der Groen, P. Piot, E. Bowen, G Platt, G. Lloyd, A. Baskerville, D. Simpson	Discovery of Ebola virus
1976	J. Bishop[a], H. Varmus[a]	Discovery of the cellular origin of retroviral oncogenes
1977	D. Henderson, F. Fenner, I. Arita, many others	Global eradication of smallpox
1978	D. Nathans[a], W. Arber[a], H. Smith[a]	Discovery of restriction enzymes and their application to problems of molecular genetics
1978	S. Harrison, M. Rossman, N. Olson, R. Kuhn, T. Baker, J. Hogle, M. Chow, R. Rueckert, J. Johnson	Atomic structure of viruses (tomato bushy stunt virus, polioviruses, rhinoviruses)
1980	P. Berg[a]	The development of recombinant-DNA technology
1980	R. Gallo, B. Poiesz, M. Yoshida, I. Miyoshi, Y. Hinuma	Discovery of human T lymphotropic viruses 1 and 2
1981	V. Racaniello, D. Baltimore	Development of an infectious recombinant clone of a virus (poliovirus)
1982	S. Prusiner[a]	Concept of the prion and their etiologic role in spongiform encephalopathies
1982	A. Klug[a]	Crystallographic electron microscopy and structural elucidation of biologically important nucleic acid–protein complexes

(Continued)

TABLE 1.1 Some Milestones in the History of Virology (Continued)

Date	Discoverer(s)	Discovery(ies)
1983	F. Barré-Sinoussi[a], L. Montagnier[a], J. Chermann	Discovery of human immunodeficiency virus 1 (HIV1)
1983	M. Balayan	Discovery of hepatitis E virus and its transmission
1985	F. Barin, F. Clavel, M. Essex, P. Kanki, F. Brun-Vézinet	Discovery of human immunodeficiency virus 2 (HIV2)
1988	G. Hitchings[a], G. Elion[a]	Discoveries of important principles for drug treatment—acyclovir
1989	M. Houghton, Q.-L. Choo, G. Kuo, D. Bradley, H. Alter	Discovery of hepatitis C virus
1993	S. Nichol, C. Peters, P. Rollin, T. Ksiazek	Discovery of Sin Nombre virus and its association with hantavirus cardiopulmonary syndrome
1994	Y. Chang, P. Moore	Discovery of human herpesvirus 8—Kaposi sarcoma herpesvirus
1995	K. Murray, P. Hooper, A. Hyatt	Discovery of Hendra virus and its reservoir host fruit bats
1996	P. Doherty[a], R. Zinkernagel[a]	Discovery of the genetic specificity of the cell-mediated immune response
1996	R. Will, J. Ironside, J. Collinge, colleagues	Discovery that bovine spongiform encephalopathy prion is the cause of variant Creutzfeldt–Jakob disease in humans
1999	K. Chua, S. Lam, W. Bellini, T. Ksiazek, B. Eaton, colleagues	Discovery of Nipah virus
1999	D. Asnis, M. Layton, W.I. Lipkin, R. Lanciotti	Extension of West Nile virus range to North America
2001	B. van den Hoogen, A. Osterhaus, colleagues	Discovery of human metapneumovirus
2003	C. Urbani, J. Peiris, S. Lai, L. Poon, G. Drosten, K. Stöhr, A. Osterhaus, T. Ksiazek, D. Erdman, C. Goldsmith, S. Zaki, J. DeRisi, others	Discovery of SARS coronavirus
2003	B. La Scola, D. Raoult, others	Discovery of mimivirus, the largest virus known at the time
2005	J. Taubenberger, P. Palese, T. Tumpey, A. Garcia-Sastre, others	1918 influenza virus genome sequenced and the virus reconstructed
2005		Beginning of global pandemic of chikungunya
2005	E. Leroy, J. Towner, R. Swanepoel, others	Discovery that the reservoir hosts of Ebola/Marburg viruses are bats
2007	T. Allander, D. Wang, Y. Chang, others	Discovery of human polyomaviruses KI, WU, MC
2008	H. zur Hausen[a]	Discovery that human papilloma viruses cause cervical cancer
2008	B. La Scola, D. Raoult, others	Discovery of virophage, Sputnik
2010	W. Plowright and the FAO Global Rinderpest Eradication Programme	Global eradication of rinderpest
2011	B. Hoffmann, M. Beer, T. Mettenleiter, colleagues	Discovery of Schmallenberg virus
2012	A.M. Zaki, R. Fouchier, W.I. Lipkin	Discovery of MERS coronavirus
2014		Beginning of an Ebola hemorrhagic fever epidemic in West Africa, the largest ever
2015		Beginning of a global epidemic of Zika virus disease—discovery of microcephaly as consequence of *in utero* infection

[a]*Scientists who were awarded the Nobel Prize for their work—cited at date of the discovery rather than the date of award.*

FIGURE 1.1 In 1881 and 1882, Louis Pasteur, Charles Chamberland, Émile Roux, and Louis Thuillier began their research toward developing a rabies vaccine. They modified Pierre-Victor Galtier's technique by inoculating nervous tissue from a rabid dog through a long series of dogs via subdural trephination. After many passages, they obtained a virus of maximum virulence and with a fixed incubation period of about 10 days. The degree of attenuation of virus recovered from each passage was measured and virus was then further attenuated in rabbits. This final attenuation procedure consisted of suspending the spinal cord of a rabid rabbit in a flask, in a warm dry atmosphere, to achieve slow desiccation. They succeeded in producing "attenuated viruses of different strengths," the weakest of which could be used to prepare the first dose of a vaccine. Inoculating dogs with a sequence of spinal cords of increasing virulence rendered the recipients resistant to inoculation with fully virulent virus. Within a year, Pasteur and his colleagues reported the results of this treatment in 350 cases of rabies exposure—only one person developed rabies, and this a child who was treated 6 days after exposure. Over the next decades many thousands of people with potential rabies exposures were immunized with ever-improving animal nervous system (brain and spinal cord) vaccines, at the Institut Pasteur in Paris, which was founded in 1888, and in other locations throughout the world. *Louis Pasteur, 1822–95. Painting by Albert Edelfeldt, 1885. From Institut Pasteur, used with permission.*

1911 Peyton Rous also showed similar properties for the etiologic agent of a sarcoma of chickens: Rous sarcoma virus was to play an essential role in determining the basic mechanism by which viruses may trigger the onset of tumors.

The realization that oncogenesis and virus infection went hand in hand was an important milestone in the early days of virology, although it took many decades for its true significance to be appreciated. In 1970 Howard Temin and David Baltimore independently were able to show that oncogenic viruses contain a reverse transcriptase enzyme,

thus explaining how an RNA virus could produce DNA copies of its genetic material.

Bacteriophages were independently discovered by Frederick Twort and Felix d'Herelle (1915) who investigated outbreaks of dysentery among troops of the First World War. Presciently, Twort foresaw that the clear plaques in plated *Micrococcus* cultures could be caused by "ultrafilterable viruses."

During the following decades of the 20th century, it was thought by many that viruses represented infectious protein particles. This was a view reinforced by Wendell Stanley's description in 1935 that crystals of pure TMV could be dissolved and transmit infection to healthy plants—he presumed that the crystals were pure protein. This was dispelled, however, when Frederick Bawden and Norman Pirie showed that TMV contained not only protein but also nucleic acid. The importance of this was shown by the classic studies of Oswald Avery, Colin MacLeod, and Maclyn McCarty (1944) and then Alfred Hershey and Martha Chase (1952), who proved DNA was linked to hereditary.

In 1933 the electron microscope was invented by Ernst Ruska and Max Knoll and in 1938 Bodo von Borries, Helmut Ruska, and Ernst Ruska published the first electron micrographs of ectromelia (mousepox) virus and vaccinia virus. It soon became clear that there was great diversity in the size and shape of the various viruses. A major advance was the development of negative-contrast electron microscopy in 1959 by Sydney Brenner and Robert Horne. Using this method, electron-dense stains surround virus particles to produce a negative image of the virus with remarkable resolution; importantly in those early days of medical virology, the method was simple to use. Figs. 1.3 and 1.4 depict the diverse spectrum of morphological shapes represented by animal viruses. By the early 1960s, the fine structure of several viruses was unraveled by Aaron Klug, Donald Caspar, and others using X-ray crystallography—they showed that many viruses are constructed from uniform subunits, following the principles of icosahedral symmetry as first understood for the Platonic solids (regular polyhedra) by the ancient Greeks. Thus through the use of several different approaches the diversity of structural detail among various viruses began to emerge.

Attempts to prevent virus disease using vaccines have paralleled the development of virology, beginning from the early pioneering days of Edward Jenner and Louis Pasteur. Notable developments included the attenuated yellow fever vaccine developed by Max Theiler in 1931, a vaccine that is still in widespread use today and has saved countless thousands of lives. Jonas Salk and Albert Sabin in 1954 and 1959 developed non-replicating (inactivated) and living attenuated virus vaccines against poliovirus, respectively, the use of which has been so extensive that poliovirus infection has all but been eradicated save for a few pockets of infection

FIGURE 1.2   In 1900, Walter Reed and his colleagues discovered yellow fever virus, the first human virus, and its transmission cycle. This is a famous allegorical painting, entitled *Conquerors of Yellow Fever* by Dean Cornwell. It depicts Walter Reed (in white uniform) and Carlos Finlay (with white hair) looking on as Jesse Lazear, who died of yellow fever a month later, applies an infected mosquito to the arm of James Carroll. The painting includes Aristides Agramonte (behind Lazear), Leonard Wood (in brown helmet), Jefferson Kean (in white helmet), and several of the volunteers who subsequently were infected in the same way. Carroll became infected as a result of this experiment—he survived, and went on to have a distinguished career as a microbiologist, but suffered from chronic illness leading to an early death, said to be a consequence of his yellow fever infection. *Purchased copy, used with permission.*

in remote parts of the world (Fig. 1.3). As is described in Chapter 11: Vaccines and Vaccination, vaccine research has often exploited novel concepts, for example, the use of plasma from chronically infected humans as a source of hepatitis B virus (HBV) envelope protein to stimulate immunity against hepatitis B virus (1976), and the use of genetically modified naked DNA preparations to induce the expression of antigens in the tissues of vaccine recipients.

In 1957, Alick Isaacs and Jean Lindemann discovered interferons, molecules that represent the initial mammalian response to infection. Great hope was placed on the use of interferons in the treatment of a wide spectrum of human virus infections: although of proven use in certain conditions, however, the use of interferons has not lived up to the earlier wide promise suggested by laboratory studies.

The sciences of immunology and cell and molecular biology have been intertwined with that of virology: landmark discoveries were made by Peter Doherty and Rolf Zinkernagel, who in 1974 discovered how the cellular immune system recognizes virus-infected cells, and Georges Kohler and Cesar Milstein, who in 1975 developed the first monoclonal antibodies.

THE VIROSPHERE

We live in what many now describe as the *virosphere*, since almost all living multicellular and unicellular organisms are susceptible to virus infection. Take as an example the oceans: every liter of seawater is populated with up to 10 billion viruses. It is estimated that there are around 5×10^{30} bacteria on planet Earth, and that viruses are numerically at least more common; this means there are more viruses in the world than all life forms. The vast majority are most likely viruses of bacteria (bacteriophages) serving to aid the recycling of organic matter, but some have a more sophisticated role in the environment, for example, determining insect behavior as an essential part of an arthropod life cycle. This staggeringly large repertoire of

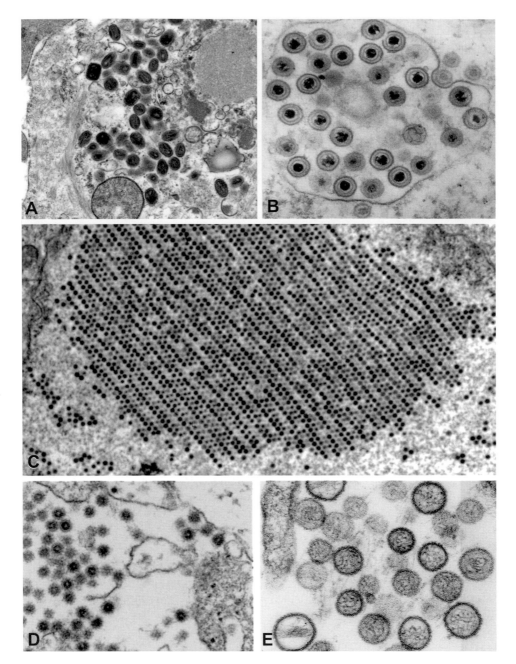

FIGURE 1.3 Thin-section electron microscopy of selected viruses. The remarkable diversity of the viruses is clearly revealed by thin-section electron microscopy of infected cells—and this technique provides important information about morphogenesis and cytopathology. (A) Family *Poxviridae*, genus *Orthopoxvirus*, variola virus. (B) Family *Herpesviridae*, genus *Simplexvirus*, human herpesvirus 1. (C) Family *Adenoviridae*, genus *Mastadenovirus*, human adenovirus 5. (D) Family *Togaviridae*, genus *Alphavirus*, Eastern equine encephalitis virus. (E) Family *Bunyaviridae*, genus *Hantavirus*, Sin Nombre virus. These images represent various magnifications; the details of the morphogenesis of the various viruses are given in the chapters of Part II of this book.

the virosphere is not restricted to inhabiting non-human life forms: we are only recently beginning to study the range of different viruses that humans appear to carry permanently (the human "virome," see Chapter 39: Viral Syndromes), yet appear to cause no harmful effects. One example is the Torque teno (TT) virus, discovered by chance in 1997 during studies of "transfusion-transmitted" infection.

THE NATURE OF VIRUSES

The unicellular *microorganisms* can be arranged in the order of decreasing size and complexity: protozoa, fungi, and bacteria (the latter including mycoplasmas, rickettsiae, and chlamydiae). These microorganisms, however small and simple, are *cells*. Such microorganisms contain DNA as the

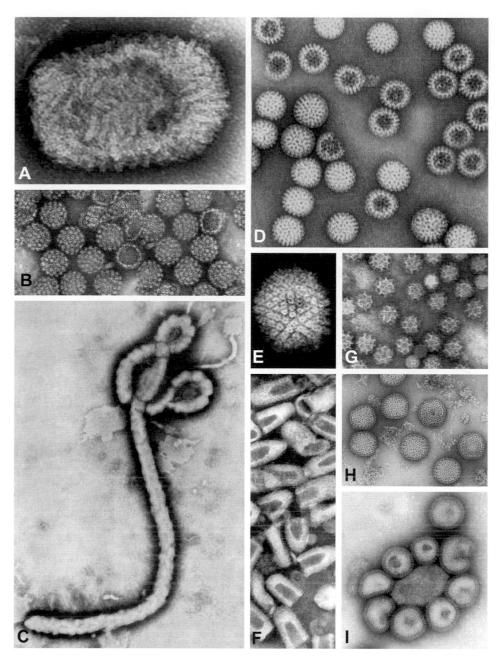

FIGURE 1.4 Negative contrast electron microscopy of selected viruses. The remarkable diversity of the viruses is revealed by all kinds of electron microscopy methods, but none better than by negative staining. (A) Family *Poxviridae*, genus *Orthopoxvirus*, vaccinia virus. (B) Family *Papovaviridae*, genus *Papillomavirus*, human papillomavirus. (C) Family *Filoviridae*, Ebola virus. (D) Family *Reoviridae*, genus *Rotavirus*, human rotavirus. (E) Family *Herpesviridae*, genus *Simplexvirus*, human herpesvirus 1 (capsid only, envelope not shown). (F) Family *Rhabdoviridae*, genus *Lyssavirus*, rabies virus. (G) Family *Caliciviridae*, genus *Norovirus*, human norovirus. (H) Family *Bunyaviridae*, genus *Phlebovirus*, Rift Valley fever virus. (I) Family *Orthomyxoviridae*, genus *Influenzavirus A*, influenza virus A/Hong Kong/1/68 (H3N2). These images represent various magnifications; the size of the various viruses is given in Chapter 2: Classification of Viruses and Phylogenetic Relationships and in the chapters of Part II of this book.

repository of genetic information, and also contain various species of RNA and most, if not all, of the machinery for producing energy and macromolecules. These microorganisms grow by synthesizing macromolecular constituents (nucleic acids, proteins, carbohydrates, and lipids), and most multiply by binary fission.

Viruses, on the other hand, are neither cellular nor microorganisms. The key differences between viruses and microorganisms are listed in Table 1.2. Viruses do not possess functional organelles (e.g., mitochondria, Golgi, chloroplasts, and endoplasmic reticulum), and thus are totally dependent on the host for the machinery of energy

FIGURE 1.5   The World Health Organization Global Polio Eradication Initiative aims for global eradication of poliomyelitis by about 2018. The Initiative is led by the World Health Organization, UNICEF, and the Rotary and Gates Foundations; it has reduced the number of cases from the many thousands per year to less than 100 (359 cases in 2014; 74 cases in 2015). Polio will be the third disease globally eradicated, after smallpox and rinderpest. The most important step in polio eradication is interruption of endemic transmission by universal infant vaccination using oral vaccine (OPV; often by organizing "national immunization days"), supplementary IPV vaccination campaigns where needed, intensive surveillance of cases of flaccid paralysis, and in some places detection of virus in sewage. Figures (clockwise from top left). An Egyptian stele (slab) thought to depict a polio victim—18th Dynasty (1403–1365 BC); patients with permanent respiratory muscle paralysis after recovery from poliomyelitis would spend the rest of their lives requiring assisted respiration (immersed in an "iron lung"); patients with permanent lower limb weakness following poliomyelitis; those remaining countries reporting cases of poliomyelitis in 2014; oral administration of polio vaccine; in 1921, 39-year-old Franklin D. Roosevelt was diagnosed with poliomyelitis and was left with permanent paralysis from the waist down, but was rarely photographed in a wheelchair. He was elected US president in 1932.

TABLE 1.2 Contrasting Properties of Unicellular Microorganisms and Viruses

Property	Bacteria	Rickettsiae	Mycoplasma	Chlamydiae	Viruses
>300 nm diameter[a]	Yes	Yes	Yes	Yes	No
Growth on non-living media[b]	Yes	No	Yes	No	No
Binary fission	Yes	Yes	Yes	Yes	No
Contain both DNA and RNA	Yes	Yes	Yes	Yes	No[c]
Infectious nucleic acid	No	No	No	No	Many
Functional ribosomes	Yes	Yes	Yes	Yes	No
Sensitivity to antibiotics	Yes	Yes	Yes	Yes	No[d]

[a]Some mycoplasmas and chlamydiae are less than 300 nm in diameter, and mimiviruses and the other new "giant DNA viruses" are greater than 300 nm in diameter.
[b]Chlamydiae and most rickettsiae are obligate intracellular parasites.
[c]A few viruses contain both types of nucleic acid, but one of these types acts as the main functional molecule and the other plays a minor role.
[d]With very few exceptions.

production and synthesis of macromolecules. Viruses contain only one type of functional nucleic acid, either DNA or RNA, never both, and differ from microorganisms in having a life cycle divisible into two clearly defined phases. Outside of the host cell, the viruses are metabolically inert and can be considered as complexes of large macromolecules; during this extracellular phase of the viral life cycle, virus transmission is dependent upon movements of air and fluid, and in some cases the life cycle of insect vectors. Once inside the host cell, however, viruses behave with many of the properties of living organisms; viruses are metabolically active in that the viral genome exploits the machinery of the host to produce progeny genome copies, viral messenger RNA, and viral proteins (often along with carbohydrates and lipids), all of which are then assembled to form new virions (*virion*, the complete virus particle). This assembly from pools of precursor molecules is in contrast to the multiplication of cellular organisms by binary fission. In contrast to any microorganism, many viruses can reproduce even if only the viral DNA or RNA genome is introduced into the host cell. These qualities have been used to argue the question, "Are viruses alive?" One answer is to envision viruses "*at the edge of life*," in some ways fulfilling the criteria we use to define life, but mostly not.

Given the unique characteristics of viruses, where might viruses have originated? There are three principal theories that have been argued for many years. First, viruses may have originated as escaped eukaryotic genes, that is nucleic acid sequences, that evolved to encode protective protein coats to allow survival outside of the environment of the cell (transposons and retrotransposons have been suggested as the progenitors of retroviruses). Second, viruses may be degenerate forms of intracellular parasites, having lost most cellular functions (bacteria have been suggested as the progenitors of mitochondria, chloroplasts, and poxviruses); and third, viruses may have originated independently along with other primitive molecules and developed with self-replicating capabilities.

In the absence of fossil remains, insight as to virus evolution relies almost entirely on sequence analyses of virus genomes. For example, the genome of a plant viroid (a subviral agent comprised of infectious naked RNA), potato spindle tuber viroid, seems to be a self-replicating RNA copy of a part of the host potato DNA. Many of the genes of poxviruses are similar to those of eukaryote hosts. In any case, it seems certain from sequence analyses of viral genomes that all presently recognized viruses did not evolve from a single progenitor; rather, different kinds of viruses likely arose independently from different origins, and then continued to diversify and adapt survival and transmission qualities to better fit particular niches by the usual Darwinian process of mutation and natural selection.

It should be stressed that the genetic blueprint of all viruses is under continuing evolutionary pressure, sometimes showing dramatic examples of genetic change and natural selection of those variants that survive the best. Some viruses have continued to evolve in long association with each associated hosts (e.g., herpesviruses, some retroviruses); others have evolved by "host species jumping" (e.g., influenza viruses), and yet others by developing zoonotic transmission schemes (e.g., rabies virus).

Several important practical consequences follow from understanding that viruses are different from microorganisms and all life forms: for example, some viruses can persist for the lifetime of the host cell by the integration of the DNA genome (or a DNA copy of the RNA genome) into the genome of the host cell, or by the carriage of viral DNA genomes by the host cell in episomal form. Since viruses use the replicative machinery of the host, virus infections present major challenges to antiviral drug development. Drugs that interfere with viral replication nearly always interfere with essential host cell functions. This is in contrast to bacteria, which have unique metabolic pathways different from those of the host, enabling these to be exploited as targets for antibiotics.

The simplest viruses consist of a DNA or RNA genome contained within a protein coat, but there are classes of even simpler infectious agents: (1) *satellites*, which are defective viruses, dependent upon a helper virus to supply essential functions such as nucleic acid replication functions or structural elements such as capsid proteins; (2) *viroids*, which as noted above consist of a naked RNA molecule that is infectious; and (3) *prions*, the agents of the spongiform encephalopathies, consisting of an infectious protein without any associated nucleic acid.

SCOPE OF THIS BOOK

From its beginning medical virology has been intertwined with many related sciences. Even though this book deals with medical virology *per se*—the viruses infecting humans and the diseases so caused—understanding the full scope of the subject requires a continuing appreciation and integration of related sciences, from cell biology to medical epidemiology and extending to human social behavior. The perspective represented by this book, of medical virology as an infectious disease science, is meant to provide a starting point, an anchor, for those who must relate the subject to clinical practice, public health practice, scholarly research, and other endeavors.

Part I of this book thus deals with the properties of viruses, how viruses replicate, and how viruses cause disease. These chapters are then followed by an overview of the principles of diagnosis, epidemiology, and how virus infections can be controlled. This first section is concluded by a discussion on emergence and attempts to predict the next major public health challenges. These form a guide for delving into the specific diseases of interest to the reader as described in Part II.

FURTHER READING

Crawford, D.H., 2011. A Very Short Introduction to Viruses. Oxford University Press, Oxford.

Flint, S.J., Racaniello, V.R., Rall, G.F., Skalka, A.M., Enquist, L., 2015. Principles of Virology, two volumes, fourth ed. ASM Press, Washington, DC, ISBN-10: 1555819516.

Kaslow, R.A., Stanberry, L.R., LeDuc, J.W., 2014. Viral Infections of Humans, Epidemiology and Control, fifth ed. Springer, New York, ISBN 978-1-4899-7447-1.

Knipe, D.M., Howley, P.M., et al., 2013. Field's Virology, sixth ed. Lippincott Williams and Wilkins, Philadelphia, ISBN 978-145-110563-6.

MacLachlan, N.J., Dubovi, E.J., 2011. Fenner's Veterinary Virology, fourth ed. Academic Press, London, ISBN 978-0-12-375158-4.

Waterson, A.P., Wilkinson, L., 1978. An Introduction to the History of Virology. Cambridge University Press, Cambridge, ISBN 978-0-521-21917-5.

Zimmer, C., 2011. A Planet of Viruses. University of Chicago Press, Chicago, ISBN 978-0-226-98335-6.

Zinsser, H., 1935, reprinted 2007. Rats, Lice and History. Transaction Publishers, New Brunswick, USA, ISBN 978-1-4128-0672-5.

Chapter 2

Classification of Viruses and Phylogenetic Relationships

Virus taxonomy brings into sharp focus the debate about the true nature of viruses. A comprehensive classification system should define boundaries within what may at first appear as a continuum of properties. This is often most challenging at the level of genome sequence analysis.

The rules and processes that have been developed are unique to the science of virology, and necessary to accommodate the astonishing variety of viruses. There is now evidence that probably all organisms in the biological world may be infected by at least one virus. Indeed it has been estimated that viruses represent the most abundant biological entities on the planet, existing as pathogens or silent passengers in humans and other animals, plants, invertebrates, protozoa, fungi, and bacteria. To date more than 4000 different viruses and 30,000 different strains and subtypes have been recognized, with particular strains and subtypes often having significant public health importance. Several hundred different viruses are known to cause disease in humans, although this is a small fraction of those viruses encountered in the surrounding environment. Since all viruses, whatever the host, share the properties described in the preceding chapter, virologists have developed a single system of classification and nomenclature that covers all viruses—this is a system overseen by the *International Committee on Taxonomy of Viruses* (ICTV). One challenge of virus classification is to define evolutionary relationships between viruses when minor changes in molecular structures may give rise to pathogens with radically different properties (Fig. 2.1).

VIRAL TAXONOMY

Although it is hierarchical and at most levels reflects evolutionary relationships, the taxonomy of viruses is deliberately non-systematic—that is, there is no intent to relate all viruses to an ancient evolutionary root—in fact, there is good evidence for several separate roots. The earliest efforts to classify viruses were based upon host organism species, common clinical and pathological properties, tropism for particular tissues and organs, and common ecological and transmission characteristics. For example, viruses that cause hepatitis (e.g., hepatitis A virus, family *Picornaviridae*; hepatitis B virus, family *Hepadnaviridae*; hepatitis C virus, family *Flaviviridae*; and Rift Valley fever virus, family *Bunyaviridae*) might have been brought together as "the hepatitis viruses." Such systems have now been superseded.

The initial principles for identifying and distinguishing different viruses involved giving equal weight to the importance of:

1. type of nucleic acid (DNA or RNA);
2. virion size, as determined by ultrafiltration and electron microscopy;
3. virion morphology, as determined by electron microscopy;
4. virion stability, as determined by varying pH and temperature, exposure to lipid solvents and detergents, etc.; and
5. virion antigenicity, as determined by various serological methods.

This approach was practicable in the era before molecular biology, as these characteristics had already been determined for a large number of viruses, and thus these properties could be used to build a taxonomic framework. Subsequently it has been necessary in most cases to determine only a few characteristics in order to place a newly described virus into an established taxon, as a starting point for further work to define its relationship with other members. For example, an isolate from the respiratory tract of a child with croup, identified by negative contrast electron microscopy as an adenovirus, might be submitted immediately for serological identification—it would certainly turn out to be a member of the family *Adenoviridae*, genus *Mastadenovirus* (the adenoviruses of mammals), and would be serologically identified as one of the >50 human adenoviruses—or perhaps, it would turn out to be a new human adenovirus!

Nowadays, the primary criteria for delineation of the main viral taxa are:

1. the type, character, and nucleotide sequence of the viral genome;
2. the strategy of viral replication; and
3. the structure of the virion.

Fenner and White's Medical Virology. DOI: http://dx.doi.org/10.1016/B978-0-12-375156-0.00002-3

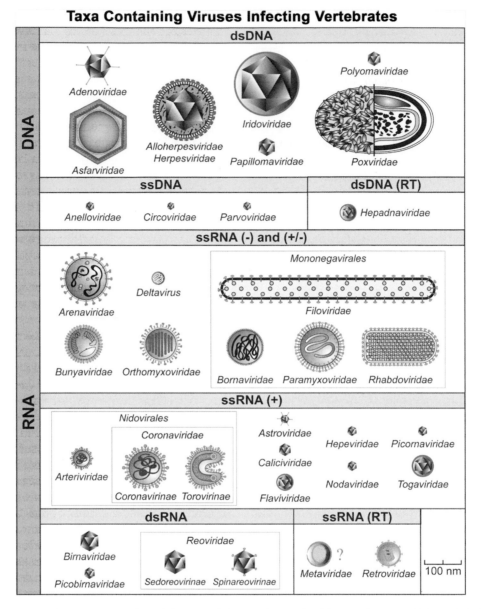

FIGURE 2.1   Diagram illustrating the shapes and sizes of viruses that infect vertebrates. The virions are drawn to scale, but artistic license has been used in representing their structure. In some, the cross-sectional structures of capsid and envelope are shown, with a representation of the genome; with the very small virions, only their size and symmetry are depicted. *Reproduced from King, A.M.Q., Adams, M.J., Carstens, E.B., Lefkowitz, E.J. (Eds.), 2011. Virus taxonomy, classification and nomenclature of viruses. In: Ninth Report of the International Committee for the Taxonomy of Viruses. Academic Press, London, with permission.*

Sequencing, or partial sequencing, of the viral genome provides powerful taxonomic information and now is often done very early in the identification process. Reference genome sequences for all viral taxa are available in public databases (e.g., GenBank, National Center for Biotechnology Information, National Library of Medicine, National Institutes of Health, Bethesda, Maryland, United States: <http://www.ncbi.nlm.nih.gov>). Such an approach in most cases allows one to immediately place a virus in a specific taxon.

The universal system of viral taxonomy recognizes five levels, namely *order*, *family*, *subfamily*, *genus*, and *species*. The names of orders end with the suffix -*virales*, families with the suffix -*viridae*, subfamilies with the suffix -*virinae*, and genera with the suffix -*virus*. The names of species also end with the term *virus*, either as a separate word or as a suffix (according to historic precedence). Lower levels, such as *subspecies*, *strains*, and *variants*, are established for practical purposes such as diagnostics, vaccine development, etc., but this is not a matter of formal classification and there are neither universal definitions nor is there any standard universal nomenclature.

As of 2015 the universal taxonomy system for viruses encompasses seven orders, four of which contain human and animal pathogens (*Picornavirales*, *Herpesvirales*, *Mononegavirales*, and *Nidovirales*), and 78 families, 27 of which contain human and/or animal pathogens, 348 genera, and 2285 species of viruses (Table 2.1). This situation is constantly changing, and the interested reader should consult the ICTV website for updates (http://www. ictvonline.org). The universal taxonomy system is nearly complete at the level of families and genera; that is, virtually

TABLE 2.1 Major Families of Viruses Infecting Vertebrates—A Subset of the ICTV Universal Virus Taxonomy System, 2015

Family	Subfamily	Genus	Type Species	Viruses Infecting Humans
Double-Stranded DNA Viruses				
Poxviridae	Chordopoxvirinae	Orthopoxvirus	Vaccinia virus	Smallpox (variola)
		Capripoxvirus	Sheeppox virus	
		Leporipoxvirus	Myxoma virus	
		Suipoxvirus	Swinepox virus	
		Molluscipoxvirus	Molluscum contagiosum virus	Molluscum contagiosum virus
		Avipoxvirus	Fowlpox virus	
		Yatapoxvirus	Yaba monkey tumor virus	Yaba monkey tumor virus; Tanapox virus
		Parapoxvirus	Orf virus	Orf virus
		Cervidpoxvirus	Deerpox virus	
Asfarviridae		Asfivirus	African swine fever virus	
Iridoviridae		Ranavirus	Frog virus 3	
		Lymphocystivirus	Lymphocystis disease virus 1	
		Megalocytivirus	Infectious spleen and kidney necrosis virus	
Alloherpesviridae		Ictalurivirus	Ictalurid herpesvirus 1	
Herpesviridae	Alphaherpesvirinae	Simplexvirus	Human herpesvirus 1	Herpes simplex viruses 1 and 2
		Varicellovirus	Human herpesvirus 3	Varicella-zoster virus
		Mardivirus	Gallid herpesvirus 2	
		Iltovirus	Gallid herpesvirus 1	
	Betaherpesvirinae	Cytomegalovirus	Human herpesvirus 5	Human cytomegalovirus
		Muromegalovirus	Murid herpesvirus 1	
		Proboscivirus	Elephantid herpesvirus 1	
		Roseolovirus	Human herpesvirus 6	Human herpesviruses 6 and 7
	Gammaherpesvirinae	Lymphocryptovirus	Human herpesvirus 4	Epstein-Barr virus
		Macavirus	Alcelaphine herpesvirus 1	
		Percavirus	Equid herpesvirus 2	
		Rhadinovirus	Saimiriine herpesvirus 2	Human herpesvirus 8 (Kaposi sarcoma-associated virus)

(Continued)

TABLE 2.1 Major Families of Viruses Infecting Vertebrates—A Subset of the ICTV Universal Virus Taxonomy System, 2015   (Continued)

Family	Subfamily	Genus	Type Species	Viruses Infecting Humans
Malacoherpesviridae		*Ostreavirus*	*Ostreid herpesvirus 1*	
Adenoviridae		*Mastadenovirus*	*Human adenovirus C*	Human adenoviruses A–G
		Aviadenovirus	*Fowl adenovirus A*	
		Atadenovirus	*Ovine adenovirus D*	
		Siadenovirus	*Frog adenovirus*	
		Ichtadenovirus	*Sturgeon adenovirus A*	
Polyomaviridae		*Polyomavirus*	*Simian virus 40*	JC polyomavirus BK polyomavirus; others
Papillomaviridae		*Alphapapillomavirus*	*Human papillomavirus 32*	Human papillomaviruses, many
		Betapapillomavirus	*Human papillomavirus 5*	Human papillomaviruses, many
		Gammapapillomavirus	*Human papillomavirus 4*	Human papillomaviruses, many
		Deltapapillomavirus	*Europeam elk papillomavirus 1*	
		Epsilonpapillopmavirus	*Bovine papillomavirus 5*	
		Zetapapillomavirus	*Equine papillomavirus 1*	
		Etapapillomavirus	*Fringilla coelebs papillomavirus*	
		Thetapapillomavirus	*Psittacus erithacus timneh papillomavirus*	
		Iotapapillomavirus	*Mastomys natalensis papillomavirus*	
		Kappapapillomavirus	*Cottontail rabbit papillomavirus*	
		Lambdapapillomavirus	*Canine oral papillomavirus*	
		Mupapillomavirus	*Human papillomavirus 1*	Human papillomaviruses 1 and 63
		Nupapillomavirus	*Human papillomavirus 41*	Human papillomavirus 41
		Xipapillomavirus	*Bovine papillomavirus 3*	
		Omicronpapillomavirus	*Phocoena spinipinnis papillomavirus*	
		Pipapillomavirus	*Hamster oral papillomavirus*	
Single-Stranded DNA Viruses				
Parvoviridae	*Parvovirinae*	*Parvovirus*	*Minute virus of mice*	
		Erythrovirus	*Human parvovirus B19*	Human parvovirus B19
		Dependovirus	*Adeno-associated virus 2*	AAV 1–5
		Amdovirus	*Aleutian mink disease virus*	
		Bocavirus	*Bovine parvovirus*	Human bocaviruses 1–4
Circoviridae		*Circovirus*	*Porcine circovirus 1*	
		Gyrovirus	*Chicken anaemia virus*	
Anelloviridae		*Alphatorquevirus*	*Torque teno virus*	TTV groups 1–5

(Continued)

TABLE 2.1 Major Families of Viruses Infecting Vertebrates—A Subset of the ICTV Universal Virus Taxonomy System, 2015 (Continued)

Family	Subfamily	Genus	Type Species	Viruses Infecting Humans
Reverse Transcribing Viruses				
Hepadnaviridae (DNA genome)		*Orthohepadnavirus*	*Hepatitis B virus*	Hepatitis B virus genotypes A–H
		Avihepadnavirus	*Duck hepatitis B virus*	
Retroviridae (RNA genome)	*Orthoretrovirinae*	*Alpharetrovirus*	*Avian leukosis virus*	
		Betaretrovirus	*Mouse mammary tumor virus*	
		Gammaretrovirus	*Murine leukaemia virus*	
		Deltaretrovirus	*Bovine leukaemia virus*	Human T-lymphotropic viruses 2 and 3
		Epsilonretrovirus	*Walleye dermal sarcoma virus*	
		Lentivirus	*Human immunodeficiency virus*	HIV-1 and HIV-2
	Spumaretrovirinae	*Spumavirus*	*Simian foamy virus*	
Double-Stranded RNA Viruses				
Reoviridae	*Sedoreovirinae*	*Orbivirus*	*Bluetongue virus*	African horse sickness virus; Kemerovo virus
		Rotavirus	*Rotavirus A*	Rotaviruses A–E; others
		Seadornavirus	*Banna virus*	Banna virus
	Spinoreovirinae	*Coltivirus*	*Colorado tick fever virus*	Colorado tick fever virus
		Orthoreovirus	*Mammalian orthoreovirus*	Mammalian orthoreoviruses 1–4
		Aquareovirus	*Aquareovirus A*	
		Cardorcovirus	*Eriocheir sinensis reovirus*	
Picobirnaviridae		*Picobirnavirus*	*Human picobirnavirus*	Human picobirnavirus
Single-Stranded Negative-Sense RNA Viruses				
Paramyxoviridae	*Paramyxovirinae*	*Respirovirus*	*Sendai virus*	Human parainfluenzaviruses 1 and 3
		Avulavirus	*Newcastle disease virus*	
		Morbillivirus	*Measles virus*	Measles virus
		Rubulavirus	*Mumps virus*	Mumps virus; Human parainfluenzaviruses 2 and 4
		Avulavirus	*Newcastle disease virus*	
		Henipavirus	*Hendra virus*	Hendra virus; Nipah virus
	Pneumovirinae	*Pneumovirus*	*Human respiratory syncytial virus*	Human respiratory syncytial virus
		Metapneumovirus	*Avian pneumovirus*	Human metapneumovirus

(Continued)

TABLE 2.1 Major Families of Viruses Infecting Vertebrates—A Subset of the ICTV Universal Virus Taxonomy System, 2015   (Continued)

Family	Subfamily	Genus	Type Species	Viruses Infecting Humans
Rhabdoviridae		*Vesiculovirus*	*Vesicular stomatitis Indiana virus*	
		Lyssavirus	*Rabies virus*	Rabies virus; others
		Ephemerovirus	*Bovine ephemeral fever virus*	
		Novirhabdovirus	*Infectious haematopoietic necrosis virus*	
Filoviridae		*Marburgvirus*	*Lake Victoria marburgvirus*	Lake Victoria marburgvirus
		Ebolavirus	*Zaire ebolavirus*	Zaire; Taï Forest; Reston; Sudan ebolaviruses
Bornaviridae		*Bornavirus*	*Borna disease virus*	Borna disease virus
Orthomyxoviridae		*Influenzavirus A*	*Influenza A virus*	Influenza A virus
		Influenzavirus B	*Influenza B virus*	Influenza B virus
		Influenzavirus C	*Influenza C virus*	Influenza C virus
		Thogotovirus	*Thogoto virus*	Thogoto virus
		Isavirus	*Infectious salmon anaemia virus*	
Bunyaviridae		*Orthobunyavirus*	*Bunyamwera virus*	Bunyamwera virus; California encephalitis virus; Oropouche virus; others
		Hantavirus	*Hantaan virus*	Hantaan virus, Sin Nombre virus; others
		Nairovirus	*Dugbe virus*	Crimean-Congo haemorrhagic fever virus
		Phlebovirus	*Rift Valley fever virus*	Rift Valley fever virus; Sandfly fever Naples virus
Arenaviridae		*Arenavirus*	*Lymphocytic choriomeningitis virus*	Old World (Lassa virus); New World (Junin virus, Machupo virus, others)
Single-Stranded Positive-Sense RNA Viruses				
Coronaviridae	*Coronavirinae*	*Alphacoronavirus*	*Alphacoronavirus 1*	Human coronaviruses 229E and NL43
		Betacoronavirus	*Murine coronavirus*	Human coronavirus HKU1 SARS-related coronaviruses; MERS coronavirus
		Gammacoronavirus	*Avian coronavirus*	
		Torovirus	*Equine torovirus*	Human torovirus
Arteriviridae		*Arterivirus*	*Equine arteritis virus*	
Roniviridae		*Okavirus*	*Gill-associated virus*	
Picornaviridae		*Enterovirus*	*Human enterovirus C*	Human enteroviruses A–D (including polioviruses)
		Rhinovirus	*Human rhinovirus A*	Human rhinoviruses A–C (>100 serotypes)

TABLE 2.1 Major Families of Viruses Infecting Vertebrates—A Subset of the ICTV Universal Virus Taxonomy System, 2015 (Continued)

Family	Subfamily	Genus	Type Species	Viruses Infecting Humans
		Erebovirus	Equine rhinitis B virus	
		Hepatovirus	Hepatitis A virus	Hepatitis A virus
		Cardiovirus	Encephalomyocarditis virus	
		Aphthovirus	Foot-and-mouth disease virus	
		Parechovirus	Human parechovirus	Human parechoviruses 1–16
		Kobuvirus	Aichi virus	Aichi virus
		Teschovirus	Porcine teschovirus	
		Sapelovirus	Porcine sapelovirus	
		Senecavirus	Seneca Valley virus	
		Tremovirus	Avian encephalomyelitis virus	
		Avihepatovirus	Duck hepatitis A virus	
Caliciviridae		Vesivirus	Vesicular exanthema of swine virus	
		Lagovirus	Rabbit haemorrhagic disease virus	
		Norovirus	Norwalk virus	Norwalk viruses
		Sapovirus	Sapporo virus	Sapporo viruses
		Nebovirus	Newbury-1 virus	
Astroviridae		Mamastrovirus	Human astrovirus	Human astroviruses 1–8
		Avastrovirus	Turkey astrovirus	
Togaviridae		Alphavirus	Sindbis virus	Equine encephalitis viruses; Ross River and Barmah Forest viruses; Chikungunya virus; others
		Rubivirus	Rubella virus	Rubella virus
Flaviviridae		Flavivirus	Yellow fever virus	Yellow fever virus; Dengue viruses 1–4; Japanese encephalitis virus group; Tick-borne viruses; others
		Pestivirus	Bovine viral diarrhoea virus type 1	
		Hepacivirus	Hepatitis C virus	Hepatitis C virus, 7 genotypes
Hepeviridae		Orthohepevirus	Orthohepevirus A	Hepatitis E viruses 1–4
Unassigned				
(single-stranded circular RNA)		Deltavirus	Hepatitis delta virus	HDV 1–8

all of the viruses mentioned in this book have been placed within a family and assigned to a genus, although there are some "floating genera" where family construction is not yet complete. Subfamilies are used only where needed to deal with very complex interrelationships among the viruses within a particular family.

Virus families are broadly divisible into those with DNA or RNA genomes respectively. Viruses within each family possess broadly similar genome structure, virion morphology, and replication strategy. *Subfamilies* are distinguished in cases where some members of a family can be grouped as possessing distinct and unique properties.

Orders are used to group together those virus families with related but distant phylogenetic properties (e.g., conserved genes, sequences, or domains). Again, since all viruses did not derive from a common ancestor, there is no intent to construct a unified viral evolutionary tree.

Genera are used to bring together viruses with clear, important evolutionary, and biological relationships, which are also usually reflected in antigenic, host range, epidemiological, and/or other relationships.

Species is the most important taxon in the systems used to classify all life forms, but it is also the most difficult to both define and use—this is especially the case with regard to viruses. In recent years the ICTV has determined criteria for defining virus species—different criteria are being used for different families. After some controversy, the ICTV recently redefined the term species:

> A species is a monophyletic ("relating to or descended from one source or taxon") group of viruses, whose properties can be distinguished from those of other species by multiple criteria. The criteria by which different species within a genus are distinguished shall be established by the appropriate Study Group. These criteria may include, but are not limited to, natural and experimental host range, cell and tissue tropism, pathogenicity, vector specificity, antigenicity, and the degree of relatedness of their genomes or genes...

Below the species level, the identification of particular lineages within an individual virus species is often extremely important because of clinical, epidemiological, or evolutionary significance. Such lineages may be designated as serotypes, genotypes, subtypes, variants, escape mutants, vaccine strains, etc. Many different conventions are used for naming at this level, depending on the virus involved—these distinctions lie outside the remit of the ICTV.

Using the above taxonomic system brings a number of practical benefits, including (1) the ability to relate a newly found virus to similar agents that have already been described and thereby to anticipate some of its possible properties, and (2) the ability to infer possible evolutionary relationships between viruses. Even though there has been little disagreement over the use of this system at the order, family, or genus level, there has been considerable confusion

at the species level, partly based in misunderstanding over the difference between the man-made taxonomic construction, the species, and the actual entity, the virus. In this book formal ICTV taxonomy and nomenclature will be cited, but virus names will be in the English vernacular.

The discovery of mimiviruses (a virus infecting the protozoan *Acanthamoeba*) in the last decade has challenged the traditional concept of *virus*. The mimivirus genome is able to direct much more than the replication of its own DNA genome, coding as it does for a large number of proteins with functions resembling some eukaryotic proteins and a large number of proteins of unknown function. At the time of writing, no mimivirus-like agent causing human illness has yet been found; still the discovery of mimiviruses has had a profound influence on our understanding of virus evolution and on our sense of what is yet to be discovered. A full discussion regarding the origin of viruses is outside the scope of this book, suffice it to say that some virologists argue that RNA viruses have evolved many aeons before the appearance of DNA viruses.

VIRAL NOMENCLATURE

Formal Usage

In formal usage, the first letters of virus order, family, subfamily, genus, and species names are capitalized and the terms are printed in italics. Further words making up a species name are not further capitalized unless they are derived from a place name (e.g., the species *St. Louis encephalitis virus*). The first letter of the names of specific viruses having the status of tentative species is capitalized, but the names are not italicized. In formal usage, the identification of the taxon precedes the name; for example: "… the family *Paramyxoviridae*" or "… the genus *Morbillivirus*." The following are some illustrative examples of formal taxonomic usage:

> Family *Poxviridae*, subfamily *Chordopoxvirinae*, genus *Orthopoxvirus, Vaccinia virus*, vaccinia virus, strain New York Board of Health Laboratories (Wyeth calf-adapted) [the strain that was used to produce smallpox vaccine in the United States].
>
> Order *Herpesvirales*, family *Herpesviridae*, subfamily *Alphaherpesvirinae*, genus *Simplexvirus, Human herpesvirus 1*, herpes simplex virus 1, strain HF [a typical laboratory strain obtainable from the American Type Culture Collection].
>
> Order *Mononegavirales*, family *Rhabdoviridae*, genus *Lyssavirus, Rabies virus*, rabies virus, strain CVS 11 [the "challenge virus standard" used in laboratories throughout the world, with passage history back to Pasteur's laboratory].

Informal Usage

In informal vernacular usage, all terms are written in lower case script (except those derived directly from place names); these are not italicized, do not include the formal suffix, and the name of the taxon follows the name. For example, "…the picornavirus family," "…the enterovirus genus," "poliovirus 1."

One particular problem in vernacular nomenclature lies in the historic use of the same root terms in family and genus names—it is sometimes difficult to determine which level is being cited. For example, the vernacular name "bunyavirus" might refer to the family *Bunyaviridae*, to the genus *Orthobunyavirus*, or perhaps even to one particular species, *Bunyamwera virus*. The solution to this problem is to add an extra word to formally identify which taxon level is being referred to; for example, when referring vernacularly to Bunyamwera virus (capitalized, because the name derives from a place name), a full vernacular description would be "Bunyamwera virus, a member of the genus *Bunyavirus* in the family *Bunyaviridae*…" For each genus there is a type species assigned that creates a link between the genus and the species.

A second problem lies in what seems to be an arbitrary incorporation of the root term, "virus," in some virus names and its separation as a detached word in others. For example, poliovirus *vs.* measles virus. The basis for this lies in history—some, but not all, of the viruses isolated early on assumed the former name style, whereas most viruses discovered more recently have been identified using the latter style. In this book, we have tried to hold to the name style used most commonly for each virus, but since this is mostly a matter of vernacular usage the reader may often find variations.

GROUPINGS OF VIRUSES ON THE BASIS OF EPIDEMIOLOGICAL CRITERIA

There are other informal categories of viruses that are practical and in common usage, distinct from the formal universal taxonomic system and the formal and vernacular nomenclature. These are based upon virus tropism and modes of transmission. Most human pathogens are transmitted by either inhalation, ingestion, injection (including via arthropod bites), close contact (including sexual contact), or congenitally.

Enteric viruses are usually acquired by ingestion (fecal–oral transmission) and replicate primarily in the intestinal tract. The term is usually restricted to viruses that remain localized in the intestinal tract, rather than causing generalized infections. Enteric viruses are included in the families *Picornaviridae* (genus *Enterovirus*), *Caliciviridae*, *Astroviridae*, *Coronaviridae*, *Reoviridae* (genera *Rotavirus* and *Orthoreovirus*), *Parvoviridae*, and *Adenoviridae*.

Respiratory viruses are usually acquired by inhalation (respiratory transmission) or by fomites (inanimate objects carrying virus contagion) and replicate primarily in the respiratory tract. The term is usually restricted to viruses that remain localized in the respiratory tract, rather than causing generalized infections. Respiratory viruses are included in the families *Picornaviridae* (genus *Enterovirus*), *Caliciviridae*, *Coronaviridae*, *Paramyxoviridae* (genera *Respirovirus*, *Rubulavirus*, *Pneumovirus*, and *Metapneumovirus*), *Orthomyxoviridae*, and *Adenoviridae*.

Arboviruses (from "*ar*thropod-*bo*rne viruses") replicate in hematophagous (blood-feeding) arthropod hosts such as mosquitoes and ticks, and are then transmitted by bite to vertebrates, wherein the virus replicates and produces viremia of sufficient magnitude to infect other blood-feeding arthropods. In all cases, viruses replicate in the arthropod vector prior to further transmission: thus, the cycle is perpetuated. The occasional passive transfer of virus on contaminated mouthparts ("the flying pin") does not constitute sufficient grounds for a virus to be identified as an arbovirus. Arboviruses are included in the families *Togaviridae*, *Flaviviridae*, *Rhabdoviridae*, *Bunyaviridae*, and *Reoviridae* (genera *Orbivirus* and *Coltivirus*).

Blood-borne viruses are those that are typically transmitted by transfusion of blood or blood products, by sharing of intravenous injecting equipment, and by other mechanisms of parenteral transfer of blood or body fluids. Some are also transmitted by sexual contact (*sexually transmitted viruses*). This group includes hepatitis B, C, and D, HIV-1 and -2, HTLV-1 and -2, and other viruses can also be transmitted occasionally by this route.

Hepatitis viruses are grouped as such because the main target organ for these viruses is the liver. Hepatitis A, B, C, D, and E viruses each belong to completely unrelated taxonomic families.

Oncogenic viruses usually cause persistent infection and may produce transformation of host cells, which may in turn progress to malignancy. Viruses that have oncogenic potential, in experimental animals or in nature, are included in the families *Herpesviridae*, *Adenoviridae*, *Papillomaviridae*, *Polyomaviridae*, *Hepadnaviridae*, *Retroviridae*, and *Flaviviridae*.

TAXONOMY AND THE CAUSAL RELATIONSHIP BETWEEN VIRUS AND DISEASE

One of the landmarks in the history of infectious diseases was the development of the Henle–Koch postulates that established the evidence required to prove a causal relationship between a particular infectious agent and a particular disease. These simple postulates were originally drawn up for bacteria, but were revised in 1937 by Thomas Rivers and again in 1982 by Alfred Evans in attempts to accommodate the special problem of proving disease causation by viruses. In many cases, virologists have had to rely on indirect causal evidence, with associations based on epidemiology and patterns of antibody prevalence among populations. The framework of virus taxonomy, again, plays a role, especially in trying to distinguish an etiological, rather than coincidental or opportunistic relationship between a virus and a given disease. Particular difficulty arises where a disease occurs in only a small fraction of infected individuals, where the same apparent disease can be caused by more than one different agent, and in various chronic diseases and certain cancers. These difficulties are confounded in many instances where diseases cannot be reproduced by inoculation of experimental animals, or where the discovered viruses cannot be grown in animals or cell culture: there may even be a "hit and run" relationship where the causative virus may no longer be present in the afflicted individual. Thus scientists have to evaluate the probability of "guilt by association," a difficult procedure that relies heavily on epidemiological observations.

The Henle–Koch postulates were reworked again in 1996 by David Relman and David Fredricks as more and more genomic sequencing criteria came to dominate the subject (Table 2.2). As a test of the value of these criteria, one can consider the level of proof that the human immunodeficiency viruses, HIV-1 and HIV-2, are the etiological agents of human acquired immunodeficiency syndrome (AIDS) (Table 2.3). Early in the investigation of AIDS, before its etiology was established, many kinds of viruses were isolated from patients and many candidate etiological agents and other theories were advanced. Prediction that the etiological agent would turn out to be a member of the family *Retroviridae* was based upon years of research on animal retroviral diseases and many points of similarity with some characteristics of AIDS. Later, after human immunodeficiency virus 1 (HIV-1) was discovered, the morphological similarity of this virus to equine infectious anemia virus, a prototypic member of the genus *Lentivirus*, family *Retroviridae*, highlighted the usefulness of the universal viral taxonomic system and of animal lentiviruses as models for AIDS.

In other examples, the causal relationships of Epstein-Barr (EB) virus to the disease infectious mononucleosis, and of Australia antigen (later known as hepatitis B surface antigen) to clinical hepatitis, were each established by matching serological evidence of acute infection with the timing of onset of clinical disease. Further, the complex role of EB virus in Burkitt's lymphoma was investigated in a large prospective study carried out by the International Agency for Research on Cancer (IARC) on 45,000 children in an area of high incidence of Burkitt's lymphoma in Africa. This showed that:

1. EB virus infection preceded development of the tumors by 7 to 54 months;
2. exceptionally high EB virus antibody titers often preceded the appearance of tumors; and
3. antibody titers to other viruses were not elevated.

In addition, it was demonstrated that the EB virus genome is always present in the cells of Burkitt's lymphomas among

---

TABLE 2.2 Fredricks and Relman's Molecular Guidelines For Causal Association

1. Strength of the association. Are viral nucleic acid sequences detected in most (all) cases of disease?
2. Specificity of the association. Are viral nucleic acid sequences localized to diseased tissues, and not to healthy tissues? Is the frequency of virus infection reduced significantly in healthy individuals?
3. Response to treatment. Does the copy number of viral nucleic acid sequences fall with resolution of illness or effective treatment, and increase if the disease relapses?
4. Temporality. Does infection with the virus precede and predict disease onset?
5. Plausibility. Do the known biological properties of the virus make sense in terms of the disease?
6. Biological gradient. Is the amount of virus higher in patients with severe disease than it is in persons with mild disease? Is the amount of virus higher in diseased tissues than in healthy tissues?
7. Consistency. Are these findings reproducible by multiple laboratories and by multiple investigators?

TABLE 2.3 Application of Fredricks and Relman's Guidelines to the Cause of Acquired Immunodeficiency Syndrome (AIDS)

1. Strength of the association. Infection with HIV is found in almost all cases that fit a clinical definition of AIDS.
2. Specificity of the association. Human immunodeficiency viruses are found preferentially in target organs (immune cells, lymphoid tissues). HIV infection is not found in healthy individuals, except for those who subsequently develop AIDS or those rare individuals considered long-term non-progressors.
3. Response to treatment. Combination therapy against HIV lowers or completely eradicates circulating virus, resulting in increased CD4 cells, improved immune function, and significant long-lasting clinical improvement.
4. Temporality. HIV infection precedes and predicts disease onset in children born to infected mothers, in medical personnel infected via needle-stick accidents, and in recipients of blood transfusions from infected persons.
5. Plausibility. HIV infects and kills CD4+ T cells and macrophages. SIV causes AIDS in experimentally inoculated macaques.
6. Biological gradient. HIV-1 RNA load is highest in lymphoid tissues and brain (diseased tissues). HIV-1 RNA load predicts the rate of disease progression.
7. Consistency. These findings are consistently reproducible, worldwide.

African children, and that a malignant lymphoma can be induced in certain primates with EB virus or EB virus-infected lymphocytes (see Chapter 9: Mechanisms of Viral Oncogenesis and Chapter 17: Herpesviruses).

Using a similar approach, Palmer Beasley and coworkers in Taiwan demonstrated unequivocally that persistent hepatitis B infection increased the subsequent risk of primary liver cancer, but not other cancers, by approximately 100-fold.

These studies are examples of important concepts now widely understood in situations where a virus has been shown to cause a specific disease, namely that not all cases of the infection may necessarily develop the clinical disease, and not all cases of the clinical disease may be caused by the particular virus in question. Thus, for many associations between a virus and a clinical disease, the concept of infection representing a "risk factor" is more appropriate than it being an absolute "cause." It also now happens frequently using modern diagnostic methods, that viruses are recovered from individuals with some ongoing disease; however, careful work is essential in such cases to distinguish a true causative role from an unrelated infection of no clinical significance occurring at the same time.

GENOME SEQUENCING AND VIRUS EVOLUTION

The breath-taking advances in genome sequencing now enable the complete genomes of many hundreds of virus isolates to be sequenced in a matter of days, if not hours. Multiple sequence alignment and the construction of phylogenetic trees are now commonplace when virologists are confronted with either a potentially new virus or an isolate with new or unexpected properties. These data are rapidly challenging previous ideas about the origin and evolution of many viruses of medical importance. Detailed phylogenetic analysis of RNA viruses in particular sometimes provides unexpected answers that in turn create more questions; for example, hepadnaviruses share a similar reverse transcriptase-based replication strategy that is common to the caulimoviruses of plants—does this reflect a common ancestor or convergent evolution?

Deep evolutionary relationships among the higher virus taxa have led to the construction of several Orders—the *Herpesvirales*, *Mononegavirales*, *Nidovirales*, and *Picornavirales*. The common conserved sequences employed here are at the lower limit of significance, but similarities in some functional and structural protein domains still appear among otherwise unrelated viruses in various taxa. Sequence analyses also suggest that it is unlikely that many more associations of diverse taxa will be found that warrant construction of further Orders.

At the other extreme, namely clarifying the phylogenetic relationships among viruses in the same taxa (i.e., families or genera), great progress is being made continually. For example, the origin of the 2009 influenza (H1N1) pandemic has been found to be complex indeed: the virus is a reassortant with genes from four different ancestral viruses—North American swine influenza, North American avian influenza, human influenza, and swine influenza virus typically found in Asia and Europe. Similarly, some member viruses of the family *Bunyaviridae* have been found to be natural reassortants with genes from known and unknown ancestors.

Thus, the development of a robust, yet flexible and continually evolving taxonomic system for viruses underpins, and gives structure to, all facets of research, management, and control of virus diseases.

FURTHER READING

Adams, M.J., Lefkowitz, E.J., King, A.M.Q., Carstens, E.B., 2013. Recently agreed changes to the International Code of Virus Classication and Nomenclature. Arch. Virol. Available from: http://dx.doi.org/10.1007/s00705-013-1749-9.

Anthony, S.J., Epstein, J.H., Murray, K.A., et al., 2013. A strategy to estimate unknown viral diversity in mammals. mBio., 4. Available from: http://dx.doi.org/10.1128/mBio.00598-13.

Fredricks, D.N., Relman, D.A., 1996. Sequence based identification of microbial pathogens: a reconsideration of Koch's postulates. Clin. Microbiol. Rev. 9, 18–33.

Holmes, E.C., 2009. The Evolution and Emergence of RNA Viruses. Oxford University Press, Oxford, ISBN 978-0-19-921112.

Virus taxonomy, classification and nomenclature of viruses. King, A.M.Q., Adams, M.J., Carstens, E.B., Lefkowitz, E.J. (Eds.), 2011. Ninth Report of the International Committee for the Taxonomy of Viruses, Academic Press, London, ISBN 978-0-12-384684-6.

Krupovic, M., Bamford, D.H., 2010. Order to the viral universe. J. Virol. 84, 12476–12479. Available from: http://dx.doi.org/10.1128/JVI.01489-10.

Chapter 3

Virion Structure and Composition

The virus particle or *virion* represents a virus in its *extracellular* phase, in contrast to the different *intracellular* structures involved in virus replication. To ensure survival of a virus, the virion must fulfill two roles: (1) protecting the genome from environmental damage, for example, from heat, desiccation, chemicals; and (2) facilitating the passage of the virus to the next host, that is, from the point of release from the original host, passage through the environment to the point of encountering a new host, followed by entry into the cells of the new host. There are many different ways that different viruses achieve these two roles, and viral genomes and virion structures show enormous variety in both size and composition—yet there are many features or principles of assembly that are shared by most viruses. Notably, many key structures within the virion are assemblies and subassemblies of a large number (usually hundreds) of identical protein subunits that lock together sterically to form a stable shell (capsid or envelope); the employment of large numbers of one or a few different primary units (structural units, capsomeres) allows the genetic coding of relatively large macromolecules by a very small number of different viral genes.

PHYSICAL METHODS FOR STUDYING VIRUS STRUCTURE

Electron Microscopy

The development of electron microscopy was pivotal in the establishment of virology as a scientific discipline. Viruses are smaller than the limit of resolution of the light microscope, which is about 0.3 μm, or 300 nm. Poxviruses were for many years considered amongst the largest of viruses, being just about visible in the light microscope using dark-field optics or certain staining techniques. In recent years, much larger viruses have been discovered but so far none of these have been shown to be human pathogens. Nevertheless, these newly discovered viruses are driving a re-examination of the limits of the concept of *virus*. For example, the virions of megavirus and mimivirus, both infectious for *Acanthamoeba*, are about 0.5 μm in diameter, have a DNA genome up to 1.26 Mb in size and

code up to 1120 proteins. The virions of pandoraviruses are 1–1.2 μm in size, have a linear DNA genome up to 2.8 Mb, and code up to 2500 proteins. An even larger virus, Pithovirus (recently isolated from Siberian permafrost), is 1.5 μm in size, the size of a small bacterium, with >2500 putative protein-coding sequences, of which only 6% have recognizable relationships with genes from known viruses, microorganisms, or eukaryotes. At the same time, viruses that are smaller than those in previously known taxa have been found, so virologists must now be prepared to work with viruses as large as bacteria and as small as large protein molecules (Table 3.1).

The first electron microscopy of viruses by Bodo von Borries, Ernst Ruska, and Helmut Ruska in 1938 employed a simple preparative method that did not show much more than the outline of virions. Later, metal shadow-casting of purified virus preparations improved the visualization of virions, but still not enough. Beginning in the 1950s, ultra-thin sectioning of virus-infected cells became widespread, providing more virion detail and also the beginning of the science of ultrastructural cytopathology—virion morphogenesis, intracellular localization of virus structures and cellular organelles and coincident damage to host cell structures. In 1959, visualization of viral ultrastructure was taken to a still higher level of resolution when Sydney Brenner and Robert Horne developed negative staining. In this method, a solution of potassium phosphotungstate (or other electron-dense salts) is added to a virus suspension on a coated specimen grid; the metal ions surround and fill the interstices in the surface of virions giving a negative image in the electron beam, thus revealing structural details not previously seen. A remarkable diversity of virus structures can be seen in negatively stained preparations. Electron micrographs of virions and infected cells from different families of viruses are shown in the various chapters of this book (see Figs. 1.3 and 1.4 for comparison of ultra-thin sectioning versus negative staining).

In the past few years, the above methods have been complemented by several new microscopy technologies, particularly scanning electron microscopy and cryo-electron microscopy, the latter using computer-based image construction of images of snap frozen, unstained virion

TABLE 3.1 Relative Sizes of Common Objects in the Biological World

Size	Equivalent Practical Units	Observations
1 m	3 ft 3 in.	Humans, adult males, are about 1.8 m tall
10 cm	4 in.	Human adult hand is about 10 cm wide
1 cm	1 cm	*Aedes aegypti*, adult, mosquito is about 1 cm long
1 mm	1 mm	*Ixodes scapularis* tick, nymphal stage, is about 1 mm long
0.1 mm	100 µm	Smallest things visible to the naked eye
0.01 mm	10 µm	Lymphocytes are about 20 µm in diameter
		Bacillus anthracis, among the largest of pathogenic bacteria, is 1 µm wide and 5 to 10 µm long
0.001 mm	1 µm	Smallest things visible in light microscope are about 0.3 µm in size
		Poxviruses, the largest of the viruses of vertebrates, are 300 nm (or 0.3 µm) in their longest dimension
0.1 µm	100 nm	Influenza viruses and retroviruses, typical medium-sized viruses, are about 100 nm in diameter
	100 nm	Flaviviruses, such as yellow fever virus, typical smaller-sized viruses, are about 50 nm in diameter
0.01 µm	10 nm	Picornaviruses, such as polioviruses, typical small viruses, are about 30 nm in diameter
		Circoviruses, the smallest of the viruses of vertebrates, are 17 to 22 nm in diameter
0.001 µm	1 nm, 10 Å	Smallest things visible in transmission electron microscope; DNA double helix diameter is 2 nm
0.1 nm	1 Å	Diameter of atoms is about 2 to 3 Å

preparations. This technique has the advantage of showing viruses in a hydrated state rather than the desiccated conditions of negative staining in electron microscopy. The resolution of these techniques is rapidly approaching that obtained by X-ray diffraction of crystallized virions and viral substructures (Fig. 3.1).

X-Ray Crystallography of Viruses

X-ray crystallography of viruses provides another technique for visualizing virion structural organization showing structural details to near atomic resolution, the location of antigenic sites on the surface of virions, and aspects of virion attachment and penetration into cells. For example, applying this technique to several picornaviruses revealed that the polypeptides of each of the three larger structural proteins are packaged to form wedge-shaped eight-stranded antiparallel β-barrel subassemblies (Fig. 3.2). The overall contour of picornavirus virions reflects the packing of these subassemblies. Relatively unstructured amino acid chains form loops that project from the main wedge-shaped domains. Some loops form flexible arms that interlock with

the arms of adjacent wedge-shaped subassemblies, thereby providing physical stability to the virion. Other loops, those involved in virion attachment to the host cell, harbor the antigenic sites (epitopes) that are the targets of the host's neutralizing antibody response against the virus.

Larger viruses are much more complex in structure, and to study the structure of these viruses it is usually necessary to separate well-defined substructures and examine crystals of these structures by X-ray diffraction. Rotaviruses are an excellent example of this approach, being composed of a core and two capsid layers, each component exhibiting unique structural details, fitting together in a precise fashion to form the complete virion.

One of the pioneering studies of viral structure was the determination by X-ray crystallography of the structure of the hemagglutinin molecule of influenza viruses and the placement and variation of neutralizing epitopes on this molecule. Today, determination of new variations in the amino acid sequence and hence the microstructure of the influenza hemagglutinin is used in the development of updated vaccines. Many individual viral proteins have been analyzed to the level of 2–3 Å resolution, revealing

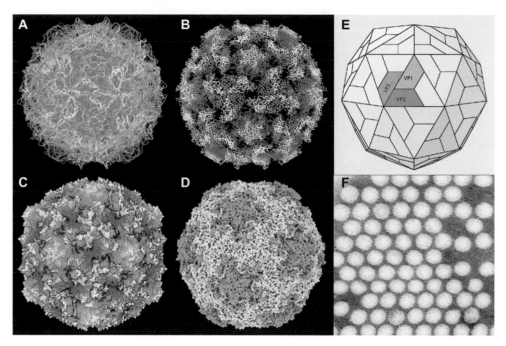

FIGURE 3.1 Picornavirus structural studies. (A–D) Computer-based virion reconstructions from cryo-electron microscopy and X-ray diffraction images. (A) Poliovirus 1, protein chain model. *From Jason Roberts and colleagues, with permission.* (B) Poliovirus 1, string model. (C) Coxsackie B3 virus, space-filling model. (D) Human rhinovirus B14, Qutemol rendering model. *(B–D) from Jean-Yves Sgro, with permission.* (E) Model of poliovirus icosahedral capsid showing location of the three proteins making up the capsid surface, VP1, VP2 and VP3 (VP4 is buried on the inner face of the capsid). (F) Poliovirus, negative contrast electron microscopy from Joseph Esposito (deceased), showing little or no virion surface detail—such was the resolution available before X-ray diffraction and cryo-electron microscopy technologies, both of which require massive computer compilation to reconstruct image data.

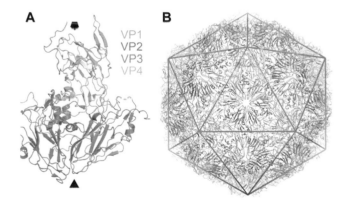

FIGURE 3.2 Structure of a typical picornavirus. (A) X-ray crystallographic structure of a native virus particle. A single substructural unit (protomer) is shown in a string-ribbon representation. The viral proteins are colored blue (VP1), green (VP2), magenta (VP3), and orange (VP4), and the approximate positions of the viral 5- and 3-fold axes are indicated by the solid black pentagon and triangle, respectively. (B) The intact, T = 3 capsid structure, again in a string-ribbon representation. The unit shown in panel (A) is highlighted in yellow, and the layout of the virion icosahedron is included as a red line overlay. *Adapted from Bakker, S.E., Groppelli, E., Pearson, A.R., Stockley, P.G., Rowlands, D.J., Ranson, N.A., 2014. Limits of structural plasticity in a picornavirus capsid revealed by a massively expanded equine rhinitis A virus particle. J. Virol. 88: 6093–6099.*

potential targets for new antiviral compounds. Notable is the development of antiviral drugs for the treatment of influenza through analysis of the detailed structure of the viral neuraminidase (Fig. 3.3).

CHEMICAL COMPOSITION OF VIRIONS

Viruses are distinguished from other macromolecular forms by a possessing rather simple, repetitive chemical composition. The *virion*, that is the complete infectious virus particle, includes a genome comprising one or a few molecules of either DNA or RNA, surrounded by a morphologically defined protein coat, the *capsid*; the capsid and the enclosed nucleic acid together constitute the *nucleocapsid*. A small number of additional proteins may be present within the virion as enzymes. The nucleocapsid of some viruses is surrounded by a lipoprotein bilayer *envelope* into which are inserted viral proteins (*peplomers*): these may or may not be glycosylated. Sometimes a matrix protein is also associated with the inner aspect of the viral envelope. The simplest virus (e.g., tobacco necrosis virus satellite, a defective virus that needs a helper virus to

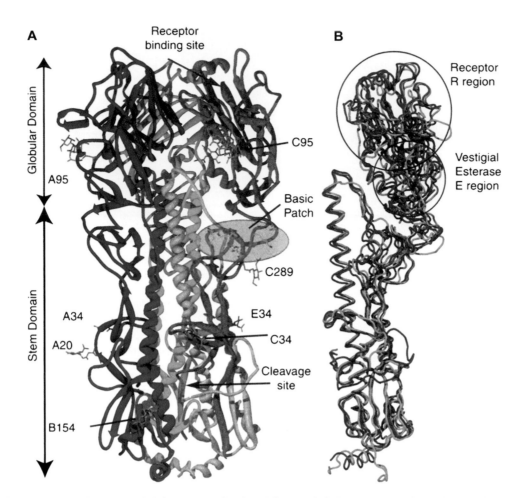

FIGURE 3.3 Crystal structure of the hemagglutinin (HA) protein of the influenza 1918 virus, and comparison with other human, avian, and swine HAs. (A) Overview of the HA0 trimer, represented as a ribbon diagram. Each of the three monomers is colored differently. Carbohydrates are colored orange and labeled with the asparagine to which each is attached. The basic patch is indicated in the light blue ellipse and consists of HAl residues. The locations of the three receptor binding sites, and cleavage sites, are shown for only one of the three monomers. (B) Structural comparison of different HA0 monomers, showing influenza 1918 HA0 (red), human H3 (green), avian H5 (orange), and swine H9 (blue). *Reproduced from MacLachlan, N.J., Dubovi, E.J., 2011. Veterinary Virology, fourth ed. Academic Press, London (Fig. 1.4), with permission.*

provide some of its functions) directs the synthesis of only one protein; many important viruses direct the synthesis of five to ten proteins; large viruses, such as the poxviruses and herpesviruses, direct the synthesis of up to 200 proteins and the recently discovered megaviruses up to 2500 proteins: still this is very few relative to the number of proteins involved in the life processes of bacteria (>5000 proteins) and eukaryotic cells (between 250,000 and 1,000,000 proteins). There are many variations on these constructions and diverse additional components are found among larger and more complex viruses (Fig. 3.4).

Viral Nucleic Acids

Viral genes are encoded in either DNA or RNA molecules; both DNA and RNA genomes can be either *double-stranded* or *single-stranded*, as well as *monopartite* (all viral genes contained in a single molecule of nucleic acid) or *multipartite*

(*segmented*) (viral genes distributed in multiple molecules or segments of nucleic acid). For example, among the RNA viruses, only viruses of the families *Reoviridae*, *Birnaviridae*, and *Picobirnaviridae* have a double-stranded RNA genome and these genomes are segmented (*Reoviridae*: 10, 11, or 12 segments, depending on the genus; *Birnaviridae* and *Picobirnaviridae*: two segments). All viral genomes are haploid, that is, they contain only one copy of each gene, except for retrovirus genomes, which are diploid. When carefully extracted from the virion, the nucleic acid of viruses of certain families of both DNA and RNA viruses is directly infectious; that is, when transfected into a cell there is sufficient genetic information to initiate a complete cycle of viral replication and produce a normal yield of progeny virions.

The sequence in which the various virus families are described in Part II of this book reflects the essential characters and diversity of viral genomes. The remarkable variety of

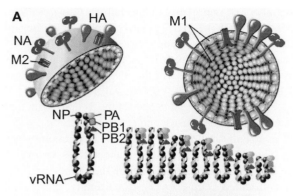

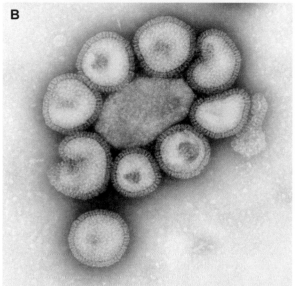

FIGURE 3.4 Structure of influenza A virions. (A) There are two major glycoproteins embedded in the lipid envelope, viz. the trimeric hemagglutinin (HA) which predominates, and the tetrameric neuraminidase (NA). The envelope also contains a small number of M2 membrane ion channel proteins. Inside the envelope lies the matrix protein (M1); the viral ribonucleoprotein which consists of the RNA genome in segments, each segment associated with nucleocapsid protein molecules; and the PA, PB1, and PB2 polymerase proteins. (B) Negative contrast EM of influenza A particles. *(A) Reproduced from MacLachlan, N.J., Dubovi, E.J., 2011. Veterinary Virology, fourth ed. Academic Press, London, Fig. 21.3, with permission.*

viruses is reflected in diverse ways in which the information encoded in the viral genome is transcribed into mRNA, then translated into proteins, and the ways in which the viral nucleic acid is replicated (see Chapter 4: Virus Replication).

Viral DNA Genomes

The genomes of all DNA viruses of vertebrates are monopartite, consisting of a single molecule that is, double-stranded except in the case of the parvoviruses, anelloviruses, and circoviruses. DNA genomes may be linear or circular, depending on the virus family. The DNA of papillomaviruses, polyomaviruses, hepadnaviruses, anelloviruses, and circoviruses is circular. Additionally, the circular DNA of the papillomaviruses and polyomaviruses is supercoiled. The DNA genome of hepadnaviruses is partially double-stranded, although the single-stranded gap is closed during replication to form a covalently closed circular DNA (cccDNA: see Chapter 22: Hepadnaviruses and Hepatitis Delta).

Most linear viral DNAs have characteristics that can facilitate a circular configuration, a requirement for replication by a rolling circle mechanism (see Chapter 4: Virus Replication). The two strands of poxvirus DNA are covalently cross-linked at each terminus (forming *hairpin ends*), so that upon denaturation, the molecule becomes a large single-stranded circle. The linear double-stranded DNA of several DNA viruses and the linear single-stranded RNA of retroviruses contain repeat sequences at the ends of the molecule that permit circularization. Adenovirus DNA contains inverted terminal repeats; these are also a feature of the single-stranded DNA of parvoviruses. Another type of terminal structure occurs in adenoviruses, hepadnaviruses, and parvoviruses (and some single-stranded RNA viruses such as the picornaviruses and caliciviruses); all of these viruses contain a protein covalently linked to the 5′-terminus (the 5′ cap) with an essential function in priming replication of the genome.

The size of viral DNA genomes ranges from 1.7 kilobases (kb) for some circoviruses to over 200 kilobase pairs (kbp) for the double-stranded DNA of herpesviruses and poxviruses—and up to 2.8 mbp for the double-stranded DNA of the giant pandoraviruses and pithovirus of amoeba. As 1 kb, or for double-stranded DNA 1 kbp, contains enough genetic information to code for about one average-sized protein, it might be surmised that viral DNAs contain anywhere between two and 200 genes, coding for some two to 200 proteins in poxviruses and more than 2500 proteins in the giant viruses of amoeba. However, the relationship between any particular nucleotide sequence and its protein product(s) is not straightforward. On the one hand, the DNA of most of the larger viruses, similar to that of mammalian cells, contains what appears to be redundant information in the form of repeat sequences, thus the coding capacity of large viral genomes may be overestimated. On the other hand, coding capacity might be underestimated: first, a given DNA or mRNA sequence may be read in up to three alternate reading frames, producing up to three proteins with different amino acid sequences; second, both strands of a double-stranded viral DNA molecule may be transcribed, each transcript yielding a different protein; third, genes may overlap, yielding various transcripts and protein products; and finally, a single primary RNA transcript may be spliced or cleaved in several different ways to yield a number of distinct mRNAs, each of which may be translated into a different protein, or a single polyprotein translation product may be subsequently cleaved by proteolysis to yield multiple discrete proteins.

Viral DNAs contain several kinds of non-coding sequences essential for genome expression and replication, some of which have been conserved throughout evolutionary time; these include DNA replication initiation sites, RNA polymerase recognition sites, sites for the initiation and termination of translation, RNA splice sites, promoters, enhancers, etc.

Viral RNA Genomes

With the exception of the reoviruses, birnaviruses, and picobirnaviruses the genomes of all vertebrate RNA viruses are single-stranded. The RNA may be monopartite or multipartite: for example, the retroviruses, paramyxoviruses, rhabdoviruses, filoviruses, coronaviruses, arteriviruses, picornaviruses, togaviruses, and flaviviruses, all have monopartite genomes. In contrast, the orthomyxoviruses, bunyaviruses, and arenaviruses have multipartite genomes. The genomes of the arenaviruses consist of two segments; the bunyaviruses three; the orthomyxoviruses six, seven, or eight (depending on the genus), the birnaviruses and picobirnaviruses two; and the reoviruses 10 to 12, again depending on the genus. Each RNA molecule in these viruses is unique, often encoding a single protein. There is no vertebrate virus RNA genome that is a covalently linked circle, with the exception of the very small circular single-stranded RNA of delta hepatitis virus which has a structure resembling that of plant viroids. However, the single-stranded RNAs of the bunyaviruses and arenaviruses appear circular owing to hydrogen bonding between complementary ends. The genomes of single-stranded RNA viruses have considerable secondary structure, regions of base-pairing causing the formation of loops, hairpins, etc., that serve as signals controlling nucleic acid replication, transcription (especially initiation and termination), translation, and/or packaging into the capsid.

Single-stranded RNA genomes can be defined according to coding *sense* (also called *polarity*). If the genomic RNA is of the same sense as mRNA, that is, it can direct the synthesis of protein, it is said to be of *positive-sense* (also called *plus sense*). This is the case with the picornaviruses, caliciviruses, togaviruses, flaviviruses, coronaviruses, and retroviruses. If, on the other hand, the genomic nucleotide sequence is complementary to that of mRNA, it is said to be *negative-sense*. Such is the case with the paramyxoviruses, rhabdoviruses, filoviruses, bornaviruses, orthomyxoviruses, arenaviruses, and bunyaviruses. All RNA virus genomes carry a gene for an *RNA-dependent RNA polymerase* (*transcriptase*), with the exception of retroviruses and hepadnaviruses which encode an *RNA-dependent DNA polymerase*. This enzyme directs the synthesis of positive-sense RNA in the infected cell using the viral genome as template. Mammalian cells do not contain an RNA-dependent RNA polymerase, and therefore this virally encoded non-structural protein has to be synthesized early in infection before rounds of replication can be completed.

Both RNA segments of the member viruses of the family *Arenaviridae* and one of the three RNA segments of members of one genus of the family *Bunyaviridae* are *ambisense*, that is, part positive-sense, part negative-sense (Chapter 29: Bunyaviruses and Chapter 30: Arenaviruses). If the viral RNA is of a positive-sense, it is usually polyadenylated at the 3′-end (in picornaviruses, caliciviruses, togaviruses, coronaviruses, and arteriviruses, but not in flaviviruses) and capped at the 5′-end (togaviruses, flaviviruses, coronaviruses, and arteriviruses).

The size of single-stranded RNA viral genomes varies from 1.7 to 32 kb and the double-stranded RNA viruses from 18 to 27 kbp—a much smaller range than found among the double-stranded DNA viruses. Accordingly, these viruses encode fewer proteins than many DNA viruses, generally less than a dozen. Most of the segments of the genomes of orthomyxoviruses and reoviruses are individual genes, each coding for one unique protein.

Anomalous Inclusion of Nucleic Acids Within Virions

Viral preparations often contain some particles with an atypical content of nucleic acid. Several copies of the complete viral genome may be enclosed within a single virion, or virions may be formed that contain no nucleic acid (empty particles) or that have an incomplete genome (*defective particles* and in some cases *defective interfering particles* by virtue of modulating replication). Moreover, host cell DNA may sometimes be incorporated into virions (for example, in papillomaviruses and polyomaviruses), while ribosomal RNA is found in orthomyxovirus and arenavirus virions. However, there is no evidence that ribosomal RNA within virus particles has any function.

Viral Proteins

Some virus-coded proteins are *structural*, that is, they are used to construct the capsid, envelope, and other components of the virion. Other proteins are *non-structural*; these are not present in the virion but are involved in various viral replication processes including virion assembly. Many non-structural proteins are enzymes, which may be involved in nucleic acid replication, transcription, and translation, as well as the shutdown of host cell functions, the inhibition of innate immunity, and the subversion of cellular machinery to viral synthetic activities. There are many kinds of viral enzymes described in this book; among these are various types of (1) *replicases* (also called *polymerases*; e.g., DNA-dependent DNA replicase or polymerase) and other enzymes involved in viral nucleic acid replication, (2) *transcriptases* that transcribe mRNA from viral DNA or RNA genomes, and (3) various *proteases*, *helicases*, and *ligases*. *Reverse transcriptase*, an enzyme that transcribes DNA from an RNA template, is found uniquely in retroviruses and hepadnaviruses. Other enzymes found only in retroviruses are involved in the integration of the DNA product of reverse transcription into cellular chromosomal DNA (*integrase*). Poxviruses, which replicate in the cytoplasm and therefore have less access to cellular machinery, duplicate a number of host enzymes that function in processing RNA transcripts and replicating the DNA genome.

Capsid Structure

The capsid is built up of identical, non-covalently linked *structural subunits*, which may be discernible by electron microscopy as protrusions or depressions on the surface of virions (Fig. 3.4). These may correspond to individual viral proteins or aggregates of proteins. Only the simplest virions are assembled from the primary products of protein synthesis; that is, individual polypeptides; in most cases virion capsids are constructed from distinct *assembly units* ("capsomeres") of several components, themselves often derived by modification or cleavage of precursors. Each subassembly may contain more than one polypeptide. One crucial step in virion assembly is the incorporation of the viral nucleic acid into the nascent virion—several different mechanisms driving this process have been recognized including the presence of *packaging signals* within the nucleic acid sequence of the genome.

Viral Envelope Lipids

In most cases the integrity of the envelope is necessary for viral infectivity. The envelope is acquired when the nucleocapsid is extruded through one of the cellular membranes—this process is known as *budding*. This process is not limited to the outer cell membrane: many viruses mature on internal, cytoplasmic membranes or even bud through the host nuclear membrane. The lipids of the viral envelope are derived directly from the cellular membrane, but the major proteins associated with the envelope are virus-coded.

Most lipids found in enveloped viruses are constructed into a typical lipid bilayer, in which the virus-coded glycoprotein peplomers and in some cases other viral proteins are embedded. As a consequence, the composition of the lipids of particular viruses differs according to the composition of the membrane lipids of the host cells in which replication takes place. The composition of the membrane lipids of viruses also varies with the particular membrane system employed for virion budding. For example, the lipids of paramyxoviruses, which bud from the plasma membrane of host cells, differ from those of bunyaviruses and coronaviruses that bud through the membranes of intracytoplasmic organelles. Lipids constitute about 20 to 35% of the dry weight of most enveloped viruses; some 50 to 60% of viral envelope lipid is phospholipid and most of the remainder consists of cholesterol.

The envelopes of some viruses also contain a variable content of host proteins, for example, MHC molecules, which may be retained after the virus buds through the cellular membrane. On occasion, the presence of host antigenic material on the viral surface can cause unexpected effects on the pathogenesis of disease and in diagnostic tests.

Viral Envelope Proteins

Most external viral envelope proteins are glycoproteins, occurring as membrane-anchored *peplomers* (*peplos, Gk= a loose outer garment*) or spikes, often assembled as dimers or trimers. These can be seen in electron micrographs extending outward from the envelope of enveloped viruses such as orthomyxoviruses, paramyxoviruses, rhabdoviruses, filoviruses, coronaviruses, bunyaviruses, arenaviruses, and retroviruses. However, the virions of some of the more complex viruses also contain glycosylated internal or outer capsid proteins. Recognition sites for cellular receptors are often located at the furthest domain from the viral envelope (distal end) whereas proximal domains interact with the lipid bilayer of the envelope (Fig. 3.5). Oligosaccharide side-chains (glycans) are attached by *N*-glycosidic, or more rarely *O*-glycosidic, linkages. Since these are synthesized by cellular glycosyl transferases, the sugar composition of these glycans is analogous to that of host cell membrane glycoproteins.

In contrast, *matrix proteins* are non-glycosylated and are found as a layer on the inside of the envelope of orthomyxoviruses, paramyxoviruses, rhabdoviruses, filoviruses, and retroviruses, but not coronaviruses, bunyaviruses, and arenaviruses. The presence of a matrix protein provides structural rigidity to a virion; for example, the helical nucleocapsid of rhabdoviruses is closely apposed to a rather rigid layer of matrix protein, and this in turn is tightly bound to the viral envelope by hydrogen bonding to the internal domains of the surface glycoprotein peplomers.

VIRION SYMMETRY

Individual protein structures are asymmetric. As discussed above, virions are assembled from multiple copies of one or a few kinds of protein subunits—the repeated occurrence of similar protein–protein interfaces leads to assembly of the subunits into symmetrical capsids. Folded polypeptide chains, specified by the viral genome, comprise *protein subunits*; assemblages of these protein subunits comprise *structural units* and in turn sets of these structural units comprise *assembly units* (capsomeres) that are the major intermediates in the formation of viral capsids. This efficiency of design depends upon principles of *self-assembly*, wherein structural units and capsomeres are brought into clusters through random thermal movement and are held in place through weak electrostatic bonds.

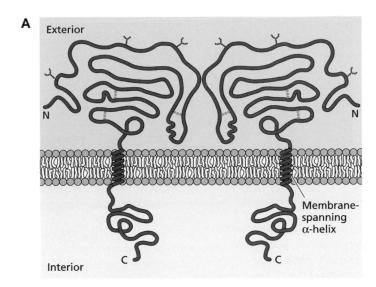

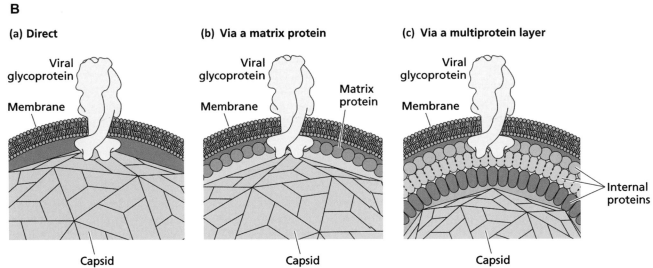

FIGURE 3.5 Glycoproteins of enveloped viruses. (A) Structural and chemical features of a typical viral glycoprotein. The protein is inserted through the lipid bilayer by a single membrane-spanning domain (transmembrane domain, TMD). There is a large external domain decorated with *N*-linked oligosaccharides, and a small internal domain involved in interactions with internal viral components. (B) Three modes of interaction between the internal domains of viral glycoproteins and viral capsids or nucleocapsids. *Reproduced from Flint, S.J., et al., 2009, Principles of Virology, third ed. ASM Press, Washington, DC, Figs. 4.20 and 4.21, with permission.*

Viruses come in a variety of shapes and sizes, depending on the shape, size, and number of protein subunits and the nature of the interfaces between these subunits; however, only two kinds of symmetry have been recognized, *icosahedral* and *helical*. Asymmetrical virus structure, such as that of poxviruses, is said to be *complex*.

Icosahedral Symmetry

In the mid-1950s Francis Crick and James Watson argued, using theoretical principles, that small spherical viruses were constructed out of identical subunits in identical environments as in true crystals. However, the maximum number of subunits that can be arranged in a sphere is 60—when more than 60 identical subunits are assembled an icosahedron is formed. The icosahedron is the optimum solution to the problem of constructing, from repeating subunits, a strong structure enclosing a maximum volume with the minimum amount of energy. The icosahedron is one of the five classical *Platonic* or *cubic solids*; the icosahedron, that is, the form with *icosahedral symmetry*, has 12 vertices (corners), 30 edges, and 20 faces, each face describing an equilateral triangle. Icosahedra have axes of two-, three-, and fivefold rotational symmetry that pass through the edges, faces, and vertices of the icosahedron, respectively (Fig. 3.6). Evidence from electron microscopy and biophysical

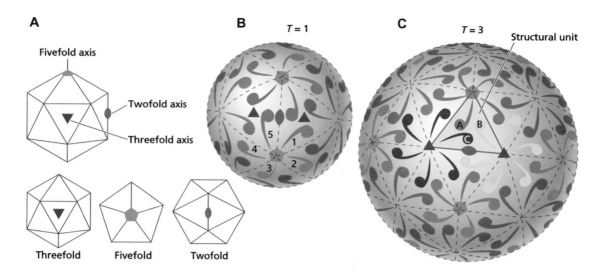

FIGURE 3.6 Icosahedral structures. (A) Different views of the icosahedron, showing threefold, fivefold, and twofold symmetry. (B) Simplest structure, where a single protein molecule (shown by a comma), forms the basic structural unit of each face. Thus there are 60 protein molecules per virion, and T, the triangulation number (defined as the number of structural units per face), is equal to 1. (C) $T=3$ structure, where there are three protein molecules per face and 180 molecules per virion. Note that the structural unit (outlined in blue and comprising one molecule each of proteins A, B, and C), is now an asymmetrical structure. *Reproduced from Flint, S.J., et al., 2009, Principles of Virology, third ed. ASM Press, Washington, DC, Fig. 4.8, with permission.*

techniques confirmed that most spherical viruses had far more than 60 protein subunits. Drawing inspiration from Buckminster Fuller's geodesic domes, Donald Caspar and Aaron Klug discovered a solution to the problem—they proposed that spherical viruses were structured like miniature geodesic domes, that is, the identical subunits are asymmetrical and are bonded together not in equivalent, but in quasiequivalent ways wherein there can be slight deformation of optimum bonding distances and angles wherein far more subunits can be accommodated to make larger structures. They also developed the idea of self-assembly after considering the viral assembly process as a type of crystallization process.

Only certain arrangements of structural units can form the faces, edges, and vertices of viral icosahedra. The structural units on the faces and edges of adenovirus virions, for example, bond to six neighboring capsomers and are referred to as *hexons* or *hexamers*; those at the vertices bond to five neighbors and are called *pentons* or *pentamers* (Fig. 3.6). In the virions of some viruses both hexons and pentons are composed of the same polypeptide(s), whereas the hexons or pentons of other viruses are formed from different polypeptides. The arrangements of structural units on the surface of virions of two icosahedral model viruses are shown in Fig. 3.6. Because of variations in the arrangement of the structural units on different viruses, some viruses appear rather hexagonal in outline and some appear nearly spherical. Even within rather smooth overall surface configurations, at higher resolution functional protrusions, bulges, and projections (often housing cellular attachment ligands and neutralizing epitopes) can be seen. There are

also depressions, clefts, and canyons that may also house attachment ligands but usually not neutralizing epitopes.

Helical Symmetry

The nucleocapsid of many RNA viruses self-assembles very differently, forming a cylindrical structure in which the protein structural units are spatially arranged as a helix, hence the term *helical symmetry*. The occurrence of identical protein–protein interfaces on the structural units promotes the symmetrical assembly of the helix. In helically symmetrical nucleocapsids, the RNA genome forms a spiral within the nucleocapsid (Fig. 3.7). Many plant viruses with helical nucleocapsids are rod-shaped, flexible or rigid, and non-enveloped. The helical structure of tobacco mosaic virus was among the first viral structures determined by negative staining electron microscopy—its detailed structure was resolved by X-ray crystallography. However, in all viruses of vertebrates with helical symmetry, the nucleocapsid is wound into a secondary coil and enclosed within a lipoprotein envelope, for example rhabdoviruses (see Fig. 27.1); and paramyxoviruses (see Fig. 26.2).

The Function of Viral Capsids and Envelopes

Viral capsids and envelopes are not just inert coverings—these must be sufficiently stable in the environment to protect the contained nucleic acid genome, and at the same time play multiple roles in the interaction between the virion and host cell. Different kinds of envelope-associated proteins are associated with at least four crucial activities:

A

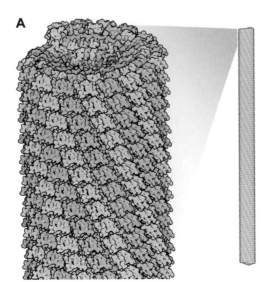

B

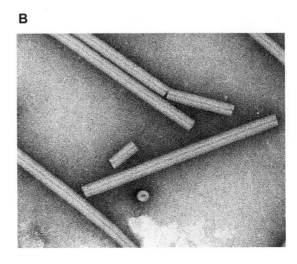

FIGURE 3.7   Helical symmetry. (A) Model of tobacco mosaic virus virion, showing interlocking capsid subunits and internal helical RNA. (B) Negative contrast electron micrograph of TMV particle. *(A) From David S. Goodsell, RCSB. (B) Reproduced from Williams, R.C. and Fisher, H.W. (1974). An electron micrographic atlas of viruses. Charles C. Thomas, Springfield, Il, with permission.*

binding to receptors, membrane fusion, uncoating, and receptor modification. For example, *fusion proteins* are involved in both viral entry and viral release, in many cases promoting the fusion of viral envelope with cellular membranes at virus entry and promoting virus "pinching off" at virus exit by budding. Moreover, before entry into the cell, viruses may be converted to a primed state to facilitate uptake and infection of target cells. This primed state usually involves conformational rearrangements of the virion surface proteins, making these structures responsive to various triggers. For example, upon entry into the host the hemagglutinin (HA0) of influenza viruses is cleaved at a specific site by the extracellular enzyme tryptase Clara, generating a primed modified structure composed of two unique subunits (HA1 and HA2). After entry of the virion into the host cell via receptor-mediated endocytosis, the primed hemagglutinin molecule then is activated when exposed to the low pH within the endosome. Activated hemagglutinin mediates endosome membrane disruption or fusion, thereby allowing the release of the viral RNA genome into the cytoplasm of the host cell.

This process of virion attachment and entry into the host cell is one of the most important stages of the virus–host relationship. In this context, the terms *receptor* and *ligand* have often been used in imprecise ways. The term *receptor* is properly used to designate specific molecule(s) or structure(s) on the surface of host cells that are involved in virus attachment. The term *ligand* is used for the molecule(s) on the surface of the virus that bind to the receptor. For example, the hemagglutinin of influenza virus is the ligand that binds to the receptor on the host cell surface, in this case a glycoconjugate terminating in *N*-acetylneuraminic acid.

Stability of Viral Infectivity

In general, viruses are more sensitive than bacteria or fungi to inactivation by physical and chemical agents, but there are important exceptions. Knowledge of specific viral sensitivity to environmental conditions and particular physical and chemical agents is therefore important for preserving the infectivity of viruses as reference reagents and in clinical diagnostic specimens. Knowledge of stability is also vital for deliberate inactivation, for example in sterilization, disinfection, and the production of inactivated vaccines.

When a virus-containing sample is being inactivated, it is important to appreciate that individual virus particles within a *population* successively will lose infectivity, at an overall rate determined by the physical conditions and the properties of the particular virus. Thus, infectivity of the sample will be lost progressively, at a specific *rate*, for example, $1 \log_{10}$ loss of titer over a specific period of time. This means that the time for complete inactivation of a sample is critically dependent on the starting titer of virus as well as physical conditions, for example, temperature. It also explains why scientists often give a qualified and even evasive answer to the common question, "How long does HIV survive outside the body?"

Temperature

The principal environmental condition that may adversely affect the infectivity of viruses is temperature. In most cases, viral envelope proteins are denatured within a few minutes at temperatures of 55 to 60°C, with the result that the virion is no longer capable of normal cellular attachment, penetration, and/or uncoating. Many viral capsid proteins are only slightly more heat resistant. At ambient temperature, the rate of decay of infectivity is slower but significant, especially in the summer or in the tropics. In order to preserve infectivity, viral preparations must therefore be stored at low temperature; 4°C (on wet ice or in a refrigerator) is usually satisfactory for a day or so, but long-term preservation requires much lower temperatures. Two convenient temperatures are −70°C, the temperature of solid CO_2 (dry ice) and of many laboratory mechanical freezers, and −196°C, the temperature of liquid nitrogen.

As a rule of thumb, the half-life of most viruses can be measured in seconds at 60°C, minutes at 37°C, hours at 20°C, days at 4°C, and years at −70°C or lower. In general, the enveloped viruses are more heat-labile than the non-enveloped viruses.

Enveloped virions, for example, respiratory syncytial virus, are susceptible to repeated cycles of freezing and thawing, probably as a result of disruption of virions by ice crystals. This poses problems for the collection and transportation of clinical specimens. The most practical way of avoiding such difficulties is to deliver specimens to the laboratory as rapidly as practicable, packed without freezing on wet ice or packed with cold (not frozen) gel packs.

In the laboratory, it is often necessary to preserve virus stocks for years. This is achieved in one of two ways: (1) rapid-freezing of small aliquots of virus suspended in medium containing protective protein and/or dimethyl sulfoxide, followed by storage at −70°C or −196°C, (2) freeze-drying (lyophilization), that is, dehydration of a frozen viral suspension under vacuum, followed by storage of the resulting powder at 4°C or −20°C. Freeze-drying significantly prolongs viability even at ambient temperatures, and is commonly used in the end-stage manufacture of attenuated virus vaccines.

In contrast to the general principles of the temperature lability of viruses, prions (which are not viruses but commonly fall within the general work domain of virologists), are amazingly stable under virtually all environmental conditions, surviving boiling, freezing, many physical and chemical insults, and even large doses of γ-irradiation (see Chapter 38: Prion Diseases).

Ionic Environment and pH

On the whole, viruses are best preserved in an isotonic environment at physiological pH, but some tolerate a wide ionic and pH range. For example, most enveloped viruses are inactivated at pH 5–6, rotaviruses and many picornaviruses survive the acidic pH of the stomach without loss of infectivity. A 1 M solution of magnesium cations has been used to stabilize enteroviruses, for example, in stocks of poliovirus vaccine.

Lipid Solvents and Detergents

Lipid solvents such as ether or chloroform or detergents such as sodium deoxycholate readily destroy the infectivity of enveloped viruses—these agents must be avoided in laboratory procedures concerned with maintaining the viability of viruses. On the other hand, mild detergents are commonly used by virologists to solubilize viral envelopes and liberate proteins for use as vaccines or serological reagents.

FURTHER READING

Harrison, S.C., 2013. Principles of Virus Structure. In: Knipe, D.M., Howley, P.M. (Eds.) Fields Virology (sixth ed.), Lippincott Williams and Wilkins, Philadelphia, PA.

Chapter 4

Virus Replication

Understanding the molecular events accompanying virus replication has been a major focus of experimental virology, and is essential for the proper understanding and control of all virus diseases. The biological "purpose" of any replication cycle is the generation of new viral genomes and proteins in sufficient quantities to ensure propagation of the viral genome (a clear example of the "selfish gene"); this requires that the extracellular viral genome is protected from enzymatic degradation and can be introduced into further target cells for further rounds of replication. Much has become known about the initial stages of attachment, and more detailed study shows that the initial recognition between virus and host is more complex than originally supposed. Temporal regulation of intracellular events is critical in all but the very simplest of viruses, with some form of suppression of the host innate immune response being common to nearly all human viruses. There are examples where the innate immune response is even used to enhance virus spread to cells otherwise unavailable to the virus. Over the next few years, the boundaries between virus-directed events and cellular processes that control specialized cell functions are likely to be even more complex; nevertheless understanding these processes will open up a range of targets for the development of novel antiviral therapies and immunotherapy.

This chapter presents an overview of the subject, indicating similarities and differences in the replication strategies adopted by viruses of each family that contains human pathogens. Major features and replication requirements of human viruses are shown in Table 4.1. More detailed information about the replication of individual viruses can be found in the relevant chapters of Part II of this book.

GROWTH OF VIRUSES

Before the development of *in vitro* culture techniques, viruses of humans could only be propagated in experimental animals. Thus much of our knowledge about virus growth and replication relied in the first instance on the study of bacterial viruses.

In 1931 it was shown that vaccinia virus and herpes simplex virus could be grown on the chorioallantoic membrane of embryonated chicken eggs. This system became routine for the study of many mammalian viruses as virtually all virus families contain viruses that can be cultured in this way. Although cell culture has long since replaced the use of eggs for this purpose, the technology is still in use for the preparation of influenza virus stocks and the manufacture of influenza vaccines.

The development of *in vitro* cell culture systems was a watershed development in virology: not only did it become possible to dissect the intracellular events accompanying virus replication in a manner similar to that of the study of bacteriophages in bacterial cells, it also provided a means of quantifying the amount of infectious virus in samples and virus stocks. Artificial medium was developed to maintain cell viability independent of the source species: these cells could be in the form of organ cultures, explant cultures, primary cell culture monolayers, or monolayer cell cultures immortalized into cell lines. Organ cultures maintain the three-dimensional structure of the tissue of origin and can be useful for short-term experiments that depend upon preserving fully differentiated cells. For example, tracheal epithelial cells attached to the cartilage matrix of the trachea during culture played a critical role in the isolation of many human respiratory viruses. The preparation of primary cell cultures uses proteases such as trypsin or collagenase to separate individual cells of a tissue such as fetal kidney or lung, and the individual cells then attach to a cell culture substrate where a limited number of divisions will occur. The limited life span of these cells requires repeated preparation of new cultures from source tissue, clearly presenting problems with reproducibility. In contrast, continuous propagation of cells is possible with two types of long-term culture:

1. "Semicontinuous," "diploid" cell strains, for example, human lung or foreskin fibroblasts (WI-38, MRC-5), in which cells eventually senesce after 20 to 30 divisions (the "Hayflick number") due to progressive shortening of telomeres, and
2. Continuous immortalized cell lines, for example, HeLa cells, BHK-21 cells. These are derived either from tumors or from primary cells that have undergone a spontaneous transformation event during cell culture, and can undergo an almost infinite number of cell divisions, thus generating consistency although often accompanied by a loss of differentiated cell functions. For example, HeLa cells were originally isolated in cell culture from a patient who died of cervical cancer

Fenner and White's Medical Virology. DOI: http://dx.doi.org/10.1016/B978-0-12-375156-0.00004-7

TABLE 4.1 Characteristics of Virus Replication Among Different Families

Family	Route of Uptake	Site of Nucleic Acid Replication	Eclipse Period (h[a])	Site of Maturation or Budding
Poxviridae	Variable	Cytoplasm	4	Golgi membrane
Herpesviridae	Variable	Nucleus	4	Nuclear membrane
Adenoviridae	Clathrin-mediated endocytosis	Nucleus	10	Nucleus
Papillomaviridae	Clathrin-mediated endocytosis	Nucleus	?	Nucleus
Polyomaviridae	Caveolar endocytosis	Nucleus	12	Nucleus
Parvoviridae	Clathrin-mediated endocytosis	Nucleus	6	Nucleus
Circoviridae	?	Nucleus	?	None
Hepadnaviridae	Clathrin-mediated endocytosis	Nucleus/cytoplasm	?	Endoplasmic reticulum
Retroviridae	Plasma membrane fusion or clathrin-mediated endocytosis	Cytoplasm/nucleus	10	Plasma membrane
Reoviridae	Clathrin-mediated endocytosis	Cytoplasm	5	Cytoplasm
Birnaviridae	Uncertain	Cytoplasm	4	None
Paramyxoviridae	Plasma membrane fusion	Cytoplasm	4	Plasma membrane
Rhabdoviridae	Plasma membrane fusion	Cytoplasm	3	Plasma membrane
Filoviridae	Plasma membrane fusion	Cytoplasm	2	Plasma membrane
Bornaviridae	Clathrin-mediated endocytosis	Nucleus	?	Plasma membrane
Orthomyxoviridae	Clathrin-mediated endocytosis	Nucleus	4	Plasma membrane
Bunyaviridae	Clathrin-mediated endocytosis	Cytoplasm	4	Golgi membrane
Arenaviridae	Clathrin-mediated endocytosis	Cytoplasm	5	Plasma membrane
Coronaviridae	Clathrin-mediated endocytosis/ plasma membrane fusion	Cytoplasm	5	Endoplasmic reticulum
Arteriviridae	Clathrin-mediated endocytosis	Cytoplasm	5	Endoplasmic reticulum
Picornaviridae	Caveolar endocytosis/plasma membrane insertion	Cytoplasm	2	Cytoplasm
Caliciviridae	Caveolar endocytosis/plasma membrane insertion?	Cytoplasm	3	Cytoplasm
Astroviridae	Caveolar endocytosis/plasma membrane insertion?	Cytoplasm	3	Cytoplasm
Togaviridae	Clathrin-mediated endocytosis	Cytoplasm	2	Plasma membrane
Flaviviridae	Clathrin-mediated endocytosis	Cytoplasm	3	Endoplasmic reticulum

[a]Differs with multiplicity of infection, strain of virus, cell type, and physiological conditions.

in 1951, but the cells remain an important system for culturing viruses to this day, and in fact proliferate so successfully that HeLa cells can become contaminants of cell cultures from other sources.

Recognition of the presence of a virus was dependent initially on observing cultures at regular intervals under the light microscope for signs of morphological change or cell death, compared to uninoculated cell cultures acting as controls. The specific appearance of the *cytopathic effect* (cpe) is often diagnostic for a certain family of viruses, for example herpesviruses can cause a distinct cytopathology often accompanied by the fusion of dying cells. However, reliance on cytopathology can give false results especially

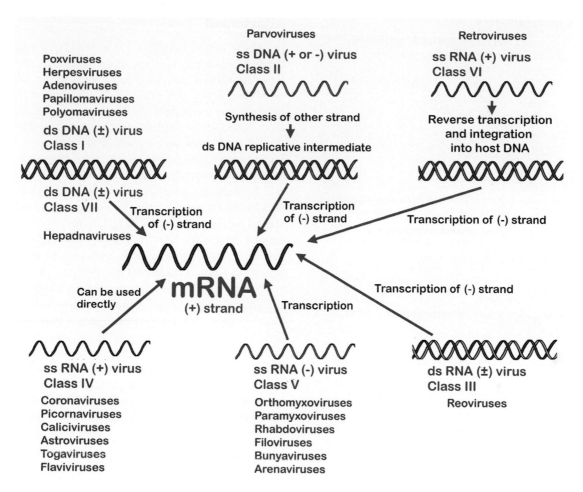

FIGURE 4.1   Virus replication strategies based on genome composition and the pathways of viral mRNA synthesis. *Originally from Baltimore, D., 1971. Expression of animal virus genomes. Bacteriol. Rev. 35, 235–241.* Then, modified from *Flint, S.J., et al., 2016. Principles of Virology, 4th ed. Washington, DC, ASM Press, with permission.*

if the sample is contaminated with bacterial toxins or other adventitious agents not related to the pathological process under study.

Using such systems, the basic mechanisms of transcription, translation, and nucleic acid replication have been characterized for all the major families of vertebrate viruses and the strategies of gene expression and regulation have been clarified. Every family of viruses employs unique replication strategies. One important unifying and simplifying concept, as originally proposed by David Baltimore in 1971, was to assign all viruses to one of six classes based on their genome composition and the pathway used to produce mRNAs for translation using host cell ribosomes (Fig. 4.1). This also reflects the parasitic property of all viruses in requiring a host cell for the synthesis of viral proteins.

THE VIRUS REPLICATION CYCLE

Most studies of the replication of viruses of humans have used cultured mammalian cells, grown either in suspension or as a monolayer adherent to a plastic surface. Classic studies

of this kind defined the *one-step growth curve,* in which all cells in a culture are infected simultaneously by using a high *multiplicity of infection,* and the increase in infectious virus over time is measured by sequential sampling and titration of infectious virus (Fig. 4.2). It is important to understand that, with synchronous infection of the whole cell population, the events observed in the total culture can be used to study events in a single cell. The time to achieve maximal yield of virus ranges from 6 hours (poliovirus) to more than 24 hours (adenovirus), compared with a generation time for *E. coli* in broth culture of 15 to 20 minutes.

Virus released into the medium can be titrated separately from virus that remains cell-associated. Shortly after infection, the inoculated virus "disappears"—infectious particles cannot be detected, even if the cells are disrupted. This *eclipse period* continues until the first progeny virions are detected some hours later. As infection progresses, non-enveloped viruses mature within the cell and may be detected as infectious, intracellular virions for some hours before being released by cell lysis. Many enveloped viruses, on the other hand, mature by budding from the plasma

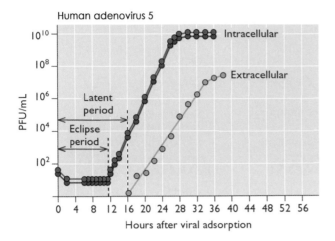

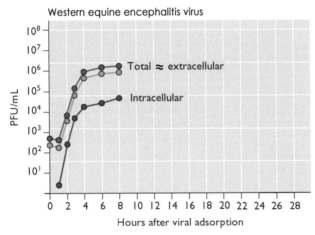

FIGURE 4.2   One-step growth curve of an enveloped and a non-enveloped virus. Attachment and penetration are followed by an eclipse period of 2 to 12 hours during which cell-associated infectivity cannot be detected. This is followed by a period of several hours when viral maturation occurs. Virions of non-enveloped viruses (e.g., adenoviruses—top) are often released late, when the cell lyses. Release of enveloped virions (e.g., the togavirus Western equine encephalitis virus) occurs concurrently with maturation by budding from the plasma membrane. *Adapted from Flint, S.J., et al., 2009. Principles of Virology, third ed. ASM Press, Washington, DC.*

membrane of the host cell and are released into the medium. Virus release via cell lysis is abrupt, whereas release via budding may continue over a protracted period of time. The eclipse period generally ranges from 2 to 12 hours for viruses of different families.

Early studies, relying on quantitative electron microscopy and the assay of infectious virions, provided an overview of the replication cycle (attachment, penetration, maturation, and release), but with little detail as to the processes involved. Investigation of the expression and replication of the viral genome became possible after the introduction of biochemical methods for the analysis of viral nucleic acids and proteins. Furthermore, imaging methods are also available nowadays that enable the tracking of viral and cellular proteins as the viral replication cycle proceeds,

Attachment

In order to initiate infection, virions must first bind to cells. Binding occurs between ligands on the surface of the virion (*viral attachment proteins*) and receptors on the plasma membrane of the cell (Table 4.2). This initial interaction between virus and host cell is sometimes complex and there is frequently a lack of correlation between attachment studies in cultured cells versus intact hosts. For example, several viral envelope glycoproteins of herpesviruses may serve as attachment proteins and several cellular receptors may be involved, in sequential order, first achieving a loose attachment via one receptor, then irreversibly binding via a second receptor and ligand. Exploitation of more than one cellular ligand provides redundancy and also may assist herpesviruses in invading both epithelial and neural tissues to support latent/recrudescent infection cycles.

While there is a degree of specificity about the recognition of particular cellular receptors by particular viruses, quite different viruses (e.g., orthomyxoviruses and paramyxoviruses) may utilize the same receptor and, conversely, viruses in the same family or genus may use different receptors. Viruses have thus evolved by making opportunistic use of a wide variety of host cell surface proteins as receptors—in most cases viruses employ as receptors host cell proteins that are crucial for fundamental cellular functions, many of which are strongly conserved over evolutionary time.

The receptor–ligand interaction between human cells and human immunodeficiency virus 1 (HIV-1) illustrates the complexity of the interaction between virus and host proteins. Attachment initially involves CD4 molecules on the surface of target cells, notably macrophages and T helper lymphocytes, via the viral gp120 envelope glycoprotein. This binding induces a conformational change exposing a high-affinity chemokine receptor-binding site on gp120, which binds one of several chemokine receptors on the cell surface. This latter binding in turn exposes a fusogenic domain of the viral transmembrane protein gp41. Direct contact between the fusogenic domain of gp41 and the cell membrane brings about the fusion of the viral envelope with the plasma membrane of the cell, permitting the viral nucleocapsid to enter the cell cytoplasm (see Fig. 23.6).

The cellular receptor for most orthomyxoviruses is the terminal sialic acid on an oligosaccharide side-chain of a glycoprotein (or glycolipid) exposed at the cell surface: the viral ligand is in a cleft at the distal tip of each monomer of the trimeric viral hemagglutinin glycoprotein (see Chapter 3: Virion Structure and Composition). Viruses can adapt to new hosts through modification of the relevant attachment ligand. For example, influenza A viruses originate in aquatic birds and recognize sialic acid residues on the surface of cells along the respiratory tract. However, influenza A viruses of birds have a greater affinity for sialic acid linked to the penultimate galactose moiety through α2–3 bonding, whereas those of humans have a greater affinity for sialic

TABLE 4.2 Examples of Cellular Receptors for Viruses of Medical Importance

Virus	Family	Receptor
Human immunodeficiency virus	*Retroviridae*	CD4; coreceptors CCR5, CXCR4, CCR3
Vaccinia virus	*Poxviridae*	Various (heparan sulfate, chondroitin sulfate glycosaminoglycans, laminin)
Poliovirus	*Picornaviridae*	PVR (CD155)—Immunoglobulin (Ig) superfamily
Human rhinovirus 14	*Picornaviridae*	Intercellular adhesion molecule-1 (ICAM-1); Ig superfamily
Echovirus 1	*Picornaviridae*	α_2/β_1 integrin VLA-2
Hepatitis A virus	*Picornaviridae*	HAVCR1; member of TIM-1 family (transmembrane, Ig, mucin)
Adenovirus	*Adenoviridae*	CAR (Coxsackie and adenovirus receptor), member Ig superfamily; coreceptor α_v integrins
Herpes simplex virus 1	*Herpesviridae*	HveA (herpesvirus entry mediator A), heparan sulfate proteoglycan
Human cytomegalovirus	*Herpesviridae*	Heparan sulfate proteoglycan
Epstein–Barr virus	*Herpesviridae*	CD21, complement receptor 2 (CR2)
Adeno-associated virus 5	*Parvoviridae*	$\alpha(2,3)$-linked sialic acid
Influenza A virus	*Orthomyxoviridae*	$\alpha(2,3)$- or $\alpha(2,6)$-αlinked sialic acid
Influenza C virus	*Orthomyxoviridae*	9-*O*-acetylsialic acid
Hendra virus	*Paramyxoviridae*	Ephrin-B2
Rotavirus	*Reoviridae*	Various integrins
Reovirus	*Reoviridae*	JAMs (junction adhesion molecules)
Dengue virus	*Flaviviridae*	Heparan sulfate proteoglycan
Rabies virus	*Rhabdoviridae*	Acetylcholine, NCAM (neural adhesion molecule)
Polyomaviruses	*Polyomaviridae*	Various gangliosides (glycosphingolipids with terminal sialic acid)

acid linked via an $\alpha 2$–6 bond; this reflects differences in the relative abundance of the two linkages in the different host species. There is considerable subtlety in these interactions, and as with many human and animal viruses a single amino acid change within a receptor site may manifest as a significant change in host range and epidemiology.

Proinflammatory reactions in the host may also affect the expression of receptors and thereby indirectly modify interactions between virus and potential host cells. For example, human adenovirus 5 recognizes both the Coxsackievirus and adenovirus receptor (CAR) and $\alpha v\beta 3/5$ integrin in order to infect cells, yet neither is normally exposed on the apical surfaces of cells until macrophages respond to infection and secrete IL-8. This in turn moves the CAR and $\alpha v\beta 3/5$ integrin receptors away from the tight junctions of polarized cells toward the apical cell surface and thus become exposed and available for virus attachment.

Penetration and Uncoating

The majority of mammalian cells are continuously engaged in *receptor-mediated endocytosis* for the uptake of macromolecules via specific receptors. Many enveloped and non-enveloped viruses use this essential cell function to initiate penetration and uncoating (Fig. 4.3). Virion attachment to those receptors clustered at *clathrin-coated* pits is followed by endocytosis into clathrin-coated vesicles. Vesicles enter the cytoplasm and, after removal of the clathrin coat, fuse with endosomes (acidic prelysosomal vacuoles). Acidification within the vesicle triggers changes in virion proteins and surface structures. For example, the acidic pH within the endosome, induces the hemagglutinin molecules of influenza viruses to undergo a conformational change. This enables fusion to occur between the viral envelope and the endosomal membrane and results in the release of the viral nucleocapsid into the cytoplasm. Many other non-enveloped and enveloped viruses undergo comparable changes.

Fusion at neutral pH is an alternative mechanism. The F (fusion) glycoprotein of paramyxoviruses causes the viral envelope to fuse directly with the plasma membrane of the cell, even at pH 7. This allows the nucleocapsid to be released directly into the cytoplasm. A number of other important enveloped viruses for example filoviruses, have

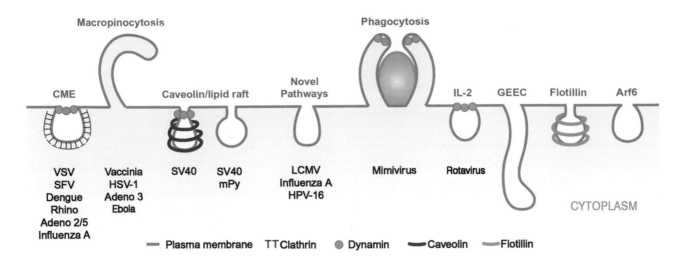

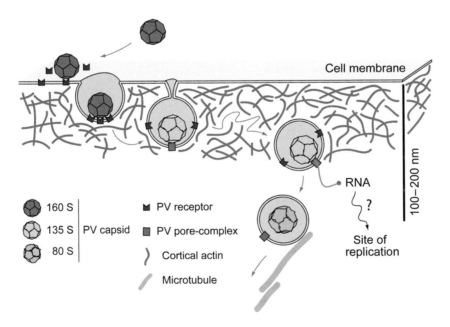

FIGURE 4.3 Endocytic mechanisms for virus entry. Endocytosis in animal cells can occur via four major mechanisms: clathrin-mediated endocytosis (CME); macropinocytosis; caveolae/lipid rafts; and phagocytosis (larger particles, >0.75 μM). Note the IL-2, GEEC (GPI-enriched early endosomal compartment), Flotillin and Arf6 (ADP-ribosylation factor 6) pathways import specific cell cargo but viruses are not currently known to use these routes. *VSV*, vesicular stomatitis virus; *SFV*, Semliki Forest virus; *HSV*, herpes simplex virus; *LCMV*, lymphocytic choriomeningitis virus; *SV40*, simian virus 40; *mPy*, mouse polyomavirus; *HPV*, human papillomavirus. *Reproduced from MacLachlan, N.J., Dubovi, E.J., 2011. Veterinary Virology, fourth ed. Academic Press, London (Fig. 2.5), with permission.*

FIGURE 4.4 Mechanism of poliovirus entry. Virus binds to specific receptors on host cell plasma membrane (CD155 = poliovirus receptor Pvr). The receptors induce conformational changes in the virion (the 160S infectious virion becomes a 135S particle), with the loss of VP4 and the externalization of the N-terminus of VP1. The 135S particles are then internalized by a clathrin- and caveolin-independent, but actin- and tyrosine kinase-dependent, mechanism. The release of the viral genome takes place only after internalization from an endocytotic compartment localized within 100 to 200 nm of the plasma membrane. Upon release of the RNA genome, the empty capsid (80S) is transported away along microtubules. *Reproduced from Brandenburg, B., et al., 2007, Imaging poliovirus entry in live cells. PLoS Biol 5 (7), e183. doi:10.1371/journal.pbio.0050183, with permission.*

the ability to fuse with the host cell plasma membrane, thereby allowing entry of viral nucleic acid into the cytosol.

A third entry mechanism ("direct entry") is used by some non-enveloped viruses (e.g., poliovirus, adenovirus).

This involves binding of virus to receptors at the endosomal membrane, which induces conformational changes in the viral capsid; these changes then expose regions that react with the endosomal membrane to induce channels for transport of the genome across the plasma membrane (Fig. 4.4).

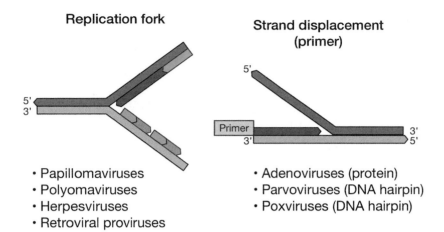

FIGURE 4.5 Two mechanisms of synthesis of DNA virus genomes. A replication fork is used by papillomaviruses, polyomaviruses, and herpesviruses. This involves initiation from an RNA primer; synthesis along the leading strand is continuous, but discontinuous along the lagging strand with formation of Okazaki fragments which are subsequently ligated. With circular DNA genomes, synthesis proceeds bidirectionally from the site of origin. Alternatively, the DNA genomes of adenoviruses, parvoviruses, and poxviruses replicate by strand displacement, using as primers either protein (adenoviruses) or DNA hairpins (parvoviruses and poxviruses). *Reproduced from Flint, S.J., et al., 2009. Principles of Virology, third ed. ASM Press, Washington, DC, with permission.*

In order for viral genes to become available for transcription it is necessary that virions are at least partially uncoated. In the case of enveloped RNA viruses entering by the process of fusion, the nucleocapsid is discharged directly into the cytoplasm and transcription commences from viral nucleic acid while still associated with the nucleocapsid protein(s). With the non-enveloped icosahedral reoviruses, only certain capsid proteins are removed and the viral genome expresses all its functions without being released from the virion core. For most other viruses, however, uncoating proceeds to completion, otherwise genome duplication cannot proceed. For some viruses, uncoating takes place in the nucleus where later stages of replication occur.

Replication of Viral Nucleic Acids

Replication of Viral DNA

Different mechanisms of DNA replication are employed by each family of DNA viruses (Fig. 4.5). These involve either synthesis of daughter strands via a *replication fork* (papillomaviruses, polyomaviruses, herpesviruses), or via *strand displacement* (adenoviruses, parvoviruses, poxviruses). Since cellular DNA polymerases cannot initiate synthesis of a new DNA strand but can only extend synthesis beginning from a short (e.g., RNA) primer, one end of newly synthesized viral DNA molecules might be expected to retain a short single-stranded region. Various DNA viruses have evolved different strategies for circumventing this problem. Viruses of some families have a circular DNA genome, others have a linear genome with complementary termini that serve as DNA primers, yet others have a protein primer covalently attached to the 5'-terminus of each DNA strand.

Several virus-coded enzyme activities are generally required for replication of viral DNA: a helicase (with ATPase activity) to unwind the double helix; a helix-destabilizing protein to keep the two separated strands apart until each has been copied; a DNA polymerase to copy each strand from the origin of replication in a 5'- to 3'-direction; an RNAase to degrade the RNA primer after it has served its purpose; and a DNA ligase to join the Okazaki fragments together. Often a single large enzyme performs two or more of these activities.

The genomes of papillomaviruses and polyomaviruses structurally and functionally resemble cellular DNA in binding to cellular histones. The viral genome utilizes host DNA polymerase α-primase to synthesize the RNA primer for genome replication. Among the polyomaviruses an early viral antigen, large T, binds to the regulatory sequence—and in the case of papillomaviruses E1 and E2—thereby initiating DNA replication. Replication of these circular double-stranded DNA molecules commences from a unique palindromic sequence and proceeds simultaneously with replication forks in both directions. Both continuous and discontinuous DNA synthesis occur (of leading and lagging strands respectively) at the two growing forks as for the replication of mammalian DNA. The discontinuous synthesis of the lagging strand involves repeated synthesis of short oligoribonucleotide primers, which in turn initiate short nascent strands of DNA (*Okazaki fragments*), each is then covalently joined by a DNA ligase to form one of the growing strands.

The replication of adenoviruses is distinct from that of other DNA viruses. The adenovirus DNA genome is linear, the 5'-end of each strand being identical to the other (terminally repeated inverted sequences) and covalently linked to a protein, the precursor of which serves as the primer for viral DNA synthesis. DNA replication proceeds from both ends, continuously but asynchronously, in a 5′- to 3′-direction, using a virus-coded DNA polymerase. It does not require the synthesis of Okazaki fragments.

Herpesviruses code for many or all of the proteins required for DNA replication, including a DNA polymerase, a helicase, a primase, a single-stranded DNA binding protein, and a protein recognizing the origin of replication. Poxviruses, which replicate entirely within the cytoplasm, are self-sufficient in DNA replication machinery. Hepadnaviruses, similar to retroviruses, utilize positive-sense single-stranded RNA transcripts as intermediates for the production of progeny DNA by a process of reverse transcription. Single-stranded DNA parvoviruses use 3′-palindromic sequences to form a double-stranded hairpin structure to provide a primer for cellular DNA polymerase binding.

Replication of Viral RNA

Transcription of RNA from an RNA template is a phenomenon unique to viruses (Fig. 4.6), and requires an RNA-dependent RNA polymerase, a virus-coded enzyme not found in uninfected cells. The replication of viral RNA requires first the synthesis of complementary RNA: this in turn serves as a template for the synthesis of further copies of viral RNA.

For viruses where the viral RNA is of negative-sense (orthomyxoviruses, paramyxoviruses, rhabdoviruses, filoviruses, bornavirus, arenaviruses, and bunyaviruses), the complementary RNA is of positive-sense and the RNA polymerase involved performs the same function as the virion-associated transcriptase used for the primary transcription of mRNAs. Most transcripts from such

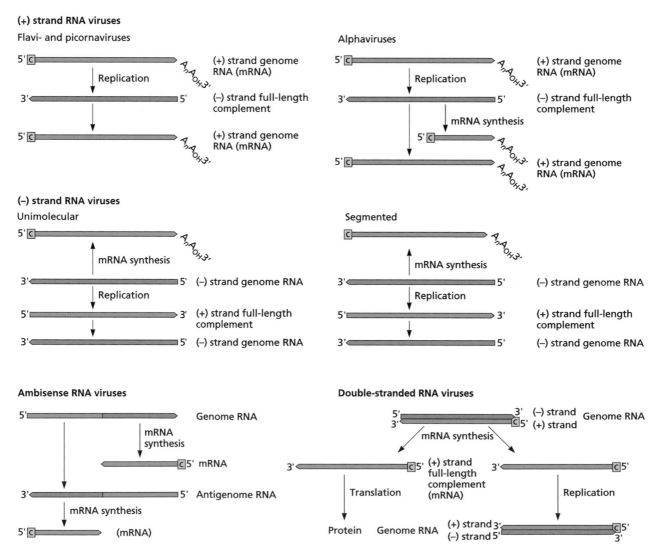

FIGURE 4.6 Strategies for replication and mRNA synthesis of RNA virus genomes are shown for representative virus families. Picornaviral genomic RNA is linked to VPg at the 5′-end. The (+) genomic RNA of some flaviviruses does not contain poly(A). Only one RNA segment is shown for segmented (±) strand RNA viruses. Cap structures are shown as blue boxes containing "C". *Reproduced from Flint, S.J., et al., 2016. Principles of Virology, 4th ed. Washington, DC, ASM Press, with permission.*

negative-sense viral RNAs are subgenomic mRNA molecules, but some full-length positive-sense strands must also be made, in order to serve as templates for full-length progeny viral RNA synthesis (replication). For some viruses there is good evidence that the RNA polymerases used for transcription and replication are distinct, in others the same enzyme performs both functions.

In the case of the positive-sense RNA viruses (picornaviruses, caliciviruses, togaviruses, flaviviruses, coronaviruses, and arteriviruses), the complementary RNA is negative-sense, and its sole purpose is to provide a template for synthesis of more positive-sense RNA. Several viral RNA molecules can be transcribed simultaneously from a single complementary RNA template, each RNA transcript being the product of a separately bound polymerase molecule. The resulting structure, known as the *replicative intermediate*, is therefore partially double-stranded with single-stranded tails. Some RNA-dependent RNA polymerases can initiate RNA synthesis *de novo*, while others require a primer with a free 3′-OH group to which further nucleotides can be added. Initiation of the replication of picornavirus and calicivirus RNA, similar to that of adenovirus DNA, requires a bound protein as primer, rather than an oligonucleotide. This small protein is covalently attached to the 5′-terminus of nascent positive and negative RNA strands in addition to virion RNA, but not to molecules used as mRNA. Little is known about what determines whether a given picornavirus positive-sense RNA molecule will be directed (1) to a *replication complex* (a structure bound to smooth endoplasmic reticulum), where it serves as template for transcription by RNA-dependent RNA polymerase into negative-sense RNA, or (2) to a *ribosome*, where it serves as mRNA for translation into protein, or (3) to a *procapsid*, with which it associates to form a virion.

Other RNA polymerases require an oligonucleotide 5′ cap structure, which may be synthesized by viral enzymes or derived from cellular mRNAs by a process of "cap-snatching" (e.g., Chapter 25: Orthomyxoviruses).

Retroviruses have a genome consisting of positive-sense single-stranded RNA. Unlike other RNA viruses, retroviruses replicate via a DNA intermediate. The virion-associated reverse transcriptase, using a transfer RNA (tRNA) molecule as a primer, makes a single-stranded DNA copy, while the same enzyme functions concurrently as a ribonuclease and removes the parental RNA molecule from most of the DNA:RNA hybrid, except for several short RNA stretches known as poly-purine tracts. The poly-purine tracts then act as primers for copying the negative-sense single-stranded DNA strand to form a linear double-stranded DNA that contains an additional sequence known as the *long terminal repeat* (LTR) at each end. This double-stranded DNA then is processed by the viral integrase and integrates into cellular chromosomal DNA. From this point on, transcription of viral RNA occurs from the integrated (proviral) DNA (see Chapter 23: Retroviruses).

Overview of Viral Gene Expression Strategies

Distinguishing the various strategies used by viruses for the production of viral proteins is fundamental to an understanding of virus replication (Fig. 4.1). A summary of the properties of viral proteins is given in Table 4.3. For many viruses, the production of viral proteins immediately after entry of the viral genome is a critical step. These early proteins act to subvert the molecular machinery of the host,

TABLE 4.3 Properties of Viral Proteins

Protein Category	Examples and Comments
Structural proteins of the virion	1. Capsid, and (for some viruses) core and/or envelope and matrix proteins 2. Virion-associated enzymes, especially polymerases (transcriptases)[a]
Non-structural proteins, mainly enzymes, required for transcription, replication of viral nucleic acid, and cleavage of proteins	DNA and RNA polymerases, helicases, proteases, etc. DNA viruses with large complex genomes, notably poxviruses and herpesviruses, also encode numerous enzymes needed for nucleotide synthesis
Regulatory proteins which control the temporal sequence of expression of the viral genome	Site-specific DNA-binding proteins (transcription factors) which bind to enhancer sequences in the viral genome, or to another transcription factor. Some may act in trans (transactivators)
Proteins down-regulating expression of cellular genes	Usually by inhibiting transcription, sometimes translation
Oncogene products (oncoproteins) and inactivators of cellular tumor suppressor proteins	Upgrade expression of certain cellular genes; may lead to cell transformation and eventually to cancer. Observed with herpesviruses, adenoviruses, papovaviruses, and retroviruses
Proteins influencing viral virulence, host range, tissue tropism, etc.	Recorded so far mainly in the more complex DNA viruses (poxviruses, herpesviruses, adenoviruses) but may be more widespread
Virokines, which act on non-infected cells to modulate the progress of infection in the body as a whole	Mainly by subverting the immune response, e.g., by inhibiting cytokines, downregulating MHC expression, blocking the complement cascade, etc.

[a]*RNA viruses of positive-sense and DNA viruses that replicate in the nucleus do not carry a transcriptase in the virion. Virions of some viruses, e.g., poxviruses, also contain many other enzymes.*

to allow the later production of new virus particles and to inhibit the activation of the host innate immunity that would otherwise prevent virus replication and spread to adjacent cells and tissues.

For most families of DNA viruses, transcription and DNA replication take place in the cell nucleus, using cellular RNA polymerase II and other cellular enzymes. For viruses such as poxviruses and herpesviruses it is possible to identify a significant number of genes by gene deletion (more than 40%) that are not essential for the replication of the virus in cultured cells. Of course, in the strict economy of viral genomes it is likely that most or all of such genes are important for virus survival in nature.

The situation is quite different for RNA viruses as these are unique having genetic information coded as RNA. Thus an effort needs to be made to understand the distinct replication and expression strategies, particularly as these have a bearing on understanding virus pathogenesis. RNA viruses with different types of genomes (single-stranded or double-stranded, positive- or negative-sense, monopartite or segmented) have necessarily evolved different routes to the production of mRNA. Since eukaryotic cells do not contain an RNA-dependent RNA polymerase, all RNA viruses need to code for a unique RNA-dependent RNA polymerase. In the case of positive-sense single-stranded RNA viruses, the incoming RNA genome can bind directly to ribosomes and be translated in full or in part without the need for any prior transcription; all other forms of incoming viral RNA must first be transcribed to produce mRNA, in order to begin the process of expression of the infecting viral genome. Thus, both negative-sense single-stranded RNA viruses and double-stranded RNA viruses need to carry an RNA-dependent RNA polymerase in the virion, originating from the preceding round of infection.

In the case of DNA viruses dependent upon the nucleus for replication, cellular DNA-dependent RNA polymerase II performs this function, while double-stranded DNA viruses that replicate in the cytoplasm carry a virus-encoded DNA-dependent RNA polymerase.

Many, but not all, viruses express different genes at different stages of the replication cycle. *Early viral genes* are first transcribed into RNA, which may then be processed in a number of ways, including splicing. The early gene-products translated from this mRNA are of three main types: proteins that shut down cellular nucleic acid and protein synthesis, proteins that regulate the expression of the viral genome, and enzymes required for the replication of the viral nucleic acid. Following viral nucleic acid replication, *late viral genes* are transcribed.

The late proteins are principally viral structural proteins used for assembly of new virions; some of these are subject to post-translational modifications before use, and are often made in considerable excess. Structural proteins for the coating of nascent viral genomes are required in multiple copies for every new nucleic acid molecule destined for encapsidation. For this reason, many viral strategies have evolved so that new copies of the viral genome can act as templates for the further specific transcription and translation of structural proteins required for new virus particles.

The existence of overlapping reading frames, multiple splicing patterns of RNA transcripts, post-translational cleavage of polyproteins, etc., mean that it is too simplistic to assume one particular gene codes for a particular protein. It is more appropriate to refer to a *transcription unit*, defined as a region of the genome beginning with the transcription initiation site and extending to the transcription termination site (including all introns and exons in between), the expression of which falls under the control of a particular promoter.

Some viral proteins serve as regulatory proteins, modulating the transcription or translation of cellular genes or down-regulating early viral genes. The large DNA viruses also encode numerous additional proteins, called *virokines*, which do not regulate the viral replication cycle *per se*, but influence the host response to infection. Included among these are viral protein homologues of cellular cytokines.

In 1978 Walter Fiers and his colleagues presented the first complete description of the genome of an animal virus, namely that of the polyomavirus SV40 (see Fig. 20.2). Analysis of the circular double-stranded DNA molecule and its transcription revealed some remarkable insights, many of which are now applicable to other double-stranded DNA viruses. First, early and late genes are transcribed in opposite directions, from different strands of the DNA. Second, certain genes overlap, the protein products generated having common amino acid sequences. Third, some regions of the viral DNA may be read in different reading frames (ORFs), so that quite distinct amino acid sequences are translated from the same nucleotide sequence. Fourth, certain long stretches of the viral DNA consist of transcribed introns that are not translated into protein as these are spliced out of the primary RNA transcript (see Fig. 20.2).

Adenoviruses exemplify some of the mechanisms that regulate the expression of viral genomes at the level of transcription. There are several adenovirus transcription units; at different stages of the viral replication cycle, "pre-early," "early," "intermediate," and "late" transcription units are transcribed in a precise temporal sequence. A product of the early region E1A induces transcription from the other early regions including E1B, but following viral DNA replication there is a 50-fold increase in the rate of transcription from the major late promoter relative to early promoters such as E1B, and a decrease in E1A mRNA levels. A second control operates at the point of termination of transcription. Early in the replication cycle transcripts terminate at a particular point in the genome, while later in infection these are read through the point of termination to produce a range of longer transcripts with different polyadenylation sites and proteins of different functions. This is but one of the many examples of the economy of viral genomes in coding for complex functions using minimal sequences of nucleic acid.

For RNA viruses, the regulation of transcription is, on the whole, not as complex as is the case for DNA viruses. In particular, the temporal separation into early genes transcribed before the replication of viral nucleic acid, and late genes thereafter, is not nearly so well defined. In the simplest examples (e.g., picornaviruses), a full-length polycistronic mRNA is translated and the resulting polyprotein is subsequently cleaved to yield equimolar amounts of all protein products. Togaviruses synthesize excess amounts of structural proteins from a separate subgenomic RNA.

Yet other mechanisms of regulation have evolved among viruses with non-segmented negative-sense RNA genomes. Once the nucleocapsid is released into the cytoplasm of an infected cell, the RNA polymerase initiates transcription from the 3′-end of the genome. With paramyxoviruses, discrete genes along the viral RNA are each separated by a consensus sequence that includes termination and start signals as well as short intergenic sequences of U residues enabling the transcriptase to generate a long poly(A) tail for each mRNA by a process of reiterative copying (also known as *stuttering*). Each discrete mRNA is released from the template while the enzyme continues to transcribe the next gene in sequence. As the polymerase moves from 3′ to 5′ along the template, decreasing amounts of each mRNA are sometimes made due to decreasing efficiency of the transcription process—thus, gene order becomes an efficient way of modulating the relative amount of each protein synthesized.

Paramyxovirus transcription also involves a process known as *RNA editing*. The P gene codes for two proteins, P and V, which share a common N-terminal amino acid sequence but differ completely in the C-terminal sequences as a result of a shift in the reading frame brought about by the insertion of two uncoded G residues into the RNA transcript by transcriptase stuttering.

Eukaryotic cells are not equipped to translate polycistronic mRNA into several individual species of protein, as mammalian cells cannot normally reinitiate translation partway along an RNA molecule. DNA viruses overcome this limitation by using cellular mechanisms for cleavage (and sometimes splicing) of viral polycistronic RNA transcripts to yield monocistronic mRNA molecules.

RNA viruses, most of which replicate in the cytoplasm, do not have access to the RNA processing and splicing enzymes of the host cell nucleus, but have developed a remarkable diversity of solutions to the difficulty of punctuating a large genome to produce multiple individual gene products. Some (e.g., orthomyxoviruses) have developed a segmented RNA genome in which each gene is, in general, expressed and duplicated as a separate molecule. Others (e.g., paramyxoviruses) have evolved a polycistronic genome but produce monocistronic RNA transcripts by termination and reinitiation of transcription. Yet other RNA viruses (e.g., coronaviruses) make use of a nested set of overlapping RNA transcripts, each of which is translated into a single gene product. Finally, some (e.g., picornaviruses) have a polycistronic genomic RNA, which is translated into a polyprotein that is later cleaved proteolytically to yield the final products.

With certain viruses, polycistronic mRNA can be translated directly to produce several gene-products as a result of initiation, or reinitiation, of translation at internal AUG start codons. Internal initiation of translation is facilitated by an upstream RNA motif known as an internal ribosomal entry site (IRES), first discovered in 1988 in poliovirus and encephalomyocarditis virus RNAs (another picornavirus). Where initiation of translation at an internal AUG takes place, a frameshift can also occur. Another mechanism, known as *ribosomal frameshifting*, occurs fortuitously when a ribosome happens to slip one nucleotide back and forth along an RNA template. This phenomenon is exploited by retroviruses, where frameshift read through leads to about 5% of Gag protein molecules being extended as a Gag-Pol polyprotein. Thus, taken together with the phenomenon of RNA splicing and RNA editing described above, it can be seen that there are several mechanisms of exploiting overlapping reading frames to maximize the limited coding potential of viruses with comparatively small genomes.

Regulatory Genes and Post-transcriptional Processing

There is considerable interest in the untranslated regions of viral genomes, particularly the numerous conserved (consensus) sequences or motifs that represent *responsive elements*. Many of the latter have a critical role to play in the regulation of transcription. For example, each transcription unit of a viral genome has near its 3′-end an mRNA transcription initiation site *(start site)*, designated as nucleotide +1. Within the hundred or so nucleotides upstream of the start site is the *promoter*, which upregulates the transcription of a particular gene or series of genes. Upstream or downstream from the start site there may be a long sequence with several, in some cases repeated, elements known as *enhancers* that amplify transcription even further. These regulatory regions are activated by the binding of either viral or cellular DNA-reactive proteins. Several such proteins may bind to adjacent responsive elements to form an interactive structure, or otherwise interact to facilitate attachment of the viral RNA polymerase. Viral regulatory genes encoding such regulatory proteins may act in *trans* as well as in *cis*, that is, these proteins may *trans*-activate genes residing on a physically separate molecule of nucleic acid.

A description of the role of the regulatory genes of human immunodeficiency virus 1 (HIV-1) serves to illustrate the sophistication of such regulatory mechanisms. A DNA copy of the viral genome is integrated into a chromosome of a resting T cell, and remains latent until a T cell mitogen or a cytokine induces synthesis of the cellular NF-κB family of DNA-binding proteins. NF-κB then binds to the enhancer present in the integrated provirus, thereby triggering transcription of the HIV-1 regulatory genes. One of these, *tat*, found in all lentiviruses, codes for a protein specific for a responsive element, TAR, within the provirus, greatly augmenting (*trans*-activating) the transcription of all viral genes (including *tat* itself). A positive feedback loop is thereby established that stimulates the production of HIV RNA transcripts. Moreover, by interacting with TAR present in all viral mRNAs as well as in the proviral DNA, *tat* also enhances translation. The HIV-1 regulatory protein Rev plays a different role to modulate gene expression by regulating the nuclear export of viral mRNAs. Although the control of lentivirus transcription may be unusually complex compared to many RNA viruses, these viruses contain only 9 genes, compared with up to 200 in the case of some DNA viruses.

Post-transcriptional Processing

Primary RNA transcripts from eukaryotic DNA are subject to a series of post-transcriptional alterations in the nucleus prior to export to the cytoplasm as mRNA. First, a cap, consisting of 7-methylguanosine (m^{7}Gppp), is added to the 5′-terminus of the primary transcript; this cap structure facilitates the formation of a stable complex with the host 40S ribosomal subunit, necessary for the initiation of translation. Second, a sequence of 50 to 200 adenylate residues is added to the 3′-terminus. This poly(A) tail acts as a recognition signal for processing and the transport of mRNA from the nucleus to the cytoplasm, and assists in the stabilization of mRNA against cytoplasmic degradation by ubiquitin. Third, a methyl group is added at the sixth position to about 1% of the adenylate residues throughout the RNA (methylation). Fourth, introns are removed from the primary transcript and the exons are linked together in a process known as *splicing*, an important mechanism for regulating gene expression in nuclear DNA viruses. A given RNA transcript can have two or more splicing sites and, moreover, be spliced in several alternative ways to produce a variety of mRNA species coding for distinct proteins; both the preferred poly(A) site and the splicing pattern may change in a regulated fashion as infection proceeds. For example, the HIV-1 protein Rev assists the nuclear export of unspliced or singly spliced (intron-containing) viral mRNA; thus, early in the replication cycle, the mRNA present in the cytoplasm is largely doubly spliced, while later in the cycle after Rev accumulates, a temporal switch in mRNA species occurs.

Special mention should be made of an extraordinary phenomenon known as *cap-snatching*. The transcriptase of influenza virus, which also carries endonuclease activity, steals the 5′-methylated caps from newly synthesized cellular RNA transcripts in the nucleus and uses these host cell RNA sequences as primers for initiating transcription from the viral genome.

The rate of degradation of mRNA provides another possible level of regulation. Not only do different mRNA species have different half-lives but the half-life of a given mRNA species may change as the replication cycle progresses.

Capped, polyadenylated, and processed monocistronic viral mRNAs bind to ribosomes and are translated into protein in the same fashion as cellular mRNAs. The sequence of events has been closely studied in reovirus-infected cells. Each monocistronic mRNA molecule binds via its capped 5′-terminus to the 40S ribosomal subunit and this then moves along the mRNA molecule until reaching the initiation codon. The 60S ribosomal subunit also binds, together with methionyl-transfer RNA and various initiation factors, after which translation proceeds.

Post-translational Modifications

Most viral proteins undergo various sorts of post-translational modification such as phosphorylation (for nucleic acid binding), fatty acid acylation (for membrane insertion), glycosylation, myristylation, or proteolytic cleavage. Newly synthesized viral proteins must also be transported to the various sites in the cell where they are needed, for example, into the nucleus in the case of viruses where the nucleus is the major site of replication. The sorting signals that direct intracellular trafficking are only now beginning to be understood, as are the polypeptide chain-binding proteins (*molecular chaperones*) that regulate folding, translocation, and assembly of oligomers of viral as well as cellular proteins.

Post-translational Cleavage

The polycistronic viral RNA is translated directly into a single polyprotein in the case of the positive-sense picornaviruses and flaviviruses. This large molecule carries protease activity that cleaves the polyprotein at defined recognition sites into smaller proteins. The first cleavage steps are carried out while the polyprotein is still associated with the ribosome. Some of the larger intermediates exist only fleetingly; others are functional for a short period but are subsequently cleaved by additional virus-coded proteases to smaller proteins with alternative functions. Post-translational cleavage occurs in several other RNA virus families, for example, togaviruses and caliciviruses, in which polyproteins corresponding to large parts of the genome are cleaved. Some viruses encode several different proteases. Most are either trypsin-like (serine or cysteine proteases), pepsin-like (aspartyl proteases), or papain-like (thiol proteases).

Cellular proteases, present in organelles such as the Golgi complex or transport vesicles, are also vital to the maturation and assembly of many viruses. For example, cleavage of the hemagglutinin glycoprotein of orthomyxoviruses or the fusion glycoprotein of paramyxoviruses is essential for virion infectivity.

Glycosylation

Viruses frequently exploit those cellular pathways normally used for the synthesis of host cell secretory glycoproteins. The amino terminus of viral envelope proteins contains a sequence of 15 to 30 hydrophobic amino acids, known as a signal sequence, which facilitates binding of the growing polypeptide chain to a receptor site on the cytoplasmic side of the rough endoplasmic reticulum and its passage through the lipid bilayer to the lumenal side. Oligosaccharides are then added in N-linkage to certain asparagine residues of the nascent polypeptide by *en bloc* transfer of a mannose-rich core of preformed oligosaccharides, and glucose residues are removed by glycosidases (called *trimming*). The viral glycoprotein is then transported from the rough endoplasmic reticulum to the Golgi complex. Here the core carbohydrate is further modified by the removal of several mannose residues and the addition of further N-acetyl-glucosamine, galactose, and the terminal sugars, sialic acid, or fucose. The completed side-chains are a mixture of simple oligosaccharides (also called high mannose oligosaccharides) and complex oligosaccharides which are

usually N-linked (to asparagine) or less commonly O-linked (to serine or threonine). A coated vesicle then transports the completed glycoprotein to the cellular membrane from which the particular virus buds.

The precise composition of the oligosaccharides is determined, not only by the amino acid sequence and tertiary structure of the proteins concerned, but more importantly by the particular cellular glycosyl transferases active in the type of cell in which the virus happens to be growing.

Assembly and Release

All non-enveloped viruses of vertebrates have an icosahedral structure. The structural proteins of simple icosahedral viruses associate spontaneously to form structural units (called capsomers when considered morphologically), which then self-assemble to form capsids into which viral nucleic acid is inserted, often accompanied by conformational changes to the nascent capsid structure. Completion of the virion may also involve proteolytic cleavage of one or more species of capsid protein.

The mechanism of packaging viral nucleic acid into a pre-assembled empty procapsid has been well elucidated for adenoviruses. A particular protein binds to a nucleotide sequence at one end of the viral DNA known as the *packaging sequence*; this enables the DNA to enter the procapsid bound to basic core proteins, after which some of the capsid proteins are cleaved to produce the mature virion.

Most non-enveloped viruses accumulate within the cytoplasm or the nucleus and are released only when the cell eventually lyses (Fig. 4.7).

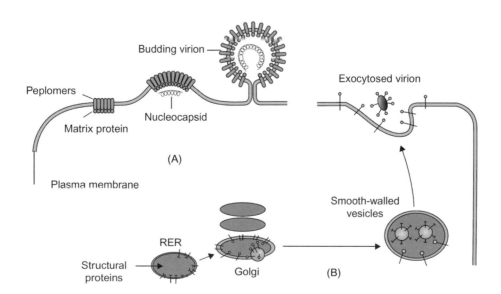

FIGURE 4.7 Maturation of enveloped viruses. (A) Viruses with a matrix protein (and some viruses without a matrix protein) bud through a patch of the plasma membrane in which glycoprotein peplomers have accumulated over matrix protein molecules. (B) Most enveloped viruses that do not have a matrix protein bud into cytoplasmic vesicles (rough endoplasmic reticulum [RER] or Golgi), then pass through the cytoplasm in smooth-walled vesicles and are released by exocytosis. *Reproduced from MacLachlan, N.J., Dubovi, E.J., 2011. Veterinary Virology, fourth ed. Academic Press, London (Fig. 2.12), with permission.*

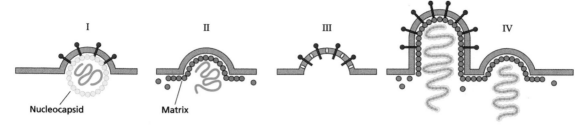

FIGURE 4.8   Four distinct strategies used by different viruses for budding. In type I budding, as with the alphaviruses, both the envelope glycoproteins and the internal capsid are essential. If altered or chimeric envelope proteins reach the plasma membrane, these do not support budding, indicating that the quaternary structures of the envelope heterodimers and capsid are involved in driving the process. Type II budding, for example Gag-dependent budding of many retroviruses, requires only the internal capsid or matrix proteins. Type III budding can be driven solely by the envelope proteins. Finally type IV budding is driven by viral matrix proteins but requires additional components; for example, internal matrix proteins alone can drive the budding of rhabdoviruses or orthomyxoviruses. but the process is inefficient or results in deformed particles unless envelope glycoproteins or the internal ribonucleoprotein are also present. *Reproduced from Flint, S.J., et al., 2009. Principles of Virology, third ed. ASM Press, Washington, DC (Fig. 13.15), with permission.*

All mammalian viruses with helical nucleocapsids, as well as some with icosahedral nucleocapsids (e.g., herpesviruses, togaviruses, and retroviruses) mature by acquiring an envelope by budding through cellular membranes.

Enveloped viruses may bud from the plasma membrane, from internal cytoplasmic membranes, or from the nuclear membrane; viruses that acquire an envelope within the cell are then transported within vesicles to the cell surface. Budding usually occurs through patches of membrane that contain viral glycoprotein(s) inserted into the lipid bilayer of the membrane. This occurs by lateral displacement of cellular proteins from that patch of membrane (Fig. 4.8). The cleaved single molecules of viral glycoprotein associate into oligomers to form typical rod-shaped or club-shaped peplomers with a hydrophilic domain projecting from the external surface of the membrane, a hydrophobic trans-membrane anchor domain, and a short hydrophilic cytoplasmic domain projecting slightly into the cytoplasm. In the case of icosahedral viruses, for example togaviruses, each protein molecule of the nucleocapsid binds directly to the cytoplasmic domain of the membrane glycoprotein oligomer, thus molding the envelope around the nucleocapsid. In the more usual case of viruses with helical nucleocapsids, a matrix protein attaches to the cytoplasmic domain of the glycoprotein peplomer, and the nucleocapsid protein recognizes the matrix protein to initiate budding. Release of each enveloped virion does not breach the integrity of the plasma membrane, hence many thousands of virus particles can be shed over a period of several hours or days without significant cell damage. Many but not all viruses that bud from the plasma membrane are only slowly cytopathic: and non-cytopathic may be associated with persistent infections.

Epithelial cells display *polarity*, that is they have an *apical* surface facing the outside world and a *basolateral* surface facing the interior of the body, the two separated by lateral cell–cell tight junctions. These surfaces are chemically and physiologically distinct. Viruses that are

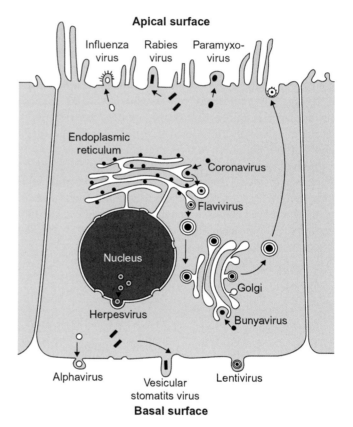

FIGURE 4.9   Polarity of virus release from infected cells. Viruses that bud from apical surfaces can be shed in respiratory or genital secretions or intestinal contents. Viruses budding from basal surfaces are available for systemic spread via viremia or lymphatics. Some viruses, for example, flaviviruses, bunyaviruses, and coronaviruses, take a more circuitous route in exiting the cell. Viruses that do not bud are usually released by cell lysis. *Reproduced from MacLachlan, N.J., Dubovi, E.J., 2011. Veterinary Virology, fourth ed. Academic Press, London (Figure 2.11), with permission.*

shed to the exterior (e.g., influenza viruses) tend to bud from the apical plasma membrane, whereas others (e.g., lentiviruses, such as human immunodeficiency virus) bud through the basolateral membrane (Fig. 4.9).

Flaviviruses, coronaviruses, arteriviruses, and bunyaviruses mature by budding through either membranes of the Golgi complex or the rough endoplasmic reticulum; vesicles containing the virus then migrate to fuse at the plasma membrane thereby releasing the virions by *exocytosis* (Fig. 4.8). Uniquely, the envelope of the herpesviruses is acquired by budding through the inner lamella of the nuclear membrane; the enveloped virions then pass directly from the space between the two lamellae of the nuclear membrane to the exterior of the cell via the cisternae of the endoplasmic reticulum.

SATELLITE VIRUSES AND VIROIDS

Satellite viruses are subviral particles that contain a DNA or RNA genome coding for a capsid protein, but that are absolutely dependent upon the presence of another virus for replication. The vast majority are found among plants but a few have an impact on medical virology and public health. Replication of the dependoviruses (family *Parvoviridae*, genus *Dependovirus*—adeno-associated viruses, now used as vectors for delivering heterologous genes of interest, e.g., vectored vaccines) for example, is dependent upon co-infection with an adenovirus—the adeno-associated virus produces coat protein but not the enzymes necessary for genome replication. Satellite viruses share little or no nucleotide sequence homology with the helper virus, yet can sometimes interfere with the replication of the helper virus.

Viroids are small, rod-like single-stranded RNA molecules with a high degree of secondary structure, approximately 250 to 450 bases in size, all sharing a common structural feature of a conserved central genomic region essential for replication. The RNA forms a hammerhead structure with the enzymatic properties of a ribozyme, an autocatalytic, self-cleaving molecule. This ribozyme function is used to cleave multimeric RNA structures produced during the course of replication (Chapter 22: Hepadnaviruses and Hepatitis Delta). Most are plant pathogens, remarkable in not coding for any protein. First described by Theodore Diener, it has been suggested that viroids may represent crucial intermediate steps in the evolution of RNA-based life forms from inorganic precursors.

Hepatitis Delta Virus

Hepatitis delta virus (HDV) is a unique example of a defective virus that requires co-infection with hepatitis B virus to provide its outer HBsAg-containing envelope. The HDV genome is a branched or rod-like, single-stranded RNA molecule similar to plant viroids in conformation, but unlike other viroids it codes for a single protein, the delta antigen. This nuclear phosphoprotein exists in two forms generated by RNA editing: the shorter form (195 amino acids) is required for HDV genome replication whereas the larger form (214 amino acids) is necessary for assembly and release from infected liver cells.

The HDV genome is thought to be replicated by the host cell RNA polymerase II enzyme via a rolling circle mechanism to produce a consecutive series of concatemers of first, antisense RNA and then, RNA of genome sense. These are immediately cleaved by the ribozyme domain in the RNA genome to generate new viral genomes: this is the only known example of a mammalian virus utilizing a ribozyme for genome replication (see Fig. 22.9).

GENERATION OF GENETIC DIVERSITY

A sample of any given virus inevitably contains a population of closely related genome sequences ("quasi-species"), replicating simultaneously at different and varying rates according to selection pressures. The *molecular mechanisms* involved in generating diversity include nucleotide substitution, insertion, or deletion; sequence duplication or deletion; genetic recombination; and genetic reassortment for viruses with segmented genomes. The error rate in nucleotide copying is approximately 1 in 10^5 nucleotides for DNA viruses, and 1 in 10^4 nucleotides for RNA viruses. This quasi-species population is then constantly varying its composition according to the most recent selection applied, providing the raw material for antigenic drift and for adaptation of the virus population to new hosts.

The selection properties that are desirable *in vivo* include the ability to replicate to high levels, the ability to be transmitted from host to host, evasion of the immune response, and resistance to host antiviral mechanisms and/or antiviral therapy. Not all of these properties are as necessary for successful replication in cell culture as they are *in vivo*, and virus stocks prepared *in vitro* are likely to include a different set of variants from those produced *in vivo*, say in experimental animal tissues. Serial passaging of wild-type virus through cell culture or a foreign host species has been a long used and successful way to generate virus strains with lowered virulence for the original host, to be used as "modified live-virus" vaccines.

Many different viral genes may be involved in such adaptations; for example, mutations in viral polymerases may affect replicative ability or resistance to antiviral drugs, changes in antigenic epitopes can lead to immune escape, changes in receptor ligands that affect receptor preference can alter many aspects of pathogenesis, and so on. Examples are discussed in Chapter 15: Emerging Virus Diseases.

QUANTITATIVE VIRUS ASSAYS

Advances in our knowledge of virus replication, pathogenesis, diagnosis, vaccine production, and many other areas would not have been possible without the tools to quantitate how much virus there is in a given sample or preparation. With the greater use of antiviral drugs, most notably in blood-borne diseases such as HIV, hepatitis B virus, and hepatitis C virus, the stage of disease and effectiveness of treatment are now routinely assessed by quantifying the level of virus (*viral load*) in the blood.

There are two approaches to measuring the titer of virus, either biological or physical. Physical assays measure actual virus particles, and include electron microscopy, hemagglutination, and serological assays for the amount of viral antigens; biological assays measure some viral function, for example, infectivity, reverse transcriptase activity. Tests performed on the same sample with different techniques will in some cases give significantly different results, and thus it is important to understand the reason for these differences.

The difference between the amount of virus detected using a physical assay such as particle counting by electron microscopy and a biological assay, for example a plaque assay (Fig. 4.10), is often referred to as the *particle to plaque-forming unit (pfu) ratio*. In virtually all instances, the number of particles exceeds the number determined in a biological assay, often with ratios greater than 1000:1. This is due to a number of reasons; first, not all virions may be capable of replication due to intrinsic defects, for example, because only a partial or defective genome is present, or the particle is empty, faulty capsid assembly, or because of lethal mutations in the genome; second, virions may have undergone environmental inactivation; third, the choice of cells for virus isolation and growth may not mimic the optimum intracellular environment of the natural target organ; fourth, the interaction between a fully viable virus particle and a permissive cell may not always successfully initiate infection and an excess of particles may be necessary to achieve a statistical chance of success.

Perhaps no other procedure has contributed as much to virology as the development of the plaque assay. The test was originally developed a century ago by Felix d'Herelle in his initial studies of bacteriophages and was subsequently adapted to mammalian viruses in 1953 by Renato Dulbecco and Marguerite Vogt. The assay is elegantly simple: serial 10-fold dilutions of a virus sample are made in a cell culture medium. These diluted samples are then added to preformed monolayer cells and incubated under agar or a carboxymethyl cellulose layer to prevent the spread of secreted virus. After 24 to 48 hours (or longer, according to the virus of interest), plaques of necrotic cells become visible under the light microscope. At an appropriate time, fixation and staining, for example, with crystal violet, reveals clearly visible holes in the monolayer, each corresponding to a focus of necrotic cells initiated by one viable virus particle. Serial dilution of the virus preparation facilitates the counting of discrete plaques so that, knowing the dilution, volume tested, and the plaque count, the concentration (titer) in the original sample can be determined. Immunohistochemical staining procedures using specific antisera can be used as an alternative for visualizing plaques formed by non-cytopathic viruses.

With the development of real-time PCR assays, the concentration of viral nucleic acid in a test sample can now be measured in virtually any context. By comparison with copy number controls, the concentration of nucleic acid in the treated sample can be determined. This type of assay does not detect empty capsids (those that do not contain viral nucleic acid), and, importantly, it does not necessarily relate to the infectivity of the sample.

Chapter 10: Laboratory Diagnosis of Virus Diseases, describes a further range of assays that are used in a diagnostic context, most of which can be also used both quantitatively and for research applications.

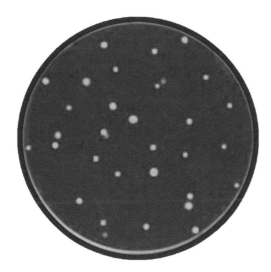

FIGURE 4.10 Plaque assay for virus quantitation. Cell monolayers are grown, typically in Petri dishes, and when confluent the virus inoculum is allowed to adsorb to the monolayer in a small volume. The monolayers are overlaid with agarose to restrict diffusion of virus particles, and incubated to allow virus replication beginning with single cells and then spreading as more cells become successively involved. The vital stain neutral red is then added; this is taken up by viable cells, staining the monolayer red and leaving clear holes. Plaque sizes and rates of development depend on the replication cycle and yield of the individual virus (see text). To determine the virus titer of the starting inoculum, serial 10-fold dilutions are inoculated on to duplicate monolayers. For those monolayers where discrete plaques can be distinguished (typically 10 to 50 plaques per dish), plaque numbers are counted and the original virus titer calculated by multiplying this plaque number by the dilution factor. By harvesting virus from a single plaque, a stock of genetically homogeneous virus can be obtained ("plaque-purifying"). Spontaneous mutants in a virus inoculum may be recognized by plaques of abnormal size or morphology, and can thus be harvested and separated for further study.

DEFECTIVE INTERFERING VIRUSES

This chapter has hitherto described how, during virus replication, non-infectious particles that lack the full genetic information essential for infectivity, are commonly produced—analogous to the assembly of defective motor vehicles from a car assembly plant. One special example involves defective interfering (DI) particles; these (1) lack the complete genome of the wild-type virus (deletion mutants), but (2) can still replicate in a new cell if the cell is also co-infected by wild-type virus which provides the missing function (by a process known as *complementation*). Such defective particles may then also (3) *interfere* with the ongoing replication of the wild-type *helper virus*, possibly due to possessing a competitive replication advantage over wild-type virus.

The production of DI particles is encouraged by infection at high multiplicity, where the chances of co-infection with full-length and defective viruses occurring in the same cell are increased. In fact, this phenomenon was first observed by Preben and Herdis von Magnus in 1944 with influenza virus; they found that very high multiplicities led to poor virus replication and the production of "interfering" virus stocks. A cyclical pattern may be observed, where DI particles increase in number as long as the culture has sufficient wild-type helper virus replication; however, as DI particles then inhibit the numbers of wild-type virus, the whole virus population diminishes, fresh rounds of wild-type virus replication can then occur and so a milieu for a return to DI replication is built up again. Despite several suggestions, it is not clear to what extent DI particles play a role in the pathogenesis of an ongoing infection in a host, for example, in modulating recovery from an acute infection or promoting chronic infection.

FURTHER READING

Bose, S., Jardetsky, T.S., Lamb, R.A., 2015. Timing is everything: fine-tuned molecular machines orchestrate paramyxovirus entry. Virology 479–480, 518–530.

Flint, S.J., Enquist, L.W., Racaniello, V.R., Rall, G.F., Skalka, A.-M., 2016. Principles of Virology, 4th ed. ASM Press, Washington, DC.

Freed, E.O., 2015. HIV-1 assembly, release and maturation. Nat. Rev. Microbiol. 13, 484–496.

Harrison, S.C., 2015. Viral membrane fusion. Virology 479–480, 498–507.

Hogle, J.M., 2002. Poliovirus cell entry: common themes in viral cell entry pathways. Annu. Rev. Microbiol. 56, 677–702.

Knipe, D.M., Howley, P.M. (Eds.), 2013. Fields Virology (sixth ed.), Lippincott Williams and Wilkins, Philadelphia, PA.

Noack, J., Bernasconi, R., Molinari, M., 2014. How viruses hijack the ERAD tuning machinery. J. Virol. 88, 10272–10275.

Ortin, J., Martin-Benito, J., 2015. The RNA synthesis machinery of negative-stranded viruses. Virology 479–480, 532–544.

Chapter 5

Innate Immunity

A virus invading the host must first breach natural barriers at the portal of entry. These may include (1) the physical barrier of the skin or the epithelial lining of the respiratory, gastrointestinal, or urogenital tract, (2) secretions at mucosal surfaces, for example, the surfactant and mucociliary functions of the respiratory tract, or the mucus, acid and detergent (bile) environment of the gastrointestinal tract.

Two major forms of immune defense then operate after an incoming virus has penetrated these physiological and anatomical barriers (Table 5.1). The first, innate immunity, is a mechanism that is continually present and operates immediately to limit tissue injury and prevent the spread of virus to adjacent, healthy cells, the so-called bystander effect. This innate response is of broad specificity, modulated largely by the secretion of an extensive array of signaling molecules (lymphokines, cytokines, etc.), and forms the basis of any immediate localized inflammatory response. In contrast, the adaptive immune response is pathogen-specific and requires more time to develop; it also generates long-term pathogen-specific memory that is the basis for a more rapid immune response and enhanced protection should the same pathogen be encountered again at a later date.

Although it is convenient to consider innate and adaptive responses as distinct responses to virus infection, many aspects of the initial innate response play a continuing role in later immune processes and thus the distinction between responses based on broad, non-specific recognition of viruses and specific recognition by immune cells is becoming increasingly blurred as more becomes known, especially in regard to cell signaling pathways. In short, the development of a robust adaptive immune response to an invading virus is intimately linked to the early innate response to infection.

Since the last edition of this book, it has become apparent that most, if not all, viruses of humans have evolved ways to circumvent the innate immune response. Some viruses can block or modify various stages in the intracellular pathways for expression of signaling molecules induced by infection. These effects may be exerted in a temporal manner during the different stages of virus replication (see below).

INNATE RESPONSES

The host responds rapidly in a matter of a few hours following virus invasion. The innate defense includes (1) phagocytic cells (macrophages, dendritic cells, and neutrophils) that engulf invading viruses, (2) natural killer (NK) cells that lyse infected cells, (3) activation of pattern recognition receptors (PRRs) that induce inflammatory mediators to stimulate the maturation of innate immune cells and their recruitment to the site of infection. This also induces the interferon response with direct antiviral activity and contributes significantly to the development of an adaptive immune response several days later, (4) small interfering RNA molecules (RNAi) that interfere with virus replication, and (5) induction of apoptosis (programmed cell death) that leads to elimination of infected cells.

MONOCYTES, MACROPHAGES, AND DENDRITIC CELLS

These cells are very much at the front line in the early host response to virus infection. Monocytes display considerable mobility and a homing capacity for sites of infection, infiltrating tissue, and thence differentiating into macrophages. Macrophages and dendritic cells occupy key locations in various tissues, for example, alveolar macrophages in the lung, Kupffer cells in the liver, and Langerhans dendritic cells in the skin. Both monocytes and macrophages are important initiators of the immune response against viral invasion. Macrophages often become the predominant cell within a focus of infection by 24 hours. Activated macrophages have increased chemotactic activity, phagocytic activity, and digestive powers. Dendritic cells carry out afferent immune functions at all body surfaces by promoting the transfer of immune signals and cells into the regional lymph nodes where together with the liver and spleen most phagocytic removal of foreign particles occurs. All three cell types bear immunoglobulin Fc and C3b receptors on their cell surface to promote the phagocytosis of immune complexes (consisting of virions coated with antibody). By serving as "professional" antigen-presenting cells these cells exercise a controlling influence over the

Fenner and White's Medical Virology. DOI: http://dx.doi.org/10.1016/B978-0-12-375156-0.00005-9

TABLE 5.1 Characteristics of Two Types of Immune Response

Property	Innate Immunity	Adaptive Immunity
Speed of response	Minutes/hours	Days—response is accelerated when the same antigen is met on subsequent occasions
Antigen specificity	No	Yes
Duration[a]	Days	Weeks
Memory	No	Yes
Effector mechanisms	(1) Complement, other serum proteins (2) Natural antibodies, prod. by B1 lymphocytes (3) Phagocytic cells (neutrophils, macrophages, dendritic cells) (4) Natural killer (NK) cells (5) Local cells (many types) that respond to PAMPs and produce cytokines including interferons (6) Apoptosis to remove infected cells (7) Small RNA molecules (RNAi) that interfere with virus replication	(1) Humoral response—different classes of antibodies produced by plasma cells which are derived from B lymphocytes (2) Cell-mediated response—mediated by cytotoxic T lymphocytes (CTLs), usually CD8+ (3) Macrophages, esp. following activation by cytokines, e.g., IFN-γ released by antigen-specific T cells and NK cells

[a]Duration is prolonged if there is continuing antigenic stimulus.

rapidity, magnitude, and dynamics of the immune response. Macrophages then also contribute to the efferent limb of the immune response: cytokines secreted by activated T cells bring more monocytes into the infection focus that then become differentiated into macrophages.

Viruses have developed a number of ways to circumvent the extreme degradative environment of the macrophage cytoplasm. Dendritic cells infected with viruses lose endocytic capacity, allowing the virus the opportunity to migrate into distant organs and tissues, for example, HIV and Ebola virus. The Ebola virus VP35 protein (see Chapter 28: Filoviruses) blocks dendritic cell activation and effectively breaks the link between innate and adaptive immunity. A few viruses, for example, poxviruses and herpesviruses, trigger dendritic cell apoptosis, with the result of a reduced capacity of the system to present viral antigens to the adaptive immune system.

THE ROLE OF NATURAL KILLER (NK) CELLS AND THE LINK WITH ADAPTIVE IMMUNITY

NK cells are another important component of the early defense system, with the activity of these cells greatly enhanced within one to two days of viral infection. Resting NK cells are found in large quantities in the spleen, uterus, liver, and blood, but can be rapidly recruited to any site in the body in response to chemokine signaling from damaged tissue, tumor cells, and cells infected with a pathogen. The NK cell surface markers include the neural cell adhesion molecule (NCAM: CD56) or the low affinity IgG receptor CD16, or both. These large, non-phagocytic cells of the

lymphoid system share the same lineage as mature T cells of the adaptive immune system, but differ in a number of important regards. The first is that NK cells lack T cell receptors (TCRs) but are activated by recognizing infected or tumor cells on which the density of molecules of the major histocompatibility complex (MHC: see Chapter 6: Adaptive Immune Responses to Infection) has been reduced. Viruses downregulate surface MHC class I molecules, in order to escape recognition by CD8+ T cells. However, cells lacking class I MHC molecules are susceptible to NK cells: the so-called missing self hypothesis. This reduced MHC density on the target cell disturbs the balance between activator and inhibitory receptor signaling at the surface of the NK cell, as the inhibitory receptors recognize class I MHC molecules on the surface of healthy cells. Second, NK cells are activated by several non-antigen-specific mechanisms to produce preformed pore-inducing (perforin) and granzyme molecules that induce apoptosis in the target cell: this is in contrast to cytotoxic, CD8+ T cells of the adaptive immune response that only produce these effector molecules once activated via recognition of class I MHC molecules in association with viral peptides. The third, and most important, difference is that NK cells do not need to proliferate and differentiate into viral-specific cells before activation. Overall, NK cells, while not displaying any immunological specificity for particular viral antigens, play an essential role in defense by mediating the death of infected cells by apoptosis, and by priming the adaptive response through secretion of several cytokines including interferon γ, tumor necrosis factor α (TNF-α), IL-4, and IL-13 (Fig. 5.1).

Several families of viruses have developed mechanisms for evading NK responses, most notably members of the families *Herpesviridae*, *Papillomaviridae*, *Poxviridae*, *Retroviridae*, and *Flaviviridae*. These fall into five broad strategies (Table 5.2). Four of these strategies are designed to interfere with the stimulation of NK cells by the presence

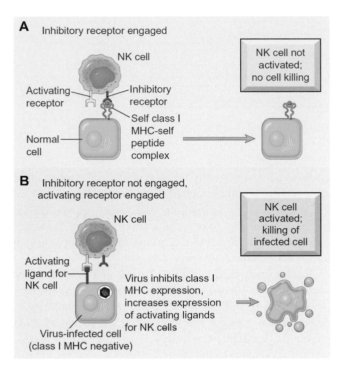

FIGURE 5.1 Activating and inhibitory receptors of NK cells. (A) Healthy cells express self class I MHC molecules which are recognized by inhibitory receptors, thus ensuring NK cells do not attack normal cells. Healthy cells may express ligands for activating receptors (not shown) or they may not express such ligands (as shown), but they do not activate NK cells because they engage the inhibitory receptors. (B) In virus-infected cells, class I MHC expression is reduced so that the inhibitory receptors are not engaged and ligands for activating receptors are expressed. The result is that NK cells are activated and the infected cells are killed *Reproduced from MacLachlan, N.J., Dubovi, E.J., 2011. Veterinary Virology, fourth ed. Academic Press, London (Figure 4.1), with permission.*

of infected cells, either by enhancing inhibitory signals or blocking stimulatory signals to NK cells. For example the herpesvirus protein UL18 mimics host cell class I MHC molecules, such that, when bound with NK inhibitory cellular receptors, cell activation is suppressed. Viruses can also have a direct effect on NK cells. Both HIV and herpes simplex virus can infect NK cells *in vitro*, and the E2 protein of hepatitis C virus binds directly to the CD81 protein on the surface of NK cells leading to inhibition of activation signals.

THE IMPORTANCE OF MOLECULAR RECOGNITION

As mentioned above, an early important line of defense is the recognition of viral components by germline-encoded pattern recognition receptors (PRRs) capable of distinguishing viral products from those of the host. PRRs are expressed in many cell types likely to be present at portals of virus entry, including macrophages, dendritic cells, neutrophils, NK cells, endothelial cells, and mucosal epithelial cells. These cellular receptor molecules recognize different classes of pathogen-associated molecular patterns (PAMPs). There are a number of classes of PRRs that are present on the majority of mammalian cells. These include the C-type lectin receptors, NOD-like receptors, the trans membrane Toll like Receptor (TLR) receptors, and the cytosolic RIG-I like receptors (RLR) such as RIG-I and MDA5. TLRs consist of amino-terminal leucine-rich repeat domains responsible for PAMP recognition and a cytoplasmic carboxyl-terminal domain that recognizes interleukin-1 and leads to downstream signal transduction. In contrast, the RLRs recognize associated PAMP via caspase activation and recruitment (CARD) domains. Common to both classes of PRR, the PRR–PAMP complex then triggers a series of signaling cascades, culminating in the production of pro-inflammatory cytokines and other

TABLE 5.2 Viral Mechanisms for the Evasion of Natural Killer (NK) Cells

Mechanism of Action	Examples	Outcome
(1) Homologues of class 1 MHC	Herpesviruses	Bind to NK cell inhibitory receptor, Inhibit NK cytotoxicity
(2) Regulation of class I MHC expression on target cell	Herpesviruses, SIV	Inhibition of NK cytotoxicity
(3) Virus-coded protein interfering with NK cell-activating receptor/ligand interactions	Herpesviruses, HIV, HTLV	Inhibition of NK cytotoxicity and IFN-γ production
(4) Inhibition of NK cell-activating cytokine by binding cytokine or producing chemokine antagonist	Herpesviruses, papillomaviruses,	Inhibition of IFN-γ production, trafficking
(5) Direct effects of virions; e.g., block an inhibitory receptor, or directly infect NK cells; HCV E2 protein binds directly to CD81 on NK cell	Herpesviruses, HIV	Reduces NK cell activity

TABLE 5.3 Toll-Like Receptors (TLRs) for PAMPs

Toll-Like Receptor (TLR) Family	Other TLR Members	Location	Ligand
1[a]	2,6,10	Cell surface	Microbial cell walls (lipoproteins, peptidoglycans)
3		Endosomes	dsRNA
4		Cell surface	Bacterial lipopolysaccharides
5		Cell surface	Bacterial lipopolysaccharides
7	8,9	Endosomes	ssRNA (TLR7,8), unmethylated CpG DNA (TLR9)
11	12, 13	Endosomes	*Toxoplasma gondii* (TLR11,12) bacterial ribosomal RNA (TLR13)

[a]*TLR2 pairs with either TLR1, 6 or 10 to form surface heterodimers.*

signaling molecules designed to limit the spread of infection, including interferons (see below). Importantly, these molecules also play an important role in the development of adaptive immune responses necessary for viral clearance and sustained, long-term immunological memory.

Most attention has been focused on the class of TLRs and RLRs. The TLRs were first discovered in studies of fruit flies (*Drosophila* spp.) and subsequently 10 homologues were found in humans. Human TLRs are classified into six major groups that recognize both bacterial and viral PAMPs (Table 5.3).

PAMPs may not necessarily be proteins and in the case of viral infection the TLRs recognize predominantly nucleic acids. One important PAMP consists of double-stranded RNA replicative intermediate structures, as will be discussed below. Although the 21 naturally occurring amino acids could theoretically assume diverse molecular structures in forming viral proteins, in reality the repertoire of secondary polypeptide folds that act as PAMPs is more limited, governed by the polar (hydrophilic) and non-polar (hydrophobic) side chains of individual amino acids. A prime illustration is a comparison of the VP1 major capsid proteins of biologically unrelated picornaviruses: sequence alignments show little homology yet these proteins fold to give the morphology characteristic of this virus family. Of course the adaptive immune response can distinguish these amino acid differences thus enabling a virus-specific response.

INTERFERONS

In 1957 Alick Isaacs and Jean Lindenmann showed that cells of the chorioallantoic membrane of embryonated hens' eggs infected with influenza virus release a non-viral protein that protects cells in culture from infection with the same or an unrelated virus. Dubbed "interferon," this discovery led in the 1980s to further research showing that interferon was in fact a large family of cell signaling molecules known as cytokines. Interferons are thus a response to the infection of any one cell, and the secreted interferon molecules induce the expression of interferon stimulatory genes (ISGs) by the surrounding cells leading to an antiviral response. The numbers of ISGs induced in the surrounding cells can number in excess of 100 (Fig. 5.2).

Interferons (IFNs) are classified into three types according to receptor usage (Table 5.4). Type I interferons (IFN-α, IFN-β, and others) are produced by a majority of cells and play a major role in limiting the spread of virus infections; type II interferon (IFN-γ) activates macrophages, recruits leukocytes to the sites of infection, and potentiates type I interferons; while type III (IFN-λ) interferons are particularly prominent in the control of infections at mucosal surfaces, for example the gastrointestinal and respiratory tracts.

All interferons bind to specific cell surface receptor complexes consisting of multiple polypeptide chains. Among the type I interferons, IFN-α and IFN-β have received much attention as potential therapeutics although other interferons are present in humans (IFN-ε, IFN-κ, and IFN-ω). IFN-α is used for the treatment of hepatitis B and hepatitis C as well as for the treatment of some cancers. IFN-β has been recommended for the treatment of multiple sclerosis.

The IFN-λ cytokine family is composed of IFN-λ1 (previously known as IL-29), IFN-λ2 (IL-28A), IFN-λ3 (IL-28B), and IFN-λ4, all of which are coded on human chromosome 19. In contrast to IFN-α and IFN-β signaling, IFN-λ receptors are specific to melanocytes, liver cells, and epithelial cells. Indeed, IFN-λ appears to have evolved specifically to protect the epithelium from virus infection. Since the induction of ISGs in surrounding cells by IFN-λ is perhaps more prolonged than with IFN-α, it is receiving considerable attention as a potential therapeutic agent. The density of receptor molecules on hepatocytes for IFN-λ3 varies according to genotype; a link has been established between IFN-λ3 genotype and the degree of responsiveness of hepatitis C patients to treatment with commercial IFN-α.

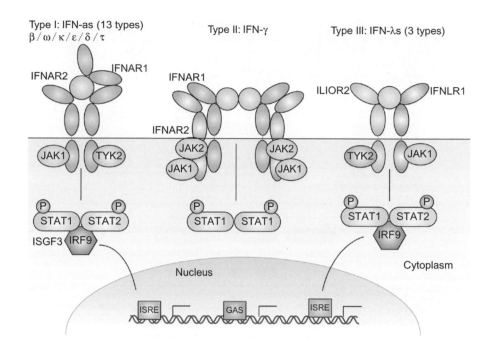

FIGURE 5.2 Pathways of action of interferons. Type I IFNs interact with IFN- (α, β, and ω) receptor 1 (IFNAR1) and IFNAR2; type II IFN interacts with IFN-γ receptor 1 (IFNGR1) and IFNGR2; type III IFNs interact with IFN-λ receptor 1 (IFNRL1; also known as IL28RA) and IL-10 receptor 2 (IL10R2; also known as IL10RB). Type II IFN-γ is an antiparallel homodimer exhibiting a twofold axis of symmetry. It binds two IFNGR1 receptor chains, assembling a complex that is stabilized by two IFNGR2 chains. These receptors are associated with two kinases from the JAK family: Janus (JAK)1 and tyrosine (TYK2) for types I and III IFNs; JAK1 and JAK2 for type II IFN. All IFN receptor chains belong to the class 2 helical cytokine receptor family, with the 200 amino acid extracellular domains usually contain the ligand binding site. IFNAR2, IFNLR1, IL10R2, IFNGR1, and IFNGR2 are classical representatives of this family, whereas IFNAR1 is atypical, as its extracellular domain is duplicated. GAS, IFN-γ-activated site; IRF9, IFN regulatory factor 9; ISGF3, IFN-stimulated gene factor 3 (refers to the STAT1–STAT2–IRF9 complex); ISRE, IFN-stimulated response element; P, phosphate; STAT1/2, signal transducers and activators of transcription 1/2. *Adapted from Borden, E.C., et al., 2007. Nat. Rev. Drug Discov. 6, 975–990 and MacLachlan, N.J., Dubovi, E.J., 2011. Veterinary Virology, fourth ed., Academic Press Fig. 4.2, with permission.*

TABLE 5.4 Three Types of Interferons

	Type I	Type II	Type III
Examples	IFN-α: Different species-specific examples IFN-β; 1 type IFN-δ, IFN-ε, IFN-κ, IFN-o, IFN-τ	IFN-γ	IFN-λ-1, IFN-λ-2, IFN-λ-3, IFN-λ-4,
Produced by	Most nucleated cell types	T cells and NK cells	Many cell types
Receptor	IFNAR, a heterodimer of INFAR-1 and IFNAR-2	IFNGR, a tetramer of 2 heterodimers of IFNGR-1 and IFNGR-2	IFNLR1, present on melanocytes, liver cells, epithelial cells
Effects of binding to receptor	Activates cascade including TYK, JAK, STAT, and IRF9 to induce IFN stimulated response elements (ISREs)	Activates pathway involving JAK and STAT to induce IFN-γ-activated site (GAS)	Activates pathway involving JAK and STAT leading to expression of IFN-stimulated genes

The induction of innate responses results in the activation of intracellular pathways that often involve overlapping signaling mechanisms. For example, transcription of IFN mRNAs is controlled by several IFN regulatory factors (IRFs), which are activated by signaling pathways that themselves become activated following interaction between different PAMPs and PRRs, particularly the TLRs 3, 7, 8, and 9 in the endosome but also RLRs. in the cytoplasm. The role of the transcription factor NF-κB is also critical for both IFN and inflammatory

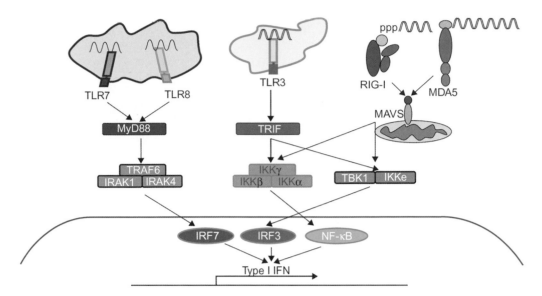

FIGURE 5.3   Pathways for virus recognition and interferon induction. In dendritic cells, TLR7 and TLR8 located in endosomal compartments recognize viral ssRNA after direct infection, autophagocytic uptake of viral material from cytoplasm, or phagocytic uptake of other infected cells or vial particles. Both TLR7 and TLR8 signal through the adapter MyD88, which through interaction with the IRAK4–IRAK1–TRAF6 complex leads to phosphorylation and activation of IRF7 and subsequent IFN transcription. TLR3 located in endosomes of DCs, macrophages, epithelial cells, and fibroblasts, is activated by encountering dsRNA. Following its activation, TLR3 signals through its adapter, TRIF, which leads to activation of non-canonical IKK kinases (TBK1/ IKKε) and subsequent phosphorylation and nuclear translocation of IRF3. Nuclear factor κ B (NF-κB) is also activated by TRIF-mediated signaling through canonical IKK kinases (IKKα, β, and γ). Cytoplasmically located RIG-1 and MDA5 are expressed in most cells and recognize 5′ppp-containing dsRNA or long dsRNA, respectively. Both of these cytoplasmic sensors upon activation interact and signal through the mitochondrially located adapter, MAVS. This signaling pathway, analogous to that of TLR3, leads to activation of the canonical and non-canonical IKK kinases and the following nuclear translocation of NFκB and IRF3. Concurrent activation of IRF3 and NFκB in turn allows for transcription of IFN genes and its synthesis and export. DC, dendritic cells; IKK, IκB kinases; IRAK, interleukin receptor-associated kinase; IRF, IFN regulatory factor; MDA5, melanoma differentiation-associated gene 5; MVAS, mitochondrial antiviral signaling protein; RIG, retinoic-acid-inducible gene; TBK, TRAF family member-associated NFκB activator binding kinase; TLR, Toll-like receptor; TRAF, tumor necrosis factor receptor-associated factor; TRIF, TIR-domain-containing adapter-inducing IFN-β. *Adapted from Baum, A., Garcia-Sastre, A., 2010. Amino Acids 38, 1283–1299 and From MacLachlan, N.J., Dubovi, E.J., 2011. Veterinary Virology, fourth ed. Academic Press (Fig. 4.3), with permission.*

responses. Normally sequestered in the cytoplasm through attachment to an inhibitory protein, the latter is degraded once virus infection has activated the relevant pathway, allowing NF-κB to translocate into the nucleus and there activate a myriad of cytokine genes, including IFN-β. Together, the IRF systems in conjunction with NF-κB are major regulatory elements controlling inflammatory gene expression and IFN production (Fig. 5.3).

It is thought that the major site of TLR recognition occurs within endosomes, where internalized viruses are detected. This endosomal location provides some specificity for recognition of viral RNA/DNA, as cellular derived nucleic acids are not normally present within endosomes. The role of each TLR may also differ between cell types. Intracellular TLR3 is particularly important as a cytoplasmic sensor of viral nucleic acid, especially dsRNA, a replicative intermediate present in many virus-infected cells. Activation of TLR3 leads to the production of both IFN types I and III and inflammatory cytokines via the TRIF-dependent pathway (Fig. 5.3): TLR3 agonists provide protection against many different viruses, for example HIV, coronaviruses, and hepatitis B virus.

Another example of endosomal recognition is the interaction between ssRNA and TLR7 and TLR8. TLR7 signaling is particularly prominent in peripheral dendritic cells, leading to the expression of type I IFNs. Viruses entering by direct membrane fusion at the cell surface may undergo autophagy, and viral ssRNA molecules are transported to endo-lysosomes where contact can be made with TLR7. A third mechanism has been uncovered whereby exosomes containing viral RNA are transferred by cell-to-cell contact to peripheral dendritic cells.

DNA viruses such as herpesviruses and poxviruses activate IFN-α production via a TLR9-dependent pathway: cooperation is required between TLR9 and TLR2 for a response to EBV and adenoviruses. The role of TLR9 in recognizing the ssDNA of parvoviruses is less clear. Also little is known regarding the role of other TLRs in responding to viral infection: although TLR2 and TLR4 sense primarily bacterial components, there is some evidence for either or both playing a role in controlling virus infection.

The action of type I interferon involves its binding to interferon receptors (IFNAR1/2), which by activating further signaling cascades (JAK/Stat pathway) leads to expression

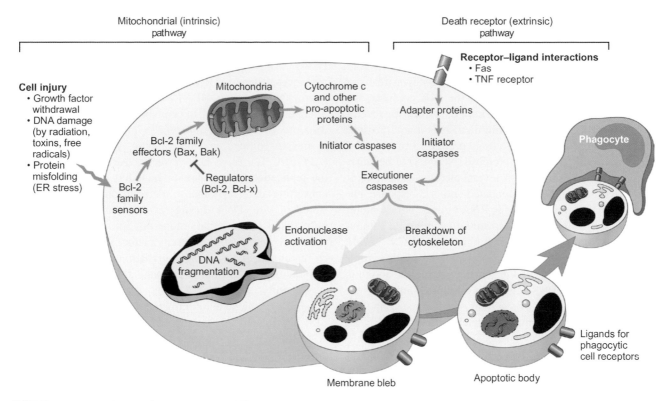

FIGURE 5.4 Two pathways of apoptosis. These differ as to mode of induction and regulation, but both culminate in the activation of "executioner" caspases that degrade cellular proteins and DNA leading to cell death. *Reproduced from MacLachlan, N.J., Dubovi, E.J., 2011. Veterinary Virology, fourth ed. Academic Press (Fig. 4.5), with permission.*

of a large number of ISGs whose main function is control of viral replication and immune modulation. Those induced ISGs that contributes to the interferon induced antiviral state are numerous. These continue to grow and include ISG15, a ubiquitin homologue that targets more than 150 proteins for degradation; MxGTPase, a hydrolyzing enzyme that affects vesicle formation and prevents virus maturation; the 2'5'oligoadenylate synthetase pathway, which activates cellular RNAse that cleaves viral RNAs; and the protein kinase (PKR) pathway responsible for the phosphorylation of elongation initiation factor eIF-2α and which therefore inhibits protein synthesis.

CELL DEATH AND APOPTOSIS

Death of an infected cell may be viewed as beneficial to the host if it impedes the production of progeny viruses and thus prevents further dissemination in the body. Various derangements of cell function and metabolism due to virus infection have long been recognized as causes of cell *necrosis*, but it is also evident that the induction of *apoptosis*, or programmed cell death, is another mechanism whereby infected cells undergo active self-destruction. Apoptosis in the context of viral infection can be induced by one of two pathways: an intrinsic mitochondrial pathway whereby cell

injury leads to increased mitochondrial permeability and leakage of mitochondrial proteins into the cytosol; or an extrinsic death domain pathway in which binding of TNF to, or interaction of cytotoxic T lymphocytes with, specific TNF cellular receptors triggers the apoptotic pathway. Once initiated by either route, the activation of host caspase enzymes degrades cellular DNA and proteins leading to cell death by apoptosis (known as the "executioner" phase) (Fig. 5.4).

As noted above, apoptosis of virus-infected target cells can also be initiated by cytotoxic T lymphocytes and NK cells using preformed mediators such as perforin and granzyme that directly activate caspases in the target cell.

EVASION STRATEGIES

Almost all viruses have developed elaborate, and often complex, strategies to circumvent host innate immunity highlighting the importance of the innate immune response; host cells have also evolved countermeasures to these evasion strategies, giving rise to a "genetic arms race" with both virus and host competing against the other. Viral evasion strategies include the degradation of TLR signaling components, interference with transcription factors, and mimicry of cellular proteins. Given its importance in dsRNA recognition, the TLR3 pathway is frequently targeted, with TRIF, the adaptor for TLR3, being particularly targeted for degradation by viral proteases. This targeting of TRIF is particularly effective, as TRIF abrogation also inhibits both NF-κB and IRF3 production. Other points downstream in the pathways may also be blocked: the IKK complex, for example, a complex of cellular kinases pivotal in several activation pathways, is often inhibited, as is ubiquitination, a regulatory process controlling protein turnover and cellular location. The serine protease (NS3/4A) of hepatitis C virus and the 3C protease of several picornaviruses can all cleave TRIF leading to an attenuated antiviral response, but targeting TRIF is not confined to RNA viruses. For example, TRIF levels are significantly downregulated during gamma herpesvirus infections. Interestingly, the HCV serine protease also cleaves IPS-1, an adaptor molecule within the RIG-I pathway involved in sensing viral RNA, and thus HCV can modulate PRR sensing at multiple points.

For many viruses, preserving and expanding the endoplasmic reticulum of the infected cell is vital for the purpose of virus assembly and maturation while preventing stress-induced apoptosis that would otherwise be a natural consequence. Dengue viruses and other flaviviruses do this by careful control of the cellular unfolded protein response, or UPR, preventing the development of cytopathicity yet allowing for the rearrangement of cellular membrane components necessary for endoplasmic reticulum expansion. Autophagy, an important element of innate immunity, is also utilized by dengue viruses. Normally autophagosomes coalesce with lysosomes but dengue viruses inhibit this process, thus allowing these viruses to sequester within autophagosomes for the purpose of RNA replication. This complements the capacity of the virus non-structural protein NS4B to block synthesis of IFN-α and IFN-β through inhibition of the JAK/STAT pathway by blocking the phosphorylation of STAT1. The result is the inhibition of interferon-inducible transmembrane protein synthesis.

Mimicry of key components of innate immunity is a strategy employed by many viruses. For example, the V protein of paramyxoviruses mimics IRF3 and acts as a non-functional substrate for IRF3 kinases, thereby inhibiting pathways initiated by TLR3 binding to viral RNA. Perhaps the best studied viruses in this respect are the poxviruses: protein A49 effectively blocks through mimicry the induction of type I IFN through abolishing NF-κB activation, a key molecule for the induction of IFN gene transcription (see above and Fig. 5.2).

Following initial infection of a host cell, many host restriction factors recognize viruses and directly inhibit their replication. HIV and simian immunodeficiency viruses (SIVs) are recognized by several host restriction factors in their respective human and non-human primate hosts. Tripartite motif-containing protein 5α (TRIM5α) is a species-specific host restriction factor that restricts the replication of HIV-1 in Old World monkeys such as rhesus and cynomolgus macaques. Rhesus TRIM5α restricts HIV-1 infection by interacting with the HIV-1 capsid at an early stage of infection, and is believed to be involved in the innate immune response to retroviral infection. Similarly, apolipoprotein B mRNA-editing, enzyme-catalytic, polypeptide-like 3G (APOBEC3G) is a host protein; APOBEC3G restricts the replication of retroviruses including HIV and SIV, by converting deoxycytidine to deoxyuridine on the minus strand of viral DNA during reverse transcription, thereby introducing (often lethal) mutations into the genome. The retroviral protein viral infectivity factor (Vif) counteracts this by binding to, and assisting degradation of, APOBEC3G via a ubiquitin-dependent proteasomal pathway.

Recently, a study of four African green monkey subspecies, which can be infected with divergent strains of SIV, highlighted that there is ongoing evolution of simian APOBEC3G in the absence of ongoing disease even in a non-pathogenic infection. In response to these changes, both natural isolates from long-term infected individuals and viruses from experimentally infected individuals also adapt to retarget the host restriction factor. These factors may contribute to the narrow species specificity of closely related retroviruses, and highlight the ongoing conflict between virus and host.

In another example, the transmembrane protein tetherin inhibits the detachment of enveloped viruses from the cell membrane; in the case of HIV, the viral protein U (Vpu) counteracts this by degrading and downregulating tetherin, thereby enhancing virus release. This action of Vpu also helps to protect cells from elimination by antibody-dependent cell cytotoxicity. Vpu also acts to enhance the degradation of newly formed CD4 molecules.

FURTHER READING

Flint, S.J., Enquist, L.W., Racaniello, V.R., Rall, G.F., Skalka, A.-M., 2016. *Principles of Virology*, 4th ed. ASM Press, Washington, DC.

Green, A.M., Beatty, P.R., Hadjilao, A., Harris, E., 2013. Innate immunity to dengue virus infection and subversion of antiviral responses. J. Mol. Biol. Available from: http://dx.doi.org/10.1016/j.mb.2013.11.023.

Jost, S., Altfield, M., 2013. Control of human viral infections by natural killer cells. Annu. Rev. Immunol. 31, 163–194. Available from: http://dx.doi.org/10.1146/annurev-immunol-032712-100001.

Lester, S.N., Li, K., 2013. Toll-like receptors in antiviral innate immunity. J. Mol. Biol. Available from: http://dx.doi.org/10.1016/j.mb.2013.11.024.

Mercer, J., Greber, U.F., 2013. Virus interactions with endocytic pathways in macrophages and dendritic cells. Trends Microbiol. 21, 380–388. Available from: http://dx.doi.org/10.1016/j.tim.2013.06.001.

Chapter 6

Adaptive Immune Responses to Infection

Two major phenomena are involved in recovery from infection: (1) the destruction of infected cells, and (2) the neutralization of the infectivity of virions. The adaptive immune response (Fig. 6.1) contributes to each of these processes. Adaptive immunity is antigen-specific, and takes at least several days to develop. Furthermore, it leads to antigen-specific memory, resulting in a more rapid antigen-specific immune response when the same agent is again encountered.

This chapter deals with the development of the adaptive immune response, how it aids in recovery from viral infection, and subsequently protects the individual against reinfection. Later chapters describe situations when the immune response can actually be harmful, becoming a significant component of disease pathogenesis. In circumstances where the virus evades the immune system a persistent infection may be established.

OVERVIEW OF THE ADAPTIVE IMMUNE RESPONSES TO VIRAL INFECTION

The adaptive immune response to viruses includes both humoral and cellular components. Humoral immunity is produced when B lymphocytes respond to an antigenic stimulus and differentiate into plasma cells that produce antibodies. Cell-mediated immunity involves T lymphocytes responding by secreting cytokines that regulate the immune response and coordinate the activities of the various types of cells involved, including antibody production by B lymphocytes. T lymphocytes also have direct effector functions such as cytotoxic functions (Fig. 6.1). These responses are triggered by highly specific receptor molecules on the surfaces of both B and T lymphocytes—these receptors recognize discrete regions of viral peptides, known as antigenic determinants or epitopes. The reader should note that many of the components involved in innate immunity and discussed in Chapter 5: Innate Immunity, for example macrophages, monocytes, dendritic cells, cytokines, also play key roles in the adaptive immune response as discussed below.

Virus infection is recognized initially by a group of molecules known as pattern recognition receptors (PRRs) on the surface of sentinel cells such as macrophages and dendritic cells. These PRRs alert both the innate and adaptive immune

systems to the presence of viruses. Some virus particles may be phagocytosed by macrophages. Except in the case of certain viruses capable of growing in macrophages, the engulfed virions are destroyed and viral proteins are cleaved and secreted as short peptides. These peptides are then ingested by dendritic cells, which migrate to nearby secondary lymphoid organs where these peptides are presented on the cell surface in association with class II MHC proteins, or in the case of highly specialized "cross-presenting" dendritic cells, additionally on class I MHC molecules (see below). This combination is recognized by naïve CD4+ or CD8+ T lymphocytes,[1] giving rise to T-helper cells (e.g., Th1), T follicular helper (Tfh) cells, and cytotoxic lymphocytes (CTLs), involving a process known as clonal selection.

Th1 lymphocytes migrate to the site of infection and secrete IFNγ that enhances the cytotoxic response against virus-infected cells. In contrast, Tfh lymphocytes assist appropriate clones of B lymphocytes, following binding of viral antigen, to divide and differentiate into plasma cells. This "help" provided to the B cells is crucial for the development of high-affinity antibodies directed against the virus. In a different pathway, cytotoxic T (Tc or CTLs) cells are activated following recognition of viral peptides in association with class I MHC on the surface of infected cells. The Tc response usually peaks at about one week after infection, compared with the antibody response being greatest later at two to three weeks.

Antibody synthesis takes place principally in the spleen, lymph nodes, gut-associated lymphoid tissues (GALT), and bronchus-associated lymphoid tissues (BALT). Viral antigens are carried directly via the blood or lymphatics to the spleen and lymph nodes, where binding to antigen receptors takes place on the surface of B cells. This ultimately results in the synthesis of antibodies mainly restricted to the IgM class early in the response and the IgG subclasses subsequently. On the other hand, the submucosal lymphoid tissues of the respiratory and digestive tracts, such as the tonsils and Peyer's patches, receive antigens directly from overlying epithelial cells, producing antibodies mainly of the IgA class (Box 6.1).

---

[1] Cellular differentiation (CD) antigens: by convention cells displaying certain CD antigens are designated for example CD4+, CD8+, etc.

Fenner and White's Medical Virology. DOI: http://dx.doi.org/10.1016/B978-0-12-375156-0.00006-0

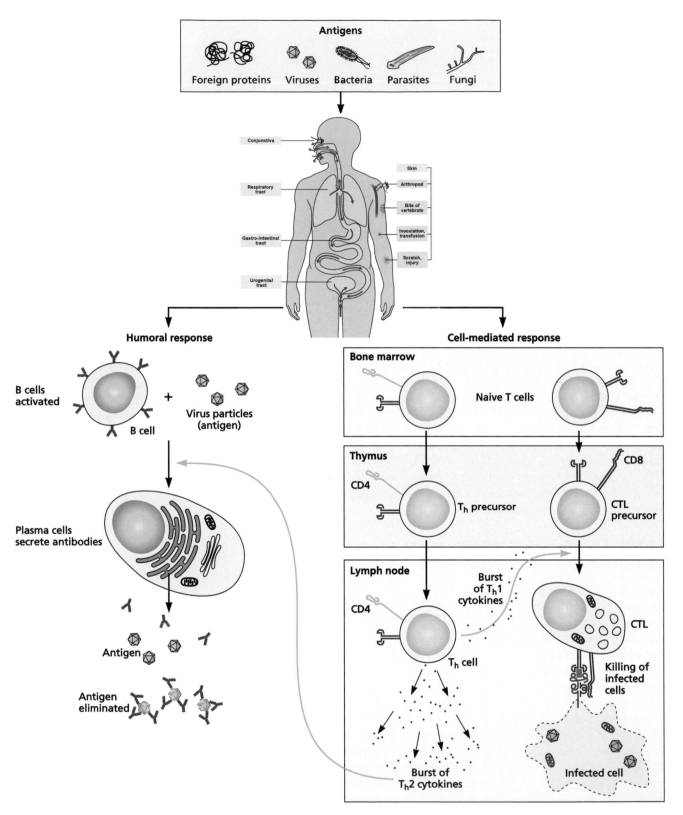

FIGURE 6.1 The humoral and cell-mediated branches of the adaptive immune system. A variety of foreign proteins and particles (antigens) may stimulate adaptive immune responses after recognition by intrinsic and innate defense systems. The humoral branch consists of lymphocytes of the B cell lineage, which produce specific antibodies (immunoglobulins). The process begins with the interaction of specific receptors on precursor B lymphocytes with antigens. Binding of antigen promotes differentiation into antibody-secreting cells (plasma cells). The cell-mediated branch consists of lymphocytes of the T cell lineage that arise in the bone marrow and are selected in the thymus. The activation process is initiated in lymph nodes when the T cell receptor on the surface of naïve T lymphocytes binds viral peptides on dendritic cells complexed with MHC class II protein. Two subpopulations of naïve T cells are illustrated: the Th-cell precursor and the CTL precursor. The Th cell recognizes antigens bound to MHC class II molecules and produces powerful cytokines that "help" activated B cells to differentiate into antibody-producing plasma cells (Th2 cytokines), or induce CTL precursors (Th1 cytokines) to differentiate into CTLs capable of recognizing and killing virus-infected cells. The Th1 or Th2 cytokines are produced by different subsets of Th cells. *Reproduced from Flint, S.J., et al., 2009. Principles of Virology: Pathogenesis and Control, third ed. ASM Press, Washington, DC, with permission.*

BOX 6.1 Immunoglobulin Classes

Immunoglobulin G (IgG)

The major class of antibody in the blood is immunoglobulin G (IgG), which occurs as IgG1, IgG2, IgG3, and IgG4 subclasses. Following systemic viral infections, IgG continues to be synthesized for many years and is the principal mediator of protection against reinfection. The subclasses of IgG differ in the constant region of their heavy chains and consequently in biological properties such as complement fixation and binding to phagocytes.

Immunoglobulin M (IgM)

IgM is a particularly avid class of antibody, being a pentamer of five IgG equivalents, with 10 Fab fragments and therefore 10 antigen-binding sites. Because IgM is formed early in the immune response and is later replaced by IgG, specific antibodies of the IgM class are diagnostic of recent (or chronic) infection. IgM is the first immunoglobulin found in the fetus as it develops immunological competency in the second half of pregnancy. Since IgM does not cross the placenta from mother to fetus, the presence of IgM antibodies against a particular virus in a newborn is indicative of intrauterine viral infection.

Immunoglobulin A (IgA)

IgA is a dimer, with four Fab fragments. Passing through epithelial cells, IgA acquires a J fragment (J, for joining, also called the secretory component) to become secretory IgA (sIgA) prior to secretion through the epithelium into the respiratory, intestinal, and urogenital tracts. Secretory IgA is more resistant to proteases than other immunoglobulins, and it is the principal immunoglobulin on mucosal surfaces and in milk and colostrum. For this reason IgA antibodies are important in resistance to infection of the respiratory, intestinal, and urogenital tracts. IgA antibody responses are much more effectively elicited by oral or respiratory than by systemic administration of antigen, a matter of importance in the design and route of delivery of some vaccines (see Chapter 11: Vaccines and Vaccination).

Immunoglobulins D and E

IgD and IgE are minor immunoglobulin species, accounting for less than 1% of total immunoglobulin levels. The majority of IgD antibodies are bound to the surface of B lymphocytes but as yet without any described function. IgE antibodies are produced by sub-epithelial plasma cells in the respiratory and intestinal tracts, and bind strongly to mast cells where they react with certain kinds of antigens (allergens). IgE stimulates the release of mediators of inflammation such as serotonin and histamine.

B Lymphocytes

Some of the pluripotent hematopoietic stem cells originating from fetal liver and later from bone marrow differentiate into B lymphocytes in the bone marrow. These are characterized by the presence on the cell surface of specific antigen-binding receptors, plus receptors for complement (C3) and receptors for the Fc portion of immunoglobulins. In the early stages of life several hundred inherited V (variable) L (light) and H (heavy) chain immunoglobulin gene segments undergo somatic recombination. There are also multiple copies of J (joining) gene segments in the case of light chains, and J and D (diversity) gene segments in the case of heavy chains that also somatically recombine with the V genes. Somatic mutation also adds to the generation of antibody diversity, to yield potentially more than 10^{12} unique specificities prior to encountering antigen.

Each individual B lymphocyte and its clonal progeny express a set of immunoglobulin genes that are specific for a single epitope. During development such cells have four possible fates: the cells (1) may react with a self-antigen and be eliminated, (2) may be non-viable and be eliminated, (3) may continue to circulate at low frequency without antigen stimulus ("naïve" cells), or (4) they may react with a foreign antigen and proliferate.

In contrast to T cells, the antibody receptors on the surface of B cells recognize native and soluble state antigens rather than peptide–MHC complexes, hence B cells interact directly with viral proteins or virions as well as with peptides. When a particular clone of B cells bearing receptors complementary to any one epitope on an antigen binds to that antigen, the cells in the clone then divide more than 1000-fold and differentiate into antibody-secreting plasma cells once the appropriate signals from Tfh cells have been received.

Each plasma cell secretes antibody of a single specificity, corresponding to the particular V (variable) region of the surface immunoglobulin (sIg) receptor it expresses. Initially, secreted antibody is of the IgM class, but somatic genetic recombination (translocation) then brings about a class switch by associating V gene segments with different H chain constant domains. Various cytokines play an important role in this isotype switching. Thus, after a few days IgG, IgA, and sometimes IgE antibodies of the same specificity begin to dominate the immune response. Early in the immune response, when large amounts of antigen are present, antigen-reactive B cells may be triggered even if the surface receptors fit the epitope with relatively poor affinity; the result is the production of antibody with a correspondingly low affinity. Later, by the time only small amounts of antigen remain, B cells have evolved by a process of hypermutation in the V region genes and produce receptors with greater affinity for the recognized epitope. This in turn leads to the secretion of antibody with a greater affinity for the antigenic determinant in question.

T Lymphocytes

T lymphocytes are so named because of a dependence on the thymus for maturation from pluripotent hematopoietic stem cells. Within the thymus there is a positive selection for those cells able to interact with MHC molecules on the surface of thymic stromal cells, and subsequently a negative selection to eliminate those T cells that recognize self-antigens presented in either class I or class II MHC. There are three possible outcomes of negative selection of T cells: (1) elimination as a result of a high affinity reaction with self-antigens, (2) differentiation into regulatory T cells (Tregs) as a result of an intermediate affinity reaction, or (3) become circulating naïve T cells in the absence of any reaction with the viral antigen. This process is designed to generate a diverse range of T cell clones while avoiding adverse autoimmune disease. Only 1% or 2% of the lymphocytes produced in the thymus populate the secondary lymphoid tissues (lymph nodes, spleen, and lymphoid follicles in peripheral organs).

Functionally, T lymphocytes are classified into two subsets: T helper (Th) lymphocytes and cytotoxic T (Tc) lymphocytes (CTLs). Th cells are generally considered to have a regulatory function and Tc cells are responsible for target cell killing. Close examination of T cell clones indicates that a single cell type can discharge a number of functions through the secretion of a range of different lymphokines (Fig. 6.2).

Antigen-Specific Receptors on Lymphocytes

Lymphocytes express surface receptors specific for particular antigens, this being the basis for immunological specificity. Individual T or B lymphocytes possess receptors specific

for a single epitope. When T or B lymphocytes bind to a corresponding antigen, the cells divide to form an expanded clone of cells (clonal expansion). B lymphocytes differentiate into plasma cells with a primary function of antibody secretion. In contrast, T lymphocytes secrete a number of cytokines, and develop regulatory or cytotoxic function. Cytokines modulate the activities of the cells involved in the immune response. Subsequently, some T and B cells revert to long-lived lymphocytes responsible for immunological memory. Antibodies and the receptors on B cells recognize epitopes present on both the conformation-dependent and conformation-independent foreign antigens. In contrast, T cell receptors recognize only small, linear peptides formed by the cleavage of viral proteins, and this only occurs when the foreign peptides are presented to T cells in association with MHC class I or II proteins.

The antigen-specific receptors on the surface of B lymphocytes are modified immunoglobulin molecules composed of four polypeptides: two light (L) and two heavy (H) chains termed surface immunoglobulins (sIg). Both are modified at the C-terminus of the H chains to form a transmembrane domain that anchors the complex into the cell membrane. Prior to primary antigen stimulation the sIg molecules are sIgM; after class switching (see below) the Ig of the class switch becomes a sIg antigen-specific receptor.

The T cell antigen-specific receptor (TCR) is quite distinct; it is a two-polypeptide heterodimer and although immunoglobulin-like in structure, it is coded by an entirely

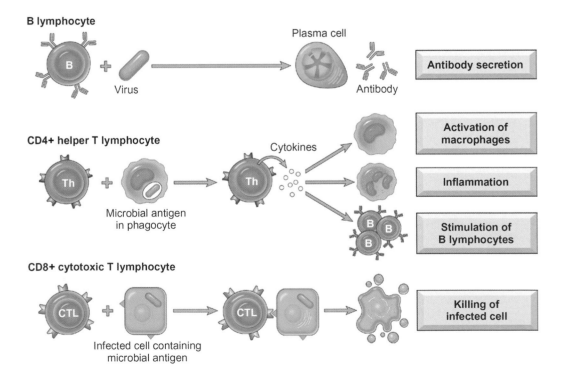

FIGURE 6.2 The principal classes of lymphocytes and associated roles in adaptive immunity. *Reproduced from MacLachlan, N.J., Dubovi, E.J., 2011. Veterinary Virology, fourth ed. Academic Press, London (Fig 4.7), with permission.*

different set of genes. The two polypeptides of the most common T cell receptors are designated α/β. A second T lymphocyte population bears a different T cell receptor designated γ/δ.

T Helper (Th) Lymphocytes

T-helper cells carry the surface marker CD4 and express a surface receptor known as the T cell receptor composed of a polypeptide heterodimer (designated e.g., α/β). T helper cells recognize viral peptides in association with class II MHC protein, usually on the surface of an antigen-presenting cell (APC). These interactions result in T helper cell activation, proliferation and differentiation, providing the affinity of binding is sufficiently high. Activated T helper cells also secrete cytokines that lead to activation of other T helper cells as well as activating Tc and B lymphocytes. The result is a population of lymphocytes cytotoxic for infected cells and antibody production.

Cytokines and receptor–ligand interactions stimulate naïve CD4+ cells to interact with dendritic cells in lymphoid tissue, differentiating into a number of functionally distinct subsets. In the case of virus infections, the two most important are Th1 and Tfh cell subsets. Th1 cells are principally involved in boosting the cytotoxic response. These cells promote the cell-mediated response to virus infection by stimulating the maturation of cytotoxic T cell precursors, partly through the secretion of the cytokines IL-2 and IFN-γ. Th1 cells also secrete tumor necrosis factor (TNF), mediate delayed-type hypersensitivity reactions, and promote the production of IgG2a antibodies. Th1 cells greatly augment the immune response by activating macrophages and other T cells at the site of the viral infection. This response is the basis for delayed-type hypersensitivity reactions that are a recognized part of the pathogenesis of many viral infections.

Tfh cells secrete the cytokine IL-21 which is essential for B cell differentiation and the development of high-affinity, isotype-switched antibody responses against viruses.

Other Th cells, including Th2 and Th17 cells may also contribute to the immune response against virus infection by promoting inflammation or the generation of specific antibody isotypes. It was previously thought that Th2 played a key role in stimulating the antibody response, but more recent work has clarified the view that Tfh cells provide the essential role described above.

Some T cells can be demonstrated to downregulate other T cell and/or B cell responses. At one time it was suggested that there may be a further class of cell, once referred to as T-suppressor cells. This area has been controversial until recently, but it is now believed to comprise a distinct subset known as regulatory T cells (T-regs). There are two basic types of T-regs; tTregs that are produced in the thymus during negative selection and are thought to be mainly involved in controlling autoimmune disease; and iTregs that are induced during immune responses and are involved in terminating immune responses and bringing the immune system back to homeostasis. T-reg cells may also help maintain the balance between protection and an immune-mediated pathology.

Cytotoxic T (Tc) Lymphocytes

Cytotoxic T (Tc) lymphocytes carry the CD8 surface marker, and possess T cell receptors that recognize viral peptides presented on the surface of virus-infected target cells in association with class I MHC molecules. Activation and subsequent killing of target cells by Tc cells require direct Tc–target cell contact in a manner reminiscent of a synapse that is also dependent on peptide–class I MHC interaction. Granules within the cytoplasm of the Tc cell polarize toward the target cell plasma membrane and the contents are then released. A monomeric protein, perforin, is secreted and polymerizes to form ~17-mer mushroom-shaped structures: these are then inserted into the target cell plasma membrane to create a pore that brings about cell lysis. Perforin is structurally and functionally very similar to C9 of the complement cascade responsible for complement-mediated lysis (see below). Both Tc and the innate equivalent, natural killer (NK) cells, release lymphocyte-specific granules with serine esterase activity (granzymes); these granules induce apoptosis in target cells.

The effector response of T cells is generally transient: in certain acute infections Th and Tc activities peak about one week after the onset of viral infection and disappear by two to three weeks. It is not yet clear whether this is attributable to the destruction of infected cells and a consequential removal of the antigenic stimulus, or whether it is due to the suppressor actions of T-reg cells.

γ/δ T Lymphocytes

An entirely different class of T cells has a different type of T cell receptor composed of polypeptide heterodimers designated γ and δ (rather than the conventional α and β chains). γ/δ T cells are found principally in epithelia such as the skin, intestine, and lungs. In humans and mice this class constitutes a small minority (about 5%) of the T cell population. The recognized antigens are not bound to cell surface MHC proteins, including both intact proteins as well as peptides. For some time, these cells received little attention because of the technical difficulties in identification, purification and analysis. However, there is emerging evidence that these T cells play a key role in the immune responses against viruses entering the body via the skin, intestine, or airways.

Monocytes, Macrophages, and Dendritic Cells

Monocytes, macrophages, and dendritic cells are important initiators of the immune response against viral invasion;

monocytes are mobile and can home to infected sites; macrophages together with dendritic cells also occupy key locations in various tissues (e.g., alveolar macrophages in the lung, Kupffer cells in the liver, Langerhans dendritic cells in the skin). All are involved early in the host's response: (1) monocytes infiltrate tissue and differentiate to become macrophages, (2) macrophages often become the predominant cell in an infection focus by 24 hours after viral invasion, and (3) dendritic cells carry out afferent immune functions at all body surfaces and in key organs such as lymph nodes, spleen, and liver, where most phagocytic removal of foreign particles occurs. Macrophages and dendritic cells bear PRRs, including virus class-specific Toll-like receptors crucial for the initial recognition of virus invasion. The surfaces of these cells also bear surface immunoglobulin Fc and C3b receptors that promote the phagocytosis of immune complexes, that is virions coated with antibody. By serving as "professional" antigen presenting cells these cells exercise a controlling influence over the rapidity, magnitude, and dynamics of the immune response.

Macrophages subsequently give expression to the efferent limb of the immune response: cytokines secreted by activated T cells bring more monocytes into the infection focus where activation occurs followed by differentiation into macrophages. Activated macrophages have increased chemotactic activity, phagocytic activity, and digestive powers.

THE MAJOR HISTOCOMPATIBILITY COMPLEX (MHC) AND ANTIGEN PRESENTATION

Antigen processing and presentation of viral proteins are intimately related to the structure and intracellular production of MHC proteins. During ontogeny, MHC molecules expressed by thymic stromal and resident hematopoietic cells play a crucial role in the development of mature T cells. Once the developing T cells undergo antigen receptor gene rearrangement, T cells are tested for an ability to interact with "self" MHC molecules in a process called positive selection. If interactions occur with a sufficiently high affinity, these cells are provided with survival signals. In the absence of these survival signals, cells die by apoptosis. This process essentially tests the developing T cells for "functional" TCRs. Once past positive selection, the developing T cells are then subjected to a second process called negative selection involving being tested for reactivity against self-peptides presented by MHC molecules. A high affinity interaction in this case, results in apoptosis. Together, these two selection processes result in the generation of mature T cells that can recognize foreign peptides, but only if the peptides are located in the peptide-binding cleft of "self" MHC protein molecules. This process does not happen if viral peptides are either

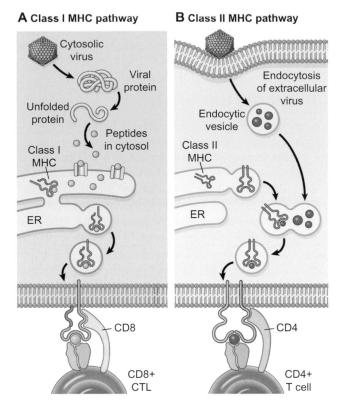

A Class I MHC pathway   **B Class II MHC pathway**

FIGURE 6.3 Antigen processing and display by MHC molecules. (A) In the class I MHC (exogenous) pathway, peptides are produced from proteins in the cytosol and transported to the endoplasmic reticulum (ER), binding to MHC class I molecules occurs. The peptide–MHC complexes are transported to the cell surface and displayed for recognition by CD8+ T cells. (B) In the class II MHC (endogenous) pathway, proteins are ingested into vesicles and degraded into peptides which bind to class II MHC molecules being transported in the same vesicles. The class II–peptide complexes are expressed on the cell surface and recognized by CD4+ T cells. *Reproduced from MacLachlan, N.J., Dubovi, E.J., 2011, Veterinary Virology, fourth ed. Academic Press (Fig 4.8), with permission.*

free in the extracellular space or if in association with non-self-MHC molecules. This phenomenon is known as MHC restriction first described by Rolf Zinkernagel and Peter Doherty.

There are two classes of MHC molecules, identified as class I and class II. The two classes of T lymphocytes, namely Tc and Th, are defined by the interactions of these cells with class I or class II MHC proteins, respectively (see Fig. 6.3). The pathways used by cells to process and present antigenic peptides to Th and Tc cells are fundamentally different: presentation is either by the exogenous pathway for those peptides presented in association with class II MHC molecules, or via the endogenous pathway for those peptides presented in association with class I MHC molecules (see below).

The MHC is a genetic locus encoding three class I MHC proteins and up to 12 class II MHC proteins, each of which occurs as 50 to 100 alternative allelic forms. Class I glycoproteins can be expressed on the plasma membrane of most types of cells with the exception of neurons. Although not constitutively expressed, class II MHC glycoproteins are expressed principally by "professional" APCs—dendritic cells, macrophages, and B cells. At the distal tip of each class of MHC protein is a cleft in which the antigenic peptide is bound and presented. Peptide binding is determined by only two or three hydrophobic amino acids, the "anchor" residues, in a particular peptide and accordingly a particular MHC protein can bind numerous different peptides: indeed some peptides can bind to several different MHC molecules. Peptides presented by class I MHC molecules are usually nine amino acids long (range 8 to 11-mers) whereas peptides binding to class II MHC proteins tend to be longer, ranging from 13 to 18 amino acids. A further difference is that the peptide binding cleft in the case of class II MHC proteins is open at each end whereas that of class I MHC proteins is closed. Specific amino acids forming depressions on the floor of the cleft of MHC proteins determine the particular range of peptides capable of binding to that molecule. The peptide–MHC complex is in turn recognized, with absolute specificity, by the T cell receptor of the appropriate clone of T cells. Amino acid residues that do not bind within the MHC cleft are hydrophobic and project outwards, inviting recognition by T cell receptors.

Although there is extensive polymorphism of MHC genes between individuals, any individual has only a limited number of different MHC proteins, and any given antigenic peptide binds only to certain MHC molecules. If certain peptide–MHC complexes are important in eliciting a protective immune response to a serious viral infection, individuals lacking suitable MHC proteins will be genetically more susceptible to that disease. A further cause of increased susceptibility lies in the possible absence from an individual's T cell repertoire of lymphocytes bearing receptors for that particular MHC–peptide complex.

The Exogenous Pathway of Antigen Presentation

Only a restricted range of cells, defined as APCs, process and present antigens to Th cells in association with class II MHC protein. APCs include dendritic cells, macrophages, and B lymphocytes. Dendritic cells, including Langerhans cells of the skin and the dendritic cells of lymph nodes and the splenic red pulp and marginal zones, are so named because these form long finger-like processes that interdigitate with lymphocytes, thereby favoring antigen presentation. Unlike dendritic cells, macrophages express relatively low levels of class II MHC protein while resting, but following activation, these levels increase particularly after exposure to IFN-γ. After primary activation, B lymphocytes become important antigen-presenting cells; B cells are especially important during the latter stages of an infection and during reinfection. Memory B cells also serve as very efficient APCs. Viral

antigen, or the virion itself, may be rapidly endocytosed by APCs. Viral protein then passes progressively through early endosomes to late (acidic) endosomes and prelysosomes, to be cleaved by proteolytic enzymes into peptides. Selected viral peptides capable of binding to class II MHC proteins are then transported to the plasma membrane to be recognized by CD4+ T cells; the result is a Th cell response (Fig. 6.3). In the case of Th1 cells, the result is generation of cells capable of producing IFNγ at the site of infection leading to enhanced macrophage function and CTL killing. In the case of humoral immunity, these peptides generally represent different epitopes from those of the same antigen recognized by the B cell for the production of antibody. CD4+ Th cells to which B cells present antigen respond by secreting the necessary cytokines for stimulating B cells to differentiate into plasma cells. Such cognate interaction, involving close physical association of T and B cells, ensures a highly efficient delivery of "helper factors" in the form of cytokines from the Th cell to the relevant primed B cell.

The Endogenous Pathway

Almost all nucleated cells constitutively synthesize class I MHC proteins. In contrast to exogenous pathway processing that makes use of specialized endosomal components, the endogenous pathway of antigenic presentation involves proteolytic degradation of viral proteins within the host cell cytosol followed by transport to the endoplasmic reticulum.

The processing of viral antigen occurs in the cytosol rather than in endosomes, through cellular mechanisms normally used by the cell to eliminate damaged or misfolded host proteins. Cytoplasmic multicomponent proteases (proteasomes) perform this essential function, producing peptides eight to ten amino acids in length. Exposure to pro-inflammatory cytokines, for example IFN-γ and TNF, upregulates a proteasome subset (immunoproteasomes) possessing a modified 20S core component, a 19S regulatory complex, and a proteasome activator, PA28. Peptides are conveyed through the cytosol to the endoplasmic reticulum through the intermediary of transporter proteins (*t*ransporter associated with *a*ntigen–*p*rocessing, TAP). This process of cytosolic transport is mediated by heat/shock proteins (HSPs), a family of chaperones that shield peptides from degradation during the transport process. Once at the endoplasmic reticulum, the peptides need to transverse across the endoplasmic reticulum membrane before complexing with MHC molecules, a process mediated by calnexin, calreticulum, and ERp57. There the peptides assemble with class I MHC protein to form a stable trimeric complex that is exported, via the Golgi complex, to the cell surface for presentation to Tc cells. Both the proteasome and transporter proteins are coded for within the MHC gene complex.

Recent work has highlighted that antigen presentation may also occur by complexing with class Ib MHC molecules and CD1, the latter being able to bind lipids as well as peptides.

While it was previously believed that these two antigen presentation pathways were distinct, it is now well established that there exists a third pathway for antigen presentation. In this pathway, peptides can be taken up by the APC by endocytosis, but instead of being presented on MHC class II molecules, the peptides are presented on MHC class I molecules. This process is known as "cross-presentation" and is carried out by a highly specialized subset of dendritic cells possessing the intracellular machinery that allows for this to occur. Cross-presentation is important for the immune response against virus infection as it allows clonal selection of CD8+ T cells without the requirement of infection of dendritic cells by the virus. In summary, peptide presentation occurs through one or more pathways in what is increasingly evident as being a complex process.

ANTIBODIES

The end result of activation and maturation of B cells is the production of antibodies that react specifically with the epitope identified initially by cell surface receptors. Antibodies fall into five main classes: IgG, IgA, IgM, IgD, and IgE (Box 6.1). All immunoglobulins of a particular class have a similar structure, but vary widely in those amino acid sequences comprising the antigen-binding site that determines the specificity for a given antigenic determinant. The commonest immunoglobulin found in serum, IgG, consists of two heavy (H) and two light (L) chains, and each chain consists of a constant and a variable domain. The chains are held together by disulfide bonds. Papain cleavage separates the molecules into two identical Fab fragments containing the antigen-binding sites, and an Fc fragment bearing the sites for various effector functions such as complement fixation, attachment to phagocytic cells, and placental or colostrum transfer (Fig. 6.4).

The immunological specificity of an antibody molecule is determined by its ability to bind specifically to a particular epitope. The binding site, that is the antibody-binding groove, is located at the amino-terminal end of the molecule. The variable regions of both L and H chains are comprised of over 100 amino acids, within which there are three hypervariable domains termed complementary determining regions, interspersed between four conserved framework regions. On folding to form the three-dimensional functional Ig structure, the six complementary determining regions (three each from L and H chains) are located in the antigen-binding groove. It is the variability of the complementary determining regions that accounts for the limitless range of different epitopes recognized by these molecules. It should be noted that similar principles underlie the generation of antigenic diversity found in T cell receptor variable regions.

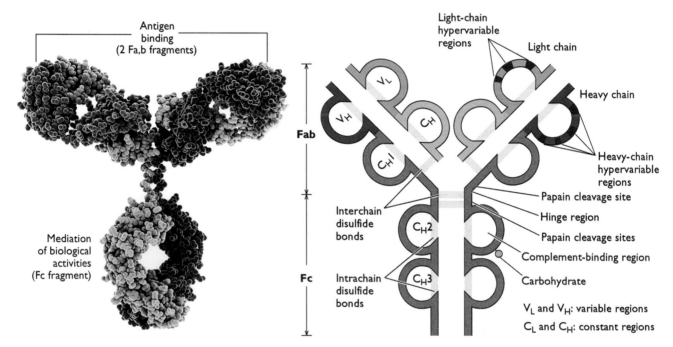

FIGURE 6.4   The structure of an antibody molecule. (Left) Model of an IgG molecule built from X-ray crystallographic data – it is a large molecule, about 150 kDa in size, composed of four peptide chains. There are two identical class γ heavy chains and two identical light chains, thus a tetrameric structure. The tetramer has two identical halves, which together form the Y-like shape. (B) The heavy (H) and light (L) chains are held together by disulfide bonds. The distal ends of the variable regions of the heavy (VH) and light (VL) chains, form the antigen binding sites. The papain protease cleavage sites are shown, as this enzyme is used to separate the Fab (containing antigen-binding sites) and Fc domains (containing sites active in biological activities such as complement binding and binding to Fc receptors on macrophages and other cells).

Antibodies directed against certain epitopes on the surface of virions neutralize infectivity. Antibodies may also act as opsonins, a process whereby antibodies bound to virions facilitates virion uptake and destruction by macrophages. In addition, antibodies may attach to viral antigens on the surface of infected cells, leading to the destruction of infected cells; this may occur either by complement following activation of the classical or alternative complement pathways, or by armed and activated Fc receptor-bearing cells such as NK cells, polymorphonuclear leukocytes, and macrophages (antibody-dependent cell-mediated cytotoxicity [ADCC]).

CYTOKINES

Cytokines are low molecular weight hormone-like proteins that stimulate or inhibit the proliferation, differentiation, and/or maturation of immune cells. These differ from true hormones in a number of ways, including production by non-specialized cells. Many are produced by T lymphocytes (known as lymphokines) or monocytes/macrophages (monokines) and serve to regulate the immune response by coordinating the activities of the various cell types involved. Many lymphokines are also known as interleukins. Thus, while cytokines are not antigen-specific, the production and actions of cytokines are often antigen-driven.

Cytokines may act either on the same cell (autocrine) or on cells in the immediate vicinity (paracrine), particularly at cell–cell interfaces, where directional secretion may occur and very low concentrations may be effective, or they may act on cells at more distant locations (endocrine). Responsive target cells carry receptors for particular cytokines. A single cytokine may exert a multiplicity of biological effects, often acting on more than one type of cell. Moreover, different cytokines may exert similar effects, though perhaps via distinct postreceptor signal transduction pathways, resulting in synergism. There is much redundancy in the actions of cytokines presumably linked to the body's need to maintain reliable defense mechanisms; it is frequently the case that knock-out mice with the deletion of a single cytokine gene do not show increased susceptibility to particular virus challenges.

Cytokines upregulate or downregulate the target cell, and different cytokines can antagonize one another. Typically, a cytokine secreted by a particular type of cell activates another type of cell to secrete a different cytokine or to express receptors for a particular cytokine, and so on in a form of cascade. Because of the intricacy of the cytokine cascade it is rarely possible to attribute a given biological event *in vivo* to a single cytokine.

Cytokines can influence viral pathogenesis in a number of ways: (1) augmentation of the immune response, for example, of cytotoxic T cells by TNF or by IFN-γ that upregulates MHC expression, (2) regulation of the immune response, for example, antibody isotype switching by a number of interleukins (e.g., IL-4, 5, 6), or IFN-γ, (3) suppression of the immune response—for example, interleukin 10 inhibits the synthesis of IFN-γ, (4) inhibition of viral replication by interferons, and (5) upregulation of viral gene expression.

RECOVERY FROM VIRAL INFECTION

Cell-mediated immunity, antibody, complement, phagocytes, and interferons and other cytokines are all involved in recovery from viral infections—in most cases several of these arms of the immune system act in concert, again depending on the particular host–virus circumstances (Box 6.2).

Lymphocytes and macrophages normally predominate in the cellular infiltration of virus-infected tissues; in contrast to bacterial infections, polymorphonuclear leukocytes are usually not at all plentiful. T cell depletion by neonatal thymectomy or antilymphocyte serum treatment increases the susceptibility of experimental animals to most viral infections; for example, T cell-depleted mice infected with ectromelia virus fail to show the usual inflammatory

---

BOX 6.2 Lessons from Natural Congenital Immunodeficiencies

One approach to understanding the mechanisms involved in recovery from viral infection that is not subject to laboratory artifact, is the clinical observation of viral infections in children suffering from congenital immunodeficiencies. Humans born with genetic deficiencies in antibody production and with little or no immunoglobulin can nevertheless control many viral infections, with the exception of some enterovirus infections. Children with Bruton's sex-linked agammaglobulinemia not only recover in the absence of detectable antibody but resist subsequent exposure to most viruses, indicating that they also have and retain immunological memory. However, in children with impaired T cell responses, such as individuals with DiGeorge's syndrome (congenital athymic aplasia), leukemia, or other lymphoreticular neoplasms, or those receiving immunosuppressive therapy, viral infections occur with increased frequency and severity. The same is the case in adults, for example in those suffering with acquired immunodeficiency syndrome (AIDS) or receiving immunosuppressive therapy. In these instances the administration of hyperimmune immunoglobulin may moderate, but often is unable to clear the viral infection.

Perhaps the most informative example is that of measles in infants with thymic aplasia. In these T cell-deficient infants there is no sign of the usual measles rash but there is an uncontrolled and progressive growth of virus in the respiratory tract, leading to a fatal pneumonia. In the normal child, the T cell-mediated immune response controls infection in the lung and also plays a vital role in the development of the characteristic skin rash.

mononuclear cell infiltration in the liver, and develop extensive liver necrosis and die as a consequence. This happens despite the production of antiviral antibodies and interferon. Virus titers in the liver and spleen of infected mice can be greatly reduced by adoptive transfer of immune T cells taken from recovered donors; this process is class I MHC restricted, implicating Tc cells.

Immune Cytolysis of Virus-Infected Cells

Specific recognition and binding of either sIg or a T cell receptor to its epitope triggers, by signal transduction across the plasma membrane of the lymphocyte, a wide range of effector processes that attack and remove the invading virus and/or virus-infected cells (Fig. 6.1). The resulting cascade of cell–cell interactions and cytokine secretion amplifies the immune response to match the scale of the virus infection and, in addition, establishes a long-lived memory that enables the immune system to respond more quickly (a secondary or anamnestic response) to reinfection with the same virus. The stimulation of immunological memory is an important goal of immunization (see Chapter 11: Vaccines and Vaccination).

Destruction of infected cells is an essential feature of recovery from viral infections, and it results from any one of four different processes, viz: (1) cytotoxic T cell action, (2) antibody-complement-mediated cytotoxicity, (3) ADCC, or (4) NK cell activity. Since some viral proteins, or peptides derived therefrom, appear in the plasma membrane before any virions have been produced, lysis of the cell at this stage brings viral replication to a halt before significant numbers of progeny virions are released. The host response is exquisitely sensitive in this regard: recent evidence suggests that a single peptide complexed to the appropriate MHC protein on a target cell is sufficient to elicit cytotoxic T cell activity against that cell.

Antibody-complement-mediated cytotoxicity is readily demonstrable *in vitro* even at very low concentrations of antibody. The alternative complement activation pathway appears to be particularly important under these circumstances. ADCC is mediated by leukocytes that carry Fc receptors: macrophages, polymorphonuclear leukocytes, and other kinds of killer cells. NK cells, on the other hand, are activated by interferon, or directly by viral glycoproteins (see Chapter 5: Innate Immunity). These NK cells demonstrate no immunological specificity, but preferentially lyse virus-infected cells. In addition, in the presence of antibody, macrophages can phagocytose and digest virus-infected cells.

One approach used to dissect the immune response of experimentally infected inbred mice is to ablate completely all immune potential (using X-irradiation, cytotoxic drugs, etc.), then to separately add back individual components. In a now classical model, virus-primed cytotoxic T lymphocytes

of defined function and specificity, cloned in culture then transferred to infected animals reduced significantly the levels of mortality among mice infected with lymphocytic choriomeningitis virus, influenza virus, and several other viruses. Generally, greater protection is conferred by CD8+ T cells than by CD4+ T cells. Moreover, transgenic mice lacking CD8+ T cells suffer higher levels of morbidity and mortality than normal mice following virus challenge. Nevertheless, CD4+ T cells have been shown to play a significant role in recovery, mediated particularly by IFN-γ and IL-2.

Although T cell determinants and B cell epitopes on surface proteins of viruses sometimes overlap, the immunodominant Tc determinants are often situated on the relatively conserved proteins located in the interior of the virion, or on non-structural virus-coded proteins that occur only in virus-infected cells. Hence T cell responses are generally of broader specificity than neutralizing antibody responses and display cross-reactivity between strains and subtypes.

Complement

The complement system consists of about 30 serum proteins that can be activated to "complement" the immune response. The classical complement pathway is activated by the interaction between antibody–antigen complexes and the complement component C1. There is also an alternative antibody-independent pathway involving component C3b. Both are important in viral infections.

Activation of complement by either pathway may lead to (1) the destruction of virions or virus-infected cells by lysis of membranes, (2) activation of inflammation and the accumulation of leukocytes, and (3) binding to virions leading to engulfment by, and destruction within, phagocytic cells (opsonization). Activation of complement via the alternative pathway appears to occur mainly after infections with enveloped viruses that mature by budding through the plasma membrane; since it does not require antibody, the alternate pathway can become active immediately after viral invasion of the body, and is one component of the innate defense.

Role of Antibody

Circulating antibody plays a significant role in recovery in generalized diseases characterized by a viremia, for example picornavirus, togavirus, flavivirus, and parvovirus infections. Antibody has been clearly shown to slow or prevent poliovirus from gaining access to the central nervous system, and it is likely to be important in slowing dissemination of many virus infections. However, it does not necessarily follow that the antibody is acting solely by neutralizing virion infectivity. Indeed it has been shown that certain non-neutralizing monoclonal antibodies can

BOX 6.3 Passive Immunity

There is abundant evidence that antibodies alone can prevent infection. For example, artificial passive immunization (injection of antibodies) temporarily protects against hepatitis A and B, rabies, measles, varicella, and several other viral infections (see Chapter 11: Vaccines and Vaccination). Furthermore, natural passive immunization protects the newborn for the first few months of life against most of the infections that the mother has experienced. In humans this occurs in two ways: (1) maternal antibodies of the IgG class cross the placenta and protect the fetus and the newborn infant during pregnancy and for several months after birth, (2) antibodies of the IgA class are secreted in the mother's milk at a concentration of 1.5 gm per liter (and considerably higher in colostrum), conferring protection against enteric infections as long as breast-feeding continues. If the infant encounters viruses when maternal immunity is waning, the virus replicates to only a limited extent, causing no significant disease but stimulating an immune response; thus the infant acquires active immunity while partially protected by maternal immunity. This "passive–active" immunity can also develop at a later age following therapeutic passive immunization. Maternally derived antibody while still present, also interferes with the efficacy of live vaccines in the very young, and must therefore be taken into account when designing vaccination schedules (see Chapter 11: Vaccines and Vaccination).

protect mice inoculated with various viruses, presumably by ADCC or antibody-complement-mediated lysis of infected cells, or by opsonizing virions for macrophages. Furthermore, antibodies that bind to the surface proteins of enveloped viruses can also help to clear virus infection from persistently infected cells; this process is not cytolytic and does not involve complement, but occurs synergistically with IFN and other cytokines (Box 6.3).

For example, infants with severe primary agammaglobulinemia recover normally from measles virus infection, but are about 10,000 times more likely than normal infants to develop paralytic disease after vaccination with attenuated poliovirus vaccine. These infants have normal cell-mediated immune and interferon responses, normal phagocytic cells, and a normal complement system, but cannot produce those antibodies essential for the prevention of poliovirus spread to the central nervous system via the bloodstream.

NEUTRALIZATION OF VIRAL INFECTIVITY BY ANTIBODIES

In contrast to T cells, B cells and antibody generally recognize conformational epitopes. These structures are often created by protein folding, which brings into close proximity critical amino acid residues normally distant on the linear polypeptide chain or other short amino sequences from adjacent polypeptides. Such B cell epitopes are generally located on the surface of the protein, often on prominent protuberances or loops, and generally represent relatively variable regions of the molecule, differing between strains of the virus.

While specific antibody of any class can bind to any accessible epitope on a surface protein of a virion, only those antibodies that bind with reasonably high affinity to particular epitopes on a particular protein of the outer capsid or envelope of the virion are capable of neutralizing viral infectivity. The key viral protein is usually that containing the ligand by which the virion attaches to receptors on the host cell or a nearby protein. Mutations in critical epitopes on such proteins allow the virus to escape from neutralization by antibody. The accumulation of mutations in the genes encoding these epitopes leads to the emergence of new viral strains and eventually to the emergence of new viruses (see Chapter 15: Emerging Virus Diseases).

Neutralization is not simply a matter of coating the virion with antibody, nor indeed of blocking attachment to the host cell. Except in the presence of such high concentrations of antibody that most or all accessible antigenic sites on the surface of the virion are saturated, neutralized virions may still attach to susceptible cells. In such cases the neutralizing antibody blocks the infection at some point following adsorption and entry. For example, in the case of picornaviruses, neutralizing antibody appears to distort the capsid, leading to loss of a particular capsid protein, thereby rendering the virion vulnerable to enzymatic attack. With influenza virus, more subtle conformational changes in the hemagglutinin molecule may prevent the fusion event preceding the release of the nucleocapsid from the viral envelope. One mechanism, demonstrated first with adenoviruses, involves the intracellular tripartite motif-containing protein 21 (TRIM21) which has a strong affinity for IgG; while virions are normally uncoated intracellularly in a controlled way to preserve viral infectivity; TRIM21 captures virion–antibody complexes, leading to ubiquitination of the complexes and destruction by lysosomal enzymes.

IMMUNITY TO REINFECTION

Although a large number of interacting phenomena contribute to recovery from viral infection, the mechanism whereby acquired immunity prevents reinfection with the same virus appears to be much simpler. The first line of defense is antibody, which, if acquired by active infection with a virus that causes systemic infections, continues to be synthesized for many years, providing solid protection against reinfection. The level of acquired immunity generally correlates well with the titer of antibody in the serum. Furthermore, the transfer of antibody alone by passive immunization or by maternal antibody transfer to fetus or newborn, provides excellent protection in the case of many viral infections. Thus it is reasonable to conclude that antibody is the most influential factor in immunity acquired either by natural infection or by vaccination. If the antibody defenses are inadequate, virus replication may commence and the mechanisms that contribute to recovery are called into play again; the principal differences on this occasion being that the interaction with antibody will reduce the effective dose of virus and that primed memory T and B lymphocytes generate a more rapid secondary response.

As a general rule the secretory IgA antibody response is short-lived compared to the serum IgG response. Accordingly, resistance to reinfection with respiratory viruses and some enteric viruses tends to be of limited duration. For example, reinfection with the same parainfluenza virus is not uncommon. Moreover, reinfection at a time of waning immunity favors the selection of neutralization-escape mutants, resulting in the emergence of new strains of viruses such as influenza virus by antigenic drift. Because there is little or no cross-protection between antigenically distinct strains of virus, repeated attacks of respiratory infections occur throughout life.

The immune response to the first infection with a virus can have a dominating influence on subsequent immune responses to antigenically related viruses, in that the second virus often induces a response that is directed mainly against the antigens of the original viral strain. For example, the antibody response to sequential infections with different strains of influenza A virus is largely directed to antigenic determinants of the particular strain of virus with which that individual was previously infected. This phenomenon, irreverently called "original antigenic sin," is also seen in infections with enteroviruses, reoviruses, paramyxoviruses, and togaviruses. Original antigenic sin has important implications for interpretation of seroepidemiology, for understanding immunopathological phenomena and particularly for the development of vaccination strategies.

IMMUNOLOGICAL MEMORY

Following priming by antigen and the clonal expansion of lymphocytes, a population of long-lived memory cells arise that persist indefinitely. These memory T cells are characterized by particular surface markers (notably CD45RO) and homing molecules (adhesins) associated with a distinct recirculation pathway. When reexposed to the same antigen, even many years later, these cells respond more rapidly and more vigorously than was the case on the first encounter. Memory B cells, on reexposure to antigen, also display an anamnestic (secondary) response, with production of larger amounts of specific antibody.

Little is known about the mechanism of the longevity of immunological memory in T or B lymphocytes in the absence of a demonstrable chronic infection. It is believed that cells are periodically restimulated by the original antigenic peptide which may be retained for long periods as peptide–MHC complexes on follicular dendritic cells in lymphoid follicles, or by a surrogate in the form of either fortuitously cross-reactive antigens or antiidiotypic antibodies. In addition, circulating in the community, memory may be occasionally boosted by later subclinical reinfection with the same agent in the case of those infections. Memory T and B lymphocytes may survive for years without dividing, until restimulated following reinfection.

FURTHER READING

Klasse, P.J., 2014. Neutralisation of virus infectivity by antibodies: Old problems in new perspectives. Adv. Biol. Vol. 2014 Article ID 157895.

Murphy, K., 2012. Janeway's Immunobiology, eighth ed. Garland Science, London and New York.

Chapter 7

Pathogenesis of Virus Infections

Pathogenesis refers to the sequence of events during the course of an infection within the host, and the mechanisms giving rise to these events. It includes entry of the virus into the body, multiplication and spread, the development of tissue damage, and the production of an immune response. It encompasses the appearance of clinical signs and symptoms, the eventual resolution of the infection and, in most cases, virus elimination. Understanding viral disease pathogenesis requires knowledge of each of the stages of infection and an awareness of the mechanisms involved. The pattern of pathogenetic events accompanying infection is remarkably consistent and specific for each individual virus. There may be variation from individual to individual in the severity and/or the duration of these events, but a sound knowledge of the typical sequence associated with each infection is crucial in both making an accurate diagnosis and recommending the appropriate treatment.

In this chapter the mechanisms underpinning the initiation, establishment, and outcome of typical virus infections are described.

For the successful initiation of an infection, three requirements must be satisfied:

1. There must be an inoculum containing sufficient viable virus to establish an infection;
2. Virus must first reach and interact with susceptible cells capable of supporting virus replication;
3. The host innate immunity and pre-existing adaptive immunity must be insufficient to immediately abort the infection.

THE INCOMING VIRUS INOCULUM

The *infective titer* of a sample of virus can be measured by titration in a defined experimental assay system (e.g., in cell culture or an animal model), by determining the highest dilution (i.e., the lowest concentration) of the sample that can still initiate an infection, for example, that will infect 50% of the cell cultures (tissue culture infective dose 50, $TCID_{50}$) or animals (lethal dose 50, LD_{50}; or infective dose 50, ID_{50}) used for measurement. It is expressed as a number per volume, for example, 100 LD_{50} per ml. One LD_{50} generally represents many more than a single virus particle as (1) usually many of the incoming virus particles are non-infective due to defective assembly, genetic errors, or inactivation caused by environmental conditions, etc., and (2) usually many or most interactions between infective particles and cells do not lead to a productive infection at the tissue/organ/individual level.

Further, the "*infective titer*" of a virus sample is not an absolute value, but varies according to the assay system used—there are examples where humans may be infected by samples assayed by a particular laboratory method as containing less than one infectious unit—the human being more sensitive than the laboratory assay. Clearly the volume of the inoculum is important. In blood transfusion, a 500 ml unit of blood may contain an infecting dose even if the virus concentration is low (and transfusion recipients are often also debilitated by a major illness or injury); hence, highly sensitive PCR-based testing of every unit of donated blood is often necessary, as for example, when West Nile virus was first introduced into the United States.

Importantly, inactivation of a virus-containing sample by environmental conditions or by a disinfectant is not an all-or-none process, but a progressive stepwise reduction in virus titer over time as more and more virions become inactivated.

ROUTES OF ENTRY—A MAJOR FACTOR GOVERNING THE SPECIFIC PATTERN OF INFECTION

The surfaces of the body are lined by an almost continuous layer of epithelium—the epidermis or skin externally, and the various kinds of mucosae internally. Any incoming virus needs to be taken up by, and/or pass through, the cells making up this physical barrier. Different viruses have evolved distinctive mechanisms and routes for crossing this first barrier. The route preferred by a particular virus is a significant determinant of the type of damage and disease seen, and also is a major factor determining epidemiological characteristics (Fig. 7.1).

Fenner and White's Medical Virology. DOI: http://dx.doi.org/10.1016/B978-0-12-375156-0.00007-2

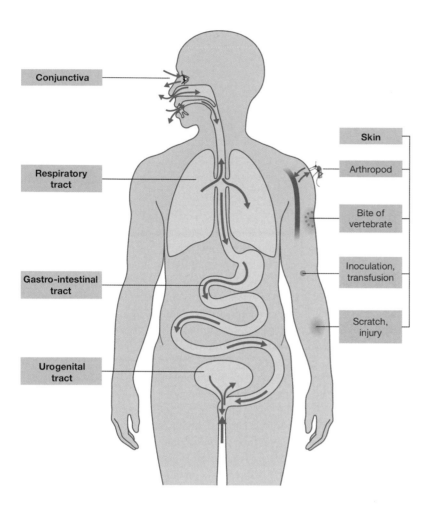

Conjunctiva

Respiratory
tract

Gastro-intestinal
tract

Urogenital
tract

Skin

Arthropod

Bite of
vertebrate

Inoculation,
transfusion

Scratch,
injury

FIGURE 7.1 Routes of virus entry—the various surfaces of the body, one of which must be broached in order for infection to occur. *Reproduced from Flint, S.J., et al., 2009. Principles of Virology, third ed., Vol, II, p. 348. ASM Press, Washington, DC, with permission.*

Following entry by either of these routes, virus replication may remain *localized* to the site of entry or may spread more widely through body compartments becoming *systemic*. Clinical signs and symptoms are often, but not always, related to the site(s) of virus replication.

Virus Entry via the Respiratory Tract

The cells lining the respiratory tract can support the replication of many viruses. The surface of the respiratory tract is protected by two cleansing systems: (1) a blanket of mucus produced by goblet cells, kept in continuous flow by (2) the coordinated beating of cilia on the epithelial cells lining the upper and much of the lower respiratory tract. Inhaled virus particles deposited on this surface are trapped in mucus, carried by ciliary action (the "*mucociliary escalator*") from the airways and nasal cavity to the pharynx, and then swallowed or coughed out. Inhaled droplets of 10 μm or more in diameter are usually deposited on the nasal mucosae covering the nasal turbinates: these project into the nasal cavity and act as baffle plates. Droplets of 5 to 10 μm in diameter are often carried to the trachea and bronchioles, where they are usually trapped in the mucus

blanket. Droplets of 5 μm or less may be inhaled directly into the lungs, and some may reach the alveoli, where contained virus particles may infect alveolar epithelial cells directly, causing viral pneumonia (interstitial pneumonia) (Fig. 7.2, Fig. 39.1).

The respiratory tract is, overall, the most important entry site of viruses into the body (Table 7.1). All viruses that infect the host via the respiratory tract do so by attaching to specific receptors on epithelial cells. Following initial respiratory tract infection, many viruses remain localized (e.g., rhinoviruses, parainfluenza, and influenza viruses), whereas others become systemic (e.g., measles, varicella-zoster, and rubella viruses).

Virus Entry via the Alimentary Tract

Many viruses are acquired by ingestion. Virus particles may either be swallowed and reach the stomach and intestine directly, or first infect cells in the oropharynx with progeny eventually carried into the intestinal tract. The esophagus is rarely infected, probably because of its tough stratified squamous epithelium and the rapid passage of swallowed material over its surface. The intestinal tract is protected by mucus, which may contain specific secretory antibodies

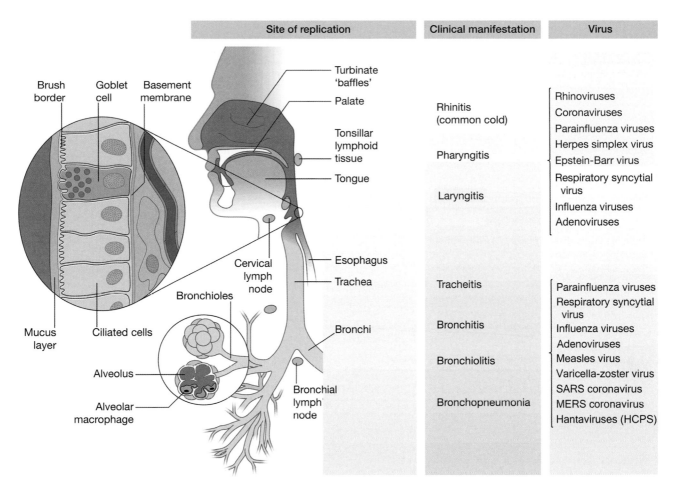

Site of replication	Clinical manifestation	Virus

FIGURE 7.2 Sites of viral entry in the respiratory tract. (Left) A detailed view of the ciliated pseudostratified columnar respiratory epithelium. A layer of mucus, produced by goblet cells, is a formidable barrier to virion attachment. Virions that pass through this layer may multiply in the ciliated cells or pass between them, reaching another physical barrier, the basement membrane. Beyond this are tissue fluids from which particles may be taken into lymphatic capillaries and reach the blood. Local macrophages patrol the tissue fluids in search of foreign particles. (Right) Different viruses replicate and cause clinical disease most prominently at different levels of the respiratory tract. *Originally from Mims, C.A., 1977–2008. Pathogenesis of Infectious Disease (through six editions). London, UK: Elsevier/Academic Press. Adapted by Nathanson, N. (ed.); and then adapted from Flint, S.J., et al., 2009. Principles of Virology: Pathogenesis and Control, third ed., ASM Press, Washington, DC.*

(IgA), but the directional peristaltic movement of gut contents provides many opportunities for virus particles to contact susceptible epithelial cells. Viruses may also be taken up by the M cells, specialized transporter cells that overlie Peyer's patches in the ileum, from where they may be passed to adjacent mononuclear cells and be carried away to draining lymph nodes and then enter the bloodstream (Table 7.2).

Many viruses entering the intestinal tract are inactivated by either acid in the stomach or the detergent activity of bile and proteolytic enzymes in the small intestine. In general, viruses causing intestinal infection, such as rotaviruses, caliciviruses, and enteroviruses are acid- and bile-resistant. However, there are acid- and bile-labile viruses that cause important intestinal infections; for example, coronaviruses. Additionally, viruses are protected during passage through the stomach of babies by the buffering action of milk. Some enteric viruses not only resist inactivation by proteolytic enzymes in the stomach and intestine, but infectivity may actually be increased by such exposure. The cleavage of a viral outer capsid protein by intestinal proteases enhances the infectivity of rotaviruses and some coronaviruses.

Rotaviruses, caliciviruses (noroviruses), and astroviruses are major causes of viral gastroenteritis and diarrhea. In contrast some of the enteroviruses (e.g., polioviruses) and hepatitis A and E viruses gain entry via ingestion and are important causes of infection but do not produce clinical signs referable to the intestinal tract. The emergence of AIDS has drawn attention to the importance of the rectum as a route of viral entry—HIV as well as other sexually transmitted agents may gain entry through damaged rectal mucosae.

TABLE 7.1 Viruses that Initiate Infection of Humans via the Respiratory Tract

With the Production of Local Respiratory Symptoms	
Picornaviridae	Rhinoviruses, some enteroviruses
Coronaviridae	SARS CoV, MERS CoV, human coronaviruses OC43, 229E, NL63
Paramyxoviridae	Parainfluenza viruses, respiratory syncytial virus, human metapneumovirus
Orthomyxoviridae	Influenza virus
Adenoviridae	Most types
Producing Generalized Disease, Usually without Initial Respiratory Symptoms	
Paramyxoviridae	Mumps, measles viruses
Togaviridae	Rubella virus
Herpesviridae	Varicella virus
Picornaviridae	Some enteroviruses
Polyomaviridae	Polyomaviruses
Parvoviridae	B19 parvovirus
Bunyaviridae	Hantaan virus
Arenaviridae	South American hemorrhagic fever viruses
Poxviridae	Variola virus (smallpox is now extinct)

TABLE 7.2 Viruses that Initiate Infection of Humans via the Alimentary Tract

Via Mouth or Oropharynx	
Herpesviridae	Herpes simplex virus, EB virus, cytomegalovirus, HHV6
Via Intestinal Tract	
Producing enteritis	
Reoviridae	Rotaviruses
Caliciviridae	Noroviruses, sapoviruses
Astroviridae	Human astroviruses (HAstV)
Adenoviridae	Some adenoviruses especially HAdV-40, HAdV-41
Producing generalized disease, usually without alimentary symptoms	
Picornaviridae	Many enteroviruses including polioviruses, hepatitis A virus
Hepeviridae	Hepatitis E virus
Usually symptomless	
Adenoviridae	Some adenoviruses
Picornaviridae	Some enteroviruses
Reoviridae	Reoviruses

Virus Entry via the Skin

The outer keratinized layer of the skin (stratum corneum) is normally impenetrable to viruses unless it is breached mechanically, either by direct trauma, by insect or animal bites, or various inoculation or transfusion procedures (Table 7.3, Fig. 7.3).

Virus replication may then remain localized either in cells of the epidermis (papillomaviruses) or in underlying dermal cells. Alternatively, virus progeny produced within the dermis may be carried by the bloodstream, lymphatics, or nerves to more distant sites. Virus may also be taken up by dendritic cells (Langerhans cells) in the skin and then be transported directly to local lymph nodes. If local virus replication remains confined to the epidermis and does not breach the basement membrane (e.g., herpes simplex virus), scarring of the skin surface is unlikely, whereas if the basement membrane is significantly damaged (e.g., human monkeypox), scarring is the likely result (Table 7.3).

Virus Entry via the Genitourinary Tract

The genitourinary tract is protected by a mucosal lining, by mucus, and by the low pH of the vagina. However, minute tears or abrasions to the vaginal, rectal, and urethral epithelium during sexual activity can facilitate virus entry (e.g., papillomaviruses). Entry by this route is also facilitated by the exchange of bodily fluids that occurs during sexual activity.

Herpes simplex virus 2 and papillomaviruses produce local lesions on the genitalia and perineum from where they may be transmitted by contact. In contrast many other viruses, for example, HIV-1 and 2, human T-lymphotropic viruses 1 and 2 (HTLV-1 and 2), and hepatitis B and C viruses, do not produce local lesions but are sexually transmitted.

Virus Entry via the Eyes

The conjunctiva, although much less resistant to viral invasion than the skin, is constantly cleansed by the flow of secreted tears and is regularly wiped by the eyelids. Infection is more likely to be introduced if abrasions to the conjunctiva or cornea are present, for example, in dusty environments. Virus can reach the eye by aerosol, by rubbing with

TABLE 7.3 Viruses that Initiate Infection of Humans via the Skin, Genital Tract, or Eye

Route	Family	Virus
Skin		
Minor trauma	*Papillomaviridae*	Many types of papillomaviruses
	Poxviridae	Molluscum contagiosum, cowpox, orf, milkers' nodes viruses
	Herpesviridae	Herpes simplex viruses
	Hepadnaviridae	Hepatitis B virus
Arthropod bite (mechanical) Arthropod bite/Mechanical	*Poxviridae*	Tanapoxvirus
Arthropod bite (with replication in the arthropod) Arthropod bite/Biological	*Togaviridae*	Many alphaviruses
	Flaviviridae	Many flaviviruses
	Bunyaviridae	La Crosse, sandfly fever, Rift Valley fever viruses
	Reoviridae	Colorado tick fever virus
Animal bite	*Rhabdoviridae*	Rabies virus
	Herpesviridae	Herpes B virus
Injection, inoculation	*Hepadnaviridae*	Hepatitis B virus
	Flaviviridae	Hepatitis C virus
	Retroviridae	HIV, HTLVs
	Herpesviridae	CMV, EBV
	Filoviridae	Ebola virus
Genital tract	*Papillomaviridae*	Genital types of papillomaviruses
	Herpesviridae	Herpes simplex viruses
	Retroviridae	HIV, HTLV-1
	Hepadnaviridae	Hepatitis B virus
	Flaviviridae	Hepatitis C virus
Conjunctiva	*Adenoviridae*	Several adenoviruses
	Picornaviridae	Enterovirus 70
	Herpesviridae	Herpes simplex viruses
	Poxviridae	Vaccinia virus

contaminated fingers, during ophthalmic procedures with improperly sterilized instruments, or from swimming pool water. Patterns of disease produced include conjunctivitis (e.g., some adenoviruses, influenza viruses, South American arenaviruses, and enteroviruses), and recurrent keratitis (inflammation of the cornea, caused by herpes simplex and several other viruses). Macular involvement is a common feature in Rift Valley fever. Rarely, infection can spread systemically following entry via the eyes, for example, paralysis following enterovirus 70 conjunctivitis. There are reports of Marburg and Ebola viruses persisting in the anterior chamber of the eye long into convalescence.

Vertical Transmission

There are three situations where a virus infection can be transmitted from the mother to her fetus or newborn infant.

1. Transmission of viral genomic DNA encoded in the germ-line or as episomes in ova or sperm may predispose

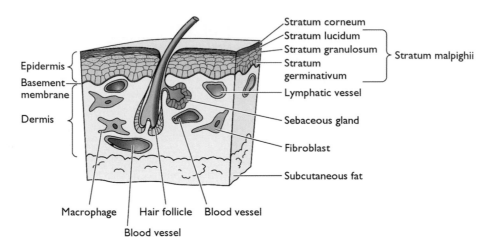

FIGURE 7.3   Cross-section of human skin, showing the different anatomical compartments with their contents. *Originally from Fenner, F.J., et al., 1974. The Biology of Animal Viruses, Academic Press, New York. Modified from Flint, S.J., et al., Principles of Virology, third ed., Vol, II. ASM Press, Washington, DC, with permission.*

the infant to disease in later life. There are well-studied examples of this in animals, and the human genome contains large tracts of integrated retrovirus DNA (endogenous retrovirus genomic DNA and fragments of such, making up about 8% of the human genome). However, no human diseases have so far been shown to be transmitted in this way.

2. Transplacental spread, where infection is passed from an infected mother to her fetus during pregnancy—this usually involves infection of the placenta. Outcomes may range from inapparent infection, development of disease in postnatal life, congenital malformations at birth, or premature labor with or without stillbirth. Well-known examples are rubella (congenital rubella syndrome, now well-controlled by vaccination in most developed countries), cytomegalovirus disease (a remaining major infective cause of congenital abnormalities) and now Zika virus encephalitis and microcephaly. With other infections such as HIV and hepatitis B, both transplacental and perinatal transmission can occur.

3. Perinatal transmission resulting from contact with infected genital secretions or bowel contents during delivery. Infection with herpes simplex viruses 1 and 2 and Coxsackieviruses acquired in this way can have severe consequences, more severe than the usual outcome of these infections in later life due to the immaturity of the newborn infant's immune system.

MECHANISMS OF VIRUS SPREAD WITHIN THE BODY

Virus replication may be localized to a body surface or, alternatively, it may become generalized or systemic following spread from entry sites via lymphatic and hematogenous routes.

Local Spread of Virus on Epithelial Surfaces

Viruses that enter the body via the respiratory or intestinal tracts can be spread rapidly through the layer of fluid/mucus that covers epithelial surfaces; consequently such infections often progress rapidly. Spread to more distant parts of the same anatomical space, for example, the sinuses, middle ear, or alveoli in the case of respiratory infections, is enhanced by sneezing, coughing, and inhalation of secretions. Infections of the respiratory tract by paramyxoviruses and influenza viruses and of the intestinal tract by rotaviruses produce little or no invasion of sub-epithelial tissues. Although these viruses usually enter lymphatics and thus have the potential to spread, they usually do not replicate well in deeper tissues.

In the skin, papillomaviruses initiate infection in the basal layer of the epidermis, but maturation of virions occurs only in cells as they move toward the skin surface and become keratinized. Since this is a slow process taking several weeks, papillomas develop slowly. Many poxviruses produce infection via the skin, but in addition to spreading from cell to cell, there is in addition localized sub-epithelial and lymphatic spread. In vaccination with vaccinia virus a few epidermal cells are infected by scarification and virus spreads locally from cell to cell, primarily in the epidermis, before spreading to the local lymph nodes. The poxviruses that cause molluscum contagiosum, orf, and tanapox remain localized in the skin and produce local lesions.

Factors that restrict an infection from spreading beyond an epithelial surface are listed in Table 7.4. One important factor is the directional shedding of viruses from infected polarized epithelial surfaces. If a virus is preferentially released from the apical end of cells either into the lumen of the respiratory or intestinal tracts or the acinar lumen of a gland, it is free to spread locally to contiguous epithelial

TABLE 7.4 Factors that Restrict Virus Spread from an Epithelial Surface

1. Directional (polarized) budding. When infecting a polarized epithelium, some viruses bud preferentially from the *apical* surface toward the lumen, while others bud from the *baso-lateral* surface toward the underlying tissues
2. Virus may be unable to cross the basement membrane unless it is damaged
3. Cell types in more distant parts of the body may lack receptors, or not be permissive for other reasons
4. There may be systemic presence of neutralizing antibody
5. The particular virus strain may be temperature-sensitive, i.e., it may grow successfully in the nasal passages at 33° but not deeper in the body at 37°
6. A fusion protein on the virion may require proteolytic cleavage for its activation; proteases capable of performing this cleavage may be restricted to a particular site, e.g., gastrointestinal or respiratory tract

surfaces and may be immediately shed from the body, but this site does not favor invasion of sub-epithelial tissues and systemic spread. Conversely, shedding from the basolateral cell surfaces of epithelial cells facilitates invasion of sub-epithelial tissues and subsequent dissemination of virus through lymphatics, blood vessels, or nerves. Paramyxoviruses, respiratory syncytial virus, and influenza viruses are released preferentially from lumenal (apical) surfaces of respiratory system epithelial cells, whereas HIV 1 and 2 are shed from basolateral surfaces of genital tract and other mucosae into sub-epithelial spaces. The polarized distribution of proteins and lipids at different regions of the cell surface is maintained by the interaction between distinct apical and basolateral sorting signals and the appropriate sorting machineries and transport carriers. Interaction between precursor virus particles and these processes facilitates the transport of viruses from sites of synthesis/assembly to their respective plasma membrane domains along the cell cytoskeleton, microtubules, and other framework constituents. Tight junctions at cell-to-cell contact points prevent the movement of proteins between the two domains and maintain the unique protein composition of each domain. In influenza and respiratory syncytial virus-infected epithelial cells, the viral ribonucleoprotein complex (RNP) is carried to a plasma membrane domain where at the same time viral surface glycoproteins are inserted into lipid rafts in the plasma membrane at the nascent budding site. Specific signaling sequences in the stem of the viral glycoprotein spikes are involved along with M-protein determinants. Much of the specificity in the pattern of spread of viruses through the host is due to such evolutionary processes.

Infections that are restricted to an epithelial surface are not always associated with less severe clinical disease.

Large areas of intestinal epithelium may be damaged by rotaviruses, causing severe diarrhea and even death from dehydration. The severity of localized infections of the respiratory tract depends upon their location; infections of the upper respiratory tract may produce severe rhinitis but few other signs; tracheitis lesions are usually repaired rapidly; infection of the bronchioles or alveoli more often produce severe respiratory distress. The entire tracheal and bronchial epithelial lining may be destroyed in influenza or parainfluenza virus infections, causing extravasation of fluids and hypoxia. Fluid and cellular debris build-up in airways may predispose to anoxia and secondary bacterial invasion.

To this point, examples have been considered where infection is typically confined to the epithelial surface of the initial entry site. Of course, in some cases, for example, in immunosuppressed individuals, such infections may become generalized. Furthermore, some superficial infections, for example, influenza, can lead to systemic symptoms including headache, malaise, and myalgia, due to the action of circulating cytokines (the extreme being called a "cytokine storm")—inflammatory mediators may act systemically even if the virus itself remains localized.

Mechanisms of Virus Spread to Distant Target Organs

The most important routes for dissemination of virus to target organs are via blood capillaries, within macrophages and dendritic cells, or via lymphatics. Circulating virus may then gain access to tissues by one of a number of mechanisms, depending on the type of blood tissue junction present. Another mechanism for virus dissemination is where virus enters local nerve endings and undergoes retrograde transit within nerve axons (see below).

Sub-Epithelial Viral Invasion and the Network of Lymphatics

Viruses can enter the network of lymphatics beneath all cutaneous and mucosal epithelia (Fig. 7.4). Virions entering the lymphatics are carried to local draining lymph nodes. As they enter the marginal sinuses of lymph nodes, virus particles are exposed to active macrophages and dendritic cells and may be engulfed, inactivated, and/or processed for viral antigen presentation to adjacent lymphocytes, thereby initiating the adaptive immune response (see Chapter 6: Adaptive Immune Responses to Infection). Some viruses, however, also replicate in monocytes/macrophages (e.g., many retroviruses, measles virus, dengue viruses, some adenoviruses, and some herpesviruses); some viruses are able to replicate in lymphocytes. Some viruses pass directly through lymph nodes to enter the bloodstream. Monocytes and lymphocytes recirculate through the blood, lymphatics,

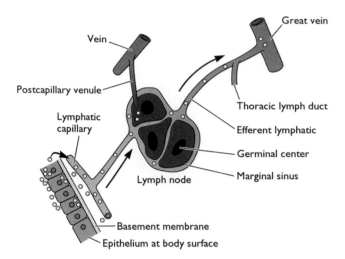

FIGURE 7.4 Subepithelial invasion and lymphatic spread of viruses. *Originally from the work of Cedric Mims 1977–2015 (e.g., Pathogenesis of Infectious Disease, through six editions, Elsevier/Academic Press), and modified many times since.*

and lymph nodes; recirculation is the key to normal "immune surveillance," but is also an effective means of disseminating some viruses throughout the body.

Normally, there is a local inflammatory response at the site of viral invasion, the extent of which depends upon the extent of tissue damage. Local blood vessels become dilated and rendered more permeable, so that monocytes and lymphocytes, cytokines, immunoglobulins, and complement components may be delivered directly into the extravascular site of infection. These events are especially vigorous once the adaptive immune response reaches its full level of activity. In some cases, viruses take advantage of these events to infect cells of the lymphoreticular system and spread locally or systemically.

Virus Spread via the Bloodstream: Viremia

The blood is the most effective and rapid vehicle for the spread of viruses through the body. Once a virus has reached the bloodstream, usually via the lymphatic system, it can localize in any part of the body within minutes. The first entry of virus into the bloodstream is referred to as the *primary viremia*. This early viremia may be clinically silent, known to have taken place only because of the invasion of distant organs. Virus replication in initial target organs leads to the sustained production of much higher concentrations of virus, producing a *secondary viremia* (Fig. 7.5), which in turn can lead to the establishment of infection in yet other parts of the body.

Virions may be free in the plasma or may be contained in, or adsorbed to, leukocytes, platelets, or erythrocytes. Enteroviruses, togaviruses, and most flaviviruses circulate free in the plasma, whereas hepatitis B and hepatitis C

viruses are complexed with different serum proteins and lipids. Viruses carried in leukocytes, generally lymphocytes or monocytes, are not cleared as readily or in the same way as viruses circulating free in the plasma; being protected from antibodies and other plasma components, virus can be transported to distant tissues even after the initiation of the immune response. Monocyte-associated viremia is a feature of measles, cytomegalovirus, and human herpesvirus 8 (Kaposi sarcoma herpesvirus) infections. In infections caused by Rift Valley fever virus and Colorado tick fever virus, virions are associated with erythrocytes— in Colorado tick fever virus infection the virus replicates in erythrocyte precursors in the bone marrow producing a viremia that lasts for the life span of erythrocytes. Colorado tick fever virus has been transmitted to blood transfusion recipients 100 days after the donor was infected by tick bite. Certain murine leukemia viruses and arenaviruses infect megakaryocytes and thereby are present in circulating platelets, the clinical significance of which is unknown. Neutrophils have a very short life span and powerful antimicrobial mechanisms; they are rarely infected despite frequently containing phagocytosed virions.

Viruses circulating in the blood encounter many kinds of cells, but two play special roles in determining the subsequent fate of infection: macrophages and vascular endothelial cells.

Virus Interactions with Macrophages

Macrophages are very efficient phagocytes and are present in all compartments of the body: they occur free in plasma, in alveoli, in sub-epithelial tissues, in sinusoids of the lymph nodes, and above all in the sinusoids of the liver, spleen, and bone marrow. Together with dendritic cells and B lymphocytes, macrophages are antigen-processing and antigen-presenting cells and therefore play a pivotal role in initiation of the adaptive immune response (see Chapter 6: Adaptive Immune Responses to Infection). The antiviral action of macrophages depends on the age and physiological status of the host and their site of origin in the body; indeed, even in a given site there are subpopulations of macrophages that differ both in phagocytic competence and in susceptibility to infection. Their state of activation is also important. Transport of viruses inside infected cells such as macrophages, has been referred to as the "*Trojan Horse*" mechanism of invasion; it is especially important in HIV infection of the central nervous system.

The various kinds of interactions that can occur between macrophages and viruses may be illustrated by the various responses of Kupffer cells, the macrophages lining the sinusoids of the liver (Fig. 7.6). Infection of Kupffer cells often contributes to acute hepatocellular damage (e.g., yellow fever, Rift Valley fever, Crimean Congo hemorrhagic fever, and Ebola and Marburg virus hemorrhagic fevers).

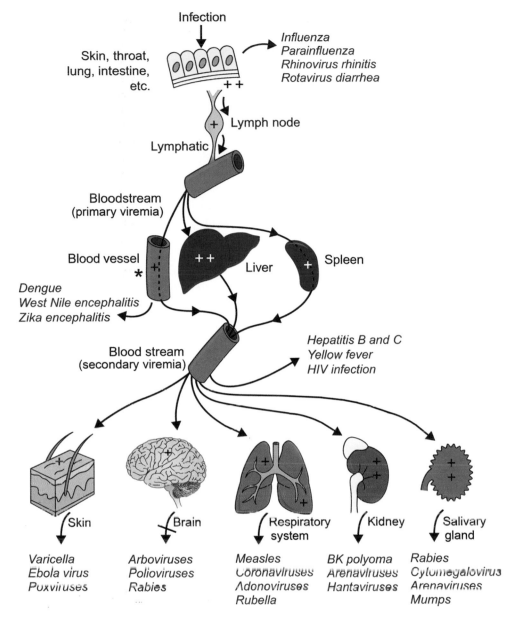

FIGURE 7.5 The role of primary and secondary viremias in the spread of viruses throughout the body, showing some, but not all, sites of replication and routes of shedding. + indicates major sites of virus replication; arrows indicate sites of shedding to the exterior; * indicates replication in vascular endothelium rather than in viscera. *Originally from the work of Cedric Mims, 1977–2015. (e.g., Pathogenesis of Infectious Disease, through six editions), Elsevier/Academic Press), and modified many times since.*

Differences in virus–macrophage interactions may account for differences in the virulence of particular virus strains and differences in host resistance. Even though macrophages are innately efficient phagocytes, this capacity is greatly enhanced during an immune response by the action of cytokines, which are released notably from T helper lymphocytes. Macrophages also have Fc-receptors and C3-receptors on their plasma membranes, which further enhance the ingestion of virus particles, especially when coated with antibodies or complement. Certain togaviruses, flaviviruses (notably dengue viruses), coronaviruses, arenaviruses, reoviruses, and especially retroviruses, are capable of replicating in macrophages— when virus uptake is facilitated by bound antibody, antibody-mediated enhancement of infection may occur. This is a major pathogenetic factor in dengue and retrovirus infections.

Virus Interactions with Vascular Endothelial Cells

The vascular endothelium with its basement membrane and tight cell junctions constitutes the blood–tissue interface and is often a barrier. Virus invasion of the tissue parenchyma

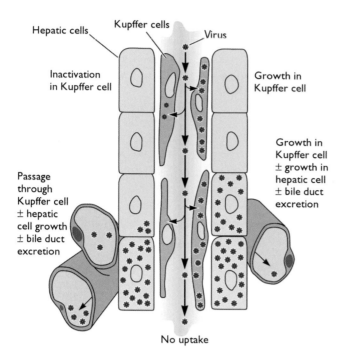

FIGURE 7.6 Different possible interactions between viruses and Kupffer cells, the macrophages that line the sinusoids of the liver. (1) Viruses may pass through the sinusoid without being phagocytosed. (2) Virions may be phagocytosed and destroyed: because the macrophage system is so efficient, viremia can only be maintained if virions enter the blood stream as fast as they are removed. (3) Virions may be phagocytosed and then transferred passively to adjacent cells (i.e., hepatocytes in the liver). (4) Virions may be phagocytosed by Kupffer cells and then replicate in them, with or without spread to adjacent hepatocytes and excretion into bile ducts or release to the bloodstream. *Originally from the work of Cedric Mims, 1977–2015 (e.g., Pathogenesis of Infectious Disease, through six editions), Elsevier/Academic Press), and modified many times since.*

by circulating virions depends upon crossing this barrier, usually exiting capillaries and venules where the blood flow is slower and the barrier is thinnest. The structure of this barrier varies from tissue to tissue (Fig. 7.7): (1) in the central nervous system, connective tissue, muscle, skin, and lungs there is a continuous lining of endothelium and basement membrane, (2) in the intestine, renal glomerulus, pancreas, endocrine glands, and choroid plexus, this lining has fenestrations or pores, (3) in liver, spleen, bone marrow, and adrenal glands, blood flows through sinusoids lined by macrophages and endothelial cells. Virions may either move passively through fenestrae between endothelial cells, or replicate in endothelial cells or macrophages and "grow" across this barrier, (4) in some cases viruses can be passively transferred through the lining cells without replicating ("*transcytosis*"), or (5) be carried within lymphocytes or monocytes trafficking between lining cells ("*diapedesis*").

The above mechanisms do not fully explain the preferential targeting of certain viruses to particular target tissues, which is likely to involve use of preferred mechanisms to exit from the vascular space in preferred organs, as well as a preference for replication in particular cell types. Much focus has been placed upon viral receptors on particular cell types, but the identification of virus receptors in cultured cells has turned out to oversimplify this complex subject. A virus may employ several different receptors and co-receptors *in vivo*, on endothelial cells, perivascular cells, parenchymal cells, etc., and these may vary between different target organs. While our knowledge of virus/receptor interactions has advanced greatly in recent years, our understanding of the full mechanisms underlying virus dissemination and tissue targeting *in vivo* is still limited.

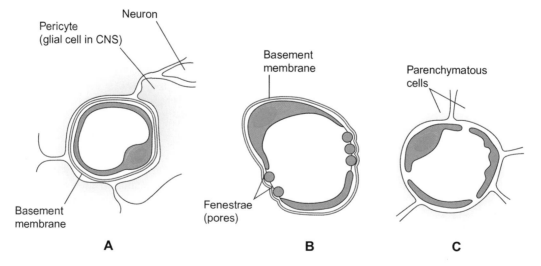

FIGURE 7.7 Three types of blood–tissue junctions, as occur in capillaries, venules, and sinusoids. (A) Continuous endothelium (central nervous system, connective tissue, skeletal and cardiac muscle, skin, lung). (B) Fenestrated endothelium (renal glomerulus, intestinal villi, choroid plexus, pancreas, endocrine glands). (C) Sinusoid (reticuloendothelial system): liver (lined by Kupffer cells), spleen, bone marrow, adrenal gland, parathyroid gland. *Originally from the work of Cedric Mims, 1977–2015 (e.g., Pathogenesis of Infectious Disease, through six editions), Elsevier/Academic Press), and modified many times since.*

Maintenance of Viremia

Viremia can be maintained only if there is a continuing introduction of virus into the bloodstream from infected tissues to counter the continual removal of virus by macrophages and other cells and the natural decay of virus infectivity over time. The balance between virus replication and removal is greatly affected if macrophage functions are impaired, as in some infections (e.g., measles). Although circulating leukocytes can themselves constitute a site of viral replication, viremia is usually maintained by replication in parenchymal cells of target organs such as the liver, spleen, lymph nodes, bone marrow, etc. In some infections, such as dengue, viremia follows infection of endothelial cells. Striated and smooth muscle cells may be an important site of replication of some enteroviruses, togaviruses, and rhabdoviruses, from where virus reaches the blood via lymphatic circulation/recirculation.

There is generally a correlation between the magnitude of viremia generated by blood-borne viruses and their capacity to invade target tissues. Conversely, the failure of some attenuated viruses to generate a significant viremia may account for their lack of invasiveness. Certain neurotropic viruses are virulent after intracerebral inoculation, but avirulent when inoculated peripherally—in such instances the level of viremia may be insufficient to favor invasion of the nervous system, or this may be due to other virus characteristics. However, the capacity to produce viremia and the capacity to invade tissues from the bloodstream are distinct viral properties. For example, some laboratory strains of Semliki Forest virus, an alphavirus, have lost their capacity to invade the central nervous system while retaining a capacity to generate a viremia equivalent in duration and magnitude to that produced by neuroinvasive strains.

In most transient systemic infections, the viremic phase is usually quite short-lived, typically a few days. As this often coincides with the period when the patient feels most unwell, and is therefore excluded from donating blood, transmission of these infections by blood transfusion or sharing of intravenous needles is often not a significant problem. One notable exception is hepatitis A; in this case there is a transient infection, but low levels of infectious virus can circulate for several months, into convalescence, and be transmitted by the blood-borne route. This situation is however quite different from the persistent systemic infections discussed below, where high-level viremia can be maintained for months or years frequently without symptoms. Transmission of these infections by unscreened donated blood can present major problems (e.g., hepatitis B virus, HIV).

Virus Spread via Nerves

Herpesvirus capsids travel centripetally from the peripheral nervous system entry site to the central nervous system in the axon cytoplasm of sensory nerves; while doing so they also sequentially infect the Schwann cells of the nerve sheath. In most instances herpes simplex and varicella-zoster viruses stop at this point, establishing persistent infection of the neurons of dorsal root ganglia. These viruses can then transit the same sensory nerves centrifugally from the ganglia to the skin/mucosa. This is what happens in the reactivation of latent infection and production of recrudescent epithelial lesions. The rate of this anterograde and retrograde transit of virus nucleocapsids approaches 200 to 400 mm a day.

Rabies virus travels to the central nervous system within the axon cytoplasm without infecting cells of the nerve sheath (Schwann cells). Following the bite of a rabid animal, virus is usually amplified by replication in striated muscle cells at the bite site. From here virus may enter the peripheral nervous system at sensory nerve end organs in muscle (neuromuscular spindles) or motor nerve end organs (motor end plates), traveling in axon cytoplasm centripetally. Rabies virus may also enter the peripheral nervous system end organs directly at the bite site: this is the likely situation in rabies cases with an exceptionally short incubation period. Rarely, the central nervous system may be invaded directly by rabies virus and some togaviruses when neurons of the olfactory end organ in the nares are directly exposed. These are the only neurons in the body that directly link the body surface with the central nervous system. In instances where virus transit uses this route, the olfactory bulb of the brain is usually seen to be infected first. This has been the likely source of infection in speleologists infected in bat caves where the rabies virus has been shown to be present in aerosols (in some areas speleologists are regularly vaccinated).

As these viruses move centripetally, they must cross cell-to-cell junctions. Rabies virus and herpesviruses are known to cross from one neuron to the next at synaptic junctions by employing structures and mechanisms normally used to transfer neurotransmitter molecules.

VIRUS INFECTION OF TARGET ORGANS

Different viruses present different unique patterns of infection (clinical signs, symptoms, laboratory data, etc.), based upon differences in their major sites of replication and damage. These sites are known as "target organs." Of course, other sites may also be infected at various stages of infection, without necessarily being clinically evident (Table 7.5).

The predilection of a particular virus for infecting a particular cell type or organ is known as its "tropism" (as distinct from "trophism," which refers to nutrients). With some viruses, the pathophysiological, molecular, and/or anatomical factors determining tropism have been partially clarified, but for many others these are not known. For example: most viruses of humans replicate optimally at 37°C, the internal temperature of the body. However some

TABLE 7.5 Target Organs in Some Acute Transient Systemic Infections

Virus enters via the gastrointestinal tract
- Hepatitis A: liver
- Poliomyelitis: anterior horn cell in spinal cord and central nervous system
- Other enteroviruses: meninges, muscle, skin, CNS

Virus enters via the respiratory tract
- Chickenpox: (sensory ganglia), skin
- Measles: conjunctiva, skin, CNS
- Rubella: skin, joints
- Mumps: parotid and salivary glands, testes, pancreas, meninges
- Smallpox: skin, mucous membranes

TABLE 7.6 Virus and Host Properties Affecting Viral Tropism

1. Distribution of virus receptors may be restricted to particular cells or expressed only at certain times or under certain physiological conditions
2. Temperature sensitivity of virus replication
3. Restricted presence of a specific cellular protease required for activation of virus attachment or fusion proteins
4. Selective presence of host innate or adaptive immune response
5. Cellular transcription factors restricted to specific cell types
6. Anatomical barriers may restrict virus spread
7. Route of inoculation or entry may dictate virus distribution

respiratory viruses, for example, rhinoviruses, replicate optimally at 33°C, reflecting the slightly lower temperature of the mucosal surfaces of the upper respiratory tract. These viruses grow less well at 37°, which restricts spread and involvement of the lower respiratory tract. This also means that rhinovirus isolation is more successful if cell cultures are maintained at 33°. Many other factors also affect virus infection patterns (Table 7.6).

Of course, the summary items listed in Table 7.6 represent necessary simplifications. For example, the poliovirus receptor (Pvr, CD155), a member of the immunoglobulin (Ig) superfamily of proteins, is present on neurons in many parts of the nervous system as well as on cells of the adrenal gland, lung, and kidney, and at low levels on skeletal muscle. Poliovirus replication in the infected host partly follows this distribution but there are a number of discordances, indicating that receptor distribution alone is not sufficient to explain the differences in virus distribution. Most striking is the selective infection and rapid destruction by polioviruses of anterior horn neurons of the spinal cord, while other neurons are often spared.

Furthermore, not only do some viruses require several cellular receptors/co-receptors to complete an infection cycle, some utilize different receptors on different host cells in different organs or tissues. For example, the glycoprotein ligand SU of HIV-1 can bind to several receptors (including CD4, CXCR4, CCR5, and others); these take part in a complex sequence of initial loose binding, tight binding, fusion, and entry of the viral RNA complex into the cell cytosol. HIV-1 can infect T lymphocytes, macrophages, and other cell types; viral strains that bind CXCR4 preferentially infect T cell lines, and strains that bind CCR5 infect monocytes/macrophages, but the full role of different receptor usage by different HIV variants in natural infection remains to be clarified. Expression of receptors can be dynamic; for example, it has been shown experimentally that animals treated with neuraminidase [the viral enzyme that digests the neuraminic acid (NA)-containing receptor] exhibit substantial protection against intranasal infection with influenza viruses, and this lasts until the neuraminidase-sensitive receptors have regenerated.

Receptors for a particular virus are usually restricted to certain cell types in certain organs, and only these cells can be infected. However, the presence of critical receptors is not the only factor that determines whether the cell may become infected—intracellular factors that exert effects subsequent to virus attachment, such as viral enhancers and cellular factors, are also required for a productive infection.

The Skin as a Target Organ

As well as being a site of initial virus entry, the skin may be invaded via the bloodstream, producing erythema and often a generalized rash. The individual lesions in a generalized rash may progress sequentially through macules, papules, vesicles, pustules, and ulcers, or alternatively the rash may resolve at an earlier stage. A lasting local dilation of subpapillary dermal blood vessels produces a *macule*, which becomes a *papule* if there is also edema and infiltration of inflammatory cells into the area. Primary involvement of the epidermis or separation of epidermis from dermis by fluid pressure leads to *vesicle* formation, characterized by the presence of clear fluid. This may subsequently be converted to a *pustule* by polymorphonuclear cell infiltration into the vesicle. Erosion or sloughing of the surface epithelium produces an *ulcer*, with or without *scab* formation. More severe involvement of the dermal vessels may lead to petechial or hemorrhagic rashes, although coagulation defects and thrombocytopenia may also be important in the genesis of such lesions. The clinical distribution, stage of evolution, and other characteristics of rashes can be distinctive for specific viruses and an aid in diagnosis (Fig. 7.8).

Virus Infection of the Central Nervous System

Because of the critical importance of the central nervous system and its vulnerability to damage by any process

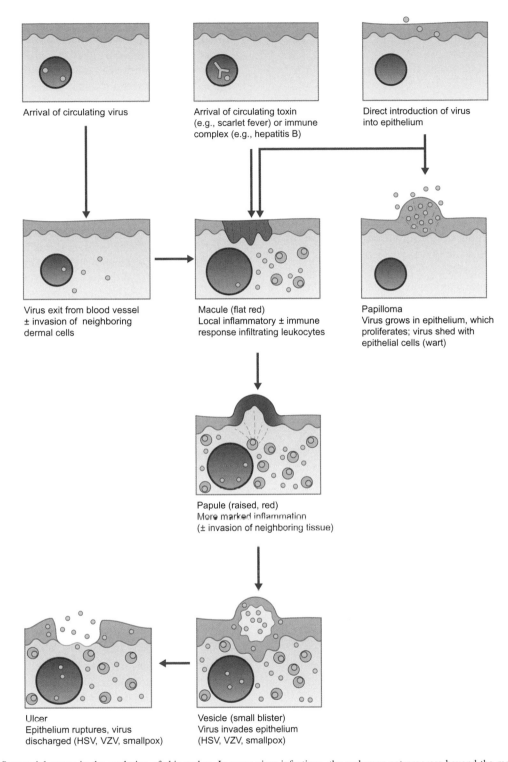

Arrival of circulating virus

Arrival of circulating toxin
(e.g., scarlet fever) or immune
complex (e.g., hepatitis B)

Direct introduction of virus
into epithelium

Virus exit from blood vessel
± invasion of neighboring
dermal cells

Macule (flat red)
Local inflammatory ± immune
response infiltrating leukocytes

Papilloma
Virus grows in epithelium, which
proliferates; virus shed with
epithelial cells (wart)

Papule (raised, red)
More marked inflammation
(± invasion of neighboring tissue)

Ulcer
Epithelium ruptures, virus
discharged (HSV, VZV, smallpox)

Vesicle (small blister)
Virus invades epithelium
(HSV, VZV, smallpox)

FIGURE 7.8 Sequential stages in the evolution of skin rashes. In some virus infections, the rash may not progress beyond the maculopapular stage (e.g., rubella), while with others a full progression through to vesicles and ulcers is usual (e.g., smallpox). *Modified from Mims, et al., 1993. Medical Microbiology, Mosby/Elsevier Europe Ltd.*

that harms neurons directly or via increased intracranial pressure, viral invasion of the central nervous system is always of serious concern.

Viruses can spread from the blood to the brain in two major ways:

A. Virus present within blood vessels in the meninges and the choroid plexus can pass through the endothelial lining (with or without replication) to circulate in the cerebrospinal fluid, infect the ependymal lining of the ventricles, and thence invade the brain parenchyma (Fig. 7.9). Some enteroviruses that cause meningitis, rather than encephalitis, may traverse endothelial junctions in the meninges and stay localized at this site without further penetration of the brain parenchyma. Others progress from this site to cause encephalitis.

B. Although the cerebral capillaries together with their underlying dense basement membranes represent a morphological blood–brain barrier, most viruses that invade the central nervous system cross these vessel walls directly. Some viruses infect vascular endothelial cells prior to infection of the cells of the brain parenchyma; others appear to be transported across the capillary walls without endothelial cell infection. In HIV infection and measles, virus may be carried across capillary walls into the brain parenchyma via infected leukocytes. Subsequent spread in the central nervous system can take place via the cerebrospinal fluid or by sequential infection of neural cells. The intercellular space of the brain parenchyma may appear completely collapsed and a barrier to virus movement, but in fact it is without barriers (e.g., tight junctions, connective tissue structures) and so local virus movement between neurons, glia, and inflammatory cells occurs causing focal parenchymal lesions.

The other important route of invasion of the central nervous system is via the peripheral nerves, as seen in rabies and varicella-zoster and herpes simplex encephalitis. This route has been described in the preceding section.

Lytic infections of neurons, whether due to polioviruses, flaviviruses, togaviruses, or herpesviruses, are characterized by the three histological hallmarks of viral encephalitis: (1) neuronal necrosis, (2) phagocytosis of neurons by phagocytic cells (neuronophagia), and (3) perivascular infiltration of mononuclear cells (perivascular cuffing), the latter reflecting an immune response. The cause of clinical neurological signs in many central nervous system infections is sometimes less clear. Rabies virus ("street virus" = wild-type strains) infection is rather non-cytocidal; infection evokes little of the inflammatory reaction or cell necrosis found in other encephalitides, yet it is lethal for most mammalian species. The basis for this is not clear but has to do with pathophysiologic functional loss. The extensive central nervous system infection of mice congenitally infected with non-cytolytic lymphocytic choriomeningitis virus, readily demonstrable by fluorescent antibody staining, has no recognizable deleterious effect—it is the following fulminant T lymphocyte response that is lethal (immunopathological disease). Characteristic pathological changes are seen in several slowly progressive diseases of the central nervous system; in prion diseases, for example, there is slow neuronal degeneration and vacuolization (Chapter 38: Prion Diseases); in progressive multifocal leucoencephalopathy (PML), JC virus infection is lytic in oligodendrocytes, producing multiple foci of demyelination and bizarre transformation of astrocytes into giant cells (Chapter 20: Polyomaviruses).

Post-infectious encephalitis occurs most frequently after measles (about 1 per 1000 cases), more rarely after rubella and varicella-zoster virus infections, and in the past was seen in 1 to 2 persons per 100,000 primary vaccinations against smallpox (see below).

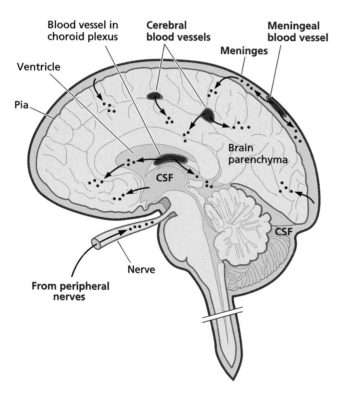

FIGURE 7.9 Viruses can gain access to the central nervous system via many routes, with specific routes seen in certain experimental models; however, in most human central nervous system diseases caused by viruses the route of entry is unknown or based upon indirect evidence. Few, such as rabies virus and herpesviruses, ascend via peripheral nerves; most employ hematogenous routes, entering via the capillaries of the brain parenchyma or surfaces (meninges, ventricles, choroid plexus). *CSF*, cerebrospinal fluid. *Originally from Mims, C.A., 1977–2008. Pathogenesis of Infectious Disease (through five editions). Elsevier/Academic Press, London, UK. Then adapted by Nathanson, N. (ed.), 1997. Viral Pathogenesis. Lippincott-Raven, Philadelphia, PA. Then adapted by Flint, S.J., et al., 2009. Principles of Virology: Pathogenesis and Control, 3rd ed. ASM Press, Washington, DC.*

Virus Infection of Other Organs

Almost any organ may be infected via the bloodstream with many viruses, but most viruses have well-defined organ and tissue tropisms. The clinical importance of infection of various organs and tissues depends in part on their role in the economy of the body. Thus invasion of the liver, causing severe hepatitis, as in yellow fever and infections with the hepatitis viruses, may be life-threatening, with the additional possibility in the case of hepatitis B and C of the establishment of a chronic carrier state that may eventually result in hepatocellular carcinoma. The critical importance of such organs as the brain, heart, and lung is equally self-evident. Thus the most dangerous viral infections tend to be those that cause encephalitis, pneumonia, carditis, hepatitis, or hemorrhagic fever.

Infection of the testis or accessory sexual organs may lead to excretion of virus in the semen and the risk of transmission during sexual activities. In some cases of hemorrhagic fever, virus continues to be shed in the semen long into convalescence. Localization of virus in the salivary glands (e.g., in mumps), mammary glands, kidney tubules, and lungs may lead to excretion in the saliva, milk, urine, and respiratory secretions. Some togaviruses and Coxsackieviruses infect muscle cells, while infection of synovial cells by rubella virus, Ross River virus, and chikungunya virus produces polyarthritis/polyarthralgia.

Virus Infection of the Fetus

Most viral infections of the mother have no harmful effect on the fetus, but some blood-borne viruses cross the placenta to enter the fetal circulation, sometimes after establishing foci of infection in the placenta. Severe cytolytic infections of the fetus cause fetal death and abortion, a pattern that was common in smallpox. Fetal death and nearly universal death of the mother is the case in Ebola, Marburg, and Lassa hemorrhagic fevers. Fetal teratogenic effects are common following rubella and cytomegalovirus infections during pregnancy.

Since Norman Gregg's initial observations in 1941, it has been recognized that maternal rubella contracted in the early months of pregnancy often leads to congenital abnormalities. A variety of abnormalities occur, of which the most severe are deafness, blindness, and congenital heart and brain defects. These defects may not be recognized until after the birth of an apparently healthy baby, or they may be associated with severe neonatal disease— hepatosplenomegaly, purpura, and jaundice—comprising the *"congenital rubella syndrome."* The pathogenesis involves infection of the placenta from maternal viremia, leading to virus spread through the developing vasculature of the fetus. Damage to fetal blood vessels and ischemia in affected organs cause a pattern of congenital defects, that are typically most prominent in those particular organs that are being formed at the gestational time when the mother acquires her infection. Classical immunological tolerance does not develop; children who have contracted rubella *in utero* exhibit high titers of neutralizing antibodies throughout their lives but there may be some diminution in cell-mediated immunity. Rubella virus is relatively non-cytocidal; few inflammatory or necrotic changes are found in infected fetuses. The retarded growth and developmental abnormalities may be due, in addition to ischemic effects, to slowing of cell division leading to the reduced numbers of cells in many fetal organs. Clones of persistently infected cells might be unaffected by the maternal antibody that develops during the first weeks after the maternal infection, even though such antibody may limit viremia in the fetus, perhaps a matter of "too little, too late."

Cytomegalic inclusion disease of the newborn results from infection acquired congenitally from mothers suffering an inapparent cytomegalovirus infection during pregnancy. The important clinical features in neonates include hepatosplenomegaly, thrombocytopenic purpura, hepatitis and jaundice, microcephaly, and mental retardation.

Apart from infection of the fetus via the placenta, germ-line transmission of retroviruses, as integrated provirus, occurs commonly in many species of animals but has not been found to be linked to disease in humans. More important in human medicine because of its epidemiological implications (see Chapter 14: Control, Prevention, and Eradication) is transovarial transmission of arboviruses, notably some bunyaviruses, togaviruses, and flaviviruses, in mosquitoes and ticks. This mechanism may be involved in virus overwintering in vector species leading to a long-term risk of human infection in certain econiches.

PERSISTENT VIRUS INFECTIONS— MECHANISMS INFLUENCING PERSISTENCE

The great majority of virus infections of humans are transient, and after a period of replication within the body and development of an immune response (with or without accompanying pathology and symptoms) the virus is eliminated from the body. In some instances what is thought to be a persistent viral infection may actually be a chronic bacterial super-infection; for example, sinusitis or bronchitis following a viral infection. In other instances, apparently long-term clinical effects may be a consequence of sequential infections with different viruses. However, more and more viruses that were usually thought only to cause acute transient clinical infections, are being shown to sometimes persist in sequestered sites in trace amounts, usually without any demonstrable effect, but sometimes causing important clinical consequences.

Establishment and maintenance of a persistent infection implies two obvious prerequisites: (1) the failure to eliminate the virus by the immune system, and (2) in the case of cytocidal viruses, a restriction of viral replication. The viral genome may persist in the absence of complete cycles of viral replication; persistent infection may only be evident after reactivation of a latent viral state at some later time in the life of the host. Whether persistence leads to disease is obviously of crucial importance to human hosts, but may be of little consequence to virus survival in nature; this depends more critically upon shedding into the environment and consequent spread to other hosts.

Long-term virus persistence is an outcome that is ultimately dependent on the balance between the virus' capacity, on the one hand, to replicate or remain within long-lived cells, and, on the other hand, the virus' ability to counter the two arms of the immune response. The outcome of this balance can vary from situation to situation.

For example, certain viruses, for example, members of the herpesvirus family and HIV, invariably produce persistent infections, whereas other viruses, for example, common respiratory viruses, almost invariably produce a transient infection. Others, for example, hepatitis B virus, can produce either a transient or persistent infection depending on the circumstances, for example, the age of the person at the time of exposure. These "typical" outcomes may also vary subject to the immune competence of the individual— immunosuppression will shift the balance in favor of virus persistence, and in immunosuppressed individuals some typically transient virus infections may persist, and other typically localized or latent infections may become generalized. The following sections describe many of the mechanisms used by different viruses in developing persistence (Table 7.7).

TABLE 7.7 Mechanisms for Ineffective Immune Responses in Persistent Viral Infections[a]

Phenomenon	Mechanism	Example[b]
Reduced or absent antigen expression	Limited viral gene expression	Herpes simplex virus (neurons)
		EBV (B lymphocytes)
		HIV-1
Evasion of immune response	Cell–cell spread by cell fusion	HIV-1
		Measles
		Cytomegalovirus
	Sequestration in sanctuaries	HIV-1 (brain)
		CMV, polyoma virus (luminal surface of kidneys, glands)
		HPV (keratinized skin)
	Antigenic drift	HIV-1, visna, HCV
Immunosuppression by infection of effector cells	Disruption of lymphocyte function, cell loss. Polyclonal B cell activation. Abrogation of macrophage function	HIV (CD4+ T cells), EB virus, HIV
	Antibody-enhanced infection of macrophage via Fc receptor	Many viruses
Induction of immunologic tolerance	Congenital or neonatal infection induces T cell non-responsiveness	Congenital rubella and CMV, LCM, HBV, HIV, parvovirus B19
Reduced antigen display on plasma membrane	"Stripping" by antibody. Down-regulation of MHC antigen. Down-regulation of cell adhesion molecules Virus-coded Fc receptor blocks immune lysis	Measles (SSPE brain). Adenoviruses. EB virus (Burkitt's lymphoma) Herpesviruses
Blocking by non-neutralizing antibodies		Many viruses
Excess soluble antigen as decoy		Hepatitis B virus
Evasion of cytokines	Interfere with IFN, TNF	Adenoviruses, HCV, others

[a]Some of these mechanisms are also operative in non-persistent infections.
[b]Speculative only; in several of these instances an association exists but no cause-and-effect relationship has been demonstrated between the immunologic phenomenon and the persistent infection listed.

Reduction or Absence of Either Viral Gene Expression or Cytocidal Activity

Restricted Expression of Viral Genes

Obviously a virus cannot persist in a cell it destroys. Therefore, long-term persistence of a potentially cytocidal virus can occur only if the viral genome remains fully or partially silent, or if a dynamic can be established whereby newly made viruses can continue to find new uninfected cells. In latent herpesvirus infections, only a few early genes are transcribed during latency. Latency is eventually overridden, often following immunosuppression and/or by the action of a cytokine or hormone that de-represses transcription of the whole viral genome, leading to reactivation of viral synthesis. In contrast, persistent HIV infection is characterized by continuing virus production in T lymphocytes leading to cell death. This is combined with massive parallel proliferation of uninfected lymphocytes which become a target for further rounds of HIV replication and damage.

Restricted viral genome expression as seen in several persistent infections naturally limits the presentation of viral antigens to the immune system, allowing such infected cells to remain immunologically invisible. For example, in herpes simplex virus infection latently infected neurons display no viral antigens. This protects these cells not only against cytotoxic T cells but also against lysis by antibodies or complement-mediated mechanisms or possibly by antibody-dependent cell-mediated cytotoxicity (ADCC).

Latency in Non-Permissive, Resting, or Undifferentiated Cells

A given virus may undergo productive replication in one cell type but non-productive latent infection of another. For example, Epstein Barr virus may replicate productively in mucosal epithelial cells but assume a latent state in B lymphocytes. Hence one cell type may serve as a repository which, following reactivation, seeds others. Even in a given cell type, permissiveness may be determined by the state of cellular differentiation or activation. For example, papillomaviruses complete a replication cycle only in fully differentiated epithelial cells. Again, HIV replicates in CD4+ T cells activated by an appropriate cytokine but remains latent in resting CD4+ T cells. Moreover, HIV enjoys a different type of association with cells of the monocyte/macrophage lineage, and other cell types in the body have been identified as further potential reservoirs of latent virus.

Non-Cytocidal Viruses

Arenaviruses and some retroviruses are examples of non-cytocidal viruses that may establish persistent infections in their reservoir hosts. In some natural pairings of virus and host, persistence is life-long with no chronic disease, while in others virus carriage declines slowly over time and transmission to further generations of reservoir hosts is required to complete the natural virus cycle. For example, Junin virus (the etiologic agent of Argentine hemorrhagic fever) produces a persistent infection, in its reservoir host, the vesper mouse, *Calomys musculinus*, with no disease but with long-term virus shedding in urine and saliva. Transmission from the mouse by aerosol and fomites to field workers, especially during harvest time, resulted for many years in outbreaks of severe hemorrhagic fever—the incidence has been greatly lowered in recent years by extensive use of vaccine.

Hepatitis B virus causes a non-cytocidal infection of hepatocytes, and the extent of liver damage is dependent on the extent of immune-mediated cell damage (see below).

Evasion of the Immune Response

Cell-to-Cell Spread of Virus by Cell Fusion

Lentiviruses (e.g., HIV), paramyxoviruses (e.g., measles virus), and herpesviruses (e.g., cytomegalovirus) can cause neighboring cells to fuse together, enabling the viral genome to spread contiguously from cell to cell without ever being exposed to virus-specific circulating antibodies or cytotoxic T cells.

Sequestration of Virus in Anatomic or Physiological Sanctuaries

A number of persistent infections involve the central nervous system. The brain is insulated from most trafficking of lymphocytes by the blood–brain barrier; further, neurons express very little MHC antigen at their surfaces, thereby conferring some protection against lysis by cytotoxic T lymphocytes and the consequent release of viral antigens that might otherwise trigger an immune response. Herpesviruses, polyomaviruses, and lentiviruses are examples of viruses that can persist in the brain by taking advantage of its separation from normal immune surveillance. HIV is relatively protected not only in the brain but also in the epididymis.

The kidney is the other major site that frequently harbors persistent viruses, for example, cytomegalovirus and JC and BK polyomaviruses. These viruses are not acutely cytopathic, and perhaps because virus is released on the lumenal surfaces of polarized epithelial cells immune surveillance is avoided.

An extreme case of inaccessibility to the immune system is employed by papillomaviruses causing warts; virus replication occurs in a stepwise fashion as skin cells mature—initial infection occurs in basal skin layers (*stratum germinativum*) but infectious virions are only completed in the fully differentiated outer layers (*stratum granulosum* and *stratum corneum*) of the skin—there is little or no immune surveillance in such sites. Nevertheless, evidence of immune control of papillomavirus infections is demonstrated by the increased occurrence of warts in immunosuppressed individuals, and the tendency for multiple warts to all undergo spontaneous regression simultaneously.

Viral Antigenic Drift

In a number of persistent infections, notably HIV, mutations emerge at an extremely high rate because of the very high replication rate, for example, $\sim 10^{10}$ new virions per day, and the high error rate of the virion reverse transcriptase (approximately 1 base error per 10^4 nucleotides transcribed). This leads to continuing generation and then selection of antigenic escape mutants which are recognized only partly or not at all by the immune response specificity at the time. Such escape mutants often also have lessened replicative fitness, that is, they do not outcompete the original virus unless the latter's replication is being retarded by the immune response.

Virus-Induced Immunosuppression

Virus infections usually lead to an immune response ensuring recovery and elimination of the virus. However, a number of infections can be associated with suppression of the immune response. This suppression may either be generalized, that is, against many antigens, or it may only involve antigens specific to the virus in question. There are three major mechanisms involved:

1. Immunosuppression due to infection of immune effector cells,
2. Immunological tolerance induced by clonal deletion of T cells that would respond to the viral antigens,
3. Interference by viral products in various stages of the immune response.

Immunosuppression due to Infection of Effector Cells

Many viruses can replicate productively or abortively in cells of the reticuloendothelial system and it is noteworthy that these cells are often implicated in persistent infections. Lymphocytes and monocytes/macrophages and dendritic cells represent tempting targets for any virus, in that they move readily throughout the body, can seed virus to any organ, and are key players in the immune response.

A dramatic example of destruction of the body's immune system is provided by HIV, which replicates in CD4+ T lymphocytes, dendritic cells, and cells of the monocyte/macrophage lineage. In the latter, cytopathic effects are minimal but the cell functions of phagocytosis, antigen processing/presentation, and cytokine production may be inhibited. DNA copies of the viral genome are integrated into the cellular DNA of CD4+ cells, and viral replication occurs only following activation of these lymphocytes by certain cytokines. CD4+ T cell death can occur by apoptosis, or following fusion with other T cells to form short-lived syncytia, or by lysis by CD8+ T cells. Clinically, immunosuppression is followed in a practical way by monitoring the number of CD4+ T cells in the blood, normally ranging from 500 to 1500/mm^3. Virtual elimination of CD4+ T lymphocytes from the body eventually results in such profound depression of the immune response that the untreated patient dies from intercurrent infections caused by opportunistic pathogens, or from cancer (see Chapter 9: Mechanisms of Viral Oncogenesis and Chapter 23: Retroviruses).

Numerous other viruses temporarily induce generalized immunosuppression by abrogating the function of one or another arm of the immune response. Many viruses are capable of productive or abortive replication in macrophages. In many chronic viral infections virus replicates in reticuloendothelial tissues; this may impair several other key immunological functions, notably phagocytosis, antigen processing, and antigen presentation to T cells.

EB virus induces polyclonal activation of B cells which diverts the immune system to irrelevant activity.

Several viruses have been shown to grow productively or abortively in activated T lymphocytes. For example, measles virus replicates non-productively in T lymphocytes and suppresses Th1 cells (DTH). It has been known since the work of von Pirquet in 1908 that measles infection depresses skin responses to tuberculin and can reactivate latent tuberculosis. Other immunological abnormalities found during the acute phase of measles include spontaneous proliferation of peripheral blood lymphocytes and increased plasma interferon-γ. The depressed delayed-type hypersensitivity response, decreased NK cell activity, increased plasma IgE level, and increased soluble IL-2 receptor can persist for up to four weeks after the onset of the rash. These and other immunological abnormalities probably account for the susceptibility to secondary infections that cause most of the mortality during outbreaks of measles.

Induction of Immunological Tolerance

The probability of an acute infection progressing to chronicity is strongly age-related. In particular, congenital virus transmission to the fetus or neonate, whether transplacental or perinatal, greatly enhances the likelihood of persistent infection—this is the case with hepatitis B virus, rubella virus, cytomegalovirus, parvovirus B19, and others. This non-responsiveness may reflect "split tolerance"—a B cell response may occur, but accompanied by a degree of T cell unresponsiveness. In the well-studied experimental model, lymphocytic choriomeningitis (LCM) virus infection, mice infected *in utero* do not mount a T cell response to the virus and the humoral immune response that is mounted does little to affect the course of the infection. This persistent infection can be reversed by adoptive transfer of sensitized CD8+ T cells, demonstrating their key role in recovery from this infection.

Interference with the Immune Response by Viral Products

Antibody-Induced Removal of Viral Antigens from Plasma Membranes of Infected Cells

Antibodies, being divalent, can bridge viral molecules on cellular membranes, bringing about "capping" followed by endocytosis of the antigen, thereby leading to a diminished display of viral antigens on the cell surface and diminishing their function as immunological targets. This phenomenon is readily demonstrable *in vitro* in "carrier cultures" of cells persistently infected with budding viruses; this phenomenon has been postulated to play a role in SSPE, where antibodies also down-regulate transcription of the measles viral genome.

Interference with MHC Antigen Expression or Peptide Display

Since CD8+ T cells recognize viral peptides bound to MHC class I antigens (and CD4+ T cells recognize them in the context of MHC class II antigens), viral persistence is presumably favored by reduction of peptide display on cell surfaces. The E1A protein of adenoviruses, the Tat protein of HIV-1, and E5 and E7 proteins of bovine papillomavirus all inhibit transcription of different genes coding for MHC components. The E7 protein of human papillomavirus 18 down-regulates the transcription of the TAP1 gene responsible for cytosolic transport of degraded peptides. Other viral proteins interfere with peptide generation (EBV, HHV-8, HIV-1) or transport (HSV, CMV, EBV), or the cell surface expression of MHC (adenoviruses, CMV, HTLV-I).

Down-Regulation of Cell Adhesion Molecules

Burkitt's lymphoma cells carrying the EB virus genome display reduced amounts of the adhesion molecules ICAM-1 and LFA-3, and therefore bind T cells with lower affinity.

Viral Evasion of Host Cytokine Actions

Interferons are induced by infection with viruses and display a wide range of antiviral as well as immunomodulatory activities, as described in Chapter 5: Innate Immunity and Chapter 12: Antiviral Chemotherapy. However, many viruses code for proteins that block or subvert one or more of the sequential steps in the interferon response (see Chapter 5: Innate Immunity). These actions are thought to facilitate persistence, for example, in the case of HCV and HIV-1, but also suppress the immune responses in some transient virus infections (for example SARS, West Nile virus, dengue, Ebola virus). Other viruses have developed the ability to counter other antiviral cytokines (e.g., an adenovirus gene-product partially protects infected cells against the action of tumor necrosis factor).

Blocking of the Immune Response by Non-Neutralizing Antibodies

Very high titers of antibodies are characteristic of many chronic viral infections, so much so that virus–antibody and antigen–antibody complexes accumulate at the basement membranes of renal glomeruli and other sites, causing a variety of immune-complex diseases. Yet the infection may not be eliminated and the complexes can remain infectious. A high proportion of these antibodies may be directed against viral proteins or epitopes that are not relevant to neutralization; by binding to virions these antibodies may block the attachment of neutralizing antibody by steric hindrance.

Production of Excess Viral Antigens

The chronic carrier state in hepatitis B is marked by the production of a huge excess of non-infectious particles of HBsAg thus mopping up neutralizing antibodies, a mechanism thought to overwhelm the body's capacity for antibody production. It is possible some other chronic infections are also sustained by this strategy.

MECHANISMS OF DISEASE PRODUCTION

The continuing survival of a virus depends on an unbroken chain of transmission to, and replication in, new non-immune hosts; whether this produces pathology in the host, and/or clinical signs and symptoms, is often of little relevance to the survival of the virus in nature. Nevertheless the clinical consequence of the interactions between virus and host, namely clinical disease, is of course at the heart of medical virology.

Damage to Tissues and Organs Caused by Virus Replication

Viruses can damage cells in a number of different ways, including by mechanisms ending in necrosis, in apoptosis, or by expression of viral antigens that serve as targets for damaging immune attack (Chapter 5: Innate Immunity). The pathological effects at the level of tissues and organs are considered. The severity of disease in humans is not necessarily correlated with the degree of cytopathology produced by the virus *in vitro*. Many viruses that are cytocidal in cultured cells do not produce clinical disease; for example, some enteroviruses that cause a striking cytopathic effect in cultured human cells cause only inapparent infections in humans. Conversely, some viruses are non-cytocidal *in vivo* and in some instances in cultured cells but cause lethal disease—rabies virus (wild-type "street virus" strains) is a dramatic example. In some instances cell and tissue damage can occur without producing obvious disease; for example, a substantial number of liver cells can be destroyed without causing significant clinical signs. Damage to a proportion of cells in some organs and tissues may be of minor importance, e.g., in striated muscle, connective tissue, and skin, whereas it may be of major consequence if it occurs in key organs such as the heart or brain. Likewise, moderate levels of tissue edema may be unimportant in most sites in the body, but may have serious consequences in the brain if it leads to an increase in intracranial pressure, or in the lung where it may interfere with gaseous exchange, or in the heart where it may disrupt nerve impulse conduction.

Damage to the Epithelium of the Respiratory Tract

Respiratory viruses initially invade and destroy just a few epithelial cells, but initiate a "daisy-chain" process that progressively damages more and more epithelial cells and eliminates the protective layer of mucus. As viral replication progresses, large numbers of progeny virions are budded into the lumen of the airways. Early in infection, the beating of cilia, the primary function of which is to cleanse the respiratory tract of inhaled particles, may actually help to move released progeny virus along the airways, thereby fostering the spread of infection. As secretions become more profuse and viscous, the cilial beating becomes less effective and ceases as epithelial cells are destroyed.

Studies of influenza and parainfluenza virus infections in experimental animals have shown virus spreading from the site of infection via contiguous expansion, not stopping until virtually all columnar epithelial cells along the major airways are infected. The result is a complete denudation of large areas of the epithelial surface (Fig. 7.10) together with the accumulation of transudates and exudates containing inflammatory cells and necrotic epithelial cell debris. A fatal outcome is invariably associated with one or more of three complications: bacterial superinfection (nurtured by the accumulation of fluid and necrotic debris in the airways), infection and destruction of the lung parenchyma and the alveolar epithelium, and/or blockage of airways that are so small in diameter that mucous plugs cannot be bypassed by forced air movements. Blockage of the airways is of most significance in the narrow-bore airways of the newborn. In all complications there is hypoxia and a pathophysiological cascade causing acidosis and an uncontrollable fluid exudation into the airways (acute respiratory distress syndrome—ARDS).

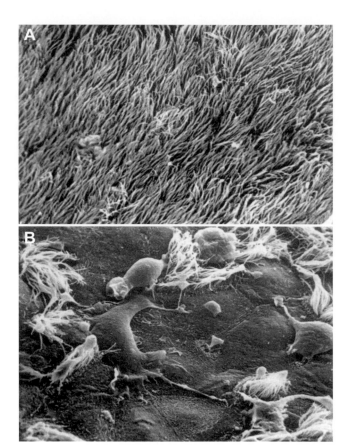

FIGURE 7.10 Organ cultures of tracheal epithelium, viewed from the luminal side by scanning electron microscopy. This has proved an excellent system to study the pathogenesis of respiratory virus infections. (A) Normal control culture; the epithelium appears as a thick carpet of fine projections—the carpet of cilia. (B) Six days after rhinovirus infection only a few tufts of cilia remain. The exposed non-ciliated sub-epithelial surface is quite smooth, consisting of flattened basal cells that will differentiate to reform the normal luminal surface. *From Reed, S.E., Boyde, A., 1972 Jul. Infection and Immunity 6(1):68–76, with permission.*

Degeneration of respiratory tract epithelial surfaces during influenza infection is extremely rapid, but so is regeneration. In studies of influenza in ferrets, for example, it has been shown that the development of a completely new columnar epithelial surface via hyperplasia and maturation of remaining transitional cells may occur within a few days. The transitional epithelium and the newly differentiated columnar epithelium that arises from it are resistant to infection, probably by virtue of interferon production and a lack of virus receptors. The role of other host defenses, including soluble factors such as mannose-binding lectins and lung surfactants, as well as macrophages, NK cells, IgA and IgG antibody, and T cell-mediated immune mechanisms in terminating the infection are discussed in Chapter 5: Innate Immunity and Chapter 6: Adaptive Immune Responses to Infection.

Damage to the Epithelium of the Intestinal Tract

The principal agents causing viral diarrhea are rotaviruses, caliciviruses, astroviruses, and certain adenoviruses (noroviruses). Infection occurs by ingestion, and the incubation period is usually very short.

Rotaviruses infect mature enterocytes at the tips of villi and cause a marked shortening and occasional fusion of adjacent villi. Infection generally begins in the proximal part of the small intestine and spreads progressively to the jejunum and ileum and sometimes to the colon. The extent of such spread depends on the initial virus dose, the virulence of the virus, and the host's immunological response. Diarrhea can occur in the absence of visible tissue damage, or, conversely, histological lesions can be asymptomatic. Factors contributing to fluid accumulation in the lumen of the gut and diarrhea include: (1) the reduced absorptive surface of the intestine as a result of the enterocyte damage, (2) the induction of chloride secretion by the rotavirus toxic protein NSP4, and (3) the impairment of intestinal disaccharidases leading to malabsorption of carbohydrates and osmotic shock. Acidosis develops due to the loss of Na^+, K^+, glucose, and water uptake, and also to the increased microbial activity associated with the fermentation of undigested milk. Acidosis can create a K^+ ion exchange across the plasma membrane of epithelial cells, affecting cellular functions essential for the maintenance of normal K^+ concentration. Hypoglycemia due to decreased intestinal absorption, inhibited glyconeogenesis, and increased glycolysis follow, completing a complex of pathophysiological changes that, if not promptly corrected by restoration of fluids and electrolytes, can result in death.

As infection progresses, the absorptive cells are replaced by immature cuboidal epithelial cells with greatly reduced capacity and enzymatic activity. However, these cells are relatively resistant to viral infection, so that the disease is often self-limiting if dehydration is not so severe as to be fatal. The rate of recovery is rapid, since crypt cells are not damaged and have a mitotic capacity.

Epithelial Damage Predisposing to Secondary Bacterial Infection

As well as having direct adverse effects, viral infections often predispose epithelia to secondary bacterial infections, increasing the susceptibility of the respiratory tract, for example, to bacteria that are normal commensals in the nose and throat (Fig. 7.10). Thus infections with influenza virus may destroy ciliated epithelia and cause exudation, allowing pneumococci and other bacteria to invade the lungs and cause secondary bacterial pneumonia, a not uncommon cause of death in the elderly. Conversely, proteases secreted by bacteria may activate influenza virus infectivity by proteolytic cleavage of its hemagglutinin spike protein. Rhinoviruses and respiratory syncytial virus damage the mucosa of the nasopharynx and sinuses, predisposing the host to bacterial superinfection that commonly leads to purulent rhinitis, pharyngitis, sinusitis, and sometimes otitis media. Similarly, in the intestinal tract, rotavirus infection may lead to an increase in susceptibility to enteropathogenic *Escherichia coli* and other bacterial pathogens; the synergistic effect can lead to more severe diarrhea.

Physiological Changes without Cell Death

In some situations infected cells may show no obvious damage, but, as discussed in Chapter 5: Innate Immunity, specialized cells may function less effectively after infection. For example, lymphocytic choriomeningitis virus infection may appear harmless in mice, but less antibody may be produced by infected than by uninfected B cells. The same virus has no cytopathic effect on cells of the anterior pituitary gland; however the output of growth hormone is reduced and as a result infected mice are runted; likewise, persistent infection of insulin-producing islet cells in the pancreas may result in a lifelong elevation of blood glucose levels. Other viruses may indirectly alter the expression of cell surface MHC molecules, leading to destruction of the infected cells by immunological mechanisms; thus enhanced class II MHC expression after infection of glial cells by mouse hepatitis virus, perhaps due to the production of interferon-γ, may render these cells susceptible to immune cytolysis by cytotoxic T cells.

Immunopathology—Cell Damage Caused by Immunological Processes

The immune response is an essential part of the pathogenesis of most virus diseases. Infiltration of lymphocytes and macrophages, with accompanying release of cytokines and inflammatory mediators, is a regular feature of viral infection that assists in recovery from infection. However these processes can also be a major cause of cell and tissue damage in some infections, and many of the common symptoms of viral disease (fever, erythema, edema, and enlargement of lymph nodes) have an immunological basis. When pathological changes are ameliorated by immunosuppression, it can be assumed that underlying immunopathology is making an important contribution to the disease process.

In particular, for non-cytolytic viruses it is likely that immune mechanisms are the main cause of disease. In most cases these mechanisms involve activated T cells, although there are some examples due to antibody or an exuberant innate immune response.

Immunopathology Resulting from the Activation of Cytotoxic T Lymphocytes

This mechanism involves both lymphocyte and macrophage accumulation at sites of virus replication, together with local secretion of proinflammatory cytokines by these cells. Of course, cell-mediated immune responses are also an important contributor to recovery from viral infections (see Chapter 6: Adaptive Immune Responses to Infection), as becomes evident if they are abrogated by cytotoxic drugs, or are absent as is the case with some immunodeficiency diseases.

The classic model of death due to a cell-mediated immune response is lymphocytic choriomeningitis in adult mice inoculated intracerebrally. The virus replicates harmlessly in the meninges, ependyma, and choroid plexus epithelium for about a week until a CTL-mediated immune response develops, causing severe meningitis, choroiditis and ependymitis, and cerebral edema, convulsions and death. Experiments with knockout mice, and with chemical immunosuppression and adoptive transfer of T cells, have clearly confirmed that CTLs are necessary for the development of disease. Normally CTLs help to control infection, but within the rigid confines of the skull this response is fatal.

A lethal pneumonia develops when some inbred strains of mice are infected with certain strains of influenza virus by the intranasal route. Adoptive transfer of influenza virus-primed CD8+ cytotoxic T cells protect mice against intranasal challenge, but CD4+ Th1 cells actually accelerate their demise. Coxsackie B virus infection of mice is manifested as myocarditis, but in transgenic mice lacking the perforin gene, the disease takes a mild course, indicating that both intrinsic virus-induced damage and cytotoxic lymphocytes contribute to the pathology.

Hepatitis B is a clear example of liver damage as a result of specific immune responses to viral proteins. Transgenic mice expressing hepatitis B envelope proteins show little liver damage unless injected with hepatitis B-specific CTLs; this causes hepatocyte damage and inflammatory cell infiltration, characteristic of acute viral hepatitis. In humans, the degree of liver damage is governed by the strength of the immune response. Patients with marked immunosuppression can express high levels of virus in the absence of significant liver damage: patients with a strong immune response clear the virus, but those with an intermediate response show continuing liver damage and viral persistence. Any restoration of immune competence can induce pathology; for example, in HBV/HIV co-infected patients there is partial recovery of immune competence with a risk of "flares" of liver damage, after patients are treated with anti-retroviral drugs.

The maculopapular rash of measles involves virus infection of endothelial cells in the superficial layers of the dermis, spread of infection to overlying epithelial cells, vascular dilatation, and infiltration of CD4+ and CD8+ T cells and macrophages into sites of replication. Individuals with deficiencies in cellular immunity may develop measles without a rash, but the disease in such patients is especially severe.

Antibody-Dependent Cell-Mediated Cytotoxicity (ADCC)

In experimental cell culture systems, when antibody combined with viral antigen is present on the surface of infected cells, there can be a sensitization in which Fc-receptor-carrying K (killer) cells, together with polymorphonuclear leukocytes and/or macrophages, cause damage. Alternatively, the binding of antibodies can activate the complement system, leading to cell lysis. While it has been clearly demonstrated that virus-infected cells *in vitro* and in some experimental animal models are readily lysed by ADCC, its importance in human viral diseases remains unclear.

Immune Complex-Mediated Disease

Antigen–antibody reactions cause inflammation and cell damage by a variety of mechanisms. If the reaction occurs in extravascular spaces the result is edema, inflammation, and infiltration of polymorphonuclear leukocytes, which may later be replaced by mononuclear cells. This is a common cause of mild inflammatory reactions. Such immune complex-mediated reactions constitute the classical Arthus response and are of major importance, especially in persistent viral infections. If these occur in the blood, the result is circulating immune complexes, which are found in most viral infections. The fate of the immune complexes depends on the ratio of antibody to antigen. When there is a large excess of antibodies, each antigen molecule is covered with antibody and removed by macrophages bearing receptors for the Fc component of the antibody molecules. Alternatively, if the amount of antigen and antibody is about equal, lattice structures develop into large aggregates that are removed rapidly by the reticuloendothelial system. However, in antigen excess, as occurs in some persistent infections, viral antigens and virions are continuously released into the blood but the antibody response is weak and the antibodies are of low avidity or are non-neutralizing. Complexes may continue to be deposited in small blood vessels and kidney glomeruli over periods of weeks, months, or even years, leading to an impairment of glomerular filtration and eventually to chronic glomerulonephritis.

A classic example is lymphocytic choriomeningitis infection of mice infected *in utero* or as neonates. Viral antigens are present in the blood, and small amounts of non-neutralizing antibodies are formed, giving rise to immune complexes which are progressively deposited on renal glomerular membranes; the end result may be glomerulonephritis, uremia, and death. Circulating immune complexes may also be deposited in the walls of the small blood vessels in skin, joints, and the choroid plexus, where attracting macrophages and activate complement. Prodromal rashes, which are commonly seen in exanthematous diseases, are probably caused in this way. A more severe manifestation of the deposition of antigen–antibody–complement complexes in capillaries is erythema nodosum (tender red nodules in the skin); when small arteries are involved, as is occasionally seen in patients with hepatitis B, the result is periarteritis nodosa.

In addition to these local effects, the mobilization of soluble mediators induced by antigen–antibody complexes may generate late systemic reactions, such as fever, malaise, anorexia, and lassitude common to most viral infections. Fever is attributed to interleukin-1 and tumor necrosis factor produced by macrophages, and possibly to interferons.

Rarely, systemic immune complex reactions may activate the enzymes of the coagulation cascade, leading to histamine release and an increase in vascular permeability. Fibrin may be deposited in the kidneys, lungs, adrenal glands, and pituitary gland, causing multiple thromboses with infarcts and scattered hemorrhages—a condition known as disseminated intravascular coagulopathy (DIC).

Systemic Inflammatory Response Syndrome ("Cytokine Storm")

The immune response to an infection is normally modulated to a level proportionate to the extent of the infection, and is down-regulated once the infection has resolved. However, in some infections there is a dramatic and large-scale release of inflammatory cytokines accompanied by stress mediators that leads to overwhelming, even lethal, damage to the host. Also known as a "cytokine storm," this is particularly seen with newly emerging or zoonotic infections, for example, SARS, Ebola hemorrhagic fever, hantavirus pulmonary syndrome, avian H5N1 influenza. It possibly was one key to the exceptional virulence of the 1918 influenza virus strain, the strain that caused the great global pandemic. The pathology associated with a "cytokine storm" includes microvascular damage, pulmonary edema, capillary leakage, hypotension, and organ failure.

Autoimmunity

Antibodies reactive against the host's own proteins (i.e., autoantibodies) can be detected frequently in viral infections, although usually only transiently and in low titer. In one study, 4% of a large panel of monoclonal antibodies raised against several viruses were found to react with normal tissues. For example, a monoclonal antibody directed against the neutralizing domain of Coxsackievirus B4 also reacted against heart muscle; this virus is known to target muscle, including cardiac muscle, and to cause myocarditis.

One likely explanation for this widespread cross-reactivity is that viral proteins share identical or near-identical sequences of 6 to 10 amino acids with cellular proteins far more frequently than would be predicted by chance. For instance, there is partial homology of amino acid sequences (*molecular mimicry*) between myelin basic protein (MBP) and several viral proteins. Alternatively, viral infections may lead to exposure to the immune system of cellular antigens normally sequestered or in some other way hidden. It has also been proposed that some T cells may carry dual T cell receptors, and as a consequence an autoreactive T cell normally depleted in the thymus may be activated should the host be exposed to a viral antigen also recognized by a second T cell receptor on the same cell.

Such mechanisms may be involved in the neurological disorders associated with animal lentivirus infections, visna, and caprine arthritis-encephalitis, and in the rare occurrence of post-vaccinial encephalitis in humans. Inoculation of a neuritogenic protein of peripheral nerve myelin can cause the experimental equivalent of Guillain-Barré syndrome. There is mimicry between the epitope involved (P2) and a sequence in the influenza virus NS2 protein. Ordinarily, this non-structural protein is removed from influenza virus vaccine during purification; failure to do this in some batches of swine influenza vaccine used in the crisis program mounted in the United States in 1976 may have accounted for the apparent increase in the incidence of Guillain-Barré syndrome at that time.

Post-infectious encephalomyelitis is a rare condition (about one in every thousand cases) that follows a few weeks after acute measles, and even more rarely following varicella, mumps or rubella infection or vaccinia vaccination. The pathology is predominantly demyelination without neuronal degeneration—unlike lesions produced by the direct action of viruses on the central nervous system. Allied with the failure to recover virus from the brain of fatal cases, this has led to the view that post-infectious encephalitis is probably an autoimmune disease. Myelin basic protein (MBP) can be found in the cerebrospinal fluid, and, at least in the case of measles, anti-MBP antibodies and T cells can be detected in the blood. Theoretically, autoimmune damage can continue long after the virus that triggered the response has been cleared from the body. Once the cross-reactive viral amino acid sequence has induced a humoral and/or cellular immune response that brings about damage to normal tissue, proteins that are normally sequestered may be exposed to the immune system. If the cellular protein is capable of immune recognition by the host, there may be a chain reaction of progressive damage. Implicit in this hypothesis is the notion of molecular mimicry being responsible for autoimmune

TABLE 7.8 Potential Mechanisms of Induction of Autoimmune Disease by Viruses

Molecular mimicry: viral protein elicits humoral and/or cellular immune response that cross-reacts with identical or similar epitope fortuitously present on a cellular protein

Activation of bystander autoreactive cells. Ongoing local inflammation may result in liberation of self-antigens that are not normally exposed to the immune system: these then become targets for bystander autoreactive T cells. This may also lead to "epitope spreading" to involve wider areas. Since such T cells are thought to have low affinity for self-antigen, viral infection itself may also contribute an adjuvant effect

Immortalization of polyclonal autoreactive effector cells or antigen-presenting cells, by a virus, e.g., EB virus

T cells expressing dual T cell receptors; an autoreactive T cell that would normally be thymically depleted might be activated if it contacts a viral antigen recognized by a second TCR that it is carrying

Induction of MHC antigens by cytokines: virus induces production of interferon γ and tumor necrosis factor, which induce expression of MHC class II protein on brain cells that do not usually express it, for example, glial cells, enabling them to present antigens such as myelin to T cells

Viral destruction or down-regulation of T cells that normally suppress immune response to self-proteins

disease while normal immunological tolerance is somehow deregulated.

Humans suffer from numerous autoimmune diseases, ranging from multiple sclerosis to rheumatoid arthritis. Most are chronic and difficult to treat, many are common, and most are of unknown etiology. Viruses are major suspects as triggers for these diseases, but definitive proof has yet to be produced, and this problem continues to be an important area for research. Current theories for which there is some evidence are summarized in Table 7.8.

VIRUSES AND IMMUNOSUPPRESSION

Immunosuppression Caused by Viral Infection

Suppression of one or more components of the immune system is seen during the course of many viral infections. In acute infections this effect is almost always transient, but it may or may not play a part in the pathogenesis of the infection in question. In a number of persistent infections, virus-induced immunosuppression may be important in the establishment and maintenance of virus persistence. This is discussed in a previous section.

Viral Infections in Immunocompromised Patients

Impairment of the immune system can exacerbate disease or predispose to superinfection with other infectious agents. An immunocompromised state may be due to genetic defects, such as in congenital agammaglobulinemia. Most commonly, however, immune dysfunction is due to some other disease or circumstance, especially lymphomas, leukemias, infection with HIV, chemotherapy, or radiotherapy employed to treat tumors or to prevent organ transplant rejection.

Patterns of viral infections in persons with immune dysfunction depend on the particular virus and which arm of the immune system is defective. Many viral infections in immunocompromised subjects follow the disease pattern seen in normal persons, differing only in severity. Occasionally, if the disease is largely immunopathological, the disease may actually be milder—for example, in immunosuppressed renal transplant patients infected with hepatitis B there may be less liver damage. Usually, however, the immunocompromised patient suffers more severe and/or prolonged disease, and sometimes relatively innocuous viruses can prove lethal. For example, measles in patients with impaired cell-mediated immunity may cause giant cell pneumonia, sometimes several months after the acute infection, and often with fatal consequences. Immunosuppressed patients are also more likely to suffer generalized infections with viruses that usually cause localized disease; for example, herpes simplex and varicella-zoster. Immunocompromised individuals are not only subject to exogenous infections but suffer increased reactivation of latent viruses, especially herpesviruses, but also adenoviruses and polyomaviruses. These situations are described in more detail in the chapters dealing with specific viral diseases in Part II of this book.

Finally, immunosuppression commonly leads to reactivation of persistent infections. We should distinguish the reactivation of virus replication in a latently infected cell, at the cellular level, and the recrudescence of clinical disease. Examples of clinical reactivation are listed in Table 7.9.

VIRAL VIRULENCE AND HOST RESISTANCE TO INFECTION

It is a common observation that different individuals in a family or community may show different levels of severity of disease when infected at the same time from a common source. Similarly, different strains of the same common virus may cause disease of different levels of severity (e.g., the H1N1 pandemic influenza strain that swept the world in 2009 was highly transmissible but of low virulence, while the H5N1 avian virus seen mostly in Asia in recent years is very poorly transmissible between humans but causes significant mortality). Influenza virus studies clearly distinguish the viral quality of *transmissibility* from that

TABLE 7.9 Examples of Clinical Reactivation of Persistent Viral Infections in Humans

Circumstances	Virus	Features
Old age	Varicella virus	Rash of shingles
Pregnancy	Polyomavirus JC, BK	Viruria
	Cytomegalovirus	Replication in cervix
	Herpes simplex virus 2	Replication in cervix
Immunosuppression in organ transplantation, autoimmune disease, malignancy, HIV	Herpes simplex virus 1, 2	Vesicular rash, potential for systemic spread
	Varicella virus	Vesicular rash (chickenpox or shingles), potential for systemic spread
	Cytomegalovirus	Fever, hepatitis, pneumonitis, GI tract disease
	Epstein-Barr virus	Increased shedding from throat, PTLD[a]
	JC polyomavirus	Viruria very common, PML[b]
	BK polyomavirus	Interstitial nephritis, hemorrhagic cystitis
	Hepatitis B virus	Viremia, hepatitis flare
	Adenoviruses	Increased shedding/disease in different organs
HIV/AIDS	Human papillomaviruses	More extensive warts (skin/anogenital)
		HPV-related cancers
	Cytomegalovirus	Retinitis, colitis, pneumonitis
	JC polyomavirus	PML[b]
	Epstein-Barr virus	Increased shedding from throat, non-Hodgkin's lymphoma
	HHV-8	Kaposi's sarcoma
	Herpes simplex virus 1, 2	Spreading vesicular rash, may be chronic
	Varicella-zoster virus	Shingles

[a]PTLD, post-transplantation lymphoproliferative disease.
[b]PML, progressive multifocal leukoencephalopathy.

of *virulence*—with other viruses this distinction is usually harder to define. Such studies also highlight the fact that the severity of an infection is influenced by both virus- and host-determined factors.

Genetic Factors Affecting Viral Virulence

Viral virulence is influenced by viral genes in four categories: (1) those that affect the ability of the virus to replicate, (2) those that affect host defense mechanisms, (3) those that affect tropism, spread throughout the body and transmissibility, and (4) those that encode or produce products that are directly toxic to the host.

It has been a longstanding goal of virologists to understand and to predict the mechanisms underpinning the virulence of pathogenic viruses. In addition to the clinical aim of a better understanding of viral pathogenesis, such knowledge allows us a rational basis for designing avirulent virus strains as attenuated live-virus vaccines. However, reductionist attempts to identify a "virulence gene" (or genes) have usually been inconclusive, as virulence has usually turned out to be based on a co-operative effect between a number of different genes (often called "*the constellation of genes*" encoding virulence and transmissibility factors). For example, after the highly virulent 1918 pandemic influenza virus (an H1N1 virus) was reassembled from permafrost-frozen cadavers and from formalin-fixed tissue samples taken during the epidemic, genetic reassortants were constructed containing genes from the epidemic virus and an avirulent H1N1 strain (TX/91); the virulence of TX/91 increased as more of its genes were replaced by genes from the pandemic strain. No single gene was found to be wholly responsible for the virulence of the epidemic virus, but the hemagglutinin gene and genes of the polymerase complex contributed most to the virulence phenotype.

Genes affecting the ability of the virus to replicate. One well-studied example comes from the comparison of wild-type and vaccine strains of poliovirus. It was clearly shown that virulent and attenuated viruses differed in only a small number of nucleotides. Infectious genomic cDNA constructs containing different combinations of wild-type and vaccine strain sequences were tested for neurovirulence in monkeys and in transgenic mice expressing the poliovirus CD155 receptor. Attenuation was shown to be determined primarily by mutations within the internal ribosome entry site (IRES) within the 5′-UTR (see Chapter 32: Picornaviruses), but there was also a contribution from mutations within the capsid region of the viral genome. The mutations in the IRES altered stem-loop structures and reduced the efficiency of translation of poliovirus RNA, while the capsid mutations may have led to impaired binding to the cellular receptor and reduced stability of the capsid.

Genes affecting virus spread or tropism. Influenza virus replication requires cleavage of the hemagglutinin precursor protein HA0 into its two subunits HA1 and HA2. Studies of avian influenza strains have shown that avirulence is associated with the presence of a monobasic cleavage site that is susceptible to cleavage by trypsin-like proteases present only in the respiratory tract (mammals) and gastrointestinal tract (birds). In contrast, highly pathogenic avian influenza virus strains contain a polybasic cleavage site that is susceptible to cleavage by furin, an enzyme widespread in the body, thereby promoting systemic infection.

Influenza virus tropism is also dependent on the distribution of appropriate sialic acid virus receptors; human viruses preferentially bind to α2,6-linked sialic acids, while avian viruses preferentially bind to α2,3-linked sialic acids. In humans, α2,6-linked sialic acid receptors are predominant on ciliated epithelial cells and goblet cells of the upper respiratory tract, whereas α2,3-linked sialic acid receptors are predominant on non-ciliated bronchiolar cells and alveolar type II cells in the lower respiratory tract. Clinical observations indicate that human seasonal influenza viruses predominantly infect the upper respiratory airways and are readily transmissible, while avian viruses acquired directly from birds (e.g., H5N1) often progress to severe lower respiratory tract disease in humans, but are not as transmissible between humans; these observations are consistent with the above patterns of receptor usage operating as virulence mechanisms.

Genes that affect host defense mechanisms. A number of viruses, particularly the large DNA viruses, encode proteins that interact with host defense systems. These include proteins that mimic cytokines or growth factors, and proteins that mimic virus receptors. Separately, several viruses down-regulate the expression of MHC molecules at the surface of infected cells, thereby interfering with the elimination of infected cells by CD8+ T cells. Hepatitis C virus is a poor inducer of interferon α, and work with the recently developed cell culture system for HCV has shown two mechanisms for this—(1) cleavage by the HCV NS3/4A protease of the cell mitochondria adapter MAVS involved in interferon induction, and (2) HCV-mediated suppression of protein translation by induction of phosphorylation of the eIF2α translation initiation factor.

Genes coding for proteins that are directly toxic. In contrast to bacterial infections, examples of this mechanism have rarely been found in viral infections, perhaps reflecting the fact that a successful virus needs to maintain the viability of its host cell long enough to allow its replication. The best documented example is the non-structural glycoprotein NSP4 of rotaviruses. This protein inhibits a Na$^+$-glucose lumenal co-transporter which is required for water resorption in intestinal cells, and also increases intracellular calcium levels by inducing a phospholipase C-dependent calcium signaling pathway. Both effects enhance fluid loss.

Host Genetic Factors Affecting Virulence

It has long been known that different animal species show different degrees of susceptibility to the same virus. The benign course of B virus (Cercopithecine herpesvirus 1, formerly called herpesvirus simiae) infection in macaques compared with the virulence of the same virus for humans is a clear example. In general, there is a tendency for low virulence to be seen with a virus that has been long established in a particular host, compared with high virulence when the same virus is introduced into a new host species; this is a frequent observation in cross-species transfer of viruses and with emerging zoonotic infections of humans.

In some examples using inbred strains of susceptible and resistant mice, susceptibility genes have been mapped and functions identified. Not surprisingly, associated susceptibility has been shown as linked to genes for MHC haplotypes or components of the interferon system, reflecting the critical role played by the *balance* between virus invasion and host defense in determining the severity of an infection. Mouse strains that are resistant to flavivirus infection were found to have a mutation in the *flv* gene, which was subsequently shown to encode 2′-5′-oligo(A) synthetase, an interferon-induced enzyme that activates RNAse L resulting in the breakdown of viral and host mRNA. One well-studied human example involves the chemokine receptor CCR5 which acts as a co-receptor for infecting strains of HIV; individuals carrying a mutation in this gene show greatly increased (but not complete) resistance to HIV infection.

In another example, humans who are predisposed to herpes encephalitis have been found to have mutations in either of the genes *TLR3* or *UNC-93B*, both of which code for proteins that affect interferon production. *TLR3* appeared not to be involved in resistance to other infections but is vital for natural immunity to herpes simplex virus infection of the CNS, implying that neurotropic viruses may contribute to maintaining TLR3 through evolutionary time.

However, in most human viral infections the reasons why different individuals respond differently to the same virus remains a mystery, very likely grounded in multifactorial subtle variations in innate and adaptive immune responsiveness.

Physiological and Other Host Factors Affecting Virulence

Host factors such as age, nutritional state, pregnancy, immunity from prior exposure, and co-infections all contribute to the outcome of virus infections. In natural infections of human populations it may be difficult to tease out the relative contribution of co-existing effects, but animal models have allowed the study of individual factors.

Age. For many years virologists used inoculation of suckling mice as a sensitive way to search for new viruses, particularly arboviruses and enteroviruses, many of which do not infect adult mice. In humans, more severe outcomes are often seen in the very young and the very old, and different outcomes in patients of different ages are well described. For example, hepatitis A infection is more likely to lead to the development of fulminant hepatitis in the elderly (Table 7.10).

An infection in early infancy with many viruses leads to more severe, sometimes devastating, disease because the maturing immune system is not yet able to respond as well as in older patients. Transplacentally acquired maternal antibody can provide a short-term protective umbrella, so such examples tend to be more severe if the mother is undergoing a primary infection and the infant is born without maternal antibodies. The tendency for very young infants with respiratory syncytial virus infection to suffer from severe bronchiolitis has also been ascribed to the narrower airways in the very young. Other viruses cause more severe infections or more complications in adults, for reasons that are not well understood. In these examples, subclinical infection acquired inadvertently in childhood can in fact give immunological protection against later adult disease.

Hepatitis B virus provides a well-studied example. Persistent infection almost invariably follows infection in infancy, but in only 5 to 10% of infections acquired by adults; intermediate rates of persistence are seen at intermediate ages. This effect again is thought to be due to immaturity of the immune system. In the duck hepatitis B model, the transition from virus persistence to virus clearance occurs in young ducklings at quite a sharply demarcated age— by increasing the dose of virus inoculated, the age point favoring persistence versus clearance is shifted to older and older ducks. This highlights the principle that the outcome of virus exposure reflects a balance between the rates of virus replication and the immune response.

Nutritional state. Malnutrition can interfere with most or all of the mechanisms that serve as barriers to the progress of virus disease. It has been repeatedly demonstrated that severe nutritional deficiencies interfere with the generation of antibody and cell-mediated immune responses, with the activity of phagocytes, and with the integrity of skin and mucous membranes. However, often it is impossible to disentangle adverse nutritional effects from other adverse factors found in developing communities. Moreover, just as malnutrition can exacerbate viral infections, so viral infections can exacerbate malnutrition, especially if repeated severe diarrhea is a feature, and especially in the first year of life.

Children with protein deficiency of the kind found in many parts of Africa are highly susceptible to measles. All the epithelial manifestations of the disease are more severe, and secondary bacterial infections cause life-threatening disease of the lower respiratory tract as well as otitis media, conjunctivitis, and sinusitis. The skin rash may be associated with numerous hemorrhages, and there may be extensive intestinal involvement with severe diarrhea, which exacerbates the nutritional deficiency. The case–fatality rate is commonly 10% and may approach 50% during severe famines, and a debilitating chronic malnutrition/malabsorbtion syndrome can occur among survivors.

Other factors. In experiments with myxoma virus in rabbits fever has been shown to increase protection against disease; blocking the development of fever with salicylates was shown to increase mortality. Pregnancy can trigger the reactivation of some persistent viruses, particularly herpesviruses and polyomaviruses, thereby adding to the risk of the neonate acquiring herpes simplex infection during birth. Certain infections are markedly more severe in pregnancy (e.g., hepatitis E, smallpox) (Box 7.1).

TABLE 7.10 Effect of Age on Severity of Infection

- Many infections are more severe in infancy than in older children: rotavirus, respiratory syncytial virus, herpesviruses, Coxsackieviruses (e.g., due to less mature immune system, no prior immunity, anatomical factors in the respiratory tract)
- Some important infections are *more* severe in older children or adults: polio, hepatitis A, measles, Epstein-Barr virus, varicella, mumps (orchitis)
- Hepadnaviruses: > 90% of infections in infancy lead to persistent infection, only 5% to 10% of adult infections become persistent
- Some chronic infections progress more rapidly in people over the age of 30 to 40: HIV, hepatitis C
- Many acute virus infections cause greater morbidity and mortality in elderly persons, in part due to declining immunocompetence and co-morbidities (influenza)
- In some infections with a very late outcome, the patient may die first from other causes (hepatitis C, human T cell leukemia virus)

BOX 7.1 Important definitions

Pathogenicity is defined as the absolute ability of an infectious agent to cause disease/damage in a host—an infectious agent is either pathogenic or not.

Virulence is defined as the relative capacity of an infectious agent to cause disease/damage in a specific host. *Virulence* must be measured relatively, comparing one infectious agent to another, or comparing our historical experience with one disease or infectious agent to another (i.e., "virus A is more *virulent* than virus B in a specified host"). The terms *pathogenicity* and *virulence* do not refer to the properties of infectiousness, infective titer, or transmissibility, only to the capacity to cause disease or damage in a host.

Virulence is dependent upon many factors including the age, sex, species, and physical condition of the host involved, the route of infection, the strain of the virus of interest, etc. For example, many viruses are relatively avirulent in a host species in which they have been endemic for a long time, but cause severe disease when introduced into a new host species (host range extension—"host species jumping"). This is an important principle in understanding the emergence of zoonotic infections in humans (Chapter 15: Emerging Virus Diseases). Furthermore, there are well-studied examples where a virus may be virulent when introduced by one route, for example, intracerebral inoculation, but harmless when given by another route, for example, intraperitoneally.

The *virulence* of a virus can be quantitated, at least in experimental terms, by titrating serial dilutions in groups of (ideally inbred) animals, and recording the virus dilution that causes an endpoint of death or disease in 50% of the animals (the lethal dose for 50% of subjects, the LD_{50}; or the infectious dose for 50% of subjects, the ID_{50}). Other endpoints that can be used are the time to death or the time to the appearance of clinical signs, weight loss, or fever (to accord with modern standards for humane care of experimental animals). Assessment of the virulence of a virus in humans cannot be made in such a controlled experimental way, but often becomes evident when observing outbreaks of acute infection. Interpretation is more complicated in chronic infections—for example, the appearance of a new strain, or a higher viral load, of HIV in a patient whose clinical condition is worsening, may indicate that a new strain or higher virus replication is *causative* of the increasing disease, or it may be a *result* of increasing immunosuppression.

CO-INFECTIONS

Co-infections with more than one virus are increasingly recognized, and are of increasing practical concern in an era of greatly improved options for antiviral drug use. At one end of the spectrum, use of better diagnostics, particularly rapid PCR-based tests for respiratory and gastrointestinal viruses, has revealed that multiple infections are not at all unusual, nor unfortunately is the nosocomial acquisition within hospital of further virus infections by a child after being admitted to hospital for one initial infection. This can create ambiguity in ascribing a patient's clinical disease to any one particular virus detected by the diagnostic laboratory, and necessitates careful case-control studies of populations at risk with and without disease before a newly discovered virus can be definitively implicated as a new pathogen.

At the other end of the spectrum, many patients have been exposed to HIV, HCV, and/or HBV, and dual or even triple infections are increasingly reported. Management of these patients poses additional challenges. In patients with HBV or HCV infection, the additional presence of HIV infection produces clinical outcomes that might be expected from a diminished immune response: with HBV, higher levels of virus replication and milder inflammatory liver disease, and with HCV a lower rate of clearance of the HCV infection. However, in both situations liver disease frequently progresses more rapidly through fibrosis to cirrhosis and brings a worsening prognosis. Conversely, the additional presence of HBV or HCV does not appear to significantly influence the course of HIV infection. Selection of optimal antiviral drug regimens requires consideration of the patient's history of prior treatment, resistance patterns of the patient's virus, and the multiple effects of some drugs against more than one virus. Successful treatment of HIV infection with anti-retroviral drugs carries the risk of "immune restoration disease," which may involve recruitment of both antigen-specific and non-antigen-specific mononuclear cells to the liver, possibly mediated by interferon-γ. With both HBV and HCV, this can lead to flares of inflammatory liver disease, which can progress to hepatic decompensation in patients with cirrhosis.

FURTHER READING

Fenner, F., 1949. Mouse-pox; infectious ectromelia of mice; a review. J. Immunol. 63, 341–373.

Griffin, D.E., 1995. Immune responses during measles virus infection. Curr. Top. Microbiol. Immunol. 191, 117–134.

Jackson, A.C., Zhen, F.U., 2013. Pathogenesis. In: Jackson, A.C. (Ed.), Rabies, Scientific Basis of the Disease and Its Management (third ed), Elsevier/Academic Press, San Diego, pp. 299–349.

Kaslow, R.A., 2014. Epidemiology and control: Principles, practice and programs. In: Kaslow, R.A., Stanberry, L.R., LeDuc, J.W. (Eds.) Viral Infections of Humans (fifth ed), Springer, Chapter 1.

Nathanson, N., Ahmed, R., Gonzalez-Scarano, F., Griffin, D.E., Holmes, K.V., Murphy, F.A. (Eds.), 1997. Viral Pathogenesis, Lippincott-Raven, Philadelphia.

Nathanson, N., Ahmed, R., Biron, C.A., Brinton, M.A., Gonzales-Scarano, F., Griffin, D.E., et al., 2007. Viral Pathogenesis and Immunity, second ed. Academic Press, London.

Racaniello, V.R., 2006. One hundred years of poliovirus pathogenesis. Virology 344, 9–16.

Taubenberger, J.K., Baltimore, D., Doherty, P.C., et al., 2012. Reconstruction of the 1918 influenza virus: Unexpected rewards from the past. mBio 3 (5). Available from: http://dx.doi.org/10.1128/mBio.00201-12.

Tisoncik, J.R., Korth, M.J., Simmons, C.P., Jeremy Farrar, J., Martin, T.R., Katze, M.G., 2012. Into the eye of the cytokine storm. Microbiol. Mol. Biol. Rev. 76, 16–32.

Virgin, H.W., 2007. *In vivo* veritas: Pathogenesis of infection as it actually happens. Nat. Immunol. 8, 1143–1147.

Chapter 8

Patterns of Infection

Viral infections can be classified as either *transient* or *persistent*, and within each of these categories an infection may be either *localized* (that is, confined to the epithelial surface where the virus initially enters), or *systemic* (that is, spreading more generally throughout body compartments, usually via the bloodstream). Under normal circumstances most common human infections typically lead to one of these four possible scenarios, transient local, systemic persistent infection, etc., depending on the virus.

Next, for each of these above four categories, the clinical impact may be *asymptomatic (subclinical)*, or it may be *symptomatic (clinical)* of varying severity ranging from mild through to severe or fatal. Once again, different viruses typically cause specific patterns of disease and severity. For example, around 80 to 90% of measles and smallpox infections are symptomatic, whereas less than 5% of cytomegalovirus (CMV) infections are symptomatic except in the very young or the immunosuppressed. In the cases of HIV and hepatitis C, a minority of individuals have symptomatic disease at the time of initial infection; subsequently, nearly all HIV-infected individuals without treatment eventually develop disease and die, while 10 to 20% of HCV-infected individuals progress to cirrhosis or liver cancer over 20 to 40 years. For polioviruses, less than 1% of childhood infections cause paralysis, although this figure is higher among older age groups. In contrast, vaccine strains of poliovirus with around 10 base changes in the genome compared to wild-type virus rarely cause paralysis at any age.

The four different infection models will now be considered, namely transient localized, transient systemic, persistent localized, and persistent systemic infections.

TRANSIENT LOCALIZED INFECTIONS

This category includes many common respiratory and gastrointestinal infections. After an initial round of replication in epithelial cells, virus infection spreads to adjacent cells and may also be carried by mucous, ciliary, and peristaltic action, coughing and sneezing, infecting more distant susceptible cells in the same anatomical compartment. The incubation period is typically 2 to 5 days; the peak of virus replication is generally just prior to, and just after, the onset of symptoms and usually declines steadily from there on. The timing of virus excretion in relation to symptoms is of practical importance for two reasons. First, it determines the period in which the patient is most likely to transmit to others and second, it indicates the stage of the infection when the detection of virus is most likely to occur. Shedding of virus into the environment occurs from the site of replication, aided by processes such as vomiting, diarrhea, sneezing, and coughing (Fig. 8.1, Table 8.1).

Influenza virus infection provides an example of this pattern (see also Chapter 7: Pathogenesis of Virus Infections). Inhaled aerosolized droplets containing virus particles become trapped on the film of mucus that covers the epithelium of the upper respiratory tract. Alternatively, virions contained on fomites may gain entry into the nares, the conjunctiva, or the oropharyngeal cavity and from there move to the respiratory tract. Immediately upon alighting, the virus is met by host defense mechanisms—if there has been previous infection or if the subject has been vaccinated with the same or a very similar strain of virus, it may be neutralized by antibody (mainly IgA) present in the mucus. Mucus also contains glycoproteins similar to the receptor molecules on respiratory epithelial cells, which may combine with virions and prevent attachment to epithelial cells. Alternatively, the viral neuraminidase may destroy enough of this host glycoprotein to allow virions to attach to and infect epithelial cells.

Very small virus-containing aerosolized droplets carried deeper into the airways face another physiological barrier, namely the cleansing action of beating cilia. Inhaled particles are normally carried in the flow of mucus generated by synchronized cilial beating to the pharynx where they are swallowed. However, initial invasion and destruction of just a few epithelial cells by influenza viruses can initiate lesions which progressively damage the mucus layer, thereby laying bare an increasing number of epithelial cells. Viral replication progresses, and large numbers of progeny virions bud into the lumen of the airways. Early in infection, cilial beating may even help to move released progeny virus along the airways, thereby spreading the infection.

The spread of the infection via contiguous expansion from initial foci may continue until virtually every columnar

Fenner and White's Medical Virology. DOI: http://dx.doi.org/10.1016/B978-0-12-375156-0.00008-4

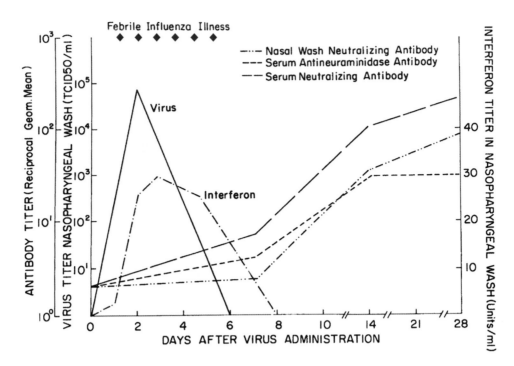

FIGURE 8.1 The time-course of influenza virus replication and the human host response. Six seronegative adult volunteers received $10^{4.0}$ TCID$_{50}$ of wild-type A/Bethesda/1015/68 virus (a Hong Kong/68-like H3N2 virus) intranasally on day 0. The time-course of the onset of clinical symptoms, IFN response, and serum and nasal wash antibody responses are shown. Virus replication peaked about 48 hours after instillation and declined slowly thereafter, coinciding with the interferon response but ahead of the appearance of antibody. There was little shedding after days 6 to 8. Peak virus titers in symptomatic adult volunteers ranged from $10^{3.0}$ to $10^{7.0}$ TCID$_{50}$/ml of nasopharyngeal wash. As in most clinical experiments, there was a positive correlation between the amount of virus shed and the severity of the clinical illness. Individuals who shed less than $10^{3.0}$ TCID$_{50}$/ml were either asymptomatic or had only minor upper respiratory tract symptoms. Even after infectious virus could no longer be recovered, viral antigen was detected for several days in cells and secretions of infected individuals. Viral antigen was isolated in conjunctival cells and secretions. In children, in naturally acquired infections, virus can be found for up to 13 days after the onset of symptoms. The higher titers and more prolonged shedding in children contribute to the important role of this population in the spread of influenza. *Reproduced from Knipe, D.M., Howley, P. (Eds.), 2001. Fields Virology, fourth ed. Lippincott Williams & Wilkins, with permission.*

TABLE 8.1 Correlation Between the Stages in the Pathogenesis of a Localized Respiratory Virus Infection, and the Evolution of Symptoms

Day Following Transmission	Symptoms	Pathology
1–3	Fever, chills, headache, watery sneezing	Peak virus replication, epithelial cell damage exposing basement membrane
4–7	Thicker yellow mucus, cough, hoarseness, blocked nose, sputum, extension to sinuses, middle ear, etc.	Mucus exudate, inflammatory cell infiltrate, dead cells, debris, commensal bacteria
>7	Clearance of mucus and sputum, improvement in blockage	Removal of mucus, dead cells, debris. Regeneration of epithelium

epithelial cell at the particular airway level is infected. Concurrently, the virus induces cytokine production; these cytokines are the cause of many of the clinical signs, and also play a central role in the induction of the immune response and in the exacerbation of the inflammatory response.

Respiratory distress follows, which is made worse by exertion. Complications that may lead to a fatal outcome include: (1) secondary bacterial infection, nurtured by the accumulation of fluid and necrotic debris in the airways, (2) infection and destruction of the lung parenchyma and alveolar epithelium (interstitial pneumonia), and/or (3) blockage of airways that are so small in diameter that mucous plugs cannot be dislodged by forced air movements. Blockage of the airways is of most significance in the very young wherein the airway diameters are smallest.

Significant spread of virus beyond the respiratory epithelium does not occur—although virus may rarely be detected in the circulation, virus is not found in the cerebrospinal fluid in cases of influenza encephalopathy, an unusual complication thought to have an auto-immune basis.

Immunity, whether evoked by natural infection or vaccination, is not long lived, and reinfection is common because of both ongoing antigenic drift of circulating viruses and because of waning immunity.

TRANSIENT SYSTEMIC INFECTIONS

These include many common infections of childhood, many of which can be prevented by vaccination. The incubation periods for such infections range from 10 up to 40 days, depending on the virus involved; for each virus a specific defined sequence of pathological events occurs, which in turn determines the usual incubation period and pattern of clinical features seen.

Virus enters the body through one of the routes discussed in Chapter 7: Pathogenesis of Virus Infections, usually via respiratory or gastrointestinal tracts, and after replication at the site of entry, virus dissemination occurs through the bloodstream, lymphatics, and/or nerve fibers. A complex set of stages then develop sequentially, the nature of which are specific to each virus. During this period clinical symptoms develop that are characteristic of the particular infection. The characteristic incubation period for the particular infection, defined by the time between the transmission of infection and onset of symptoms in the target organ, is determined by the time required for each of these sequential steps.

The successive stages in the pathogenesis of an acute transient systemic infection were first demonstrated by Frank Fenner in 1948, in pioneering experiments using mousepox (ectromelia) infection of mice as a model. Groups of mice were inoculated in the footpad of a hind limb and at daily intervals organs were removed and extracts titrated to determine the amount of virus present. Fenner showed that during the incubation period, infection spread through the mouse body in a stepwise fashion (Fig. 8.2). Virus first replicated locally in tissues of the footpad and then in local lymph nodes. Virus produced at these sites gained entry into the bloodstream, causing a primary viremia which brought virus to its ultimate target organs, especially lymphoreticular organs and the liver. This stage of infection was accompanied by the development of focal necrosis, first in the skin and draining lymph nodes in the inoculated hind limb, thereafter in the spleen and liver. Within days there was massive necrosis in the spleen and liver leading to death. However, this was not the whole pathological sequence—to complete the viral life cycle, shedding and infection of the next host had to be explained. Fenner found that the virus produced in the major target organs, that is the spleen and liver, caused a secondary viremia, seeding larger amounts of virus back to the skin. Infection in the skin caused a macular and papular rash, from which large amounts of virus were shed, the result being contact exposure of other mice (Table 8.2).

In the example of chickenpox (varicella), initial virus replication in the respiratory tract and regional lymph nodes

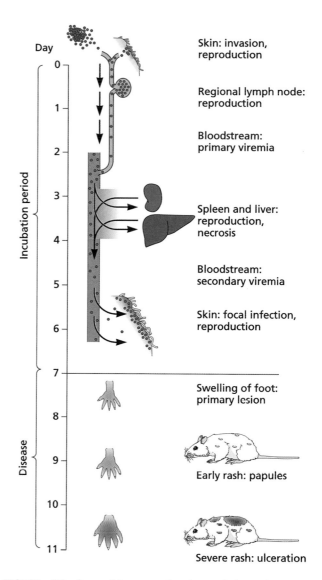

FIGURE 8.2 Sequential stages in the evolution of mousepox infection. This is the original classic figure from Fenner's 1948 paper in *The Lancet*—the first quantitative virus pathogenesis study, the model for many that have followed. *Originally from Fenner, F., 1948. The pathogenesis of the acute exanthems; an interpretation based on experimental investigations with mousepox; infectious ectromelia of mice. Lancet 2 (6537), 915–920. Redrawn many times since.*

TABLE 8.2 Successive Stages in a Transient Systemic Infection (Mousepox)

1. Initial entry and replication at local site of entry ± involvement of regional draining lymph nodes
2. Primary viremia, leading to replication in liver and spleen
3. Secondary viremia, with localization in target organ(s)
4. Replication in target organ, leading to damage and symptoms
5. Adaptive immune response, clearance of infected cells, elimination of virus, repair of damage

leads to a primary viremia, followed by seeding in the liver, spleen, and other organs and further replication. The secondary viremia disseminates virus widely, to produce lesions in the skin, conjunctiva, and sometimes other mucosal sites including the lung. The major complications include bacterial superinfection of lesions, encephalitis, and pneumonia. During the acute infection, latent infection of sensory ganglia is established which provides a lifelong site of potential reactivation (Fig. 8.3).

The clinical severity of transient systemic infections varies from individual to individual. In patients with *subclinical* or *asymptomatic* infection, there may be significant replication at the site of entry without leading to symptoms—such individuals will still develop a specific immune response to the infection, and can act as an unsuspected source of infection to others. Second, in *abortive* infection, there may be non-specific symptoms typical of the early *prodromal* phase (e.g., headache, fever, muscle aches), coinciding with the phase of extensive replication and viremia. However, if there is insufficient involvement of the target organ to cause specific symptoms characteristic of the viral syndrome in question, the diagnosis of the infecting agent will not be evident from the patient's clinical presentation. However, such patients will also be infectious for others and will mount an immune response.

Poliovirus provides a direct stepwise model of pathogenesis that illustrates how severity can vary between individuals (Table 8.3, Fig. 8.4).

There is evidence that specific antiviral antibody limits poliovirus access to the CNS and prevents paralytic disease. Thus the infection can be viewed as a race in time between virus replication and spread on one hand, and the development of the immune response on the other. For any individual, the point seen on the clinical spectrum of disease depends on the balance between these two forces, as well as other factors such as the age of the patient, virulence of the particular virus strain, and other factors. Overall, up to 95% of infections are asymptomatic, approximately 4 to 8% show prodromal symptoms only (URTI, gastrointestinal disturbances or an influenza-like illness), and paralytic disease occurs in 1 to 20 per 1000 infections, more frequently among older individuals.

In general, virus shedding occurs usually from the primary site of replication, for example, poliovirus is excreted in the feces (and to a lesser extent from the nasopharynx), and measles virus from the respiratory tract. Significant shedding is usually delayed until a large number of infected cells have built up.

PERSISTENT INFECTIONS

Many viruses, especially those that remain localized in the respiratory tract or the intestinal tract, cause only acute (self-limited) infections and rarely result in death. Recovery

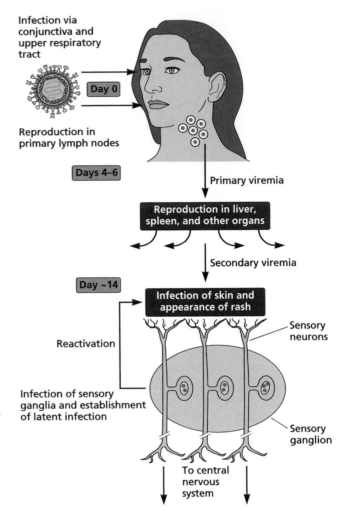

FIGURE 8.3 Sequential steps in the evolution of acute varicella (chickenpox) infection. The progression of many other acute virus infections follows similar phases. However, herpesvirus infections are unique in having a subsequent latent phase with later reactivation and recrudescent disease. Virus enters via the mucosa of the upper respiratory tract or conjunctiva from where virus spreads to regional lymph nodes. After 4 to 6 days, infected T cells enter the bloodstream, causing a primary viremia. These infected cells invade the liver, spleen, and other organs, initiating a second phase of infection. Virus is released from these sites into the bloodstream (as free virions or cell-associated)—the secondary viremia. About 2 weeks post-infection, infected skin-homing T cells invade the skin and initiate the third phase of infection. The characteristic vesicular rash results from local virus replication with cell damage and an inflammatory response. Virus produced in the skin infects sensory nerve terminals and spreads to spinal dorsal root ganglia, where a latent infection is established. Later in life, viral reactivation can occur, in which another infectious cycle is initiated. In this case, virus spreads from the peripheral nerve endings and infects skin epithelium. The recurrent disease, called zoster or shingles, is often accompanied by a long-lasting painful condition called post-herpetic neuralgia. The recurrent infection restimulates the immune system, as does adult vaccination, preventing further disease episodes. *Modified from Flint, S.J., et al., Principles of Virology, third ed., vol. II. ASM Press, with permission.*

TABLE 8.3 Successive Stages in the Pathogenesis of Poliovirus Infection

Approx. Timing	Stages in Pathogenesis	Clinical Outcome
Day 1	Ingestion via fecal–oral route. Virus replicates in pharynx, binds to M cells of intestine, crosses intestinal wall, and replicates in lymphoid cells	Asymptomatic
Days 1–3	Primary viremia→infection of secondary sites	Asymptomatic
Days 4–6	Replication in secondary sites→secondary viremia	Prodromal symptoms, abortive infection
Days 4–6	Access to neurons within CNS by (1) crossing blood–brain barrier, (2) via peripheral nerves	Prodromal symptoms, abortive infection
Days 6–18	Destruction of lower motor neurons	Focal lower motor neuron paralysis
	Control and elimination of virus replication	Progression of paralysis
6+ months	Accessory muscle function improves to substitute for muscle fibers directly lost	Regaining of some motor function
20+ years	Slow muscular atrophy of unknown cause	Post-polio syndrome

follows with elimination of the virus from the body, as was discussed in Chapter 6: Adaptive Immune Responses to Infection and Chapter 7: Pathogenesis of Virus Infections. However, there are other infections that have the capacity to establish infections and these may persist for years or even for life.

Persistent viral infections are of practical significance for a number of reasons: (1) These are often epidemiologically successful, since carriers of the virus serve as a source of infection for other persons, thereby enabling such viruses to persist in small populations for very long periods, even if virus infectivity is low, (2) Viral genomes may be reactivated and cause acute episodes of disease, (3) Chronic or progressive disease may develop, (4) Virus is sometimes associated with neoplasms, (5) Virus may become established even in vaccinated persons unless the vaccine has elicited a strong memory T cell response, (6) Virus may be very difficult to eradicate by antiviral chemotherapy,

particularly if the infectious cycle includes a latent phase when the genome is not replicating and hence infection is not susceptible to antiviral agents, (7) The presence of a persistent virus infection may confuse the diagnosis of another acute disease, for example if an asymptomatic persistent infection is detected as an incidental finding during infection with another agent. In recent years, persistent infections with hepatitis B, hepatitis C, HIV, and human papillomavirus (HPV) have been the subject of much public anxiety, significant health impact, and present major challenges to virologists (Table 8.4).

Having made this neat distinction, the reader should also be aware that, as in so much of biology, gray areas and exceptions can be expected. For example, after the apparent resolution of hepatitis B with "clearance" of virus from the body accompanied by a negative assay for HBV, residual HBV genomes can frequently be found persisting in the liver for years and occasionally can reactivate. Poliovirus genomes can sometimes be found in cerebrospinal fluid samples from patients with the post-polio syndrome decades after recovery from acute poliomyelitis; measles virus must persist in at least some patients for years after recovery from acute measles, to explain the late development of subacute scerosing panencephalitis. Thus, very long-term persistence in tissue reservoirs of viral genomes or genome fragments may not be unusual after apparent virus elimination in many common virus infections. While this phenomenon warrants further study, it is apparent that accompanying clinical consequences are rare.

PERSISTENT LOCALIZED INFECTIONS

The most important example is papillomatosis, caused by any of the many HPVs. Papillomas (skin warts), long considered of minor significance, are now recognized as important when occurring at certain sites such as the larynx, genital tract, and conjunctiva. More so, the most important aspect of certain oncogenic HPV types is squamous cell carcinoma, particularly of the uterine cervix and other parts of the genital tract of both sexes. German virologist Harald zur Hausen clarified the central etiological role of certain HPV types in cervical carcinoma, and was awarded the 2008 Nobel Prize in Physiology or Medicine for this work.

HPVs fall into different subgroups with a predilection for either mucosal surfaces or the skin. Infection is established in the basal cell layers of the epithelium but this involves only limited expression of the viral genome. Differentiation of cells as they move externally leads to increasing expression of the viral genome. In particular the activation of a gene cascade including the E6 and E7 viral proteins leads to cell division and inhibition of apoptosis; full replication with production of mature virus only occurs in the differentiated, more superficial layers of the epidermis. The normal cellular process of differentiation to superficial keratinocytes is also suppressed by the virus. Clinical lesions typically last for

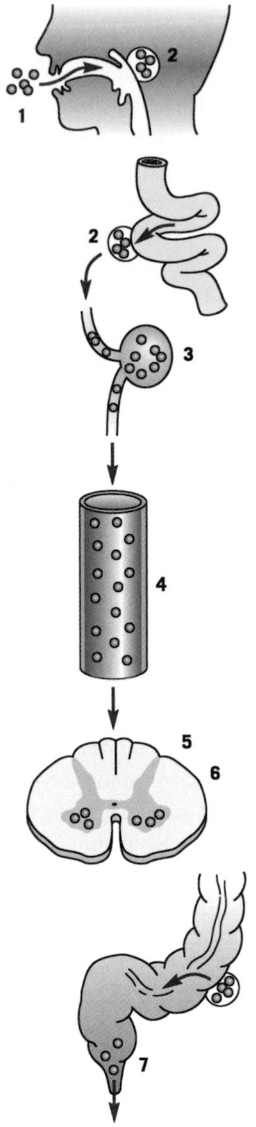

1. Virus ingested		Day 0
2. GUT ASSOCIATED LYMPHOID TISSUE		Day 0–3
• tonsils, Peyer's patches		
• virus invades (via M cells?)		
• replicates in monocytes		
3. REGIONAL LYMPH NODES		Day 3–5
• virus replicates in monocytes		
4. BLOOD		Day 5–15
• plasma viremia		
5. BLOOD BRAIN BARRIER		Day 8–12
• virus crosses endothelium		
6. SPINAL CORD		
• virus replicates in anterior horn cells		Day 10–30
• cell destruction		
• paralysis		Day 12–30
7. GUT		Day 5–45
• virus excreted in feces		

FIGURE 8.4 Sequential steps in the evolution of poliovirus infection. Virus enters via the alimentary tract (1) and multiplies locally (2) at the initial sites of virus implantation (tonsils, Peyer's patch) or the lymph nodes that drain these tissues (3), and virus begins to be excreted in the throat and the feces into the environment. Virus spread then occurs via the bloodstream (4) to other susceptible tissues—namely, other lymph nodes, brown fat, muscle, and the CNS, and amplification in these sites contributes to the secondary viremia. Virus can spread into the CNS by crossing the blood–brain barrier (5), and also by means of peripheral or cranial nerve retrograde axonal flow—for example, from muscle. If a high level of virus replication occurs within the CNS, motor neurons die and paralysis ensues (6). *Originally drawn by Joseph Melnick and used in several publications; then redrawn by Neal Nathanson: Nathanson, N., 2007. Viral Pathogenesis and Immunity, second ed. Academic Press, with permission.*

TABLE 8.4 Most Important Persistent Virus Infections of Humans

	Herpes Simplex	Hepatitis B	Hepatitis C	HIV	Human Papillomavirus
Global total infected	4+ billion	350 million	300 million	35 million	Up to 500 million
Viremia	No	Yes	Yes	Yes	No
Long-term mortality	Low	25–30%	~20%	>95%	~500,000 cases cancer p.a.
Vaccine licensed	No	Yes	No	No	Yes
Viral genome integrates	No	Late	No	Essential for repl'n	In many but not all tumors

months before resolving, but infection of basal cells may be much more long-lived. Although this intraepidermal site is to some extent protected from the immune system, some degree of immunological control must operate because lesions reactivate more often and are more extensive in immunosuppressed individuals (see Fig. 19.4).

Papillomavirus infections always remain localized to the epithelium, usually in sites adjacent to the site of entry, although self-inoculation can produce further crops of lesions at sites distant from the initial infection. In most patients the lesions, and viral DNA detected by PCR, are cleared from the genital tract or the skin within 12 to 18 months, in which case persistence is apparently not lifelong. However, the emergence of multiple skin warts in some patients following immunosuppression, and the detection by PCR of unusual HPV genotypes in samples of healthy skin, both suggest that much longer subclinical persistence of HPV genomes may be common.

PERSISTENT SYSTEMIC INFECTIONS

For practical convenience, persistent systemic infections may be subdivided into three categories (Table 8.5):

1. *Acute infections with rare late complications*;
2. *Latent infections with reactivation*, in which infectious virus is not demonstrable except when reactivation occurs, and disease is usually absent except during some or all of such recurrences;
3. *Chronic infections with ongoing viral replication*, in which infectious virus is usually demonstrable and often shed. Disease may be absent, or slowly progressive, or may develop late, and often involves immunopathological or neoplastic mechanisms.

These three categories are defined primarily in terms of the extent and continuity of viral replication in the body during the long period of persistence. The presence or absence of shedding, and of disease, are secondary issues as far as this categorization is concerned. Some persistent infections in all three categories are associated with disease, some are not, but the *carrier* status of all such persons offers a potential source of infection for others.

Acute Infections with Rare Late Complications

Subacute Sclerosing Panencephalitis

The paradigm of this type of persistent infection is subacute sclerosing panencephalitis (SSPE), an invariably fatal complication occurring in apparently normal children 1 to 10 years after recovery from measles. It has an incidence of approximately 4 to 10 cases per 100,000 cases of measles per year, and has become very rare following the global introduction of childhood vaccination. Nevertheless, it remains of considerable interest to virologists owing to the questions it raises about pathogenesis of persistent infections of the CNS (see Chapter 26: Paramyxoviruses).

SSPE patients harbor very little measles virus in their brains but have an exceptionally high titer of neutralizing antibodies in their cerebrospinal fluid. The RNA viral genome is detectable in neurons by nucleic acid hybridization, and nucleocapsids are demonstrable by electron microscopy and immunofluorescence. Studies of infected neurons in culture indicate that the presence of high-titer neutralizing antibody brings about a pronounced inhibition of viral genome transcription. Even in the absence of antibody it is clear that there is something unique to neurons, and/or their state of differentiation that down-regulates transcription of measles virus RNA. Transcription of paramyxovirus genomes characteristically displays "polarity," that is, a progressive diminution in the efficiency of transcription as the polymerase progresses from the 3′ to the 5′ end of the genome (see Chapter 26: Paramyxoviruses). This gradient is more pronounced in neurons infected with measles virus, with the result that the production of the envelope proteins M, F, and H is reduced, and the L protein is almost non-existent. Thus, virtually no infectious virions are assembled but there is sufficient transcription of the 3′ non-structural genes, N and P/C, to permit viral RNA replication and transfer of nucleocapsids at a very slow rate from cell to cell. Further, during the long incubation period of SSPE, numerous mutations accumulate, seen particularly in the M gene transcripts. All in all, SSPE illustrates beautifully how peculiarities in the infection by particular viruses of particular host cells may afford enabling conditions for the persistence of viral genome—in this case, to the detriment of the host.

Progressive Multifocal Leukoencephalopathy

Progressive multifocal leukoencephalopathy (PML) is a rare, lethal manifestation of reactivation of an almost universal latent infection due to human JC virus (a polyomavirus). The acute infection is spread by the respiratory route, usually in childhood, and is usually subclinical or there may be a mild respiratory disease. The virus then persists for life in the kidneys, and perhaps also in the brain and/or elsewhere. The virus is shed in urine from time to time, especially during pregnancy or immunosuppression. Almost nothing is known of the mechanism of establishment and maintenance of latency. PML is a demyelinating disease of the brain, seen only in severely immunocompromised individuals, especially AIDS patients, organ transplant recipients, and those with advanced disseminated lymphoid malignancies. The target cell is the oligodendrocyte rather than the neuron (see Chapter 20: Polyomaviruses).

TABLE 8.5 Major Persistent Virus Infections of Humans

Virus	Site(s) of Persistence	Chronic Disease
Localized Infection		
Papillomavirus	Skin, mucosal cells	Warts; genital, and other cancers
Adenoviruses	Adenoids, tonsils, lymphocytes	None known
Systemic Infection		
1. Acute infection with late rare complications		
Measles	Central nervous system	SSPE (panencephalitis)
Rubella	Central nervous system	Progressive panencephalitis
JC polyomavirus	Kidney, CNS	Progressive multifocal leukoencephalopathy (PML)
BK polyomavirus	Kidney	Hemorrhagic cystitis
2. Latent infection with reactivation		
Herpesviruses		
• Herpes simplex 1 and 2	Sensory, autonomic ganglia	Cold sores, genital lesions
• Varicella-zoster virus	Sensory, autonomic ganglia	Herpes zoster (shingles)
• Cytomegalovirus	Kidney, salivary gland, bone marrow progenitor cells	Pneumonia, retinitis
• Epstein-Barr virus	B cells, nasopharynx	Lymphomas, carcinomas
• HHV6, HHV7	Salivary glands, CNS	Graft rejection? multiple sclerosis
• HHV8	B lymphocytes, vascular	Kaposi's sarcoma, endothelium
3. Chronic infection with ongoing viral replication		
Hepatitis C virus	Liver	Cirrhosis, liver cancer
HIV	CD4 T cells, macrophages	AIDS
Human T cell leukemia virus	T cells	Leukemia
Rubella	Many organs	Congenital rubella syndrome
Hepatitis B virus	Liver, lymphocytes	Cirrhosis, liver cancer

Latent Infections with Reactivation

This pattern is characteristic of infections caused by all eight human herpesviruses. Viral carriage is lifelong, and there may be spontaneous reactivations at any time. In patients with profound immunosuppression, for example, those with AIDS or those immunosuppressed for tissue/organ transplantation, reactivation of a latent herpesvirus infection may be lethal. Paradoxically, although all herpesviruses share the common characteristic of evoking persistent latent infection, each has evolved a different strategy for achieving that end. The key distinction is that herpes simplex and varicella-zoster (VZV) viruses persist in neurons, cells that are non-dividing but long-lived, whereas, cytomegalovirus (CMV), Epstein Barr virus (EBV), and human herpesvirus 6 persist in lymphocytes that actively divide but are short-lived (Table 8.5, Table 8.6).

The state of the genome is a central consideration for understanding latency. With some viruses the genome must become integrated into a cellular chromosome, whereas for others it survives satisfactorily as a free plasmid (episome) in the cytoplasm or nucleus. Of course, it is essential that

TABLE 8.6 Examples of Latent Infections: Acute Infection Followed by Viral Latency and Reactivation With or Without Recurrent Clinical Episodes

Virus	Disease in (1) Acute Infection (2) Reactivation	During Acute Infection	During Latency	Virus Shedding[a]
Herpes simplex viruses	1. Primary oral or genital herpes 2. Recurrent herpes simplex or genitalis	Epithelial cells Neurons then epithelium	As DNA, in neurons of sensory ganglia	Plentiful from vesicles in acute phase Sporadically in saliva or genital secretions between attacks Plentiful in recurrent herpes vesicles
Varicella-zoster virus	1. Chickenpox 2. Shingles	Widespread Neurons then epithelium	As DNA, in neurons of sensory ganglia	From throat and skin lesions; no shedding after recovery from acute disease From skin lesions; contacts may contract chickenpox
Cytomegalovirus	1. Usually subclinical except in fetus or immunocompromised 2. Reactivation may lead to a range of syndromes	Epithelial cells	As DNA, in salivary glands, kidney epithelium, leukocytes	Sporadically throughout life in saliva and urine, especially pregnancy
Epstein-Barr virus	1. Glandular fever (mononucleosis) 2. Burkitt's lymphoma, nasopharyngeal carcinoma, other lymphomas	Epithelial cells and lymphoid tissue	As DNA, in B cells	In saliva during the acute phase

[a]All reactivated by immunosuppression.

any latent viral genome, whatever its physical state or location, must code for all the genetic information required for virus replication, if the virus is ever reactivated. This contrasts with the status of many defective viruses that may induce cancer following genome integration even though a full replication cycle is not possible. In any case, expression of the latent viral genome is, by definition, repressed, either wholly or partially. Generally, only a restricted range of viral genes is transcribed during latency, but of course all are derepressed during reactivation. Often the particular genes that are expressed during latency fulfill in a myriad of ways a vital role in the maintenance of latency.

Herpes Simplex Virus Infection

Primary infections with herpes simplex virus 1 (HSV-1) may be subclinical, or may present as an acute stomatitis in early childhood, or tonsillitis/pharyngitis in adolescence. At intervals of months or years after recovery from this primary infection, the characteristic vesicular lesions may reappear, usually on the lips. Herpes simplex virus type 2 (and sometimes type 1) causes comparable initial and recurrent lesions on the male and female genitalia (see

Chapter 17: Herpesviruses). During primary infection with herpes simplex viruses, viral nucleocapsids gain access to neurons of cranial or spinal sensory ganglia, where the viral genome persists indefinitely. Periodically, reactivation of these latent infections can be triggered by a variety of stimuli, such as "stress," ultraviolet light, fever, nerve injury, or immunosuppression (Fig. 8.5).

The mechanism of establishment and maintenance of latency, and of subsequent reactivation, have fascinated virologists and been the object of study for many years.

Following replication in epidermal cells of the skin or mucous membranes, some virions access the peripheral nervous system by fusion of the viral envelope with sensory nerve endings. Viral nucleocapsids are released into the axoplasm, and ascend within axons to reach the nucleus of a small minority of the neurons in the corresponding sensory ganglion in the brain stem (e.g., trigeminal ganglion following oral infection) or spinal cord (sacral ganglion following genital infection). Productive infection leads to the destruction of some neurons but in many the viral genome is coated with nucleosomes and silenced, most likely after the productive-cycle gene expression pathway has begun. Studies in mice suggest that these infected neurons are

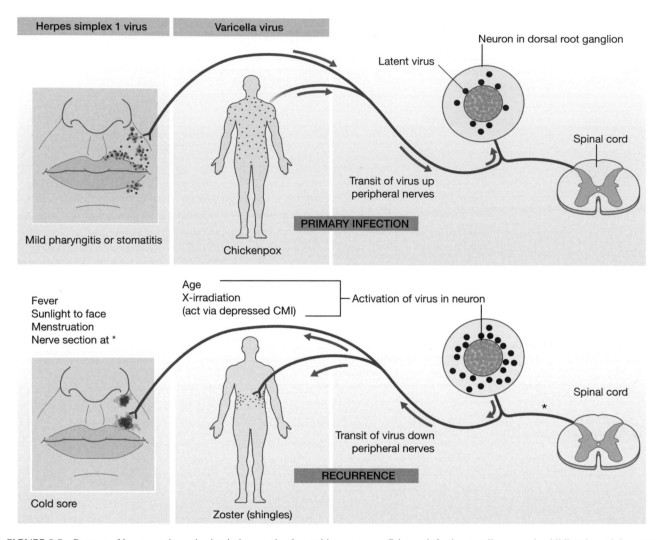

FIGURE 8.5 Patterns of latency and reactivation in herpes simplex and herpes zoster. Primary infection usually occurs in childhood or adolescence. Latent infection becomes established in cerebral or spinal dorsal root ganglia by retrograde axonal spread from the primary infection site (herpes simplex) or following viremia (varicella-zoster). Reactivation of herpes simplex virus causes recrudescence of lesions around the primary site (e.g., cold sores); reactivation of varicella-zoster virus causes recrudescence in a dermatomal distribution (shingles) related to the ganglia involved in reactivation. *Reproduced from White, D.O., Fenner, F.J., Medical Virology, fourth ed., Academic Press, with permission.*

protected by antiviral cytokines secreted by CD8+ T cells that have been activated by exposure to viral peptides presented on the surrounding capsular cells (analogous to microglia) but that cannot lyse infected neurons because neurons do not express significant amounts of MHC class I or II glycoproteins. Interferons reduce expression of HSV "immediate-early" (α) genes, which are absolutely required for the expression of all other HSV genes. Surviving neurons harbor 10 to 100 copies of the viral genome indefinitely, in the form of non-integrated, circular, extended concatemers. No virions are made because transcription of the standard mRNAs is absent or tightly restricted, and viral protein is virtually absent. However, a proportion of the latently infected neurons produces thousands of molecules of a family of unusual overlapping non-polyadenylated antisense

RNA transcripts known as LATs (latency-associated transcripts). The function of these LATs remains to be clarified. The 2 kb LAT appears to be a non-coding intron, complementary in sequence to the mRNA for a key regulatory protein, RL2, which is required for the transactivation of all later HSV genes. Other suggestions are that LATs may be precursors of micro-RNAs that interact with host cell mRNAs, or that they may block apoptosis of neurons or assist latency by altering chromatin structure.

Following reactivation, with or without triggering by the factors noted above, replication of virus is induced in a small proportion of the latently infected neurons. Virus is transported down the axon to the periphery where it multiplies once again in epithelial cells in the same general locality as those infected originally. A secondary CD4+ Th1 cell-mediated inflammatory response, perhaps supported by a CD8+ Tc response, quickly leads to the elimination of infected epithelial cells, and pre-existing antibody mops up any free virions to bring the recurrence of disease to a halt.

Neurons, being non-dividing cells, offer certain unique survival advantages to a viral genome: (1) in the absence of MHC antigens these cells are shielded from lysis by cytotoxic T lymphocytes, and because the latent viral genome expresses no protein they remain a safe haven from lysis by antibody plus complement, or ADCC, (2) because the neuron does not divide there is no corresponding need for the viral genome to divide to maintain a fixed number of copies per cell, (3) the axon of a sensory neuron, which may be many centimeters long, provides a direct pathway to the periphery. Long-term survival of the virus requires that sufficient productive replication in susceptible epithelial cells must occur, despite the presence of pre-existing immunity, to ensure dissemination of infectious virus to other susceptible hosts. Regular throat swabbing of asymptomatic individuals has shown that asymptomatic *reactivation* with peripheral shedding occurs more frequently than true clinical *recrudescence*, thereby explaining anecdotes of primary herpesvirus infection having been acquired from someone without any visible lesions.

Herpes Zoster

This disease, familiarly known as zoster or "shingles," is characterized by a painful vesicular rash that is usually limited largely to an area of skin innervated by a single sensory ganglion (see Chapter 17: Herpesviruses). It occurs among older people who had varicella in childhood; its occurrence is more likely if accompanied by immunosuppression. Zoster results from the reactivation of VZV (see Fig. 8.5). VZV latency is more tightly maintained than that of HSV, as indicated by (1) lack of evidence of asymptomatic shedding of virus between attacks, (2) reactivation triggered only by declining immunity to the virus, not by fever, ultraviolet light, etc., (3) most latent VZV infections never reactivate, and in the remaining 15% only a single recurrence (herpes zoster) occurs: this is largely confined to the elderly. On the other hand, herpes zoster typically involves a whole dermatome, with much more extensive lesions than generally occur with recurrent herpes simplex, and much greater pain, sometimes including protracted neuralgia ("a belt of roses from Hell"). A major distinction from herpes simplex infection is that the primary infection (chickenpox) is a systemic infection, acquired by the respiratory route and disseminated via the bloodstream. The epithelial lesions (rash) are widely distributed, and hence the distribution of latently infected ganglia can be widely distributed in the body.

These clinical observations are consistent with the following findings arising from limited investigations of sensory ganglia taken at autopsy from people with asymptomatic VZV infection. Viral replication cannot be reactivated by explantation or co-cultivation of neurons as it can with HSV, but *in situ* hybridization, cell separation, and other studies have shown that a proportion of neurons contain viral genomes as well as at least five species of RNA transcripts and corresponding proteins, but no equivalent to the LATs of HSV. Unlike HSV, VZV-infected ganglia do not show T cell infiltration, suggesting that maintenance of VZV latency involves different mechanisms, in contrast to HSV infections where neurons containing latent virus are frequently surrounded by a corona of CD8+ T cells expressing markers of antigen activation. When reactivation is accompanied by clinical disease, the shingles rash typically involves one dermatome on one side of the body, suggesting that the virus spreads within one reactivating ganglion before being transported to the periphery. Despite superficial similarities between latency in HSV and VZV infection, the establishment and maintenance of latency in VZV is still incompletely understood, yet thought to involve different processes.

Epstein-Barr Virus Infection

The mechanism of Epstein-Barr virus (EBV) persistence is fundamentally different from that of HSV and VZV. Because the site of long-term latent infection is B lymphocytes, which unlike neurons are short-lived dividing cells, the evolution of the capacity for latency has involved the development of two special capabilities: (1) stimulation of the lymphocyte to divide, and (2) maintenance of the genome in the form of an episome (plasmid) while controlling the viral genome copy number in synchrony with the increase in cell number.

In developing countries, EBV usually occurs in infancy and is subclinical, whereas infection in higher socioeconomic groups usually occurs later, for example, in adolescence, and has a much higher rate of clinical disease. The clinical disease may take the form of infectious mononucleosis (glandular fever). EBV replicates productively in epithelial cells of the nasopharynx, for example, the tonsils, and salivary glands, and also spreads to resting B cells in lymphoid tissues. This is followed by a striking proliferation of B lymphocytes which is usually quickly contained by a T cell-mediated immune response. However, if the patient suffers from any severe underlying congenital or acquired T cell immunodeficiency, various types of severe EBV-induced progressive lymphoproliferative diseases may prove lethal. Moreover, in certain parts of the world EBV infection in childhood may lead decades later to death from either a B cell malignancy known as Burkitt's lymphoma or nasopharyngeal carcinoma (see Chapter 9: Mechanisms of Viral Oncogenesis and Chapter 17: Herpesviruses).

The usual outcome of acute EBV infection is viral persistence as a latent infection in approx. 0.001% of non-proliferating memory B cells. The latter express one viral mRNA known as LMP-2 but no infectious virus. These cells lodge in lymphoid organs and bone marrow. The EBV genomes are maintained as circular, self-replicating episomes in B cell nuclei, in association with nucleosomes and DNA methylation. Reactivation of virus replication can occur in response to a range of signal transduction cascades. This process is normally held in check by surveillance and removal of virus antigen-expressing B cells by CTLs specific for these antigens.

The availability of EBV-transformed lymphoblastoid cell lines has greatly extended our understanding of the mechanism of EBV latency, in B cells at least. The entire EBV genome persists in the nucleus as multiple copies of a supercoiled covalently closed circular DNA molecule, but only a very small part of this large viral genome is transcribed. Three levels or programs of latency are recognized. Latency III is normally seen when transformation of a naïve resting B cell is first established, and with time this may progress to Latency II and Latency I. Viral gene products transcribed in Latency III include EBNA-1 and -2, EBNA-LP, EBNA-3A, -3B and 3-C, LMP-1, -2A and -2B, and the small RNA molecules EBER-1 and -2, while fewer genes are expressed in Latency II and Latency I.

Less is known about reactivation of EBV from the latent state. This can be triggered *in vitro* by treatment with various agents including cytokines, and *in vivo* it is likely that differentiation of B cells into plasma cells, and differentiation of epithelial cells, are important mechanisms.

Cytomegalovirus Infection

Infections with CMV and EBV are characterized by initial prolonged excretion of virus followed by a state of latency. CMV infections are normally subclinical, but severe disease occurs in patients with immunodeficiency as a result of AIDS, chemotherapy for malignancy, or immunosuppression for organ or tissue transplantation. In such patients generalized CMV infection may result from reactivation of an endogenous latent infection, or from an exogenous primary infection resulting from the organ graft or from a blood transfusion, reflecting the widespread occurrence of symptom-free carriers in the general population. Indeed, surveys show that at any time about 10% of children under the age of 5 years are excreting CMV in their urine. Transplacental infection of the fetus when a primary CMV infection or reactivation of a latent infection occurs during pregnancy may induce devastating congenital abnormalities in the neonate.

CMV establishes latent infection in both salivary glands and kidneys, as well as in monocytes and CD34+ bone marrow progenitor cells. Cellular differentiation of CD34+ cells into mature myeloid cells including dendritic cells, is concomitant with the induction of a full lytic transcription program, viral DNA replication, and production of progeny virus. Virus is shed, intermittently or continuously, mainly into the oropharynx, from which it may be transmitted via saliva; virus is also shed into the urine.

Chronic Infections with Ongoing Viral Replication

Since the 1980s the medical profession has recognized the widespread prevalence and late-onset disease associated with hepatitis B and hepatitis C, and the appearance in humans of a completely new infection, HIV. Diagnostic testing, antiviral drug treatment, epidemiological monitoring, and control of these three infections now play a large role in the practice of medical virology. There are also a large number and variety of chronic persistent infections in domestic animals associated with continuous virus production (Table 8.7). In a number of these examples, progressive organ damage or late disease may develop, often with fatal consequences (Table 8.8).

Ongoing replication provides a complex substrate for virus evolution by Darwinian natural selection. The population of viruses present in a host at any time is known as a "quasi-species," a mixture of genetically similar viruses all replicating concurrently. At the point of transmission a genetic "bottleneck" operates, leading to selection of a particular variant or variants most fit to achieve transmission. Once a new persistent infection is established, errors in genetic copying during ongoing virus replication produce a never-ending source of new genetic variants. Continuing selection of the most fit variants produces a constantly evolving mixture of different variants, the quasi-species. In the case of HBV, the long-term persistence of covalently closed circular ("ccc") viral DNA within hepatocytes provides some genetic continuity for this infection, while with HIV the continuing turnover of infected cells means that viral DNA templates also turn over rapidly and are subject to genetic drift. HCV is assumed to maintain chronic infection by a combination of continuing release from infected hepatocytes together with ongoing recruitment of newly infected cells, but the details and dynamics remain to be clarified.

Hepatitis B

During acute HBV infection the virus replicates in the liver and enters the blood, usually together with a great excess of smaller subviral envelope particles composed of viral surface antigen (HBsAg). In most infected individuals HBsAg and virions are then cleared from the circulation, but in 5 to 10% of adults and over 90% of those infected during infancy, a persistent infection is established which typically extends for many years, often for life. The carrier state is characterized by continuous production of HBsAg and usually infectious virions. These are plentiful in the

TABLE 8.7 Examples of Chronic Infections, with Virus Always Demonstrable, With or Without Late Development of Disease

Virus/Genus	Host	Site of Persistent Infection	Late Disease[a]
Human immunodeficiency virus-1 and -2 (Lentivirus)	Human	T-lymphocytes, macrophages	Opportunistic infections due to profound immunodeficiency
Visna/maedi (Lentivirus)	Sheep	Macrophages, brain, lung	Slowly progressive pneumonia or encephalitis
Caprine arthritis/encephalitis virus (Lentivirus)	Goat	Macrophages, brain and joints	Arthritis, encephalitis
Equine infectious anemia virus (Lentivirus)	Horse	Macrophages	Anemia, vasculitis, glomerulonephritis
Human T cell lymphotropic virus-1 and -2 (Deltaretrovirus)	Human	T-lymphocytes	Adult T cell leukemia, tropical spastic paraparesis
Avian leukosis virus[b] (Alpharetrovirus)	Chicken	Widespread	Occasionally leukemia
Hepatitis B virus (Hepadnavirus)	Human	Hepatocytes	Chronic active hepatitis, cirrhosis, hepatocellular carcinoma
Hepatitis C virus (Flavivirus)	Human	Hepatocytes	Chronic active hepatitis, cirrhosis, hepatocellular carcinoma
Aleutian disease virus (Parvovirus)	Mink	Macrophages	Hyperglobulinemia, arteritis, glomerulonephritis

[a]Sometimes occurring despite presence of high titers of antibody which fail to eliminate the virus.
[b]Also occurs as latent infection with integrated provirus, transmitted congenitally and not causing disease.

TABLE 8.8 Different Pathological Changes That May Develop During the Course of a Persistent Virus Infection

Process	Example	Result
Gradual recovery of immunocompetence	Hepatitis B	Elimination of infection, either with or without flares of disease (hepatitis) immune restoration disease
Emergence of escape mutants	Equine infectious anemia	Flares of viremia, with or without disease (anemia, vasculitis)
Immunosuppression	HIV-1 and -2	Increased viral replication Opportunistic infections and malignancies
Progressive tissue damage	HCV, HIV	Organ failure
Virus-directed cell transformation	HTLV-1, HPV	Development of cancer
Cell damage and increased cell proliferation, increased mutation	HBV, HCV	Development of cancer

bloodstream but less so in semen and saliva, hence the danger represented by such persons as blood donors and sexual partners. Some carriers develop chronic hepatitis and cirrhosis; hepatitis B virus (HBV) infection is also an important cause of primary hepatocellular carcinoma (see Chapter 22: Hepadnaviruses and Hepatitis Delta).

In acute hepatitis B and in the initial "high replicative phase" of the chronic carrier state the viral DNA genome exists in hepatocytes as unintegrated covalently closed circular (ccc) DNA, and infectious virions are produced reaching levels in the plasma as high as 10^8/ml. In the subsequent "low replicative phase" of the chronic carrier state, more copies of the genome become progressively integrated into the chromosomes of the hepatocyte. Transcription from this integrated DNA is restricted largely to subgenomic mRNA, and the major protein translated is HBsAg. No full-length RNA transcripts are produced from the integrated genome, therefore it cannot replicate as it requires faithful complete RNA copies for the viral reverse transcriptase to utilize as a template. The fact that the number of circulating non-infectious HBsAg particles can exceed the number of virions by a factor of up to 10^5 suggests that this may have evolved as a survival strategy whereby the abundant HBsAg serves as a decoy by absorbing most or all of the neutralizing antibody. Certainly, no free antibody against HBsAg can be detected throughout the many years of the chronic carrier state, but extensive deposition of antigen–antibody complexes in kidneys and arterioles commonly causes immune complex disease. Cytotoxic CD8+ T lymphocytes, which are principally responsible for clearance of the acute infection by cytolysis of infected hepatocytes, recognize class I MHC associated peptides derived from HBcAg and HBeAg; however, in the low replicative phase of the chronic carrier state, integrated HBV DNA generally produces no RNA transcripts encoding these two antigens, and further, only productively infected hepatocytes express substantial amounts of MHC class I protein. In addition, there is some evidence that chronic carriers display a degree of HBV-specific immunosuppression, mediated by suppressor T cells and/or tolerization of B cells.

Liver damage is produced by the action of CTLs against infected hepatocytes. During the high replicative phase, damage is less severe because CTL activity is less effective, while in the low replicative phase damage is also less because the expression of target antigens on hepatocytes is more restricted. However, as patients evolve from the former to the latter they pass through an "intermediate" phase when liver damage can be at its greatest, sometimes fluctuating in the form of "flares." Long-term sequelae of this process include cirrhosis and hepatocellular carcinoma.

Hepatitis C

Although acute HCV infection is clinically milder than hepatitis B, persistent infection follows approximately 60 to 80% of hepatitis C infections, and careful long-term studies have clarified that 10 to 20% of these individuals will progress to chronic hepatitis and cirrhosis over the ensuing 20 to 40 years (see Chapter 36: Flaviviruses).

Individuals with persistent HCV infection usually have virions circulating in their plasma as immune complexes, together with an excess of virus-specific antibodies. Virus replication *per se* is non-cytopathic, and the outcome of infection and degree of liver damage is the result of complex interactions with many arms of the immune system. A strong anti-HCV immune response is associated with severe liver damage, often with virus clearance, whereas a mild immune response allows virus persistence but also facilitates liver cell damage and inflammatory cell infiltrates. HCV triggers an innate immune response through recognition by RIG-I receptors, toll-like receptors, and the pattern recognition receptor PKR, but this virus has multifunctional gene products that counteract this. HCV also induces endogenous interferon-α, but responds poorly. Natural killer (NK) cells are also involved in acute and chronic infection, and may contribute to the regulation of hepatic fibrosis in chronic HCV infection via cytotoxicity against hepatic stellate cells. Both humoral and cellular arms of the adaptive immune response play important roles; a strong initial CTL response is associated with viral clearance, while chronic HCV is characterized by a progressive functional exhaustion and eventually loss of HCV-specific CD4+ and CD8+ cells. Regulatory T cells also play multiple roles, for example, by suppressing inflammatory responses, facilitating HCV persistence, and contributing to the regulation of hepatic fibrosis.

Human Immunodeficiency Virus (HIV) Infection/AIDS

The pathogenesis of HIV infection demonstrates many principles learned in the study of chronic viral infections (see Chapter 23: Retroviruses). The major targets in HIV infection are cells bearing the CD4+ receptor used by the virus for entry—activated CD4+ lymphocytes in particular are highly permissive for infection. Following the most common mode of transmission to mucosal surfaces via sexual contact, free virus and infected cells migrate to both draining and other lymph nodes. Massive virus amplification then occurs, detectable as a viremia and sometimes an associated acute glandular fever syndrome, after which a prolonged phase ensues where the individual is largely asymptomatic (clinical latency). However it is important to recognize that, during this phase, circulating virus is turning over rapidly (half-life 6 hours), and productively infected CD4+ lymphocytes are being destroyed and rapidly replaced (half-life 1.1 days). Meanwhile, the host mounts a cell-mediated and humoral immune response but fails to eliminate the virus. Lifelong persistent infection is promoted by a number of different attributes of the infection (Table 8.9). Virus is continuously released not only into the bloodstream but also into various body secretions, hence carriers can be a source of infection to sexual and blood contacts, and survival of the virus in the community is ensured.

Throughout the prolonged asymptomatic phase averaging about 10 years, there is an eventual decline in the number of circulating CD4+ cells, coinciding with a steady disruption of function of the various arms of the immune response. The patient becomes progressively susceptible to infection with a succession of viruses, bacteria, and fungi,

TABLE 8.9 Attributes of HIV that Assist Its Long-Term Survival in the Host

1. Inducing the formation of syncytia allows HIV to spread from cell to cell without encountering antibody
2. Once viral cDNA is integrated into the host cell chromosome, it can persist indefinitely as a latent infection to be reactivated later
3. The low fidelity of reverse transcription leads to an inexhaustible production of different HIV mutants which, after selection, enable the virus to alter its cellular tropism, gain in virulence, become resistant to antiviral drugs, and evade neutralization by the prevailing antibodies
4. The principal pathologic effect of the virus is to suppress the host's immune system by destruction of helper cells and macrophages
5. Widespread dissemination within the body, including into sites less accessible to the immune system

infections that add to the heavy demands on a weakened immune system. Cancers also commonly develop, in particular forms of cancer where other oncogenic viruses are involved—Kaposi's sarcoma (human herpesvirus 8), B cell lymphomas (EBV and HHV8), and anogenital carcinomas (HPVs). The progressive effect of debilitating opportunistic infections is accompanied by an ever-increasing decline in CD4+ T cells and death inevitably follows unless interrupted by antiviral therapy.

In common with all retroviruses (see Chapter 4: Virus Replication and Chapter 23: Retroviruses), HIV replication requires reverse transcription of its RNA genome to produce a dsDNA copy (cDNA) which becomes integrated permanently into a chromosome of the host cell as a provirus. During the long preclinical phase of HIV infection an increasing proportion of CD4+ T cells carry HIV cDNA, either integrated or unintegrated into the host genome; this HIV genome may remain latent until the T cell becomes activated by an appropriate cytokine or by transactivation by another virus (see Chapter 23: Retroviruses). Such latently infected T cells do not express viral proteins and

therefore are not susceptible to immune elimination. These infected T cells and macrophages are able to transfer infection to nearby uninfected cells by fusion—this involves interaction of the HIV envelope protein with the CD4+ receptor on lymphocytes and macrophages, thus evading neutralizing antibody. During its long sojourn in the body, HIV undergoes an unparalleled degree of genetic drift. The reverse transcriptase is notoriously error-prone, so numerous nucleotide substitutions accumulate in the genome as the years progress. Clearly the stage is set for the selection of any virus variant possessing a replication advantage. In particular, three phenotypes tend to be regularly selected. First, escape mutants with amino acid changes in key antigenic determinants recognized by antibody or by T cells (antigenic drift) progressively evade the immune response. Second, there is progressive selection for mutants with a predilection for using different co-receptors (e.g., the change from macrophage-tropic virus using chemokine receptor CCR5 as co-receptor, to T cell-tropic virus using CXCR4). Third, because the use of antiviral drugs is now widespread in such patients, drug-resistant mutant viruses are continually being selected.

Thus, HIV is a remarkable and devastating example of a persistent viral infection with continuing extensive virus replication and the ultimate development of end-stage disease and death.

FURTHER READING

Flint, S.J., Enquist, L.W., Racaniello, V.R., Rall, G.F., Skalka, A.-M., 2016. Principles of Virology, 4th ed. ASM Press, Washington, DC.

Heise, M.T., Virgin, H.W., 2013. Pathogenesis of viral infection. In: Knipe, D.M., Howley, P. (Eds.), Fields Virology, sixth ed., Wolters Kluwer Health, Philadelphia, PA.

Kaslow, R.A., Stanberry, L.R., LeDuc, J.W., 2014. Viral Infections of Humans, fifth ed Springer, New York, NY

Nathanson, N , 2007. Viral Pathogenesis and Immunity, 2nd ed. Academic Press, Amsterdam.

Racaniello, V.R., 2005. One hundred years of poliovirus pathogenesis. Virology 344, 9–16.

Chapter 9

Mechanisms of Viral Oncogenesis

It is now recognized that 15 to 20% of human cancers are attributable to certain viruses, and it has been suggested that infection with known oncogenic viruses is second only to the use of tobacco as a risk factor for cancer. Viral oncology originated from observations in the early part of the 20th century, when the transmission of avian leukemia and Rous sarcoma were described by Vilhelm Ellermann and Olaf Bang in 1908 and Peyton Rous in 1911 respectively. The 1966 Nobel Prize in Physiology or Medicine was awarded to Peyton Rous nearly 50 years after his initial discovery of this avian sarcoma virus. These early findings were regarded as insignificant and intellectual curiosities until the 1950s when the experimental transmission of murine tumor viruses to laboratory mice was described. Since the 1950s there has been a flood of discoveries implicating many viruses of the family *Retroviridae* in a variety of benign and malignant tumors in animals. Finally, after years of false hopes, two retroviruses, human T lymphotropic viruses 1 and 2 (HTLV-1 and HTLV-2), were implicated as causes of human cancers.

Since then, studies have progressed rapidly on exogenous and endogenous retroviruses, as well as on the functions of viral oncogenes and insertion mutations in association with tumor suppressor genes, and many new tumor viruses in animals have been recognized. The discovery of tumor viruses in animals not only made it possible to study viral oncology in well-controlled experiments, but also provided a research tool for exploring the role of viruses in human cancers.

In addition to the role of retroviruses, oncogenesis by small DNA viruses was explored by studying transformation of cells *in vitro*, beginning with SV40, a simian polyomavirus. This work was extended to adenoviruses, human papillomaviruses (HPV), and other polyomaviruses. Subsequently, certain large DNA viruses, such as human herpesvirus 8 (HHSV 8) and Epstein-Barr virus (EBV), hepatitis B virus, and an RNA virus (hepatitis C virus) have also been shown to be associated with tumors. Experimental, clinical, and epidemiological studies in this interesting and challenging field have already led to the development of vaccines against two virus-induced human tumors (HBV-induced hepatocellular carcinoma and HPV-induced cervical carcinoma), and have opened the prospect of improved prevention, management, or control of others (Table 9.1).

DEFINITIONS AND OVERVIEW OF VIRUS TRANSFORMATION

It is useful to clarify a few commonly used terms. *Oncology* is the study of tumors. A *benign tumor* is a growth produced by abnormal cell proliferation that remains localized and does not invade adjacent tissue; a *malignant tumor*, in contrast, is usually locally *invasive* and may also be *metastatic*, that is, it spreads through lymphatic and blood vessels to other parts of the body. Such malignant tumors are often referred to as *cancers*. Malignant tumors of epithelial cell origin are known as *carcinomas*, those arising from cells of mesenchymal origin as *sarcomas*, and those from leukocytes as *lymphomas* (if solid tumors) or *leukemia* (when circulating cells are involved). The process of development of tumors is termed *oncogenesis*, synonyms for which are *tumorigenesis* and *carcinogenesis*.

The study of oncogenesis at a molecular level was greatly facilitated after it became possible to reproduce those genetic changes in cultured cells that led to a phenotypic behavior resembling cancer—this process is known as *cell transformation*. This then allowed the study and manipulation of those genetic changes responsible. Transformation by DNA viruses is usually non-productive (i.e., the transformed cells do not yield infectious progeny virus, but may express virus-encoded antigens); cells transformed by non-defective retroviruses often express the full range of viral proteins and antigens, and new virions bud from cellular membranes. Viral (or proviral) DNA in transformed cells is usually integrated into the cell DNA, except in the case of papillomaviruses and herpesviruses in which viral DNAs usually remain episomal.

Transformed cells, like cells in malignant tumors, continue to divide under circumstances in which the normal cellular counterparts do not. Transformed cells differ from normal cells in several specific ways (Table 9.2, Fig. 9.1):

1. Transformed cells escape from the normal regulation of cell growth, division, differentiation, and lose the characteristic of cell–cell contact inhibition—the cells often continue to grow, piling up into dense aggregates. Normal primary cells reach a crisis phase and die after about 50 generations in culture, whereas transformed cells are "immortalized" and may be passaged indefinitely.

Fenner and White's Medical Virology. DOI: http://dx.doi.org/10.1016/B978-0-12-375156-0.00009-6

TABLE 9.1 Viruses Associated with Malignant Tumors in Humans

Virus Family	Virus	Type of Tumor
Retroviridae	HTLV-1, HTLV-2 viruses	Adult T cell leukemia
Papillomaviridae	Human papillomaviruses 5, 8	Cutaneous squamous cell carcinoma
	Human papillomaviruses 16, 18, others	Genital carcinoma
Hepadnaviridae	Hepatitis B virus	Hepatocellular carcinoma
Herpesviridae	Epstein-Barr virus	Burkitt's lymphoma
		Nasopharyngeal carcinoma
		B cell lymphomas
	Human herpes virus-8	Kaposi's sarcoma
Flaviviridae	Hepatitis C virus	Hepatocellular carcinoma

TABLE 9.2 Characteristics of Cells Transformed by Viruses

Viral DNA sequences are present, integrated into cellular DNA or as episomes

Greater growth potential *in vitro*, demonstrated by:
- Formation of three-dimensional colonies of randomly oriented cells in monolayer culture, usually due to loss of contact inhibition
- Capacity to divide indefinitely in serial culture
- Higher efficiency of cloning
- Capacity to grow in suspension or in semi-solid agar (anchorage independence)
- Reduced serum requirement for growth

Altered cell morphology

Altered cell metabolism and membrane changes

Chromosomal abnormalities

Virus-specified tumor-associated antigens; some at the cell surface behave as tumor-specific transplantation antigens

Capacity to produce malignant neoplasms when inoculated into isologous or severely immunosuppressed animals

2. The normal dependence on exogenous growth factors is lost, and transformed cells often acquire different, usually reduced nutritional requirements.

3. Anchorage-dependent growth characteristics are also often lost—many normal cells grow only when in contact with a specific substrate, but transformed cells may grow in "colonies" in agar.

4. Transformed cells grow in experimental animals—for example, some transformed cells produce tumors in athymic mice with defective T cell immunity or in transgenic knockout mice lacking certain immune response genes, whereas such mice do not support the growth of normal foreign cells.

5. A number of chromosomal abnormalities are found in transformed cells, such as aneuploidy (the chromosomal number is other than normal), whereas normal cells retain their usual chromosomal number and morphology.

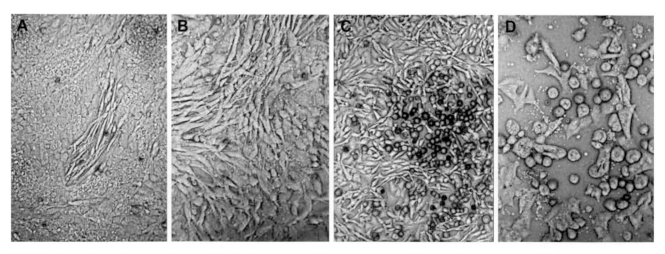

FIGURE 9.1   Foci formed by avian cells transformed with two strains of Rous sarcoma virus. Differences in morphology are due to genetic differences in the transduced *src* oncogene. (A) A focus of infected cells with fusiform morphology seen on a background of flattened, contact-inhibited uninfected cells. (B) Higher magnification of a fusiform focus showing lack of contact inhibition of the transformed cells. (C) A focus of highly refractile infected cells with rounded shape and reduced adherence. (D) Higher magnification of rounded infected cells, showing tightly adherent normal cells in background. *Originally from Peter Vogt, The Scripps Research Institute. Reproduced from Flint S.J., et al., Principles of Virology third ed., Vol. II, ASM Press, Washington, DC, with permission.*

6. Transformed cells may express distinctive antigens, called *tumor-associated antigens*, usually arising from restricted expression of the viral genome. Some tumor-associated antigens are located in the plasma membrane and as a result present potential targets for immunological attack: these antigens are sometimes referred to as *tumor-specific transplantation antigens*. A failure of immunological surveillance mechanisms, that is, a failure to eliminate cells expressing on their surfaces tumor-associated antigens is clearly important in the development of some tumors.

The cellular cancer phenotype results from mutations in, or aberrant expression of, one or more genes that encode transcripts, proteins involved in the regulation of cell growth, division, differentiation, and/or death. These mutations may result from chemical agents (carcinogens, mutagens), radiation, or viruses. There are two primary steps leading to oncogenic mutation: initiation and promotion. Initiation refers to the damaging event to the host cell DNA that is perpetuated in the form of mutation; promotion pertains to the effect of repeated exposure to the initiating factor (and cofactors), thereby preventing damaged cells from self-repair. Many initiators have been identified and respective risks assessed; however, promoters have been much more difficult to identify and evaluate. In this context viruses may be seen as initiators and also promoters. Viruses can act as mutagens, directly, either by introducing an oncogene into a cell or by inserting viral DNA into the genome of the host cell in ways that disrupt normal cellular regulatory functions. Viruses can also act as promoters because of a capacity to cause persistent infection, increased cell division, secondary genetic damage, and immunosuppression, thereby interfering with normal immune surveillance and repair functions. Study of the infections caused by oncogenic viruses has provided major insights into the nature of the pathways that regulate cell growth—and in turn, study of these pathways has provided major insights into the molecular cell biology of viral infections.

ONCOGENES AND TUMOR SUPPRESSOR GENES

An important element in our present understanding of oncogenesis has come from the discovery of *oncogenes*. These were originally found as discrete genes of retroviruses and collectively referred to as v-*onc* genes (by convention, v-*onc* genes, such as *src*, are designated in lower case and in italics; the resulting oncoproteins, such as Src protein, are designated in Roman script, with the first letter capitalized).

For each of the more than 60 v-*onc* genes so far identified there is either a corresponding normal cellular gene, which is referred to as a c-*onc* gene, or a *proto-oncogene*,

a cellular gene that is capable of functioning as an oncogene after mutation or altered expression. In normal cells, c-*onc* genes are involved in the regulation of cell growth, division, and differentiation. At some time in the past, retroviruses are believed to have acquired cellular genes (c-*onc*) possessing growth promoting or transforming activities. These have then continued to be propagated along with the virus as v-*onc* genes but are not essential for virus replication. This was discovered by Harold Varmus and J. Michael Bishop, in work that led to the award of the Nobel Prize in Physiology or Medicine (Table 1.1). The term oncogene is now applied broadly to any genetic element associated with cancer induction, including some cellular genes not known to have a viral homolog, and relevant genes of the oncogenic DNA viruses not having a cellular homolog.

The v-*onc* genes differ from c-*onc* genes in several significant ways:

1. v-*onc* genes usually contain only that part of the corresponding c-*onc* gene that is transcribed into messenger RNA—in most instances introns are lacking.

2. v-*onc* genes are separated from the cellular context that normally controls gene expression, including the normal promoters and other sequences regulating c-*onc* gene expression.

3. v-*onc* genes are under the control of the viral *long terminal repeats* (LTRs) that are not only strong promoters but are also influenced by cellular regulatory factors. For some retrovirus v-*onc* genes, such as *myc* and *mos*, the presence of viral long terminal repeats is all that is needed for tumor induction.

4. v-*onc* genes may undergo mutations (deletions and rearrangements), which alter the structure of oncogene protein products; such changes can interfere with normal protein–protein interactions leading to escape from normal regulation of cell division.

5. v-*onc* genes may be joined to other viral genes in such a way that associated functions are modified. For example, in Abelson murine leukemia virus the v-*abl* gene is expressed as a fusion protein with the viral gag protein; this arrangement directs the fusion protein to the plasma membrane where the Abl protein is functional. In feline leukemia virus the v-*onc* gene v-*fms* is also expressed as a fusion protein with the viral gag protein, thus allowing the insertion of the Fms protein into the plasma membrane.

As with the c-*onc* genes and gene products in normal cells, the mechanisms of action of the various v-*onc* genes and the expressed proteins, called *oncoproteins*, act in four major ways (Table 9.3).

1. Growth factors. One example is the oncogene v-*sis* which encodes one of the two polypeptide chains of platelet-derived growth factor (PDGF).

TABLE 9.3 Some Retroviral Oncogenes and Functions of Encoded Oncoproteins

Oncogene	Retrovirus	Subcellular Location of Oncogene Product	Nature of Oncoprotein
Growth Factors			
sis	Feline leukemia virus	Secreted and cytoplasm	Platelet-derived growth factor β chain
	Simian sarcoma virus		
Receptors			
Growth Factor and Hormone Receptors			
erbB	Avian erythroblastosis virus[a]	Plasma membrane	Truncated epidermal growth factor receptor
fms	Feline sarcoma virus	Plasma membrane	Mutated CSF-1 receptor
Intracellular Receptors			
erbA	Avian erythroblastosis virus[a]	Nuclear	Thyroxine receptor; activated form prevents differentiation
Intracellular Transducers			
Ras Proteins			
H-ras	Harvey murine sarcoma virus	Plasma membrane	Guanine nucleotide-binding proteins with GTPase activity
Ki-ras	Kirsten murine sarcoma virus	Plasma membrane	
Protein-Tyrosine Kinases			
src	Rous sarcoma virus	Cytoplasm	Protein kinases that phosphorylate tyrosine residues
fps[b]	Fujinami avian sarcoma virus	Cytoplasm	
fes[b]	Feline sarcoma virus	Cytoplasm	
abl	Feline and murine leukemia viruses	Cytoplasm	
Protein Serine/Threonine Kinases			
mos	Moloney murine sarcoma virus	Cytoplasm	Cytoplasmic serine kinase (cytostatic factor)
Nuclear Transcription Factors			
myc	Avian myelocytoma virus	Nuclear matrix	Binds to DNA; regulates transcription
myb	Avian myeloblastosis virus	Nuclear matrix	Binds to DNA; regulates transcription
fos	Murine osteosarcoma virus	Nucleus	Transcription factor AP-1
jun	Avian sarcoma virus	Nucleus	Transcription factor AP-1

Modified from Darnell, J.E., Lodish, H., Baltimore, D., "Molecular Cell Biology," second ed., p. 986. Scientific American Books, New York, 1990.
[a]Transducing retrovirus with two oncogenes.
[b]fps and fes are the same oncogene derived from avian and feline genomes, respectively.

2. Growth factor receptors or hormone receptors. In normal cells, growth factor receptors bind particular growth factors, thereby sending a growth signal to the cell nucleus. For example, the v-*erbB* oncogene product is a modified epidermal growth factor (EGF) receptor that retains tyrosine kinase activity. In normal cells this kinase becomes activated only after the binding of circulating epidermal growth factor to the receptor. However, in the presence of the v-*erbB* gene product, ErbB, the enzyme is permanently activated, phosphorylating any intracytoplasmic protein in the vicinity, thereby initiating a cascade of events culminating in transmission of conflicting signals to the nucleus. In some cells this results in uncontrolled growth.

The product of the v-*erbA* gene mimics the intracellular receptor for thyroid hormone. This oncoprotein, ErbA, competes with the natural receptor for the hormone, causing uncontrolled cellular growth. In avian erythroblastosis virus, v-*erbA* and v-*erbB* oncogenes act synergistically to cause uncontrolled cell growth.

3. Intracellular signal transducers. Most retrovirus v-*onc* genes mediate oncogenesis by altering signal transduction pathways. Signal transduction pathways link the cell surface to synthetic and regulatory functions in the nucleus. Taken together, the various v-*onc* genes encode proteins that may interfere at virtually any step along these pathways.

   A typical growth signal arrives at the surface of the cell in the form of a polypeptide growth factor (ligand) that binds to its specific receptor. The receptor, often an integral membrane protein with tyrosine kinase activity, is activated by binding the ligand; that is, its kinase activity is switched on, resulting in auto-phosphorylation of specific residues in the tail of the molecule protruding into the cytoplasm. The signal is further propagated by specific sequential protein/protein interactions (the signal-transduction cascade) involving numerous different intracytoplasmic proteins and eventually the signal reaches the nucleus. The ultimate recipients of the propagated signal are transcription factors that upregulate specific sets of genes and start a sequence of synthesis steps leading to cell growth.

   The best understood of the v-*onc* genes that act as signal transducers are Ha-*ras* and Ki-*ras*, the first non-viral oncogenes to be discovered. Most Ras protein molecules exist in an inactive state within the resting cell bound to guanosine diphosphate (GDP). Following a physiological stimulus via a transmembrane receptor the *ras* oncogenes are temporarily activated, leading to the synthesis of guanosine triphosphate, one of the key ingredients in DNA synthesis. The *ras* genes acquire transforming properties by mutation, mostly point mutations at specific sites, the consequence of which is to stabilize Ras proteins in an active state. This causes continuous signal transduction, leading to malignant transformation.

4. Nuclear transcription factors. By one mechanism or another, the activity of oncogenes eventually results in a change in gene expression in the cell nucleus. In most cases this effect is indirect, as noted above; however, some oncogenes encode proteins that bind to DNA or directly affect transcription. The v-*jun* gene product, Jun protein, can bind tightly to another nuclear oncoprotein, Fos, to form a heterodimer that is homologous to AP-1, an important transcription factor. Components of mitotic signal transduction cascades switch on v-*jun*, leading to binding to AP-1 sites in the host DNA within promoter and enhancer regions of target genes.

Tumor Suppressor Genes

In 1989 a completely different category of cellular genes called *tumor suppressor genes* was discovered; these genes play an essential regulatory role in normal cell division through negative regulation of growth. This regulatory role may be ablated by mutation in a tumor suppressor gene. However, because mutations in genes exerting a negative effect are recessive, both copies must be inactivated for excessive activity of a c-*onc* gene to occur and cancer to develop. Although it currently appears that at least half of all cancers may be associated with altered tumor suppressor genes, these genes may be only indirectly involved in most cancers caused by viruses, and then only at a late stage in the multi-step process that leads to full-blown malignancy. Examples are the retinoblastoma (Rb) and p53 tumor suppressor proteins, each of which blocks the progression of the cell cycle at G1. The p53 protein, in response to various signals, also plays a role in triggering programmed cell death (apoptosis).

Some early proteins of DNA viruses have a dual role in viral replication and cell transformation. With a few possible exceptions, the oncogenes of DNA viruses have no homologs or direct ancestors (c-*onc* genes) among cellular genes of the host. The protein products of these oncogenes are multifunctional, having particular functions which mimic functions of, or bind to, normal cellular proteins. Adenovirus E1B oncoprotein binds specifically to the N-terminal activation domain of p53, thereby converting the cellular protein from an activator to a suppressor of transcription. Thus, it inhibits the action of the p53 tumor suppressor protein and thereby allows an otherwise resting (in G1) cell to enter S-phase. The expression of the viral E1B gene in a cell is the equivalent of a mutation in the p53 gene—it renders p53 protein inactive and this leads to uncontrolled cell growth. Similarly, the E6 and E7 proteins of human papillomaviruses bind to p53 and Rb respectively, leading to reduced activity of these tumor suppressor proteins.

Tumor suppressor genes and proteins are likely to be as important in the pathogenesis of cancer as oncogenes. There has been a long-standing wish that despite the heterogeneity of cancer cells and the triggers that lead to oncogenic transformation, there might be common mechanisms involved—tumor suppressor genes and proteins may represent this common ground. Whereas oncogenic transformation may follow the *activation* of a particular oncogene, the *inactivation* of tumor suppressor genes might be a more general pathway associated with oncogenic transformation.

Multi-Step Oncogenesis

The development of full malignancy requires multiple steps. A potentially neoplastic clone of cells must bypass apoptosis (programmed death), circumvent the need for growth signals from other cells, escape from immunological surveillance, organize a blood supply, and possibly metastasize. Thus, tumors other than those induced by rapidly transforming retroviruses, for example, Rous sarcoma virus, generally do not arise as the result of a single event, but by a series of steps leading to progressively greater loss of regulation of cell division. Significantly, the genomes of some retroviruses carry two different oncogenes and polyomaviruses three. Two or more distinct oncogenes are activated in certain human tumors (e.g., Burkitt's lymphoma).

Co-transfection of normal rat embryo fibroblasts with a mutated c-*ras* gene plus the polyomavirus large PyT gene, or with c-*ras* plus the *E1A* early gene of oncogenic adenoviruses, or with v-*ras* plus v-*myc*, converts fibroblasts into tumor cells. It should be noted that, whereas v-*ras* and v-*myc* are typical v-*onc* genes, originally of c-*onc* origin, the other two had been assumed to be typical viral genes before an oncogenic role was identified. Furthermore, a chemical carcinogen can substitute for one of the two v-*onc* genes; following immortalization of cells *in vitro* by treatment with a carcinogen, transfection of a cloned oncogene can convert the cloned continuous cell line into a tumor cell line. To achieve full malignancy, mutations in tumor suppressor genes may also be needed.

These points resurrect earlier unifying theories of cancer causation that viewed viruses as analogous to other mutagenic carcinogens, any one of which is capable of initiating a chain of two or more events leading eventually to malignancy. If viruses or oncogenes are considered as co-carcinogens in a chain of events culminating in the formation of a tumor, it may be important to determine whether these function as initiator or promoter, or both. The most plausible hypothesis may be that (1) c-*onc* genes represent targets for carcinogens such as chemicals, irradiation, and other tumor viruses, and (2) the full expression of malignancy may generally require the mutation or enhanced expression of more than one class of oncogene, and perhaps also mutations in both copies of critical tumor suppressor genes.

ONCOGENESIS BY RETROVIRUSES

Although retroviruses are a major cause of leukemias and lymphomas in many species of animals, including cattle,

cats, non-human primates, mice, and chickens, only two retroviruses, human T lymphotropic viruses 1 and 2 (HTLV-1 and 2), are known to cause leukemia in humans.

There are two types of retroviruses, endogenous (carried within the host genome) and exogenous (transmissible) retroviruses. Most endogenous retroviruses never produce disease, cannot transform cultured cells, and do not contain an oncogene. Many exogenous retroviruses, on the other hand, are oncogenic; some characteristically induce leukemias or lymphomas, others sarcomas, and yet others induce carcinomas, usually displaying a predilection for a particular target cell.

The replication cycle of retroviruses is described in Chapter 23: Retroviruses, but certain aspects associated with the integration of the DNA copy of the RNA genome into the cellular DNA are described here in the context of oncogenesis. Upon viral entry into the cytoplasm, the single-stranded RNA genome is converted to double-stranded DNA by the virion reverse transcriptase. This is then integrated into the chromosomal DNA as provirus. The expression of mRNA from the provirus is under the control of the viral transcriptional regulatory elements (promoter and enhancer elements) located in the long terminal repeats. In some ways, the proviral DNA behaves similarly to other chromosomal genes, segregating into daughter cells during mitosis and in some cases gaining access to the germ line of the host.

Based on the presence or absence of an oncogene (v-*onc*) in the viral genome, and on the cellular sites where v-*onc* becomes integrated, retroviruses fall into three general categories in terms of their capacity to produce tumors (Table 9.4)—transducing retroviruses, *cis*-activating retroviruses, and *trans*-activating retroviruses.

Transducing Retroviruses

The viral genomes of *transducing retroviruses* carry one or more v-*onc* genes and upon infection introduce these genes into the chromosome of the cell. Cell transformation follows rapidly both *in vitro* and *in vivo*. Transducing retroviruses are rapidly oncogenic; Rous sarcoma virus is the most rapidly acting—transforming cultured cells in a day or so and causing death in chickens within two weeks

TABLE 9.4 Oncogenesis by Retroviruses—Three Categories

	Transform Cells in Culture	Replication Competent	Time to Tumor Development	v-*onc* Gene Present	Mechanism of Action
Transducing retroviruses	Yes	Usually not	Rapid (weeks)	Yes	Direct v-*onc* gene action
Cis-activating retroviruses	No	Yes	Moderately slow (months)	No	Regulate c-*onc* gene expression
Trans-activating retroviruses	No	Yes	Slow (years)	No	Viral protein modulates cell gene expression

after infection. These properties are attributable to the v-*src* gene carried as part of the viral genome.

Retrovirus v-*onc* genes are not essential for viral replication; rather, these genes have been acquired over time by the viruses and have been selected for, most likely because these genes cause cellular transformation, which in turn favors viral growth and perpetuation in nature. Because the oncogene is usually incorporated into the viral RNA genome in place of part of one or more of the normal viral genes (*gag, pol,* or *env*), such viruses are usually defective, that is, the transforming retroviruses are dependent upon co-infection with non-defective helper retroviruses for replication, the latter supplying missing functions, such as an environmentally stable envelope. The advantage to both viruses is presumably that when together the duality can infect a wider range of cells and produce more progeny of both viruses than would be the case independently. The exception is Rous sarcoma virus, which has a viral oncogene (v-*src*) in addition to complete functioning *gag, pol*, and *env* genes.

Cis-Activating Retroviruses

The *cis-activating retroviruses* lack a v-*onc* gene but become integrated into the host cell DNA close to a c-*onc* gene and take over the regulation of this gene. These retroviruses are replication-competent, and induce tumors after a much longer incubation period by causing over expression or inappropriate expression of a c-*onc* gene, for example, in the wrong cell or at the wrong time. They do not transform cells in culture. For example, avian leukosis viruses produce a lifelong viremia with no associated clinical disease. Proviral insertional mutagenesis is an important mechanism whereby replication-competent retroviruses can initiate tumors. In infected animals, a large number of cells acquire new proviral DNA inserts and each insertion constitutes a somatic mutation; thus, retrovirus infection can be regarded as similar to a massive exposure to a potent mutagen. Though most of the insertions are harmless, on the rare occasion when a provirus integrates near a gene that controls growth, it can lead to inappropriate expression of a host gene; the cell can then proliferate and ultimately form a clonal tumor in which all cells contain the provirus integrated at the same site. Many cellular genes have been identified as potential targets for insertional activation in retrovirus-induced tumors.

These include transcription factors c-myc, N-myc, c-myb, and Spi1 (PU.1); a number of secreted growth factors, such as Wnt1 (Int1), Wnt3 (Int4), Int2 (Fgf3), and Fgf8; growth factor receptors, including c-erbB, Int3 (Notch4), Mis6 (Notch1), c-fms (Fim2), the prolactin receptor, and genes implicated in intracellular signal transduction pathways, such as the serine or threonine kinases Pim1 and Pim2.

There are at least four mechanisms whereby cellular oncogene activity may be modified by *cis*-activating retroviruses:

1. Oncogene Activation via Insertional Mutagenesis

    The presence upstream from a c-*onc* gene of an integrated provirus, with its strong promoter and enhancer elements, may greatly amplify the expression of the c-*onc* gene. This is the likely mechanism of tumor induction by the weakly oncogenic avian leukosis viruses lacking a v-*onc* gene. The viral genome is generally found to be integrated at a particular location immediately upstream from a c-*onc* gene when avian leukosis viruses cause malignant tumors. The integrated avian leukosis provirus increases the synthesis of the normal c-*myc* gene product, Myc protein, 30- to 100-fold. Experimentally, only the long terminal repeat needs to be integrated to cause this effect; furthermore, by this mechanism c-*myc* may be expressed in cells in that it is not normally expressed in.

2. Oncogene Activation via Transposition

    Transposition of c-*onc* genes may result in an enhanced expression of c-*onc* genes by bringing the sequences under the control of strong promoter and enhancer elements. For example, the 8:14 chromosomal translocation that characterizes Burkitt's lymphoma (a tumor associated with Epstein-Barr herpesvirus infection: Chapter 17: Herpesviruses) brings the c-*myc* gene into position immediately downstream of a strong immunoglobulin promoter; v-*onc* genes may be transposed from the initial site of integration in a similar way, thereby causing uncontrolled cell proliferation.

3. Oncogene Activation via Gene Amplification

    Amplification of oncogenes is characteristic of many tumors; for example, a 30-fold increase in the number of copies of the c-*ras* gene is found in one cell line derived from a human carcinoma and the c-*myc* gene is amplified in several human tumors. The increase in gene copy number leads to a corresponding increase in the amount of oncogene product, thus producing cancer.

4. Oncogene Activation via Mutation

    Direct mutation within a c-*onc* gene, for example, c-*ras*, may alter the function of the corresponding oncoprotein. Such mutations can occur either *in situ* as a result of chemical or physical mutagenesis, or in the course of recombination with integrated retroviral DNA. Given the high error rate of reverse transcription, v-*onc* gene homologs of c-*onc* genes will always carry mutations, and the strongly promoted production of the viral oncoprotein will readily exceed that of the normal cellular oncoprotein. The result can be uncontrolled cell growth.

It is worth noting that insertional activation of a proto-oncogene by a provirus is not sufficient to fully transform a cell, but merely represents one step in a progression to neoplasia. Other point mutations in other proto-oncogenes, or loss of function mutations in tumor suppressor genes, are generally also needed. In addition, gene inactivation is also an important event in some tumors. Retrovirus insertions also can frequently disrupt gene expression. When the target gene is a tumor suppressor gene, this insertion will promote tumorigenesis.

Trans-Activating Retroviruses

The *trans-activating retroviruses* contain genes that code for regulatory proteins which modulate transcription of virus and host cell genes; these retroviruses either do not possess oncogenic activity or induce tumors very late through modulating cellular transcription.

Human T lymphotropic virus 1 (HTLV-1), associated with adult T cell leukemia, provides an example of this third mechanism. HTLV-1 is a replication-competent exogenous retrovirus that is spread horizontally as a typical infectious agent and acts as a *trans*-activating retrovirus.

This was the first human retrovirus discovered (in 1980); its geographical distribution in Japan, the Caribbean, and central Africa was found to correspond with that of an uncommon adult T cell leukemia-lymphoma (ATLL). It is now clear that HTLV-1 is an ancient virus and has been maintained successfully in rather isolated communities in various parts of the world. It is transmitted from mother to baby across the placenta and through breast milk, and between adults by sexual contact. Virions are not abundant, being mainly cell-associated, and the efficiency of transmission is apparently too low for long-term viral survival in any but such communities. In recent years, however, viral spread has been augmented by blood transfusion and needle-sharing by drug users as well as by broader sexual contact. The virus persists for life, generally causing no disease. However, after a remarkably long incubation period (20 to 40 years) up to 4% of individuals develop adult T cell leukemia-lymphoma (Chapter 23: Retroviruses).

Tumor cells carry the HTLV-1 provirus integrated in a monoclonal fashion, that is, into the same site within a particular chromosome in every tumor cell. However, the integration site is different in each infected individual. The virus can also be demonstrated to infect and immortalize CD4+ T cells *in vitro*.

In addition to the standard three retroviral genes, *gag*, *pol*, and *env*, HTLV-1 carries several different regulatory genes, two of which, *tax* and *HBZ*, also play a crucial role in transformation. Elucidating the underlying mechanism of action of these genes has proved difficult, partly due to the blending of primary and secondary activation events observed. *HBZ* is consistently expressed in all ATL cell lines and ATL patient T cells, suggesting it plays an essential role, while continued expression of *tax* is not necessary for maintenance of malignancy. HBZ inhibits tax-mediated 5′ LTR transcription, and modulates cell signaling pathways involved in cell growth, T cell differentiation and immune responses. HTLV-2 infection has not been associated with ATL, and the protein of HTLV-2 homologous to HBZ (known as APH-2) has certain differences in its activities (see also Chapter 23: Retroviruses). The protein product of *tax* is a transcriptional activator that promotes transcription not only under control of the proviral long terminal repeat (LTR) but also controlled by the regulatory sequences of cellular genes. By acting on the enhancer for the gene encoding the T cell growth factor interleukin-2, and also on the enhancer for the IL-2 receptor gene, the tax protein establishes an autocrine loop promoting lymphocyte proliferation. The *tax* gene also upregulates two cellular oncogenes, *Fos* and *PDGF*, thereby initiating a cascade of processes favoring cellular proliferation. As with many oncogenic viruses, it is still not fully understood why it takes up to 40 years for malignancy to become manifest *in vivo*, and why in such a small proportion of those known to be infected with the virus. Is immunosurveillance initially effective, or is there an unknown co-factor required for the full flowering of malignancy?

Recently, another unique mechanism of retroviral tumor induction has been revealed with Jaagsiekte sheep retrovirus (JSRV), the causative agent of a contagious lung cancer in sheep that shares similarities with human *in situ* pulmonary adenocarcinoma (AIS, previously called bronchioloalveolar carcinoma, BAC). The envelope *(env)* gene of JSRV can transform cells in culture and induce tumors in animals. The phosphatidylinositol 3-kinase (PI3K)–Akt–mTOR and H/N-Ras–MEK–mitogen-activated protein kinase (MAPK) pathways have been shown to be critical targets for transformation by the Env protein of this virus.

ONCOGENESIS BY SMALL DNA VIRUSES

Although retroviruses are the most important oncogenic viruses, certain DNA viruses are also important as known causes of cancers (Table 9.1). Small DNA tumor viruses interact with cells in one of two ways: (1) productive infection, in which the virus completes its replication cycle, resulting in cell lysis, or (2) non-productive infection, in which the virus transforms the cell without completing its replication cycle. During such non-productive infection, the viral genome, or a truncated version of it, is integrated into the cellular DNA. Alternatively, the complete genome persists as an autonomously replicating plasmid (episome). The genome continues to express early gene functions.

Interest in small DNA viruses as oncogenic agents can be traced back 50 years to work on the polyomavirus SV40, followed by studies of the molecular basis of oncogenesis by other polyomaviruses, papillomaviruses, and adenoviruses,

TABLE 9.5 Oncogenes of Adenoviruses, Papillomaviruses, and Polyomaviruses and Associated Outcomes

Virus	Gene Product	Molecular Target	Activities
Human adenoviruses	E1A	Binds to Rb protein	Induces inappropriate entry of cell into S-phase
	E1B	Binds to p53 and inactivates it	Prevents p53-induced cell cycle arrest and apoptosis
HPV 16 & HPV 18	E6	Binds to p53 protein and cell ubiquitin protein ligase	Prevents p53-induced cell cycle arrest and apoptosis
	E7	Binds to Rb protein	Induces inappropriate entry of cell into S-phase
Bovine papillomavirus 1	E5	Binds to receptor for platelet-derived growth factor	Transforms fibroblasts in culture; induces fibropapillomas
Murine polyomavirus	LT	Binds and sequesters p53 and Rb proteins	Induces inappropriate entry of cell into S-phase
	mT	Binds to c-Src protein, increasing its protein kinase activity Binds to protein phosphatase 2a	Transforms rodent cell lines; induces endotheliomas in transgenic animals
Simian virus 40	LT	Binds and sequesters p53 and Rb proteins	Induces inappropriate entry of cell into S-phase
	sT	Binds to protein phosphatase 2a	Stimulates synthesis of cyclins D1 and A; required for cell transformation

all of which contain genes that behave as oncogenes in a variety of ways. Table 9.5 summarizes several of these genes and respective proteins. This field is complex, as many of these proteins are multifunctional, with targets frequently involving complex pathway cascades, similar protein products from related viruses may have evolved different mechanisms of action, and determining which exact activity is responsible for oncogenesis is not always straightforward.

With a few possible exceptions, these oncogenes of DNA viruses have no homologs or direct ancestors (c-*onc* genes) among cellular genes of the host.

In every case the relevant viral oncogenes encode early proteins with a dual role in both viral replication and cell transformation. The protein products of these oncogenes are multifunctional, with certain functions that mimic functions of normal cellular proteins due to the presence of common domains in the folded protein molecules. These may interact with host cell proteins at the plasma membrane or within the cytoplasm or nucleus. For example, polyoma middle T protein (Py-mT) interacts with c-Src protein, resulting in an increased level of the protein kinase activity of src protein.

A most striking mechanism of oncogenesis by small DNA tumor viruses is the ability of some viral oncogenes to bind to and functionally compromise tumor suppressor genes. This was first shown for the interaction of the EIA oncoprotein of adenoviruses with the retinoblastoma gene product (pRb). The binding of large T protein of SV40 to another tumor suppressor gene (p53), and the discovery that the 55 kDa product of the adenovirus EIB gene and the E6 protein of high-risk HPV types, also bind to p53 disabling its function, are further examples of oncogenes targeting and inactivating cellular tumor suppressor gene products leading to tumor development. In addition, the viral oncoproteins are multifunctional and may target other cellular proteins; numerous other cellular signal transduction signals can be altered to create neoplastic phenotypes. Other interactions between virus and host cells have been discovered, for example, studies of SV40 and murine polyomavirus transcription led to the discovery of DNA-based transcriptional enhancer elements. Studies with E6 and p53 identified p53 as the first mammalian substrate of the ubiquitin-dependent degradation system.

Cellular transformation *in vitro* has been employed to study the potential role of small DNA viruses in carcinogenesis. It should be stressed that integration of viral DNA does not necessarily lead to transformation. Many or most episodes of integration of polyomavirus or adenovirus DNA have no recognized biological consequence. Transformation by these viruses in experimental systems is a rare event, requiring that the viral transforming genes be integrated in the location and orientation needed for efficient expression. Even then, many transformed cells revert and are considered as having undergone abortive transformation. Furthermore, cells displaying the characteristics of transformation in cell culture do not necessarily produce tumors in animals.

Tumors Induced by Papillomaviruses

Most human papillomaviruses produce benign papillomas (warts) on the skin and mucous membranes. These benign tumors are hyperplastic outgrowths and generally regress spontaneously. However, certain types of human papillomaviruses initiate changes that may progress to malignancy. There is evidence that cofactors may be required for this process.

The etiological association between human papillomaviruses and carcinoma of the cervix has been established by a number of lines of evidence, including (1) epidemiological studies, (2) demonstration of plausible molecular mechanisms, and (3) the reduction in cancer incidence following introduction of vaccination against papillomaviruses. Certain human papillomaviruses, notably papillomaviruses 16 and 18, induce cervical dysplasia, which may progress to invasive carcinoma. The viral genome in the form of an unintegrated, autonomously replicating episome is regularly found in the nucleus of pre-malignant cells in cervical dysplasia. In contrast, invasive cervical cancers reveal chromosomally integrated DNA of human papillomaviruses 16 or 18. Each cell carries at least one, and sometimes up to hundreds, of monoclonal incomplete copies of the human papillomavirus genome (see Chapter 19: Papillomaviruses).

Integration disrupts one of the early viral genes, E2, and other genes may be deleted, but the viral oncogenes E6 and E7 remain intact and are expressed efficiently. Transfection of cultured cells with the human papillomavirus 16 E6/E7 genes alone results in immortalization; co-transfection with the *ras* oncogene allows full expression of malignancy. Further, the proteins encoded by the E6/E7 genes from the highly oncogenic human papillomaviruses 16 and 18, but not from non-oncogenic human papillomaviruses, have been demonstrated to bind to the protein products of the human tumor suppressor genes p53 and Rb. In HPV-positive patients with certain p53 genetic polymorphisms, more rapid *progression* from pre-malignant to malignant cervical disease has been demonstrated, but this effect was not seen in HPV-negative disease nor with *initiation* of pre-malignant cervical lesions.

Most cutaneous human papillomaviruses cause benign skin warts that do not turn malignant. However, there is a rare autosomal recessive condition, epidermodysplasia verruciformis, caused by mutation of two adjacent genes EVER1/TMC6 and EVER2/TMC8 that codes for the transmembrane proteins within the endoplasmic reticulum. Affected children develop multiple disseminated red scaly patches on the skin; over 20 different rare HPV types have been isolated from these lesions. Later, some of these lesions tend to undergo a malignant change. These squamous cell carcinomas, but generally not those found in immunocompetent persons, are often found to carry as episomes the genome of human papillomaviruses 5 or 8, the E6 gene of which displays greater transforming activity than that of other dermatotropic types. The fact that malignant change occurs much more commonly in warts on exposed areas of skin strongly suggests that ultraviolet light is a co-carcinogen in a succession of events culminating in malignancy.

Tumors Induced by Polyomaviruses and Adenoviruses

During the 1960s and 1970s two members of the family *Polyomaviridae*, murine polyomavirus and simian virus 40 (SV40), as well as certain human adenoviruses (types 12, 18, and 31), were found to induce malignant tumors following inoculation into baby hamsters and other rodents. Although, with the exception of murine polyomavirus, none of these viruses induce cancer under natural conditions in a natural host, these viruses transform cultured cells of certain other species and provide good experimental models for analysis of the molecular events in cell transformation. However, despite extensive search for human tumors caused by these viruses it was not until 2008 that Merkel cell polyomavirus (MCPyV) was isolated from Merkel cell carcinoma tissues and became the first candidate virus in this group etiologically implicated in a human tumor.

Polyomavirus- or adenovirus-transformed cells do not produce virus. Virus can be rescued from polyomavirus-transformed cells; that is, virus can be induced to replicate by irradiation, treatment with certain mutagenic chemicals, or by co-cultivation with certain types of permissive cells. This cannot be done with adenovirus-transformed cells, as the integrated adenovirus DNA contains substantial deletions.

Polyomavirus and adenovirus DNA is integrated at multiple sites in the chromosomes of transformed cells. Most of the integrated viral genomes are complete in the case of polyomaviruses, but defective in the case of adenoviruses. Only certain early viral genes are transcribed, albeit at an unusually high rate. By analogy with retrovirus genes, these are now also referred to as oncogenes. The products of these genes, demonstrable by immunofluorescence, were previously known as *tumor (T) antigens*. A great deal is now known about the role of these proteins in transformation. For example, the polyomavirus "large-T" (Py-T) and "middle-T" (Py-mT) antigens, which are required for several stages in virus replication (DNA replication, transcription, and virus assembly), also act on host cell functions including cell proliferation, cell death, and inflammation. Py-mT enables the cells to grow in suspension in semi-solid agar medium as well as on solid substrates (anchorage independence); Py-T inactivates the p53 and Rb tumor suppressor genes (as do the E6 and E7 genes of human papillomaviruses), and is responsible for the reduction in dependence of the cells on serum and an enhanced life span of cells in culture.

ONCOGENESIS BY LARGE DNA VIRUSES

Herpesviruses of the subfamily *Gammaherpesvirinae* cause lymphomas and carcinomas in hosts ranging from amphibians, through birds, to primates including humans.

The human herpesvirus Epstein-Barr virus (EBV) is found in nearly all Burkitt's lymphoma tumors from patients in central Africa where the disease is most common, but in only 15 to 20% of Burkitt's lymphomas in other parts of the world. EBV is also involved in various B cell lymphomas and Hodgkin's disease. EBV-associated nasopharyngeal carcinoma (NPC), while uncommon generally in other countries, is the commonest cancer in the densely populated regions of southern China and in Cantonese populations who have settled in other parts of the world, as well as in Arctic Inuit people and the peoples of East and North Africa. Another human herpesvirus, namely herpesvirus-8 or Kaposi sarcoma (KS)-associated herpesvirus, is linked to Kaposi's sarcoma and two other rare lymphoproliferative disorders, primary effusion lymphoma and multicentric Castleman's disease. Another herpesvirus, Marek's disease virus, transforms T lymphocytes, causing T cells to proliferate and produce a generalized lymphomatosis in chickens. Although some poxviruses are regularly associated with the development of benign tumor-like lesions in animals, there is no evidence that these can ever become malignant, nor is there evidence that poxvirus DNA is ever integrated into cellular DNA.

Epstein-Barr Virus and Burkitt's Lymphoma

Burkitt's lymphoma (Fig. 9.2) is a malignant B cell lymphoma of high prevalence among children in tropical Africa at elevations below 1500 m where malaria is endemic—it also occurs in other regions of the world, notably in a broad band across the tropics and subtropics, but at much lower prevalence. Epstein and his colleagues succeeded in isolating from cultured tumor cells the herpesvirus now known as Epstein-Barr virus (EBV), which more commonly causes infectious mononucleosis in the developed world.

The regular association between EBV and Burkitt's lymphoma suggested a cause-and-effect relationship. Yet the virus is not associated with most of the sporadic cases of Burkitt's lymphoma found outside Africa or New Guinea, whereas in areas of high prevalence it is a ubiquitous virus that causes subclinical infection in most children. Hence, it was conceivable that the association was fortuitous. However, a long-term prospective seroepidemiological survey in Uganda firmly established that EBV infection very early in life predisposes to Burkitt's lymphoma. In particular, a high antibody titer against EBV capsid antigen (VCA), arising early in childhood, indicates a high risk of subsequently developing Burkitt's lymphoma.

The EBV genomic DNA is present in multiple copies in each cell of most African Burkitt's lymphomas, in the form of closed circles of the complete viral DNA molecule, found free as autonomously replicating episomes; however the cells do not produce virus unless induced to do so following cultivation *in vitro*. Burkitt lymphoma cells *in vivo* express the EBV nuclear antigen EBNA1 (latency I;

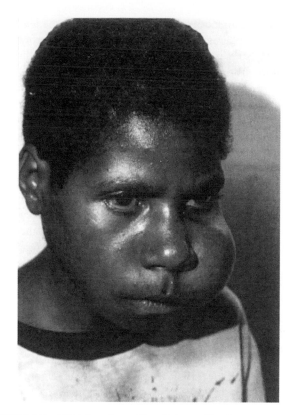

FIGURE 9.2 Child from Papua New Guinea with a jaw tumor that on biopsy was confirmed as a case of Burkitt's lymphoma. *Reproduced from Cooke, R.A., Infectious Diseases, McGraw-Hill, with permission.*

see Chapter 17: Herpesviruses), while when cultured many such cells progress to express all six EBNAs and the latent membrane proteins LMP1 and LMP2 (latency III). Notably, the two major antigens seen by cytotoxic T cells on the surface of cells acutely infected with EBV, namely EBNA 2 and LMP, are suppressed in Burkitt's lymphoma cells *in vivo*, thus suppressing induction of, and recognition by, CTLs. Unlike asymptomatic carriers, children with Burkitt's lymphoma make antibodies to the viral antigens EA and MA, as well as unusually high titers of antibodies against VCA.

The malignant cells also contain a characteristic 8:14 chromosomal translocation. The human c-*myc* oncogene, located on chromosome 8, is transposed to one of three chromosomes that contain genes for immunoglobulins (usually chromosome 14, sometimes 2 or 22), leading to enhanced transcription of c-*myc*. Transgenic mice with the rearranged c-*myc* locus also develop B cell lymphomas. Some Burkitt's lymphomas also have mutations in the cellular tumor suppressor gene p53. Although the details are unclear, Burkitt's lymphoma may develop as a consequence of differentiation arrest due to this translocation, and subsequent growth stimulation resulting from c-*myc* deregulation. In parallel, host defense mechanisms may be weakened by suppression of EBV membrane antigen expression and by blocking programmed cell death mediated by tumor necrosis factor. Moreover, the tendency for tumors to arise in non-lymph nodal tissue, particularly the jaw, suggests additional clonal selection through migratory properties or local growth factors.

George Klein proposed that Burkitt's lymphoma develops in three stages: (1) chronic Epstein-Barr virus infection arrests B cell differentiation and stimulates division, thereby enhancing the probability of chromosomal damage, (2) an environmental cofactor, postulated to be infection with the malaria parasite *Plasmodium falciparum*, impairs the capacity of cytotoxic T cells to control this proliferation of EBV-immortalized B cells, (3) chromosomal translocation leads to constitutive activation of the c-*myc* oncogene, resulting in Burkitt's lymphoma. There is also evidence that short non-coding RNAs (miRNAs) have important functions in B cell maturation and the biology of lymphomas; the regulation of cellular miRNAs by EBV expression, and the role of EBV miRNAs, are areas of ongoing investigation.

Nasopharyngeal Carcinoma and Epstein-Barr Virus

The anaplastic form of nasopharyngeal carcinoma that occurs in certain Asian and African populations (but not the well-differentiated type more commonly seen in other parts of the world) contains multiple copies of EBV DNA, in episomal form, in every cell. As in the case of Burkitt's lymphoma, these malignant epithelial cells do not synthesize virus except either on transplantation into athymic mice, or co-cultivation *in vitro* with B lymphocytes. Interestingly, EBV normally fails to replicate in or transform cultured epithelial cells, which seem to lack receptors for the virus, but it can if EBV or its genomic DNA is introduced into epithelial cells artificially. It has been postulated that EBV produced by B lymphocytes trafficking through the copious lymphoid tissue in the pharynx binds to secretory IgA antibody, allowing infection of mucosal epithelial cells via the IgA transport pathway. Nasopharyngeal carcinoma cells produce EBNA1 and LMP2A/B antigens, and EBER RNAs. There is evidence that inappropriate expression of LMP antigen in an undifferentiated epithelial cell may be important in the development of nasopharyngeal carcinoma.

The host age at the time of primary infection with EBV does not seem to be as critical in nasopharyngeal carcinoma as it is in Burkitt's lymphoma, and the interval between infection and development of nasopharyngeal carcinoma can be up to 40 years. Reactivation of a latent EBV infection in the nasopharynx, with a consequent rise in anti-VCA antibodies of the IgA class, frequently heralds the development of the tumor. The striking ethnic restriction of EBV nasopharyngeal carcinoma, retained when Cantonese people emigrate, suggests either a genetic predisposition or a cultural (e.g., dietary) cofactor. Around 5 to 15% of gastric adenocarcinomas contain EBV DNA and express EBNA1, LMP2A, and EBERs, as do some carcinomas of other sites (salivary gland, lung, thymus).

Lymphoproliferative Disorders (LPD) and Other Tumors

B cell lymphomas in immunocompromised patients, clinically very different from Burkitt's lymphoma, sometimes follow EBV primary infection or reactivation; for example, in transplant patients, in AIDS patients (typically non-Hodgkin's lymphomas), or in children with the rare X-linked lymphoproliferative syndrome. The molecular pathogenesis of these connections has not been as comprehensively analyzed, although the gene responsible for X-linked lymphoproliferative syndrome (SH2DIA) is known to code for a 128-amino acid protein involved in signal transduction leading to apoptosis in T lymphocytes.

EBV DNA and a latency II pattern of genome expression are also present in about 35 to 50% of cases of Hodgkin's lymphoma. This disease occurs most commonly in young adults, particularly those whose childhood social environment would foster susceptibility to late (post-adolescent) infection with EBV. The mechanism of EBV involvement may involve infected cells inappropriately surviving from apoptosis in germinal centers through expression of LMP antigens.

Human Herpesvirus 8

Kaposi's sarcoma associated herpesvirus (KSHV) or HHV-8 is associated with Kaposi's sarcoma in patients both with and without human immunodeficiency virus (HIV) infection. Evidence for KSHV has also been found in body-cavity-based lymphomas, suggesting that the virus may play a role in causing this unusual subgroup of malignant lymphomas. Body-cavity-based lymphomas are effusions of lymphocytes that occur in the lung cavity, in the pericardium, or the abdominal cavity of some patients with advanced HIV disease. A number of HHV-8-encoded proteins including LANA 1, LANA 2, V-cyc, VFLIP, and vGPCR, have been shown to have complex activities leading to cell proliferation, enhanced immune evasion, and the inhibition of tumor suppressor proteins. Furthermore, KSHV encodes at least 12 microRNAs, some of which have been linked to regulatory roles in cellular processes such as adhesion, migration, and angiogenesis.

Tumors Induced by Poxviruses

Although some poxviruses are regularly associated with the development of benign tumor-like lesions in animals, there is no evidence that these ever become malignant, nor is there any evidence to suggest that poxvirus DNA is ever integrated into cellular DNA. A very early viral protein produced in poxvirus-infected cells displays homology with epidermal growth factor, and is probably responsible for the epithelial hyperplasia characteristic of many poxvirus infections. For some poxviruses (e.g., fowlpox, orf, and rabbit fibroma viruses), epithelial hyperplasia is a dominant clinical manifestation and may be a consequence of a more potent form of the poxvirus epidermal growth factor.

OTHER IMPORTANT VIRUSES ASSOCIATED WITH CANCERS

Hepadnaviruses

Although primary hepatocellular carcinoma (HCC) is relatively uncommon in most developed countries (<5 cases per 100,000 per year), it ranks as one of the commonest cancers (>20 cases per 100,000 per annum) in China, Southeast Asia, and sub-Saharan Africa, and is currently the third leading cause of cancer deaths (>500,000 per annum) worldwide. In high incidence parts of the world five to 20% of the population are persistently infected with hepatitis B virus (HBV) and perinatal transmission to offspring ensures perpetuation of the pool of chronic carriers, now totaling 250 to 300 million—this will change, but quite slowly, as more and more people are vaccinated across the developing world. Landmark prospective studies by Palmer Beasley and co-workers in Taiwan showed that virus carriers are approximately 100 times more likely to develop liver cancer than uninfected individuals, that the virus infection precedes the development of cancer, that the risk does not persist in individuals who clear the virus and develop antibody, and that the cancer risk is specific to the liver only. HBV infection early in life typically leads to chronic hepatitis, cirrhosis of the liver, and eventually to liver cancer 20 to 50 years post-infection.

In over 80% of cases of hepatocellular carcinoma, part of the HBV genome is found integrated into host cell chromosomes—the viral genome is integrated at five to seven sites scattered throughout the host cell genome. As HBV does not code for an integrase, cellular enzymes such as topoisomerase I are presumably involved. Integration generally occurs first during chronic infection, but decades before cancer becomes apparent. The HBV genome does not contain an oncogene; however, it does contain a gene, HBx, which may facilitate transformation, as it stimulates transcription of many growth-activating host cell genes (e.g., c-*myc* and c-*fos*) and possibly inhibits cellular growth suppressor proteins. Deregulated over-expression of HBx and viral surface proteins is often found in the early stages of hepatocellular carcinoma.

The integrated HBV DNA usually contains deletions, duplications, inversions, and mutations. Later, further recombination events occur, producing inverted duplications of flanking cellular sequences, often accompanied by further deletion of viral sequences. This suggests that HBV integration somehow promotes genetic instability in the cell. Most integrated viral sequences, although truncated, retain at least part of the X gene, including the X-promoter and the viral enhancer. It has been postulated that specific activity of X gene products, together with a chromosomal instability due to long-standing infection, could deregulate expression of nearby cellular oncogenes. In addition, active hepatocellular regeneration associated with cirrhosis of the liver, which frequently develops in persistent HBV infection, has been shown to increase the sensitivity to environmental carcinogens and also promote the development of tumors.

Environmental carcinogens such as dietary aflatoxins that have been shown to increase the risk of liver cancer, and mutagenic oxidants generated by phagocytic and other cells during chronic inflammation, may also play a role.

Hepatitis C Virus

It is a paradox that chronic infection with the flavivirus hepatitis C virus (HCV), an RNA virus with no DNA intermediate in its life cycle, no known oncogenes, and without any evidence of integration, is at least as important a precursor to hepatocellular carcinoma as is HBV in many developed countries. Currently a molecular hypothesis is absent other than the cellular proliferation that enables the liver to regenerate remarkably following the destruction of hepatocytes, also maximizes opportunities for the sequential development of chromosomal abnormalities leading to cancer; this sequence may be initiated and/or promoted by genetic abnormalities arising from long-standing chronic infection with HCV, and may be aggravated by alcoholism, a known risk factor, carcinogens, and oxidants as described for hepatitis B.

Human Immunodeficiency Virus (HIV)

HIV has not been regularly isolated from tumor cells, but people with advanced disease, for example, AIDS, often develop characteristic tumors. Kaposi's sarcoma, once common but now less so in HIV disease, is thought to arise through a combination of the effects of herpesvirus 8 infection and immunosuppression. Patients with very low CD4+ cell counts may also develop other types of malignancy, notably aggressive B cell lymphoma or non-Hodgkin's lymphoma, as well as anogenital warts and genital cancers, particularly cervical cancer. These are thought to be due to weakened immunological surveillance of cancer precursor cells arising in the context of latent infection with other oncogenic viruses (EBV or HPV).

THE PREVENTION AND MANAGEMENT OF VIRALLY INDUCED TUMORS

The above knowledge has allowed the realization of one of the aspirations of medical research—vaccines against cancer. Indeed, the introduction of hepatitis B vaccine in the early 1980s provided the scientific tool to prevent the great majority of newly transmitted hepatitis B infections, and thereby eliminate the etiological factor in a large number of liver cancers worldwide. Proof of this dream was soon demonstrated—in Taiwan the rate of primary liver cancer in childhood has fallen markedly since the adoption of universal infant HBV immunization in 1984, and similar findings have been observed elsewhere. This incidentally also provides the ultimate evidence for causality, namely that reducing the rate of an infection results in a reduction in the associated disease.

More recently, vaccination against human papillomaviruses has quickly led to a major reduction in pre-cancerous lesions of the female cervix. However, applying these successes on a worldwide scale will take time—there is the matter of cost, public health infrastructure and organization, and societal/political will (Table 9.6).

It has proved more problematic to use this knowledge to develop a satisfactory treatment for established cancers. In a majority of examples, active virus replication is not relevant to maintenance of the cancer, and antiviral drugs are therefore of no value. However, attempts are being made to develop treatments, for example, for cervical carcinoma, by vaccinating patients against individual viral oncoproteins that have been shown to be involved in maintenance of the malignant state.

TABLE 9.6 Oncogenic Viruses: How This Knowledge Can Be of Clinical Benefit

Virus	Prevention of Cancer	Screening	Treatment of Precursor Condition
HBV	Vaccine	If HBsAg positive, α-fetoprotein testing and ultrasound	Antivirals
	Reduce transmission		
HCV	Reduce transmission	If cirrhosis present, α-fetoprotein testing and ultrasound	Antivirals
HPV	Vaccine	Yes—Pap smear and HPV DNA	Yes—surgery
	Reduce transmission		
EBV	None routine	No	No
HTLV-I	Screen blood donations	Not routine	No

FURTHER READING

Flint, S.J., Enquist, L.W., Racaniello, V.R., Skalka, A.M., 2009. Principles of Virology, third ed. Transformation and Oncogenesis, Vol. II. ASM Press.. Ch 7.

Rickinson, A.B., 2014. Coinfections, inflammation and oncogenesis; future directions for EBV research. Sem Cancer Biol 26, 99–115.

Chapter 10

Laboratory Diagnosis of Virus Diseases

Arriving at an accurate virus diagnosis on clinical information alone is unreliable in most cases. Exceptions include the setting of known epidemics where large numbers of cases may present with a similar clinical picture; during the smallpox eradication campaign, photographs of the distinctive rash of smallpox were distributed to allow field workers to institute immediate containment measures. However, accurate virus diagnosis nearly always requires laboratory testing. The past few decades have seen a major revolution in the operation of virus diagnostic laboratories, and their role in the clinical management of patients has become vital. This has been driven by a number of factors:

1. The development of molecular technologies and ways to make these accessible for diagnostic laboratories;
2. The plethora of new antiviral agents and a more sophisticated understanding of how these should be used; and
3. An increased awareness of the clinical value of, and demand for, prompt information about viral loads, viral sequence data, and antiviral resistance information.

The modern virus diagnostic laboratory is characterized by high test throughputs, rapid turnaround times, and a close liaison with clinical staff. Many of the older and slower diagnostic approaches such as animal inoculation, virus isolation in cell culture, and serological demonstration of a four-fold rise in antibody titer, are now of minor importance, if practiced at all.

RATIONALE FOR PERFORMING LABORATORY VIRUS DIAGNOSIS

1. The appropriate management of the patient depends on knowledge that follows from the diagnosis. Specific management does not always include, or stop at, chemotherapy. A rapid, specific, and accurate diagnosis can often avoid the need for further unnecessary tests, limit unnecessary or continuing antibiotic use (e.g., in cases of meningitis), ensure quicker management of patients, allow a more informed prognosis, and save on costs. Furthermore, specific active management measures may be indicated. The management of

needlestick injuries to medical staff, screening of donated organs before transplantation, and application of blood isolation precautions for particular surgical procedures, are all highly dependent on the testing for blood-borne viruses. Cesarean section may be advisable if a woman has primary genital herpes at the time of delivery. Special care and education are required for a baby with congenital defects attributable to rubella or cytomegalovirus. Pregnancy termination is recommended if rubella is diagnosed in the first trimester of pregnancy.

2. *Use of antiviral agents* usually requires accurate laboratory identification of the infecting virus. In some chronic diseases, for example, those caused by HIV and hepatitis B virus, decisions about when to initiate therapy require a full understanding of clinical and laboratory findings. Drug combinations appropriate for each patient are selected and later modified based on the drug resistance profile of the infecting virus strain. Results of virus load testing are used to monitor the response to therapy and guide changes to therapy.

3. Infections may demand *public health measures* to prevent spread to others. For instance, blood banks routinely screen for HIV and hepatitis B and C viruses that may be present in blood donated by symptomless carriers. Since herpes simplex virus type 2 is readily transmissible to sexual partners, in some settings contact tracing helps protect sexual partners. Nosocomial infections (e.g., varicella, measles), often in epidemic form, may create havoc in a leukemia ward of a children's hospital, unless hyperimmune IgG is promptly administered to potential contacts following diagnosis of the sentinel case. Documentation of a novel strain of influenza virus may herald the start of a major epidemic against which vulnerable older members of the community should be immunized. Positive identification of a particular arbovirus in a case of encephalitis enables authorities to promulgate warnings and initiate appropriate mosquito control measures. Introduction of a dangerous exotic disease demands containment and surveillance, and so on.

4. *Surveillance* of viral infections may shed light as to the significance, natural history, and prevalence of a virus in the community, allowing control measures to

Fenner and White's Medical Virology. DOI: http://dx.doi.org/10.1016/B978-0-12-375156-0.00010-2

be designed, control priorities to be established, and immunization programs to be monitored and evaluated.

5. Continuous surveillance of a community may provide evidence of *new epidemics, new diseases, new viruses*, or *new virus–disease associations*. New viruses and new virus–disease associations continue to be discovered every year. It should be stressed that over 90% of all the human viruses known today were completely unknown at the end of World War II. Opportunities are legion for astute clinicians as well as pathologists, virologists, and epidemiologists to be instrumental in such discoveries.

The traditional approaches to laboratory diagnosis of viral infections have been (1) *direct detection* in patient material of virions, viral antigens, or viral nucleic acids, (2) *isolation of virus* in cultured cells, followed by identification of the isolate, and (3) detection and measurement of antibodies in the patient's serum (*serology*). In recent decades direct detection methods, capable of providing a definitive answer in less than 24 hours, have undergone major advancement, whereas virus serology has become restricted to particular purposes only. Virus isolation is now used infrequently outside research or specialist areas. A summary of the major strengths and limitations of the several alternative approaches to the diagnosis of viral infections is given in Table 10.1.

The best diagnostic methods should satisfy the criteria of: speed, simplicity, sensitivity, specificity, and cost. Standardized diagnostic reagents of reliable quality are widely

TABLE 10.1 Advantages and Disadvantages of Various Diagnostic Methods

Diagnostic Method	Advantages	Disadvantages/Problems
Virus isolation	Produces further material for study of agent	Slow, time-consuming, can be difficult and expensive
	Usually highly sensitive	Selection of cell type, etc., may be critical
	"Open-minded"	Useless for non-viable virus or for non-cultivable agents
Direct observation by electron microscopy	Rapid	Relatively insensitive
	Detects viruses that cannot be grown in culture	Cumbersome for large numbers of samples
	Detects non-viable virus	Limited to a few infections
	"Open-minded"	
Serological identification of virus or antigen, for example, EIA	Rapid and sensitive	Not applicable to all viruses
	Provides information on serotypes	Interpretation may be difficult
	Readily available, often as diagnostic kits	Not as sensitive as PCR
		Targeted to a specific agent
Detection of viral genomes by PCR	Rapid, very sensitive	High sensitivity may lead to detection of non-relevant co-infections
	Potentially applicable to all viruses incl. non-cultivable	Risk of DNA contamination
	Reagents (primers) for additional viruses easily made	Needs good quality control
	Good quantitation of load	Targeted to a specific agent
	Can be multiplexed	
Antibody seroconversion (acute and convalescent sera)	Useful if appropriate samples for direct detection cannot be obtained, or to exclude a particular infection retrospectively	Slow, late (retrospective)
		Interpretation may be difficult
		Targeted to a specific agent
IgM serology	Rapid	False positives may occur
		Targeted to a specific agent

available commercially, assays have been miniaturized to conserve reagents, and instruments have been developed to automate the barcoding and dispensing of samples, diluting and rinsing, reading the tests, and computerized analysis and delivery of the results. Moreover, a veritable cascade of commercial kits for nucleic acid tests, antigen and antibody assays is now available, together with the relevant equipment to allow automated high-throughput assays and ongoing quality assurance. Solid-phase enzyme immunoassays (EIAs) and real-time PCR technology, in particular, have revolutionized diagnostic virology and are now methods of choice for a wide range of indications.

COLLECTION, PACKAGING, AND TRANSPORT OF SAMPLES

The chance of successful diagnosis by virus isolation or direct detection of virus depends critically on the attention given by the attending physician to the collection of the sample. Clearly, the clinical sample must be taken from the right place at the right time. The right time is as soon as possible after the patient first presents, because virus is usually present in maximum concentration at about the time symptoms first develop, and then declines during the ensuing days. Samples taken as a last resort after days or weeks of empirical treatment (e.g., antibiotic therapy) are almost invariably useless.

The site from which the specimen is collected will depend on the clinical symptoms and signs, together with knowledge of the pathogenesis of the suspected disease. As a general rule the epithelial surface that constitutes the portal of entry and the primary site of viral replication is usually the best site for obtaining samples (Table 10.2). The key specimen in respiratory infections as well as in many generalized infections, is a nasal or throat swab, or in the case of a young child a nasopharyngeal aspirate in which mucus is drawn from the back of the nose and throat into a mucus trap using a vacuum pump. A sample of feces is also essential in enteric and many generalized infections of the gastrointestinal tract.

Swabs may be taken from the genital tract, from the eye, or from vesicular skin lesions. Some viruses responsible for systemic infections can be isolated from blood (plasma or leukocytes). Cerebrospinal fluid (CSF) may yield virus in cases of meningitis or encephalitis. Biopsy or autopsy specimens may be taken by needle or knife from any part of the body for virus isolation, or snap-frozen for immunofluorescence. Obviously, tissue taken at autopsy or biopsy for the purpose of virus isolation must not be placed in formalin or any other fixative. In the case of many generalized viral diseases it may not be obvious as to what specimen is required. As a rough working rule it can be said that at a sufficiently early stage in the disease, virus can usually be isolated from a throat swab, feces, or blood. In

TABLE 10.2 Specimens Appropriate for Laboratory Diagnosis of Various Clinical Syndromes

Syndrome	Specimen
Respiratory	Nasal or throat swab; nasopharyngeal aspirate; sputum
Enteric	Feces
Genital	Genital swab, urine
Eye	Conjunctival (and/or corneal) swab
Skin	Vesicle fluid/swab/scraping; biopsy solid lesion
Central nervous system	Cerebrospinal fluid; feces (enteroviruses)
Generalized	Throat swab[a]; feces[a]; blood leukocytes[a]
Autopsy/biopsy	Relevant organ
Any	Blood for serology[b]

[a]Depending on known or presumed pathogenesis.
[b]Blood is allowed to clot, then serum kept for assay of antibody.

all cases, it is essential to capture sufficient patient material on the (preferably moistened) swab; negative diagnoses are to be expected if a medical attendant fails to collect enough material from the patient.

Because of the lability of many viruses, specimens intended for virus isolation must always be kept cold and moist. Immediately after collection the swab should be swirled around in a small screw-capped bottle containing virus transport medium. This medium consists of a buffered balanced salt solution, to which has been added protein (e.g., gelatin or albumin) to protect the virus against inactivation, and antibiotics to prevent the multiplication of bacteria and fungi. (If it is at all probable that the specimen will also be used for attempted isolation of bacteria, rickettsia, chlamydia, or mycoplasmas, the collection medium must not contain antibiotics—the portion used for virus isolation can be treated with antibiotics later.) The swab stick is then broken off aseptically into the fluid, the cap is tightly fastened and secured with adherent tape to prevent leakage, and the bottle is labeled with the patient's name, date of collection, and nature of specimen. This is then dispatched immediately to the laboratory, accompanied by a completed laboratory request form which must include an adequate clinical history, a provisional diagnosis, request for a particular test, and the date of onset of the illness.

If a transit time of more than an hour or so is anticipated the container should be sent refrigerated (but not frozen), with cold packs (4°C) or ice in a thermos flask or styrofoam box. International or interstate transport of specimens, particularly in hot weather, generally requires that the container be packed in dry ice (solid CO_2) to maintain the virus in a frozen state. Governmental and International Air Transport Association (IATA) regulations relating to the transport of biological materials require precautions such as double-walled containers with absorbent padding in case of breakage. Permits must be obtained from the appropriate authorities for interstate and international transportation. Especially, hazardous samples are subject to additional regulations, differing in different countries.

Nucleic acid tests have revolutionized diagnostic virology. The sensitivity of nucleic acid assays means that previously unsuitable specimens can now be used (e.g., deep nasal swabs or throat swabs can be examined for respiratory viruses instead of requiring nasal pharyngeal aspirates or tracheal aspirates); in addition, the quality of the collection is less important (self-collection is suitable for human papilloma virus testing) and transport is less critical as the nucleic acid in viral particles is relatively stable (e.g., respiratory samples in VTM can be sent through the mail, and dried blood spots can be used). Nucleic acid tests have increased the range of detected viruses, but the down side is that testing is specific for the suspected virus, with the consequence that variants may be missed unless the assays are designed to be inclusive.

DIRECT IDENTIFICATION OF VIRUS, VIRAL ANTIGEN, OR VIRAL GENOME

Direct detection of virus material in a patient sample has become the method of choice for a very large number of different infections. These methods have the advantage of speed, as they do not rely on virus culture or a rise in antibody titer. The sensitivity and specificity of antigen detection methods have increased greatly through use of high-quality reagents and solid phase reagent supports allowing ease of washing and signal detection; the sensitivity of nucleic acid detection has advanced from early dot blot and Southern blot assays to the widespread and reliable use of PCR.

Direct Detection of Virions by Electron Microscopy

The morphology of most viruses is sufficiently characteristic to allow assigning many viruses to the correct family by appearance in the electron microscope. Moreover, viruses that have never been cultured may nevertheless be recognized. During the 1970s electron microscopy was the means to the discovery in feces of several new groups of previously non-cultivated viruses. The human rotaviruses, caliciviruses, astroviruses, hepatitis A virus, and previously unknown types of adenoviruses and coronaviruses were all initially identified in this way.

The most widely used procedure is negative staining, where virus-containing fluid is placed on a carbon or formvar-coated grid usually with prior clarification and concentration by ultracentrifugation; virions adhere to the surface and become "negatively stained" when an electron-dense fluid such as sodium phosphotungstate is added and surrounds the virions. There are several variations on the method, mainly to remove excess salts and proteins which affect the translucency of the specimen in the microscope. Characteristic virion size and structural details provide useful information toward virus identification when routine diagnostic approaches fail. Pre-incubation of the virus specimen with specific antibody can be used to assist in virus identification—antibody-coated virions appear fuzzier than controls (see Fig. 33.1), and the procedure (immune electron microscopy) can also be used for quantitating specific antibody.

A second procedure is thin-section electron microscopy of fixed tissue sections. This is used particularly in a research setting, for example, where viral-like inclusions suggest a viral etiology or where the pathology of a particular disease is being investigated.

The biggest limitation of electron microscopy as a diagnostic tool is its low sensitivity; specimens need to contain at least 10^6 virions per milliliter or milligram for there to be any chance to detect the virus by electron microscopy. Such levels are often surpassed in feces and vesicle fluid, but not in respiratory mucus. The method can be quick for one or several samples only, but is impractical for large batch testing.

Detection of Viral Antigens

A great variety of different immunological assays can be used to detect viral antigens in patient samples, using well-characterized reagents containing known antibody specificities. Where the antigen-containing reagent is defined, these same assay techniques can be used similarly to detect unknown antibodies (see below).

Enzyme Immunoassay (EIA)

The introduction of EIA, also known as ELISA (for enzyme-linked immunosorbent assay), revolutionized diagnostic virology in the days before the development of polymerase chain reaction (PCR), and still has widespread specific uses. EIAs can be designed in different formats to detect antigen or antibody. The high sensitivity of the method means that less than 1ng of viral antigen per milliliter can be detected in specimens taken directly from a patient.

The development of solid-phase assays was a major technical advance; the available test configurations are almost limitless, and a wide variety of direct, indirect, and reversed assays have been developed for diagnosing viral infections. The critical principle is that all the components making up the layers of the reaction are known and defined,

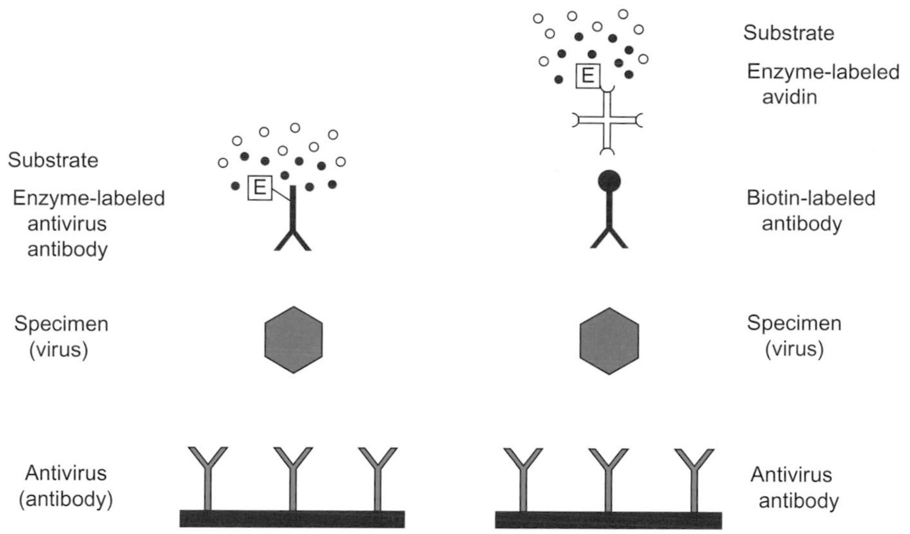

FIGURE 10.1   Enzyme immunoassays (EIA or ELISA) for detection of virus and/or viral antigen. *Left*: Direct method. *Right*: Indirect method, using biotinylated antivirus antibody, followed by enzyme- (e.g., peroxidase)- labeled avidin. In each case an enzyme substrate is then added to develop a color reaction. Note the immobilization of the capture antibody on a solid support to facilitate subsequent washing steps. *Reproduced from MacLachlan, N.J., Dubovi, E.J., 2011. Veterinary Virology, fourth ed., Academic Press, with permission.*

with the exception of the "unknown" component that may be present in the specimen. For the solid-phase format, the "capture" antibody can be attached (by simple adsorption or by covalent bonding) to a solid substrate, typically the wells of polystyrene or polyvinyl microtiter plates, or polystyrene beads, this format aiding considerably the rinsing of the solid surface between applications of reagents.

The simplest format is the direct EIA (Fig. 10.1). Virus and soluble viral antigens from the specimen are allowed to adsorb to the captured antibody. After unbound antigen has been washed away, an enzyme-labeled antiviral antibody (the "detector" antibody) is added. (Various enzymes can be linked to antibody; horseradish peroxidase and alkaline phosphatase are the most commonly used.) After a final washing step, readout is based on the color change that follows addition of an appropriate organic substrate for the particular enzyme. The colored product of the action of the enzyme on the substrate should be clearly seen by eye. The test can be made quantitative by serially diluting the antigen to obtain an endpoint, or by using spectrophotometry to measure the amount of enzyme-conjugated antibody bound to the captured antigen. Sensitivity can be enhanced using the high binding affinity of avidin for biotin; antibody is conjugated to biotin, a reagent of low M_r that gives reproducible labeling and does not alter the antigen-binding capacity of the detector antibody. The antigen–antibody complex is identified by adding avidin-labeled enzyme, followed by enzyme substrate (Fig. 10.3, right). To further increase sensitivity, high-energy substrates are available which release fluorescent, chemiluminescent, or radioactive products that can be identified in very small amounts.

Indirect immunoassays are widely used because of a somewhat greater sensitivity and the avoidance of the need to label each antiviral antibody in the laboratory repertoire. In one protocol, the detector antibody is unlabeled, and a further layer, labeled (species-specific) anti-immunoglobulin, is added as the "indicator" antibody; of course, the viral antibodies constituting the capture and detector antibodies must be raised in different animal species. Alternatively, the detector antibodies can be labeled with biotin and a positive reaction indicated using enzyme-labeled avidin. In another method, labeled staphylococcal protein A, which binds to the Fc moiety of IgG of most mammalian species, can be used as the indicator in indirect immunoassays.

Monoclonal viral antibodies (mAbs) are frequently used in immunoassays as detector antibodies, but the use of mAbs as capture antibodies is limited because of the restricted epitope repertoire recognized by such cloned antibody molecules and thus antigen capture is less stable. The obvious advantages of mAbs are that these represent the well-characterized, purified, monospecific antibody of a defined class, recognizing only a single epitope, as well as being free of "natural" and other extraneous antibodies against host antigens or adventitious agents. Monoclonal antibodies also make a vital contribution as reference reagents, being easily manufactured in large quantities. It is important to select mAbs of high affinity, but not of such high specificity that some strains of the virus being sought might be missed in the assay. Indeed, the specificity of the assay can be pre-determined by selecting a mAb directed at an epitope that is either confined to a particular viral serotype or common to all serotypes within a given species or genus.

Radioimmunoassay and Time-Resolved Fluoroimmunoassay

Radioimmunoassay (RIA) predates EIA. The only significant difference is that the label is not an enzyme but a radioactive isotope such as ^{125}I, and the bound antibody is measured in a gamma counter. RIA is a highly sensitive and reliable assay that lends itself well to automation, but the cost of the equipment and the health hazard of working with radioisotopes argue against its use in small laboratories.

Time-resolved fluorescence immunoassay (TR-FIA) is a non-isotopic immunoassay, in which the indicator antibody is labeled with a fluorophore (a europium chelate). Following excitation by light, the fluorophore emits fluorescence of a different wavelength that can be measured in a time-lapse fluorometer. The method is as sensitive as RIA and has the advantage of a stable fluorophore, but it requires costly equipment that can be contemplated only by laboratories with a large throughput of specimens.

Latex Particle Agglutination

Perhaps the simplest of all immunoassays is the agglutination by antigen of small latex beads previously coated with antiviral antibody. The test can be read by eye within minutes. Not surprisingly, diagnostic kits based on this method have become popular with small laboratories and with medical practitioners. However, these assays suffer from both low sensitivity and low specificity. Thus, false negatives occur unless large numbers of virions are present, and therefore this assay for antigen tends to be restricted to examination of feces. False positives, on the other hand, can occur commonly with fecal specimens.

Immunochromatography

Immunochromatography or lateral flow tests involve the migration of an antigen, or antigen–antibody complexes, through a support, for example, nitrocellulose film, filter paper, or agarose. In a typical format, a labeled antibody is allowed to react with the unknown antigen, the antigen–antibody complex migrates by capillary action through the solid support, and a second antibody embedded in the solid support allows detection of the complexes if antigen is present. Positive and negative controls are included to ensure that individual tests are valid. Immunochromatography tests are available for measuring viral antigens such as HIV p24, dengue NS1, influenza A and B, RSV, etc. and are of particular value for rapid point-of-care testing where rapid results are needed and access to equipment is limited.

Detection of Antigen in Tissues

Viral (or non-viral) antigens can be detected in fixed or frozen tissue sections, or in exfoliated cells, using the same principles as above, where the cells or tissues on a glass slide fill the role of the solid support. Commonly used indicators include fluorescein, horseradish peroxidase, and alkaline phosphatase. The approach can be valuable in virus diagnosis, for example, by detecting viral antigens in exfoliated cells or tissue samples, and also in studying disease pathogenesis should the presence of viral material be matched to particular organs or histopathology.

Once antibody is labeled with a fluorochrome, the antigen–antibody complex emits light of a particular longer wavelength when excited by short wavelength light. This can be visualized as fluorescence in an ordinary microscope after light of all other wavelengths is filtered out. The sensitivity of the method is generally too low to detect complexes of fluorescent antibody with virions or soluble antigen; hence, the antigen in the test typically takes the form of virus-infected cells. There are two main variants of the technique.

Direct Immunofluorescence

For direct immunofluorescence, a frozen tissue section, or an acetone-fixed cell smear, or monolayer on a coverslip, is incubated with fluorescein-tagged antiviral antibody (Fig. 10.3, A, left). Unbound antibody is then washed away, and the cells are viewed by light microscopy using a powerful ultraviolet/blue light source. The apple-green light emitted from the specimen is revealed (against a black background) by incorporating filters into the eyepieces so that all the blue and ultraviolet incident light is adsorbed (Fig. 10.2).

Indirect Immunofluorescence

Indirect ("sandwich") immunofluorescence differs in that the antiviral antibody is untagged, but fills the role of the meat in the sandwich. It binds to antigen and is itself recognized by fluorescein-conjugated anti-immunoglobulin (Fig. 10.3, A, right). The high affinity of avidin for biotin can also be exploited in immunofluorescence, by coupling biotin to antibody and fluorescein to avidin.

In the diagnostic setting, immunofluorescence has proved to be of great value in the early identification of viral antigens in infected cells taken from patients with diseases known to have a relatively small number of possible etiological agents. There is little difficulty in removing partly detached infected cells from the mucous membrane of the upper respiratory tract, genital tract, eye, or from the skin, simply by swabbing or scraping the infected area with reasonable firmness. Cells are also present in mucus aspirated from the nasopharynx; aspirated cells must be extensively washed by centrifugation to remove mucus before fixation and staining. Respiratory infections with paramyxoviruses, orthomyxoviruses, adenoviruses, and herpesviruses are particularly amenable to rapid diagnosis by immunofluorescence.

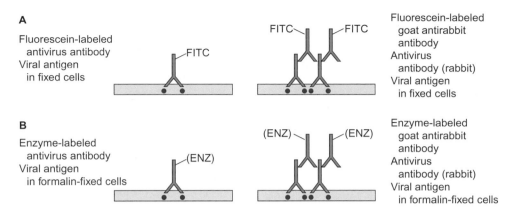

FIGURE 10.2 Detection of viral antigens in tissues or cell smears by immunofluorescence (A) and immunohistochemistry (B). In each case, the direct method is shown on the left, and the indirect method on the right. *Reproduced from MacLachlan, N.J., Dubovi, E.J., 2011. Veterinary Virology, fourth ed., Academic Press (Figure 5.6), with permission.*

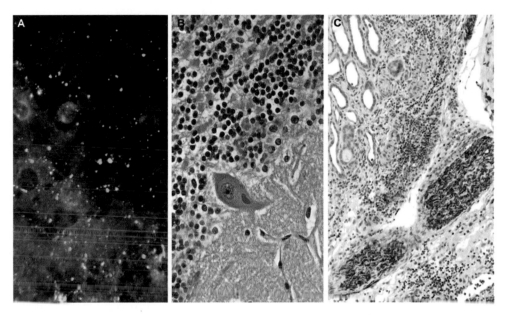

FIGURE 10.3 Diagnosis of rabies. (A) Direct immunofluorescence on impression smear of fox brain (possible human contact), showing prominent brilliant apple-green masses of viral antigen (Negri bodies of light microscopy) and punctate "dust," against a dark background. FITC-labeled anti-rabies globulin, UV microscopy. (B) Histopathology on human postmortem brain section. Finding Negri bodies (shown here in the cytoplasm of a Purkinje cell in the cerebellum) often requires careful searching. H&E. (C) Immunohistochemistry used in rabies research and in special diagnostic situations, for example, for postmortem diagnosis in patients infected via organ transplants. Here viral antigen fills the axoplasm of peripheral nerves. Human rabies, biotin-conjugated anti-rabies globulin, avidin-alkaline phosphatase, and naphthol fast-red method.

Immunofluorescence can also be applied to infected tissue, for example, brain biopsies for the diagnosis of such lethal diseases as herpes simplex encephalitis or measles SSPE, or at necropsy, for the verification of rabies in the brain of animals captured after biting humans (Fig.10.3).

Immunoperoxidase Staining

An alternative method of locating and identifying viral antigen in infected cells is to use antibody coupled to horseradish peroxidase; subsequent addition of hydrogen peroxide together with a benzidine derivative forms a colored insoluble precipitate. The advantages of the method are that the preparations are permanent and only an ordinary light microscope is needed. Endogenous peroxidase, present in a number of tissues, particularly leukocytes, can produce false positives, but this problem can be circumvented by proper technique and adequate controls. Other enzymes, for example, alkaline phosphatase, can also be used as indicator systems (Fig. 10.4).

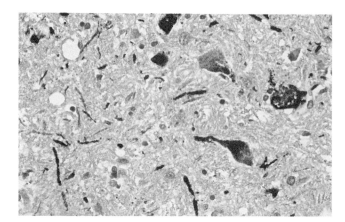

FIGURE 10.4 Immunohistochemistry showing West Nile virus antigens in a brain tissue section from a fatal case of encephalitis. Cortical neurons are particularly involved. Paraffin-embedded section. Biotin-conjugated West Nile virus specific hyperimmune globulin conjugated with avidin-alkaline phosphatase; naphthol fast-red method. Counterstained with hematoxylin. *Reproduced from Sherif Zaki, U.S. Centers for Disease Control and Prevention, with permission.*

Detection of Viral Nucleic Acids

Nucleic acid detection is now widely applied by use of polymerase chain reaction (PCR) assays, combined with advances in oligonucleotide synthesis, standardized automated procedures for nucleic acid extraction, and real-time detection of PCR products. Rapid advances in nucleic acid sequencing technology and the availability of sequence databases have greatly enhanced analysis of the results obtained. Isothermal RNA and DNA amplification assays are also readily available. Signal amplification assays were used previously but have been superseded.

Nucleic Acid Hybridization

From the time of the development of the first nucleic acid hybridization techniques, a variety of test formats were applied to viral nucleic acid detection. Early approaches usually involved a hybridization reaction between an immobilized nucleic acid target and a labeled probe, washing away unbound probe, followed by subsequent detection of bound probe. These include dot-blot assays using nucleic acid-containing samples immobilized onto filters; Southern blot hybridization, where viral nucleic acids are separated by electrophoresis according to molecular weight, blotted to a filter and detected by hybridization; and *in situ* hybridization applied to infected tissue sections or exfoliated cells. Real-time PCRs nowadays often use hybridization in solution as part of the product detection strategy.

Traditionally, radioactive isotopes such as ^{32}P and ^{35}S were used to label nucleic acids or oligonucleotides used as probes for hybridization tests, with the signal being read by counting in a spectrometer or by autoradiography.

These have been largely replaced by non-radioactive labels. Some of these (e.g., fluorescein or peroxidase) produce a signal directly, whereas others (e.g., biotin or digoxigenin) act indirectly by binding another labeled ligand that then emits the signal. Biotinylated probes can be combined with various types of readouts, for example, an avidin-based EIA. Chemiluminescent substrates, such as luminol, have also been widely exploited.

Polymerase Chain Reaction

PCR (Fig. 10.5) constitutes one of the greatest advances in molecular biology. It enables a single copy of any gene sequence to be enzymatically amplified *in vitro* at least a million-fold within a few hours. Thus viral DNA extracted from a very small number of virions or infected cells can be amplified to the point where it can be readily identified. PCR can also be used to detect viral RNA by including a preliminary step in which reverse transcriptase is used to convert RNA to DNA. It is not necessary or usual to amplify the whole genome, but it is necessary to know at least sufficient nucleotide sequence in order to synthesize two oligonucleotide primers, usually about 20 residues in length, that hybridize to opposite strands of the target DNA and flank the region one chooses to amplify. The two primers (sometimes known as "forward" and "reverse" primers) provide the DNA polymerase with an initiation point to which additional nucleotides can be attached, and also attach the reaction to the specific DNA target region. Primers can be synthesized containing attached ligands or molecular tags, thereby generating tagged DNA product molecules to facilitate further analysis. Computer programs can be used to design optimum primer sets and to predict optimal PCR reaction conditions, for example, time/temperature/ionic conditions. Where the target might be expected to show variability at a particular site, "degenerate" primers containing different bases at that site can be synthesized to ensure all variants are detected.

The process is carried out under carefully controlled conditions of temperature, ionic strength, primer concentration, and nucleotide concentration. There are three main steps: (1) melting the target DNA at 95°C, (2) cooling to around 50–60°C to allow binding of two oligonucleotide primers, and (3) synthesis by extension from the oligonucleotide primers of two DNA strands located between and including the two primers, thus generating complementary copies of the original (plus and minus) target strands. The synthesis reaction is catalyzed by DNA polymerase added to the reaction. The primer extension products from the first cycle act as templates in a second reaction, and the cycle of melting, primer binding, and primer extension are repeated many times. The number of DNA copies increases exponentially. Thus after 30 cycles the number of DNA copies, beginning with a single copy of the target sequence, is theoretically 2^{30} or 1 billion, and

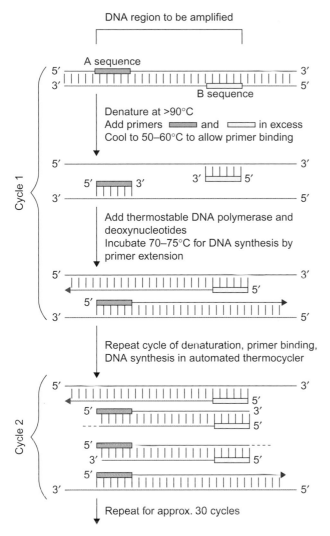

DNA region to be amplified

Cycle 1

A sequence

5′ 3′
3′ 5′

B sequence

Denature at >90°C
Add primers and in excess
Cool to 50–60°C to allow primer binding

Add thermostable DNA polymerase and deoxynucleotides
Incubate 70–75°C for DNA synthesis by primer extension

Repeat cycle of denaturation, primer binding, DNA synthesis in automated thermocycler

Cycle 2

Repeat for approx. 30 cycles

FIGURE 10.5 Principle of DNA sequence amplification using the polymerase chain reaction. The method relies on thermal cycling, that is cycles of repeated heating and cooling to melt and re-anneal DNA along with enzymatic replication of the targeted melted strands of DNA. Primers (short DNA fragments) containing sequences complementary to each end of the target region of the DNA of interest must first be synthesized. Then, the two strands of the DNA double helix are physically separated at a high temperature in a process called DNA melting. In the second step, the temperature is lowered and the two DNA strands become templates for DNA synthesis. In the presence of heat-resistant DNA polymerase and deoxynucleotide triphosphates, two new copies of the desired region of the DNA are produced. As PCR progresses through cycles of melting, annealing, and extension, the DNA generated is itself used as a template for replication, setting in motion a chain reaction in which the DNA template is exponentially amplified. After the first few cycles, virtually all the templates consist of just the short region chosen for amplification by the choice of primers. After 30 cycles, this region has been amplified many million fold. Many different variations of the PCR method have been developed to perform a wide array of genetic manipulations. *Courtesy of Holmes, I.H., Strugnell, R., reproduced from MacLachlan, N.J., Dubovi, E.J., 2011. Veterinary Virology, fourth ed., Academic Press, with permission.*

in practice over 1 million can be achieved. Forty cycles can be completed in less than an hour depending on reaction volumes and method, and by a suitable automated thermal cycling device. The amplified DNA may be detected either as a stained band of correct molecular weight on agarose gel electrophoresis, or by hybridization with labeled DNA or RNA probes culminating in any of a wide variety of readout systems. However the development of real-time detection systems (see below) has revolutionized the speed and quantitative accuracy of the method and greatly helped reduce false-positive reactions due to contamination.

Selection of the most suitable pair of primers is a matter of critical importance. Each may be chosen as highly specific for a particular virus strain or, alternatively, to represent consensus sequences within a gene known as conserved within a given family or genus. By probing for a conserved gene, such as that for RNA polymerase, it is even possible to discover a previously unknown agent. Reactions must be carried out under very carefully controlled conditions of ionic strength, temperature, primer concentration, and nucleotide concentration. Deviations can result in non-specific amplification.

PCR technology has undergone numerous major developments since its first description in 1983. These include:

1. Use of the heat-stable DNA polymerase (*Taq*) of *Thermus aquaticus*, an organism naturally found in hot springs. Since this makes it no longer necessary to replenish the enzyme between cycles, the reaction tubes can remain sealed, and it also allows the use of higher annealing temperatures thereby increasing reaction specificity by reducing false-positive reactions due to mismatching;
2. Use of "nested" PCR reactions, whereby a second amplification is carried out with primers internal to the initial target after the initial PCR amplification of a target sequence. This can greatly increase sensitivity, albeit at the potential cost of increasing the risk of false-positive reactions;
3. Due to the detection of trace amounts of DNA sequence with extreme sensitivity, early PCR results suffered from regular problems of environmental contamination with DNA products leading to false-positive results. The procedure developed a reputation for unreliability, and laboratories went to heroic efforts to prevent reaction contamination, including stringent physical segregation of the different steps in the procedure. This has become far less of a problem with recent advances, in particular the use of real-time detection methods;
4. Real-time PCR testing. This method for detecting PCR products required the development of a thermocycler with an inbuilt fluorimeter allowing measurement of newly synthesized PCR products in the reaction tube as were accumulating, that is, in real time. The reaction products can be detected by dyes that fluoresce when

bound non-specifically to any double-stranded DNA, for example, SYBR Green; or by sequence-specific reporter probes (EgTaqMan) containing a fluorescent reporter at one end and a fluorescence quencher at the other, that bind specifically to newly synthesized target sequences leading to loss of quenching and emission of a quantitative fluorescence signal. As the reaction progresses, the intensity of the fluorescent signal is continually monitored and plotted, generating curves the position of which reflects the original concentration of target. This approach has largely replaced other methods for product detection because it does not require the reaction tube to be opened once the reaction has started, thereby minimizing contamination, it achieves excellent quantitative measurements, is more sensitive than earlier detection systems, and if a sequence-specific reporter probe is used it has improved specificity because only the target sequence and no other non-specific DNA molecules is detected;

5. Development of more accurate quantitation. Real-time detection methods allow generation of standard curves the position of which corresponds to the original concentration of target, and an unrelated copy number control can be included in the reaction to detect any non-specific inhibition of the enzyme reaction. With a number of infections, the clinical interpretation and management are now dependent on regular quantitation of the patient's "viral load";

6. Use of multiplex PCR. Once different fluorescent probes became available for PCR product detection, it became possible to include different sets of primers bearing different fluorescent labels for different targets in the same reaction, and thereby to detect a number of different products simultaneously. This is exploited in a number of standard protocols for specific sites, for example, respiratory virus infections (Fig. 10.6).

Isothermal amplification methods are variations of PCR that do not require high (95°C) temperature cycling to melt the newly synthesized DNA product to allow binding of a new round of primers. These methods rely on enzymes to displace the strands and include Loop Mediated Isothermal Amplification (LAMP), Strand Displacement Amplification (SDA), Helicase-Dependent Amplification (HAD), and Nicking Enzyme Amplification Reaction (NEAR). These methods have the potential advantage in being extremely fast and do not require thermal cycling equipment. Several examples are available as approved diagnostic tests.

Due to the speed, sensitivity, and versatility of PCR methods for detecting and quantifying viral genomes, this approach is now a significant part of the routine work of the diagnostic laboratory, and many commercial test systems are available. Methods such as antigen detection and virus isolation tend to require different reagent preparation expertise and methods specific for each individual virus, whereas PCR technology is more universally applicable because its specificity relies primarily on the chosen primer sequences. The procedure is invaluable when dealing with (1) viruses that cannot be cultured satisfactorily, (2) specimens that contain predominantly inactivated virus, as a result of prolonged storage or transport, (3) where quantitative results are needed, or (4) to avoid exposing laboratory staff to culturing larger quantities of a biohazard category 3 or 4 agent. Furthermore, it is easy to make new PCR reagents for new viruses, and thereby rapidly expand the capacity for testing.

Problems that may be encountered with PCR include (1) some clinical samples that may contain factors that inhibit the activity of the DNA polymerase, leading to falsely low or negative results. This can be detected by testing in parallel a sample control that contains a standard unrelated DNA target or "copy number control," (2) the possibility of contamination of the reaction with extraneous DNA not originating from the patient although nowadays it is much less of an issue, (3) as with all methods that are specific for a particular virus or viruses (in contrast to "open-minded" procedures such as virus isolation or electron microscopy), PCR will only detect those agents that are recognized by the primer set and test conditions. Thus, variant viruses may be missed if these sequences are not amplified or the sequences amplified are not detected in the product assay system used, (4) the extreme sensitivity of PCR means that low levels of chronic, latent, or concurrent infecting genomes unrelated to the patient's clinical condition, may be detected.

As with all diagnostic procedures, before finally ascribing a laboratory finding to a patient's condition, the patient's history and physical findings, knowledge of disease pathogenesis, and clinical acumen, must all be brought into play. For example, varicella zoster virus DNA has been detected as a contaminant in a genital herpes simplex virus infection—the PCR giving a strong HSV2 but weak VZV reaction because of the primers used.

Microarray Technologies

Another technological advance that is impacting the field of diagnostics is the use of microarrays or microchips for nucleic acid detection. The microchip is a solid support matrix onto which have been "printed" spots, each containing one of several hundred to several thousand unique oligonucleotides. These oligonucleotides can represent conserved sequences from virtually all viruses represented in the various genetic databases, or can be customized to represent only viruses from a given specific disease syndrome, such as acute respiratory disease in children. The basis of the technology is the capture by these oligonucleotides of randomly amplified labeled nucleic acid

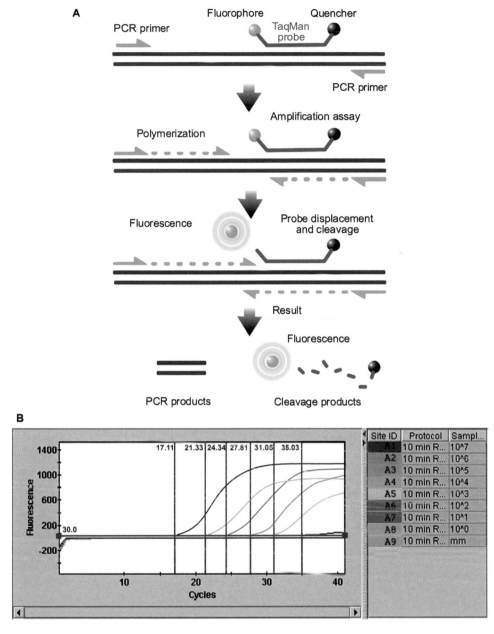

FIGURE 10.6 (A) TaqMan probe chemistry mechanism. These probes rely on the 5′-3′ nuclease activity of Taq DNA polymerase to cleave a dual-labeled probe during hybridization to the complementary target sequence. (B) Real-time quantitative PCR data. Reaction curves for a test run are used to assess assay conditions, using dilutions of an RNA transcript (copy number control) of a cloned segment of canine pneumovirus. The vertical lines represent the Ct value, which is the number of PCR cycles required for the fluorescent signal to cross the threshold value. TaqMan1 probe was labeled with FAM (6-carboxy fluorescein; the reporter dye) at the 5′ end and BHQ (Black Hole Quencher; the quencher) at the 3′ end. *Reproduced from MacLachlan, N.J., Dubovi, E.J., 2011. Veterinary Virology, fourth ed., Academic Press, with permission.*

sequences from clinical specimens. The binding of a labeled sequence is detected by laser scanning of the chip and software programs that assess the strength of the binding. From the map position of the reacting oligonucleotides, the software identifies the virus in the clinical sample. This type of test was used to initially determine that the virus responsible for severe acute respiratory syndrome (SARS) was a coronavirus. With knowledge of the oligonucleotide sequences that bind an unknown virus, primers can be made to eventually determine the entire nucleotide sequence of the virus. The low cost of oligonucleotide synthesis, development of laser scanning devices, nucleic acid amplification techniques, and software development have made this technology available in specialized reference laboratories, increasingly more common in national diagnostic laboratory networks.

Microarray technology has been used as a research tool for some years, for example, for studying comprehensive gene expression in cells and tissues. It is now being applied increasingly for monitoring host responses to infections and for assessment of the development of antiviral drug resistance. It has been used to map immune response gene expression in hepatitis C-infected liver tissue, and changes in leukocyte gene expression in patients with sepsis. It is now also being directly applied in the diagnostic arena. For example, commercial microarrays are available for the simultaneous detection of a large panel of respiratory viruses, by reacting the product of a multiplex PCR and RT-PCR reaction with a solid chip or liquid bead platform carrying oligonucleotide probes for multiple viruses. Similar approaches have been described for the simultaneous detection and typing of human papillomavirus sequences in routine clinical specimens. The transfer of this technology to the routine diagnostic laboratory still presents challenges, not least due to the requirements for a rapid turn-around, high-throughput, quality-assured procedure for the diagnostic setting, and the unpredictability and heterogeneity of clinical samples. Nevertheless it is likely that application of microarray technology will eventually lead to another transformation of the diagnostic laboratory.

Next-Generation Deep Sequencing

"Deep" sequencing, or "next-generation" sequencing refers to novel techniques of DNA sequencing directly from DNA fragments without the need for cloning in vectors, allowing the generation of enormous amounts of sequence data at high speed and low cost from a single run. A number of systems are commercially available, the approach has yielded enormous benefit in various research fields, and its application to diagnostic virology is now being explored. Different platforms have different advantages and pitfalls, for example, differences in read length and read depth, and access to a capable bioinformatics resource is essential to ensure that data analysis can be done appropriately. For an unbiased or broad-based analysis, all nucleic acids in a clinical sample can be sequenced on a multiplex platform; however, because host sequences can frequently outnumber infecting virus sequences by 10^5 or more, prior selective enrichment of virus sequences can be achieved, for example, by hybrid capture using PCR-derived capture probes, by prior PCR amplification of suspected virus sequences in the unknown target sample, or by physical methods of partial purification, for example, based on the nuclease resistance of encapsidated viral genomes.

This approach is already being applied to many purposes in virology, for example:

1. Discovering new pathogens in undiagnosed illness or outbreaks (e.g., the coronavirus causing Middle East respiratory syndrome [MERS] first reported in Saudi Arabia in 2012), the novel highly divergent rhabdovirus Bas-Congo virus (BASV), and the novel polyoma viruses HPYV9 and Merkel cell polyomavirus (MCPyV);
2. Retrospective diagnosis of undiagnosed illness, for example, encephalitis, using stored autopsy samples;
3. Screening vaccines for contaminants;
4. Analysis of the quasi-species sequence composition of the viruses in a clinical sample, including detection of minor variants with new pathogenic implications, for example, drug resistance;
5. Investigation of the diversity and evolution of particular viral genomes;
6. Studies of the human virome in health and disease.

Full application of deep sequencing technology to virus diagnosis has not yet been achieved, and deep sequencing methods are likely to complement existing methods rather than replace them. Costs have plummeted enormously, now allowing the technology to move from large institutions to the more widespread diagnostic setting. However many development challenges remain, including the huge computing and data analysis requirements that are created by the generation of enormous amounts of data.

VIRUS ISOLATION

In recent years, virus isolation has changed from its role as the "gold standard" of the diagnostic laboratory, to occupying more of a research or historical role. Newer techniques, particularly sensitive antigen EIAs and PCRs, do not suffer from the drawbacks of virus isolation that included high cost, slowness, false negatives due to virus being inactivated during transport or complexed with antibody, cell lines losing sensitivity, and increased concerns about live virus transmission to staff. Many diagnostic laboratories no longer try to isolate viruses in cell culture; virus isolation in cell culture is mostly limited to research laboratories or where there is a need to grow up large amounts of virus. Even bulk antigen production is now often done preferentially by expression of recombinant antigens, and unknown agents can be sought using molecular techniques such as random-priming PCR.

Similarly, techniques for animal inoculation, for example, suckling mice, and egg inoculation procedures, are only performed where a laboratory has a specialist need, for example, influenza virus to be used in vaccine production is grown in embryonated hens' eggs.

For successful virus isolation in cell culture, the following steps are followed.

1. Care must be given to maintaining a cold chain at 4°C from the patient to the laboratory, but with care not to freeze the sample as some viruses are labile to freezing.

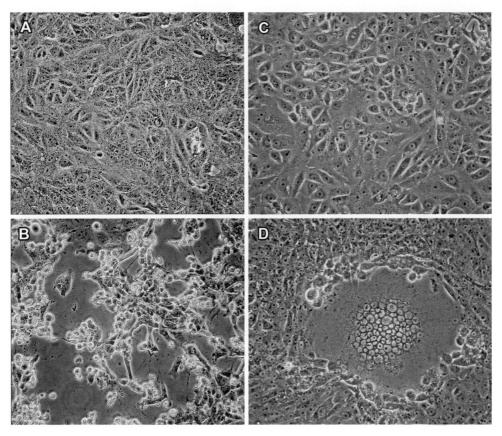

FIGURE 10.7 Cytopathic effect of viruses in cell cultures. (A) Confluent monolayer of uninfected Vero cells. (B) Vero cells infected with the SARS coronavirus (SARS-CoV) 24 hours post-infection. Massive cytopathic effect (CPE), with cells rounded up and detaching from the substrate. (C) Confluent monolayer of uninfected Vero cells. (D) Vero cells infected with herpes simplex virus 1 (HSV-1) at 24 hours post-infection. A single large syncytium of at least 50 cells (massed nuclei) which is unstable and later progresses to a necrotic/apoptotic CPE. Since most diagnostic specimens contain a much lower titer of infectious virus, CPE is usually more asynchronous, beginning with few foci involving few cells and taking longer to achieve widespread infection. Phase contrast microscopy (C, D) monochromatic green light provides slightly better resolution. *(A), (B) from Institute for Medical Virology, University Hospital, Frankfurt, with permission.*

2. Samples may need preparation by treatment with antibiotics, clarification of debris, or disruption of mucus.
3. Appropriate cell lines for the virus(es) sought are chosen and inoculated, taking care that the cells are metabolically healthy; that cells are being grown in an appropriate cell culture medium and serum concentration, and the cultures are in the appropriate state for the virus(es) being sought, that is, confluent or actively dividing; at the appropriate temperature, that is, 37° or 33° (for respiratory viruses); and that the laboratory periodically checks the cell lines being used for their continuing sensitivity to support growth of reference viruses. Specially designed cell lines may express particular virus receptors to broaden specificity or indicator systems for easier detection of the presence of virus.
4. Evidence that a new virus has been isolated may come from routine reading of monolayers for cytopathic effects (CPEs), which may be distinctive for a particular virus. Alternatively, monolayers may be screened by a variety of techniques including antigen staining, hemadsorption, electron microscopy, or PCR. Blind passage of material not showing a CPE may be advisable to identify slow-growing agents, and regular comparison with uninoculated control cells is essential to distinguish viral CPE from degeneration of uninfected cell monolayers.
5. Finally, a variety of methods may be needed to obtain final identification of the virus that has been isolated, sometimes including detailed strain identification or sequencing.

For further detailed description of virus isolation, the reader is encouraged to consult more detailed sources, for example, national reference laboratories or information distributed by the US Centers for Disease Control and Prevention (http://www.cdc.gov) and the World Health Organization (http://www.who.int) (Fig. 10.7).

Further Characterization of Viruses Detected

For most routine diagnostic purposes it is usually not necessary to determine the exact strain or sequence of the

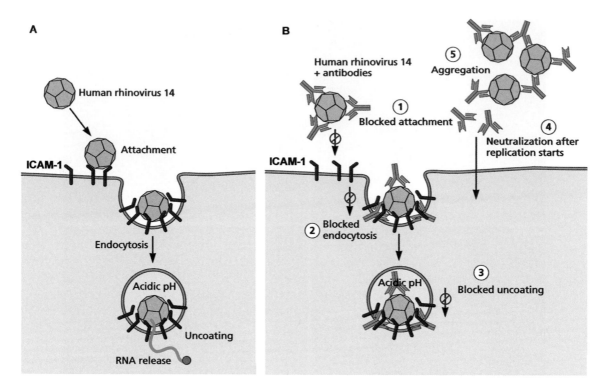

FIGURE 10.8 Virus neutralization, as exemplified by the interactions of neutralizing antibodies with human rhinovirus 14. (A) The normal route of infection. The virus attaches to the ICAM-1 receptor and enters by endocytosis. As the internal pH of the endosome decreases, the particle uncoats and releases its RNA genome into the cytoplasm. (B) Possible mechanisms of neutralization of the virus by antibodies. At least five mechanisms of neutralization have been supported by experimental evidence: (1) blocked attachment—binding of antibody molecules to virus results in steric interference with virus-receptor binding, (2) blocked endocytosis—antibody molecules binding to the capsid can alter the capsid structure, affecting the process of endocytosis, (3) blocked uncoating—antibodies bound to the incoming particle fix the capsid in a stable conformation so that pH-dependent uncoating is not possible, (4) blocked uncoating inside the cell—antibodies themselves may be taken up by endocytosis and interact with virions inside the cell after infection starts, (5) aggregation—because all antibodies are divalent, they can aggregate virus particles, facilitating their destruction by phagocytes. Any or all of these mechanisms may operate in quantitative neutralization assays. Virus neutralization assays may be used to measure the titer of neutralizing antibody in clinical samples, by reacting a standard virus inoculum with different dilutions of serum. There are many other variations, for example, known antisera to different virus serotypes may be used to identify the serotype of a new virus isolate; tests may be run in cell cultures with cytopathic effect or plaque endpoints, or in animals when necessary, and many steps may be automated. *Adapted from Smith, T.J., et al., 1995. Semin Virol 6:233–242. Then from Flint, S.J., et al., 2016. Principles of Virology: pathogenesis and control, fourth ed., Washington, DC: ASM Press, with permission.*

virus that has been detected. On the other hand, there are certain situations when this information can be crucial for patient management, or for epidemiological or public health purposes. Serological methods including EIAs and neutralization of virus growth can be used to identify different serotypes, assisted by the availability of appropriate monoclonal antibodies (Fig. 10.8). Different influenza virus isolates are characterized and compared by comparing hemagglutination-inhibition titers using standard antisera.

Complete nucleotide sequencing of a viral genome is not often attempted for diagnostic purposes at present although it probably will be increasingly widespread over the next few years. However, obtaining a partial sequence from PCR products is widely used and has replaced earlier molecular approaches such as restriction mapping of DNA products and oligonucleotide fingerprinting. For example, the resistance of a patient's circulating HIV to

the commonly used anti-retroviral drugs is predicted from the pattern of resistance-associated point mutations in the sequence of the viral polymerase. This can be determined from the sequence of the patient's HIV PCR product, which is then compared with commercially available sequence databases to correlate polymerase sequence with drug resistance phenotype. The result is used to plan the patient's individualized drug regimen or to modify a drug regimen to which the patient's virus has become resistant. Because different rabies virus sequence variants have segregated according to geographical region and animal host species, sequencing of rabies virus recovered from a patient can be used to determine the origin of the infection. In order to investigate an apparent cluster of HIV patients who may or may not have acquired their infection from a common source, for example, a single medical practice or surgeon or a common sexual partner, partial sequences of their viruses

TABLE 10.3 Serological Procedures Used in Virology

Technique	Principle
Enzyme immunoassay	Patient antibody binds to antigen; enzyme-labeled anti-Ig binds to antibody; washing; added substrate changes color
Radioimmunoassay	Patient antibody binds to antigen; radiolabeled anti-Ig binds to antibody; washing and counting
Western blot	Virus is disrupted, proteins separated by gel electrophoresis and transferred (blotted) onto nylon membrane; antibodies in test serum bind to viral proteins; labeled anti-Ig binds to particular bands; revealed by EIA or autoradiography
Latex particle agglutination	Patient antibody agglutinates antigen-coated latex particles
Virus neutralization	Patient antibody neutralizes virus infectivity; readout involves CPE inhibition or plaque reduction in cell culture, or protection of animals
Hemagglutination inhibition	Patient antibody inhibits agglutination of red blood cells by virus or antigen
Immunofluorescence	Patient antibody binds to intracellular antigen; fluorescein-labeled anti-Ig binds; fluoresces by UV microscopy
Immunodiffusion	Antibodies and soluble antigens produce visible lines of precipitate in a gel
Complement fixation	Antigen–antibody complex binds complement, which is thereafter unavailable for lysis of sheep RBC in the presence of antibody to RBC

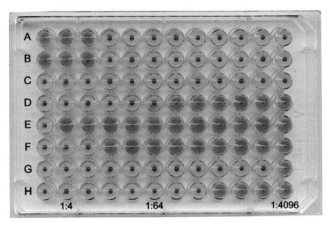

FIGURE 10.9 Hemagglutination inhibition (HI) test for detecting antibodies specific for influenza virus (H3N8, one of the highly pathogenic viruses that are of concern should these gain increased human-to-human transmissibility capability). Sera are (A) treated to remove non-specific agglutinins and non-specific inhibitors of agglutination. (B) Treated sera are serially diluted (twofold) in buffered saline in rows of wells of 96-well microtiter plates. (C) A volume of influenza virus containing 4 hemagglutinin units is added to each well and plates are incubated for 30 minutes. (D) A standard volume of turkey red blood cells (0.5% suspension) is added to each well. (E) The plates are incubated and read when control wells show complete red blood cell settling (button). Where virus agglutinates the erythrocytes, these form a shield pattern; however, where enough antibody is present to coat the viral HA, hemagglutination is inhibited, and the erythrocytes settle to form a button on the bottom of the cup. Rows (A), (B): Titration of the influenza virus used in the test, confirming that the correct amount (4 units) has been added to test wells. Row (C): Red blood cell control. Rows (D–H): Serially diluted test sera, with the first well in each row serving as a serum control for non-specific hemagglutination. Results: HI titers are the reciprocal of the last dilution of the serum showing inhibition of agglutination by test virus. Row (D): HI titer = 64; row (E): HI titer <4; row (F): HI titer = 8; row (G): HI titer = 2048; row (H): HI titer = 256. *Reproduced from MacLachlan N.J. and Dubovi E.J. (2011), Veterinary Virology, fourth ed., Academic Press, with permission.*

can be compared with the usual random sequence variation seen in that population, to assess the likelihood that this truly represents common source infection.

MEASUREMENT OF SERUM ANTIBODIES

The identification of unknown viruses or viral antigens using panels of antibodies of known specificity has already been discussed. Any of the above serological techniques may be employed the other way around, using panels of known antigens to identify unknown antibodies. The basic principles of those most widely used today are set out in Table 10.3 (Fig. 10.9).

Diagnosis of Acute Infection by Demonstrating a Rise in Antibody

Following the traditional approach, "paired sera" are taken from the patient, the "acute-phase" serum sample as early as possible in the illness and the "convalescent-phase" sample at least 10 to 14 days later. Blood is collected in the absence of anticoagulants and given time to clot, and then the serum is separated. Acute and convalescent serum samples should be tested simultaneously. For certain tests that measure inhibition of some biological function of a virus, for example, virus neutralization or hemagglutination-inhibition, the sera must first be inactivated by heating at 56°C for 30 minutes and sometimes treated by additional methods, in order to destroy various types of non-specific inhibitors of infectivity or hemagglutination, respectively. Prior treatment of the serum is not generally required for assays that simply measure antigen–antibody binding, such as EIA, RIA, Western blot, latex agglutination, immunofluorescence, or immunodiffusion.

The paired sera are then titrated for antibodies using any of a wide range of available serological techniques. Demonstration of a significant rise in antibody titer (conventionally, a four-fold rise when titrating the test serum as two-fold dilutions) is taken as proof of "seroconversion," that is, acute infection with the agent in question. In practice, there are many subtleties in the interpretation of results. For example, the timing of the collection of the paired sera in relation to the date of onset of illness must be carefully assessed to allow for inappropriate sample timing. Moreover, an individual with immunity due to prior infection or immunization may undergo a second or reinfection, often of a subclinical nature, but showing a rise in antibody titer. It is obvious that in reality, two well-timed paired sera to allow demonstration of a four-fold rise in titer, are often not available. Attempts are sometimes made to infer that a recent infection has occurred, on the basis of one serum sample containing a relatively high antibody titer; indeed laboratories may issue guidelines about what titer of antibody in a single serum may imply recent infection. In practice this is often unreliable, as the peak antibody titer seen in the convalescent stage of an infection can vary significantly from person to person.

Immunoglobulin M Class-Specific Antibody Assays

A rapid diagnosis can be made on the basis of a single acute-phase serum by demonstrating virus-specific antibody of the IgM class. Because IgM antibodies appear early after infection but decrease to low levels within 1 to 2 months and generally disappear altogether within 3 months, the presence of these antibodies is indicative of either recent or chronic infection. Moreover, if present in a newborn baby, these antibodies are diagnostic of intrauterine infection, because maternal IgM antibodies, unlike IgG, do not cross the placenta. All of the immunoassays described above can readily be rendered IgM class-specific; EIA and immunofluorescence are the most generally useful. Typical indirect assays for virus-specific IgM antibodies are depicted in Fig. 10.10.

A particular problem with IgM antibody assays is interference by the so-called *rheumatoid factor*, which is antibody, mainly of the IgM class, directed against the constant region (Fc) of normal IgG. Though first described in rheumatoid arthritis and other autoimmune diseases, rheumatoid factor is in fact prevalent in many infectious diseases, and it is found in the majority of congenitally infected neonates. Rheumatoid factor produces false positives in IgM immunoassays, because it binds to antiviral (as well as normal) IgG antibodies in human serum, forming IgM–IgG complexes. Thus, any antiviral IgG antibodies in the sample, which will also bind to the immobilized antigen, may then capture rheumatoid factor present, which will in turn be detected by the anti-human IgM antibodies employed as detector antibody in the assay format depicted in the left-hand side of Fig. 10.10.

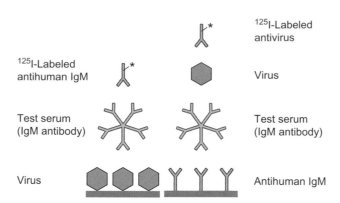

FIGURE 10.10 Radioimmunoassay for IgM class antibody. In the simpler configuration (left), virus-specific IgM antibody in the test serum reacts with viral antigen immobilized on the solid phase, and the IgM antibody is then detected by labeled anti-IgM antibody. This configuration can lead to false-positive results due to rheumatoid factor, or false-negative results due to blocking by competing antiviral IgG antibody in the test serum. In the alternative configuration (right), IgM antibody is captured from the test serum by immobilized anti-IgM, and any specific antiviral IgM is detected by progressive incubation with viral antigen and then labeled antiviral antibody. Enzyme-labeled antibodies have now largely replaced I^{125}-labeled antibodies as a preferred indicator system.

In order to minimize the impact of rheumatoid factor, it is advisable to employ a *reverse* (or *capture*) *IgM assay* (Fig. 10.10, right). In its simplest form monoclonal anti-human IgM is used as the capture antibody and labeled virus as the detector/indicator. More commonly, unlabeled virus is the detector, then labeled monoclonal antiviral IgG antibodies, or perhaps better still F(ab')$_2$ antibody fragments, are added as indicator.

Another pitfall with IgM antibody assays is the potential for cross-reaction between closely related viruses. This is because antibodies produced early in an immune response, particularly IgM antibodies, are typically more broadly reacting until B cell affinity maturation has taken place. For example, infection with a dengue virus may produce IgM antibodies that cross-react with a number of related flaviviruses, and it may be necessary to titrate the serum against a panel of antigens, or test later IgG samples, to identify the particular flavivirus involved. False-negative IgM results are occasionally encountered in infants, in some respiratory infections, and during reactivation of latent herpesviruses. Despite these pitfalls, IgM antibody capture assays are an important first-line diagnostic approach for some infections, for example, hepatitis A. Class-specific immunoassays can also be designed to measure specific antiviral antibodies of the IgA, IgE, or IgG class, or of any given IgG subclass.

Immunoblotting ("Western Blotting")

Western blotting allows simultaneous but independent detection of antibodies against a number of different proteins present in a particular virus preparation. There

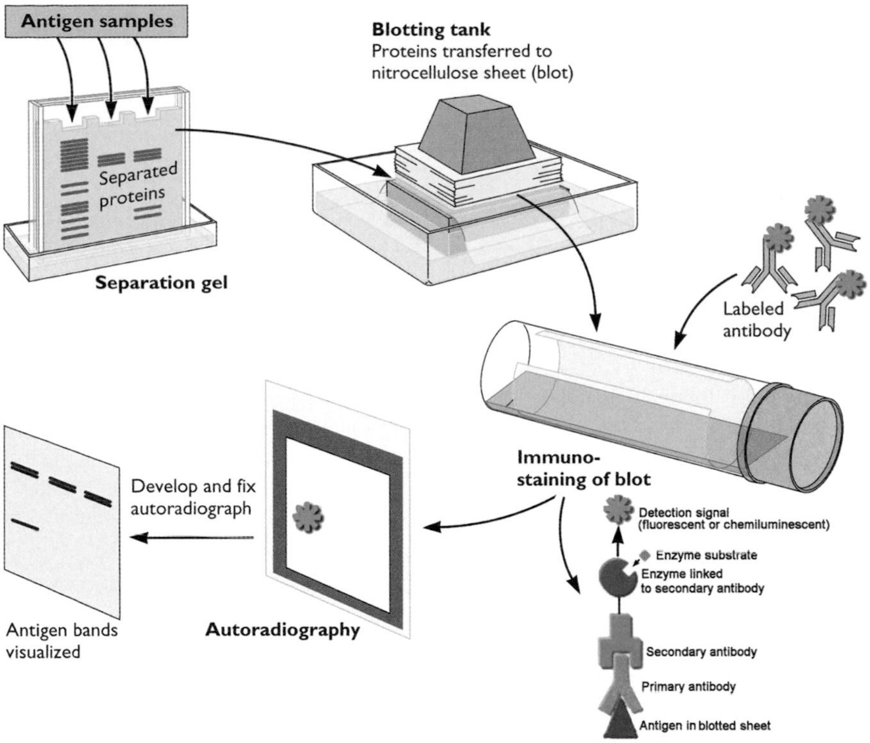

FIGURE 10.11 Principles of western blotting for identification of antibodies or antigens. (1) A virus-containing sample is digested with the anionic detergent sodium dodecylsulfate (SDS) and electrophoresed on a polyacrylamide slab gel (PAGE), which separates the different viral proteins according to M_r. (2) The bands of viral protein are then transferred ("blotted") onto a nitrocellulose membrane by capillary transfer or usually by electrophoresis in a different plane to immobilize the polypeptides. (3) The unoccupied areas of the membrane are blocked ("quenched") by saturation with a suitable protein, then washed, dried, and cut into strips which can be used to test individual patient sera. Each test serum or plasma ("primary" antibody) is then incubated with one strip to enable antibodies to bind to the individual viral proteins. Following rinsing, bound antibody is detected by the addition of enzyme-labeled anti-human immunoglobulin ("secondary" antibody). Following another wash, the bands are revealed by the addition of a substrate chosen to produce an insoluble colored product. This technique may be used with a preparation of known antigens, for example, an infected cell lysate, having been separated on a gel, to examine a serum sample containing unknown antibodies; alternatively, a known antiserum can be used to detect those antigens in an unknown sample that react after running on a gel. *Reproduced from Flint, S.J., et al., Principles of Virology, second ed., ASM Press, with permission.*

are four key steps to western blotting (Fig. 10.11). First, concentrated virus is solubilized and the constituent proteins separated into discrete bands according to molecular mass (M_r) by sodium dodecylsulfate-polyacrylamide gel electrophoresis (SDS-PAGE). Second, the separated proteins are electrophoretically transferred ("blotted") onto a nitrocellulose sheet in order to immobilize the separated polypeptides and to make these available for reaction with antibodies. Third, the test serum is allowed to react with the viral proteins on the nitrocellulose sheet and unbound components are washed away. Finally, any bound antibody is demonstrated using an enzyme-labeled anti-species antibody. Thus, immunoblotting permits the demonstration of antibodies to some or all of the proteins of any given virus preparation, and can be used to monitor the development of antibodies to different antigens at different stages of infection. The technique has played a crucial role as the ultimate confirmatory test for samples giving an initial reaction in screening EIAs for anti-HIV antibody; experience has taught which western blotting patterns are

likely to represent true anti-HIV antibodies, and which represent likely false-positive reactions in the initial EIA.

Western blotting is also an important research tool for characterizing viral antigen mixtures, using the antibody reagent rather than the antigen as the "known" component in the reaction.

Applications of Serology

Serological tests are used for many important purposes apart from diagnosing or excluding acute or chronic infection (see Box 10.1). These assays are valuable in many contexts because these provide indications of both clinical and subclinical infections, thereby giving a truer record of *total* number of infections. The finding of antibodies to a virus in a single sample carries very different clinical implications depending on the virus (see Box 10.2). The screening for immunity is very useful to document successful immunization in an individual, and to check the coverage and efficacy of vaccination in a population. In addition, testing the susceptibility of the close

BOX 10.1 Applications of Serological Tests

1. Diagnose acute infection by demonstrating a fourfold rise in titer of viral antibodies
2. Diagnose acute infection by demonstrating the presence of specific IgM antibodies
3. Estimate *prevalence* of a persistent infection in a population, for example, what percentages have persistent CMV infection
4. Estimate *incidence* of an infection over a time period, for example, the rate of new hepatitis C infections
5. Evaluate immunization programs, for example, measles, hepatitis B
6. Assess level of *herd immunity*, for example, influenza
7. Assess exposure resulting from occupation, lifestyle—whether certain activities are associated with increased rates of infection?
8. Use age-specific prevalence patterns to analyze transmission efficiencies and possible transmission routes, for example, age-specific acquisition of EBV antibody
9. Determine whether an apparently new infection in man has actually been already circulating subclinically
10. Search for an animal reservoir of an apparently new zoonotic infection, for example, which non-human species show antibodies to SARS or Ebola viruses

BOX 10.2 Significance of a Single Positive Result for Antibody with Different Viruses

- Anti-HBs = immune, not infectious
- Anti-HIV = infected and infectious
- Anti-HCV = past infection, and ~60–80% chance of being *still* infected
- Anti-HSV = still infected, intermittently infectious by salivary route but not blood transfusion
- Anti-HSV type 2 = as above, + the site of primary infection is *most likely* to be genital
- Anti-CMV, anti-EBV = still infected, potentially infectious (by salivary contact or blood transfusion)
- Antibody to rubella virus = immune to second clinical attack; may undergo asymptomatic reinfection but the fetus is *not* at risk
- Antibody to an exotic virus (e.g., Lassa fever, Ebola) in an acutely sick, recently returned traveler = likely to be acute infection

contacts of an individual with a potentially dangerous infection allows for the protection of at-risk (non-immune) individuals by segregation or immunization, and provides a baseline for subsequent monitoring as to the course of infection.

Sensitivity and Specificity

The interpretation and performance of any diagnostic assay are judged by two essential criteria, specificity and sensitivity. The *sensitivity* of a given test is a measure of the percentage of those with the disease (or infection) in question who are identified as positive by that test. For example, a particular EIA used to screen a population for HIV antibody may display a sensitivity of 98%, that is, of every 100 infected people, 98 will be correctly detected and 2 will be missed (the *false-negative* rate equals 2%). In contrast, the *specificity* of a test is a measure of the percentage of those without the disease (or infection) who yield a negative result. For example, the same EIA for HIV antibody may have a specificity of 97%, that is, of every 100 uninfected people, 97 will be correctly diagnosed as clear but 3 will be incorrectly scored as infected (the *false-positive* rate equals 3%).

Whereas sensitivity and specificity are fixed percentages intrinsic to the particular diagnostic assay, the predictive value of an assay is greatly affected by the prevalence of the disease (or infection) involved. Thus, if the above EIA is used to screen a high-risk population with an HIV prevalence of 50%, the predictive value of the assay will be high; however, if it is used to screen blood donors with a known HIV prevalence of

0.1%, the great majority of the 3.1% who register as positive will in fact be false positives and will require follow-up with a confirmatory test of a much higher specificity. This striking illustration draws attention to the importance of selecting diagnostic assays with a particular objective in mind. A test with high sensitivity is required when the aim is to screen for a serious infection, the diagnosis of which must not be missed; a test with high specificity is also required for confirmation that the diagnosis is correct.

The *sensitivity* of a given immunoassay is really a measure of its ability to detect small amounts of either antibodies or antigen. For instance, EIA, RIA, and neutralization assays generally display substantially higher sensitivity than immune electron microscopy, immunofluorescence, or the older serological methods of complement fixation and immunodiffusion. Improvements in sensitivity are achieved by miniaturization of assays, use of purified reagents, sensitive instrumentation, and signal amplification. On the other hand, the *specificity* of an immunoassay is a measure of the test's capacity to discriminate; thus it is influenced mainly by the purity of the key reagent, that is, antigen when testing for antibodies, or antibodies when testing for antigen. Use of cloned antigens or synthetic peptides will usually improve the specificity of antibody detection. Note that use of signal amplification to increase the sensitivity of a test is very likely to also increase the signal due to background or false-positive reactions, that is, the signal:noise ratio may not be improved.

INTERPRETATION OF LABORATORY RESULTS

The isolation and identification of a particular virus, or demonstration of infection by serological methods, in a patient with a given disease, are not necessarily relevant to the

BOX 10.3 Information that May Be Sought from Laboratory Virus Testing

1. Is the patient infected with a particular virus?
2. If so, which strain or serotype is involved?
3. What is the level of replication or virus load?
4. Is the patient infectious to others?
5. Which site(s) of the body are involved?
6. What is the stage or duration of the infection?
7. What is the type or extent of the immune response?
8. What are the implications for therapy or prevention of infection of others?
9. Has the patient been infected in the past?
10. Is the patient now immune?

TABLE 10.4 Laboratory Hazards

Route of Hazard	Procedure or Source of Hazard
Aerosol	Homogenization (e.g., of tissue in blender)
	Centrifugation
	Ultrasonic vibration
	Broken glassware
	Pipetting
Ingestion	Mouth pipetting
	Eating or smoking in laboratory
	Inadequate washing/disinfection of hands
Skin penetration	Needle stick injury
	Hand cut by broken glassware
	Leaking container contaminating hands
	Animal bite
	Pathologist handling infected organs
	Receptionist handling blood-stained request form
Splash into eye	Break in pressurized tubing
	Sudden liquid spillage

patient's immediate clinical problem. Concurrent subclinical infection with a virus unrelated to the illness in question is not uncommon. In attempting to interpret whether a virus that has been isolated is related to a patient's disease, one must be guided by the following considerations. First, the site from which the virus was isolated is important. For example, one would be quite confident about the etiological significance of herpes simplex virus isolated from a sample of brain tissue, or of mumps virus isolated from the CSF of a patient with meningitis, because these sites are usually sterile without any normal bacterial or viral flora. On the other hand, an echovirus recovered from the feces, or herpes simplex virus from the throat, may not necessarily be causing any pathology as such viruses are often associated with unapparent infections. Second, knowledge that the virus and the disease in question are often causally associated engenders confidence that the isolate is significant.

In addition to the obvious issue of diagnosing the cause of a patient's illness, it is very important to remember that laboratory diagnostic methods are used to answer many other significant clinical and public health questions (see Box 10.3).

A number of very important practical issues should not be forgotten as these can seriously interfere with the reliability of laboratory results for a given clinical situation. These include:

1. Contamination of the sample with extraneous viruses or nucleic acids, unrelated to the disease being investigated. This may occur at any stage in the diagnostic chain, from collection of the sample through to the final stage of the test procedure;
2. False-positive and false-negative results may occur with any test, for numerous reasons including sensitivity and specificity;
3. Mix-up of samples, labeling errors. Laboratories frequently build in system checks to detect such errors.

Frequently, laboratories will include a requirement for confirmation by repeat testing in the case of results that are subject to particular significance or unreliability.

LABORATORY SAFETY

Although virology can be considered one of the less hazardous human occupations compared with building, mining, or driving a car, it is also true that many cases of serious illness and over 700 deaths from laboratory-acquired infection have been recorded over the years, particularly from togaviruses, flaviviruses, arenaviruses, and filoviruses. Potentially hazardous procedures are described in Table 10.4. In many countries, including the United States, European Union, Canada, and Australia, pathogens are classified into categories 1 to 4 based on properties of pathogenicity, transmissibility, host range, the availability of treatment and vaccination, etc. These categories are endorsed by the World Health Organization. An appropriate level of containment and handling practice is specified for each risk category, ranging from Biosafety level 1 (BSL 1)—standard microbiological practice—for category 1 organisms, through to BSL 4—maximum security facilities and containment protocols—for category 4 viruses and organisms. In research laboratories, work usually involves known agents where the hazards are predictable and safety procedures appropriate to the level of hazard can be applied. However, diagnostic laboratories accept human samples of unknown infective status that are in a practical sense an extension of the patient. These may contain HIV, hepatitis

B or C virus, or other common or exotic infective agents. The usual approach is to regard all unknown biological specimens as potentially infectious, and to apply a level of containment appropriate to the majority of agents likely to be present, for example, BSL2; individual samples from known or suspected cases of greater hazard, for example, Ebola, are handled at a higher containment level once these are identified. Similarly, material from laboratory or wild animals, particularly primates, may contain unknown pathogens with a danger to humans.

For all containment levels, good laboratory technique is an essential aspect for avoiding laboratory hazards. Use of elaborate safety equipment, negative air pressure management, etc. will not compensate for sloppy or unsafe practices. Rigorous aseptic technique must be practiced. Mouth pipetting is banned. Laboratory coats must be worn at all times, gloves must be used for handling potentially infectious materials, and special care taken with sharp objects. Rigorous attention must be given to sterilization, where possible by autoclaving, of all potentially infectious waste as well as used equipment. Special arrangements for the disposal of "sharps" are essential. Spills are cleaned up with an appropriate chemical disinfectant. Staff should be immunized against such diseases as hepatitis B and poliomyelitis as a matter of routine and, when vaccines are available, against more exotic agents in those special laboratories handling such viruses. Limitations should be placed on the type of work undertaken by pregnant or immunosuppressed employees.

It is also important to inactivate the infectivity of any particularly dangerous viruses employed as antigens in serological tests other than virus neutralization (e.g., arenaviruses, rhabdoviruses, togaviruses, or flaviviruses). This can be done, without destroying antigenicity, by γ-irradiation, or by photodynamic inactivation with ultraviolet light in the presence of psoralen or related compounds. Viral proteins produced by recombinant DNA technology provide a simple, standardized, and safe alternative to whole virus as antigen in many serological tests.

FURTHER READING

Jerome, K.R. (Ed.), 2010. Lennette's Laboratory Diagnosis of Viral Infections (fourth ed.), CRC Press, Boca Raton, Florida, ISBN-10: 142008495X.

Kaslow, R.A., Stanberry, L.R., LeDuc, J.W., 2014. Viral Infections of Humans, Epidemiology and Control, fifth ed. Springer, New York, ISBN 978-1-4899-7447-1.

Chapter 11

Vaccines and Vaccination

Vaccination is a highly efficient way of preventing viral diseases and remains one of the most effective healthcare measures introduced into medical practice. Indeed the control of so many diseases by vaccination is probably the single most outstanding achievement of medical science over the past century after the introduction of clean water. The eradication of smallpox is rightly hailed as a major achievement of modern medicine. Poliomyelitis is likely to be consigned to history also within the next few years. Morbidity and mortality due to measles virus have declined precipitately over the past decades, despite unwarranted public concerns in some countries resulting in a temporary drop in vaccination uptake and the re-emergence of local epidemics. Other vaccines should make a greater impact than is currently the case but there is frequently a lack of political will to implement effective and sustained vaccination programs.

Newer technologies are expected to bring improvements to existing vaccines, both in terms of delivery and efficacy, for example, the development of multi-seasonal influenza vaccines. Despite considerable effort, however, control of HIV and hepatitis C by vaccination remains out of reach for the foreseeable future, due to a number of scientific and logistic obstacles (see Chapter 23: Retroviruses and Chapter 36: Flaviviruses). Considerably more optimism surrounds the development of a quadrivalent dengue vaccine, one of the most important unmet medical needs worldwide.

Edward Jenner introduced vaccination in 1798 to protect people against smallpox using fluid from cowpox pustules. Indeed, the term vaccine is derived from the Latin word for cow, *vacca*. Nearly a century later, the concept was shown to have wider applications by Louis Pasteur, including his use of an experimental killed rabies "vaccine" to protect a boy bitten by a rabid dog. In the 1930s Max Theiler developed the 17D yellow fever vaccine, arguably one of the most significant advances in the control of an arthropod-transmitted disease (Box 11.1).

With the advent of cell culture techniques in the 1950s, a second era of vaccine development began, and many attenuated virus and inactivated virus vaccines were developed. Notably, Albert Sabin and Jonas Salk independently developed live and killed poliovirus vaccines respectively—"oral" (OPV) and "inactivated" (IPV). These illustrate two traditional strategies for the development of viral vaccines, either by attenuating virulent virus sufficiently to abrogate its disease potential while retaining immunogenicity (live attenuated vaccines), or by chemical inactivation of virus infectivity without destroying those antigens necessary to elicit a protective immune response. Each has distinct advantages and disadvantages as to efficacy, duration of immunity, delivery, cost, and safety.

Today, the field of vaccinology has entered a new era through the application of advanced technologies in a myriad of different ways. Some have produced significant new vaccines, most notably against hepatitis B and chimeric vaccines against Japanese encephalitis virus (JEV). Although other candidate vaccines produced by recombinant DNA technology are yet to realize their full potential, it is clear that these will provide major advances in developing vaccination against contemporary diseases such as dengue and Zika viruses. Despite such developments, our understanding of the mechanisms by which vaccines prevent disease is still incomplete in many instances.

VACCINE DESIGN AND PRODUCTION

The rational design of any vaccine takes account of efficacy, safety, and practical usage, including cost. Vaccine efficacy is most often measured by the prevention of clinical disease rather than an absolute prevention of infection, the latter sometimes referred to as "sterile immunity." The long-term duration of immunity is important, especially among communities where exposure is likely to occur throughout life. The number of doses required and the requirement for a cold chain become critical issues in regions where healthcare facilities are either minimal or difficult to access (Box 11.2).

Considerable attention is paid to the safety of vaccines. New products undergo rigorous safety testing prior to licensure. Although care often needs to be exercised among vaccine recipients known to have natural or acquired immunodeficiency or be pregnant, heightened concerns by the public as to the safety of individual vaccines are largely unfounded.

Vaccine design also requires knowledge of the virus, its antigens, and to what extent phenotypic and antigenic

Fenner and White's Medical Virology. DOI: http://dx.doi.org/10.1016/B978-0-12-375156-0.00011-4

BOX 11.1 Edward Jenner and Vaccination Against Smallpox

Edward Jenner (1749–1823), an English country physician and naturalist, is credited as the pioneer of vaccination. As with many physicians of his day, he practiced variolation whereby small quantities of smallpox virus were deliberately scratched into the skin of people in order to stimulate protection against naturally acquired disease. However, Jenner was intrigued by country lore that those who contracted cowpox from milking and cows appeared protected from smallpox. In 1798 one such milkmaid, Sarah Nelmes, consulted Jenner as to a rash that had developed on a hand through handling a cow (named Blossom!) with evidence of cowpox on her udders. Jenner realized this was an opportunity to test the protective effect of cowpox for someone yet to catch smallpox. Thus he chose the eight-year-old son of his gardener, James Phipps, to test his hypothesis. Not only was Jenner able to transmit cowpox from Sarah Nelmes to James Phipps, he then found James was now protected from smallpox by subjecting the boy to variolation. To Jenner's relief, James was fully protected. Although there are records of others who had considered previously the use of cowpox in humans, the challenging of James Phipps with smallpox to demonstrate immunity is the seminal contribution Jenner made, almost a century before our understanding as to the nature of infectious agents and the early days of immunology.

Portrait of Edward Jenner by an unknown artist, Wellcome Library, London.

BOX 11.2 The Objective of Vaccination

The object of vaccination is to stimulate immunological memory, thereby protecting the recipient against future disease. Any subsequent infection with wild-type virus is likely to be subclinical as a result. Vaccination should ideally also totally prevent infection in the recipient after future exposure, and prevent transmission of virus to contacts. At a population level, this will increase herd immunity and reduce the circulation of virus in the community.

TABLE 11.1 The Characteristics of an Ideal Vaccine

High efficacy in target populations
Few or no adverse reactions
Manufacture is inexpensive and easy to translate into developing countries
Safe in immunocompromised individuals and pregnant women
Easy and inexpensive to deliver
Stable during transport and storage
Induces life-long immunity
Stimulates both humoral and cell-mediated immune responses
Reproducible manufacturing process
Delivery informed by adequate and detailed epidemiological knowledge of the disease and its outcomes

variation influences the quality of protective immunity. Of these factors, antigenic drift, for example, among influenza viruses, or the presence of multiple variants as is the case with human immunodeficiency virus (HIV), presents real challenges (Table 11.1).

Attenuated Virus Vaccines

Attenuated virus vaccines have dramatically reduced the incidence of several important diseases and the spread of wild-type virus. The vaccine virus replicates in the recipient and in so doing, amplifies the amount of antigen available for presentation to the host's immune system. There are important benefits in this, since the replication of the vaccine virus mimics that of wild-type virus, the host immune response resembling what occurs after natural infection. This is not the case with either inactivated or subunit vaccines (see below). Most attenuated vaccines are injected either subcutaneously or intramuscularly, but some are delivered orally, and a few intranasally (Table 11.2, Figs. 11.1–11.3).

Attenuated virus vaccines are derived from several sources. The first vaccine, introduced by Jenner in 1798 for the control of human smallpox, utilized cowpox virus, a

TABLE 11.2 Methods for Production of Viral Vaccines

Live-Virus Vaccines

Related virus from a heterologous host species (e.g., cowpox)
Naturally occurring attenuated virus strain
Serial passage in a heterologous host animal
Serial passage in cultured cells
Selection of cold-adapted mutants

Non-replicating Native Antigen Vaccines

Vaccines produced from inactivated whole virions
Vaccines produced from native virion subunits
Vaccines produced from purified native virion proteins

Vaccines Produced by Recombinant-DNA and Other Technologies

Attenuation of viruses using recombinant DNA technology
for gene deletion or site-directed mutagenesis
Expression of viral proteins in eukaryotic (yeast, mammalian,
insect) or bacterial cells by recombinant DNA technology
In vitro expression of viral proteins that self-assemble into
virus-like particles
The expression of viral antigens by viral or bacterial vectors
delivered *in vivo*
Use of reassortants or of viral chimeras (viruses with the
replicative machinery of one virus and the protective
antigens of another)
Chemical synthesis of viral peptides
The production and administration of anti-idiotypic
antibodies
DNA immunogens

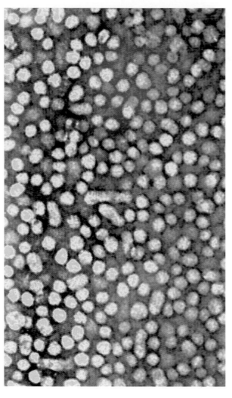

FIGURE 11.1 Hepatitis B surface antigen (HBsAg) purified from human plasma was introduced as a human vaccine in the late 1970s, but has subsequently been replaced in most countries by HBsAg particles produced by recombinant DNA technology.

natural pathogen of cattle. This virus produced only a mild lesion in humans, but it conferred protection against the severe disease because it is antigenically related to smallpox virus. The derivation of vaccinia virus subsequently used for over 150 years to eradicate smallpox is unknown, but is thought to have arisen by cross-contamination with other poxviruses during the early decades of the 19th century.

Second, serial passage in a heterologous host was a classic means of empirically attenuating viruses for use as vaccines before the advent of cell culture. For example, in the 1930s yellow fever virus was adapted to grow in mouse brain and in embryonated hens' eggs: the 17D yellow fever vaccine virus was developed by such empirical serial passage of virulent Asibi strain virus, first in mice by intracerebral inoculation, then in mouse embryo cells, and finally in chick embryo cell culture. Attenuation was laboriously assessed at every few passages by inoculation of experimental animals. At the point when the virus became sufficiently attenuated it was formulated into the 17D yellow fever vaccine that remains in extensive use today. The 17D yellow fever vaccine also offers attractive opportunities as a vector for vaccines against other flavivirus diseases, such as and dengue, diseases for which new or improved vaccines are urgently needed. These chimera

vaccine products generated by heterologous recombination are likely to offer much needed products in the next few years.

Third, the majority of attenuated virus vaccines in present use were derived empirically by serial passage in cultured cells. Adaptation of virus to more vigorous growth in cultured cells is fortuitously accompanied by progressive loss of virulence for the natural host. Loss of virulence may be demonstrated initially in a convenient laboratory model, often a mouse, before being confirmed by clinical trials. Because of the practical requirement that the vaccine must not be so attenuated that it fails to replicate satisfactorily *in vivo*, it is sometimes necessary to compromise by using a strain that may induce mild clinical signs among a few recipients.

It has been shown that for most viruses, a number of different genes contribute to virulence and tropism and do so in several different ways (see, e.g., Chapter 7: Pathogenesis of Virus Infections). Vaccine candidate viruses accumulate numerous point mutations in one or more viral genes during multiple passages in cultured cells, leading cumulatively to attenuation. In recent years, sequencing of virus genomes has brought considerable understanding about the basis of attenuation and has allowed for better predictions as to vaccine efficacy and safety.

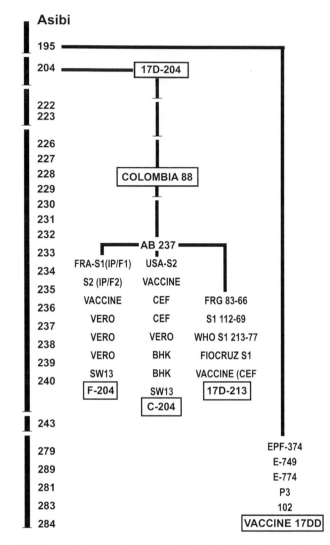

FIGURE 11.2 Passage history of the yellow fever 17D vaccine. The original Asibi monkey isolate was subcultured in embryonic mouse tissue and minced whole chicken embryo, yielding the parent 17D strain at passage 180 and the 17D-204 strain at passage 204. Further subcultures were carried out in embryonated chicken eggs, different cell lines and chick embryo fibroblasts to produce experimental vaccine batches. *Adapted from Bonaldo, M.C., et al., 2000. Mem. Inst. Oswaldo. Cruz. 95 (Suppl. 1), 215–223.*

The observation that temperature-sensitive (*ts*) mutants (mutants unable to replicate satisfactorily at temperatures greater than normal body temperature) generally display reduced virulence, suggested that such mutants might make satisfactory live vaccines. Unfortunately, candidate vaccines containing one or more *ts* mutations have displayed a disturbing tendency to revert toward virulence as a result of replication within recipients. Attention accordingly has moved to cold-adapted mutants, derived by adaptation of virus to grow at suboptimal temperatures. In general, these mutants have proven to be more stable. One rationale for the use of vaccines containing a cold-adapted mutant virus is that they may be suitable for intranasal administration—the mutant virus replicating at the lower temperature of the nasal cavity (about 33°C), but not at the temperature of the more susceptible lower respiratory tract. In 1997, after extensive clinical trials, cold-adapted influenza vaccines containing mutations in almost every gene were licensed for use in the United States.

Inactivated Virus Vaccines

Inactivated virus vaccines are usually made by exposure of virulent virus to chemical or physical agents, for example, formalin or β-propiolactone, in order to destroy infectivity while retaining immunogenicity. Initially, virus for this purpose was often obtained from infected animal sources, for example, mouse brain, but infected cell cultures provide cleaner starting material.

The need to use large amounts of antigen to elicit an adequate antibody response is a major disadvantage. Generally with such vaccines, the primary vaccination course comprises two or three injections; further "booster" doses may be required at intervals to maintain protective immunity. The chemical or physical treatment used to eliminate infectivity of inactivated virus vaccines may be damaging enough to modify immunogenicity, especially of antigens needed to elicit cell-mediated immune responses. The result is an immune response shorter in duration, narrower in spectrum for viral antigens, weaker cell-mediated and mucosal immune responses, and possibly less effective in preventing viral entry. The most commonly used inactivating agent, formalin, is known to induce irreversible changes in many viral antigens; its continued use derives from the conservative attitude of regulatory agencies and vaccine manufacturers and a paucity of research in this area. The use of β-propiolactone in the manufacture of some human rabies vaccines has advantages in that proteins are not damaged and the inactivating agent is completely hydrolyzed within hours to non-toxic products.

Non-ionic detergents such as the polyoxylene ethers are used in the case of enveloped viruses to solubilize virions and release glycoprotein peplomers and other envelope proteins. Differential centrifugation or ultrafiltration is used to semi-purify the solubilized glycoproteins prior to formulation for use as so-called "split" vaccines. Some commonly used influenza vaccines are produced by this procedure.

Vaccines Produced by Alternative Methods

Hepatitis B vaccines were the first to prevent a human cancer of viral etiology, namely hepatocellular carcinoma. The first HBV vaccines consisted of purified hepatitis B surface antigen (HBsAg) particles isolated from the blood of chronically infected carriers of the virus. The concentrated products were treated with two or three inactivating regimens to ensure that no infectious hepatitis

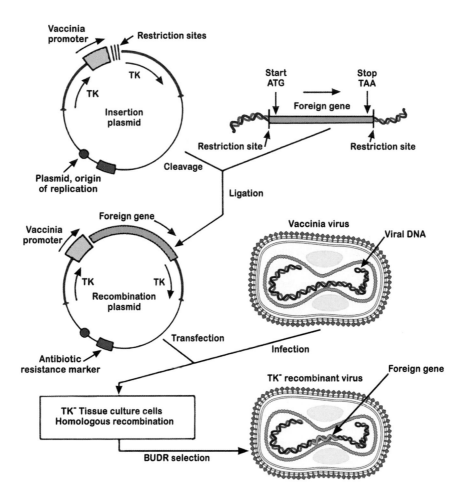

FIGURE 11.3 Method of constructing a vaccinia virus vector carrying a selected gene from another virus. This approach has been used to develop several successful veterinary vaccines, but as yet no human vaccine has been licensed. It has also been extended by using other viruses as vectors (e.g., adenoviruses, paramyxoviruses, parvoviruses), and is used for gene expression in virus research and for delivery of many kinds of therapeutic genes. As shown in the figure, a vaccinia promoter is inserted into the plasmid just upstream of the intended insertion site, vaccinia virus DNA sequences (e.g., from an interrupted thymidine kinase gene) are ligated upstream and downstream of the site, and the foreign gene is inserted. Tissue culture cells are infected with vaccinia virus and transfected with the insertion plasmid, and the vaccinia TK sequences direct homologous recombination insertion into the vaccinia virus genome. Stable recombinant viruses are then selected; in this example, the recombinant virus has an interrupted thymidine kinase gene and can be selected for by use of 5-bromodeoxyuridine in the culture medium. *TK*, thymidine kinase gene of vaccinia virus; *BudR*, bromodeoxyuridine. *Courtesy of B. Moss and J. Esposito, with permission.*

B virus remained in the final product. The initial doubt that impurities may cause unwanted autoimmune reactions proved unfounded. More recently, these vaccines have been largely but not completely replaced by HBsAg expressed in yeast cells using recombinant DNA technology.

A raft of new technologies is being explored commercially. These are directed toward solving problems of low yield or purity, or by exploiting unrelated carrier organisms to improve routes of delivery, for example, by delivery of relevant immunogens to lymphoid tissue in the gut. However, up till now most of the vaccines in widespread human use are based on well-established, conventional technologies (Table 11.2).

The problem of back mutation (that is, a mutation by which the vaccine virus regains virulence) may be largely circumvented by completely deleting those genes that are essential for virulence. The large DNA viruses, in particular, carry a significant number of genes that are required for replication *in vivo* but not for replication in cultured cells. Genetic manipulation may be used to construct deletion mutants that are attenuated and stable over many passages. Experimental herpesvirus vaccines have been constructed using this strategy; one such vaccine, a pseudorabies vaccine for swine with a deletion of the whole thymidine kinase (TK) gene of the virus, is in wide use by veterinary practitioners in many countries. This vaccine is the model for herpes simplex virus vaccines currently under development for human use.

Site-directed mutagenesis permits further variations in this strategy—not only are large deletions easy to identify, construct and test, but prescribed nucleotide substitutions and insertions are now feasible. As more becomes known about particular genes determining viral virulence and immunogenicity, these genes can be modified with the aim of improving the balance between immunogenicity and virulence.

Expression of Viral Proteins in Recombinant Systems

Recombinant-DNA technology provides a means of producing large and consistent amounts of viral proteins, that are easy to purify and can be readily formulated into vaccines. Once the critical viral protein conferring protection has been identified, the gene (or, in the case of an RNA virus, a cDNA copy of the gene) may be cloned into one of a wide choice of expression plasmids and expressed in any number of cell systems. If the immunogenic viral protein of interest is glycosylated, eukaryotic expression systems must be used to ensure the expressed protein is also glycosylated and secreted in its native conformation.

Useful eukaryotic expression systems include yeast cells (*Saccharomyces cerevisiae*), insect cells (*Spodoptera frugiperda*), and various mammalian cells deemed free of adventitious agents, for example, MRC-5 cells. Yeast offers the advantage that there is extensive experience with scale-up for industrial production; the first vaccine produced by expression of a cloned gene, hepatitis B vaccine, was produced in yeast. Insect cells offer the advantage of simple technology derived from the silk industry: moth cell cultures (or whole caterpillars) may be made to express very large amounts of viral proteins through infection with baculoviruses carrying the gene(s) of the virus of interest. The promoter for the gene encoding the baculovirus polyhedrin protein is so effective that, when the heterologous gene is inserted within the baculovirus polyhedrin gene, the viral protein may comprise up to half of the mass of the infected moth cells or caterpillars. However, mammalian cells offer the advantage over cells from lower eukaryotes in that they are more likely to possess the machinery for correct post-translational processing of viral proteins, including glycosylation and secretion.

Virus-Like Particles

Capsid proteins may assemble into virus-like particles (VLPs) when the genes coding for the production and cleavage of capsid proteins of certain non-enveloped icosahedral viruses are cloned into plasmids and co-expressed. These particles may be used as a vaccine. Examples of viruses for which VLPs have been made and shown to be immunogenic include papillomaviruses, picornaviruses, caliciviruses, rotaviruses, and reoviruses. VLPs may be equated to so-called naturally occurring "empty virus particles" in that VLPs are totally devoid of nucleic acid and therefore non-infectious. The VLPs are similar to inactivated whole virus vaccines, but avoid the need for the potentially epitope-damaging step of chemical inactivation. The first human VLP-based vaccine in wide commercial use was the vaccine against human papillomavirus, which was introduced in many countries around 2007.

Use of Live Vectors and Reassortants

Recombinant DNA technology allows the expression of viral epitopes on the surface of non-pathogenic bacteria, which can be directly administered to non-immune individuals. The general approach is to insert the DNA encoding a protective viral antigen into a region of the genome of a bacterium or one of its plasmids that encodes a prominent surface protein. Provided that the added viral protein structures do not seriously interfere with the transport, packaging, stability, or function of the bacterial protein, the bacterium can multiply and present the viral epitope to the immune system of the recipient. Many consider bacteria that multiply naturally in the intestinal lumen as ideal expression vectors for presenting protective antigens of virulent enteric viruses to the gut-associated lymphoid tissue (GALT). Vaccines under development employ attenuated strains of *Escherichia coli*, *Salmonella* spp. and *Mycobacterium* spp. for immunization against enteric organisms and/or for the preferential stimulation of mucosal immunity.

Recombinant DNA techniques allow the introduction of heterologous genes coding for protective antigens of another virus into the genome of an avirulent virus acting as a vector: the foreign gene is then expressed in cells susceptible to the avirulent carrier. Cells in which the vector virus replicates *in vivo* will express the heterologous protein, with the recipient mounting both a humoral and a cell-mediated immune response to the relevant viral antigen. Viral vectors are also used to deliver into target cells genes of interest for gene therapy, cancer therapy, degenerative disease therapy, etc.

Although vectors such as adenovirus have shown considerable promise in the laboratory, an adenovirus-vectored HIV vaccine proved unsuccessful early on in clinical trials. However, success may be in part dependent upon the particular heterologous antigen being expressed, as a similar experimental vaccine against Ebola virus has shown total protection in non-human primates and has been deployed in an attempt to control the 2014 epidemic of Ebola virus in West Africa.

Another widely used vector is vaccinia virus. Vaccination of experimental animals with these recombinant vectored vaccines has, in nearly every case, produced a good antibody response. For example, vaccinia-virus vectored rabies vaccines have been shown to protect foxes and raccoons when incorporated into orally administered baits. Since the large vaccinia genome can accommodate at least a dozen foreign genes and still be packaged satisfactorily within the virion, it would theoretically be possible to construct a vector capable of protecting against several different viral diseases. Fowlpox virus and canarypox viruses have been tested as substitutes for vaccinia virus: these viruses do not undergo a complete infection cycle in mammals, including humans, but can undergo an aborted infection cycle sufficient to allow the expression of any inserted heterologous gene, with the result that antibodies to the heterologous gene product(s) are produced. Research suggests that adenoviruses, herpesviruses, and parvoviruses also have value as vectors, and indeed may have certain advantages in terms of long-term antigen expression.

Chimeras and Reassortants

Virus strains with proven, useful vaccine properties (e.g., safety, high yield) may be altered to substitute relevant antigen genes or gene segments from wild-type pathogenic strains. The value of such viral chimeras was well established with influenza viruses even before the advent of recombinant DNA technology. Reassortant influenza viruses were produced by co-cultivation of a vaccine virus with a new isolate to encourage genome segment exchange. Viruses with the desired growth properties of the vaccine virus but with the immunogenic properties of the recent isolate were selected, cloned, and used in inactivated vaccines. This approach was adapted for the H5N1 avian influenza virus that was causing severe illness and deaths in Hong Kong in 1995. However, the H5N1 virus grows poorly in both cultured cells and embryonated hens' eggs—because it is so rapidly lethal in these systems. In this case, the virus that was used for vaccine production was a reassortant virus containing most of the genes of the high-yielding existing PR8 vaccine strain, plus the hemagglutinin and neuraminidase genes of the H5N1 virus. Furthermore, the multiple basic amino acids in the HA cleavage site that are associated with virulence in H5 avian viruses were changed to a sequence characteristic of low-virulence avian strains.

Synthetic Peptides

Techniques have been developed for locating protective epitopes on viral proteins and chemically synthesizing peptides corresponding to these epitopes (Table 11.2). Such synthetic peptides have been shown to elicit neutralizing antibodies, but in general this approach has been disappointing. One limitation is that most epitopes eliciting humoral immunity are conformational, that is, are not composed of linear arrays of contiguous amino acids, but rather are assembled from amino acids separated in the primary sequence and brought into close apposition by the folding of one or more polypeptides. An effective antigenic stimulus requires that the three-dimensional shape of an epitope present in either the native protein molecule or virus particle is maintained in a vaccine. Since short synthetic peptides lack any tertiary or quaternary conformation, most antibodies raised against these molecules are incapable of binding to whole virions. When neutralizing antibody is present after injection of a peptide, the titer is often orders of magnitude lower than that induced by inactivated whole virus vaccines or purified intact proteins.

In contrast, epitopes recognized by T lymphocytes are mostly composed of short linear peptides. Some of these epitopes are conserved between strains of virus and therefore elicit a broad T cell response. Thus attention moved toward the construction of artificial heteropolymers of T cell epitopes and B cell epitopes, in some cases also coupled to peptides facilitating fusion with cell membranes to enhance uptake. Such constructs prime T cells to respond more vigorously when either boosted with an inactivated whole-virus vaccine or when challenged by infection.

DNA Vaccination

Perhaps the most revolutionary new approach to vaccination stems from the discovery that naked viral DNA can be used as an immunogen. In the early 1990s, it was observed that a plasmid construct expressing β-galactosidase gene after inoculation into mouse skeletal muscle expressed the enzyme for up to 60 days post inoculation. In a few short years, this approach that would previously have been ridiculed by grant-giving bodies became an area of great promise, and DNA vaccines were extensively explored. Experimental DNA vaccines exist for a wide range of applications including viral, bacterial, and parasitic infectious agents, as well as for anti-tumor and anti-allergic therapies. Experimentally, the methodology can also be applied to the screening of protective antigens.

It was shown for many viruses that whole viral DNA, RNA, or cDNA of viral RNA when transfected into cells could undergo a full cycle viral replication. In principle, the technique is relatively simple; most notably, it eliminates the step of using an infectious viral or bacterial vector, or the need to prepare large amounts of antigenic protein. Recombinant plasmids are constructed to contain genes encoding viral antigens. The DNA insert in the plasmid, upon injection, is transcribed and the mRNA is translated. Antigen-presenting cells of bone marrow origin infiltrate the site and take up the antigen, from where it is transported to a regional lymph node. Direct transfection of antigen-presenting cells, particularly Langerhans cells of the skin, with the inoculated DNA also occurs at the site of injection. The expressed protein elicits an immune response similar to that during an infection. Both humoral and cellular immune responses can be measured. Most importantly, for a growing number of viruses the immune response following injection has been shown to confer protection.

An improvement in the delivery of DNA vaccines has been achieved by coating the plasmid DNA onto microparticles, commonly 1 to 3 μm diameter gold particles, and injecting an aerosol of particles using a gas-driven gun-like apparatus (the *gene gun*); less than 1 μg of DNA delivered subcutaneously by this means is often immunogenic. As with other recombinant vaccines, the incorporation of genes encoding immunostimulatory proteins including IL-2, IL-12, and GM-CSF adds to the immunogenicity of the viral DNA.

Advantages of DNA vaccines include purity, physiochemical stability, simplicity and relatively low cost of production, distribution, and delivery. Multiple antigens can be included in a single plasmid, new antigens can be added easily, and antigens are expressed in a native conformation resembling that on the virion, thereby facilitating both processing and presentation to the immune system. Repeated injection may be given without interference and T cell immunity as well as antibody responses are elicited. One interesting aspect of DNA immunization is that immunity can develop in the presence of maternal antibodies. However, for reasons not fully understood, the efficacy of DNA vaccines demonstrated in mice is not always reproduced in non-human primates or humans.

Methods for Enhancing Immunogenicity

The immunogenicity of inactivated vaccines, and especially that of purified protein vaccines and synthetic peptides, usually needs to be enhanced; this may be achieved by mixing the antigen with an *adjuvant*, by incorporating the antigen in *liposomes*, or incorporating the antigen in immunostimulating complexes (ISCOMs). Adjuvants are selected for the property of recruiting additional signaling pathways involved in the activation of the innate immune system and antigen-presenting cells, as well as promoting the persistence of the immunizing antigen through the establishment of an immunogenic depot.

Adjuvants

Adjuvants are substances that, when mixed with vaccine immunogens, potentiate the immune response resulting in the need for either a lesser quantity of antigen or fewer doses, or both. Considerable variation exists between adjuvants as to chemical composition and mode of action. Mechanisms of action include:

1. prolongation of release of antigen;
2. activation of macrophages, leading to secretion of lymphokines and attraction of lymphocytes; and
3. mitogenicity for lymphocytes.

Until recently alum was the only adjuvant licensed for use in humans, and has been widely used for many years. However, alum is not particularly effective, as cell-mediated immunity is often poorly stimulated. Mineral oil (e.g., Freund's adjuvants) and other adjuvants are used in animals but generally these are far too reactogenic to be acceptable for human use. There has been a clear requirement for better adjuvants for some considerable time, preferably chemically defined and of known mode of action.

Antibody responses can be improved by combining alum with an additional immunostimulatory compound, for example, a modified synthetic form of monophosphoryl lipid A (MPL) that is a ligand for Toll-like receptor 4 (TLR-4) on the surface of antigen-presenting cells. Such formulations also stimulate IFN-γ to produce additional primed CD4+ helper T cells. Cervarix, a VLP preparation of human papilloma virus, is so formulated and is the first vaccine to contain an additional TLR-4 ligand approved for human use. An alternative is MF59, a squalene oil-in-water emulsion successfully used to adjuvant seasonal influenza vaccines leading to greater cross-reactive antibodies with higher avidity.

Liposomes and ISCOMs

Liposomes are artificial spherical vesicles composed of a lipid bilayer, into which viral proteins can be incorporated. When purified viral envelope proteins are included, the resulting "virosomes" (or "immunosomes") resemble somewhat the original envelope of the virion. This enables not only a reconstitution of virus envelope-like structures lacking nucleic acid and other viral components, but also allows the incorporation of non-pyrogenic lipids with the properties of adjuvants.

A spherical cage-like structure 40 nm in diameter is formed when viral envelope glycoproteins or synthetic peptides are mixed with cholesterol plus a glycoside known as Quil A. These ISCOMs (immunostimulating complexes) display significantly enhanced immunogenicity. A combined influenza/hepatitis A liposome vaccine is in limited use.

VACCINE SAFETY, EFFICACY, AND PRACTICAL USAGE

The practical advantages and limitations of different vaccines can be illustrated by comparing the properties of live, non-replicating and DNA vaccines (Table 11.3). Attenuated virus vaccines delivered orally or intranasally depend critically for efficacy on vaccine replication in the intestinal or respiratory tract respectively. Interference can occur between the vaccine and other enteric or respiratory viruses incidentally co-infecting the person at the time of vaccination—a common situation among infants, particularly in poor socio-economic conditions. Interference can occur between different live viruses contained in vaccines delivered via the natural route, for example, between the three virus components of the Sabin oral polio vaccine, if

TABLE 11.3 Advantages and Limitations of Attenuated, Inactivated, and DNA Vaccines

Property	Attenuated Virus Vaccine	Inactivated Virus Vaccine	DNA Vaccine
Route of administration	Injection, inhalation, or oral	Injection	Injection, gene gun
Amount of virus in vaccine dose	Low	High	Nil
Number of doses	Single, generally	Multiple	Single, or in prime-boost regimen
Need for adjuvant	No	Yes	No
Duration of immunity	Many years	More limited, for example, several years	Not known
Antibody response	IgG, IgA (if via mucosal route)	IgG	IgG
Cell-mediated response	Good	Generally modest	Good
Heat lability	Yes, for most viruses	No	No
Interference by prior antibody	Yes	Usually no	Apparently no
Side-effects	Occasional, local, or systemic	Occasional, local	Uncertain
Use in pregnant women	Usually not advised	Yes	Yes
Reversion to virulence	Rare	No	No
Cost	Low	High	Low
Potential hazards	Contaminating viruses	Vaccine-enhanced disease	Not yet known

respective concentrations are not appropriately balanced. It is for this reason that the Sabin oral polio vaccine is routinely administered on three occasions, separated by at least two months. Interference is not a problem with systemically administered polyvalent vaccines, for example, the live attenuated measles/mumps/rubella vaccine.

IgA is the most important class of immunoglobulin relevant to the prevention of infection of mucosal surfaces, such as those of the intestinal, respiratory, genitourinary, and ocular epithelia. One of the advantages of an oral attenuated virus vaccine, for example, the Sabin polio vaccine, is that replication of the attenuated virus in the intestinal tract leads to prolonged synthesis of local IgA antibodies. By preventing infection, circulation of virus in the community is suppressed and the prospect of eradication of the virus from the local population has become possible.

The duration of immunity after vaccination with live attenuated vaccines is usually many years or decades, particularly against those viruses spreading via the blood to reach target organs. Re-infections, if they occur, are usually asymptomatic and short-lived, and lead to a boost in immunity. This solid immunity is attributable to systemic antibody of the IgG class successfully neutralizing the infectivity of any incoming virus. In contrast, protection afforded by mucosal IgA antibodies tends to be much shorter in duration, for example, with respiratory and enteric viruses where pathogenicity is largely restricted to the site of body entry. As each new vaccine is introduced, its duration of protection needs to be documented by long-term field studies.

There are particular difficulties in producing vaccines against viruses known to establish persistent infections, such as herpesviruses and retroviruses; a vaccine must be effective in inducing sterilizing immunity if it is to prevent not only the primary disease, but also the establishment of lifelong latency. As already stated, attenuated virus vaccines are generally found to be much more effective in eliciting cell-mediated immunity than inactivated viruses. Further, cell-mediated immunity is known to be the most effective arm of the immune system in modulating, if not eliminating, latent/persistent infections.

Attenuated vaccines are susceptible to inactivation by high ambient temperatures, a particular problem in the tropics. Formidable problems can be encountered in maintaining the "cold chain" from manufacturer to the point of administration in remote, often hot, rural areas. To some extent the problem has been alleviated by the addition of stabilizing agents to some vaccines (e.g., polio vaccine), selection of vaccine strains that are inherently more heat stable, and by preparing freeze-dried products for reconstitution immediately before administration (e.g., vaccinia). Portable refrigerators for use in vehicles and temporary field laboratories have proven invaluable in recent years to overcome these difficulties.

FACTORS AFFECTING VACCINE SAFETY

To be accepted, a vaccine must be safe as well as efficacious. Concerns about safety and unwarranted effects of vaccines

FIGURE 11.4 The Cow-Pock—or—the Wonderful Effects of the New Inoculation!—Publications of the Anti-Vaccine Society Print (color engraving) published June 12, 1802 by H. Humphrey, St. James's Street. In this cartoon, the British satirist James Gillray caricatured a scene at the Smallpox and Inoculation Hospital at St. Pancras, showing cowpox vaccine being administered to frightened young women, and cows emerging from different parts of people's bodies. The cartoon was inspired by the controversy over inoculating against the dreaded disease, smallpox. Opponents of vaccination had depicted cases of vaccinees developing bovine features and this is picked up and exaggerated by Gillray. Although the central figure is often assumed to be Edward Jenner, circumstantial evidence suggests this may not be so. Although the director of the Smallpox Hospital William Woodville had originally supported Jenner, he and his colleague George Pearson, were in dispute with Jenner by the time the caricature was published. *From Wikimedia Commons.*

have been expressed ever since the earliest use of cowpox as a vaccine (Fig. 11.4). Licensing authorities are extremely vigilant and insist on rigorous safety tests since residual live virulent virus in certain batches of inactivated vaccines caused a number of tragedies in pioneering days. Live attenuated vaccines present a different set of challenges that must be met before any particular product is released onto the market; these are discussed below. Vaccines are made under a broad set of guidelines, termed Good Manufacturing Practices (GMP). After successfully completing phase III studies, the release of a new vaccine is accompanied by firm data on the recorded side effects and the frequency and severity of each.

Some excellent human viral vaccines in routine use, such as rubella and measles vaccines, can produce some symptoms—in effect, a very mild case of the disease—in a minority of recipients. Attempts to attenuate virulence further by additional passages in cultured cells have been accompanied by a decline in the capacity of the virus to replicate in humans, with a corresponding loss of immunogenicity. Such trivial side effects as do occur with current human viral vaccines are of limited consequence and do not prove to be a significant disincentive to immunization, provided that parents and recipients are adequately informed in advance.

Spread of a live vaccine to contacts is not usually a problem if it occurs, but is not a preferred outcome. Some vaccine strains may revert toward virulence during replication in the recipient, or in those close contacts to which the vaccine virus has spread. Most vaccine viruses are incapable of such spread, but in those that do there may be an accumulation of back mutations that gradually leads to a restoration of virulence. The principal example of this phenomenon is the reversion to virulence of Sabin poliovirus type 3 oral vaccine, which can lead to paralytic disease in one of up to 3 million vaccine recipients.

The 17D yellow fever vaccine is widely regarded as one of the safest vaccines yet a problem arose when, during World War II, 17D yellow fever vaccine was administered to US servicemen along with human immune serum. Among approximately 2.5 million troops vaccinated, 28,600 cases of icteric hepatitis occurred, leading to 62 deaths. Careful epidemiological studies led to the conclusion that some of the lots of immune serum were contaminated with a hepatitis virus now known to be hepatitis B virus. When the use of immune serum along with the yellow fever vaccine was discontinued, the problem ceased.

Since vaccine viruses are grown in cells derived from humans or animals, there is always a possibility that a vaccine will be contaminated with another virus, derived from those cells or from the medium (especially the serum) in which those cells are cultured. For example, primary monkey kidney cell cultures, once widely used for the manufacture of polio vaccines, have, at one time or another, yielded over 75 simian viruses, some of which are considered pathogenic for man. Some early batches of inactivated poliovirus vaccine were contaminated with SV40 virus but there has been no evidence of any adverse consequences among recipients. This danger has led to a move away from primary cell cultures and toward well-characterized continuous cell lines that can be subjected to comprehensive screening for endogenous agents before being certified as safe, for example, MRC-5 cells.

Another important source of viral contaminants is bovine fetal calf serum, universally used in cell cultures; all batches must be screened for contamination with bovine viruses. Likewise, porcine viruses have been contaminants of crude preparations of trypsin prepared from swine pancreases, which is commonly used in the preparation of cell cultures. The risk of contaminating viruses is obviously greatest with live-virus vaccines, but may also occur with inactivated whole-virus vaccines, since some viruses are more resistant to inactivation than others.

Vaccination During Pregnancy

Ideally women planning a pregnancy should be up to date with immunizations at least one month before conception. These should include as a minimum influenza, measles, mumps, rubella, and chickenpox in addition to diphtheria, whooping cough, and tetanus. As high fever is linked to a slight increase in risk of fetal defects and possibly miscarriage, the possibility of vaccine-associated febrile reactions means that immunization is generally avoided during pregnancy, particularly with live vaccines. For this reason, and for the theoretical possibility of infection of the fetus, it is recommended that rubella vaccination should not be undertaken within 28 days of the intention to become pregnant, or during pregnancy, although in the limited instances where this has happened inadvertently, no harmful consequences to the fetus have been seen.

However, where a particular virus disease represents a significant risk to both a pregnant woman and the unborn child, then the benefits of immunization are usually considered as outweighing the risk of vaccine reactions.

Influenza, for example, can pose serious risks during pregnancy and thus vaccination with the standard inactivated vaccine is recommended at any time.

Vaccination of Immunocompromised Persons

Most estimates of the efficiency of immunization among immunocompromised recipients are based upon the measurement of antibodies. The titer of the antibody response may be reduced or even absent, but this is not necessarily a direct measure of protection. Limited evidence suggests that inactivated vaccines have the same safety profile in immunocompromised individuals as in those fully immunocompetent. Live attenuated vaccines are not generally recommended, although there may be exceptions, and prolonged excretion of vaccine virus by immunocompromised vaccinees is well described. Nevertheless, MMR and varicella vaccines have been successfully administered to children infected with HIV and displaying only mild or moderate immunodeficiency. Any decision to administer a vaccine has to be considered in the context of risk and benefit, for example, likely exposure to a vaccine-preventable disease and the risk of developing adverse reactions to the vaccine. In the case of immunocompetent patients facing the prospect of immunosuppressive medication, live vaccines should be given at least four weeks, and inactivated vaccines two weeks, before the start of medication.

An important precautionary measure is the immunization of household and close contacts of the immunocompromised individual in order to reduce their risk of exposure to vaccine virus undergoing reversion. Care needs to be exercised, however, in giving live attenuated vaccines, for example, rotavirus and oral poliovirus vaccines may be shed by vaccinated contacts.

PASSIVE PROTECTION

As an adjunct to the above strategies for stimulating the body to induce *active* immunity, short-term *passive* protection can be achieved by the intramuscular or intravenous administration of preformed antibody. The available preparations and policies for their use are described in Chapter 14: Control, Prevention, and Eradication.

VACCINATION POLICY AND PROGRAMS

The major human viral vaccines in widespread usage are set out in Table 11.4. Many are recommended to be given routinely to all infants or older individuals (see Table 14.2). Others are used against viruses of restricted geographical distribution or for individuals at particular risk, such as Japanese encephalitis, yellow fever, tick-borne encephalitis, rabies, and hepatitis A. There are many more vaccines in various stages of development or clinical trial, for example,

TABLE 11.4 Vaccines in Use for Routine Immunization

Bacterial	Viral—Live	Viral—Non-replicating
BCG	Polio (Sabin, OPV)	Polio (Salk, IPV)
Diphtheria (toxoid)	Measles	HBV–recombinant
Tetanus (toxoid)	Mumps	Human papillomavirus
Pertussis (acellular)	Rubella	Japanese encephalitis
Hemophilus influenzae b	Rotavirus	Tick-borne encephalitis
Pneumococcal polysaccharide	Yellow fever	Hepatitis A
Typhoid	Varicella-zoster	Influenza
Cholera	Vaccinia	Rabies
Meningococcal C conjugate		

vaccines against HIV, HCV, several herpesviruses, and several respiratory viruses. The subjects of vaccination policy and eradication of diseases by vaccination are discussed in Chapter 14: Control, Prevention, and Eradication, together with a typical schedule for the immunization of infants against common viral diseases (Table 14.2). Detailed discussion of current immunization practices and/or vaccine development prospects for individual viral diseases is provided in the relevant chapters of Part II.

Recently Introduced or Under-Utilized Viral Vaccines

Cervical Cancer Due to Human Papillomavirus (HPV) Infection

This is the second most common cancer in women, accounting for 13% of all female cancers in developing countries. There are an estimated 500,000 new cases each year. Virtually all cervical cancer cases are linked to persistent genital human papillomavirus (HPV) infection, the most common viral infection of the female reproductive tract. Hence the target group for immunization is adolescent females, although some programs now also include males. Available HPV vaccines consist of VLPs made up of the viral L1 capsid protein. These are highly efficacious in preventing infection with, and cancer due to, HPV types 16 and 18, which together are responsible for approximately 70% of cervical cancer cases worldwide. One product is also highly efficacious in preventing genital warts caused by types 6 and 11 (see Chapter 19: Papillomaviruses).

Rotaviruses

Rotaviruses are the most common cause of severe diarrheal disease in infants and young children across the developing world, accounting for around 530,000 deaths annually, with more than 85% of these deaths occurring in low-income countries in Africa and Asia: over 2 million infants are hospitalized each year suffering from dehydration, adding a further burden to often very limited healthcare resources. Over a third of diarrhea cases among hospitalized children less than five years of age have been estimated as due to rotavirus infection. Children between six months and two years of age are the most susceptible to severe gastroenteritis.

Vaccination is an important measure in the control of rotavirus diarrheal disease, along with improvements in hygiene and sanitation, oral rehydration, and zinc supplementation. Two vaccine products became available in the mid-2000s, both consisting of living attenuated strains and both compatible with national childhood vaccination schedules and the recommendations of the WHO Expanded Program on Immunization (EPI). Rotavirus vaccination is strongly recommended for developing countries where rotavirus mortality is important, and presently an increasing number of the more developed countries are adding rotavirus vaccine to their childhood vaccination schedules to prevent common non-fatal diarrheal disease. An early vaccine was withdrawn because of an increased risk of intussusception, but this concern has not been seen with current vaccines. One vaccine, RV5, is administered orally in a 3-dose series, with doses administered at ages two, four, and six months; a second vaccine, RV1, is administered orally in a 2-dose series, with doses administered at ages two and four months.

Japanese Encephalitis Virus

Japanese encephalitis virus is the main cause of encephalitis in many Asian countries. Transmission is difficult to halt, given the virus exists in a transmission cycle between mosquitoes, pigs, and/or aquatic birds. Humans become infected incidentally when bitten by an infected mosquito and the disease is usually prevalent both in rural and periurban areas (see Chapter 36: Flaviviruses). As the disease has spread over the past half-century as agricultural development and intensive rice cultivation have been expanded in many Asian countries, the need to employ effective vaccination policies has become ever greater. Indeed the disease is most likely under-reported, with annual mortality estimated at 10 to 15,000 deaths. Recent estimates put the number of clinical cases at 60,000, a third of whom develop long-term neurological impairment.

Vaccines have been available for many years, although price and inconvenient vaccination schedules together with a lack of governmental awareness as to the extent of infection have hampered immunization programs. For many years vaccines were prepared as purified, inactivated products made from either the Nakayama or Beijing strains of virus grown in mouse brain tissue. An alternative, a live attenuated SA 14-14-2 vaccine produced in China, has been extensively used both in China and in other Asian countries.

More recent cell-derived, inactivated vaccines have come onto the market, one designed to protect travelers from outside of endemic areas and the second for childhood immunization in Japan. In addition, a live recombinant product based on the yellow fever vaccine has been introduced in some Japanese encephalitis virus-endemic countries.

Therapeutic Immunization—a Goal for the Future?

Some attention is now directed to the possibility of inducing the body to clear established persistent infections (HIV, HCV, HBV) or even virus-related cancers (cervical carcinoma associated with HPV) by using a vaccine strategy to boost immunity. A number of possible endpoints could be adopted for success—a lowering of viral load, suppression of viral replication, total elimination of virus, or just slowing disease progression. These attempts, which have had limited but tantalizing results, would of course be helped by better understanding of how these viruses avoid immune elimination.

Commercial Aspects of Vaccine Manufacture

Vaccine development and manufacture are not as financially attractive to the private sector as are antiviral drugs. The development costs, including the need to carry out large-scale phase III trials to ensure safety, are considerable, and many of the populations in greatest need of new vaccines are in parts of the world where health expenditure per person is extremely low. Vaccines are used at most several times only during the life of the individual, so revenues are limited, and there is detailed community scrutiny for rare adverse events

TABLE 11.5 Commercial Attractiveness of Vaccines Compared to Antiviral Drugs

Vaccines	Antivirals
Used 1 to 3 times only, then maybe an occasional booster	Dosing is repeated, e.g., several times every day, sometimes for months
Used *before* the need arises, similar to insurance	Used *after* the subject has acquired the disease or high risk
Community scrutinizes rare adverse events critically, often does not appreciate risk	Community accepts risk/benefit ratio
Suspicion of authorities, big pharmaceutical companies	Similar

that may lead to legal ramifications. Thus, the enormous global benefit of vaccination against common serious infections is offset by the reticence of some manufacturers to take the necessary commercial risks. Efforts by government and public–private partnerships such as the vaccine alliance GAVI and the Bill and Melinda Gates Foundation, are helping to address this difficulty (Table 11.5).

FURTHER READING

Atkinson, W., Wolfe, S., Hamborsky, J. (Eds.), 2012. Epidemiology and Prevention of Vaccine-Preventable Diseases (twelfth ed.), Centers for Disease Control and the Public Health Foundation, Washington, DC.

Centers for Disease Control Health Information for International Travelers. The Yellow Book. <http://www.nc.cdc.gov/travel/yellowbook>.

Public Health England, 2013. Immunisation Against Infectious Diseases ("the Green Book"). The UK Government Stationery Office, London, UK, <https://www.gov.uk/government/collections/immunisations>.

Rubin, L.G., Levin, M.J., Ljungman, P., Davies, E.G., Avery, R., Tomblyn, P., et al., 2014. Infectious Diseases Society of America. 2013 IDSA clinical practice guideline for vaccination of the immunocompromised host. Clin Infect Dis. 58: 309–318.

Chapter 12

Antiviral Chemotherapy

Over the past two decades, antiviral therapy has undergone a revolution. After a long era of hope tempered by disappointment when there were only one or two encouragingly successful agents, many viral infections can now be treated by one or more antiviral regimens, and our understanding of chemotherapeutic strategies and drug resistance has become considerably more sophisticated. A majority of the agents in Table 12.1 have been developed in the past 20 years (Box 12.1).

In broad terms, there are three classes of chemical agents used to combat infection: (1) *Antiviral drugs* act by suppressing or preventing viral replication in infected cells, and are sufficiently non-toxic that they can be used to treat infected patients, (2) *Disinfectants* are designed to destroy the infectivity of free virus particles, but are mostly toxic for the host cell and have no significant clinical effect against established infection, (3) The many *antibiotics* now available to fight bacteria have no activity against viruses. The only circumstances when it may be appropriate to prescribe antibiotics in viral infections are (1) to prevent or treat serious bacterial superinfection of a viral disease (e.g. bacterial pneumonia complicating influenza), or (2) to play safe where there is a real possibility of a serious bacterial illness, as in meningitis or pneumonia, until laboratory identification of the etiological agent is available. However, the widespread practice of prescribing antibiotics as a knee-jerk response to any infection leads to a number of undesirable consequences and is irresponsible medical practice.

The development of antiviral drugs has lagged behind the success with antibacterial agents for a number of reasons (Box 12.2). However, three major advances in the 1980s and 1990s provided the pharmaceutical industry with considerable optimism. These were (1) the discovery and increasing use of acyclovir, (2) the production of interferon by recombinant DNA technology, and (3) the introduction of an increasing array of effective drugs against HIV coupled with a new understanding as to how these weapons may be applied in clinical practice. We are now seeing a plethora of new agents against many significant human viral diseases reaching the stage of clinical trials, combined with a greatly expanding knowledge about how to manage such treatments and associated side effects. It is noteworthy that many of the currently available antiviral drugs are for treatment of *chronic* infections, where the timing of drug administration is not as critical as for the treatment of acute infections. Also most drugs only *suppress* replication but do not by themselves eradicate infection, and true therapeutic success and eradication of infection also requires an effective immune response, otherwise rebound of replication may be seen when the administration of the drug ceases.

STRATEGIES FOR THE DEVELOPMENT OF ANTIVIRAL AGENTS

In theory, any of the steps in the viral replication cycle represent possible targets for selective chemotherapeutic attack. Steps unique to virus replication are potential targets, as are those for which an inhibitor can be devised that has a greater selective activity against a virus-specific process than against a corresponding host process. Examples are shown in Table 12.1. A further refinement of this approach is illustrated by the nucleoside analog acycloguanosine (acyclovir), which preferentially inhibits the herpesvirus DNA polymerase essential for replication of viral DNA, rather than host cell polymerases. Acyclovir is in fact an inactive prodrug that requires another herpesvirus coded enzyme, thymidine kinase, to phosphorylate it to its active form. As the viral enzyme occurs only in infected cells, such prodrugs are non-toxic for uninfected cells; the specificity of its action against viral rather than host replication is based on two separate selective steps.

New antiviral chemotherapeutic agents can come from a number of different sources:

1. Large pharmaceutical companies house large banks of compounds with possible biological activities. These are screened for a desired function, for example, inhibition of the *in vitro* replication of a particular virus, either by direct testing of a candidate drug, or as part of sophisticated very high-throughput screening protocols. An example is AZT (zidovudine), the first successful nucleoside reverse transcriptase inhibitor (NRTI) used against HIV. This was synthesized in 1964 as a possible anti-cancer drug but was found to be ineffective. Much later it was reported as having anti-retrovirus activity in 1974, and then screened and found to be active against HIV in the mid-1980s (Box 12.3).

Fenner and White's Medical Virology. DOI: http://dx.doi.org/10.1016/B978-0-12-375156-0.00012-6

TABLE 12.1 Antiviral Drugs in Clinical or Experimental Use

Process	Target	Viruses	Agent
Virus attachment to receptor	Ligand on virus capsid	Picornaviruses	Receptor analogs, disoxaril, pleconaril
Virus interaction with co-receptor	CCR5 co-receptor	HIV	Maraviroc
Conformation change during fusion of virion with cell	HR1 region of gp41	HIV	Enfuvirtide (Fuzeon)
Uncoating and release	M2 ion channel	Influenza A	Amantadine, rimantadine
Replication of viral DNA	Viral DNA polymerase	Herpesviruses	Nucleosides; acyclovir, valacyclovir, famciclovir
		Herpesviruses (CMV)	Ganciclovir, valganciclovir, cidofovir
		Herpesviruses (CMV)	Pyrophosphate analog phosphonoformic acid (foscarnet)
		Hepatitis B	Adefovir, entecavir, lamivudine, telbivudine, emtricitabine, tenofovir disoproxil fumarate
Replication of viral RNA	NS-5B polymerase	HCV	Sofosbuvir (nucleoside) dasabuvir (non-nucleoside)
Reverse transcription	Reverse transcriptase (RT)	HIV	*Nucleoside RT inhibitors*; zidovudine, didanosine, zalcitabine, stavudine, lamivudine, abacavir, emtricitabine (FTC)
Reverse transcription	Reverse transcriptase (RT)	HIV	*Non-nucleoside RT inhibitors*; nevirapine, delavirdine, efavirenz
Reverse transcription	RT, DNA polymerase	HIV, HBV	Tenofovir disoproxil fumarate
Integration of viral genome	Viral integrase	HIV	Raltegravir, elvitegravir
Cleavage of viral polyprotein precursors	Viral protease	HIV	Saquinavir, ritonavir, indinavir, nelfinavir, amprenavir, lopinavir
	NS3-NS4A serine protease	HCV	Boceprevir, telaprevir, simeprevir, asunaprevir, paritaprevir
RNA virus replication	Multiple targets	HCV (with interferon) Respiratory syncytial virus	Ribavirin
	NS5A	HCV	Daclatasvir, ledipasvir
Release of progeny virions from cell	Viral neuraminidase	Influenza A and B	Zanamavir, oseltamivir
General antiviral and immunomodulatory effects	Multiple IFN-stimulated genes	HBV, HCV (with ribavirin), AIDS-related Kaposi's sarcoma	Interferon-α

BOX 12.1 Major Landmarks in the History of Antiviral Chemotherapy

- 1962—idoxuridine, adenine arabinoside; these were relatively toxic, and their efficacy and value limited to specific situations, for example, topical use (for idoxuridine)
- 1983—acyclovir was a major breakthrough. It provided proof that highly effective, safe antiviral drugs were possible, and showed an example of how to do it
- 1980s—interferon-α became used more widely following its cloning, allowing its benefits and toxicity to be defined better

- 1990—AZT (zidovudine); the first RT inhibitor of major benefit in HIV
- Mid-1990s—protease inhibitors and Highly Active Anti-retroviral Therapy (HAART) for HIV infection—interferon/ribavirin for hepatitis C
- Early 2000s—lamivudine and other RT inhibitors for HBV, neuraminidase inhibitors for influenza
- 2011—directly acting antivirals against the HCV protease and RNA-dependent RNA polymerase

BOX 12.2 Why Has Development of Antiviral Drugs Lagged Behind the Introduction of Antibiotics?

1. Unlike bacteria, replication of viruses is carried out within host cells and largely by host biochemical processes. Hence it is more difficult to find inhibitors of virus replication that are not also damaging to host cells or their functions, leading to side effects
2. With many acute viral infections, the stage of peak virus replication is finished by the time the patient seeks treatment
3. For many agents, there are problems in attaining adequate intracellular drug concentrations, or adequate distribution of drug to the site of infection
4. Screening and testing candidate drugs can be more difficult, because, for example, there are no certified animal models
5. Until recently, accurate laboratory diagnosis has often taken longer for viral than for bacterial infections
6. Drugs shown to be effective *in vitro* are often ineffective *in vivo* for a variety of other reasons

BOX 12.3 Sources for New Candidate Antiviral Drugs

1. Blind screening of banks of randomly synthesized compounds
2. Selection or synthesis of compounds known or predicted to inhibit a target process in virus replication, for example, by binding to an active site or intermediate
3. Modification of compounds already known to have antiviral effects ("leads"), to improve efficacy or overcome drawbacks
4. Naturally occurring substances or products reported to be of therapeutic benefit

2. Knowledge of the detailed structure of an active site critical for virus replication, for example, a viral enzyme, ligand for a receptor, or a regulatory target, allows for the creation of specific inhibitors through rational design. The neuraminidase inhibitors, a class of potent inhibitors of influenza replication, were developed by Mark von Itzstein after the fine structure of the neuraminidase protein was solved by Peter Colman using x-ray crystallography; this knowledge allowed the rational synthesis of chemical derivatives that bind tightly to the active pocket of the enzyme. A similar approach can be used to develop drugs that act by binding directly to the capsid (or envelope) of the virion itself, thereby blocking early steps in the replication cycle, for example, attachment, penetration, or uncoating. Once the three-dimensional structure of the whole surface of isometric virions such as picornaviruses is known, the receptor-binding site (the ligand) on the critical capsid protein

can be characterized in atomic detail. Complexes of viral proteins with purified cell receptors or receptor mimics can be crystallized and examined directly. The receptor-binding site on the virion has generally turned out to be a "canyon," cleft, or depression on the external surface of the protein. Further steps are taken to analyze the structure of the viral protein after binding to a compound known to neutralize infectivity, to map the particular amino acid residues found to be substituted in resistant mutants of virus, or to use site-specific mutagenesis in order to identify critical residues. This information can then be exploited to design better synthetic drugs, using computer modeling to optimize the fit and the binding energy of the drug–virus interaction.

3. Once an agent is found which displays a degree of specific inhibition for a viral enzyme or other process integral to viral replication ("lead" molecule), related analogs (congeners) of the prototype are then synthesized with a view to enhancing its activity, solubility, and bioavailability, or reducing its toxicity.

4. It is often proposed that natural products (e.g., "folk medicines") that are reported to be of benefit against particular diseases may contain active ingredients from which new classes of potent antivirals can be developed. This area of discovery continues to hold hope, but all too often these products turn out to be complex mixtures and the apparent therapeutic benefit seems to wane as the material is fractionated into its components.

Other approaches based on similar principles include the targeting of regulatory genes crucial for optimal virus replication, such as the TAR region of nascent HIV RNA transcripts, to which the product of the HIV *tat* gene binds. Potentially, the replication of such viruses could be blocked by agents that bind either to the protein product of such a viral regulatory gene, or to the recognition site in the regulatory region of the viral genome with which that protein normally interacts. Antisense RNAs, micro-RNAs (miRNAs), and small interfering RNAs (siRNA) are other approaches that are yet to be fully exploited.

A successful new antiviral drug typically goes through three stages of evaluation—initial *in vitro* screening in cell culture to determine its therapeutic index against a particular virus, followed by animal studies to determine pharmacokinetics, toxicity, and (if a suitable model exists) efficacy, followed finally by successive human phase I, II, and III studies. For any one successful drug this process can take 10 years or longer and can cost many millions of dollars. Many thousands of other potential candidates usually need to be discarded along this route, having failed one of the successive stages listed in Box 12.4.

The classical first test of any putative chemotherapeutic agent is, of course, inhibition of viral replication. In the presence of dilutions of the agent, the multiplication of

BOX 12.4 Sequential Stages in the Development of New Antiviral Drugs

I.   Screening *in vitro* for antiviral inhibitory activity
     Infected and control cell cultures are incubated with dilutions of the compound to be tested.

   Chemotherapeutic index = (1) the minimum drug concentration toxic to cells (e.g., measured by trypan blue staining, reduced plating efficiency, or rate of cell division), divided by (2) the minimum drug concentration that inhibits virus replication (e.g., measured by virus yield or cytopathic effect). A promising drug would have a C.I. >100

II. Animal studies
   • *Pharmacokinetics*; includes mechanism and routes of absorption; tissue distribution; metabolism, detoxification, excretion; half-life $(t\frac{1}{2})$ in different tissue compartments
   • *Toxicity*—acute (biochemistry, hematology, immuno-suppression)—*chronic* (allergenicity, mutagenicity, carcinogenicity, teratogenicity)
   • *Efficacy* in experimental infection *in vivo*, if such a model is available

III. Human studies—Phase 1, 2, and 3
   • *Route*; oral versus parenteral (intramuscular, intravenous, subcutaneous), topical (e.g., nasal, aerosol, skin, eye) versus systemic
   • *Dose and timing*
   • *Efficacy*; this requires
     1. careful selection of population to be studied (e.g., enlist individuals already being exposed or infected)
     2. comparison with (usually) a control group receiving current standard treatment (withholding treatment from a placebo control group is not ethically acceptable)
   • *Toxicity*; recording and evaluation of major and minor adverse events in control and treated group

suitable indicator viruses in cultured cells is measured by a reduction in the yield of virions or of some convenient viral marker. The toxicity of the drug for uninfected human cells may be measured crudely by the cytopathic effect or, more sensitively, by reduction in cell plating efficiency or cell doubling time. In general, only those agents displaying a therapeutic index of at least 10 and preferably 100 to 1000 are worth pursuing further (Box 12.4).

Ideally, the drug should be water-soluble, chemically and metabolically stable, moderately apolar, and satisfactorily taken up into cells. Pharmacokinetic studies, first in animals and then in humans, address such questions as the mechanism and rate of absorption following various routes of administration, tissue distribution, metabolism, detoxification, and excretion of the drug. Tests for acute toxicity encompass comprehensive clinical surveillance of all the body systems, biochemical tests (e.g., for liver and kidney function), hematology, tests for immunosuppression, and so on. Longer-term investigations screen for chronic toxicity, allergenicity, mutagenicity, carcinogenicity, and teratogenicity.

CLINICAL APPLICATION

Formulation and Methods of Delivery

The route of administration of an antiviral agent is a prime consideration in assessing its general acceptability. The oral route is, naturally, by far the most convenient for the patient. Nasal drops or sprays may be acceptable for upper respiratory infections but can be irritating, whereas continuous delivery of aerosols through a facemask or oxygen tent is generally appropriate only for very sick and hospitalized patients. Topical preparations (creams, ointments, etc.) are satisfactory for superficial infections of skin, genitalia, or eye, provided the infection is relatively localized; penetration of drugs through the skin can be enhanced by mixing with substances such as polyethylene glycol. Parenteral administration is the only option in the case of some drugs and may, in any case, be required for serious systemic infections; intravenous infusion usually necessitates hospitalization.

Some drugs have to be used at very high, potentially toxic concentrations because of poor solubility or poor penetration into cells. Delivery of antiviral concentrations of compounds into cells can sometimes be achieved either by incorporating the drug into liposomes or by conjugating the compound to a hydrophobic membrane anchor. Sophisticated chemistry may also be required to modify potential antivirals, such as synthetic peptides or oligonucleotides, which are otherwise rapidly degraded intra- or extracellularly. Some experimental antiviral drugs are being conjugated to antiviral antibodies, or incorporated into liposomes coated with such antibody, to direct these to virus-infected cells.

Emergence of Drug-Resistant Mutants

With almost every new antiviral agent, drug-resistant mutants soon emerge *in vitro* and *in vivo*, especially during long-term therapy of chronic infections and among immunocompromised patients (Box 12.5). In its simplest form, resistance may be due to a single point mutation in the gene encoding the particular viral protein that is the target of the compound. Stepwise increases in the degree of resistance may occur as further nucleotide substitutions accumulate, often in a particular order. Drug-resistant strains often show less replication fitness and/or less virulence, and may be replaced by wild-type virus in cell culture or in a patient unless maintained by the selection pressure of the continuing presence of the drug. However,

BOX 12.5 Resistance to Antiviral Drugs

- During replication of virus nucleic acids, errors (base changes) continually occur. Error rates range from 1 in 10^{-4} or less. Selection then occurs, and any base change with a replication advantage will be amplified. We will then see this as a "mutation"
- Drug resistance arises by single or multiple base changes, that may affect a key viral protein or nucleic acid motif or sequences affecting this, for example, folding, leading to reduced susceptibility to the action of the drug
- Resistant strains are often less *virulent* and/or *replication competent* than wild-type virus, but are given a selective advantage and maintained by the selection pressure of drug presence
- The drug may still be clinically useful even after resistant strains appear
- May get *rebound* of wild-type virus replication when the drug is withdrawn
- *Compensatory mutations* may develop that enhance the replication competence of virus containing a drug resistance mutation
- Assays for drug resistance may be
  (1) Clinical, that is, the reappearance of virus, or an increase in the virus level in the patient during therapy
  (2) Phenotypic *in vitro*, that is, virus replication occurring in cell culture in the presence of higher concentrations of drug
  (3) Genotypic, that is, detection of base changes in the virus sequence that are known to be associated with drug resistance

as a demonstration of the power of selection, such resistant strains may also develop further compensatory mutations that enhance replication fitness.

Clinical virus isolates may be tested for drug sensitivity by growth in cultured cells in the presence of serial dilutions of the agent (a "phenotypic" assay). For virus/drug combinations where resistance is regularly associated with particular mutations, the genome sequence of the patient's isolate is matched against a library of known sequences; this allows prediction of the resistance pattern of the new isolate and choice of optimal antiviral therapy. This approach is most highly developed in the case of HIV. Of course, often the first indication during long-term treatment of a persistent infection that resistance has developed may be clinical, that is, a deterioration in the patient's symptoms or an increase in viral load.

Resistance mutations can only develop while a virus is replicating, a situation most likely to develop during monotherapy where the drug concentration is intermediate, that is, sufficient to create selection pressure but not high enough to prevent replication completely. The principles needed to minimize this problem are well understood from experience with antibiotic resistance in bacteria and cancer chemotherapy. Antiviral agents should be administered in sufficiently high dosage and without interruption, so that inhibitory drug levels are maintained. Combination therapy with two or more agents (preferably with distinct modes of action) is theoretically attractive and has been shown with HIV to greatly delay the appearance of resistance and improve the therapeutic response. Furthermore, if this allows one or both drugs to be given at a lower dose, combination therapy can reduce the incidence of toxic side effects. Patient compliance became a major difficulty for the treatment of HIV infections requiring the maintenance of strict drug regimens, as patients eventually grow tired of maintaining strict drug regimens necessitating the taking of 20 pills at different times each day, particularly when they start to feel better! It was a major advance with the licensing in 1997 of a single tablet containing two drugs (Combivir) and later (2006) three drugs, taken once a day (e.g., Atripla, containing emtricitabine, tenofovir, and efavirenz); six more combination treatments have been approved by the United States Federal Drug Agency in the past 4 years.

CLINICAL STRATEGIES

Antiviral drugs are used successfully with a number of different strategies in mind, including:

1. *Long-term therapy in chronic infection*, particularly HIV, HBV, and HCV. It is important to realize that different possible therapeutic goals may be targeted, for example (a) eradication of virus from the body, (b) suppression of virus replication so that viral load in the plasma is greatly reduced or undetectable, (c) delaying or preventing clinical disease, disease progression, or death, or (d) improving laboratory markers of disease, for example, liver function tests (hepatitis), and CD4 cell count (HIV). These options should be clearly understood both in the planning of clinical treatment and in the design of drug trials. The outcome of a long-term treatment course is designated as either IR (initial response), ETR (end-of-treatment response, i.e., possibly followed by rebound), or SR (sustained response, i.e., no rebound occurring).

2. *Immediate treatment of acute infection*, for example, neuraminidase inhibitors for influenza, acyclovir or a derivative for acute herpes simplex infection. As in many of these infections the virus replication peaks early in the clinical disease and wanes thereafter: a rapid initiation of treatment has been shown to be very important to achieve optimal benefit.

3. *Prophylaxis against disease progression or recrudescence, or to prevent viral shedding.* Patients suffering from frequent recrudescence of genital

herpes can have their lives transformed by long-term daily maintenance therapy with acyclovir, famciclovir, or valaciclovir. Similarly, use of valganciclovir in transplant patients, either prophylactically to all those with pre-existing CMV antibody (i.e., persistent infection), or pre-emptively for those found to have laboratory evidence of CMV replication, has led to a major reduction in CMV disease among these patients.

4. *Prevention of virus transmission.* Without intervention, approximately 25% of infants born to HIV-infected mothers become infected. This can be reduced to approximately 2%, a dramatically important result, by an appropriate combination of three antiviral drugs given to the mother in late pregnancy and to the infant after birth, combined with Caesarean section if viral suppression has not been achieved by the time of labor, and combined with avoidance of breastfeeding. Other regimens include intravenous zidovudine, or nevirapine as a short course or single dose to mother and infant at delivery.

Antiviral drug regimens are also used for post-exposure protection against HIV transmission to individuals following known sexual exposure to HIV. Recently, daily *pre*-exposure prophylaxis has been approved for individuals thought to be at on-going high risk of acquiring HIV, using the combination tablet Truvada (tenofovir and emtricitabine). Wider adoption of this approach in the future could even bring some similarity between HIV prevention and malaria prophylaxis.

MECHANISMS OF ACTION AND ROLE OF INDIVIDUAL ANTIVIRAL DRUGS

Interferons

The interferons (IFNs) have offered potential advantages as ideal antiviral agents since being discovered in 1957. These are natural cellular products of viral infection and display a broad spectrum of activity against essentially all viruses (see Chapter 5: Innate Immunity, Chapter 7: Pathogenesis of Virus Infections, and Chapter 8: Patterns of Infection). Early clinical trials were conducted with inadequate amounts of semi-purified interferons produced either by treating cultured human leukocytes or fibroblasts with a paramyxovirus, or with a synthetic double-stranded RNA. However, the cloning and expression of the gene for human interferon α (IFN-α) in *Escherichia coli* in 1980 allowed the production of therapeutic quantities of IFN-α at a greatly reduced cost, allowing its true clinical role to be explored. Since then, the genes for all known subtypes of human IFN-α as well as IFN-β and IFN-γ have been cloned in prokaryotic and/or eukaryotic cells.

Interferons are not effective by mouth and must therefore be administered by injection. IFN-α is much more active *in vivo* than IFN-β or IFN-γ, probably because the latter do not achieve or maintain the required blood levels after intramuscular administration. Toxic side effects are regularly observed and may be significant with doses in excess of 10^7 units per day, even when highly purified cloned IFN subtypes are employed. Fever regularly accompanies high doses but lasts only a day or so. Severe fatigue is the most debilitating symptom and may be accompanied by malaise, anorexia, myalgia, headache, nausea, vomiting, weight loss, erythema, and tenderness at the injection site, partial alopecia (reversible), dry mouth, reversible peripheral sensory neuropathy, or signs referable to the central nervous system. Depression is a common and sometimes severe side effect. Various indicators of myelosuppression (granulocytopenia, thrombocytopenia, and leukopenia) and abnormal liver function tests, both reversible on cessation of therapy, are regularly observed should high-dose interferon administration be prolonged.

Despite its promise, there are only a limited number of situations where interferon-α has become standard therapy. These include hepatitis C (in combination with ribavirin), hepatitis B, and AIDS-related Kaposi's sarcoma. Genital warts have been successfully treated, and juvenile laryngeal papillomatosis, a severe condition calling for repeated surgical removal following recurrences, can be arrested by local injection of interferon; however, the tumors reappear when therapy is withdrawn (Chapter 19: Papillomaviruses). Other non-viral conditions that may undergo temporary remission or partial regression with vigorous interferon therapy include hairy cell leukemia, malignant melanoma, chronic myelocytic leukemia, and in the case of multiple sclerosis, interferon-β. In these situations interferons may be acting not as antivirals but as cytokines exerting immunomodulatory effects. Because of the above pattern of side effects, long-term systemic treatment with interferon is often uncomfortable or unpleasant, and it is not uncommon for patients undergoing long-term treatment for hepatitis C to withdraw from their treatment course. The recent development of very effective interferon-free antiviral regimens for treatment of hepatitis C is a major advance that has been hailed by patients and clinicians alike.

Blocking Attachment or Fusion

One theoretical option for antiviral therapy is to inhibit the first step in the viral replication cycle, namely, attachment of the virion to its specific receptor on the host cell membrane. Substances designed to mimic either the cell receptor or the viral ligand should theoretically accomplish this. For example, attachment of HIV to its receptor (CD4) can be blocked by soluble CD4 (which binds to the virion), or by a synthetic peptide corresponding to the ligand on the HIV envelope glycoprotein gp120 that binds to the cell receptor. Another example of the latter is the blockage of the HIV co-receptor CCR-5, using either a ligand mimic or an antibody that binds to the site. Maraviroc is a CCR5 co-receptor antagonist now approved for treatment of HIV; when co-administered with standard treatment it has been shown to lead to an improved outcome. A major problem with ligand mimics is that saturation of cell receptors may occur, and therefore interfere with the normal physiological function of that membrane glycoprotein. Maraviroc has been reported to cause allergic reactions and hepatotoxicity. Receptor mimics may be safe but would need to be confirmed as not eliciting an autoimmune response.

X-ray crystallography has provided detailed information about the binding site of a wide range of antiviral agents that block the uncoating of picornaviruses. Many of the studies to date have used human rhinovirus type 14 (HRV-14) as a model. Most of the drugs, in spite of a diversity in chemical structure, bind to the same site on HRV-14, namely, a hydrophobic pocket that lies immediately beneath the floor of the canyon that comprises the ligand (receptor-binding site) on the viral capsid protein VP1. Following binding of the drug, hydrophobic interactions result in deformation of the canyon floor; this inhibits virion attachment to cell receptors, but, more importantly, also inhibits uncoating of the virion. This is thought to occur by locking VP1 into a conformation, thus preventing the disassembly of the virion that normally occurs in the acidic environment of the endosome. When administered prophylactically rather than therapeutically, antivirals of this nature have been claimed to reduce the symptoms of the common colds induced by certain rhinovirus serotypes but not others. A promising example of such a drug was Pleconaril, which was shown in randomized double-blind clinical trials to reduce the symptoms in patients with self-diagnosed colds: resistant virus strains have been reported. However, its efficacy was not fully established, and as a consequence the Food and Drug Administration of the United States declined to approve its use in 2003. While this approach has been slow to bear fruit, it does engender some optimism that antiviral drugs may eventually provide a better answer to the virologists' age-old challenge, "when will you produce a cure for the common cold?"

The anti-HIV drug enfuvirtide (T-20, Fuzeon) represents the first example of the novel approach of fusion inhibitors. During fusion between the HIV envelope and the host cell membrane, a necessary step is binding between the HR1 and HR2 regions of the gp41 envelope protein which leads to gp41 refolding. The short peptide enfuvirtide mimics the HR2 segment of the gp41 envelope protein and binds to the HR1 region of gp41, thereby preventing HR1–HR2 binding and virus–cell fusion. Being a 36-amino acid peptide, enfuvirtide has presented a special challenge in drug synthesis and delivery. It is administered subcutaneously, being used in combination with other drugs where other treatments have failed. Local side effects at the injection site, and a range of systemic side effects, are not uncommon.

Blocking Uncoating—Ion Channel Blockers

In the 1960s the simple three-ringed symmetrical amine amantadine was synthesized and shown to inhibit the replication of influenza A viruses (but not influenza B viruses). The principal target of amantadine is the protein M2, which is a minor component of the influenza viral envelope. Influenza B lacks an M2 protein, and thus amantadine's specificity for influenza A can be understood. M2 forms a tetrameric transmembrane ion channel that reduces the pH gradient across the envelope of incoming virions within acidic endosomes. It is also thought to play a role in assisting the transport of newly synthesized HA to the plasma membrane by reducing transmembrane gradients across the trans-Golgi cisternae. Thus, amantadine acts at these two distinct steps in the replication cycle as an ion channel blocker. First, by raising the pH of the endosome, amantadine prevents the pH 5-mediated conformational change in the HA molecule required for fusion of viral envelope with endosomal membrane. Second, later in the replication cycle, by disturbing the ionic environment within the exocytic pathway, amantadine prevents newly synthesized HA from assuming the correct conformation for incorporation into the envelope of budding virions.

Therapeutically, amantadine has been reported to reduce the severity of symptoms in about 50% of cases, but only if given both within the first 24 to 48 hours and at high doses. Administered prophylactically, it can significantly reduce the incidence of clinical influenza (50% to 90% in various trials). However, in practice, its use as a prophylactic demands the ingestion of 200 mg daily for 1 to 2 months from the commencement of an influenza epidemic in the community, and it is clear that vaccination presents a safer and cheaper alternative.

Amantadine has a narrow therapeutic window, with side effects commonly occurring at doses that do not greatly exceed the therapeutic dose. Such side effects relate mainly to the central nervous system (loss of concentration, insomnia, nervousness, light-headedness, drowsiness, anxiety, confusion) but are generally reversible and mild. In 1969 the drug was found to relieve symptoms of Parkinson's disease and other extrapyramidal syndromes, and today is more widely used for this indication than as an antiviral.

Rimantadine (Fig. 12.4), is a methylated derivative that shows less central nervous system side effects and is the drug of choice in most cases. Both drugs are given orally, but either may be delivered to hospitalized patients by aerosol spray. As a majority of recent isolates of H3N2 and pandemic H1N1 show resistance to adamantanes, the newer neuraminidase inhibitors have largely replaced the use of both drugs.

Although the above compounds are relatively specific for influenza A virus, different short hydrophobic transmembrane proteins that function as ion channels are encoded by a number of different viruses; these include the vpu protein of HIV, the E protein of SARS-CoV, and the p7 protein of HCV. Drugs that act as ion channel blockers against some of these are under active development, and may represent the prototype of a new class of compounds active against many other viruses.

Inhibitors of Viral DNA Polymerase

Many of the successful inhibitors of viral replication are nucleoside analogs, with antiviral activity particularly against the herpesviruses or HIV. The early prototypes, such as adenine arabinoside, are relatively undiscriminating inhibitors of both cellular and viral DNA synthesis. Understandably, these early compounds often produced toxic side effects, directed especially at dividing cells in the bone marrow and gastrointestinal tract.

Acycloguanosine (Acyclovir) and Homologs

A major breakthrough in antiviral chemotherapy occurred in 1977 when Elion and colleagues developed a prodrug that depends on a viral enzyme to convert it to its active form. Acycloguanosine, now commonly known as acyclovir, is a guanine derivative with an acyclic side chain, the full chemical name being 9-(2-hydroxyethoxymethyl) guanine (Fig. 12.1). Its unique advantage over earlier nucleoside derivatives is that the herpesvirus-encoded enzyme, thymidine kinase (TK), which has broader specificity than cellular TK, is required to phosphorylate acycloguanosine intracellularly to acycloguanosine monophosphate (ACG-P); a cellular GMP kinase then completes the phosphorylation to the active agent, acycloguanosine triphosphate (ACG-PPP) (Fig. 12.2). Further, ACG-PPP inhibits the herpesvirus-encoded DNA polymerase at

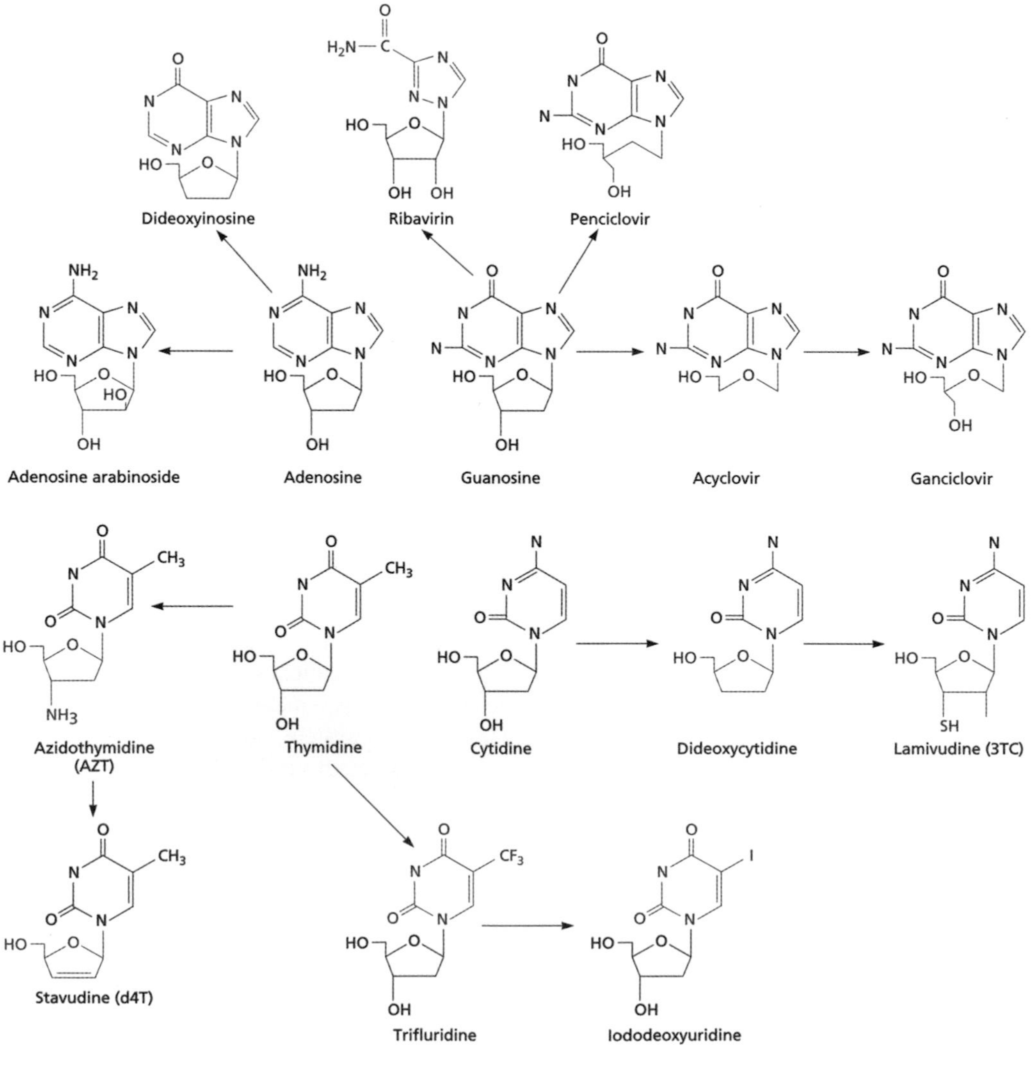

FIGURE 12.1 Many successful inhibitors of viral replication are nucleoside or nucleotide analogs. The four natural deoxynucleotides are highlighted in the yellow central box, and arrows connect these to related antiviral drugs. Chemical modifications giving rise to each antiviral drug are highlighted in red. *Reproduced from Flint, S.J. et al., 2009. Principles of Virology, third ed., vol. II, p. 292, ASM Press, Washington, DC, with permission.*

least 10 times more effectively than it does cellular DNA polymerase α. It acts as both inhibitor and substrate of the viral enzyme, competing with GTP and being incorporated into DNA: chain termination is the result as acyclovir lacks the 3'-hydroxyl group required for chain elongation. Since activation of the prodrug needs the viral TK, acyclovir is essentially non-toxic to uninfected cells but is powerfully inhibitory to viral DNA synthesis in infected cells, giving it much greater selectivity than is the case with the earlier nucleoside analogs.

Herpes simplex viruses types 1 and 2 (HSV-1 and -2) are both highly sensitive to acyclovir; varicella-zoster virus (VZV) is susceptible at somewhat higher concentrations of the drug. Other human herpesviruses, lacking a gene coding for TK, are susceptible only at much greater doses; this results from the limited production of ACG-P by cellular GMP kinase. The relative sensitivity of different herpesviruses seems to depend on a rather complex interplay of at least three variables: (1) the efficiency of the virus-coded TK (if any) in converting acyclovir to ACG-P, (2) the efficiency of cellular kinases in converting this intermediate to ACG-PPP, and (3) the susceptibility of the viral DNA polymerase to ACG-PPP. The use of acyclovir for treatment of various herpesvirus diseases is discussed in Chapter 17: Herpesviruses. Acyclovir (Zovirax) may be delivered orally, by slow intravenous infusion, or topically as an aqueous cream. As anticipated from

in vitro studies, the drug is essentially non-toxic. Acyclovir levels must be carefully monitored in patients with dehydration or renal impairment, as the drug, which is excreted unchanged through the kidneys, is relatively insoluble, and crystalluria may occur. Nevertheless, its use has revolutionized treatment and suppression of severe muco-cutaneous herpes simplex infections, herpes simplex encephalitis, and disseminated infections with HSV, varicella-zoster, and CMV.

Acyclovir-resistant mutants of HSV can be recovered *in vivo* and in cell culture. The mutation is usually located in the gene coding for the viral thymidine kinase, but more rarely is seen in the DNA polymerase gene. There are two kinds of TK mutants: (1) those failing to produce appreciable levels of TK (TK$^-$), (2) those in which the enzyme is produced but has an altered substrate specificity such that it can no longer satisfactorily phosphorylate acyclovir (TK^a). TK$^-$ mutants, the most common, may contain a mutation, deletion, or insertion leading to premature termination of translation or the production of a non-functional enzyme, whereas TK^a mutants result from a point mutation causing a more subtle alteration in substrate specificity so that the enzyme no longer phosphorylates acyclovir. Nearly all TK$^-$ mutants, while able to establish latent infection of ganglia, have reduced virulence and reduced ability to reactivate; these strains most commonly arise, and cause severe disease, in immunocompromised hosts.

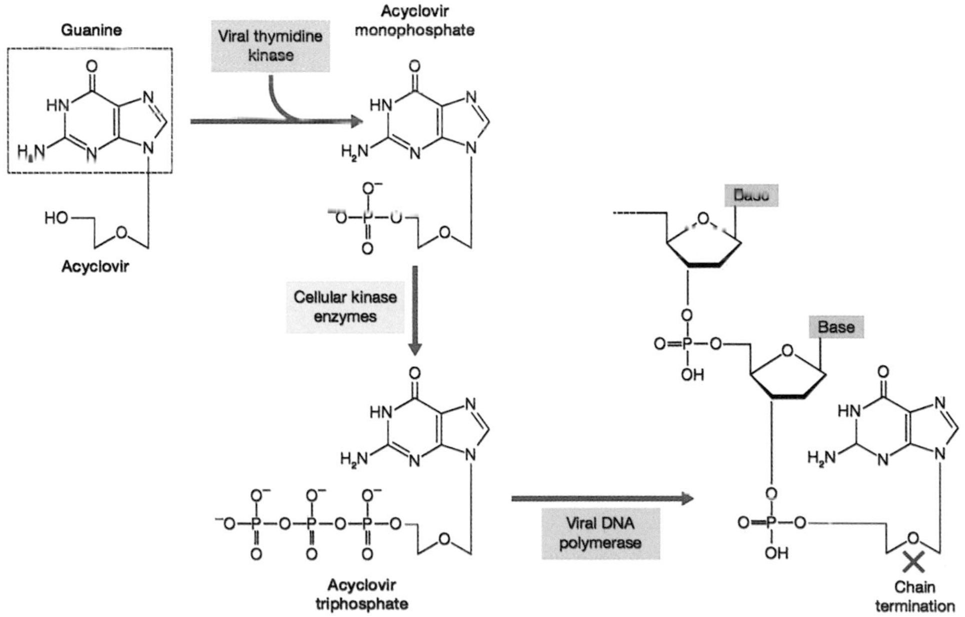

FIGURE 12.2 Mechanism of selective inhibition of herpesvirus replication by acyclovir. Acyclovir is an inactive prodrug, and must be phosphorylated to the active triphosphate compound within the cell. The first step, to the monophosphate is carried out by the viral thymidine kinase but not the host cell kinases, so this step can only proceed in infected cells. The monophosphate is then converted to the active triphosphate by cellular kinases. ACV-TTP then inhibits the viral DNA polymerase 10 times more than the host DNA polymerase α; furthermore, when it is incorporated into the growing DNA, it acts as a chain terminator because it has no 3' hydroxyl group of the sugar ring. *Reproduced from Flint, S.J. et al., 2009. Principles of Virology, third ed., p. 694, ASM Press, Washington, DC, with permission.*

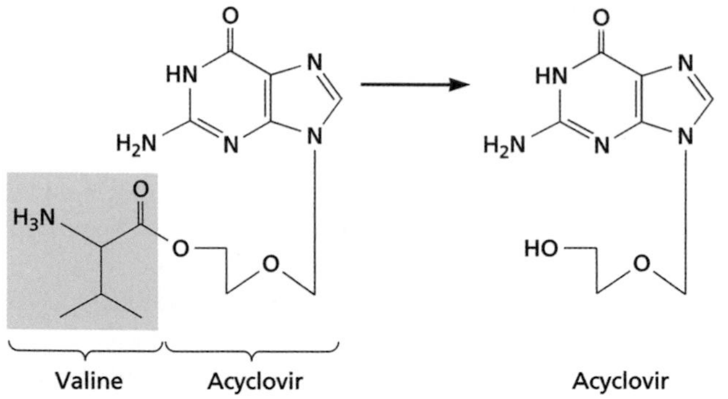

FIGURE 12.3 Structure of valacyclovir. Valacyclovir, the valine ester of acyclovir, is absorbed from the gut into the circulation three to five times faster than acyclovir. Once taken up by the cell, host enzymes cleave off the valine side chain. Its longer half-life in the body means that doses do not need to be as frequent as for the parent compound acyclovir. *Reproduced from Flint, S.J. et al., 2009. Principles of Virology, third ed., vol. II, p. 289, ASM Press, Washington, DC, with permission.*

Acyclovir has the limitations of low oral absorption and a short half-life, typically needing to be taken five times a day. More recent derivatives of acyclovir are valacyclovir (a valine ester of acyclovir), which is absorbed three to five times better than acyclovir after oral administration, and famciclovir, which is also well absorbed and is rapidly metabolized to the active compound penciclovir. Both these latter drugs have longer half-lives and require less frequent dosage than acyclovir. As expected, TK⁻ mutants resistant to acyclovir have a cross-resistance to related nucleoside analogs, but not to foscarnet or the nucleotide analog cidofovir which therefore can be used to treat infected patients (Fig. 12.3).

Ganciclovir

Acyclovir is much less potent against CMV as much less phosphorylated ACV accumulates in CMV-infected cells than in HSV- or VZV-infected cells. A derivative of acyclovir, 9-(1,3-dihydroxy-2-propoxy) methylguanine (DHPG), known as ganciclovir (GCV) (Fig. 12.1), has much greater inhibitory activity against CMV than does ACV, and was the first drug approved for use against CMV. GCV is phosphorylated in CMV-infected cells, not by TK but by an unusual virus-coded protease, the product of the UL97 gene; further phosphorylation by cellular kinases yields the active triphosphate, which inhibits the viral DNA polymerase. Resistance in some mutants maps to the phosphorylation gene, in others to the DNA polymerase gene.

Ganciclovir has been used principally to treat severe CMV infections such as retinitis, colitis, and pneumonia in AIDS patients and in transplant recipients. Given intravenously for some weeks, ganciclovir may produce a temporary remission in a proportion of cases, but unfortunately the condition generally recurs following its withdrawal. The antiviral activity of GCV is not as selective as ACV against HSV, and accordingly the drug is relatively toxic. Severe neutropenia, anemia, and thrombocytopenia are common side effects, and the drug also is associated with gastrointestinal symptoms including diarrhea, renal problems, and hypersensitivity. The valine ester valganciclovir is a prodrug that is rapidly converted to GCV in the body. It is more available orally, and can be used as daily oral prophylaxis against CMV in HIV and transplant patients.

Acyclic Nucleoside Phosphonates

Cidofovir is an acyclic cytosine analog that contains a phosphonate group. This group mimics a monophosphate group, and accordingly cidofovir is poorly taken up into cells, but has a long intracellular half-life. Its activity does not require a viral TK or protein kinase to synthesize the monophosphate, and it therefore has a broad antiviral spectrum against herpesviruses, adenoviruses, papillomaviruses, and poxviruses. It is used to treat CMV retinitis in HIV patients by intravenous administration. Drugs with similar structure include adefovir (used against hepatitis B) and tenofovir (used in HIV and hepatitis B).

Trisodium Phosphonoformate (PFA, Foscarnet)

Trisodium phosphonoformate, known also as phosphonoformic acid (PFA) or foscarnet (Fig. 12.1), is a non-nucleoside inhibitor of the DNA polymerases of herpesviruses and hepatitis B, as well as the reverse transcriptase of HIV. It acts through a non-competitive inhibition of the pyrophosphate-binding site on the enzyme. Resistance maps to the DNA polymerase gene. PFA may be given intravenously to treat CMV retinitis and severe herpesvirus infections, particularly those resistant to other antiviral drugs. The drug also displays some activity against hepatitis B *in vitro* but has been ineffective *in vivo*.

Although foscarnet displays some selectivity in that it inhibits cellular DNA polymerase α only at higher concentrations than the level required to inhibit viral DNA polymerase, it accumulates in bone and can lead to renal toxicity, electrolyte disturbances, and genital ulceration. Accordingly, it tends to be reserved for life-threatening conditions.

Inhibitors of RNA Virus Replication

A rather unusual nucleoside analog, 1-β-D-ribofuranosyl-1,2,4-triazole-3-carboxamide, known as ribavirin (Fig. 12.1), was first synthesized in 1972. Despite the fact that it inhibits the growth of a wide spectrum of RNA and DNA viruses in cultured cells and experimental animals, it is still only approved in most countries for a limited range of indications. These are chronic HCV infections in combination with interferon (although this role has now been largely superseded by the use of direct acting antiviral drugs); and severe respiratory syncytial virus infections, where it is delivered using a nebulizer to generate a small-particle aerosol administered via a mask or oxygen tent for 3 to 6 days. When given by oral administration at the usual dosage of about 1 gram per day, a substantial minority of recipients develop a reversible anemia with increased reticulocyte numbers and elevated serum bilirubin levels: immunosuppression and teratogenic effects have been demonstrated in animals. Its mechanism of action is not clear. The fact that ribavirin monophosphate inhibits the cellular enzyme IMP dehydrogenase, decreasing the pool of GTP, as well as inhibiting guanylyltransferase-mediated 5′-capping of mRNA, suggests that it may be acting on cellular pathways that are somewhat more critical to the virus than to the cell. It has also been proposed that it may inhibit viral RNA polymerases, and/or increase the mutation rate of the viral polymerase to a "catastrophic" level where few functional genomes are produced. The 3-carboxamidine derivative viramidine (taribavirin), a prodrug of ribavirin, has a similar spectrum of activity but may have slightly less toxicity.

Oral or intravenous ribavirin has been used in viral hemorrhagic fevers including Lassa, Crimean-Congo hemorrhagic fever, and Hantavirus infection, and there is some evidence of benefit.

Inhibitors of Reverse Transcriptase

Nucleoside RT Inhibitors

The advent of AIDS in the early 1980s stimulated a major search for new approaches to antiviral therapy. Few in the medical and scientific community had experience with lentiviruses and there were no effective anti-retroviral agents. Enormous strides were thenceforth made rapidly, and a new era of viral chemotherapy was born.

The first compound to display sufficient antiviral activity *in vivo* to be licensed for human use was 3′-azido-2′,3′-dideoxythymidine, otherwise known as azidothymidine, AZT, or zidovudine (Fig. 12.1), an inhibitor of reverse transcriptase. Unlike acyclovir, AZT is phosphorylated by cellular kinases to AZT triphosphate (AZT-PPP), which exerts its antiviral effect against HIV by inhibiting HIV reverse transcriptase, being as it is accepted by the enzyme in preference to TTP. AZT-PPP binds to the reverse transcriptase approximately 100 times more efficiently than it does to the cellular DNA polymerase α. AZT-PPP is incorporated into the growing HIV DNA chain, leading to premature chain termination. In addition, AZT monophosphate (AZT-P) competes successfully for the enzyme thymidylate kinase, resulting in depletion of the intracellular pool of TTP. Since AZT's activity does not depend on phosphorylation by viral enzymes, it is less selective and more toxic than acyclovir. Clearly, AZT suppresses replication but does not eliminate proviral DNA from infected cells. The ultimate goal of eradication of HIV from the body can only be achieved if recruitment of infection to new cells is prevented and if cells already infected (the HIV "reservoir") eventually die.

Toxicity of AZT is a serious issue, including marrow suppression (neutropenia, macrocytic anemia), reversible wasting of proximal muscles, nausea, and headaches. The newer NRTIs (see Table 12.1) have different patterns of side effects, and resistance to many of these agents maps at different sites in the reverse transcriptase molecule from that of AZT (see also Chapter 23: Retroviruses).

Non-nucleoside Reverse Transcriptase Inhibitors

In the late 1980s several classes of inhibitors ("TIBO" and "HEPT") were developed and found to be highly specific and potent inhibitors of the reverse transcriptase of HIV-1. These drugs bind non-competitively to a hydrophobic pocket 10Å distant from the enzyme catalytic site, leading to conformational changes in the p66 thumb domain of the reverse transcriptase, so impairing the function of the reverse transcriptase catalytic site. The first non-nucleoside reverse transcriptase inhibitor to be approved (in 1997) was nevirapine, followed by delavirdine and efavirenz, while etravirine (approved by the United States Food and Drug Agency in 2008) has a slightly different structure and is of value in patients with virus resistant to the earlier non-nucleoside reverse transcriptase inhibitors. These are well absorbed and the associated long half-lives in the body means once or twice daily dosage is effective. Side effects include hepatotoxicity and psychiatric manifestations. Since non-nucleoside reverse transcriptase inhibitors have a different molecular target and mechanism of action from non-nucleoside reverse transcriptase inhibitors, these are of value in combination with non-nucleoside reverse transcriptase inhibitors in delaying the emergence of drug resistance, and indeed the two together along with protease inhibitors are a mainstay of current therapy.

Emergence of resistance to all of the above drugs is a continuing issue that requires careful management. Further details can be found in Chapter 23: Retroviruses.

Nucleoside Analogs in Hepatitis B

In addition to tenofovir and adefovir, several other analogs have become useful for the treatment of hepatitis B virus infection. The first of these, lamivudine (3TC), was first synthesized in 1988 and licensed for use against HIV in 1995. Later it was found to be active against the HBV DNA polymerase at lower dosages, and has since been used widely for the treatment of chronic hepatitis

B infection. It has a low toxicity and a long half-life, but drug-resistant HBV variants regularly develop when used alone, and its preferred use is in combination therapy. Emtricitabine (FTC) is a similar drug, and is often used in combination with tenofovir in patients co-infected with HIV and hepatitis B virus. Entecavir is a more recently introduced nucleoside analog that is particularly active against the DNA polymerase of hepatitis B virus but not against the HIV reverse transcriptase. A 48-week course of entecavir treatment in chronic hepatitis B has been shown to give suppression of serum hepatitis B virus DNA and normalization of liver function in a greater proportion of patients than with lamivudine treatment, although the safety profiles of the two drugs are similar. Other drugs to which hepatitis B virus responds in a proportion of patients include telbivudine and pegylated interferon.

The efficacy and accompanying drawbacks in treating hepatitis B patients have been documented for each of these drugs as monotherapy, but evidence for or against their use in combination therapy is only now being explored, despite the obvious theoretical appeal of combination therapy.

Inhibitors of Viral Proteases

Cleavage of viral proteins by proteases is essential at several stages in many viral replication cycles: examples include activation of envelope fusion glycoproteins, activation of some viral enzymes, post-translational cleavage of the polyprotein product of polycistronic mRNA, and maturation of the virion. As many of the proteases are virus-coded, it should be possible to find agents that specifically inhibit a viral protease without interfering with essential cellular proteases.

The development of HIV protease inhibitors is the first example of this approach. The three-dimensional structure of the HIV protease was solved by X-ray crystallography. The active site was identified and found to accommodate seven amino acids, using enzyme produced in large quantities by recombinant DNA technology. The enzyme was found to cleave sequences containing the dipeptides Tyr-Pro or Phe-Pro (Fig. 12.4). When a hydroxyethylene linkage replaces the peptide bond, cleavage does not occur; viral proteins continue to be made, but the viral particles budding from cells are immature and non-infectious. A range of different inhibitors that bind to, and block, the active site are now in routine use, usually as part of combination therapy (Table 12.1; Fig. 12.5). As would be expected, resistant strains of HIV may arise, many of which are unique to a particular inhibitor. Side effects, which can be troublesome, include nausea, vomiting, diarrhea, hyperlipidemia, insulin resistance, redistribution of body fat to the trunk regions, and abnormal liver function.

Direct-acting Antiviral Drugs for Hepatitis C Virus Infection

Antiviral treatment of hepatitis C infection has undergone a radical change since 2011, with the progressive introduction of three classes of direct-acting antiviral drugs (DAAs). This development has been possible due to the synthesis of specific inhibitors of the active sites of three key viral gene products: the NS3-4A protease, the NS5A RNA-binding protein, and the NS5B RNA-dependent RNA polymerase. For example, sofosbuvir is a nucleotide monophosphate analog that, after conversion to the active triphosphate within the cell, becomes incorporated into the growing RNA chain and acts as a chain terminator. The NS3-NS4A protease of hepatitis C virus complex offers an attractive target. The peptidomimetic drugs boceprevir and telaprevir were licensed for use in the United States in 2011, but have since been largely replaced by second-generation protease inhibitors (simeprevir, asunaprevir, paritaprevir) with improved safety and dosage profiles. When used in combination with pegylated interferon and ribavirin the rate of sustained virological response nearly doubled and they have been of particular benefit in more difficult-to-treat patients such as those with cirrhosis or those infected with HCV genotype 1. Sustained viral responses were achieved with shorter courses of treatment, and more recent trials have demonstrated success with interferon-free regimens in most patients. These drugs represent the first of a large number of direct-acting antivirals at different stages of development, ushering in an exciting new phase in the treatment of hepatitis C virus. Six different combinations of these drugs are now recommended for use against different HCV genotypes, and permanent cure can be achieved for most patients using a shorter and completely interferon-free regimen. A truly remarkable advance.

Inhibitors of Viral Integration

Integration of the HIV genome through the action of the virus-coded enzyme integrase is an essential stage in HIV replication. The development of an assay for the DNA strand transfer step in the integration process allowed screening of compounds for inhibition of this step. The first compound to reach clinical use was raltegravir in 2008, followed by elvitegravir and dolutegravir. A number of clinical trials have shown that combination regimens inclusive of an integrase inhibitor were equally effective or superior to standard regimens, and more detailed information about toxicity and guidelines for use are being developed in the light of further experience.

Neuraminidase Inhibitors

The development of highly effective neuraminidase inhibitors became feasible when analysis of the three-dimensional

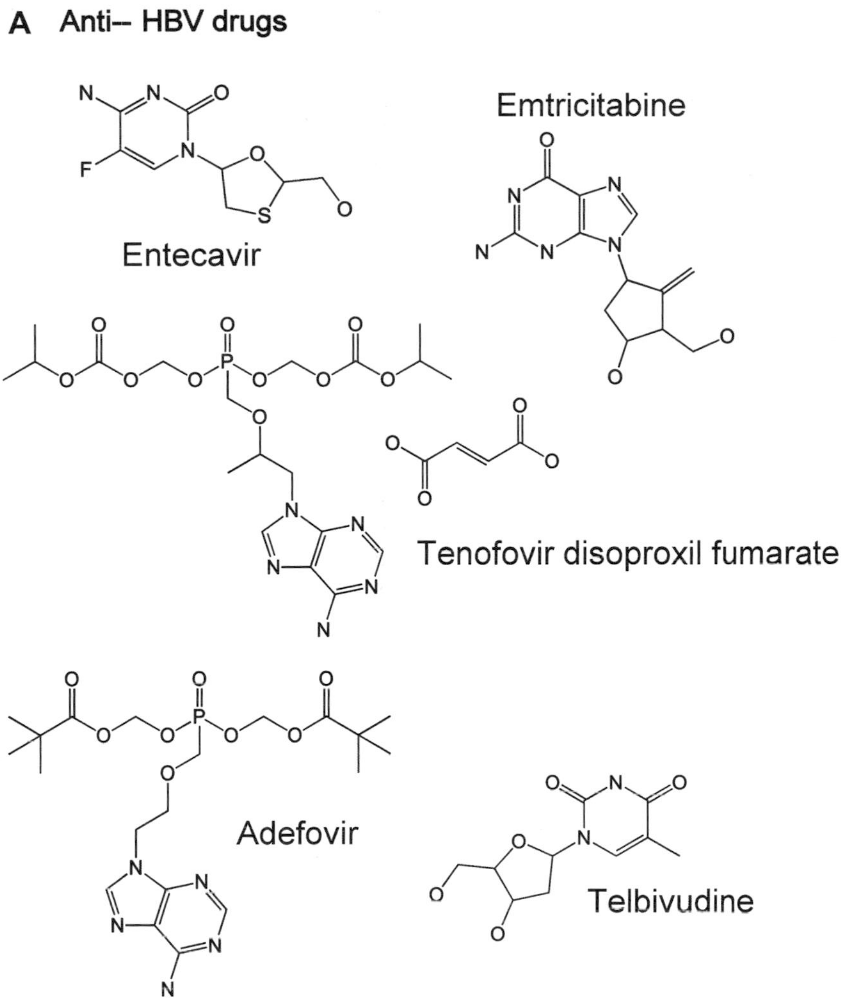

A Anti-- HBV drugs

Entecavir

Emtricitabine

Tenofovir disoproxil fumarate

Adefovir

Telbivudine

FIGURE 12.4 Chemical structures of some antiviral agents used against hepatitis B, hepatitis C and influenza viruses. (A) Anti-HBV shopped; (B) Anti-HCV and Influenza shopped. *Courtesy of Steven Polyak.*

structure of influenza neuraminidase revealed the location and structure of the catalytic site (see above). Two successful agents zanamavir (delivered intranasally) and oseltamivir (delivered orally) have become important in influenza treatment and prophylaxis. Both bind strongly to the enzyme active site and prevent release of progeny virions, thereby preventing infection of new cells and the further spread of infection. When used within 24 to 30 hours of the onset of symptoms, either drug can shorten the duration of symptoms by one to three days, and these are also useful as prophylactic treatment. In contrast to the adamantanes, zanamavir and oseltamivir cause very little toxicity and are associated with less problems of drug resistance. Each has a broad spectrum of action across influenza strains of both type A and type B, and a low rate of drug resistance makes each an important component in planning for future pandemics. Stockpiles of drugs are maintained to provide rapid prophylaxis and treatment, if required in the lead-up period before a specific vaccine against a new pandemic strain becomes available. However, cost and logistics will of course limit how

extensively and for how long such treatment can be sustained in the face of a new pandemic. More recently developed neuraminidase inhibitors include peramivir, an intravenous agent authorized for emergency use in hospitalized patients suffering from 2009 H1N1 influenza, and laninamivir, which is active against oseltamivir-resistant viruses and has such a long-lasting effect that a single inhalation on the first day of treatment appears to be effective.

NEWER APPROACHES UNDER DEVELOPMENT

Virus-Specific Oligonucleotides

Theoretically, short synthetic antisense oligo-deoxynucleotides, complementary in sequence to viral mRNA, could inhibit viral gene expression with high specificity. For example, hybridization to viral mRNA may prevent the splicing, transport, or translation of that mRNA, or render it susceptible to degradation by

FIGURE 12.4   (Continued)

B   Anti-HCV drugs

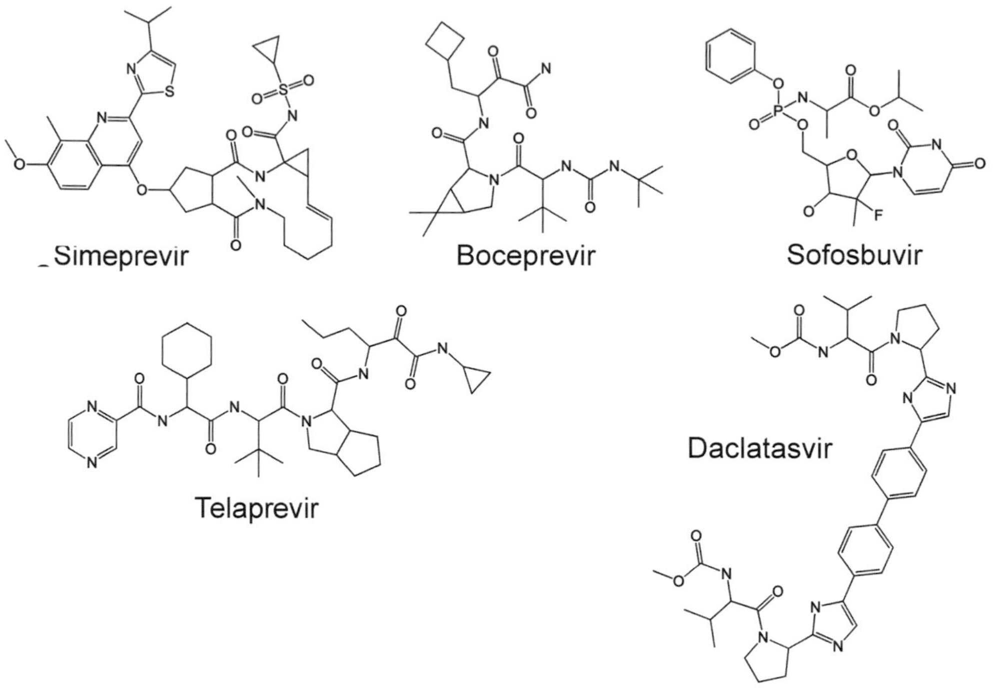

Simeprevir

Bocepravir

Sofosbuvir

Telaprevir

Daclatasvir

Anti-Influenza virus drugs

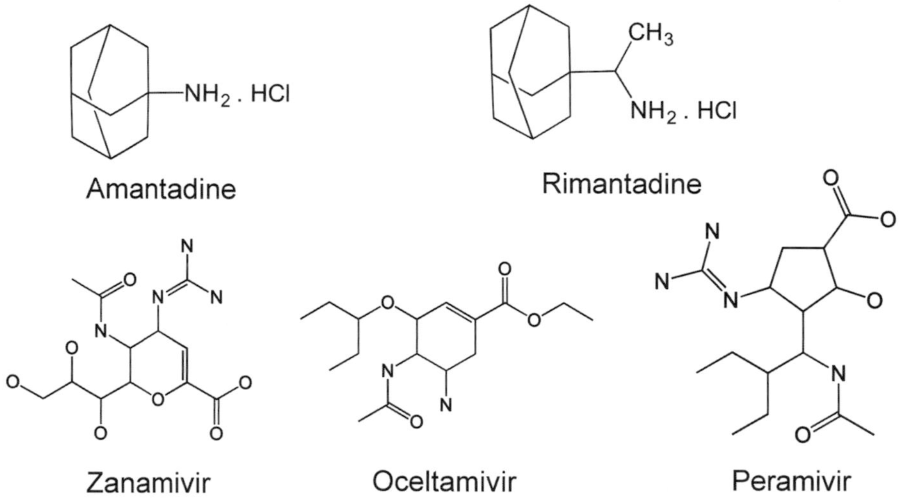

Amantadine

Rimantadine

Zanamivir

Oceltamivir

Peramivir

RNase H. On the other hand, hybridization to either viral DNA or cDNA might block transcription or replication, or block the attachment of DNA-binding regulatory proteins. Short single-stranded DNA sequences of this nature have displayed antiviral activity against HIV, herpes simplex virus, and influenza viruses in cultured cells. Moreover, the ability of naturally occurring micro-RNAs (miRNAs) to regulate gene function has provided encouragement for such an approach against viruses. In practice, major problems with the specificity, stability, and uptake of oligodeoxynucleotides have hindered this work. Oligonucleotides are rapidly degraded by extracellular and intracellular nucleases, unless the backbone of the molecule is modified, for example, to phosphorothioate linkages, to circumvent this problem. Moreover, the uptake of oligonucleotides into cells is very inefficient; attempts are being made to facilitate entry, such as the coupling of miRNAs to a hydrophobic peptide or lipid.

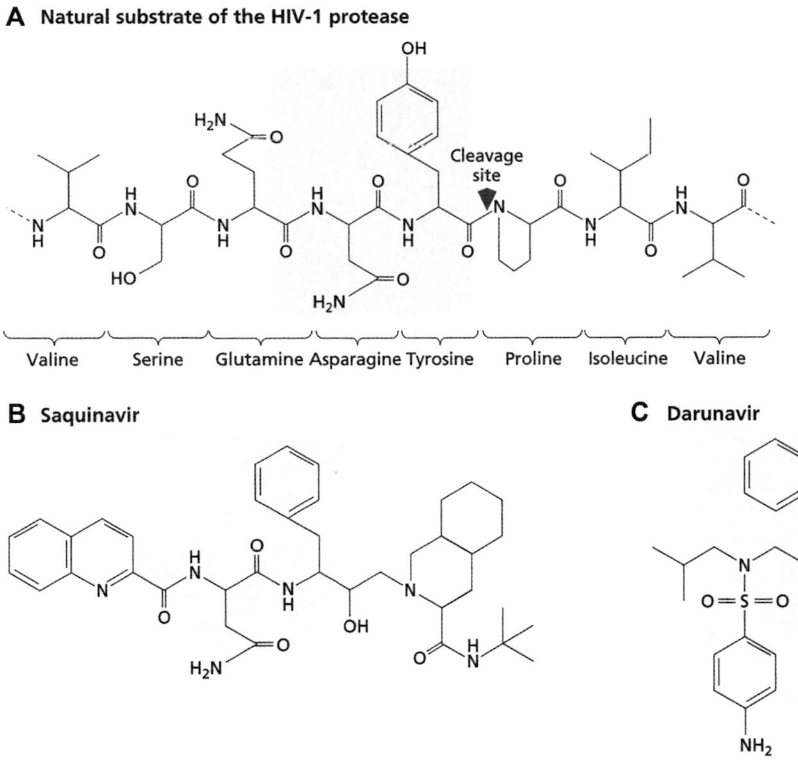

FIGURE 12.5 Comparison between one of the sites cleaved by the HIV protease and the protease inhibitor saquinavir. (A) Amino acid sequence within the Gag-Pol protein showing one cleavage site (between tyrosine and proline, shown by the red arrow). (B and C) The corresponding structures of the protease inhibitor saquinavir (Ro 31-8959) and darunavir (WHO/ATC code J05AE10). *Reproduced from Flint, S.J. et al., 2009. Principles of Virology, third ed., vol. II, p. 303, ASM Press, Washington, DC, with permission.*

Inhibitors of Regulatory Proteins

Many of the genes of HIV are regulatory genes, the sole or principal function of which is to control the expression of other genes. For example, the Tat protein binds to a specific responsive element called TAR which is present in both the integrated HIV cDNA and all HIV mRNAs. This binding results in enhanced expression of all the HIV genes, including *tat* itself, thereby constituting a positive feedback loop that enables production of large numbers of progeny virions. Clearly, an agent capable of binding either to the Tat protein or to the TAR nucleotide sequence should be a most effective inhibitor of HIV replication. Screening based on inhibition of the Tat–TAR interaction has identified a number of compounds active in cell culture, but none of these is currently used clinically, partly because these agents are not easily delivered. As other regulatory elements are being discovered for many other viruses, this novel approach to antiviral chemotherapy has considerable appeal.

Microbicides

Efforts have increased recently to develop effective topical microbicides—ointments or creams designed to prevent virus uptake at the site of entry into the body. Such agents might act by inactivating extracellular virus, by blocking attachment to cells, or by preventing the first round of virus replication at the entry site. Prevention of HIV and genital herpes simplex virus infection has received most attention, and the attraction of this approach includes the fact that it can be applied by the patient at the time when it is most needed. Ideally it should prevent infection before it is initiated, and the agents are likely to be relatively cheap. Agents that have been investigated include nonoxynol-9, a commercially available spermicide that unfortunately was found to damage epithelia and possibly enhance HIV transmission; various polymers, for example, polystyrene sulfonate that binds to herpes simplex virus glycoprotein B and inhibits virus uptake and spread; specific blockers of virus entry; and classical antiviral drugs, for example, reverse transcriptase inhibitors, applied locally. This area is under active research.

FURTHER READING

Mousseau, G., Valente, S., 2012. Strategies to block HIV transcription: Focus on small molecule tat inhibitors. Biology 2012 (1), 668–697. Available from: http://dx.doi.org/10.3390/biology1030668.

Richman, D.D., Whitley, R.J., Hayden, F.G. (Eds.), 2009. Clinical virology (third ed.), ASM Press

Chapter 13

Epidemiology of Viral Infections

Epidemiology is the study of the distribution, the dynamics, and the determinants of diseases in populations. The risk of virus infection and/or disease in a human population is determined by the characteristics both of the virus, and of susceptible individuals and of the host population such as innate and acquired resistance. In addition, virus transmission is affected by behavioral, environmental, and ecological factors. Virus epidemiology aims to meld these factors using quantitative measurements to provide a rational basis for explaining the occurrence of virus diseases and for directing disease-control measures, in particular the identification of outbreak sources and how best to implement prevention strategies. Epidemiology can also help to clarify the role of viruses in the etiology of diseases, understanding the interaction of viruses with environmental determinants of disease, determining factors affecting host susceptibility, clarifying modes of transmission, and the testing of vaccines and therapeutics on a large scale.

Epidemics are peaks in disease incidence that exceed the *endemic* baseline or expected rate of disease. The size of the peak required to constitute an epidemic is arbitrary and is related to the background endemic rate and the rate of clinical to sub-clinical infection. Sometimes a few cases of a disease that arouse anxiety because of their severity, for example encephalitis, will be loosely termed an "epidemic" whereas a few cases of influenza will not, but the term strictly implies unusually wide and rapid spread of infection within the population.

MECHANISMS OF VIRUS SURVIVAL

Because viruses, unlike most bacteria, cannot replicate outside of living cells, perpetuation of a virus in nature depends on the maintenance of serial infections, that is, a chain of transmission; the occurrence of disease is neither required nor necessarily advantageous. Indeed, although clinical cases may often produce more infectious virus than inapparent infections, the latter are generally more numerous and, because these do not restrict the movement of infectious individuals, they can provide a major mechanism of viral dissemination. Epidemiologists recognize three different patterns of virus survival in mammalian hosts, distinguished through use of virus reservoirs: acute self-limiting infections with no reservoir, persistent infections with a reservoir in humans, and involvement of an animal reservoir.

Most viruses have a principal mechanism for survival, but if this mechanism is interrupted, for example, by a sudden decline in population of the host species due to another disease or a short-term climate change, alternate mechanisms, previously less apparent, may emerge. This should be remembered when relating the epidemiology of a specific disease to a particular mechanism of survival, as proposed in Table 13.1.

An appreciation of the pathogenesis and clinical features of a particular infection is valuable in designing and implementing control programs. For example, the knowledge that variola virus caused an acute self-limiting infection in which the vast majority of infected individuals showed clinical disease, and that it had no animal host, was important in the successful eradication of smallpox.

The majority of human viral infections fall into the category of acute self-limiting infections. Optimum transmissibility is crucial, and with viruses that cause systemic infections with lifelong immunity, perpetuation is possible only in large, relatively dense populations. Viruses that cause superficial mucosal infections with short-lived immunity may survive in somewhat smaller populations, and a capacity to survive in a circumscribed population may be enhanced by antigenic drift (see below).

VIRAL SHEDDING AND ROUTES OF TRANSMISSION

Transmission cycles require virus entry into the body, replication, and shedding with subsequent spread to another host. Molecular and cellular aspects of entry and shedding were described in Chapter 4: Virus Replication; here only those aspects that are relevant to epidemiology are discussed.

Fenner and White's Medical Virology. DOI: http://dx.doi.org/10.1016/B978-0-12-375156-0.00013-8

TABLE 13.1 Different Survival Mechanisms Used by Viruses in Nature

Features of Infection	Survival Mechanism	Virus
Acute self-limiting infection—lifelong immunity	No reservoir; need large population with continuing chain of transmission	Measles, mumps, rubella, polio[a], hepatitis A[a], enteroviruses[a,c], dengue
Acute self-limiting infection—immunity more short-lived	No reservoir; reinfections occur, virus can survive in smaller-sized population	Respiratory syncytial virus, rotavirus[b], influenza[b], coronaviruses, rhinoviruses[c]
Persistent infection-intermittent replication +/− shedding	Human reservoir; infected individuals can provide lifelong source of virus	Herpes simplex, varicella-zoster, CMV, EBV, other herpesviruses
Persistent infection—continuous replication	Human reservoir; infected individuals can provide lifelong source of virus	HIV, HBV, HCV, HTLV-1, human papillomavirus
Zoonoses—no human–human spread	Survival depends on enzootic infection in animal reservoir and transmission to humans	Most arboviruses except dengue, yellow fever (urban cycle). Avian influenza, rabies, Hendra
Zoonoses—human–human spread also significant	Survival depends on enzootic infection in animal reservoir and transmission to and between humans	Marburg/Ebola, Hantaan, Nipah, dengue, yellow fever (urban cycle)

[a]*Virions are resistant and can survive in the environment for days.*
[b]*Antigenic shift and drift helps new infections overcome preexisting immunity.*
[c]*Existence of many serotypes allows new infections with related serotypes.*

Virus transmission (Fig. 13.1) may be horizontal or vertical; however, most transmission is horizontal, that is, between individuals within the population at risk. More so than in the case with bacteria, different viruses tend to use specific defined transmission routes that are ultimately determined by factors such as the physical properties of the virion, route of shedding, and aspects of pathogenesis such as cell tropism. These routes are primarily defined by the different ways used by different viruses to broach the continuous external epithelial lining of the body. How the particular route of entry can be an important determinant in pathogenesis is discussed in Chapter 7: Pathogenesis of Virus Infections; equally, the route of entry is a major factor in determining the patterns of occurrence, mode of spread, and risk populations for each virus infection.

Shedding of virus usually occurs from one of the body openings or surfaces also involved in the entry of viruses. With localized infections the same body openings are involved in both entry and exit (see Fig. 7.1); in generalized infections a greater variety of modes of shedding is recognized, and some viruses are shed from multiple sites, for example, hepatitis B virus, HIV, and cytomegalovirus in semen, cervical secretions, milk, and saliva. The amount of virus shed in an excretion or secretion is important in relation to transmission. Very low concentrations may be irrelevant unless very large volumes of infected material are transferred; on the other hand some viruses occur in such high concentrations that a minute quantity of material, for example, less than 1 μl, can transmit infection.

Respiratory and Oropharyngeal Route

Many different viruses causing localized disease of the respiratory tract are shed as aerosols in the mucus or saliva expelled from the respiratory tract during coughing, sneezing, and talking. Viruses are also shed from the respiratory tract in several systemic infections, such as in measles, chickenpox, and rubella. A few viruses, for example, the herpesviruses, cytomegalovirus, and Epstein-Barr virus, are shed into the oral cavity, often from infected salivary glands or from the lung or nasal mucosa, and are transmitted by salivary exchange through kissing and other social activities.

Aerosols are most infectious early in a respiratory infection at the peak of virus replication, but there is also variation from individual to individual, and some patients, for reasons not well understood, seem to be more infectious than others ("super shedders"). There are three distinct components of spread of respiratory viruses. Small-droplet aerosols (<10 μm diameter particles) can create rapid explosive outbreaks and spread to more distant contacts (>1.8 m). These particles are more likely to settle lower down the respiratory tract. Second, large-droplet aerosols (10 to 100 μm diameter) sink to the ground more quickly, and therefore transmission needs closer contact between source and recipient (<~0.9 m); spread may be slow, intermittent, and without clustering. These particles are mostly trapped in the upper airways. Third, spread occurs when fomites, including tissues, ward equipment, and household objects become contaminated with respiratory secretions or aerosol droplets and contacts transfer this material to their own

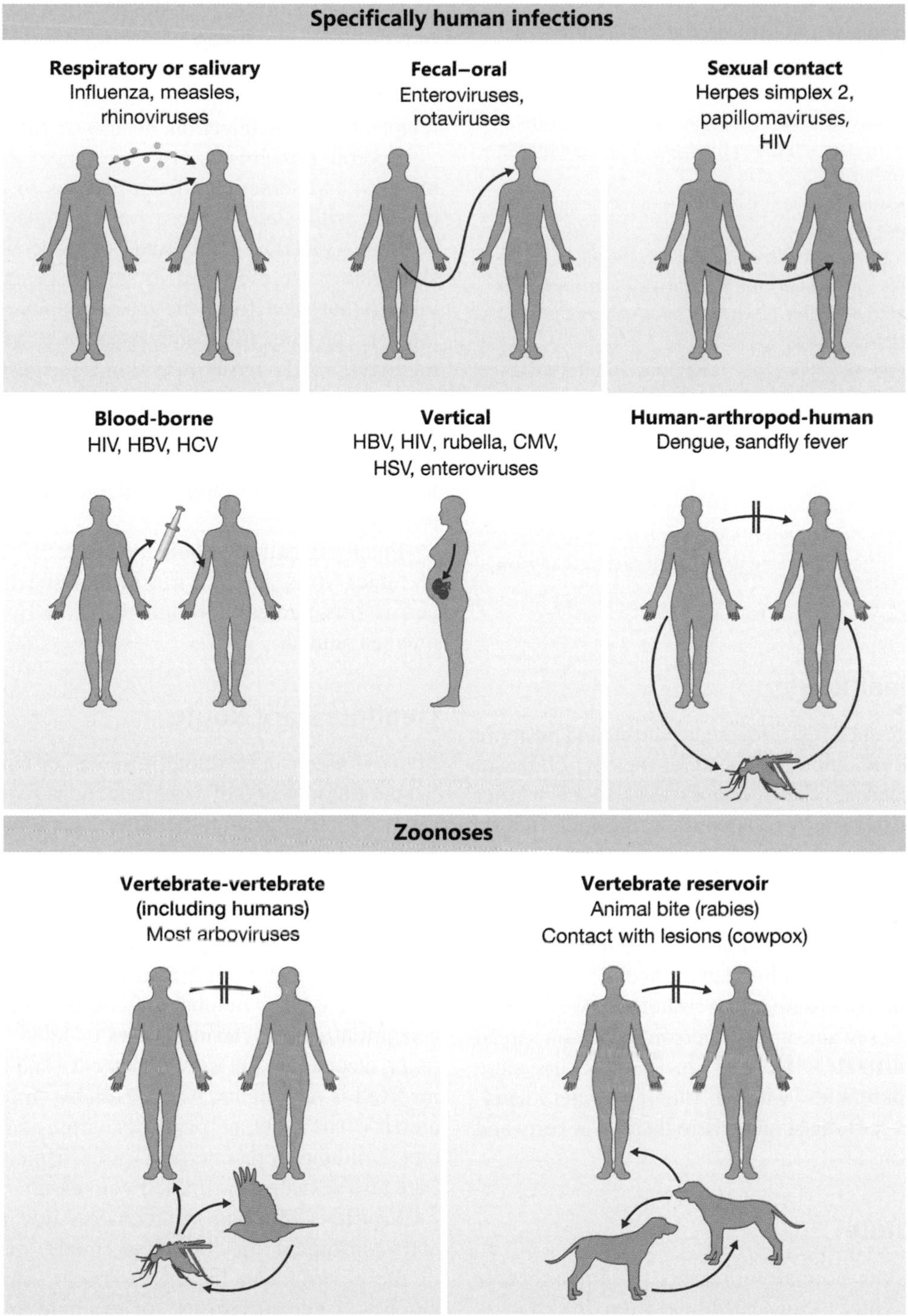

FIGURE 13.1   Modes of transmission of human viral diseases. *Modified from Mims, C.A., 1982. "The Pathogenesis of Infectious Disease," second ed., Academic Press, London.*

respiratory tracts; this mode of spread particularly involves those in close contact with others under conditions of poor hygiene (Box 13.1).

Environmental sources of airborne transmission include virus-contaminated dust, thought to be the source of arenavirus infections, and aerosols of infected urine from rodents (arenaviruses) or bats (rabies). However, most respiratory viruses, being enveloped, are relatively labile and do not survive long outside the body unless kept moist in secretions.

Respiratory transmission is a very efficient way to quickly infect a large number of contacts and spread virus globally (compare influenza with HIV). There are two mechanisms:

1. *Inhalation of aerosol.* Humans filter ~600 liters of air per hour when breathing at rest

   Aerosols are created especially by sneezing and coughing, less by talking.

   Large droplets (>10 μm) soon sink to the floor, while small droplets can spread further but dry out and virus becomes inactivated; this is why the atmosphere does not remain infective for long

   Seasonality of respiratory infections is affected by increased virus survival in cooler temperatures, seasonal differences in social activities, school attendance patterns, and many other factors

2. *Contaminated objects.* Infected respiratory secretions contaminate tissues, environmental surfaces and objects, hands. Shaking hands, handling objects etc. provides a pathway for virus to the nose and mouth of a new person

Gastrointestinal Route

Enteric viruses are shed in the feces and vomit, and the more voluminous the fluid output the greater is the environmental contamination caused. These viruses tend to be hardier and able to survive environmental conditions longer outside the body than the enveloped respiratory viruses. Two epidemiological patterns are seen: (1) a point source outbreak occurs when many people ingest contaminated food or water, for example at weddings or other functions. This particularly occurs with salads, uncooked shellfish, or through drinking unsafe river water or well water contaminated with sewage, and (2) person-to-person spread by the fecal–oral route is more gradual and occurs more efficiently in households without running water, hand-washing facilities, or toilets, and where there is poverty and a lack of education.

Cutaneous Route

The unbroken skin normally presents an impermeable barrier to virus entry. Virus shedding from the skin is usually so insignificant that individuals with systemic blood-borne infections, for example, should be reassured that their unbroken skin does not present a continuing risk to their social contacts. However, minor skin abrasions can be an important source of virus for diseases in which transmission is by direct contact, for example, molluscum contagiosum and warts. Blood-borne infections (see below) can be shed through bleeding from broken skin, and hepatitis B has been shown to spread by inapparent horizontal transmission between children, particularly in poor socio-economic conditions with overcrowding and/or prevailing skin disease. However, blood-borne infections are of course typically spread by one of a number of ways of directly introducing infected blood parenterally into a recipient (e.g., by injection, needle-sharing, or transfusion).

Several poxviruses may be spread from animals to humans, and sometimes from humans to animals through contact with skin lesions, for example, the viruses of cowpox, vaccinia, orf, and pseudocowpox. Although skin lesions are produced in several generalized diseases, virus is not shed from the maculopapular skin lesions of measles, or from the rashes associated with picornavirus, togavirus, or flavivirus infections. Herpesvirus infections, on the other hand, produce vesicular lesions in which virus is plentiful in the fluid of the lesions. Even here, however, virus shed in saliva and aerosols is much more important as far as transmission is concerned, than that shed via the skin lesions.

Finally, mention should be made of the transmission of rabies virus and B virus (Macacine herpesvirus 1 or herpesvirus simiae) through the skin by the bite of an infected animal.

Genitourinary Route

Many viruses can be found in semen or vaginal secretions. Sexual transmission of virus infections by mucosal contact can be efficient because the virus is kept moist and does not need to survive long outside the body; however dissemination through risk populations by this route is usually slower than by respiratory spread because involvement of multiple contacts is usually slower. Studies particularly with HIV have shown that sexual transmission is enhanced when there is a greater number of consecutive partners, when concurrent genital mucosal tears or intercurrent infections (e.g., ulcerating STDs) are present, and when the male involved is not circumcised. The most important examples are HIV, HBV, human papillomavirus, and herpes simplex type 2, although other herpesviruses, hepatitis B, and HTLV I are also sexually transmitted with ease.

Viruria is life-long in arenavirus infections of rodents and constitutes the principal mode of contamination of the environment by these viruses. However, while a number of human viruses, for example, mumps virus and cytomegaloviruses, replicate in tubular epithelial cells in the kidney and are shed in the urine, this is not a major source of transmission from human to human.

Blood-Borne Route

Just as viremia is an important route of virus dissemination within individual hosts, so it is between hosts. Hepatitis B, C, and D viruses, HIV, and HTLV were once commonly spread by blood transfusion, but the risk has been obviated by comprehensive testing of donated blood. Highly sensitive tests, for example polymerase chain reaction (PCR), are often used because of the extra risk caused by the large volume of blood transfused and the usual compromised health of recipients. Today in most countries, blood-borne transmission of these viruses is more a problem among intravenous drug users due to contaminated needles and injecting paraphernalia. Of course, blood is the usual source from which arthropods (e.g., mosquitoes, ticks, sandflies) acquire viruses during the course of taking a blood meal. Less commonly, some arthropods (e.g., horseflies, other biting flies) transmit viruses passively by contamination of their mouth parts and interrupted blood feeding on multiple hosts.

Despite this, in most instances blood-borne viruses are not shed from the unbroken skin of an infected person, and transmission by normal skin contact is negligible.

Ophthalmic Route

Virus infection can be introduced into the eye from the patient's fingers (herpes simplex, vaccinia), from swimming pools (adenoviruses), from inadequately sterilized ophthalmic equipment (adenoviruses, prions), via aerosols (enterovirus 71), or via the blood stream in a systemic infection (measles).

Milk-Borne Route

Several viruses, for example, cytomegalovirus, HIV-1, and HTLV-1, are excreted in milk, which may serve as a route of transmission to the newborn infant. In some situations, the additional risk of transmission from an infected mother by breastfeeding may be much smaller than the risk of vertical transmission already incurred during the birth process; breastfeeding may still be recommended where infectious disease or malnutrition is a common cause of death in infancy, despite the additional risk of transmission incurred.

Vertical Transmission

Transmission of virus from the mother to the embryo, fetus, or newborn, is an important instance of cross-generational transmission that facilitates survival of some viruses in nature. The three situations where this occurs are described in Chapter 7: Pathogenesis of Virus Infections, namely (1) via the integration of proviral DNA directly into the DNA of the germline of gametes and fertilized eggs, (2) transplacental spread during pregnancy, and (3) perinatal or postnatal spread via saliva, milk, or other secretions. Vertical transmission of a virus may be lethal to the fetus and cause abortion, it may be associated with congenital disease or congenital abnormalities, or it may cause a sub-clinical infection.

In the case of HIV and hepatitis B, vertical transmission introduces infection to a new generation of infants who are then capable of transmitting infection to succeeding birth cohorts for many years to come. In addition, vertical transmission in arthropod vectors is an important mode of perpetuation of some arthropod-borne viruses.

No Shedding

Many sites of viral replication are a "dead end" in regard to transmission to other hosts, for example, no virus is shed from the brain or other organs that do not communicate with a body opening or the body surface. One might question how this could benefit the long-term survival of the virus in nature, but the stepwise augmentation of virus titer by replication in cells located in internal organs is often an important pre-requisite for shedding from another site, or for infection of blood-sucking arthropods.

In animals such as mice and chickens, many retroviruses are not shed, but are transmitted directly to the next generation via the germline, or by transplacental spread, as described above.

FACTORS AFFECTING THE DYNAMICS OF VIRAL INFECTIONS

Transmissibility

Transmissibility is affected by physical properties of the virus, the extent and nature of shedding from the body, and social interactions between hosts. Obviously, the shedding of high titers of infectious virus enhances human-to-human transmission. Respiratory viruses tend to be shed over a relatively brief period of a few days but are expelled at high concentration as an aerosol generated by explosive sneezing or coughing, thus ensuring transmission to close contacts. The complex usage of different receptors by different influenza variants can determine virus transmissibility between humans (see Chapter 25: Orthomyxoviruses). Enteric viruses are also shed in large numbers but usually for a longer period (a week or more) in feces. These viruses may contaminate hands, fomites, food, and water. Enveloped respiratory viruses are relatively labile, especially during summer or in the tropics year-round. In contrast, many enteric viruses are non-enveloped and may survive for several days or weeks in water or dust, or on fomites, as is also true for poxviruses, adenoviruses, papillomaviruses, and hepatitis A and B viruses.

Improvements in socio-economic conditions, sanitation, and education have slowed the transmission of a number of common childhood infections, resulting in many infections being acquired at an older age. Some infections cause more clinical disease in older age groups, and thus these

BOX 13.2 Effect of Sanitation on Enteric Disease Transmission

As standards of sanitation and hygiene improved in many parts of the world during the 20th century, transmission of enteric virus infections became less efficient and thus the typical age at which infection occurred became older. As a result, most people no longer had acquired immunity by the time they reached adolescence. For reasons still unknown, primary infection in older children and adults with poliovirus or hepatitis A is more likely to cause *clinical* disease, compared to infection at a young age. The consequence was a paradoxical *increase* in clinical disease following this improvement in living standards. This was exemplified most strikingly in "virgin soil" epidemics occurring, long before the vaccination era, in isolated communities with no prior experience of the virus; most of the deaths occurred in adults. Finally, countries with advanced living standards began to see a reduction in the total numbers of infections and of clinical disease, as virus circulation declined even further.

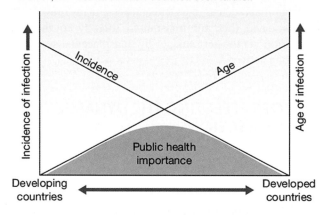

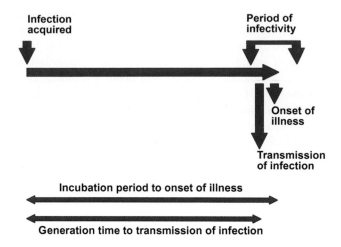

FIGURE 13.2 Sequence in an acute infection showing various time points and intervals. NB. If contact involved one event only, then the date of transmission can be inferred. However, if contact was continuous, the date of transmission is unknown. *Adapted from Neal Nathanson. In Knipe, D.M., Howley, P.M., 2013. Fields Virology, seventh ed. P. Lippincott, Williams and Wilkins, New York, NY.*

improvements may result in a paradoxical *increase* in cases of clinical disease (Box 13.2).

Fig. 13.2 defines the different time intervals relevant to a typical acute transient infection, and Table 13.2 sets out these parameters for some common human viral diseases. In many infections, such as in measles and chickenpox, persons become contagious a day or so before they themselves become ill.

The zoonoses, whether involving domestic or wild animal reservoirs, usually occur only under conditions where humans are engaged in activities involving close contact with animals (Table 13.3) or if the viruses are transmitted by arthropods (Table 13.4).

Seasonality

Many viral infections show a pronounced seasonal variation in incidence. In temperate climates, arbovirus infections transmitted by mosquitoes or sand-flies occur mainly during the summer months when the vectors are most numerous and active. Infections transmitted by ticks occur most

commonly during the spring and early summer months. More interesting, but also more difficult to explain, are the variations in seasonal incidence of infections in which humans are the only host.

Table 13.5 shows the season of maximal incidence of several human respiratory, enteric, and generalized infections. In temperate climates most respiratory infections are mainly prevalent in winter or to a lesser extent in spring or autumn. Annual winter outbreaks of severe respiratory syncytial virus infections in infants are a feature in temperate climates (Fig. 13.3); epidemics of influenza also occur almost exclusively in the winter, but these may vary greatly in extent from year to year. Many of the rash diseases of childhood transmitted by the respiratory route reach a peak in the spring. Among the enteric virus infections, seasonality varies with the etiologic agents: the incidence of a number of enterovirus infections (in common with most enteric bacterial infections) is greatest in the summer, but caliciviruses show no regular seasonal patterns and rotaviruses tend to be more prevalent in the winter months. Infections with the herpesviruses (HSV, cytomegalovirus, and EB virus), all transmitted by intimate contact with saliva and other bodily secretions/excretions, show no seasonal variations in incidence; neither do other sexually transmitted diseases. The patterns shown in Table 13.5 are found in both the northern and southern hemispheres. Different factors probably affect seasonality in the tropics, where wet and dry seasons tend to replace summer and winter. The peak incidence of measles and chickenpox is late in the dry season, with an abrupt fall when the rainy season begins, whereas influenza and rhinovirus infections reach a peak during the rainy season

TABLE 13.2 Epidemiological Features of Important Human Viral Diseases

Disease	Mode of Transmission	Incubation Period[a] (days)	Period of Infectivity[b]	Clinical:Sub-clinical Ratio[c]
Influenza	Respiratory	1–2	Short	Moderate
Common cold	Respiratory	1–3	Short	Moderate
Bronchiolitis	Respiratory	3–5	Short	Moderate
Dengue	Mosquito bite	5–8	Short	Moderate
Alphavirus encephalitis	Mosquito bite	4–10	Short	Low
Herpes simplex	Saliva, sexual	5–8	Long	Moderate
Enteroviruses	Enteric, respiratory	6–12	Long	Low
Poliomyelitis	Enteric	5–20	Long	Low
Measles	Respiratory	9–12	Moderate	High
Smallpox	Respiratory	12–14	Moderate	High
Chickenpox	Respiratory	13–17	Moderate	Moderate
Mumps	Respiratory, saliva	16–20	Moderate	Moderate
Rubella	Respiratory, congenital	17–20	Moderate	Moderate
Infectious mononucleosis	Saliva, parenteral	30–50	Long	Low
Hepatitis A	Enteric	15–40	Long	Low[d]
Hepatitis B	Parenteral, sexual, perinatal	50–150	Very long	Low[d]
Hepatitis C	Parenteral (perinatal, sexual)	40–60	Very long	Low in acute infection
Rabies	Animal bite	30–100	Nil	High
Warts	Contact, sexual	50–150	Long	High
HIV/AIDS	Sexual, parenteral, perinatal	1–10 years	Very long	High[e]
SARS	Respiratory	2–10 days	Moderate	Moderate

[a]Until first appearance of prodromal symptoms. Diagnostic signs, e.g., rash or paralysis, may not appear until 2–4 days later.
[b]Most viral diseases are highly transmissible for a few days before symptoms appear, as well as after. Long, >10 days; short, <5 days.
[c]High, >90%; low, <10%.
[d]If acquired when young. Moderate if acquired as an adult, as occurs with higher socioeconomic conditions.
[e]Eventually, after a long incubation period.

Both biological and sociological factors may play a role in these seasonal variations. Measles, influenza, and vaccinia viruses survive in air better at low rather than high humidity, whereas polioviruses, rhinoviruses, and adenoviruses survive longer at high humidity. All survive longer in aerosols, and at lower temperatures. These situations correspond with the conditions prevalent during those seasons when infections due to these viruses are most prevalent. It has also been suggested that there may be seasonal changes in the susceptibility of the host, perhaps associated with changes in nasal and oropharyngeal mucous membranes, such as drying as a result of smoke, central heating, or air conditioning.

Second, seasonal differences in social activities also markedly influence the opportunities for transmission of viruses, especially by the respiratory route. Although experiences in the Arctic and Antarctic show that cold weather alone is not enough to influence the incidence of common colds and other respiratory infections, the crowding into restricted areas and ill-ventilated vehicles and buildings that occurs in temperate climates during the winter months promotes the transmission of respiratory viruses. In places subject to monsoonal rains, the onset of the rains early in summer is accompanied by a greatly reduced movement of people, both in daily life and to fairs and festivals. While this may reduce the opportunity for exchange of viruses with those from other villages, confinement to smoke-filled dwellings maximizes the opportunity for transfer of respiratory viruses within family groups.

In urban communities young children appear to be particularly important as the persons who introduce viruses into families from school and from neighbors' children, because they have not yet acquired the immunological memories of past infections, and often shed larger amounts of virus compared to adults.

TABLE 13.3 Non-Arthropod-Borne Viral Zoonoses

Virus Family	Virus	Reservoir Host	Mode of Transmission to Humans
Herpesviridae	B virus	Monkey	Animal bite
Poxviridae	Cowpox virus	Rodents, cats, cattle	Contact, through abrasions
	Monkeypox virus	Squirrel, monkeys	
	Pseudocowpox virus	Cattle	
	Orf virus	Sheep, goats	
Rhabdoviridae	Rabies virus	Various mammals	Animal bite, scratch, respiratory
	Vesicular stomatitis virus	Cattle	Contact with secretions[a]
Filoviridae	Ebola, Marburg virus	Fruit bats	Human-human blood-borne contact[b]
Orthomyxoviridae	Influenza A virus[c]	Birds, pigs	Respiratory
Bunyaviridae	Hantaviruses	Rodents	Contact with rodent urine, feces
Arenaviridae	Lymphocytic choriomeningitis, Junin, Machupo, Lassa viruses	Rodents	Contact with rodent urine, feces
Paramyxoviridae	Hendra, Nipah viruses	Pteropid fruit bats	Contact with infected horses (Hendra), pigs (Nipah)[d]

[a]May also be arthropod-borne.
[b]Human outbreaks involve human-human spread, after initial introduction presumably from infected bats or amplifying infected primate species.
[c]Usually maintained by human-to-human spread; zoonotic infections occur only rarely, but reassortants between human and avian influenza viruses (perhaps arising during coinfection of pigs) may result in human pandemics due to antigenic shift.
[d]Human–human spread is also seen.

TABLE 13.4 Major Arthropod-Borne Viral Zoonoses

Virus Family	Virus Genus	Disease	Reservoir Hosts	Arthropod Vector[a]
Togaviridae	*Alphavirus*	Chikungunya	Monkeys, humans	Mosquitoes
		Eastern equine encephalitis	Birds	
		Western equine encephalitis	Birds	
		Venezuelan equine encephalitis[b]	Mammals, horses	
		Ross River polyarthritis[b]	Macropods, other mammals	
Flaviviridae	*Flavivirus*	Japanese encephalitis	Birds, pigs	Mosquitoes
		St. Louis encephalitis	Birds	
		West Nile fever	Birds	
		Murray Valley encephalitis	Birds	
		Yellow fever[b]	Monkeys, humans	
		Dengue[c]	Humans, monkeys	
		Zika	Humans, monkeys	
		Kyasanur Forest disease	Mammals	Ticks
		Tick-borne encephalitis	Mammals, birds	
Bunyaviridae	*Phlebovirus*	Rift Valley fever	Mammals	Mosquitoes
		Sandfly fever[a]	Mammals	Sandflies
	Nairovirus	Crimean–Congo hemorrhagic fever	Mammals	Ticks
	Bunyavirus	California encephalitis	Rodents, rabbits	Mosquitoes
		La Crosse encephalitis	Chipmunks, squirrels	Mosquitoes
		Tahyna virus infection	Hares, rabbits, hedgehogs, rodents	
		Oropouche fever	Monkeys, birds, sloths	Mosquitoes, midges
Reoviridae	*Coltivirus*	Colorado tick fever	Mammals	Ticks

[a]Arbovirus transmission often requires a very specific insect species for a particular virus (see Part II for examples). On the other hand, if a transmission cycle changes to involve a different vertebrate host species, or to a different habit (e.g., saltwater marshes or freshwater marshes), different insect species may then become involved.
[b]In certain episodes, transmitted from person to person, by insects.
[c]Usually transmitted from person to person, by mosquitoes.

TABLE 13.5 Season of Maximal Incidence of Specifically Human Viral Infections in Temperate Climates[a]

Type of Infection	Winter	Spring	Summer	Autumn	None in Particular
Respiratory					
Adenoviruses		+			
Rhinoviruses		+		+	
Influenza	+				
Coronaviruses	+				
Respiratory syncytial virus	+				
Parainfluenza 1 and 2				+	
Parainfluenza 3					+
Enteric					
Enteroviruses			+		
Rotaviruses	+				
Caliciviruses					+
Generalized					
Rubella		+			
Measles		+			
Mumps		+			
Varicella		+			
Hepatitis B					+
Herpes simplex 1 and 2					+
Cytomegalovirus					+
Epstein-Barr virus					+
Most arboviruses			+		

[a]Seasonality is often different in tropical climates, where there is little temperature fluctuation between summer and winter, but the occurrence of some infectious diseases is influenced by "wet" and "dry" seasons.

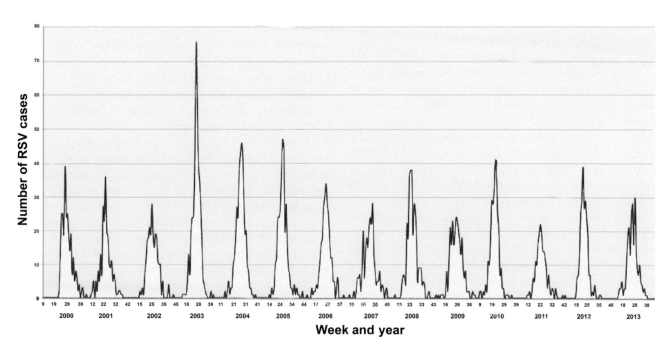

FIGURE 13.3 Epidemic occurrence of respiratory syncytial virus infections admitted to a children's hospital in Buenos Aires between 2001 and 2013. Virological diagnosis was performed by indirect immunofluorescence using nasopharyngeal aspirates. Note the sharp clustering of cases occurring each southern hemisphere winter. *Adapted from Lucion, M.F., et al., 2014, Arch. Argent. Pediatr. 112, 397–404.*

Critical Community Size

The survival of viruses that produce acute self-limiting infections requires a large and relatively dense susceptible host population. Such viruses may disappear from a population if the potential supply of susceptible hosts becomes exhausted as individuals increasingly acquire immunity to reinfection. Persistent viruses, on the other hand, may survive in very small populations, sometimes by spanning generations. Depending on the duration of immunity and the pattern of virus shedding, the critical community size varies considerably with different viruses. The principle can be exemplified by a comparison of measles and chickenpox.

Measles is a cosmopolitan disease that is characteristic of the generalized viral infections of childhood, similar to rubella, mumps, and poliomyelitis. Persistence of the virus in a community depends on a continuous supply of susceptible subjects. With an incubation period of about 12 days, maximum viral excretion for the next six days, and solid immunity to reinfection, between 20 and 30 susceptible individuals would need to be infected in series to maintain transmission for a year. For a variety of reasons, nothing like such precise one-to-one transmission occurs, and thus many more than 30 susceptible persons are needed to maintain endemicity. Analyses of the incidence of measles in large cities and among island communities have shown that a population of about 500,000 persons is needed to ensure a large enough annual input of new susceptible individuals, provided by the annual birth cohort, to maintain measles indefinitely as an endemic disease. For this reason it is believed that measles is a relatively new infection of humans, as communities of this size would not have been present prior to the development of settled agrarian societies around 10,000 to 4000 BC.

Because infection depends on close contact, the duration of measles epidemics is inversely correlated with population density. If a population is dispersed over a large area, the rate of spread is reduced and the epidemic can last longer, and therefore the number of susceptible persons needed to maintain endemicity is reduced. On the other hand, in such a situation a break in the transmission cycle is much more likely. If a large proportion of the population is initially susceptible, the intensity of the epidemic builds very quickly.

Attack rates were almost 100% in an epidemic of measles in southern Greenland in 1951, spreading through the entire population in about 40 days before running out of susceptible individuals and disappearing completely (see Box 13.3). Such virgin-soil epidemics in isolated communities may have devastating consequences, likely more due to a high proportion of the community being affected, a lack of adequate medical care, and the disruption of social life, rather than to a higher level of genetic susceptibility of the people or an exceptionally virulent virus strain.

BOX 13.3 Measles Epidemic in Greenland

Until 1951 measles had not been seen in Greenland, mainly due to the relative isolation of the population and the slow transport from surrounding countries. However in April 1951, a Greenlandic seaman traveled home from Copenhagen after being in contact with someone in the prodromal phase of infection. Unfortunately in his case the incubation period was unusually long (19 days before appearance of the rash, longer than the duration of the voyage), and second, while in the prodromal phase of infection he took part in a dancing festival involving several hundred people. The resulting epidemic infected all but 5 of the 4262 theoretically susceptible individuals in southern Greenland, with a mortality of 1.8% (greatest in those over 55 years of age) and encephalitis in slightly more than 1 in 1000. Since then, epidemics have recurred in other parts of Greenland following further introductions.

Bech, V. Measles epidemics in Greenland. Am J Dis Child. 1962;103(3): 252–253. doi:10.1001/archpedi.1962.02080020264013.

The peak age of incidence of measles depends on local conditions of population density and the chance of exposure. In large urban communities, before the days of vaccination, epidemics occurred every two to three years, exhausting the available susceptible cohort of children. Epidemics on a continental scale occurred annually in the United States and European countries. Although the newborn population (the birth cohort) provides the input of susceptibles each year, the age distribution of cases in unvaccinated communities is primarily that of children just entering school, with a peak of secondary cases at about two years of age through family contacts. The cyclical nature of measles outbreaks is determined by several variables, including the build-up of susceptibles, introduction of the virus, and environmental conditions that promote viral spread. Both the seasonality of infectivity (Table 13.5) and the occurrence of school holidays affect the epidemic pattern. Following the widespread introduction of immunization programs the epidemiology of measles has changed dramatically (see Chapter 26: Paramyxoviruses).

Although chickenpox is also an acute exanthem in which infection is followed by lifelong immunity to reinfection, it requires a dramatically smaller critical community size for indefinite persistence of the disease: less than 1000, compared with 500,000 for measles. This is because varicella virus causes a persistent infection (see below) that, after being latent for decades, may be reactivated and cause zoster (shingles) (see Chapter 17: Herpesviruses). Although zoster is not as infectious as chickenpox itself (secondary attack rates of 15%, compared to 70% for chickenpox), it can, in turn, produce a new cycle of chickenpox in susceptible children or grandchildren, often many decades after the source individual suffered his or her primary infection.

TABLE 13.6 Viral Infections of Humans Associated with Persistent Viral Secretion

Virus Family	Virus	Comments
Herpesviridae	Epstein-Barr virus, cytomegalovirus, HHV6	Intermittent shedding in saliva, genital secretions, milk
	Herpes simplex virus types 1 and 2	Recurrent excretion in saliva, genital secretions, for years
	Varicella-zoster virus	Shed from zoster lesions many years later
Adenoviridae	All adenoviruses	Intermittent excretion from throat and/or feces
Polyomaviridae	BK, JC polyomaviruses	Excreted in urine
Arenaviridae	All arenaviruses	In rodents, intermittent shedding in urine
Hepadnaviridae	Hepatitis B virus	Persistent viremia; shedding may occur in semen, saliva
Togaviridae	Rubella virus	Persistent only in congenitally infected children; excreted in urine
Flaviviridae	Hepatitis C virus	Persistent viremia; shedding may occur in semen
Retroviridae	Human immunodeficiency viruses	Persistent viremia, excretion in genital secretions for years
Papillomaviridae	Human papillomaviruses	Shed from warts; latent genomes may persist after warts regress

Effects of Immunity

Immunity acquired as a consequence of either prior infection or vaccination plays a vital role in the epidemiology of viral diseases. In most generalized infections, acquired immunity, manifested largely by circulating IgG antibody, appears to be lifelong. This occurs even in the absence of repeated sub-clinical infections, as evidenced by studies of measles and poliomyelitis in isolated populations. In a classic paper on measles in the Faroe Islands written in 1847, Peter Panum, a Danish physician, demonstrated that the attack rate was almost 100% among exposed susceptibles, but that immunity conferred by an attack of measles experienced during an epidemic 65 years earlier remained solid in spite of no further introduction of the virus to the islands in the interim.

The situation is different with viral infections that are localized to mucosal surfaces such as the respiratory tract, since mucosal immunity is relatively short-lived. A large number of serotypes of rhinoviruses and a few serotypes of both coronaviruses and enteroviruses can produce superficial infections of the upper respiratory tract. The seemingly endless succession of common colds suffered by urban communities reflects a series of minor epidemics, each caused by a different serotype of virus—or a different virus. Protection against reinfection is due mainly to antibodies in the nasal secretions, primarily IgA antibodies. Although short-lived type-specific immunity does occur, there is no intertypic cross-immunity, hence the convalescent individual is still susceptible to all other rhinoviruses or coronaviruses. Most persons contract between two and four colds each year. Shedding of most respiratory viruses is short-lived (three to seven days after the onset of symptoms), but rhinoviruses can show prolonged shedding of virus for up to three weeks, long after the acute symptoms have subsided.

Epidemiological observations within isolated human communities illustrate the need for a constant supply of susceptible subjects or antigenically novel viral serotypes to maintain respiratory diseases in nature, and repeated, often sub-clinical, infections are important in maintaining herd immunity. Explorers, for example, are notably free of respiratory illness during their sojourns in the Arctic and Antarctica, despite the freezing weather, but invariably contract severe colds upon establishing contact once more with other humans.

The more radical change known as antigenic shift, attributable to genetic reassortment in influenza A virus, occurs much less frequently than antigenic drift (attributable to small mutations) but often leads to widespread epidemics since there is no background population immunity to the new virus. As genetic reassortment can also occur in rotaviruses and there are rotaviruses affecting several animal hosts, antigenic shift may also occur with these viruses, although this appears to be infrequent.

Persistent Infections

Persistent viral infections (see Chapter 7: Pathogenesis of Virus Infections, and Chapter 8: Patterns of Infection), whether or not associated with episodes of clinical disease, provide an enhanced mechanism for perpetuation of viruses. Individuals with persistent infection may shed infectious virus intermittently or continuously (Table 13.6). In the extreme case, this can reintroduce virus into a population in which many individuals have been born since the last clinically apparent episode of disease (thus is an immunologically naive population). This transmission pattern is important for the survival of herpesviruses in small populations (see above for varicella-zoster).

Persistence of infection, production of disease, and transmission of virus are not necessarily linked. Thus persistent arenavirus infections have little adverse effects on associated rodent reservoir hosts but are efficient in continuing the infection chain. On the other hand, the persistence of viruses in the central nervous system, as with measles virus in subacute sclerosing panencephalitis (SSPE), is lethal but of little epidemiological significance since no infectious virus is shed. It is reasonable to postulate that herpesviruses and retroviruses, so well adapted to lifelong persistence and transmission within small isolated populations, were among the important human viruses found in our hominid ancestors.

Involvement of Non-human Reservoirs

The regular re-introduction of infection from a non-human reservoir, as in the case of zoonoses, both assists the perpetuation of a virus in human populations and governs the distribution and extent of these infections. Examples include many arboviruses (which are discussed in more detail below), rabies, and hantaviruses. The extent of human infection depends on the degree of contact with the animal reservoir and the prevalence of infection in the reservoir. The existence, and possible extent, of an animal reservoir is of fundamental importance in considering plans for the regional elimination or global eradication of any human viral disease.

Arthropod Transmission

Arthropod transmission is ecologically the most complex of all virus transmission modes. The term arbovirus (*ar*thropod-*bo*rne virus) refers to a virus whose life cycle involves alternating replication stages in a vertebrate host and in a blood-feeding arthropod (usually mosquitoes or ticks). "Arbovirus" is an epidemiologically based term that includes viruses of many different families sharing this common mode of spread. The arthropod vector acquires virus by feeding on the blood of a viremic animal or person. The ingested virus replicates, initially in the arthropod gut and then in the salivary gland, over several days (the extrinsic incubation period); this period varies with different viruses and is influenced by ambient temperature. Virions in the salivary secretions of the vector are injected into new vertebrate hosts when the arthropods subsequently take a blood meal. In addition to the above, there are also certain situations where arthropod transmission may occur mechanically by contamination of the insect's biting parts ("flying pin").

Diseases caused by arboviruses tend to fall into several distinct forms:

1. asymptomatic infection or a non-specific syndrome of malaise and fever;
2. encephalitis, usually diffuse, in contrast to the usual focal pattern of herpes simplex encephalitis;
3. a syndrome of fever, arthralgia, myalgia with or without rash, and;
4. hemorrhagic fever, with or without hepatitis and jaundice.

Symptomatic dengue virus infection usually involves fever, arthralgia, and rash, but severe dengue (formerly called dengue hemorrhagic fever or dengue shock syndrome) involves hemorrhagic manifestations, hypotension, and shock that may be fatal.

Arthropod transmission provides a way for a virus to cross species barriers, since the same arthropod may bite birds, reptiles, and mammals that rarely or never come into close contact with each other. Vertebrate reservoir hosts are usually wild mammals or birds, which generally sustain sub-clinical infections producing an ongoing vertebrate-arthropod-vertebrate cycle. Humans are rarely involved in this typical maintenance cycle (enzootic cycle), unless they venture into a site where contact occurs with the infected. However there are important exceptions—urban yellow fever and dengue—where human–arthropod–human cycles are the usual pattern. Another pattern involves spread of infection from the primary maintenance cycle in the wild, to a cycle involving an amplifying host (e.g., a domestic animal species and/or a different arthropod species) and finally to a human host (Fig. 13.4).

Transmission of some arboviruses from one vertebrate host to another can also occur by additional mechanisms not involving an arthropod. Thus, in central Europe a variety of small rodents and ticks are reservoir hosts of tick-borne encephalitis virus. Goats, cows, and sheep are incidental hosts and are sub-clinically infected via tick bites; however, they excrete virus in milk, and the drinking of virus-laden milk may infect new-born animals. Humans may be infected either by being bitten by a tick or by drinking unpasteurized milk from an infected animal (see Fig. 36.10).

In contrast to the arthropod, which carries the virus throughout its short life, infected vertebrates usually recover rapidly, eliminate the virus, and develop a lasting immunity to reinfection. To be an efficient reservoir host, the vertebrate must be abundant and have a rapid turnover rate, and after infection it must maintain a high level of viremia for an adequate period. In turn, to be an efficient vector the arthropod must (1) be easily infected even when feeding on vertebrate host with a low titered viremia (called a low infection threshold), (2) be supportive of virus replication to a titer sufficient to infect its next vertebrate victim, (3) be able to deliver virus from the productive infection in its

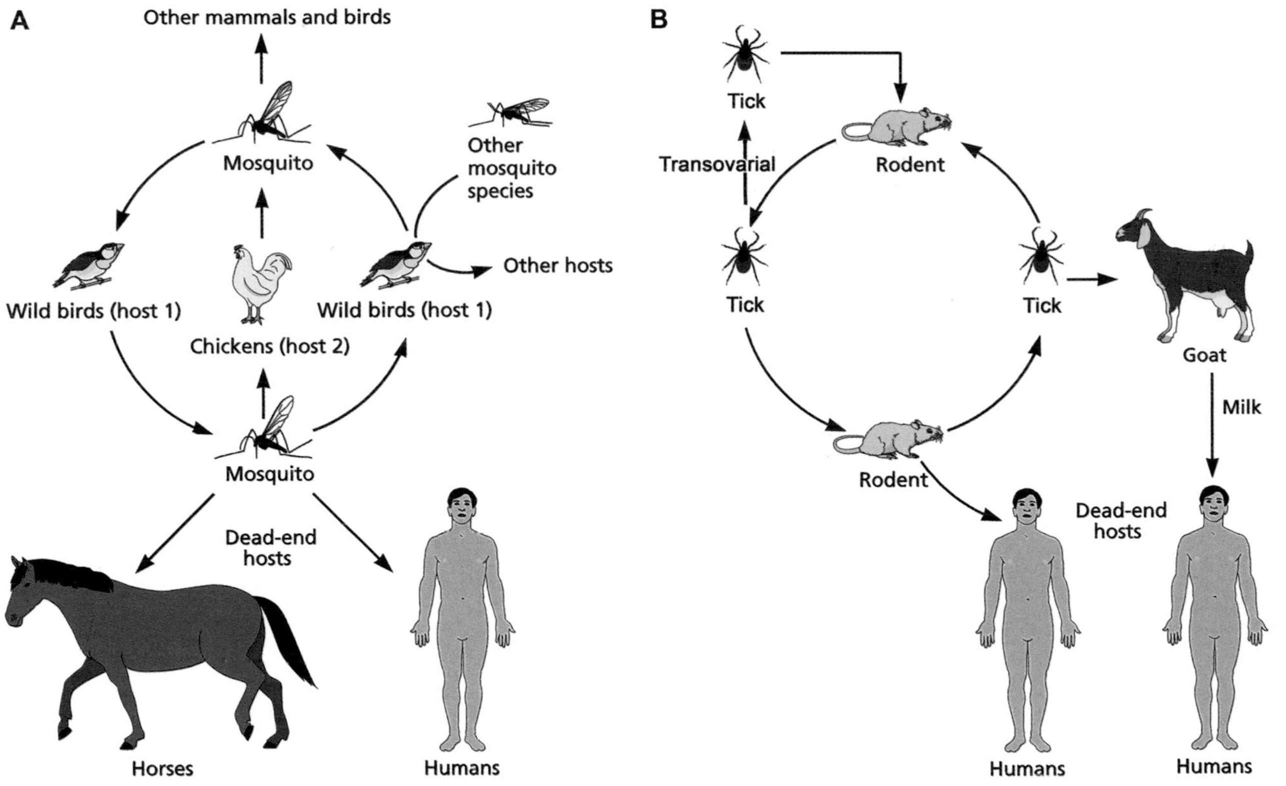

FIGURE 13.4 Patterns of arbovirus life cycles. For many infections, a reservoir is maintained by a natural cycle between a particular arthropod species and a wild vertebrate species, for example a bird or mammal. Domestic species may become involved, with or without showing disease, and these may help amplify the infection and bring it closer to human habitation. Further species including humans may then become infected, but if they do not provide a source for further arthropod infection, they function as dead-end hosts. Finally, for some viruses a human-arthropod-human cycle occurs, either continually or in situations of high arthropod and human population density. *Reproduced from Flint, S.J. et al., 2009. Principles of Virology, third ed., ASM Press, Washington, DC, with permission.*

salivary gland into its saliva and then into its vertebrate host's blood/tissues, and (4) be able to continue this sequence for its lifespan without adverse pathological effects of the infection. The arthropod must also have a distribution pattern, a flight range, a longevity, and biting habits adapted to the habitat and behavior of its vertebrate host(s). Under these circumstances, the virus flourishes indefinitely in cycles of co-existence with vertebrate and arthropod hosts. Given the multiplicity of different parameters affecting transmission and survival, arboviruses can be vulnerable of dying out, but also capable of causing rapid large human epidemics in different circumstances.

Humans living in regions where a particular arbovirus is enzootic are vulnerable to infection. A proportion of those infected may suffer severe, even fatal disease. Visitors such as tourists, soldiers, or forest workers are at greater risk as, unlike the indigenous population, they will not have acquired immunity from sub-clinical infection in childhood. In tropical countries where the arthropod vector is plentiful year-round, the risk is always present and human disease is endemic (e.g., jungle yellow fever). In regions subject to monsoonal rains, epidemics of mosquito-borne diseases

may occur toward the end of the wet season, for example, Japanese encephalitis in parts of Southeast Asia. In some temperate countries, and particularly in arid areas, human epidemics of mosquito-borne arbovirus disease occur following periods of exceptionally heavy rain.

A puzzle that has concerned many investigators has been to understand what happens to the viruses during the winter months in temperate climates when the arthropod vectors are inactive. One important mechanism for overwintering is transovarial transmission from one generation of arthropods to the next. Arthropods such as ticks have several larval stages, and this is necessarily associated with transstadial transmission (across stadia or life-stages). Transovarial infection occurs in most tick-borne arbovirus infections and is often sufficient to ensure survival of the virus independently of a cycle in vertebrates; as far as virus survival is concerned, vertebrate infection is only important in amplifying the population of infected ticks.

Other possible mechanisms for overwintering are still unproven or speculative. For example, hibernating vertebrates have been thought to play a role. In cold climates, bats and some small rodents, as well as snakes and frogs, hibernate during the winter months. The low body temperature of these species has been thought to favor persistent infection, with recrudescent viremia occurring when normal body temperature returns in the spring. Although demonstrated in the laboratory, this mechanism has never been proved to occur in nature. Finally, there is always the possibility that infection may be re-introduced from another more favorable geographical region where the vector, and the virus, have been able to survive.

Examples of the complexity of the life cycles of arboviruses are given in Chapter 39: Bunyaviruses, Chapter 35: Togaviruses, and Chapter 36: Flaviviruses. Many ecological changes produced by human activities disturb natural arbovirus life cycles and have been incriminated in the geographical spread or increased prevalence of the diseases they cause. These include:

1. Population movements and human intrusion into new arthropod habitats, notably tropical forests; for example, the epidemic yellow fever among those working on the Panama Canal in 1904 set back construction for a number of years, both because of the high rates of sickness and death and because of desertion of the remaining workers through fear of infection;
2. Deforestation, with development of new forest–farmland margins and exposure of humans and their livestock to new arthropods and their viruses;
3. Irrigation, especially primitive irrigation systems, which pay no attention to arthropod control;
4. Uncontrolled urbanization, with vector populations breeding in accumulations of water and sewage;
5. Increased long-distance air travel, with potential for carriage of arthropod vectors and of persons incubating diseases such as dengue and yellow fever. The introduction of West Nile virus from the Middle East into the Americas may have been due to carriage of an infected mosquito—or human—by air transport; alternatively, a long-distance migration of infected birds may have been responsible;
6. New routing of long-distance bird migrations brought about by recent man-made water impoundments; and
7. Climate change, affecting sea level, estuarine wetlands, fresh water swamps, and human habitation patterns, may be affecting vector–virus relationships throughout the tropics and sub-tropics.

Nosocomial and Iatrogenic Transmission

Nosocomial transmission refers to transmission while a person is in a hospital or clinic, whereas iatrogenic transmission refers to transmission "by the hand of the doctor." The lethal Ebola virus outbreak in Zaire in 1976, is a classic example of an iatrogenic and nosocomial infection. More common examples of nosocomial virus infections spreading by the respiratory route are chickenpox, influenza, and respiratory syncytial virus infections in hospital settings. Hepatitis B and C viruses, and to a lesser extent HIV, can also be transmitted by doctors, dentists, acupuncturists, tattooists, etc., and there is also a risk to attending staff and laboratory personnel, via needle stick and similar injuries. The risk of nosocomial transmission is exacerbated by the fact that infectious patients may congregate together in health-care facilities, and invasive procedures and blood exposure may occur. Understandably, health professionals exercise particular care to prevent such events.

EPIDEMIOLOGICAL INVESTIGATIONS

Definitions of Disease Activity: Incidence and Prevalence

The comparison of past disease experience, and expected future risk, in different populations is expressed in the form of rates, that is the number of events in a standard population size, for example 1000, 100,000, 1,000,000, etc. Two rates are widely used: incidence and prevalence. In all cases the denominator (total number of persons at risk) may be general, that is the total population in a state or country, or it may be a specific cohort of individuals known to be susceptible or at risk. The latter is often equated with the number of persons in a specified population who lack antibodies to the virus of interest ("susceptibles"). In each situation it is imperative to be clear about the nature of the denominator. All rates may be affected by various attributes that distinguish one person from another: age, sex, genetic constitution, immune status, nutritional status, and various behavioral parameters. The most widely applicable attribute is age, which may be linked with immune status as well as various physiological variables.

The *incidence*, or attack rate, is a measure of the number of events over time, for example, per month or per year, and is especially useful for acute diseases of short duration. The denominator includes both the population size and time frame, and incidence rates are usually expressed as cases per standard population size (e.g., 100,000) per standard time (e.g., one year). It will be immediately apparent that two additional aspects are important. First, it is usually the case that not all members of a population are susceptible, for example because of prior infection leading to immunity. Thus, an incidence rate recorded in a total population may produce a lower figure than a more targeted incidence rate based on those who are truly susceptible. Second, with nearly all viruses, not all infected individuals develop clinical disease, and many infections go unrecognized. Thus the incidence rate of clinical disease is invariably lower than the incidence rate of total infections. The ratio of clinical to sub-clinical, or inapparent, infections varies widely between different viruses. For example, measles infections are almost always clinically apparent, whereas less than 1% of those infected with encephalitogenic arboviruses or with polioviruses develop encephalitis or poliomyelitis, respectively (Table 13.2).

The proportion of a population becoming infected during the course of a season or a year may fluctuate considerably, depending on factors such as the season, changes in human behavior, emergence of a new virus strain, etc. The *secondary attack rate*, applied to comparable relatively closed groups like households or classrooms, is a useful measure of the "infectiousness" of viruses transmitted by aerosol or droplet spread. It is defined as the number of persons in contact with the primary or *index* case who become infected or ill within the maximum incubation period, expressed as a percentage of the total number of susceptible persons exposed to infection.

It is difficult to measure the incidence of chronic diseases, especially where the onset is insidious, and for such diseases it is customary to determine the *prevalence*, that is, the ratio, at a particular point in time, of the number of cases currently present in a population divided by the size of that population. Prevalence is a snapshot of the frequency that prevails at a given time, and it is thus a function of both incidence and duration of the disease. It may be expressed as a percentage, or as number of cases per population unit, for example 100,000. *Seroprevalence* refers to the frequency of individuals with antibody to a particular virus in a population, and as neutralizing antibodies often last for many years, or even for life, seroprevalence rates usually represent the cumulative experience within a studied population.

Deaths from a disease can be categorized in two ways: the *cause-specific mortality rate* (the number of deaths from the disease in a given year, divided by the total population at mid-year), usually expressed per 100,000;

or the *case-fatality rate* (the percentage of persons with a particular disease who die from the disease).

Laboratory Approaches

Seroepidemiology

Traditional surveillance is based on the reporting of clinical disease. However, examination of sera for antibodies gives a more accurate measure of the true prevalence of a particular virus in a given population. By detecting antiviral antibodies in various age groups it is possible to determine how effectively viruses have spread, or how long it has been since the last appearance of non-endemic viruses. Correlation of serological tests with clinical observations also makes it possible to determine the ratio of clinical to sub-clinical infections (Fig. 13.5).

Seroepidemiology is extremely useful in support of public health policy and research (Box 13.4). Advantage is often taken of a wide range of sources of human sera, such as prior population surveys, entrance examinations for military and other personnel, blood banks, hospitals, and public health laboratories. Such sera can be used in order to determine the prevalence of particular infections, to evaluate eradication and immunization programs, and to assess the impact, dynamics, and geographical distribution of new, emerging, and reemerging viruses. For example, serological surveys have been essential to determine the prevalence and geographical distribution of HIV, HBV, and HCV. More recently, concerns about the possibility of breaches of privacy have led to the

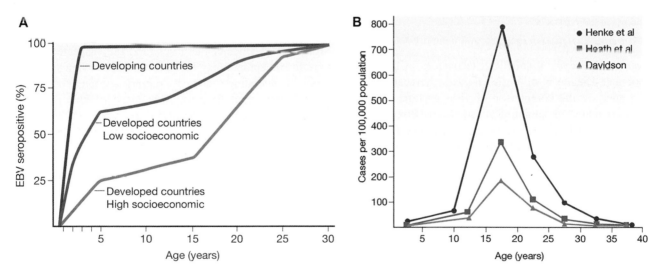

FIGURE 13.5 (A) Seroepidemiology of Epstein-Barr infection in relation to socioeconomic conditions. The y-axis shows the prevalence of EBV antibody at different age groups in three different populations. Note, in developed countries with a high SE status, peak rates of transmission (steepest curve increases) correspond to <1 to 5-year-old and 15 to 25-year-old cohorts, the two ages where salivary contact is likely to be greatest. (B) Incidence of clinical cases of infectious mononucleosis at different age groups in three different studies. The relatively small number of cases of clinical disease in children is because infections in the young carry a low rate of clinical disease, while the decline in clinical cases in older age groups is because fewer new infections occurred in this age group. *Modified from Fields, B.N., et al. (Eds.), 1996. Fields Virology. Lippincott-Raven, Philadelphia, PA.*

BOX 13.4 Uses of Serological Studies in Investigating Virus Epidemiology

1. Measure the *total* number of infections, that is clinical and sub-clinical
2. Estimate the *prevalence* of total infections in a population, for example what percentage of individuals have persistent CMV infection
3. Estimate the *incidence* of total infections over a time period, for example the rate of new hepatitis C infections in a given period
4. Age-specific prevalence patterns allow analysis of transmission efficiencies and possible transmission routes
5. Evaluate the efficacy of immunization programs, for example measles, hepatitis B, or the level of herd immunity, for example influenza
6. Assess the exposure resulting from occupation, lifestyle—do particular activities truly lead to increased rates of infection?
7. Investigate the natural reservoir of a new infection, for example SARS
8. Is an apparently new infection in man, truly new or a pre-existing infection that has just been recently recognized?

BOX 13.5 Role of Virus Genome Sequencing in Epidemiology

1. Predict the likely source of the virus in a newly diagnosed human case of infection, for example geographic/animal source of rabies, country of acquisition of HIV
2. With a cluster of cases of an infection, distinguish a common source origin from a group of sporadic unrelated cases all occurring about the same time, for example HIV transmission at a doctors' surgery
3. Distinguish a vaccine strain of poliovirus isolated from a patient or sewage (a common occurrence) from a wild-type strain (a cause for alarm, given the near global eradication of polio)
4. Detect emergence of a new strain of influenza
5. Investigate the possible origins of a virus newly circulating in humans, for example assign it to a particular virus family, assess its relationship to previously described viruses, predict whether it is from human or animal sources
6. Quickly determine the sensitivity of a new virus isolate to antiviral drugs, and monitor the spread of drug-resistant viruses in the community

imposition of constraints on the use of anonymous stored human sera as an epidemiological research resource.

Sentinel studies in animals are widely used for assessing the seasonal prevalence of arbovirus infections; for example, sentinel chickens are used for the early detection of eastern equine encephalitis and St. Louis encephalitis viruses in the southern United States and for the detection of Murray Valley encephalitis virus in Australia.

Molecular Epidemiology

Advances in rapid sequencing techniques (Chapter 10: Laboratory Diagnosis of Virus Diseases) have allowed the ready analysis and comparison of virus genome sequences to answer many epidemiological questions (Box 13.5). An excellent example is the use of partial genome sequencing in order to distinguish poliovirus vaccine strains from wild strains of poliovirus, and also to identify patient isolates containing one or more of the successive base changes that occur when each of the three vaccine strains progressively reverts to virulence. For many other viruses where different sequence variants are known to be endemic in different regions of the world, sequencing can identify the likely geographical origins of a new virus isolate. For example the strain of West Nile virus that first appeared in North America in 1999 was matched to a strain then circulating in Israel, suggesting the latter country as its likely source. The genome sequence of a rabies virus isolate from a human case is used to determine its geographical origin and reservoir animal species (Box 13.5).

Whenever an unusual cluster of cases occurs suggesting the appearance of a new or previously unrecognized virus, recovery of the virus and genome sequencing are done as soon as possible. From such information, the virus family, and often the genus, can be immediately identified by comparison with sequences in available databases. Occasionally it is possible to predict virulence attributes. Conversely, when a cluster of cases from around the same place or time are suspected of having a common source, sequence comparison between different samples can be used to ascertain whether the cases (and the suspected source) are linked epidemiologically, for example when successive patients attending a doctor's surgery are later found to be HIV-positive and the question is raised whether or not this is due to iatrogenic transmission.

Routine Surveillance

The collection of accurate data about the occurrence of disease often requires considerable resources and ingenuity. Data on the population (denominator) are usually available, but it is difficult to obtain accurate information about the number of cases. Where such information is regarded as essential for public health purposes, cases may be notifiable by law. In practice, physicians tend not to be sufficiently conscientious about reporting, and of course some infected individuals do not consult a physician. To help overcome this problem, many public health authorities enlist selected practitioners into a network of "sentinel practices," and engage hospital and public health diagnostic laboratories to provide integrated information about clinical cases, virus identification, and serology. Local and international data on infectious diseases are available in electronic and printed formats in order to disseminate information and maintain the interest of both practitioners and laboratories, examples being the Morbidity and Mortality Weekly Report (MMWR) published by the Centers for Disease Control and Prevention (CDC) in the United States, and the Weekly Epidemiological Record of the World Health Organization (WHO). Public health departments of most governments also maintain websites containing information, advice, and warnings about recent and current infectious diseases (see also Chapter 14: Control, Prevention, and Eradication).

Special arrangements for surveillance are set up to address high-priority problems. For example, HIV/AIDS surveillance programs supported by national and international bodies provide detailed data. Other examples are the WHO influenza-reporting network aimed at rapidly identifying new epidemic strains of influenza. Yet another is the clinical/laboratory surveillance system for acute flaccid paralysis (AFP) that is an integral component of the global polio eradication campaign. WHO, and national public health authorities such as CDC in the United States, respond promptly to the occurrence of unusual outbreaks of disease by setting up special "task forces" of appropriately experienced epidemiologists and laboratory scientists including virologists and pathologists to investigate such problems. Recent examples include the investigation of severe acute respiratory syndrome (SARS) in 2003, H5N1 avian influenza (since 2003), and recurring outbreaks of Ebola virus disease in Uganda, South Sudan, the Democratic Republic of Congo, and the countries of West Africa.

Cross-Sectional, Case-Control, Cohort, and Long-Term Population Studies

A *cross-sectional study* to measure the prevalence of a given marker in a population can be carried out relatively quickly. One caveat is that populations are often not homogeneous; a small subgroup at high risk may contribute most of the cases, and a major determinant of prevalence in any one study may be the size of such a subgroup in the total sample. For instance, blood donors have long been a carefully selected group where individuals with risk factors for blood-borne diseases are excluded, and, depending on the thoroughness with which this exclusion is done, blood donors may not be representative of the general population. Age-specific prevalence rates go further and give considerable information about historical changes and mechanisms of virus transmission (Fig. 13.5).

A *case-control study* is initiated after the disease has occurred, the purpose being to identify the cause by comparing cases and controls. It is thus a retrospective study,

going back in time to determine causative events; it requires careful selection of the control group, matched to the test group so as to avoid bias, together with careful selection of the questions and tests used. It can be used, for example, to answer whether a particular enteric virus can cause disease, by comparing the rate of excretion of the virus in a cohort of children hospitalized with gastroenteritis, with a group of age-matched children hospitalized for some other reason.

Cohort studies, usually carried out prospectively, start with a presumed cause or possible future risk, for example a new treatment or vaccine. A population exposed to infection is followed over time in order to identify significant correlates, such as outcome of disease, reinfection, vaccine efficacy, etc. This type of study requires the recording of new data, and the careful selection of a control group that is as similar as possible to the exposed group, except for the absence of contact with the presumed causative influence or treatment. Such studies do not lend themselves to rapid conclusions, as groups must be followed until disease is observed. Such studies are invariably expensive; however, successful cohort studies can provide strong evidence for cause–effect relationships and are essential for defining the safety of new vaccines. Long-term studies of families or larger groups, for example the population of a city, can yield much useful information about the natural history of diseases and the long-term effects of, for example, interventions, chronic infections, or environmental factors, but such studies are expensive and require long-term dedication of both personnel and resources.

The discovery as to the cause of congenital defects by rubella virus provides examples of both retrospective and prospective studies. Norman Gregg, an ophthalmologist working in Sydney, Australia, was struck by the large number of cases of congenital cataract he saw between 1940 and 1941, and by the fact that many of the children also had cardiac defects. By interviewing the mothers he found that the majority had experienced rubella early in pregnancy. His hypothesis that there was a causative relation between maternal rubella and congenital defects quickly received support from other retrospective studies, and prospective studies were then organized. Groups of pregnant women were sought who had experienced an acute exanthematous disease during pregnancy, and the subsequent occurrence of congenital defects among their children was compared with that among women who had not experienced such infections. Gregg's predictions were thus confirmed and the epidemiology of congenital rubella syndrome more precisely defined.

Human Volunteer Studies

Many major discoveries that have led to the control of viral diseases were possible only with the use of human volunteers. Early work with yellow fever, viral hepatitis, the common

One notable study of hepatitis viruses was carried out by Saul Krugman and Joan Giles among children with intellectual disabilities in the Willowbrook State School on Staten Island in New York. Hepatitis infections were endemic in the 1960s among the institutionalized children despite all medical and nursing measures to prevent transmission. Krugman and Giles studied these infections, isolated two prototype infective sera that we now recognize contained hepatitis A and B respectively. Their work represented pioneering advances including defining the clinical characteristics of HAV and HBV infections, demonstrating that there were two distinct agents with no cross-protection, and ultimately demonstrating the feasibility of passive and active immunization against HBV. This work laid the foundation for HBV vaccine development. In subsequent years Krugman and Giles were severely criticized because this work was considered by many commentators as unethical. Krugman and Giles had gone to great lengths to obtain proper informed consent and had fully described their sensitivity to the interests of their subjects. It can now be said that the work was carried out in full accordance with the ethical standards and practices of the 1960s. However, given the ethical standards and regulatory practices of today, it can also be said that despite the importance of the subject and the results obtained, this work would not now be approved in any country with a human subjects regulatory system.

cold, and a range of other respiratory infections involved human volunteers because of a lack of relevant animal models. An absolute requirement has been that investigators obtain informed consent from the subjects or, in the case of minors, from their parents (Box 13.6). In most countries today, human subject research is highly regulated by governmental agencies; at the heart of review and oversight are local institutional review boards [IRBs, also known as independent ethics committees (IECs) and human subject committees (HSCs)]. It is essential in such oversight that short- and long-term risks be carefully assessed, including the possibility of transferring adventitious agents present in an inoculum as a contaminant; subjects may need to be isolated for the duration of such studies, thus reducing the risk of secondary transmission to contacts.

Outbreak Investigation

Epidemiologists, virologists, and other professionals are frequently involved in investigating new outbreaks of disease. These may involve a recognized virus infection that for some reason has changed its characteristics or usual prevalence or distribution; alternatively, it may potentially involve a new variant of a known virus, or a completely new agent. Some of the approaches followed are outlined in Chapter 15: Emerging Virus Diseases; this work includes

some of the more challenging, glamorous, and sometimes dangerous roles for our profession.

MATHEMATICAL MODELING

Ever since Daniel Bernoulli attempted in 1760 to model the likely effect of variolation on the spread of smallpox, there have been attempts to develop and refine mathematical models of disease transmission. Most of these studies have focused on identifying how single source, local outbreaks may be prevented from spreading, but the complexities of modern societies call for more models that recognize how individuals incubating emerging or unknown infections can quickly seed infections in multiple localities, for example as a result of air travel. Ease of air travel was the major determining factor in the spread of SARS virus to three continents in 2003 within days of the index case arriving in Hong Kong.

As modeling has evolved in recent years, it has provided more and more insights that inform decisions regarding control. First, in some instances it has uncovered patterns in the evolution of an outbreak in terms of rate of spread and numbers of susceptible individuals likely to come into contact with the pathogen. This allows resources to be focused in such a way that an outbreak can be contained within the shortest period of time and the number of contacts is minimized. Second, in other instances modeling has allowed for a variety of control measures to be simulated and compared. Third, in yet other instances modeling has supported comprehensive studies of the epidemiology of the disease under study, the manner of its transmission, peak of infectiousness, and even if new variants are emerging as the outbreak progresses. Modeling in this way was invaluable in controlling the 2001 foot and mouth disease outbreak in the United Kingdom. Modeling has been the key in refining global guidelines for dealing with a smallpox bioterrorism event—three models have confirmed that large-scale ring vaccination using vaccines stores under WHO auspices would extinguish spread very quickly. Such methods proved useful also in predicting how monkeypox spread in the United States in 2003 and in understanding the reemergence of arthropod-borne diseases such as dengue. As with all modeling approaches, it is essential both to be clear as to the questions being asked of the model and to take into account all known epidemiological data. Finally, it is essential that overgeneralizing predictions drawn from modeling, often by scholars not regularly involved in disease control, not be allowed to influence the public through the press or media nor influence political decision makers.

Epidemiological Parameters

The following epidemiological parameters are required for modeling: the basic reproductive number (R_0), the degree of variation in infectivity between cases, the time for infections to recycle (the generation time, T), and the proportion of transmissions occurring before the onset of symptoms (θ).

The basic reproductive number (R_0) is an important parameter, representing the average number of secondary infections produced by an infected individual among a population of susceptible individuals. If the value of R_0 is less than 1, then the chances of an infection generating new cases is insufficient for an outbreak to be maintained. If R_0 exceeds 1, however, the number of secondary cases multiplies the infection and an epidemic ensues until the proportion of susceptible individuals declines. In cases where some of the population have pre-existing immunity, the effective reproduction number (R) is modified by applying a fraction representing this proportion of the population ($R = fR_0$, with f representing the fraction of population susceptible to infection). Control measures look to reduce the value of R to below 1.

The value of R_0 for a given virus in a given host population is determined by a number of different factors including the transmissibility of the virus, the period over which an infected host is infectious, the population density of hosts, and where appropriate, the density of relevant arthropod vectors and the capacity of such vectors to transmit the virus. The details of these relationships depend on the mode of transmission [e.g., direct contact transmission, indirect contact transmission (fomite transmission), sexual contact transmission, vector-borne transmission]. Equations used to calculate R_0 have been developed which take into account the impact of life expectancy, duration of protection due to maternal antibody, the duration of protection afforded by a vaccine, etc. Many active immunization policies and programs take such calculations into account.

One of the difficulties with modelling is chance variation early in an epidemic. This was a particular difficulty in estimating the value of R_0 during the SARS epidemic as a small number of individuals spread the virus to a disproportional greater number of secondary cases. Why this happened remains unclear, but it is known that some persons infected with other respiratory tract infections can shed larger than average amounts of virus. Importantly, this initial variation can be considerable, and greater variation can lead to more severe outbreaks. Fortunately, the value of R_0 for SARS was around 3, somewhat more than for Ebola but considerably less than figures computed for smallpox and measles. Interestingly, SARS is also similar to smallpox in having a generation time (T) of approximately one week and θ with a value of 0.11. The significance of this is that the quarantine and isolation of SARS cases were predicted as effective, as indeed was the case. Quarantine measures also have the benefit of allowing for more accurate estimates of θ as the time from contact to onset of symptoms can be accurately defined.

Models have inherent difficulties in predicting just how large an epidemic will be. This in the main is due to the enormous difficulty in measuring how individuals interact with each other within their communities. Most models assume a uniform mix of people, each having the same probability as all others in terms of coming into contact with an infected person. This might be true at the local level, but becomes increasingly invalid as individuals move between communities, cities, and national boundaries. One individual may travel from one subpopulation to another, introducing the pathogen to a completely new population of susceptible individuals. In this case, the size of an outbreak is determined by the behaviors of a relatively small number of infected individuals. In the end, the final size of an epidemic and its duration is determined by the structure of the population as much as it is by R_0, and estimates of epidemic size need to recognize a series of subpopulation structures through which an infected individual may pass. Therefore epidemics are the result of smaller local outbreaks within subpopulations where most transmission occurs, accompanied by much broader spreading by a small number of infected individuals: increasingly over the past 50 years, this process of seeding has been the result of long-distance air travel.

The usefulness of modeling is critically dependent upon accurate and rapid diagnosis of an infection within days of the first cases being recorded. Modeling shows that the effectiveness of control measures can fall off rapidly unless the correct control policy is implemented: crucially this can often be when variability of secondary transmission is greatest and statistical treatment of small case numbers least reliable. Once robust data are available, however, predictions as to the progression of an epidemic, its economic impact, and the likely consequences on healthcare resources can work wonders in galvanizing political support to ensure adequate resources are allocated.

FURTHER READING

Bloom-Feshbach, K., Alonso, W.J., Viboud, C., 2013. Latitudinal variations in seasonal activity of influenza and Respiratory Syncytial Virus (RSV): A global comparative review. PLoS One 8, e5444.

Kaslow, R.A., 2014. Epidemiology and control: Principles, practice and programs. In: Kaslow, R.A., Stanberry, L.R., LeDuc, J.W. (Eds.) Viral Infections of Hhumans (fifth ed.), Springer, Chapter 1.

Nathanson, N., 2007. Epidemiology. In: Knipe, D.M., Howley, P.M., Griffin, D.E., Lamb, R.A., Martin, M.A., Roizman, B., Straus, S.E. (Eds.) Fields Virology (fifth ed.), Raven, New York, pp. 267.

Panum, P.L., 1847. Observations Made during the Epidemic of Measles on the Faroe Islands in the Year 1846. American Publ., New York, reprinted 1940.

Chapter 14

Control, Prevention, and Eradication

Great strides have been made over the past quarter of a century in the prevention and control of viral diseases, largely through a combination of measures to minimize transmission and vaccination of populations at risk. More recently, the third pillar of a control strategy, the use of specific antiviral drugs, has also begun to be significant in reducing the dissemination and impact of a number of viral diseases, most notably for viral hepatitis and HIV (see Chapter 12: Antiviral Chemotherapy). Improved socioeconomic living conditions also play an important general role: less overcrowding leads to reduced disease transmission, and improved nutrition and physical well-being lessen the clinical impact of many infections. However, the ultimate step for the control of any infectious disease is global eradication, as was achieved for smallpox in 1977 and is hoped soon for poliomyelitis.

Control measures are applied at many different levels (Table 14.1). Government public health authorities have national responsibility for the control of infectious disease threats, although they are often constrained by a lack of scientific knowledge, inadequate budgets, expediency and political will. International bodies such as the World Health Organization (WHO) provide expert guidelines, trained personnel, and disease surveillance, but practical constraints often limit what can be achieved. At a local level, hospital infection control committees and workplace health and safety units promote policies for protecting patients and staff from infectious diseases. Exotic diseases may be excluded by quarantine, and different levels of barrier isolation are practiced for individuals who may present different degrees of risk as sources of infection. Hygiene and sanitation are important methods of controlling enteric infections, and vector control is clearly important for the control of arbovirus diseases. However, the most generally useful control measure is vaccination.

Outbreaks of severe diseases such as Ebola hemorrhagic fever and Lassa fever can be inflamed by delayed and/or inadequate control procedures. The outbreak of Ebola hemorrhagic fever in Zaire (now DRC) and the Sudan in 1976 was controlled largely by closing the hospitals at the center of the outbreaks combined with the meticulous tracking and isolation of contacts and family members. Recent experience with SARS has shown vividly both how applying classical infection control measures can stop outbreaks, and that nosocomial outbreaks of disease may still occur even in the best-equipped and staffed hospital settings. Traditional isolation methods consisted of barrier nursing, use of disposables, and quarantine, in the early hours of containing a disease outbreak. Recent experience has shown these need to be supplemented by additional disease-specific measures.

To the above could be added the necessity to involve veterinarians, especially where zoonotic disease is suspected. The 1999 West Nile outbreak in the United States showed how valuable time was lost when the first signs of disease incursion were seen in wild and domestic animals, but these observations were not communicated efficiently to the relevant local public health authorities. Traditionally there has been little integration of human and veterinary public health, yet the principles and practice of disease control are broadly the same regardless of the target species. Just how important this can be is shown by new pathogens that have come to light almost annually since the early 1990s, almost all of which are pathogens with animal reservoirs.

SURVEILLANCE AND MODELING OF VIRUS DISEASES

Epidemiological surveillance is the foundation for immediate and long-term strategies for combating infectious diseases. Such monitoring is usually the responsibility of national authorities and includes assessing individual cases, identifying the causative organisms, and compiling population-based data that inform public health policy (see also Chapter 13: Epidemiology of Viral Infections).

Infections know no boundaries and thus rapid communication at an international level is essential. The SARS outbreak of 2003 and new pandemics of influenza in particular have done much to strengthen international efforts to ensure better integration of national and international reporting systems. The focus must be on collecting background data and discerning trends, with the lead being taken by international agencies such as WHO and others. The

Fenner and White's Medical Virology. DOI: http://dx.doi.org/10.1016/B978-0-12-375156-0.00014-X

TABLE 14.1 Four Essential Components for Successful Surveillance and Control of Viral Diseases

Alerting clinicians to recognize the unusual; ensuring they have easy access to local laboratories and specialist expertise

High-standard local diagnostic facilities, backed by national reference laboratories with experienced staff

Involving epidemiologists and communicable disease specialists early whenever unusual events are suspected

The authority and resources to deliver effective and prompt control measures

difficulty is that outbreaks of emerging diseases frequently arise in regions lacking both clinical and epidemiological expertise in infectious disease. Many national laboratories—especially in Africa—are often poorly equipped and lack adequately trained personnel for recognizing the unusual and being able to react appropriately.

In the longer term, disease monitoring should also be correlated with data and data interpretation of climatic variation and other determinants of disease activity, particularly if use is to be made of the data in predicting the emergence of zoonoses and vector-borne diseases in regions free of these diseases. Satellite surveillance of vegetation growth can give an early warning of escalating vector numbers.

Passive surveillance. Historically this has often been achieved by a system of "notifiable" diseases. These are specified clinical conditions that must be reported to health authorities by attending clinicians and pathologists. This provides a window into the understanding of infections with clinical impact; importantly, notification allows detection in the community of any new or more frequent occurrence of disease. It is, however, notoriously inefficient as typically only a minority of infections are reported. It also fails to detect the size of any disease occurrence where most infections are asymptomatic, for example, as in the spread of poliovirus infection through a community which is only recognized after sporadic cases of paralysis become apparent.

Laboratory data add an important dimension, through routine reporting of positive diagnoses to local and national health authorities. These data will include some asymptomatic infections and thus add value to any incidence estimate; furthermore, laboratory information about particular viral strains may be crucial. However, most systems suffer from the drawback that often only a minority of clinical cases are actually tested for markers of infection.

Active surveillance. Specific policies are in place for the reporting of certain human infectious diseases. For example, influenza surveillance is a coordinated international effort that aims to recognize as rapidly as possible the emergence of virus strains exhibiting new antigenic

profiles or increased virulence. Local surveillance may include use of sentinel clinical practices that report numbers of influenza-like illnesses in real time; hospital emergency departments and large employers also report trends in numbers of respiratory infections. New virus isolates are sent by local laboratories to a member of an international network of WHO influenza reference laboratories, where analyses as to antigenic composition and sensitivity to common antiviral drugs are tested.

Surveillance of HIV/AIDS is an example of a successful international operation coordinated by WHO; test data arising from the surveillance programs of most countries are assembled and analyzed to give a remarkable picture of the global prevalence, pattern of spread, and demographics of HIV/AIDS.

In some specific situations, active case identification and contact tracing are important to track the spread of infection, for example, following identification of a case of measles in a measles-free community, contacts may be followed up and offered prophylaxis, advice, and diagnostic testing. Persons suspected of having rabies or exotic infections such as Ebola hemorrhagic fever or Lassa fever, suspected because of their clinical presentation and travel history, may be isolated and contacts managed appropriately. Cases of acute flaccid paralysis are reported and investigated virologically wherever poliovirus infection is a possibility.

Much surveillance information is easily and rapidly available on excellent websites; for example, from the WHO (www.who.int/csr) and the US Centers for Disease Control and Prevention (www.cdc.gov/mmwr), and in specialist publications and reports.

Role of Modeling in Control of Virus Infections. Mathematical modeling can provide important guidance about the likely progression of an infection and the impact of proposed control measures. It is based on certain assumptions, for example that the host population is homogeneous with respect to the epidemiology of infection and the consequences of vaccination. All hosts are considered to be equally exposed and equally susceptible (unless vaccinated) to infection, the assumption being that the populations of infectious and susceptible hosts are perfectly mixed. While these assumptions are rarely satisfied in practice, the approach has proven sound for many vaccination programs. The predicted consequences of vaccination at the population level, to the interpretation of the results of vaccine trials, all contribute to the design of optimal vaccination programs.

Mathematical theory can integrate this information to address such questions as to whether it is possible to eliminate an infection; what proportion of individuals must be vaccinated to achieve this; at what age should individuals first be vaccinated; and at what interval, if at all, should individuals be revaccinated. As one might expect for diseases that are transmitted directly by contact or aerosol, vaccination has advantages not only for the vaccinated individuals, who are directly protected, but also for unvaccinated individuals. The latter are indirectly protected because the opportunities for the transmission of the virus in the population as a whole are reduced.

Computer modeling has also provided useful insights into the effects of different vaccination regimens and different levels of acceptance of vaccination. One effect of vaccination programs is to increase the average age at which unimmunized individuals contract the infections against which that particular vaccine protects. Up to a certain level, increasing herd immunity may actually increase the risk of disease if older individuals are more vulnerable, by increasing the age of first exposure to the virus. A second conclusion from such modeling is that because of the cyclic incidence of the viral rash diseases, with intervals between peaks that increase as vaccination coverage improves, the evaluation of vaccination programs that stop short of countrywide elimination must be carried out over a prolonged period of time.

MEASURES TO MINIMIZE TRANSMISSION

Quarantine and Isolation

Quarantine is a measure first used by the Venetian Republic in the 14th century for the control of plague. Literally meaning "40 days" (*quaranta giorni*), the quarantine of shipping was also used by the English colonists in North America in 1647 to try to prevent the entry of yellow fever and smallpox. Quarantine proved very effective in keeping Australia free of endemic smallpox, and in delaying the entry of pandemic influenza into that country in 1919. However, with the onset of air travel and the consequent arrival of passengers before the end of most incubation periods, quarantine as a sole measure of control has become less effective. It was replaced, for smallpox, by the widespread requirement that international travelers were required to have a valid certificate of smallpox vaccination, a requirement that is no longer necessary since eradication of the virus. Currently, a similar provision operates for travelers who come from or pass through countries where yellow fever is endemic. Attempts were made during the recent H1N1 swine flu pandemic to quarantine new or suspected cases on arrival into a new country: thermal scanners became commonplace at airports around the world. While not permanently preventing introduction of pathogens, it is argued that such precautions may have slowed the subsequent spread of H1N1 influenza virus, thereby gaining time for the implementation of other control measures.

The isolation of individual patients suspected of presenting an infection risk is routine and essential practice within hospitals; the measures adopted, for example, "enteric" or "blood" precautions, and the stringency of the isolation, are governed by the characteristics of the virus involved. The recent Ebola hemorrhagic fever epidemic in West Africa has demonstrated that much more stringent precautions, special training, and personal protective equipment (PPE) may be required in some instances.

Measures to Reduce Enteric Spread

Hygiene and sanitation have had a profound effect on the incidence of enteric infections, both viral and bacterial. Viruses that infect the intestinal tract are shed in feces, and in many human communities the recycling of fecal material back into humans through the oral route following fecal contamination of food or water is common. A more voluminous and more fluid output (diarrhea) increases environmental contamination. Hands contaminated at the time of defecation and inadequately washed may transfer viruses directly or indirectly to food, a particular problem among those responsible for the preparation of meals to be eaten by others. Education about adequate hand washing, and provision of running water for toilet purposes, are vitally important measures.

In many densely populated parts of the world there are no reticulated sewerage systems, and sewage may seep into wells, streams, or other drinking water supplies, particularly after heavy rain. Explosive outbreaks of many different gastroenteritis viruses occur from time to time when sewerage mains burst or overflow and contaminate drinking water supplies.

Raw sewage contains 10^3 to 10^6 infectious virus particles per liter, mostly enteroviruses, caliciviruses, adenoviruses, and rotaviruses. Titers drop 100-fold, typically to 10 to 100 pfu per liter, following treatment in modern activated sludge plants, because virions adsorb to the solid waste and this sediments as sludge. The primary sludge is generally subjected to anaerobic digestion, leading to a significant reduction of virus titer. Some countries require that treated sludge be inactivated by pasteurization prior to being discharged into rivers and lakes or prior to use as landfill or fertilizer in agriculture.

In countries where wastewater is recycled for drinking and other domestic purposes, the treated effluent is further treated by coagulation with alum or ferric chloride, adsorption with activated carbon, and finally chlorination. Evidence that by-products of chlorination are toxic to fish and may be carcinogenic for humans has encouraged several countries to turn instead to ozonation. Ozone is a very effective oxidative disinfectant, for viruses as well as bacteria, provided that most of the organic matter containing adsorbed viruses is first removed.

There is also a good case for the chemical disinfection of recycled wastewater not used for drinking purposes, such as agricultural irrigation by sprinklers, public decorative fountains, and industrial cooling towers, as such procedures disseminate viruses in aerosols. However, the lability of viruses to heat, desiccation, and ultraviolet light ensures that the virions remaining in wastewater from which most of the solids have been removed will be inactivated within a few weeks or months, depending on environmental conditions, without further intervention. Even during a cold northern winter, the number of viable enteroviruses in standing water drops by about 10-fold per month; during a hot dry summer the rate of decay is as high as 100-fold per week. Hence storage of the final effluent in an oxidative lagoon for one to two months is an inexpensive and effective way of inactivating viruses.

Measures to Reduce Respiratory Spread

Increased understanding about how respiratory viruses are transmitted via both aerosol droplets and contaminated fomites (Chapter 13: Epidemiology of Viral Infections) has led to improved guidelines to reduce respiratory virus transmission, both between hospital patients and staff and in the community. These guidelines include advice about hand washing after contacting respiratory secretions, care in disposal of used tissues, and advice to minimize transfer of contaminated objects (including stethoscopes!) between individuals. Commercially available facemasks have varying grades of performance specifications with respect to fluid resistance, breathability, and efficiency in filtering microparticles. Masks should be chosen appropriate to the degree of risk, the purpose required, and closeness of fit. Attempts to achieve "air sanitation" by filtration and/ or ultraviolet irradiation in public buildings have proved to have only a marginal effect, although these measures are an important feature of the biosafety cabinets widely used in virus laboratories.

Respiratory viral infections remain very common, and the congregation of large numbers of people in crowded public places and the frequency of air travel enhance the rapid spread of respiratory viruses to non-immune populations. This is seen most clearly with newly emerging infections such as SARS and the 2009 H1N1 pandemic influenza. The population of the world now constitutes a single ecosystem for human respiratory viruses, although seasonal and climatic differences between the northern and southern hemispheres affect the incidence of particular diseases, like influenza, at any particular time.

Measures to Reduce Spread of Sexually Transmitted Infections

The emergence of HIV/AIDS, together with increases in genital warts and chlamydial infections, led in the 1980s to major efforts to understand sexual behavior and factors affecting sexual transmission of viruses. This was assisted in many Western countries by increased acceptance of homosexuality and frankness in discussing sexuality. Psychologists and behavioral scientists have provided

BOX 14.1 Principles of Safer Sex—Example of Guidelines for Teenagers

- Seriously consider whether you want to have sex
- Get tested. Be sure you know your and your partner's HIV status
- Prevent exposure to blood, semen, vaginal and other bodily fluids
- Cover up body parts that could be infectious
- Always use a condom, and use a new one every time
- Be in a monogamous relationship, or limit your number of sexual partners
- Get screened for other STDs
- If you or your partner have/has contracted an STD, receive treatment immediately and do not have sex until your doctor says it's okay
- Don't abuse alcohol or drugs, which are linked to sexual risk-taking
- Talk openly with your partner. Remember these discussions are easier with your clothes on

notable guidance to public health campaigns regarding motivation, cultural barriers, and other reasons why particular target groups may or may not accept advice about the risks of sexually transmitted infections (STIs). This has led to increased understanding and promotion of "safe sex" principles in many societies (see Box 14.1), and increased acceptance of testing for STIs, including HIV, as a normal part of routine health care. Another fundamental aspect of public health education is the understanding of the importance of individuals' rights over their own sexuality and sexual choices, and the awareness that health messages may have little effect unless they also address the cultural, religious, and educational context of those receiving the messages. For example, the guidelines in Box 14.1 are targeted to a developed Western community, and very different language and specific advice may be more appropriate in other cultures.

Campaigns based on the above have had different degrees of impact in many parts of the world. In a number of developed countries, HIV remains largely confined to more promiscuous or risk-taking groups. In many other countries HIV is more widespread among the general community, despite vigorous control campaigns that include blood transfusion screening, improved access to HIV testing and antivirals, marketing of condoms, self-treatment kits for STIs, and safe-sex education programs. HIV prevention campaigns can be impeded by "AIDS-fatigue" among carers, complacency, and/or less practice of safe sex following on from greater antiviral drug use, new laws against homosexual practices, and condemnation of condoms by religious bodies. On the other hand, individuals with successful suppression of viral replication by antiviral therapy now pose a very low risk of transmitting infection to others ("Treatment as Prevention"), and post-exposure prophylaxis and pre-exposure prophylaxis with antiviral drugs are becoming more widely used to restrict transmission (see Chapter 23: Retroviruses). This demonstrates the complex and unpredictable outcomes in the interaction between health advice, human behavior, government policy, and health funding, against a background of cultural and religious issues. There is an enormous challenge to apply medical and social advances, in an appropriate manner, more widely to affected populations in a global context.

Measures to Reduce the Spread of Blood-Borne Infections

In the past three decades there has been major recognition of the importance of blood-borne infections including HIV and hepatitis B and C, and far greater understanding of the mechanisms and social settings for transmission of these agents. This has led to widely adopted guidelines and public health policies in order to reduce the risks of transmission of these infections. Within health and treatment settings, the risks of transmission can be considerably reduced, if not eliminated, by sterilization of all invasive medical instruments, treating all blood and secretions as potentially infective, avoiding the reusage of equipment, for example needles, wherever possible, the screening of donated blood, blood products and organs, the management of needle-stick injuries by established protocols, and by adopting strict protocols for acupuncturists and tattooists. The more difficult goal of reducing transmission between injecting drug users involves the promotion of needle exchange programs, safe injecting rooms, education, and drug rehabilitation centers. Establishing such measures and gaining co-operation amongst intravenous drug abusers can be challenging and sometimes can meet with political and legal opposition. However a significant restriction of the spread of infection, for example, of HIV among intravenous drug abusers, has been achieved where such measures are promoted (Box 14.2).

Vector Control

Because the control of non-human vertebrate reservoir hosts is usually difficult to achieve, control of arboviruses may be approached both (1) by directly reducing the numbers of arthropod vectors, and (2) by minimizing human exposure to these vectors. Guidelines and advice are available from many government public health authorities, environmental protection agencies, international bodies such as WHO, and local councils. The philosophy of an "Integrated Pest Management" is widely used: it consists of a combination of approaches selected for the particular problem and with concern for the full environmental, health, and economic context.

The targeting of vectors includes the elimination of breeding sites and the direct destruction of adult mosquitoes

BOX 14.2 Control of Blood-Borne Infections

1. Mechanisms of transmission
   - transfusion of infected, unscreened blood, or blood products, organs
   - use or reuse of unsterilized medical equipment
   - needlestick injury to health care or laboratory worker
   - sharing injecting equipment by IVDUs
   - tattooing, bodypiercing, acupuncture, initiation rites
   - contact sport with bleeding
2. Control measures to prevent transmission
   - screening of donated blood and blood products
   - strict guidelines for medical procedures, instrument sterilization
   - single use for needles and other equipment
   - protocol for management of needlestick injuries
   - strict guidelines and regulation of tattooists, bodypiercing, acupuncture
   - IVDUs—needle exchange, drug rehabilitation centers, education
   - safe injecting rooms (NB effect of legality, needle policy in prisons)

or larvae. The flight range of many vector mosquitoes is so limited that much can be achieved by concentrating on the immediate vicinity of human settlements, particularly in the case of species attracted to human habitation such as *Aedes aegypti*. Any still water constitutes a potential breeding site. Swamps and ditches should be drained, and water-collecting refuse such as discarded tires, tin cans, and plastic containers destroyed. Domestic pot-plants, wet shower floors, toilet water tanks, and rainwater tanks pose special problems. Larvicidal chemicals can be placed in domestic water jars, and kerosene or diesel oil layered on the surface of non-potable water.

The biological control of mosquito larvae is an alternative, albeit less developed, control strategy. Approaches include the introduction of fish species that ingest mosquito larvae and the introduction of sterilized male mosquitoes in order to reduce insect breeding. The use of *Bacillus thuringiensis*, which releases crystals toxic to mosquito larvae, has also been attempted. The introduction of the bacterium *Wolbachia* has also been found to reduce the capacity of *Aedes aegypti* mosquitoes to transmit dengue, chikungunya, and yellow fever viruses. However, the growth of cities in developing countries, largely by the growth of urban slums, has greatly accentuated the difficulties of control of domesticated *Aedes* mosquitoes. The "Asian tiger" mosquito *Aedes albopictus*, an aggressive daytime biting species and important vector of dengue and chikungunya viruses, has spread widely since the 1950s and now covers large areas of the world. Once established it is difficult to eradicate. For most arbovirus infections other than urban dengue and yellow fever, the vectors breed over too wide a geographical area to make vector control feasible save on a local scale, and then only in the face of a threatened epidemic.

The use of insecticides is a controversial issue, because there are frequent objections from environmentalists and mosquitoes eventually develop resistance. Any policy should be decided on the basis of a risk–benefit analysis on a situation-by-situation basis. Some countries have based their arbovirus control programs on aerial insecticide spraying, but most retain this approach only for emergency control in the event of an epidemic, specifically aimed at the rapid reduction of the adult female mosquito population. Organophosphorus insecticides such as malathion or fenitrothion are delivered as an ultra-low-volume (short-acting) aerosol generated by spray machines mounted on backpacks, trucks, or low-flying aircraft. Spraying of the luggage bays and passenger cabins of aircraft with insecticides may reduce the chances of intercontinental transfer of exotic arthropods, whether infected or not. Whether this is effective is questionable, however. Some experts believe the introduction of West Nile virus into North America in 1999 occurred as a result of infected insects being carried into the United States on a commercial flight from the Middle East.

Avoidance of exposure to the bite of arthropods is the other mainstay of control. Personal protection against mosquito bites can be achieved by the use of screens on doors and windows, nets over beds, long-sleeved protective clothing, especially at dusk, and repellents. In developed countries the major barrier has been whole-house air conditioning. Insect repellents containing the chemical N,N-diethyl-meta-toluamide (DEET) are particularly effective and do not present a health concern when used as directed. However, in many economically disadvantaged parts of the world, many of such measures are beyond reach and infections like dengue continue to thrive.

IMMUNIZATION

Each of the foregoing methods of control of viral diseases is focused on reducing the transmission of the causative virus. The second major method of control, immunization, is directed primarily at making the individual resistant to infection, and in many cases by reducing the incidence of new cases it can also substantially reduce the source of further spread. Immunization may be active, that is, induction of an immune response by administration of antigen (vaccine), or passive, that is, by administration of antibody (immune serum or immunoglobulin).

As outlined in Chapter 11: Vaccines and Vaccination, and discussed at length in the chapters of Part II, there are effective vaccines for many common viral diseases. These vaccines are especially effective in diseases with a viremic phase, such as poliomyelitis, yellow fever, and the acute exanthems. The dramatic success of vaccination programs in reducing the incidence in the United States of measles, mumps, and rubella is illustrated in Fig. 14.1. Similar declines have occurred in the incidence of poliomyelitis. In the United States, endemic circulation of polio ceased in 1979 and of measles around 2000.

Active Immunization Policy

Although the principal aim of vaccination is to protect the vaccinated individual, vaccination on a sufficiently wide enough scale will also enhance *herd immunity* to such an extent that transmission can be restricted, or even arrested altogether, within a given community or country. In different examples, vaccines may be used with different goals in mind, for example, to protect *individuals* at risk, to protect particular *groups* within a community or in specific parts of the world, to manage outbreaks, or to eradicate the causative virus.

The optimal design of a vaccination program varies according to the characteristics of the vaccine and the epidemiology of the virus infection. Relevant vaccine characteristics are the proportion of those vaccinated that achieve protection after vaccination, the duration of protection, and the coverage achieved by the vaccination program.

Many important aspects need to be considered in developing an effective strategy for a particular vaccine (see Box 14.3). For example, vaccination against hepatitis A is most needed for protection against *clinical* hepatitis A among older children and adults, at an age when clinical cases are more likely. In economically developing countries where hepatitis A infection is endemic and occurs in childhood, many see little clinical disease and therefore may choose to fund alternative health priorities. Many of the currently used vaccines, both bacterial and viral, are aimed at preventing diseases the risks of which are greatest in infancy. Live vaccines, for example, measles, are less reliable if given while maternal antibody is present but need to take effect before the typical age at which natural infection occurs; thus the recommended age of vaccination needs to fit a window that matches the dynamics of infection within each community.

Even in the case of a disease as feared as poliomyelitis, it has been difficult to maintain enthusiasm for a program of universal immunization after the disease has become rare, localized to only a few remote sites in Asia and Africa. Measles quickly reemerges when immunization programs wane. With many virus diseases it is essential to continue routine vaccination after the threat of epidemic has declined. The short-term absence of wild virus in the population leaves any unvaccinated people uniquely susceptible to further introductions of virus as the protective effect of herd immunity wanes. For these reasons it is essential that all countries continue to maintain highly organized and robust health services, paying particular attention to unimmunized cohorts, such as those living in urban ghettos, immigrants, and certain religious minorities.

A

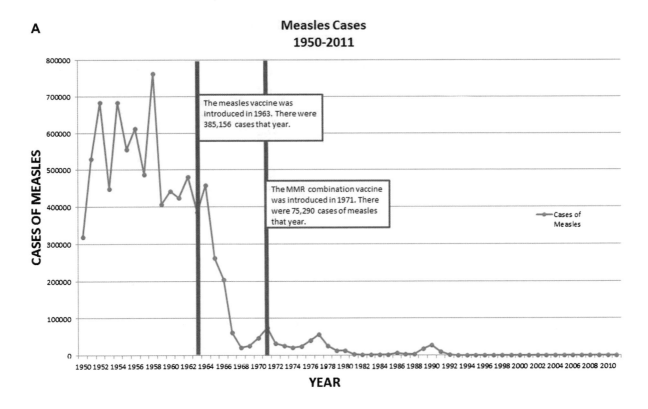

Measles Cases
1950-2011

The measles vaccine was introduced in 1963. There were 385,156 cases that year.

The MMR combination vaccine was introduced in 1971. There were 75,290 cases of measles that year.

B

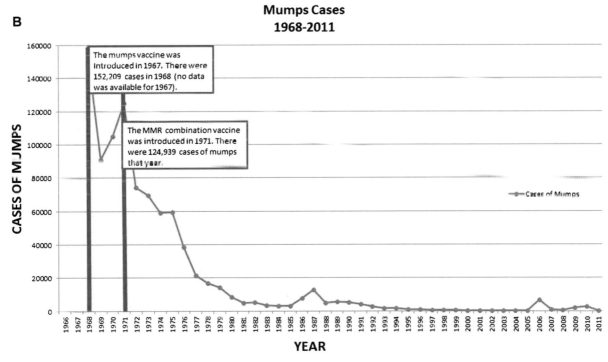

Mumps Cases
1968-2011

The mumps vaccine was Introduced in 1967. There were 152,209 cases in 1968 (no data was available for 1967).

The MMR combination vaccine was introduced in 1971. There were 124,939 cases of mumps that year.

FIGURE 14.1 (Continued)

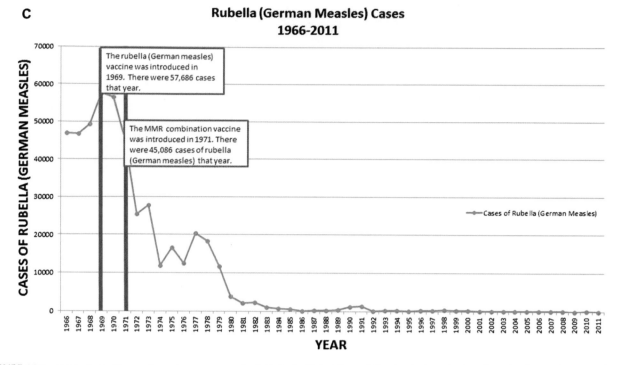

FIGURE 14.1 Fall in the incidence of measles, mumps, and rubella in the United States following the introduction of vaccination against these diseases. A similar dramatic decline occurred earlier with poliomyelitis – in recent years there has been no circulation of wild-type polioviruses anywhere in the Western Hemisphere – the same is the case with these other viruses, however sporadic reintroductions occur every year. Note: (1) a major resurgence of measles in 1989–91 (with a total of 55,622 cases and 123 deaths) was attributed to declining vaccine coverage in many cities, and (2) a resurgence of mumps in 1986–87 was attributed to waning immunity in adolescents at a time when only a single vaccine dose was recommended (subsequently two doses were recommended and from 1989 forward mumps cases steadily declined). *Data from U.S. Centers for Disease Control and Prevention and Pro.con.org.*

BOX 14.3 Issues to be Considered in Planning Vaccination Strategy for a Particular Virus

- Is the medical aim to reduce total infections, or to protect against clinical disease?
- Is the strategy aimed to protect high-risk individuals, to protect certain communities, to contain outbreaks, or to eradicate the virus?
- How should usage be targeted to achieve these goals, that is, by selective use, in response to outbreaks, mass vaccination programs?
- What is the true cost versus benefit of the program? How can this be presented to politicians and the public?
- What is the optimal dose, timing, and route of administration?
- What is the optimal age for vaccination?
- Can it be incorporated into existing programs?
- Are there validated laboratory markers indicating protection?
- Is pre- and/or post-vaccination testing useful and, if so, practical?
- How often is vaccination required—is there a need for boosters?
- What post-vaccination surveillance is necessary to monitor continuing efficacy and safety?
- How can acceptance by different communities be encouraged? What terminology is or is not appropriate, for example, "living" and "dead"? How can possible mistrust of authorities be managed?

Acceptability of a vaccine by a community is affected by many issues: balancing vaccine efficacy against safety, balancing fear of disease against fear of needles and side effects, balancing trust of public health authorities and the medical profession against trust in misinformed anti-vaccine activist groups, and balancing personal complacency and inertia against the usual deep-seated understanding as to what is correct for protecting the health of communities.

If the disease is lethal or debilitating, both the public and vaccine-licensing authorities will accept a risk of even moderately serious consequences of vaccination in a tiny minority of recipients. If, on the other hand, the disease is perceived as trivial, no side effects will be countenanced. Where more than one satisfactory vaccine is available, considerations such as cost and ease of administration tip the balance. Many other mechanisms are used to help improve compliance, including minimizing the number of clinic visits using polyvalent vaccines such as that available for measles–mumps–rubella (MMR vaccine), simplified vaccine schedules, in some instances promoting "national immunization days," and requiring proof of vaccination before allowing enrollment in school.

In designing and implementing vaccination campaigns other issues also need to be considered: these include religious sensitivities (e.g., use of the terms "live" or "dead" to characterize a vaccine may discourage its acceptance) and political sensitivities such as the country where the vaccine was manufactured. In recent years, vocal anti-vaccination lobbies have become prominent, and in some cases have seriously undermined vaccine uptake rates leading to localized disease outbreaks. Arguments raised by these groups are often ill-informed or based on misinterpretations of data, and not based on any quantitative risk assessment. For example, the introduction of a human papillomavirus vaccine was opposed by some groups on the grounds it would encourage promiscuity. Health professionals have a responsibility to provide accurate, balanced information.

Table 14.2 shows schedules for the use of the major human viral vaccines in common use. The vaccines used in particular diseases are discussed in the relevant chapters of Part II.

Passive Immunization

It is possible to confer short-term protection through the intramuscular or intravenous inoculation of preformed antibody, either using immune serum or immune (serum) globulin. Human immunoglobulin has replaced animal-derived (e.g., horse) preparations, because heterologous protein may provoke serum sickness or anaphylaxis. Pooled normal human immunoglobulin contains sufficiently high titers of antibody to protect against measles and hepatitis A. Specific high-titer immunoglobulin collected from individuals who have been recently immunized or have recovered from a particular infection is available to protect against hepatitis B, varicella-zoster, respiratory syncytial virus, human cytomegalovirus, and vaccinia. Specific high-titer immunoglobulin collected from paid donors who are vaccinated, revaccinated, and tested for high antibody titer is used for post-exposure rabies treatment. Humanized monoclonal antibodies against particular viruses, for example, respiratory syncytial virus (Palivizumab), can be used to prevent infection in high-risk infants (Chapter 26: Paramyxoviruses).

These preparations give immediate passive protection that then diminishes over the next two to four months, and can also confer "passive–active" immunity if a partially protected individual subsequently undergoes a subclinical infection with wild-type virus. Passive immunization should be regarded as an emergency procedure to achieve immediate protection of unimmunized individuals exposed to special risk. It can be used as a *pre*-exposure measure to give short-term protection against hepatitis A for travelers to developing countries. It is also important for *post*-exposure management of close contacts of a hepatitis A index case, for hepatitis B in new-born babies of infected mothers, or in unimmunized laboratory or health workers following a needle stick or comparable accident, for measles in unimmunized close contacts of a patient, and for varicella to protect new-born babies of mothers suffering from chickenpox at the time of delivery, or to protect non-immune contacts who are immunosuppressed.

TABLE 14.2 Schedules for Immunization against Human Viral Diseases[a]

Vaccine	Primary Course	Subsequent Doses
Live vaccines		
Poliomyelitis	2, 4, 6 months[b]	School entry (5 years)
Rotavirus	2, 4, (6) months	4–6 years
Measles	12, 18 months[c,d]	4–6 years
Rubella	12, 18 months[d]	4–6 years
Mumps	12, 18 months[d]	4–6 years
Varicella	18 months	10–15 years
Yellow fever	Before travel to endemic area	Booster after 10 years
Japanese encephalitis	>9 months, in endemic regions	1–2 years later
Inactivated vaccines		
Influenza	Autumn annually[e]	Annual booster
Rabies	Pre-exposure: 0, 7, 28 days[f]	Every 2 years[g]
	Post-exposure: 0, 3, 7, 14, 28, 90 days[h]	
Hepatitis A	For routine infant vaccination—2 doses at least 6 months apart, commencing after 12 months of age	Booster after 10 years
	If living in or traveling to endemic area—2 doses at least 6 months apart	
Hepatitis B	Birth[i], 2, 4, 6 months	Booster 10–15 years
	When at risk, then 1 and 6 months later	
HPV	12–13 years (girls and/or boys)—3 doses at 0, 1 and 6 months	
Japanese encephalitis	2 doses 28 days apart, in endemic regions or for travellers	
Tick-borne encephalitis	2–3 doses (see local schedules)	

[a]*Schedules vary from country to country. This table is to be taken only as a guide.*
[b]*Two or three doses spaced 2 months apart, commencing between 2 and 6 months of age, conveniently timed to coincide with diphtheria pertussis–tetanus (DPT) vaccine. The third dose can be delayed to 15 months, at the time of MMR vaccine.*
[c]*Given shortly after first birthday in most developed countries, but at 9 months in developing countries where measles death rate is high in the second 6 months of life.*
[d]*Usually as combined measles–mumps–rubella (MMR) vaccine.*
[e]*Vulnerable groups only, especially the aged and chronic cardiopulmonary invalids.*
[f]*Veterinarians, animal handlers, etc.*
[g]*If in rabies-endemic country or at high risk.*
[h]*Plus rabies immune globulin.*
[i]*Ideally within 24 hours.*

Specific antibody can also occasionally be used as therapy for an established viral disease, for example, among immunocompromised individuals with disseminated herpes zoster or vaccinia. Immune plasma together with ribavirin has been shown to reduce the mortality of Lassa fever in monkeys and has been used in exposed humans, but lack of availability in target areas is unresolved. Immune plasma is also effective in the treatment of Argentine hemorrhagic fever caused by Junin virus—its use has largely been replaced by vaccination of at-risk populations using the live attenuated vaccine Candid #1.

The WHO Expanded Program on Immunization and Other Initiatives

In developed countries both public health agencies and private medical practitioners carry out immunization. However, many developing countries lack adequate health services, political will, and funds to provide for the majority of the population. To capitalize on the health infrastructure that had been developed to support the Intensified Smallpox Eradication Program, the WHO in 1974 established the Expanded Programme on Immunization (EPI), with the specific goal of immunizing the world's children against six diseases: Bacillus Calmette-Guérin (BCG) for tuberculosis, diphtheria–tetanus–pertussis (DTP), oral polio, and measles. In 1985 the WHO program was greatly strengthened by the participation of the United Nations Children's Fund (UNICEF) as a provider of vaccines and augmented funding. Most countries have now added hepatitis B, *Haemophilus influenzae* type b (Hib) and meningococcal group A vaccines, and an increasing number are adding pneumococcal conjugate vaccine and rotavirus vaccines. By 2010 it was estimated that 85% of children under 1 year of age globally had received at least three doses of DPT vaccine. Between 2000 and 2008, measles deaths dropped worldwide by over 78%, and maternal and neonatal tetanus had been eliminated in 20 of 58 high-risk countries. Other vaccines, for example, yellow fever, hepatitis A, tick-borne encephalitis, and Japanese encephalitis, are offered in areas where these diseases are prevalent.

Other recent international initiatives include the Global Alliance for Vaccines and Immunization (GAVI), a consortium founded in 2000 that aims to deliver new vaccines to all children of the developing world. GAVI initially focused on vaccines against yellow fever and hepatitis B, but more recently has also included measles, rubella, HPV, rotavirus, and Japanese encephalitis. The Bill and Melinda Gates Foundation has also provided substantial support for vaccine research and delivery.

The adoption of vaccines into regional and national childhood immunization programs requires effective integration of vaccine delivery into primary healthcare systems, and long-term commitment to ensure sustainability coupled with effective monitoring for disease. Long-term planning is often hindered by a lack of data about the burden of disease in any particular region and effective communication to those groups most at risk.

ERADICATION

So far, global eradication has been achieved for only two viral diseases, the human disease smallpox and the animal disease rinderpest. The last naturally occurring case of smallpox was reported in Somalia in October 1977. Smallpox eradication was achieved by an intensive effort that involved a high level of international co-operation and used a potent, inexpensive, and easily administered stable vaccine. However, mass vaccination alone could not have achieved eradication of the disease from the densely populated tropical countries where it remained endemic in the 1970s, as it was impossible to achieve the necessary very high level of vaccine coverage. The effective approach was to combine vaccination with *surveillance and containment* in a strategy of ring vaccination; individual cases and pockets of infection were actively sought out (health workers were financially rewarded for finding new cases), isolated, and their contacts vaccinated, initially in the household and then at increasing distances from the index case.

The global smallpox eradication campaign was a highly cost-effective operation. The expenditure by the WHO between 1967 and 1979 was US$81 million, to which could be added about US$32 million in bilateral aid contributions and some US$200 million in expenditures by the endemic countries involved in the campaign. Against this expenditure of about US$313 million over the 11 years of the campaign could be set an *annual* global expenditure of about US$1000 million for vaccination, airport inspections, etc., made necessary by the existence of smallpox. This equation takes no account of the deaths, misery, and costs of smallpox in afflicted people, or of the medical complications of untoward reactions to vaccine.

In the years since smallpox eradication, most laboratory stocks of the virus have been destroyed, except for those held in two designated, tightly regulated laboratories, one in Russia, and the other in the United States. Research on smallpox has been discontinued except in these laboratories, and arguments for and against the total destruction of all virus stocks are regularly discussed. Nevertheless, the possibility of a related zoonotic poxvirus becoming a new human problem, or of smallpox virus being obtained for bioterrorism purposes, has led to the maintenance of large supplies of smallpox vaccine (vaccinia virus) under the auspices of the WHO and some national governments.

The achievement of smallpox eradication gave rise to discussions as to whether other diseases could also be eradicated, in particular measles and poliomyelitis. The biological characteristics of the three diseases that affect the ease of eradication are set out in Table 14.3. These diseases share several essential characteristics:

1. The absence of an animal reservoir;
2. No long-term persistent infectivity in the human host;
3. One or few stable serotypes; and
4. The availability of an effective vaccine.

However, other characteristics of smallpox are not typical of measles or poliomyelitis: smallpox cases present with such a typical clinical picture that rather accurate diagnosis was made in the field by physical examination (the most common confounder being chickenpox). The absence of virus transmission during the prodromal stage of infection made surveillance and isolation/containment feasible. This strategy was very important in the eradication of smallpox from tropical countries. The natural history of poliomyelitis and measles is quite different and eradication concepts and strategies must therefore also be different.

There are three antigenically distinct polioviruses requiring three vaccines, there is a preponderance of subclinical infections (from around 100 to 1000 to 1), there can be rather long-term shedding by infected people and virus may survive in sewerage systems for some time. Despite these concerns, in 1988 WHO, UNICEF, and other international organizations agreed on a plan for global eradication. Good progress was made in the early years, utilizing novel techniques for surveillance (flaccid paralysis notification) and for vaccination (national vaccination days, mopping up vaccination around recognized cases). As a result, paralytic poliomyelitis due to wild poliovirus was eliminated from the Americas in August 1991 (see Chapter 32: Picornaviruses).

Measles presents different problems. Because cases become infectious before the subject becomes ill, it is not possible to control measles by vaccination supplemented by surveillance and containment, but rather only by attaining very high levels of herd immunity, estimated at 96% of the population. Such a level is very difficult to achieve, because it is almost impossible to reach this proportion of the population with a reliably potent vaccine. In addition, maternal antibody inhibits replication of the standard Schwarz vaccine until some time between 9 and 12 months after birth; hence, vaccination is not recommended before the infant is 12 months old. However, in developing countries where measles is still a common disease, many infections occur in children aged 9 to 12 months. It has also become apparent that once the circulation of virus becomes rare and the basic immunity provided by vaccination is not being boosted by later subclinical infections, vaccination in infancy does not provide a certainty of protection throughout life. It is therefore recommended that there should be a booster inoculation of vaccine at school entry.

Three actions were recommended: (1) a one-time national campaign to bring children between 1 and 14

TABLE 14.3 Comparison of Features Influencing the Feasibility of Eradication of Measles and Poliomyelitis, Compared with Smallpox, in Which All Features were Favorable

	Smallpox	Measles	Poliomyelitis
Biological features			
Animal reservoir	No	No	No
Persistent infection occurs	No	Yes[a]	No
Number of serotypes	1	1	3
Antigenically stable	Yes	Yes	Yes
Infectivity in prodromal stage	No	Yes	Yes
Subclinical cases occur	No	No	Majority of infections
Early containment possible	Yes	No	No
Vaccine			
Effective	Yes	Yes[b]	Yes
Cold chain necessary	No	Yes	Yes
Number of doses	1	2	4
Sociopolitical features			
Country-wide elimination achieved	Yes[c]	No	Yes[d]
Financial incentive for assistance	Strong	Weak	Weak[e]
Records of vaccination required	No—scar	Yes	Yes

[a]As subacute sclerosing panencephalitis, but since no shedding occurs in this disease it is epidemiologically irrelevant.
[b]Vaccination is ineffective in the presence of maternal antibody.
[c]Before the Intensified Smallpox Eradication Programme commenced, in many countries.
[d]Before global eradication proposed, in several countries.
[e]But since 1985 considerable help provided by UNICEF, the World Bank, and others.

years of age up to date with measles vaccination, (2) the strengthening of routine vaccination to reach a minimum of 95 per cent of children every year, and (3) to undertake massive follow-up campaigns every 4 years, to reach a minimum of 95 per cent of children aged 1 to 4 with a 2nd dose of vaccine.

Following this strategy, the last indigenous measles outbreak was registered in Venezuela in 2002. However, some countries in the Americas still notified imported cases. Between 2003 and 2014, 5077 imported measles cases were registered in the region.

After declaring the elimination from the Western Hemisphere of rubella and congenital rubella syndrome in

2015, an International Expert Committee sought evidence for the interruption of a measles outbreak in Brazil, which had begun in 2013 and lasted for more than a year. After a year of targeted actions and enhanced surveillance, the last case of measles in Brazil was registered in July 2015.

With this achievement the elimination of measles from the Americas was declared.

BIOTERRORISM

Several viruses have been identified as having the potential as biological weapons. The former Soviet Union, the United States, and other countries have had bioweapon programs in past years. The Russian research and production program ran from 1926 to 1992 and was massive in scale. The production program of the United States ended in 1943 and its offensive research program ended in 1969. In 1975, 22 countries ratified the Biological and Toxins Weapons Convention that outlawed offensive biological weapons, and by 2013, 170 countries had ratified or acceded to the Convention. Research aimed at defensive capability (infectious agent detection, identification, diagnosis, and potentially dual-use research and development of vaccines and drugs) is carried out in many countries. Such research is mindful of the potential to make biological threat agents more dangerous by genetic engineering technologies.

Biological weapons share along with nuclear and chemical weapons the label of weapons of mass destruction. The Biological and Toxins Weapons Convention stipulated that all biological threat agent stockpiles were to be destroyed and new work on offensive biological and chemical weapons abandoned. Unfortunately, it became clear in the 1990s that a number of countries—including Russia—had continued such programs after having ratified the terms of the Convention. One ominous sign of this was the sudden announcement by Russian president Boris Yeltsin in 1992 that an anthrax outbreak, which occurred in 1979 in Sverdlovsk (Yekaterinburg), was the result of an accidental release of spores from a factory producing large amounts of anthrax spores for use in warfare. In the following years many international inspections and agreements have led to the shutdown of known facilities, but there are still countries where bioweapon research, development, and production facilities are operational and protected from view by strict security measures.

FURTHER READING

Barnighausen, T., Bloom, D.E., Cafiero-Fonseca, E.T., O'Brien, J.C., 2014. Valuing vaccination. Proc. Natl. Acad. Sci. U.S.A. 111 (34), 12313–12319.

Conway, M.J., Colpitts, T.M., Fikrig, E., 2014. Role of the vector in Arbovirus transmission. Annu. Rev. Virol. 1, 71–88. Available from: http://dx.doi.org/10.1146/annurev-virology-031413-085513.

Fenner, F., Henderson, D.A., Arita, L., Ježek, Z., Ladnyi, I.D. (1988) Smallpox and its Eradication. World Health Organisation, Geneva, Switzerland. ISBN 92-4-156110-6. This is the definitive account of the WHO Smallpox Eradication Programme, detailing both the background of the disease, early efforts at its control, and a region by region account of how the virus was eliminated.

Kaslow, R.A., 2014. Epidemiology and control: principles, practice and programs. In: Kaslow, R.A., Stanberry, L.R., LeDuc, J.W. (Eds.) Viral Infections of Humans (fifth ed.), Springer, Chapter 1.

Knipe, D.M., Howley, P.M. (Eds.), 2013. Fields Virology (sixth ed.), Lippincott, Williams and Wilkins, Philadelphia, PA.

Chapter 15

Emerging Virus Diseases

Emerging disease is a term used to describe the appearance of a previously unrecognized infection in a particular host species, or a previously known infection that has expanded into a new ecological niche or geographical zone, often accompanied by a significant change in pathogenicity. Prominent examples in the last two decades include the emergence of SARS in 2003, the sudden crossing of H5N1 avian influenza virus into humans in 1995, and the unexpected West African outbreak of Ebola virus in 2014. In the last few years Zika virus, thought for decades as causing only a mild febrile illness in East Africa, has emerged to cause significant morbidity in regions as far apart as Polynesia and South America. Difficult to predict, emerging threats to public health are the direct manifestation of a number of factors, singly or collectively, that cause significant changes in the constantly evolving relationship between pathogen and host. There is general agreement that factors include population growth, migration into large conurbations, climatic changes, ease of transportation and changes in human behavior.

Among 1400 pathogens of humans over 50% have originated from animal species, with over 300 being recognized as having emerged in the years 1940 to 2004, coinciding with the rapid changes in agricultural practice plus the rapid growth in urbanization over this period. Thus emerging or re-emerging pathogens are more likely to be zoonotic, that is, naturally transmitted from animals to humans. Furthermore, viruses are over-represented in this group, with RNA viruses accounting for a third of all emerging and re-emerging infections.

Recent interest in emerging virus disease has focused on understanding three key areas. First, how the interplay of climate, environment, and human societal pressures can trigger emergence. Second, how viruses can transmit between an established reservoir species and a new host species and what determines pathogenicity in each species. And third, which aspects of these processes offer opportunities for therapy and prevention. To these must be added a broader understanding of how viruses evolve over time; further clues to this are now being uncovered by studying host genetic elements responsible for resisting virus invasion.

INCREASED RECOGNITION AND CHANGES IN DISEASE PATTERNS

The perception that a new disease has recently emerged does not always reflect the situation that a new virus is actually circulating among humans. For example, the identification of a virus for the first time and the development of new diagnostic tests can create the impression of a new or increasingly prevalent disease even if the virus is already endemic, through allowing a better definition of the true distribution of infection within the community. In the years following the introduction of new tests for hepatitis B, hepatitis C, rotavirus, and human papillomaviruses, more accurate recognition of the true extent of infection contributed to a sense of new risks of disease. Of course, some true increase in prevalence had also occurred with some of these examples.

Furthermore, changes in human behavior can alter the epidemiology of a virus infection that is already endemic, leading to the appearance of new diseases. The slowing in circulation of polioviruses that followed improved sanitary conditions in the early 20th century led to infection being acquired at older ages with a greatly increased rate of paralytic disease. The introduction of more widespread parenteral medical procedures (e.g., parenteral therapy for syphilis, blood transfusion, renal dialysis) led to dramatic outbreaks of hepatitis B caused by a virus that was already endemic in the community. This transmission route has now been largely eliminated by rigorous screening of donated blood and infection control measures, while widespread dissemination of blood-borne infections among intravenous drug users has led to new patterns of disease. Changes in sexual practices and in numbers of sexual partners can affect an individual's exposure to sexually transmitted infections including HIV, and if this applies to groups of individuals or a population, changes in the disease prevalence will be seen.

The appearance of the new disease dengue hemorrhagic fever/dengue shock syndrome in Manila in the 1950s is thought to be due to the increase in sequential superinfection with a second related strain of dengue virus, the result of increased virus circulation that occurred following the migration of

Fenner and White's Medical Virology. DOI: http://dx.doi.org/10.1016/B978-0-12-375156-0.00015-1

people from rural areas into towns and cities. A regular driver of the reappearance of infectious diseases is the social and economic dislocation caused by war. For example, Syria had been free of poliomyelitis from 1999 until the outbreak of war in 2011; in 2013, the appearance of 35 cases was reported, leading to an intensified polio vaccination campaign.

Finally, for completeness, mention should be made of the spectrum of previously unknown diseases and syndromes, caused by common or endemic human viruses when they infect immunosuppressed individuals.

Theoretically, a new human virus disease might emerge purely by mutation of a currently endemic virus to a more lethal or transmissible phenotype. In practice, although viruses in circulation are constantly evolving by mutation and recombination and are under constant selection pressure, these changes do not seem to be a common basis for new diseases or increased virulence. The new genetic material responsible for most of the newly emerging diseases is in most cases derived from infections of non-human species.

FACTORS AFFECTING EMERGENCE

Some of the factors affecting the emergence and distribution of virus diseases are shown in Box 15.1. These include properties of the particular virus, factors affecting transmission, and aspects of host resistance, and many of these operate concurrently or cooperatively to affect the occurrence of disease.

Given that many emerging diseases are zoonotic, it is perhaps not surprising (and indeed fortunate) that many do not appear capable of sustaining human-to-human transmission, and thus R_0 approaches zero (see Chapter 13: Epidemiology of Viral Infections). This means that for an outbreak of such infections to occur, there needs to be repeated exposure to the animal reservoir. This may be the case with Lassa fever, for example, where there is good evidence showing that local communities in West Africa are continually exposed to the virus. Where human-to-human transmission occurs more readily, R_0 approaches 1 and therefore multiple outbreaks may occur. These pathogens generate the most concern, as comparatively small change to R_0 as a result of adjustments to the host–pathogen relationship may have led to a significant escalation as to the risk of outbreaks.

Once passage between humans becomes the sole route of transmission, an infection can no longer be regarded as zoonotic. At this juncture a balance will have been established between the evolving viral genome and the ability of the human immune response to limit the infective process. Thus at any one moment in time, many emerging diseases can be viewed as being in this process of adaptation prior to reaching the ultimate host–parasite balance between virus maintenance and survival of the host.

The emergence of viruses from animals to humans can be considered as progressing through five key stages, although the boundaries are often indistinct (Table 15.1). The rate by which viruses move through these stages inevitably

---

BOX 15.1 Factors Affecting the Emergence or Distribution of Virus Diseases[a]

A. Involving the virus
  Mutation and selection
  New genetic material—zoonosis or recombination/reassortment
B. Involving transmission
  Climate and weather
  Overcrowding
  Rapid air travel
  Changes in sexual activities or numbers of partners
  Intravenous drug use
  Introduction of new medical interventions
  War and famine
  Humans venturing into new environments
  Vector density and exposure
  Occupational exposure
C. Involving host resistance
  Immunosuppression
  Nutritional state
  Herd immunity

[a]More than one of these factors will often be interacting together simultaneously; for example, genetic changes in a virus frequently coincide with changes in transmission patterns or host species affected, and it may be of less importance which change was a true cause and which was a result.

TABLE 15.1 The Key Stages in the Emergence of a Zoonotic Infection

Stage	Description	Transmission to Humans	Examples
I	Agent only in animals	None	
II	Primary transmission	Only from animals	Rabies, West Nile virus, Hendra, Avian influenza H5N1
III	Limited human outbreak	From animals or a few cycles in humans	Ebola, Marburg, monkeypox
IV	Long human outbreak	From animals or many cycles in humans	Yellow fever, dengue, influenza A, HIV-1[a]
V	Exclusively human agent	Only between humans	Hepatitis C, measles, smallpox, mumps, rubella

[a]HIV-1 is now maintained as an exclusively human-to-human infection, although its initial emergence in humans is thought to have resulted from a limited number of recent animal–human transmissions.

slows as environmental barriers become progressively less favorable and host responses adapt to meet virus challenge. Simultaneously, new strains carrying mutations that favor survival in the new host are continually selected.

CLIMATE CHANGE

Our environment is changing on an unprecedented scale. Climate change needs to be distinguished from climate variation: change is where there is statistically significant variation from the mean state over a prolonged period of time.

The most notable manifestation has been the climatic conditions initiated by changes in sea surface temperatures in the Pacific Ocean, known as the El Niño Southern Oscillation (ENSO). For example, in the summer of 1990 an El Niño event occurred, which in turn led to a period of prolonged drought in many regions of the Americas: this led eventually to the emergence of hantavirus pulmonary syndrome caused by Sin Nombre virus (see Chapter 29: Bunyaviruses). Conversely a sudden reversal in sea temperature in the summer of 1995 resulted in heavy rainfalls, especially in Columbia, resulting in resurgence of mosquito-borne diseases such as dengue and Venezuelan equine encephalitis.

The geographical distribution of vector-borne diseases is particularly sensitive to climatic conditions, especially temperature. Even a small extension of a transmission season may have a disproportionate effect, as transmission rates rise exponentially rather than linearly with increasing temperature. Climatic change can also alter vector distributions if suitable areas for breeding become newly available. Again, the effect may be disproportionate, particularly if the vector transmits disease to human or animal populations that lack pre-existing levels of acquired immunity. The result is that clinical cases are more numerous and potentially more severe. Increased temperatures and seasonal fluctuations in either rainfall or temperature favor the spread of vector-borne diseases to higher elevations and to more temperate latitudes. *Aedes aegypti*, a major vector of dengue, is limited in distribution by the 10°C winter isotherm, but this isotherm is shifting, thereby threatening an expansion of disease ever further from the equator.

The relentless change inflicted by humans on the environment in the name of progress has also had a marked effect on rodent habitats. Outbreaks of Bolivian hemorrhagic fever (BHF) in Bolivia and hantavirus pulmonary syndrome (HPS) in the United States have been linked to abnormal periods of drought or rainfall, leading to unusually rapid increases in rodent numbers. Of all species of mammals, rodents are among the most adaptable to comparatively sudden changes in climate and environmental conditions. Small climatic changes can bring about considerable fluctuations in population size, particularly with those rodents inhabiting desert and semi-desert areas. Such variations are directly related to oscillations in food quantity and quality. A prolonged drought in the early 1990s in the Four Corners region of the United States led to a sharp decline in the numbers of rodent predators, such as coyotes, snakes, and birds of prey. But at the end of the drought heavy rainfall resulted in an explosion in piñon nuts and grasshopper populations, which in turn resulted in a rapid escalation of rodent numbers, among them deer mice carrying hantaviruses; the result was the emergence of hantavirus pulmonary syndrome.

Arthropod-borne infections such as Congo-Crimean hemorrhagic fever (CCHF) and Rift Valley fever (RFV) virus (see Chapter 29: Bunyaviruses) could pose a substantial risk to both humans and livestock in Europe should climatic conditions raise further the ambient spring temperature. Immature ticks infected with CCHF virus carried on migratory birds would molt in much greater numbers, although such an enhancement in molting might be offset by a significant reduction in the number of migratory birds. Many experts consider that there are already competent mosquito species in the northern hemisphere to enable the spread of RFV virus should this be introduced into Europe.

Deforestation has accelerated enormously since the beginning of the 20th century, and in the Amazonian basin and parts of Southeast Asia this has had a profound effect on local ecosystems, particularly by constraining the range of natural predators that normally keep rodents, insects, and other potential carriers of infectious disease under control. The reduction in biological diversity can trigger the invasion and spread of opportunistic species, facilitating the emergence of disease through increased contact with local human populations.

EASE OF TRAVEL

Air travel represents a major risk factor for the global spread of a new infectious agent. It is estimated that over 100 million passenger journeys are made by air every year. This is in marked contrast to just 50 years ago when many people rarely if ever traveled any distance from their place of residence. Frequent air travel is now regarded as a major contributing factor to the spread of emerging diseases as evidenced by the rapid spread of the SARS virus in 2003, when the infection was disseminated from China to at least 17 countries in less than a week. Mathematical modeling can be used to predict those regions most at risk in the event of any future SARS epidemic. Were a vaccine available, the initial spread of virus might be contained if only a third of the population were immunized in the regions where the outbreak is focused. This assumes an index case made a single air journey. However, the risk increases substantially in the event of an index case making two journeys, with the whole population requiring vaccination if the same passenger made three trips. Analysis of air traffic from

Mexico at the start of the 2009 influenza H1N1 pandemic suggests the risk of spread is particularly great when the volume of air traffic is high and the resources to report and trace diseased individuals are very restricted.

Ground transport offers another major route for transmission. Approximately 17% of all travel in Europe, for example, is by public ground transport, while air travel represents less than 0.2% of all passenger kilometers traveled. In contrast to airliners, public trains, buses, etc. are rarely fitted with high-efficiency particulate air (HEPA) filters.

It is not only humans who travel: the International Air Transport Association (IATA) estimates that around 80,000 wild-caught animals are air freighted each year, many being placed in holding facilities close to populated areas whilst in transit. Even mosquitoes may be carried: one theory is that West Nile virus (WNV) entered the United States as a result of an infected mosquito surviving the air journey from the Middle East to New York City in 1999 (see Chapter 36: Flaviviruses).

The incursion of WNV into North America is an excellent example of a virus expanding into an ecological niche where transmission-competent vectors were already present. Once established in and around the New York area, the availability of vertebrate hosts, most notably birds of the crow family, together with optimal climatic conditions for vector populations, enabled the rapid spread of WNV across the United States. Epizootic outbreaks of WNV have occurred frequently, with an escalating number of human neurological cases among the immunocompromised and the elderly (see Chapter 36: Flaviviruses) (Fig. 15.1).

ANIMALS AS A SOURCE OF HUMAN DISEASES

The advent of agriculture around 10,000 years ago was pivotal in giving rise to many of the human infections we know today. Agricultural-based societies led to humans living in close proximity both to each other and to livestock. In turn, human settlements provided fertile ground for inter-species transmission between farm animals, rodents, dogs, cats, and insects. Once established in humans, the diseases could be maintained indefinitely if the numbers of susceptible individuals remained above a certain threshold and in frequent contact with diseased persons. It is widely thought that measles emerged at this time, probably from rinderpest in cattle, and diverged into an exclusively human pathogen as human centers of population grew to a level where an animal reservoir was no longer necessary. Similarly, smallpox may have evolved about 4000 years ago from camel pox, its closest phylogenetic relative.

Wild Animal Populations

The emergence of HIV/AIDS provides an interesting example. The different major groupings of HIV-1 strains (M, N, and O) probably arose through separate chains of events involving (1) recombination between different simian immunodeficiency virus (SIV) strains in different monkey species and (2) transmission of these strains to other primates and/or chimpanzees and thence to humans. Many details remain conjectural, including the likely frequency of such events, and the different factors (viral, ecological, human behavioral, etc.) that led to a global, human-to-human pandemic becoming established in the 1980s and not before (Fig. 15.2).

Rodents

Rodents constitute an important part of the Earth's biomass. Among all species of mammals, members of the family *Muridae* have been the most successful and are found in almost all habitats. This family contains species that are the natural hosts of almost all arenaviruses and hantaviruses. As noted above, rodents are highly susceptible to climate

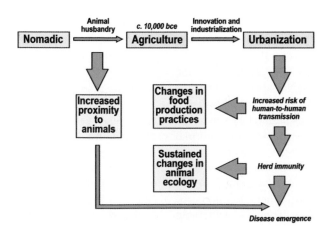

FIGURE 15.1 Historical changes in agriculture and human urbanization. Some of the impacts of these changes on infectious disease transmission (human-to-human, and animal-to-human) are shown. *Adapted from Howard, C.R., Fletcher, N.F., 2012. Emerg. Microbes Infections 1, e46; doi 10.1038/emi 2012.47.*

and ecological change, resulting in variable population numbers. Among the fastest reproducing mammals, field voles can have over 15 broods per year, each with an average of 6 pups. This in turn considerably increases the risk of human exposure to any pathogens they may carry, as well as stimulating such pathogens to undergo mutational adaptations to a changing ecosystem, with rodents thriving on potentially contaminated food and water.

The Four Corners outbreak of HPS in the United States described above instigated intensive research as to how fluctuations of rodent populations precipitate outbreaks of human disease. If the environment can suddenly sustain a rapidly expanding number of animals, population sizes explode, and the chances of rodents encroaching into peri-domestic areas and households also increase, especially when the overabundance of food comes to an end. As a consequence there is a rise in the incidence of human illness, as individuals have a much greater chance of coming into contact with excreta from persistently infected animals. The chance of a virus switching into other rodent species also becomes a greater possibility as rodent territories expand and overlap.

Switching to a new rodent host can have a profound effect on virus evolution. Adaptation of hantaviruses to new hosts can stimulate the development of new virus phenotypes and hence expansion into new ecological niches. Examples of this in Europe include the divergence of Saaremaa virus from Dobrava virus, as a consequence of Dobrava virus switching from the yellow-striped field mouse (*Apodemus flavicollis*) to *A. agrarius*, the striped field mouse (see Chapter 29: Bunyaviruses). The result is a virus with presumed reduced pathogenicity for humans.

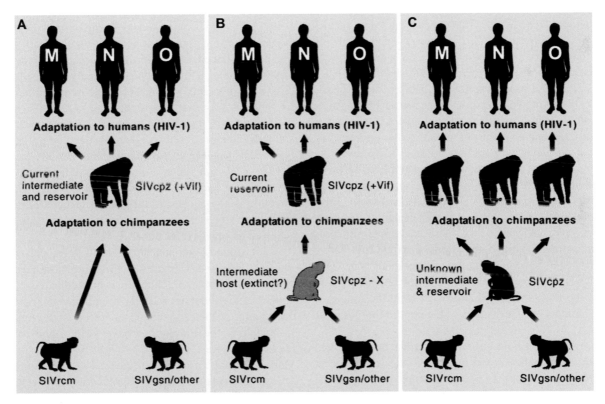

FIGURE 15.2 Simian immunodeficiency virus (SIV) strains and cross-species transmission to great apes and humans. Old World monkeys are naturally infected with over 40 different strains of SIV. These strains are species-specific and hence are denoted with a suffix to indicate their species of origin. The three major types of HIV (N, M, and O) each derived from a separate transfer event. Cartoons showing three possible alternative routes of cross-species transmissions giving rise to chimpanzee SIV (SIVcpz) as a recombinant of different monkey-derived SIVs illustrate the possible complexity of the steps leading to the introduction of viruses into a new host. + Vif indicates the presence of an HIV-like Vif, which is required to overcome the virus inhibitory effects of APOBEC3B. (A) *Pan troglodytes troglodytes* as the intermediate host. Recombination of two or more monkey-derived SIVs (likely SIVs from red-capped mangabeys [SIVrcm] and the greater spot-nosed monkeys [SIVgsn] or related SIVs) and possibly a third lineage requiring co-infection of an individual monkey with one or more SIVs. Chimpanzees have not been found to be infected by these viruses. (B) The SIVcpz recombinant develops and is maintained in a primate host that has yet to be identified, giving rise to the ancestor of the SIVcpz/HIV-1 lineage. *P. t. troglodytes* functions as a reservoir and was responsible for each of the human introductions. (C) Transfer through an intermediate host (yet to be identified) that is the current reservoir of introductions of SIVcpz into current communities of *P. t. troglodytes* and *P. t. schweinfurthii* as a potential source of diverse SIVcpz variants that are each found in limited geographic regions of Africa. *Reproduced from Heeney, J.L., Dalgleish, A.G., Weiss, R.A., 2006. Science 313, 462–466, with permission of AAAS.*

Bats

During an investigation of the 1998 Hendra virus outbreak in Queensland, Australia, it was noticed that grazing horses often sought shelter under trees containing bat roosts. Wild fruit bats in such roosts were found positive for virus and neutralizing antibodies found in otherwise healthy bats. Similarly, the related Nipah virus found in Malaysia and Bangladesh has also been associated with *Pteropus* bats: youngsters had been exposed to the secretions of fruit bats when picking fruit or processing date palm oil from bat-infested trees.

Bats have long since been known as the principal hosts of lyssaviruses, and distinct phylogenetic differences exist between rabies virus strains circulating in bats and terrestrial mammals, such as foxes, raccoons, and dogs. The link between genetic variability and spatial epidemiology among the lyssaviruses gives a particularly good insight as to how viruses of wildlife can adapt and emerge into different animal populations. Rabies virus in Europe has switched host many times over the past century, adapting rapidly to new animal hosts as the virus expands into new species with time. Rabid bats exhibit abnormal behavior, losing their natural fear of people and thus present a greater risk of transmission to humans (see Chapter 27: Rhabdoviruses).

Bats are found in most terrestrial habitats, with species distribution varying widely, some being restricted to a single island, others being found across continents. The fruit bats have evolved along a very different path to the insect-eating species. Fruit-eating bats are not normally cave dwelling, preferring to form roosts in treetops or crevices in decaying trees, and thus present opportunities for spread to humans. Many bats travel distances of several kilometers for food, especially fruit-eating species which respond to ever-varying supplies of food and which must compete with birds and other animals. While both insectivorous and fruit-eating bats have been shown to harbor zoonotic viruses, fruit-eating bats represent the biggest risk for human contact: most of the flesh of fruit is discarded from the mouth of feeding animals, thus providing ample opportunity for virus spread.

Asymptomatic Ebola virus infection has been reported in insectivorous bats trapped in Central Africa and recent exposure to fruit bats has been a feature of at least one outbreak. *Rousettus aegyptiacus* is one species of bat in which antibodies to both Marburg and Ebola viruses have been found. *Rousettus* species are the exception: despite being fruit eaters, these bats form roosts deep within caves. Marburg virus sequences have also been found in wild-caught *Rhinolophus eloquens* and *Miniopterus inflatus*. Intriguingly, filovirus genetic elements have been found in some mammalian species as diverse as shrews and South American marsupials, leading to the suggestion that filoviruses have co-evolved with mammals over many millennia. This latter observation suggests that other filoviruses are yet to be discovered in the New World, or that South American species harbored ancestral filoviruses that gave rise to present-day Ebola and Marburg viruses.

Domestic Livestock and Poultry

Pigs have been implicated in several outbreaks of emerging infections. Starting in September 1998, clusters of human cases of encephalitis were reported from the Malaysian states of Perak and Negri Sembilan. Almost all of the cases had a direct link to the local piggeries, and coincided with accounts of illness among pigs 1 to 2 weeks beforehand. Initially these outbreaks were believed due to Japanese encephalitis virus but a number of cases had been vaccinated previously against Japanese encephalitis virus and there was no evidence of virus antibodies among the remainder. The link with Hendra virus soon followed after the isolation of virus from an infected pig farmer. The new agent, now named Nipah virus after the locality it was first reported, shares 80% sequence homology with Hendra virus, with both viruses now classified as henipaviruses within the *Paramyxoviridae* family. It is clear that Nipah virus is widely distributed across Northeast India, Bangladesh, and Southeast Asia, with phylogenetic analyses revealing the virus to be diverging within specific geographical localities.

Since it was first shown that pigs could be infected with hepatitis E virus (HEV) there has been interest in the zoonotic potential of this agent. In developing countries types HEV-1 and -2 are restricted to humans and are spread by the fecal–oral route, usually via contaminated water. On the other hand, in developed countries HEV-3 and -4 infection is a zoonosis, with pigs and some poultry as the main reservoir, leading to sporadic human cases often acquired by ingestion of poorly cooked meat. Pigs become infected around 3 months of age but suffer only a mild, transient infection. Hepatitis E virus infection may be the commonest form of viral hepatitis worldwide; it is probably underdiagnosed but has recently been increasingly reported due to improvements in diagnostic testing and screening.

Swine in the Philippines have been found to act as reservoirs for Reston virus, a filovirus related to Ebola and Marburg viruses. This was discovered during an unusually severe outbreak of porcine reproductive and respiratory syndrome (PRRS). Previously, Reston virus had been first identified in 1990 among non-human primates that had been imported from the Philippines to several primate-handling facilities in the United States and Europe. In contrast to its African relatives, however, Reston virus does not appear to cause human illness, although there is ample evidence of Reston viral antibodies in humans working in primate-holding facilities or working with swine. Thus, Reston virus appears to be transmitted by the respiratory route but to possess low virulence, while other Ebola species are highly virulent but do not undergo airborne transmission. Fortunately, no viruses have yet emerged bearing the undesired characteristics of both.

Pigs are susceptible to human, avian, and swine influenza viruses, and thus play an important role in the epidemiology of human orthomyxoviruses. Influenza A virus is one of the comparatively few viral respiratory pathogens of pigs. Currently three subtypes circulate in swine: H1N1, H1N2, and H3N2. Domesticated pigs have often been regarded as a mixing vessel for influenza viruses by allowing the reassortment of the seven viral gene segments presenting an opportunity for new human strains to arise. Until 2009, however, swine influenza was not regarded as a significant cause of serious disease in humans.

However, cases of human infection began to emerge toward the end of April 2009 in what is normally regarded as the influenza season in the northern hemisphere. Beginning first in Mexico, the new virus subtype often referred to as "swine flu" by the popular press, spread rapidly throughout the world in a matter of weeks (see Chapter 25: Orthomyxoviruses). Analyses of human isolates quickly showed the unusual nature of this swine-origin influenza virus (S-OIV) as being a triple reassortment virus containing genes from avian, human and "classical" swine influenza viruses. The ancestors of this virus had probably been circulating in pig populations for over 10 years but had remained undetected. At the time, there was considerable uncertainty as to the pathogenic potential of this virus but data soon showed the severity for humans to be less than that seen with the 1918 pandemic but on a par with the 1957 "Hong Kong" pandemic. Transmissibility appeared higher than is normally the case for seasonal influenza with a higher than normal attack rate. Importantly, younger age groups appeared more susceptible, possibly due to partial immunity among older cohorts as a result of being infected during previous pandemics.

Of more recent concern is the spread of H7N9 influenza: this subtype was previously known to circulate in poultry, but from March 2013 the first human cases were reported, most of these in southern China. By March 2015 there had been over 220 deaths among 640 infected persons. Nearly all cases had been exposed to live poultry or visited live poultry markets, but several very limited family clusters were also reported, suggesting that very limited human-to-human transmission may also be possible. H7N9 influenza virus may become a serious public health threat should it become established among wild birds.

Companion and Captive Animals

Frequent contact with companion animals such as dogs, cats, and horses provides additional opportunities for the transmission of animal diseases to humans. Although companion animals have been kept within households over the centuries, the number of known emerging infections from such sources is remarkably few. There are a number of canine homologs of human flaviviruses. The discovery of a canine flavivirus distantly related to human hepatitis C virus (HCV) raises some intriguing questions as to the origin of hepatitis C virus in human populations. Although evidence was found for virus in the canine liver, there is as yet no evidence of this canine hepatitis C-like virus causing liver disease in dogs. Whether or not hepatitis C virus first emerged from dogs remains speculative, but the finding of virus in the respiratory secretions of infected dogs indicates a potential route of transmission to humans. A related flavivirus has been described among seropositive horses in the State of New York. There was no supporting evidence of clinical disease among all the animals tested but this does not preclude a pathogenic potential for humans.

There is an increasing trend, particularly in more affluent economic countries, to keep more exotic wild animals as pets. It is estimated that approximately 350,000 wild-caught animals are traded around the world each year, adding to the risk of potentially zoonotic infections crossing the species barrier into humans. The finding of a new arenavirus in boa constrictors (*Boa constrictor*) suffering from snake inclusion body disease has raised questions as to how common such viruses might be among captive wild animals. Intriguingly, sequence data from this arenavirus showed diversity compatible with a pre-existing relationship between host and virus over time. Moreover, sequences were found homologous to those of arenaviruses causing severe hemorrhagic fever, e.g., Lassa virus (see Chapter 30: Arenaviruses), but surprisingly the snake arenavirus also shared glycoprotein sequences with filoviruses. This would suggest that there has been recombination along one of the two RNA genome segments at some point in time with an ancestral filovirus that subsequently evolved to become the Ebola and Marburg viruses of today.

The keeping of small rodents and mammals has been linked to zoonotic disease for many decades, a prime example being lymphocytic choriomeningitis virus (LCMV) transmitted as a result of handling persistently infected hamsters. The keeping of prairie dogs is common in the United States, and indirectly led in 2003 to an outbreak of monkey pox in the State of Wisconsin (see Chapter 16: Poxviruses). This totally unexpected occurrence was the result of housing prairie dogs intended for sale in close proximity to small rodents imported from the African continent, most notably rope squirrels (*Funisciurus* spp.) and Gambian giant rats (*Cricetomys* spp.). Although there were no fatalities among the reported cases, it presented an opportunity for the spread of monkey pox into the feral mammal population of North America. It remains to be seen if wild animals become a source of monkey pox outbreaks in years to come.

A worrying complication is the emergence of mild human infections due to vaccinia virus. This virus, successfully used in the past for the eradication of smallpox, has been transmitted from herds of dairy cattle in Brazil and in buffaloes in India. These instances of "feral" vaccinia may have originated from human vaccines being inadvertently introduced into livestock from whence the virus has been re-introduced among agricultural workers to cause a disease resembling cowpox. There is also evidence for vaccinia virus infection among black howler (*Allouata caraya*) and capuchin monkeys (*Cebes apella*) inhabiting the Amazonian rainforest.

PREVENTION AND CONTROL

Outbreaks of emerging diseases vary widely in duration, frequency, and case numbers. Some can be predicted to occur regularly, for example new strains of influenza, whereas many decades may elapse between episodes, as is the case with Marburg virus. Yet others, as exemplified by Zika virus, have been known for decades but until recently thought to not represent a threat to public health. Thus planning a single, integrated strategy against all eventualities is therefore almost impossible. The task may be compounded by the emergence of escape mutants in populations vaccinated against known diseases, the emergence of strains resistant to antiviral therapy, or even the recycling through livestock of attenuated vaccines designed for use exclusively in humans. It is instructive to compare three major emerging disease outbreaks in recent years—HIV/AIDS, pandemic H1N1 swine influenza 2009, and Ebola in West Africa. For each of these, the transmission routes, transmissibility and R_0 together with the case-fatality rates were very different: the rates of spread also varied, and achieving control presented very different problems requiring specific control measures.

Improved epidemiological surveillance of infectious diseases is the foundation for combating emerging diseases. Integration with the veterinary community is essential: several "One Health" programs have been set up to help developing nations strengthen their overall capacity to react quickly and effectively. Valuable time was lost in 1999 when the first cases of West Nile virus occurred in New York City among both humans and birds.

Technology can play a major role in predicting disease emergence, for example, the use of satellite imagery to detect changing patterns of vegetation in response to rainfall. The use of satellite maps taken over East Africa accurately predicted the outbreak of Rift Valley fever among livestock as a consequence of increased vector activity. The use of the Internet is essential in allowing rapid dissemination of serological, clinical, and molecular sequencing data. Such rapid communications played a vital role in combating the SARS outbreak in 2003 and also in identifying the spread of swine-origin H1N1 influenza virus in 2010.

Crucial aspects for controlling any new outbreak include early recognition of the outbreak, good access to diagnostic

TABLE 15.2 The Major Requirements for Recognition and Response to Emerging Infections

- Alerting clinicians to quickly recognize and react to any unusual clinical presentation
- Access to a high standard of diagnostic capabilities plus availability of reference laboratory expertise
- Formulating practical case definitions for field use
- Instituting isolation and infection control for cases
- Involvement of epidemiologists and communicable disease specialists at the earliest opportunity to assess the crucial characteristics of the outbreak
- Quarantining contacts, travel restrictions where appropriate
- Mobilization of local health system to handle the increased number of sick patients
- Enhanced vaccination programs, access to antivirals where available
- Enhanced education for healthcare staff, families, and general communities

facilities, and dissemination and analysis of surveillance data (Table 15.2). As early as possible, an assessment needs to be made of the likely origin of the new agent (in terms of both its virological ancestors, and its likely source or reservoir); its transmission routes; the case-fatality rate; its transmissibility and R_0 (see Chapter 13: Epidemiology of Viral Infections). Based on this information, appropriate control measures need to be quickly activated. These often include instituting appropriate isolation and infection control for cases; quarantining of contacts; travel restrictions; enhanced vaccination programs; provision of antivirals; and mobilization of the local health system to manage the predicted increase in hospitalized patients and community anxiety. Time is of the essence, as delays lead inevitably to an escalation in numbers of cases that can overwhelm locally available manpower and capacity. The immediate closure of hospitals was pivotal in limiting the nosocomial spread of Ebola virus in the original outbreaks in Sudan and Zaire in 1996. The importance of early recognition and the availability of local expertise were highlighted in Uganda, where the discovery of the Bundibugyo strain of Ebola virus in 2000 led to the strengthening of local capacity. Cases of Ebola virus in July and August 2012 have been rapidly diagnosed as a result of this regional investment in infrastructure, thus preventing its spread to Kampala, the Ugandan capital. However, outbreaks may spread even in countries fully equipped to deal with infectious disease outbreaks, unless the clinical and epidemiological data can be reviewed quickly and critically and the appropriate control measures instigated.

Emergence of new infectious diseases has undoubtedly been ongoing for millennia. Although the rate at which new infections are being discovered has accelerated in the past half-century, it is some comfort that newly identified emerging viruses fall invariably within well-characterized virus families. However this may change as we discover vast numbers of hitherto uncharacterized viruses in what is now commonly referred to as the virosphere.

Viruses can evolve faster than mammals by many orders of magnitude, being near instantaneous compared to the scale of mammalian adaptation over years and decades. This enormous capacity for adaptation is offset by the generally very demanding survival requirements of viruses in their precarious existence between hosts. Emergence of vector-borne diseases can represent a major threat in the short term once conditions for adaptation result in emergence and an extension of host range as a consequence, as was the case for chikungunya virus in 2005 (see Chapter 35: Togaviruses).

How can we ensure we are prepared to combat emerging diseases? A capacity to respond quickly and effectively with appropriate measures is essential (Table 15.2). Many governments now recognize and take into account the likely impact of environmental developments on ecosystems and disease emergence, extending environmental impact studies beyond conservation of natural habitats and species. Responding to the threat of disease emergence, the International Health Regulations sponsored through WHO were substantially revised in 2005, requiring countries to develop national preparedness capabilities, conduct meaningful surveillance, and to promptly report internationally significant events.

Considerable international effort is now being made to collect and record samples of viruses and microorganisms from wild animal populations and potential arthropod vectors in order to expand the current molecular databases. A leading example is the establishment of the WHO-led Global Influenza Surveillance and Response System (GISRS), an international program serving as an alert mechanism for the emergence of influenza viruses with pandemic potential through its continual monitoring of isolates and risk assessments. By making available extensive catalogs of genome sequences across international and political boundaries it is hoped that newly emerging agents would in future be more readily identified and thus control measures put in place much more rapidly.

FURTHER READING

Institute of Medicine, 2014. Emerging Viral Diseases: The One Health Connection. National Academy of Sciences, Washington, DC.

Institute of Medicine (US), 2003. Committee on Emerging Microbial Threats to Health in the 21st Century. In: Smolinski, M.S., Hamburg, M.A., Lederberg, J. (Eds.) Microbial Threats to Health: Emergence, Detection, and Response, National Academies Press, Washington, DC, (US).

Kilpatrick, A.M., Randolph, S.E., 2012. Drivers, dynamics, and control of merging vector-borne zoonotic diseases. Lancet 380, 1946–1955.

Morens, D.M., Fauci, A.S., 2013. Emerging infectious diseases: threats to human health and global stability. PLoS Pathogens 9 (7), e1003467.

Parrish, C.R., Holmes, E.C., Morens, D.M., et al., 2008. Cross-species virus transmission and the emergence of new epidemic diseases. Microbiol. Mol. Biol. Rev. 72, 457–470.

Wolfe, N.D., Dunavan, C.P., Diamond, J., 2007. Origins of major human infectious diseases. Nature 447, 279–283.

World Health Organisation (2009) International health regulations. http://www.who.int/about/FAQ2009.pdf.

World Health Organisation Influenza Pandemic Preparedness Framework. http://www.who.int/influenza/pip/en.

Part II

Specific Virus Diseases of Humans

Chapter 16

Poxviruses

The family *Poxviridae* includes several viruses that can infect humans: variola (smallpox) virus (now extinct, but remains a threat with regard to bioterrorism), vaccinia virus (including a strain called buffalopox virus), monkeypox virus, cowpox virus, molluscum contagiosum virus, tanapox virus, orf virus, and milker's nodule virus. Variola and molluscum contagiosum viruses are specifically human viruses; the others are zoonotic.

The history of the poxviruses has been dominated by smallpox. Use of the vaccine that traces its ancestry to Edward Jenner and the cowsheds of Gloucestershire in England led to the eradication of this disease, once a worldwide and greatly feared disease. Although Jenner's first vaccines probably came from cattle and contained cowpox virus, the origins of vaccinia virus, the smallpox vaccine virus, are unknown. Jenner, in his *Inquiry* published in 1798 (*An Inquiry into the Causes and Effects of the Variolae Vaccinae, or Cow-Pox*), described the clinical signs of cowpox in cattle and humans, and how human infection provided protection against smallpox. At the time in England smallpox was responsible for more than 10% of all deaths among children. Jenner's discovery quickly led to the establishment of vaccination programs all around the world. However, it was not until Pasteur's work nearly one hundred years later that the principle was used again. Other landmark discoveries came from early work on poxviruses: myxoma virus, an important cause of disease in domestic rabbits, described first by Giuseppe Sanarelli in 1896, was the first viral pathogen of a laboratory animal. Rabbit fibroma virus, described in 1932 by Richard Shope, was the first virus proven to cause tissue hyperplasia (which at the time was considered related to neoplasia). Fenner used ectromelia virus infection of mice as a model to elucidate the stepwise dissemination of virus during an acute systemic infection—this was the first systematic, quantitative study of viral pathogenesis.

With the global eradication of smallpox in 1977, one of the greatest achievements of humankind, vaccination was discontinued throughout the world except for military personnel in some countries. However vaccines using vaccinia virus as a vector for delivery of a wide range of viral, microbial, and eukaryotic antigens have been the subject of intensive research. Although no human vaccine employing this strategy has been licensed, a number of animal vaccines have been developed (e.g., a vaccine for rabies in wildlife, a vaccine for rinderpest in cattle). Less virulent strains of vaccinia virus than those used for smallpox vaccination have been investigated for use as vectors—these are less likely to produce serious complications than the strains previously used for smallpox vaccination. Canarypox virus and other poxviruses have also been developed as vaccine vectors (see Chapter 11: Vaccines and Vaccination).

PROPERTIES OF POXVIRUSES

Classification

The family *Poxviridae* is divided into two subfamilies: *Chordopoxvirinae* (poxviruses of vertebrates) and *Entomopoxvirinae* (poxviruses of insects). The subfamily *Chordopoxvirinae* is divided further into nine genera, four of which (*Orthipoxvirus, Parapoxvirus, Molluscipoxvirus,* and *Yatapoxvirus*) contain viruses that cause human infections. Smallpox and molluscum contagiosum are specifically human diseases; the other two are zoonoses. The genus *Leporipoxvirus* includes myxoma virus, which causes the fatal disease myxomatosis in European rabbits and was used successfully to control rabbit plagues. There are other poxviruses that have not yet been classified (Table 16.1).

Virion Properties

Most poxvirus virions are brick-shaped, about $250\times200\times200$ nm in size; in contrast, the virions of the members of the genus *Parapoxvirus* are cocoon-shaped with dimensions of 260×160 nm (Fig. 16.1). The isometric nucleocapsid conforming to either the icosahedral or helical symmetry found in most other viruses is absent among poxviruses (see Chapter 3: Virion Structure and Composition); hence poxviruses are said to have a "complex" structure.

Virions of most poxviruses are composed of an outer layer of tubular structures, arranged rather irregularly to give a characteristic appearance; in contrast, the virions of

Fenner and White's Medical Virology. DOI: http://dx.doi.org/10.1016/B978-0-12-375156-0.00016-3

TABLE 16.1 Human Poxvirus Infections

Genus/Virus	Disease	Features and Epidemiology
Orthopoxvirus/variola virus	Smallpox (variola)	Globally eradicated: Narrow host range
	Variola major	Generalized infection with pustular rash; case-fatality rate 10 to 25%
	Variola minor	Generalized infection with pustular rash; case-fatality rate less than 1%
Orthopoxvirus/vaccinia virus	Vaccinia infection as a complication of vaccination	Usually, a local pustule, slight malaise. Rarely, eczema vaccinatum or generalized vaccinia (low mortality); progressive vaccinia (high mortality in immunocompromised vaccinees); postvaccinial encephalitis (high mortality): frequently used as a vaccine vector
Orthopoxvirus/monkeypox virus	Human monkeypox	Generalized infection with pustular rash; case-fatality rate in humans 15%: numerous animal hosts, e.g. squirrels, anteaters: found in Central and West Africa
Orthopoxvirus/cowpox virus	Human cowpox infection	Localized ulcerating lesion on the skin, usually acquired from cats or cows: found in Asia and Europe
Parapoxvirus/milker's nodule virus	Milker's nodule	Trivial localized nodular infection on the hands acquired from cows
Parapoxvirus/orf virus	Orf	Localized papulo-vesicular lesion on the skin acquired from sheep: worldwide distribution
Molluscipoxvirus/molluscum contagiosum virus	Molluscum contagiosum	Multiple benign nodules in skin
Yatapoxviru/Yabapox virus	Yabapox	Localized skin tumors acquired from monkeys (rare): narrow host range: Found in West Africa
Yatapoxvirus/Tanapox virus	Tanapox	Localized skin lesions probably from arthropod bites; common in parts of West Africa

the members of the genus *Parapoxvirus* are covered with long thread-like surface tubules, which because of the superimposition of features on the apical and basal surfaces appear arranged in criss-cross fashion resembling a ball of yarn (Fig. 16.1B). The virions of some ungrouped poxviruses are brick-shaped but have a surface structure similar to that of the parapoxviruses. Interestingly, cryoelectron microscopy studies have shown that all these virion surface structures, which have been used to identify poxviruses ever since the development of negative-contrast electron microscopy, may actually represent dehydration artifacts. The outer layer encloses a dumbbell-shaped core and two lateral bodies. The core contains the viral DNA together with several proteins. Nascent virions that are released from cells by budding, rather than by cellular disruption, have an extra envelope that contains cellular lipids and several virus-specified proteins.

The poxvirus genome consists of a single molecule of linear double-stranded DNA varying in size from 130 kbp (parapoxviruses), to 280 kbp (fowlpox virus), and up to 375 kbp (entomopoxviruses). The genomes of vaccinia virus (191,636 bp) and many other poxviruses have been sequenced. Poxvirus genomes have covalent linkage joining the two DNA strands at both ends; the ends of each DNA strand have long inverted tandem repeated nucleotide sequences forming single-stranded loops (Fig. 16.2).

Poxvirus genomes have the capacity to code for up to 200 proteins, as many as 100 of which are contained within virions. The genome can be divided functionally into a conserved central region coding for those proteins essential for virus replication (e.g., nucleic acid synthesis, virion structural component synthesis). Examples include a DNA-dependent DNA polymerase, DNA ligase, DNA-dependent RNA polymerase, enzymes involved in capping and polyadenylation of messenger RNAs and thymidine kinase. The two flanking regions toward the termini code for a large number of proteins that determine host range, virulence, and immunomodulation. In the case of vaccinia virus, these regions account for nearly half the genome.

Poxviruses are transmitted between individuals by several routes: by aerosol and droplets (variola virus), by introduction of virus into small skin abrasions after direct or indirect contact with an infected animal (orf virus, milker's nodule virus), and in the case of some animal poxviruses mechanically by biting arthropods. The viruses generally have narrow host ranges. Poxviruses are resistant to degradation in the environment at ambient temperatures and may survive many years in dried scabs or

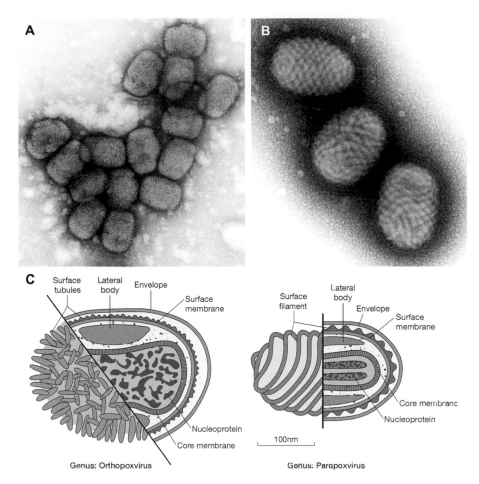

FIGURE 16.1 (A) Negatively stained vaccinia virus virions, showing surface tubules characteristic of member viruses of all genera except the genus *Parapoxvirus*. (B) Negatively stained orf virus virions, showing characteristic surface tubules of the member viruses of the genus *Parapoxvirus*. (C, left) Schematic diagram, genus *Orthopoxvirus* (and all other vertebrate poxvirus genera except the genus *Parapoxvirus*), showing biconcave core and lateral bodies. (C, right) Schematic diagram, genus *Parapoxvirus*. Part of the two diagrams shows the surface structure of an unenveloped virion, the other part shows a cross-section through the center of an enveloped virion.

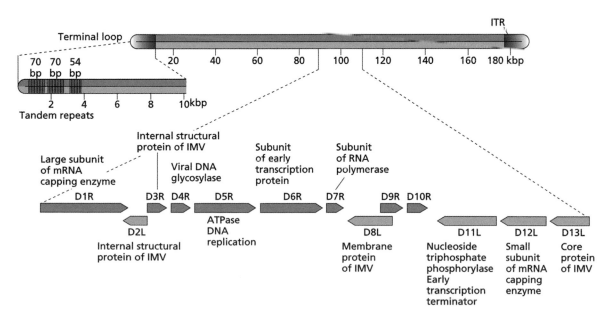

FIGURE 16.2 Structure and organization of the genome of vaccinia virus. ITR, inverted terminal repetitions; IMV, immature virion. *Reproduced from Flint, S.J. et al., 2009. Principles of Virology, third ed., ASM Press, Washington, DC, with permission.*

other virus-laden material, a property much exploited in the early days of vaccination when vaccines were transported by casual means over long distances.

All poxvirus infections are associated with lesions of the skin, which may be either localized or widespread. The lesions associated with many diseases are pustular, but lesions due to molluscum contagiosum virus, parapoxviruses, and yatapoxvirus are proliferative. Generalized poxvirus infections have a stage of leukocyte-associated viremia, which leads to localization of virus in the skin and to a varying extent in internal organs. Immunity to such infections is prolonged. However in some localized poxvirus infections, notably those produced by parapoxviruses, immunity is short-lived and reinfection is common.

Viral Replication

Most poxviruses, except for parapoxviruses and molluscum contagiosum virus, grow readily in cell culture. All produce pocks on the chorioallantoic membrane of embryonated hens' eggs, the appearance of which was used to differentiate orthopoxviruses from each other before the introduction of molecular methods. Unusually for a DNA virus, replication of poxviruses occurs entirely in the cytoplasm. To achieve this total independence from the host cell nucleus, poxviruses, unlike other DNA viruses, have evolved to code for all of the enzymes required for transcription and replication of the viral genome; several of these enzymes are carried in the virion itself. Virus entry into the target cell is by fusion of the virion with the plasma membrane or by endocytosis; after entry the viral core is released into the host cell cytoplasm (Fig. 16.3).

Transcription is characterized by a cascade in which transcription of each temporal class of genes ("early," "intermediate," and "late" genes) requires the presence of specific transcription factors that are made by the preceding temporal class of genes. Intermediate gene transcription factors are coded for by early genes, and late transcription factors by intermediate genes. Transcription is initiated by the viral transcriptase and other factors carried in the core of the virion—this enables the production of mRNAs and viral proteins within minutes after infection. These proteins complete the uncoating of the viral core and are involved in the transcription of about 100 early genes; all this occurs before viral DNA synthesis begins. Early proteins include a DNA polymerase, thymidine kinase, and several other enzymes required for genome replication.

Poxvirus DNA replication involves the synthesis of long concatemeric intermediates that are subsequently cut into unit-length genomes. With the onset of DNA replication there is a dramatic shift in gene expression. Transcription of "intermediate" and "late" genes is controlled by binding of specific viral proteins to promoter sequences in the viral genome. Some early gene transcription factors are made late in infection, packaged in virions and used early in the next round of infection.

Since poxviruses are composed of a very large number of proteins, it is not surprising that viral assembly is a complex process. Virion assembly and maturation occur by coalescence of DNA within crescent-shaped immature core structures and the progressive addition of outer coat layers. Replication and assembly occur in discrete sites within the cytoplasm (referred to as viroplasm or viral factories); a majority of mature virions remain in the cytoplasm and are released after cell lysis, but some migrate to the Golgi, become wrapped in a cell-derived envelope and are released from the plasma membrane by exocytosis ("extracellular enveloped virions"). Both forms of virus are infectious, but enveloped virions are rapidly taken up by cells and appear to be important in virus spread through the body.

During replication numerous viral gene products block innate immunity. These inhibit the production of interferon, secreted interferon, and proinflammatory cytokines such as TNF-α, IL-1, and IL-18, these being required to amplify the innate response and determine the nature of adaptive immunity (Table 16.2).

DISEASES CAUSED BY MEMBERS OF THE GENUS ORTHOPOXVIRUS

Vaccinia Virus Infection

Modern molecular analysis has established that vaccinia virus is different from the cowpox virus. Likewise, it is neither an attenuated form of variola (smallpox) virus nor a recombinant between cowpox and smallpox viruses. Today, vaccinia virus is considered a virus derived from an unknown original natural host.

For vaccination against smallpox, vaccinia virus was inoculated into the superficial layers of the skin of the upper arm by a "multiple puncture" technique. Severe complications occurred occasionally in children with eczema—this occurred when eczematous children were vaccinated mistakenly or were infected by contact. Eczema vaccinatum was rarely fatal, especially if treated with human vaccinia-immune globulin (VIG). Other very rare but more serious complications were progressive vaccinia, which occurred only in persons with defective cell-mediated immunity, and post-vaccinial encephalitis.

With the eradication of smallpox, routine vaccination of the general public ceased and the requirement that international travelers possess a valid vaccination certificate was abolished. Vaccination of military personnel has been continued in some countries.

The primary diagnosis of vaccinia virus infection, in common with other poxvirus infections in humans, is based firstly upon electron microscopy. The morphology of the virions is unique and the concentration of virus in vesicle

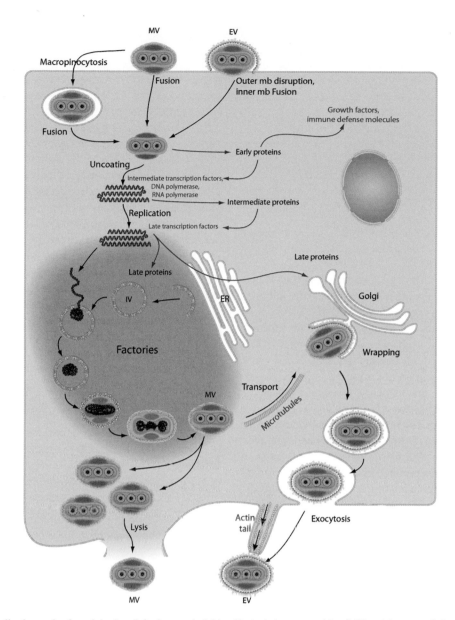

FIGURE 16.3 The replication cycle of vaccinia virus. Infection may be initiated by both the mature virion (MV) and the extracellular (enveloped) virion (EV).

1. Viral proteins attach to host cell membrane glycosaminoglycans (GAGs), and virions are endocytosed into the host cell. Alternatively, fusion with the plasma membranes can release cores into the host cytoplasm.

2. *Early phase*: Early genes are transcribed in the cytoplasm by viral RNA polymerase. Early expression begins at 30 minutes postinfection.

3. The core is completely uncoated as early expression ends; viral genome is now free in the cytoplasm.

4. *Intermediate phase*: Intermediate genes are expressed, triggering genomic DNA replication at approximately 100 minutes postinfection.

5. *Late phase*: Late genes are expressed from 140 minutes to 48 hours postinfection, producing all the structural proteins.

6. Assembly of progeny virions starts in cytoplasmic viral factories, producing a spherical immature particle (IV). This virus particle matures into the brick-shaped intracellular mature virion.

7. Intracellular MVs can be released after cell lysis, or can acquire a second double membrane from the *trans*-Golgi and bud as external enveloped virions (EV).

Reproduced from ViralZone, Swiss Institute of Bioinformatics, with permission.

fluid is so high that negative contrast electron microscopy is usually sufficient. Confirmatory diagnosis, used when deemed necessary by circumstances, is performed by PCR in national reference laboratories.

Human Buffalopox

This disease has been described in water buffalo (*Bubalis bubalis*) in Egypt, the Indian subcontinent, and Indonesia,

---

TABLE 16.2 Properties of Poxviruses

Virions in most genera are brick-shaped virions, 250×200×200 nm, with an irregular arrangement of surface tubules. Virions of members of the genus *Parapoxvirus* are ovoid, 260×160 nm, with regular spiral arrangement of surface tubules

Virions have a complex structure with a core, lateral bodies, outer membrane, and sometimes an envelope

Genome is composed of a single molecule of linear double-stranded DNA, 165 to 210 kbp (genus *Orthopoxvirus*); 280 kbp (genus *Avipoxvirus*), or 130 kbp (genus *Parapoxvirus*) in size. The molecule has covalently closed termini and inverted terminal repeats

Genomes have the capacity to encode approximately 200 proteins, as many as 100 of which are contained within virions. Poxviruses, unlike other DNA viruses, encode all of the enzymes required for transcription and replication

Cytoplasmic replication, enveloped virions released by exocytosis; non-enveloped virions released by cell lysis

---

and still occurs in India. The causative virus has been shown by restriction enzyme mapping to be vaccinia virus, although most strains differ in some biological properties from common laboratory strains of vaccinia virus used for smallpox vaccination. The disease is characterized by pustular lesions on the teats and udders of milking buffalo; occasionally, especially in calves, a generalized disease is seen. Human infection, seen usually among dairy workers who are no longer protected by vaccination against smallpox, is acquired by contact with infected animals. Infection is evident by the appearance of typical poxvirus lesions on the hands and face, sometimes associated with a febrile illness. Further cases of "feral" vaccinia have been reported in cattle and dairy workers in Brazil, most likely originating from vaccinia virus originally used for the smallpox eradication program in that country.

Human Monkeypox

Human infections with monkeypox virus, a distinct member of the genus *Orthopoxvirus*, were recognized in West and Central Africa (especially in the Democratic Republic of Congo) in the 1970s, after smallpox had been eradicated from the region. The clinical signs are similar to those of smallpox, with a generalized pustular rash, fever, and toxemia (Fig. 16.4). Human monkeypox occurs as a zoonosis; it is acquired by direct contact with wild animals killed for food, especially squirrels and monkeys. Monkeypox infection has been found in many animal species: rope squirrels, tree squirrels, Gambian rats, striped mice, doormice, and monkeys; which of these serve as reservoir hosts is unknown. A critical issue is the potential risk of sustained human–human transmission; a few cases of person-to-person transmission have been documented, but the secondary attack rate has been too low for the virus to become established endemically.

The human disease has hitherto been rare; only 400 or so cases were diagnosed in the 1980s. However, in the 1990s there were outbreaks involving more and more people. More than 500 human cases were reported in the Democratic Republic of Congo in 1996–97, the largest outbreak ever reported. Vaccination with smallpox vaccine (vaccinia virus) protects against monkeypox; vaccination has been reintroduced in some villages in Africa to deal with zoonotic monkeypox infections. An outbreak also occurred in the United States in 2003 in 81 individuals who had close contact with prairie dogs: these pet animals had been housed with exotic rodents originating in Africa before being distributed through the pet animal trade (see Box 16.1).

Human Cowpox

Humans can acquire three different poxviruses from cattle: cowpox virus (a member of the genus *Orthopoxvirus*), vaccinia virus (in the days of smallpox vaccination; also a member of the genus *Orthopoxvirus*), and milker's nodule virus (a member of the genus *Parapoxvirus*). In each case, infection is usually seen in individuals who have been milking cows with lesions on the teats or udders.

Cowpox lesions in humans usually appear as single maculopapular eruptions on the hands or the face; infection is accompanied by systemic signs such as nausea, fever, and lymphadenopathy. Children are often hospitalized; while human deaths have been rare, these have been reported.

Cowpox virus has been found only in Europe and in western regions of Russia. Despite the name, the reservoir hosts of cowpox virus are rodents, from which the virus occasionally spreads to cows, humans, domestic cats, and several zoo animal species. During one outbreak at the Moscow Zoo, the virus was isolated from laboratory rats used to feed large felids; a subsequent survey demonstrated infection in wild susliks and gerbils in Russia. In the United Kingdom, the reservoir hosts are bank voles (*Clethrionomys glareolus*), field voles (*Microtus agrestis*), and wood mice (*Apodemus sylvaticus*). In Western Europe, infection with cowpox virus is most commonly seen in domestic cats, with only occasional transmission to humans.

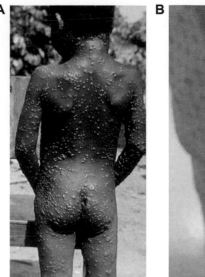
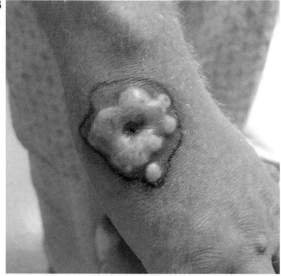

FIGURE 16.4 Human monkeypox. (A) Generalized vesiculo-pustular rash similar to that seen in smallpox, in a case in 1996 in the Democratic Republic of the Congo. *Reproduced from US Centers for Disease Control and Prevention, with permission.* (B) One of the skin lesions in a 30-year-old female who was infected during the 2003 monkeypox outbreak in the United States. She developed approximately 150 lesions over her entire body. *Reproduced from Cooke, R.A., Infectious Diseases, McGraw Hill, with permission.*

BOX 16.1 The 2003 Outbreak of Monkeypox in the Midwest, USA

In 2003, a widely publicized outbreak of monkeypox virus infection occurred in the United States. It began when a 3-year-old child in Wisconsin was bitten by a prairie dog purchased from a local pet shop. The child developed fever, swollen eyes, and a vesicular rash, and the child's parents also subsequently developed a rash. The history of animal bite alerted the local health department, and laboratory testing of both the child and the prairie dog confirmed the diagnosis of monkeypox. Over the following five weeks, a total of 81 human cases occurred in 5 Midwestern states. Most had had direct contact with prairie dogs, and there was no human-to-human transmission. The disease tended to be milder than that previously described in African cases, and there were no deaths. The origin of the virus was traced to imported African rodents [*Funisciurus* spp. (rope squirrel), *Cricetomys* spp. (giant pouched rat), and *Graphiurus* spp. (African dormouse)] that were co-housed with prairie dogs (*Cynomys* spp.). Infected prairie dogs then transmitted the virus to humans, which subsequently resulted in a ban on the importation of African rodents into the United States.

This episode reinforces the lesson that unusual or geographically restricted infectious agents are likely to appear with increasing frequency in new or unexpected settings due to the increase in global travel, trade, and military activity.

Di Giulio, D.B., Eckburg, P.B., 2004. Human monkeypox: an emerging zoonosis [review]. Lancet Infect. Dis. Jan;4(1):15–25.

DISEASES CAUSED BY MEMBERS OF THE GENUS *MOLLUSCIPOXVIRUS*

Molluscum Contagiosum

Molluscum contagiosum is specifically a human disease, but it is often confused with zoonotic poxvirus infections. Attempts to transmit the infection to experimental animals have failed, and reported growth of this virus in cultured human cells has not been reproduced. Infection is characterized by multiple discrete nodules 2 to 5 mm in diameter, limited to the epidermis, and occurring anywhere on the body except on the soles and palms. The nodules are pearly white or pink in color and painless. At the apex of each lesion there is an opening through which a small white core can be seen. The disease may last for several months before recovery occurs. Cells in the nodule are greatly hypertrophied and contain large hyaline acidophilic cytoplasmic masses called molluscum bodies. These consist of a spongy matrix divided into cavities in each of which are clustered masses of viral particles with the same general structure as those of vaccinia virus.

The disease is most commonly seen in children and occurs worldwide, although it is more common in some localities, for example the Democratic Republic of Congo and Papua New Guinea. It is also recognized as a sexually transmitted disease of adults. The virus is transmitted by direct contact, through minor abrasions and sexual contact. In developed countries communal swimming pools and gymnasia have been sources of infection.

DISEASES CAUSED BY MEMBERS OF THE GENUS *YATAPOXVIRUS*

Human Yabapox and Tanapox

Yabapox and tanapox occur naturally only in tropical Africa. Yabapox virus was discovered because it produced large benign tumors on the hairless areas of the face, on the palms and interdigital areas, and on the mucosal surfaces of the nostrils, sinuses, lips, as well as palate, in African green monkeys (*Chlorocebus aethiops*) kept in a laboratory in Nigeria. Subsequently cases occurred in primate colonies in California, Oregon, and Texas. Yabapox is believed to be endemic in African and Asian monkeys. The virus is zoonotic, spreading to humans through contact with infected monkeys, causing similar lesions in humans as in affected monkeys.

Tanapox is a relatively common skin infection of humans in parts of Africa extending from eastern Kenya to the Democratic Republic of Congo. It appears to be spread mechanically by biting insects from an unknown wild animal reservoir, probably a species of monkey. In humans, skin lesions start as papules and progress to umbilicated vesicles, although progression to pustules does not occur. There is usually a febrile illness lasting for three to four days, sometimes accompanied by severe headache, backache, and prostration.

DISEASES CAUSED BY MEMBERS OF THE GENUS *PARAPOXVIRUS*

Several parapoxviruses are zoonotic; farmers, sheep shearers, veterinarians, butchers, and others who handle infected livestock or their products can develop localized lesions, usually on the hands. Lesions begin as an inflammatory papule, and then enlarge to become granulomatous before regressing. Lesions may persist for four to six weeks. The infection acquired from milking cows is known as milker's nodule (pseudocowpox), while that from sheep is orf (see below).

Milker's Nodule

Milker's nodules (or nodes) occur on the hands of humans, the virus coming from lesions on cows' teats. The infection occurs worldwide. Lesions are small non-ulcerating nodules. Immunity following infection does not last long and second attacks may occur at intervals of a few years. The disease is trivial and no measures for prevention or treatment are warranted.

Orf

Orf is an occupational disease associated with handling of sheep or goats (shearing, docking, drenching, slaughtering); the virus also occurs in related wild species.

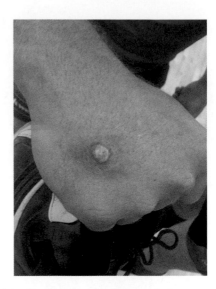

FIGURE 16.5 Human lesion caused by orf virus, a member of the *Parapoxvirus* genus. This is a zoonotic infection acquired by handling infected sheep, and occurs usually in shearers and farm workers.

Orf is an Old Saxon term applied to the infection and is also known either as contagious pustular dermatitis or scabby mouth of sheep and goats. This disease of sheep occurs throughout the world and is found particularly in lambs during spring and summer. Lesions are papulovesicular eruptions that are usually confined to the lips and surrounding skin.

In humans the incubation period is two to four days. Lesions, as a rule, occur on the hands or forearms or occasionally on the face; these are solitary, although multiple lesions have been described (Fig. 16.5). The following stages in the development of lesions may be observed: (1) macular lesions, (2) slowly developing papular lesions, and (3) rather large flat nodules, becoming papillomatous in some cases. Lesions persist for four to nine weeks. Healing takes place without scarring, but secondary infections may retard this process. Severe complications, such as fever, regional adenitis, lymphangitis, or blindness when the eye is affected, are seen only rarely.

FURTHER READING

Fenner, F., Henderson, D.A., Arita, I., Jezek, Z., Ladnyi, I.D., 1988. Smallpox and Its Eradication. World Health Organization, Geneva.

Moss, B., 2013. Poxvirus DNA replication. Cold Spring Harbor Perspectives in Biology. Available from: http://dx.doi.org/10.1101/cshperspect.a010199.

Smith, G.L., Benfold, C.O., Maluquer de Motes, C., Mazron, M., Ember, S.W.J., Ferguson, B., 2013. Vaccinia virus immune evasion: Mechanisms, virulence and immunogenicity. J. Gen. Virol. 94, 2367–2392.

Chapter 17

Herpesviruses

Virtually every vertebrate species supports at least one host species-specific herpesvirus, and even a herpesvirus of a non-vertebrate species (oysters) is known. Most are well-adapted to each natural host and usually cause asymptomatic infections except in the very young, in immunosuppressed individuals or when infecting a species other than the natural host. This implies that herpesviruses have co-evolved with vertebrate species over millennia. All herpesviruses persist indefinitely, nearly always in an episomal form within the nucleus of the infected cell.

Herpesviruses of different subfamilies occupy different anatomical niches within the human body in a non-competitive state. Varicella-zoster virus (chickenpox) and herpes simplex virus both establish latent infections in neurons. On reactivation, varicella-zoster virus precipitates an attack of herpes zoster (shingles), whereas herpes simplex virus type 1 typically causes recurrent attacks of labial herpes, (and herpes simplex virus type 2 is mainly responsible for genital herpes. Epstein-Barr (EB) virus and human herpesvirus 6 (HHV-6) persist in lymphocytes, and cytomegalovirus is latent in granulocyte/macrophage precursors in the bone marrow. Cytomegalovirus is now the major infectious cause of mental retardation and other congenital defects following the control of rubella by immunization and the consequent near elimination of congenital rubella syndrome. EB virus is the etiological agent of infectious mononucleosis and is associated with a number of human cancers including carcinomas and lymphomas. In contrast, HHV-6 and HHV-7 cause a common exanthem in children and HHV-8 is associated with several late-onset tumors including Kaposi's sarcoma in immunocompromised patients.

The patterns of disease caused by some of these human pathogens have changed somewhat as a result of developments in modern medicine and changing sexual practices. Herpesviruses are frequently reactivated in AIDS (see Chapter 23: Retroviruses) and following immunosuppressive therapy for organ and hematopoietic stem cell transplantation, and in cancer. Under these circumstances, together with infections of the fetus or newborn infant, lethal disseminated disease may occur. Although most herpesviruses pose unsolved problems for vaccine development, some respond well to antiviral chemotherapy.

PROPERTIES OF THE VIRUSES

Classification

The order *Herpesvirales* contains three families (*Herpesviridae*, *Alloherpesviridae*, and *Malaco-herpesviridae*), the first of which contains viruses of humans. Division of the family *Herpesviridae* into three subfamilies was at first based on biological properties, and the subfamilies were then divided into genera based mainly on antigenic cross-reactivity, genome size, and structure. The subfamily *Alphaherpesvirinae* includes the human pathogens herpes simplex virus types 1 and 2 (varicella-zoster virus). The alphaherpesviruses all grow rapidly, lyse infected cells, and establish latent infections in sensory nerve ganglia. The subfamily *Betaherpesvirinae* includes human cytomegalovirus, HHV-6 and HHV-7. In contrast, the replication cycle of betaherpesviruses is slow and accompanied by the production of large, often multinucleate cells (cytomegalia). The viral genome remains latent in lymphorcticular tissue, secretory glands, kidneys, and other tissues. The subfamily *Gammaherpesvirinae* contains EB virus and HHV-8. Both viruses replicate in B lymphoid and endothelial cells and may also be cytocidal for epithelial cells. Latency is frequently demonstrable in lymphoid tissue. Since herpesvirus genomes have been sequenced, these assignments have generally been confirmed by extensive phylogenetic analyses (Fig. 17.1).

All herpesvirus species have now been assigned systematic names incorporating (1) a term derived from the natural host, (2) the word "herpesvirus," and (3) a number reflecting the historical order in which it was recognized, for example, "human herpesvirus 3" (HHV-3), "equine herpesvirus 4" (EHV-4). However, common names in long-standing use, for example, herpes simplex virus type 2 (HSV-2), EB virus, etc., are still most widely used and will be used here (Table 17.1).

Virion Properties

The herpesvirus virion comprises four concentric layers: an inner *core*, surrounded by an icosahedral *capsid*, an amorphous *tegument*, and an *envelope* (Fig. 17.2). The

Fenner and White's Medical Virology. DOI: http://dx.doi.org/10.1016/B978-0-12-375156-0.00017-5

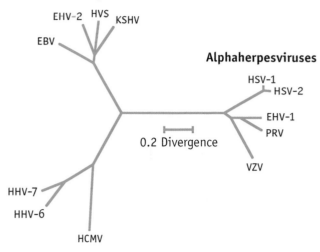

Gammaherpesviruses

EHV-2 HVS KSHV

EBV

Alphaherpesviruses

HSV-1

HSV-2

EHV-1

PRV

0.2 Divergence

VZV

HHV-7

HHV-6

HCMV

Betaherpesviruses

FIGURE 17.1 Phylogenetic relationships of the human herpesviruses (and a few closely related animal viruses) within the very large family *Herpesviridae*, order *Herpesvirales*. The tree is based on the amino acid sequences of the major capsid protein of each virus. Classification and abbreviations: subfamily *Alphaherpesvirinae*, genus *Simplexvirus*, herpes simplex viruses 1 and 2 (human herpesviruses 1 and 2) (HSV-1, -2), B virus (Macacine herpesvirus 1, Cercopithecine herpesvirus 1, herpesvirus simiae); genus *Varicellovirus*, varicella-zoster virus (human herpesvirus 3) (VZV), pseudorabies virus (PRV), equine herpesvirus 1 (EHV-1); subfamily *Betaherpesvirinae*, genus *Cytomegalovirus*, human cytomegalovirus (human herpesvirus 5) (HCMV); genus *Roseolovirus*, human herpesviruses 6 and 7 (HHV-6, -7); subfamily *Gammaherpesvirinae*, genus *Lymphocryptovirus*, Epstein-Barr virus (human herpesvirus 4) (EBV); genus *Rhadinovirus*, Kaposi sarcoma herpesvirus (human herpesvirus 8) (KSHV), equine herpesvirus 2 (EHV-2), *herpesvirus saimiri (HVS)*.

DNA genome is wound like a ball of wool and is associated with a protein core in the shape of a doughnut suspended by fibrils anchored to the inner side of the surrounding capsid. The capsid is an icosahedron, 100 nm in diameter, composed of 162 hollow capsomeres: 150 hexamers, and 12 pentamers (Fig. 17.2). Surrounding the capsid is a layer of globular material, the tegument, and this is enclosed by the lipoprotein envelope containing numerous small embedded glycoprotein peplomers. The envelope is fragile and the enveloped virion is somewhat pleomorphic, with a diameter ranging from 120 to 200 nm.

The component polypeptides within the virion vary between different herpesviruses. For example, the herpes simplex type 1 virion contains over 30 proteins, of which about 6 are present in the nucleocapsid, 10 to 20 in the tegument, and 10 in the envelope; a smaller number is associated with the DNA in the core. The envelope proteins are mainly glycoproteins, most but not all of which contribute to the peplomers. Antigenic relationships are complex. There are some shared antigens within the family, but different species have distinct envelope glycoproteins (Table 17.2).

The herpesvirus genome consists of a linear dsDNA molecule that is infectious under appropriate experimental conditions. There is a remarkable degree of variation in the composition, size, and structure of herpesvirus DNA genomes. The genomes of the herpesviruses, together with those of the poxviruses, are among the larger human viral genomes, ranging from 125 to 295 kb and coding for about 70 to around 200 proteins. The genomes of the alphaherpesviruses appear to be mainly colinear, that is, the presence and order of the individual genes are similar.

TABLE 17.1 Herpesviruses of Humans

Subfamily	Genus		Common Name	Biological Properties
Alphaherpesvirinae	*Simplexvirus*	HHV-1	Herpes simplex virus 1	Fast-growing, cytolytic, latent in neurons
	Simplexvirus	HHV-2	Herpes simplex virus 2	Fast-growing, cytolytic, latent in neurons
	Varicellovirus	HHV-3	Varicella-zoster virus	Fast-growing, cytolytic, latent in neurons
	Simplexvirus	CeHV-1	B virus (previously Macacine herpesvirus 1 or Cercopithecine herpesvirus 1)	Fast-growing, cytolytic, latent in neurons
Betaherpesvirinae	*Cytomegalovirus*	HHV-5	Cytomegalovirus	Grows in many cell types. Latent in myeloid lineage cells, shed from kidney and salivary gland
	Roseolovirus	HHV-6		Grows in T lymphocytes, salivary gland; latent in macrophages, lymphocytes
	Roseolovirus	HHV-7		Grows in T lymphocytes, salivary gland; latent in macrophages, lymphocytes
Gammaherpesvirinae	*Lymphocryptovirus*	HHV-4	Epstein-Barr virus	Grows in epithelial cells. Latent in B lymphocytes
	Rhadinovirus	HHV-8	Kaposi's sarcoma herpes virus	Infects B lymphocytes, epithelial cells

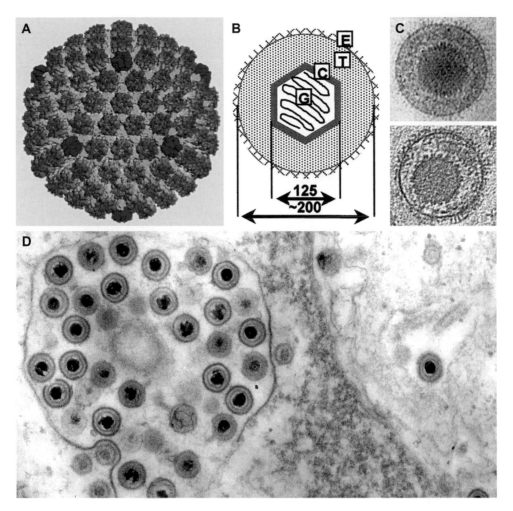

FIGURE 17.2   Herpesvirus morphology and structure. (A) Reconstruction of a herpes simplex type 1 virus icosahedral capsid generated from cryo-electron microscopy images, viewed along three-fold axis of symmetry. The hexons are shown in blue, the pentons in red and the triplexes in green. (B) Schematic representation of a herpes simplex virion with diameters shown in nm: *G*, genome; *C*, capsid; *T*, tegument; *E*, envelope. (C) Cryo-electron microscopy of herpes simplex virus capsids, showing the nucleocapsid, tegument, envelope, and surface projections (peplomers). (D) Thin-section electron microscopy of herpes simplex virus showing enveloped virions in various stages of maturation in a cytoplasmic invagination into the nucleus of an infected cell in culture. Several capsids are about to bud through the outer lamella of the nuclear envelope. *(A–C): Reproduced from King, A.M.Q., et al., 2012. Virus Taxonomy. In: Ninth Report of the International Committee on Taxonomy of Viruses, Academic Press, p. 100, with permission.*

TABLE 17.2 Properties of Herpesviruses

Virions are enveloped and roughly spherical, 120 to 250nm in diameter. An internal icosahedral nucleocapsid with 162 capsomers is surrounded by amorphous tegument and then a lipid envelope containing numerous different glycoproteins, some forming peplomers

Linear dsDNA genome 125 to 229kbp; terminal reiterated sequences are characteristic

Virus DNA replicates in the nucleus by a rolling circle mechanism; sequential transcription and translation of immediate early (α), early (β), and late (γ) genes, producing α, β, and γ proteins respectively; the earlier genes and their gene products successively regulate the transcription of the later ones

Encapsidation of viral DNA occurs in the nucleus; nucleocapsids acquire an envelope by budding through the nuclear membrane; enveloped virions accumulate in the ER and are released by exocytosis

Productive infection in permissive cells is cytocidal; intranuclear inclusions are seen, and sometimes cytomegalic cells or syncytia

Latent infection is established following initial acute infection; genome persists in the nucleus in neurons or lymphocytes. Reactivation triggers virus replication and recurrent or intermittent shedding of virus, sometimes accompanied by recrudescence of clinical disease

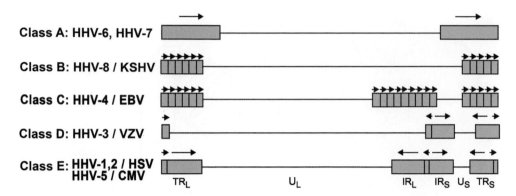

FIGURE 17.3 Simplified genome structure of human herpesviruses. Herpesvirus genomes consist of linear, double-stranded DNA molecules that range in size from about 125 to 240 kbp. The genomes characteristically contain direct or inverted repeats; the reasons for this are not known but are presumed to contribute to the complex replication cycle. In this highly simplified graphic, the various arrangement of reiterated sequences [direct or inverted (arrows), at the termini (TRL and TRS) or internally (IR)] results in a number of different genome structures that have been divided into several classes. Class A: Genomes consist of a unique coding sequence flanked by direct terminal repeats. Examples: human herpesviruses 6 and 7 (HHV-6, HHV-7). Class B: Genomes have directly repeated sequences at the termini, but these consist of variable copy numbers of a tandemly repeated sequence. Example: Human herpesvirus 8/Kaposi sarcoma herpesvirus (HHV-8; KSHV). Class C: Genomes have an internal set of direct repeats that are unrelated to the terminal set. Example: Epstein-Barr virus (EBV). Class D: Genomes contain two unique regions, each flanked by inverted repeats. Example: Varicella-zoster virus (VZV). Class E: Genomes are the most complex of all, with terminal and internal repeats that are much larger and inverted giving rise to four equimolar genome isomers. Examples: Herpes simplex viruses 1 and 2 (HSV-1, HSV-2) and human cytomegalovirus (HCMV). There is a Class F, but it does not contain any human virus. *From Bernard Roizman and Phillip Pellett, with permission.*

Herpesvirus genomes display some unusual features and generally follow one of four distinct organizational patterns (Fig. 17.3). Reiterated DNA sequences generally occur at each end of the genome and, in some viruses, internally also, dividing the genome into two unique components, designated large (U_L) and small (U_S). When these reiterated sequences are inverted in orientation, the unique L and S components can become inverted relative to each other during replication, giving rise to two or four different isomers of the genome, each present in equimolar proportions. Further, intragenomic and intergenomic recombinational events can alter the number of any particular reiterated sequence, creating *polymorphism*.

Viral Replication

Herpesvirus replication has been mostly studied using herpes simplex viruses types 1 and 2 (HSV-1 and -2); betaherpesviruses and gammaherpesviruses replicate more slowly and exhibit certain significant differences but generally follow a similar pattern of genome expression and replication. Unlike other DNA viruses such as papovaviruses and parvoviruses that use the cellular DNA synthetic machinery, herpesviruses encode most of the enzymes required to increase the pool of deoxynucleotides necessary for the replication of viral DNA. This facility is vital for viral replication in resting cells such as neurons, which throughout most of the life of the host never divide and thus never synthesize DNA. Interestingly, about half of the 73 genes of herpes simplex virus are not essential for viral replication in cultured cells, and it is likely that a similar ratio applies in the case of other herpesviruses; presumably many of these additional genes encode regulatory proteins and virokines important for the optimization of growth, dissemination, and pathogenicity *in vivo*, and by such means extend viral tissue tropism, establish and maintain latency, and suppress the host immune response.

The HSV virion attaches via its envelope glycoproteins gC and gB to the heparan sulfate moiety of cellular proteoglycans, and then forms a firmer association between the envelope gD glycoprotein and one of further cellular receptors—herpesvirus entry mediator (HEM) or nectin-1. Entry into the cytoplasm requires viral glycoproteins gB, gD, and gH and occurs by pH-independent fusion of the virion envelope with the plasma membrane in some cells (e.g., neurons) and via endocytosis in others (keratinocytes): the gB protein is responsible for fusion to the cell. Tegument proteins are released, one of which (the UL41 gene product, an endonuclease) shuts down cellular protein synthesis. The capsid is transported by retrograde microtubule transport to a nuclear pore, where viral DNA is released, enters the nucleus, and is circularized.

Viral gene expression is tightly regulated, with three classes of mRNA, α, β, and γ, transcribed by the host RNA polymerase II in a strictly ordered progression (Fig. 17.4). Immediate early (α) genes can be transcribed in the absence of *de novo* protein synthesis, while early (β) gene transcription is dependent upon the expression of immediate early proteins and late (γ) gene expression is dependent on viral DNA synthesis. One of the released tegument proteins (the UL48 gene product) transactivates transcription of the five "immediate early" (α) genes. This viral protein associates with two cellular proteins to form a multi-protein complex that specifically recognizes a nucleotide sequence in the promoter region of the viral

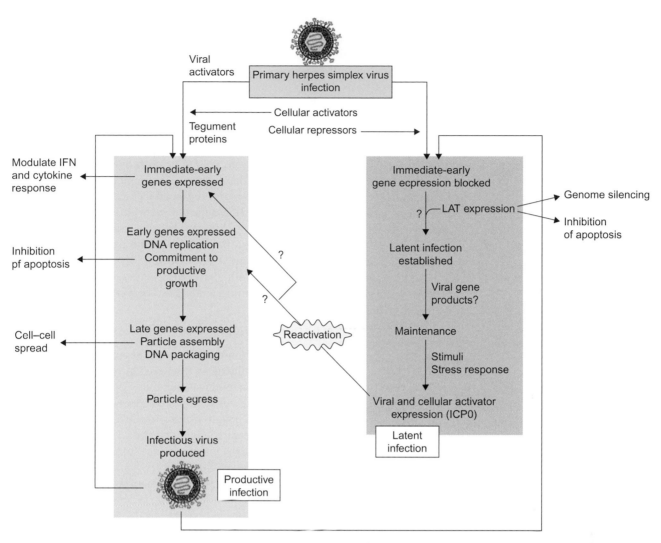

FIGURE 17.4 General strategies for the establishment of productive or latent infection with herpes simplex virus. The productive infection is shown by the pathway on the left, and the latent infection by the pathway on the right. Infectious particles produced by the productive pathway may infect other cells and enter either the productive or latent pathways. Infection can also spread from cell to cell without release of particles. Apoptosis induced by infection is inhibited by viral gene products. Reactivation is indicated by the diagonal arrow from the latent state to the start of the productive infection. There is debate whether reactivation requires "return to go" (immediate early gene expression) or "start in the middle" (expression of early genes required for DNA replication). *Reproduced from Flint, S.J., et al., Principles of Virology, Vol. II, p. 151. ASM Press, Washington DC, with permission.*

DNA, triggering transcription by a cellular polymerase. The α mRNAs are transported to the cytoplasm and translated to the several α proteins: these are regulatory proteins controlling the expression of all late viral genes. One is a protein that initiates transcription of the "early" (β) genes. The β proteins include enzymes required to increase the pool of nucleotides (e.g., thymidine kinase, ribonucleotide reductase) and others needed for viral DNA replication (e.g., a DNA polymerase, primase–helicase, topoisomerase, single-strand and double-strand DNA-binding proteins).

The viral genome replicates by a rolling circle mechanism from one or more origins of replication. In the case of HSV-1, this involves at least seven virus-coded proteins—an origin-binding protein, an ssDNA-binding protein, a DNA polymerase composed of two subunits, and a helicase–primase complex composed of three gene products. Following DNA replication, certain β proteins induce the program of transcription to switch templates once more, and the resulting "late" (γ) mRNAs are synthesized and translated into the γ proteins, most of which are structural proteins required for morphogenesis of the virion. Capsid proteins assemble to form empty "immature" capsids in the nucleus. Unit-length viral DNA cleaved from newly synthesized DNA concatemers is packaged to produce mature nucleocapsids: these then associate with patches of nuclear membrane to which specific tegument proteins are bound. This triggers envelopment by budding through the inner nuclear membrane, but the precise details of nucleocapsid exit from the nucleus are a matter of controversy. Removal of the envelope occurs at the outer nuclear membrane and the tegument proteins are added in at least two steps in the cytoplasm. These nucleocapsids are then enveloped through interactions with glycoproteins embedded in the Trans-Golgi network. The mature virions are transported in vesicles to the plasma membrane and released there by exocytosis. Virus-specific proteins in the plasma membrane may serve as Fc receptors and may be targets for immune cytolysis.

Such productive infections (as distinct from latent infections) are lytic, as a result of virus-induced shutdown of host protein and nucleic acid synthesis. Major changes are obvious microscopically, notably the margination and pulverization of chromatin and the formation of large eosinophilic intranuclear inclusion bodies. These inclusion bodies can usually be found both in herpesvirus-infected tissues and in appropriately stained cell cultures.

An alternative outcome to the above is latent infection, a process important for virus persistence in the infected host but more difficult to study. This is generally considered to involve a failure of immediate early (α) gene expression leading to persistence of the input genome(s) as circular episomal elements. However, some evidence also suggests that productively infected cells may on occasions then progress to latency. At a later time, reactivation of immediate early gene expression and entry into the productive replication cycle may be triggered by external stimuli affecting the transcription factor milieu within the latently infected cell.

HERPES SIMPLEX VIRUS INFECTION

Herpes simplex virus infections of the lips, mouth, and genital tract were described in early Sumerian and Greek literature: the first successful, relatively non-toxic antiviral drug, acyclovir, was developed to treat herpes simplex infections, while genital herpes still continues to cause a disabling problem for some sufferers.

Clinical Features

In considering the various clinical presentations it is important to distinguish between primary and recurrent infections (Table 17.3). Primary infections with HSV-1 and -2 are generally inapparent, but when clinically manifest, primary infections tend to be more severe than recurrences in the same dermatome. Since immunity to exogenous reinfection is long-lasting, nearly all second clinical episodes with the same HSV type are reactivations of an endogenous latent infection. However, because cross-immunity is only partial, *de novo* infection with the heterologous type can occur (e.g., genital herpes caused by HSV-2 in an HSV-1 immune person); these cases are referred to as "initial disease, non-primary infection" and are usually mild.

Oropharyngeal Herpes Simplex

Primary infection with HSV-1 most commonly involves the mouth and/or throat. In young children the classic clinical presentation is gingivostomatitis. The mouth and gums exhibit varying numbers of vesicles which soon rupture to form ulcers. Though febrile, irritable, and suffering from the pain of bleeding gums, the child recovers uneventfully. In adults, primary infection more commonly presents as a pharyngitis or tonsillitis.

Following recovery from primary oropharyngeal infection the individual retains HSV DNA in neurons of the trigeminal ganglion for life and has at least a 30 to 40% chance of suffering recurrent attacks of herpes labialis (otherwise known

TABLE 17.3 Diseases Caused by Herpes Simplex Viruses

Disease	Primary (P) or Recurrent (R)	Age	Frequency	Severity	Type
Gingivostomatitis	P	Young children	Common	Mild	1
Pharyngotonsillitis	P	Adults	Common	Mild	1 > 2
Herpes labialis	R	Any	Common	Mild	1 > 2
Genital herpes	P, R	>15 years	Common	Mild/moderate	2 > 1
Keratoconjunctivitis	P, R	Any	Common	Mild/moderate	1
Skin infection[a]	P, R	Any	Rare	Mild/moderate	1, 2[b]
Encephalitis	P, R	Any	Rare	Severe[c]	1 > 2[d]
Neonatal herpes	P	Newborn	Rare	Severe[c]	2 > 1
Disseminated herpes	P, R	Any	Rare	Severe[c]	1 > 2

[a]Including herpes simplex virus infection of burns, eczema herpeticum, etc.
[b]Skin above waist, 1 > 2; below waist, 2 > 1; arms, either 1 or 2.
[c]Often fatal.
[d]HSV-2 in neonates.

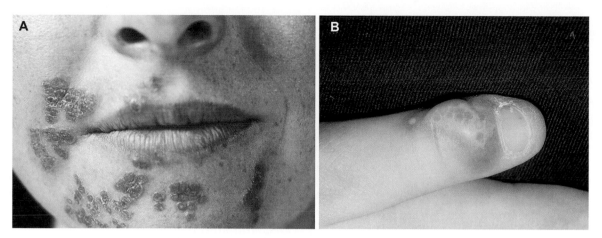

FIGURE 17.5 (A) Lesions of recurrent labial herpes simplex, in the healing and crusting phase. (B) Herpetic whitlow on the right thumb, occurring in a nurse who attended the laryngeal airway of a patient with a tracheostomy. Wearing gloves is now essential for such procedures. *Reproduced from Cooke, R.A., Infectious Diseases, McGraw Hill, with permission.*

as herpes facialis, herpes simplex, fever blisters, or cold sores) from time to time throughout the remainder of life. However, almost all infected people shed virus asymptomatically in saliva periodically throughout life which allows transmission. A brief prodromal period of hyperesthesia heralds the development of a cluster of vesicles, generally around the mucocutaneous junction on the lips (Fig. 17.5).

Genital Herpes

Most primary genital herpes is still caused by HSV-2, despite an increasing proportion of cases attributable to HSV-1, especially in adolescent and young women probably as a result of orogenital sexual practices. As a sexually transmitted disease it is seen mainly in young adults, but it is occasionally encountered following accidental inoculation at any age, or in young girls who have been victims of sexual abuse. Though the majority of primary infections are subclinical, disease may occasionally be severe, particularly in females. Ulcerating vesicular lesions develop on the vulva, vagina, cervix, urethra, and/or perineum in the female (Fig. 17.6), and the penis, including foreskin, in the male, or the rectum and perianal region in male homosexuals. Local manifestations are pain, itching, redness, swelling, discharge, dysuria, and inguinal lymphadenopathy; systemic symptoms, notably fever and malaise, are often marked, especially in females. Spread may occur to the central nervous system, causing mild meningitis in about 10% of cases. Initial disease is less severe in those with immunity resulting from previous infection with HSV-1 or previous subclinical infection with HSV-2. Asymptomatic shedding from the anogenital mucosa is very frequent (at least daily in some) and the commonest cause of sexual transmission.

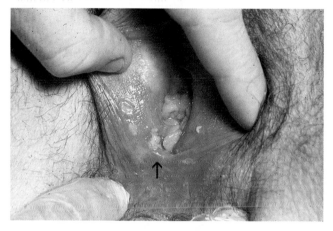

FIGURE 17.6 Female with genital herpes showing vesicular lesions on the labia. *Reproduced from Cooke, R.A., Infectious Diseases, McGraw Hill, with permission.*

Even though the vast majority of recurrent genital herpes cases are less severe than was the primary attack, the resultant pain, sexual frustration, and sense of guilt can have psychological effects as well as consequences to sexual behavior.

Keratoconjunctivitis

Primary infection of the eye with HSV-1 may occur with exogenous virus or may result from autoinoculation. Involvement of the cornea (keratitis) often leads to a characteristic "dendritic ulcer" which may progress to involve the underlying stroma. Immunopathological mechanisms including autoimmunity are thought to be involved. The corneal scarring that results from repeated

HSV infections is a common cause of blindness; such recurrences are usually unilateral.

Skin Infections

Primary and recurrent HSV infections may also involve the skin on any region of the body. Rarely this occurs by direct traumatic contact, for example, in wrestlers or rugby players—called "herpes gladiatorum." An occupational hazard for dentists and nurses is herpetic paronychia ("herpetic whitlow"—painful cutaneous infection of the distal aspect of the finger caused by HSV) (Fig. 17.5). Much more dangerous are disseminated skin infections with HSV that complicate burns or eczema. HSV-1 can also cause and complicate erythema multiforme.

Encephalitis

Although encephalitis is a rare manifestation of HSV infection, it is nevertheless the most common sporadic cause of this disease. The virus may spread to the brain during primary or recurrent HSV infection, with the temporal lobes most commonly involved. Vesicles are generally not present on the body surface. HSV-1 is usually responsible except in the case of neonates. The presentation can include fever, headache, behavioral changes, seizures, vomiting, and focal neurological signs, with no reliable pathognomonic signs. Untreated, the case–fatality rate is 70%, and the majority of survivors suffer permanent neurological sequelae. The prognosis has improved significantly with the current policy of immediate pre-emptive treatment with intravenous acyclovir (the drug of choice), once herpes simplex encephalitis is suspected. Because of the risk of relapse often within the first three months, a follow-up prolonged course of an oral antiviral agent, for example, valacyclovir, is generally recommended. HSV is also a significant cause of meningitis and more rarely, associated with a range of other neurological conditions.

Neonatal Herpes

Herpes neonatorum is a serious disease acquired by babies, usually from their mothers at delivery. The majority are HSV-2 infections acquired during passage through an infected birth canal, but some may be acquired postnatally, and a very few prenatally by viremic transmission across the placenta or by ascending infection from the cervix. If virus is present in the maternal genital tract at the time of delivery the risk of neonatal herpes ranges from three to four percent with recurrent maternal infection, to 30% to 40% in primary infection. This difference is thought due to the protective effect of pre-existing antibodies acquired transplacentally from a mother undergoing reactivation. Neonatal herpes simplex infection occurs in 1 in 5000 to 10,000 live births.

Neonatal herpes may present as (1) disseminated disease, with a case–fatality rate of 80%, most of the survivors being left with permanent neurological or ocular sequelae, (2) encephalitis, which also carries a high risk of death or sequelae, or (3) disease localized to mucocutaneous surfaces such as skin, eye, and mouth (SEM disease—which may, however, progress to disseminated disease if not treated promptly). Once again, intravenous acyclovir (20 mg/kg every 8 hours) has been shown to reduce morbidity and mortality compared to lower dosages. Because this dosage can be associated with neutropenia, the patient's white cell count should be monitored.

Disseminated Herpes in Immunocompromised Hosts

Patients particularly at risk of potentially lethal disseminated HSV infections are those who are already compromised by congenital or acquired immunodeficiency, for example, following immunosuppression for organ transplantation, HIV/AIDS, or malignancy. Other risk groups include those suffering from severe malnutrition with or without concomitant measles, or those with severe burns, eczema, and certain other skin conditions.

Pathogenesis, Pathology, and Immunity

The initial site of replication for primary herpes simplex virus is usually epithelial cells of mucosal surfaces, typically oropharyngeal or genital mucosa. Focal areas of epithelial cell necrosis are seen, involving the ballooning of cells and sometimes typical intranuclear herpesvirus inclusions. An intense inflammatory cell response and accumulation of fluid rapidly lead to vesicle formation with progression to ulcers (see Chapter 7: Pathogenesis of Virus Infections). Virions are released from the basal surface of infected cells, and fusion between viral envelopes and adjacent sensory or autonomic nerve endings releases nucleocapsids into the axoplasm of nerve endings. Internalized nucleocapsids can then be transported on microtubules by dynein motors over long distances to reach the body of the nerve cell in the corresponding ganglion (a process known as "retrograde axonal transport").

When the viral DNA enters the nucleus of the neuron, productive or latent infection may ensue. Despite intensive study, many details of the maintenance of the latent state are not clear. The viral genomes within neurons (and some other non-neuronal cells within ganglia) become coated with nucleosomes and exhibit very limited transcription. In the case of neurons latently infected with either HSV-1 or HSV-2, latently infected neurons synthesize a limited number of specific RNA molecules termed the "latency-associated transcripts" (LATs). Possible functions for LATs include serving as mRNAs for critical proteins, as antisense inhibitors of translation or as regulatory micro-RNA precursors, but evidence supporting any of these roles is scant. While the regulatory viral and cellular factors that maintain latency are not well understood, it is clear that reactivation of productive replication is a very frequent event, probably in a tiny fraction of infected cells, leading to transmission of virions to the periphery via the axonal transport system. On some but not all such occasions, productive infection in epithelial cells of the area of epithelium innervated by the neuron undergoing reactivation occurs and lesions are seen. A distinction is made between "reactivation," that is, renewed virus replication, and "recrudescence," that is, reappearance of clinical lesions (Fig. 17.4).

Although pre-existing neutralizing antibody directed against envelope glycoproteins, notably gB and gD, may successfully prevent primary infection and limit spread of herpes simplex virus from epithelial cells to nerve endings, cell-mediated immunity is the key to recovery from primary infection and maintenance of latency. Following lytic infection of epidermal cells, viral antigens on dendritic cells and macrophages are presented to CD4+ Th$_1$ lymphocytes, and these initiate viral clearance by secreting cytokines such as interferon γ (IFN-γ) that recruit and activate macrophages and natural killer (NK) cells. CD4+ and CD8+ T cells and antibody-dependent cell-mediated cytotoxicity (ADCC), lyse infected cells. The overt epithelial infection is cleared, but some virus ascends the local sensory neurons by retrograde axonal transport and establishes lifelong latency in the corresponding spinal or cerebral ganglion (see Fig. 8.5). The mechanism of establishment, maintenance, and reactivation of latency is discussed in Chapter 8: Patterns of Infection. Experimental studies in animal models, as well as the clinical observation that immunocompromised humans are much more prone to severe HSV infections and to reactivation, make it clear that CD8+ T cell-mediated immunity is the key to recovery from primary infection. HSV-specific CD8+ T cells may also suppress the full expression of HSV DNA during the establishment of latency in sensory ganglia. It is not yet understood as to the link among the apparently disparate events known to trigger reactivation, for example, immunosuppression, "stress," trauma, ultraviolet irradiation, fever.

Recurrences of disease are typically less severe than primary disease, and the frequency and severity of such recurrences diminish with time. HSV-1 and -2 display a degree of selectivity in tissue tropism. HSV-2 replicates to a higher titer than does HSV-1 in genital mucosa, and is more likely to lead to encephalitis and severe mental impairment in neonates. HSV-2 is twice as likely as HSV-1 to establish reactivatable latent infection, and recurs almost 10 times as frequently in the anogenital region. The converse applies to orolabial infections, where HSV-1 predominates. The severity of primary HSV infection is influenced by three major factors: (1) age, premature infants being particularly vulnerable, (2) site, systemic and CNS infections being much more serious than infections confined to epithelial surfaces, and (3) immunocompetence, T cell-mediated immunity being crucial in the control of infection.

Two of the HSV glycoproteins, gE and gI, form a receptor for the Fc domain of IgG. This Fc receptor is found on the surface of both virions and infected cells and can protect both against immunological attack, by steric hindrance resulting from binding of normal IgG, or from "bipolar bridging" of HSV antibodies that can attach to gE/gI by its Fc end and simultaneously to another HSV glycoprotein via one Fab arm. Moreover, gC is a receptor for the C3b component of complement and may protect infected cells from antibody–complement-mediated cytolysis. Also HSV ISP47 downregulates MHCI expression on infected target cells making these invisible to (cytotoxic) CD8 T cells, until MHCI is restored by interferon gamma secreted by CD4+ T cells.

Laboratory Diagnosis

Some of the clinical presentations of HSV, such as recurrent herpes labialis, are so characteristic that laboratory confirmation is not required. Others are not so evident, for example, if the lesions are atypical or if vesicles are not visible at all, as in encephalitis, keratoconjunctivitis, or herpes genitalis infections confined to the cervix. Specimens include vesicle fluid, cerebrospinal fluid (CSF), or swabs or scrapings from the genital tract, throat, eye, or skin, as appropriate. Speed is important in situations where the patient would benefit from early commencement of antiviral therapy. Rapid diagnostic methods include detection of HSV DNA in detergent-solubilized cells or mucus from the site of lesions by real-time polymerase chain reaction (PCR). Alternative approaches include demonstration of HSV antigen in cells scraped from lesions, genital tract, throat, or cornea, by immunofluorescence or immunoperoxidase staining (using type-specific monoclonal antibodies if desired), or by enzyme immunoassays (EIAs) on CSF or detergent-solubilized cells and mucus. Diagnosis of encephalitis is particularly difficult, and chemotherapy is now commenced pre-emptively before laboratory confirmation of the diagnosis. Brain biopsy has been used in the past but is unnecessarily invasive. PCR is now the method of choice for detection of HSV DNA in CSF.

Virus isolation in cell culture is the traditional method for diagnosing HSV infection, today perhaps only used when an isolate is needed for research purposes, for determining drug resistance where lesions are refractory to treatment in immune-compromised patients; the specimen should be taken early, placed in an appropriate transport medium, kept on ice, and transferred to a laboratory without delay. Human fibroblasts or Vero cells are both equally sensitive, but HSV replicates rapidly in many mammalian cell lines.

Distinctive foci of swollen, rounded cells appear within one to five days (Fig. 20.4A). The diagnosis can be confirmed within 24 hours by immunofluorescent staining of the infected cell culture. Differentiation of HSV-1 from HSV-2 is not usually relevant to the acute management of the case but is required for counseling of patients as to the likelihood of recurrence of genital disease (much greater with HSV-2); it simply involves selection of appropriate monoclonal antibodies that distinguish the two viruses.

When HSV encephalitis is suspected EIA can be used to demonstrate anti-HSV IgM in CSF; an abnormally high ratio of total HSV antibody in CSF compared to the titer in blood may also aid in diagnosis. In most other HSV infections, however, serology is not widely used, except again for epidemiological and research purposes. Most HSV antibodies react with both serotypes, but type-specific antigens (e.g., baculovirus-cloned external domain of the gG glycoproteins of HSV-1 and HSV-2) are available for EIA or "immunodot" screening assays. Type-specific serology for HSV-2 antibodies can be used to identify latent genital herpes infections in between recrudescences, but its clinical value is limited by inability to distinguish recent and remote infection and also to distinguish HSV-1 oral from genital disease.

Epidemiology, Prevention, Control, and Therapeutics

Herpes simplex viruses spread principally by close person-to-person contact with lesions or mucosal secretions. HSV-1 is shed principally in saliva, and virus may be transmitted to others directly, for example, by kissing, or indirectly, via contaminated hands, eating utensils, etc. Contaminated fingers may occasionally spread HSV-1 to the eye or genital tract but viral survival on the surface of skin is short. HSV-2, on the other hand, is transmitted mainly, but not exclusively, by sexual intercourse. Neonates can be infected with HSV-2 during passage through an infected birth canal, or with HSV-1 by postnatal contact with an infected parent or hospital staff member. Transplacental transmission during pregnancy has been reported rarely, and may be associated with skin lesions, chorioretinitis, microcephaly, hydranencephaly, and microphthalmia.

The probability and age of acquisition of HSV are closely linked to socioeconomic circumstances. In the developing world and in the poorer communities within developed nations, most children first become infected with HSV-1 within a few years after losing the protection of maternal antibody, and over 80% are seropositive by adolescence. Doubtless this reflects the influence of such factors as overcrowding, poor hygiene, and patterns of human contact including contact associated with social mores. In contrast, the majority of children in more affluent communities escape infection until adolescence, when there is a peak of infection as a result of exchange via kissing. The prevalence of infection in adolescence rarely exceeds 50% but there is considerable variation in adult seroprevalence in countries ranging from 50% to 80% (e.g., the United States 50%, Australia 80%). The worldwide seroprevalence of HSV-2 is even more variable ranging from 10% to 80% (Australia 12%, parts of sub-Saharan Africa 80%). In the United States the overall seropositivity rate is 15%, although it is four times higher in African-Americans than in whites. Because almost all the transmission of HSV-2 is sexual, most initial HSV-2 infections in African-Americans are subclinical, as they occur in adolescents and young adults with pre-existing antibody against HSV-1, whereas those in Caucasians tend to be more severe and to recur more frequently.

Asymptomatic shedding in genital secretions (typically HSV-2) or saliva (typically HSV-1) occurs frequently and is not limited to recrudescences with overt symptoms; for example, studies indicate that HSV-1 can be isolated from the saliva of up to 20% of children and 1% to 5% of adults at any given time whereas HSV-2 shedding can occur as frequently as 12-hourly in short bursts or for less frequent, longer durations. At least 25% of people suffer from recurrent herpes labialis and perhaps 5% from recurrent genital herpes. Importantly prior HSV-2 infection also enhances HIV acquisition 3- to 7-fold through an ulcerated and inflamed anogenital mucosa.

Prevention and Control

The very substantial risk of contracting genital herpes from a partner who is a known carrier can be reduced by the diligent use of condoms irrespective of whether or not lesions are present. The risk of herpetic paronychia for dentists and nurses handling oral catheters can be eliminated by the wearing of gloves. Patients with active herpes simplex virus infections, particularly heavy shedders such as babies with neonatal herpes or eczema herpeticum, should be isolated from patients with immunosuppressive disorders.

The prevention and management of neonatal herpes is an important challenge, because timely antiviral treatment significantly improves prognosis. Cesarean section has been shown to reduce transmission of infection to the infant if carried out within 4 hours of membrane rupture, but is of unproven value if delayed beyond four hours. The identification of pregnancies at risk has been attempted by establishing a series of cultures from genital specimens taken at least weekly through the final four to six weeks of gestation from "high-risk" women, usually defined as those with a clinical history of genital herpes. However, this costly procedure is a poor predictor of viral shedding at the time of labor and therefore of fetal risk, and therefore is no longer advocated except when primary genital herpes is diagnosed or suspected. Currently, the preferred policy is to await the onset of labor, then carefully examine the cervix and vulva, and collect swabs for testing. Culture or PCR should be done, as a reliable indicator of infection. However, if genital lesions are present at labor, the baby should be delivered by Cesarean section as soon as possible, usually before a laboratory result is at hand and not more than four hours after the membranes have ruptured. If no lesions are present, vaginal delivery is appropriate; however, should the vaginal swabs subsequently prove to be HSV positive, evidence of neonatal infection should be sought by taking samples from the baby's eyes, nose, throat, umbilicus, and anus. Prompt treatment with high-dose (60 mg/kg/day) intravenous acyclovir has led to major improvements in infant survival. Neonatal transmission occurs in approximately 33% of cases of initial genital herpes during pregnancy, but in <5% of cases of recurrent herpes.

Therapeutics

The drug of choice for the treatment of alphaherpesvirus infections is acycloguanosine (acyclovir), the mechanism of action of which is discussed in Chapter 12: Antiviral Chemotherapy. Acyclovir as an ointment formulation is used topically for the treatment of ophthalmic herpes simplex. Intravenous acyclovir (20 mg/kg 3 times daily) should be commenced promptly and continued for two to three weeks in the case of the life-threatening HSV diseases, encephalitis, neonatal herpes, and disseminated infection in immunocompromised patients. Oral acyclovir (200 mg 5 times daily for 10 days) is appropriate treatment for primary herpes genitalis and may be of value for unusually severe primary orofacial herpes, for example, with atopic dermatitis. Acyclovir-containing creams applied early in recurrences of herpes labialis have been shown in large clinical trials to shorten by around 12 hours the mean duration of clinical episodes. Acyclovir is used cautiously during pregnancy for appropriate indications, for example, severe primary herpes infection, and has not been associated to date with maternal complications or fetal damage.

A number of derivatives of acyclovir have found particular uses. Valaciclovir is more effectively absorbed orally than acyclovir and is rapidly converted to acyclovir *in vivo*, leading to plasma levels that are 8 to 10 times higher than can be achieved with acyclovir. Other acyclovir derivatives include penciclovir and famciclovir. Older antiviral agents such as adenine arabinoside (vidarabine) and phosphonoformate (foscarnet) are too toxic to be recommended for general use, but the latter can be used to treat acyclovir-resistant life-threatening HSV infections, for example, in AIDS patients.

Prolonged oral prophylaxis with valaciclovir 500 mg twice daily (or less frequently used, aciclovir 200 mg 4 times daily) is used to suppress recurrences in patients with a history of more than 10 attacks of genital herpes per annum. It also markedly reduces transmission to uninfected partners. Chemoprophylaxis is also justified in patients receiving an organ transplant, or in immunocompromised patients with any clinically apparent mucocutaneous HSV infection. It is also indicated for the prevention of neonatal herpes in babies delivered vaginally from mothers with proven primary herpes genitalis at the time of delivery. Renal function monitoring is not usually required in patients receiving prophylaxis except for immunocompromised patients on high doses of valacyclovir. In none of these situations is valaciclovir or acyclovir a panacea, and it does not prevent the establishment of latency; at best it can be claimed that early administration of the drug will restrict the progression of the disease and ameliorate symptoms. However, prolonged chemoprophylaxis actually reduces the risk of drug resistance by keeping replicating virus populations low.

Vaccines

The development of herpes simplex virus vaccines has proved difficult. Evidence from human studies and murine models has demonstrated roles for innate, humoral, and cell-mediated immunity in controlling both primary and latent infection. The ultimate goal of total prevention of primary infection and latency may be difficult because virus gains access to nerve ganglia by retrograde axonal transport and not by viremic spread. Nevertheless, a therapeutic vaccine that reduces recrudescence of disease in latently infected subjects may in itself be valuable, particularly with a view to preventing genital and neonatal herpes and is preferable to constant chemoprophylaxis. The delivery of such a vaccine could be delayed until puberty, when a significant proportion of recipients would already have been primed by natural infection with HSV-1. Various HSV glycoproteins elicit neutralizing antibodies, but the most important appears to be gD. HSV 1 and HSV 2 vaccines consisting of either gD or gD plus gB produced by recombinant DNA technologies combined with an adjuvant, have been shown in some trials as significantly reducing the risk of primary genital herpes in women, but not in men, but the recurrent disease in seropositive women has not been reduced. A DNA plasmid vaccine expressing gD, and a recombinant human HSP-70/gB HLA-restricted epitope, have also been tested in phase I trials.

VARICELLA-ZOSTER VIRUS INFECTION

The single herpesvirus known as varicella-zoster virus (VZV) is responsible for two almost universal human diseases: varicella (chickenpox), one of the exanthems of childhood, and herpes zoster (shingles), a disabling disease, most common among aging persons and immunocompromised patients.

Clinical Features

Varicella

The rash of chickenpox appears suddenly, with or without a prodromal fever and malaise. Erupting first on the trunk, then spreading centrifugally to the head and limbs, crops of vesicles progress successively to pustules then scabs. Though painless, the lesions are very itchy, tempting the child to scratch, and this may lead to secondary bacterial infection and permanent scarring. Painful ulcerating vesicles also occur on mucous membranes such as in the mouth and vulva. The disease tends to be more severe in adults, in whom potentially life-threatening varicella pneumonia occurs more frequently.

Primary varicella pneumonia can be a significant complication, accounting for many of the deaths associated with the infection. In one study of military recruits with varicella, 16% of cases had x-ray evidence of pneumonia although only one quarter of these had symptoms. Neurological complications are uncommon but potentially serious. In about 1 case in 1000, encephalitis develops a few days after the appearance of the rash; this can be lethal, particularly in adults. Rarer neurological complications include Guillain-Barré syndrome and Reye's syndrome (see Chapter 39: Viral Syndromes).

Varicella is a particularly dangerous disease in immunocompromised persons and in non-immune neonates. Children with deficient cell-mediated immunity are especially vulnerable, whether congenital or induced by malignancy (e.g., leukemia), anticancer therapy, or steroid therapy. Not only may the skin manifestations be necrotizing and hemorrhagic, but the disease becomes disseminated, involving many organs including the lungs, liver, and brain.

If women have not been infected as children, varicella tends to be more serious in pregnancy and may affect the fetus. Infections occurring in the few days immediately before or after parturition in a non-immune woman can be particularly dangerous, since the baby does not have maternal antibodies and may die from disseminated varicella. Very rarely, maternal infection in the first half of pregnancy has been associated with congenital malformations in the fetus (cutaneous scarring, limb hypoplasia, and eye abnormalities).

Herpes Zoster

Herpes zoster results from reactivation of virus that has remained latent in one or more sensory ganglia following an attack of chickenpox many years earlier. Vesicles are usually distributed unilaterally and confined to the area of skin innervated by a particular sensory ganglion (zoster, Gk=girdle), usually on the trunk or on the face involving the eye (Fig. 17.7); scattered lesions outside the primarily affected dermatome may also occur. The accompanying

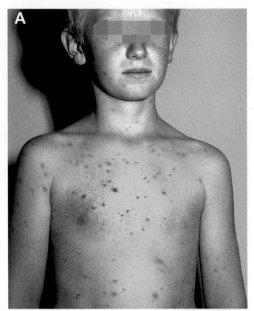
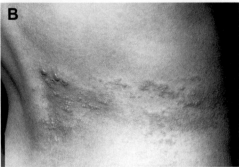
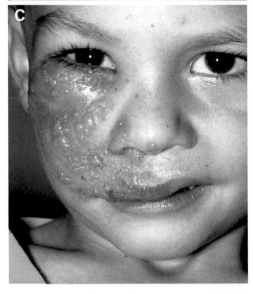

FIGURE 17.7 (A) Varicella in a child. Vesicles appear in recurring crops every 24 to 48 hours, mainly on the trunk. Note lesions at various stages of development, some of which have become secondarily infected as a result of scratching. (B) Typical rash of herpes zoster, distributed along the right first lumbar nerve dermatome, and (C) herpes zoster lesion distributed along the second branch of the right fifth cranial nerve dermatome. *Reproduced from Cooke, R.A., Infectious Diseases, McGraw Hill, with permission.*

pain is often very severe for up to a few weeks, and postherpetic neuralgia, which occurs in half of all patients over 60 years of age, may persist for many months. Motor paralysis and encephalomyelitis are rare complications. Ophthalmic zoster can result in uveitis and occasionally blindness. Disseminated (visceral) zoster is sometimes seen in cancer patients or those otherwise immunocompromised.

Pathogenesis, Pathology, and Immunity

Varicella-zoster virus enters by inhalation and replicates initially in the mucosae of the respiratory tract and oropharynx. Its progression during the 10- to 20-day incubation period (typically 14 days) is presumed to be comparable to that seen in other generalized exanthems. Dissemination occurs via lymphatics and the bloodstream, and the virus multiplies in mononuclear leukocytes and capillary endothelial cells. Eventually the rash results from multiplication of virus in epithelial cells of the skin; keratinocytes show ballooning and intranuclear inclusions, and virus is plentiful within the characteristic vesicles. At this time, virus ascends the axons of various sensory nerves to localize in sensory ganglia, where it becomes latent for life until reactivated by immunosuppression, as discussed in Chapter 7: Pathogenesis of Virus Infections. Unlike HSV, where the distribution of latently infected ganglia is usually tightly confined to the site of primary infection, ganglia latently infected with VZV may be widely distributed. The primary infection is generally rapidly controlled by T cell-mediated immunity. Varicella infection usually induces prolonged immunity, but subclinical reinfections may occur more frequently than once thought, and mild disease is sometimes seen, particularly in immunocompromised individuals.

Herpes zoster occurs when varicella virus in a sensory ganglion is reactivated and is transported anterogradely in the axon of a sensory nerve. In contrast to herpes simplex, VZV latency is rather tightly suppressed by host immunity; herpes zoster occurs in only 10 to 20% of those previously infected with VZV, it is mainly confined to the elderly, and more than a single attack is exceptional. Zoster annually affects about one percent of persons aged 50 to 60 years, with the incidence climbing rapidly thereafter to the point where most 80-year-olds have suffered an attack. The condition is particularly common in patients suffering from Hodgkin's disease, lymphatic leukemia, or other malignancies, following treatment with immunosuppressive drugs, particularly for organ or hematopoietic stem cell transplantation or irradiation, in AIDS, or following injury to the spine. A decline in the level of cell-mediated immunity is thought to precipitate attacks of herpes zoster, and the protracted course of the disease in older or immunocompromised patients is presumably due to a weakened cell-mediated immune response.

Laboratory Diagnosis

The clinical picture of both varicella and herpes zoster is so distinctive that the laboratory is rarely called on for assistance. Rapid diagnostic methods include fluorescent monoclonal antibody staining of fixed smears from the base of early skin lesions or sections from organs taken at autopsy. Alternatively, EIA can be used to demonstrate VZV antigens in vesicle fluid. Finally, PCR can be used to amplify DNA extracted from virions in vesicle fluid.

The virus can be isolated from early vesicle fluid in cultures of human embryonic lung fibroblasts; however, virus tends to remain cell-associated, very little being released, and hence the cytopathic effect (CPE) develops slowly and only in distinct foci over a period of 2 or more weeks. VZV antigen can be demonstrated in nuclear inclusions by immunofluorescence before the end of the first week.

Recent infection can also be confirmed by detecting a rising titer of antibodies, or by IgM serology preferably using EIA. The immune status, for example, of potentially vulnerable leukemic children following contact with a person with chickenpox, can be determined rapidly using EIA. Specific tests for cell-mediated immunity are also used both for research purposes and vaccine assessment.

Epidemiology, Prevention, Control, and Therapeutics

Prior to the widespread use of vaccine, varicella occurred throughout the year but was most prevalent during late winter and spring. Epidemics occurred among groups of susceptible children, for example, in schools or children's hospitals. The annual number of cases was similar to the birth cohort size, and most children became infected during their first years at school. Spread probably occurs via airborne respiratory droplets generated from vesicles on oropharyngeal mucosa, as well as by contact with skin lesions or fomites. Children are highly contagious and should be excluded from school for as long as moist vesicles are present on the skin; a week is normally sufficient.

Passive immunization with zoster immune globulin obtained originally from convalescent zoster patients has an important place in the prevention of varicella. Zoster immune globulin should be administered to non-immune pregnant women who have come into close contact with a case of varicella within the preceding three days but is ineffectual if delayed further. Should a pregnant woman contract varicella within the few days before or soon after delivery, the probability of disseminated disease in the baby may be reduced by treating the mother with valacyclovir and the baby (at birth) with zoster immune globulin. Administration of zoster immune globulin is also indicated for immunocompromised patients who become exposed to the risk of infection; a typical crisis situation would be the occurrence of a case of varicella in the leukemia ward of a children's hospital.

Chemotherapy

Varicella in the normal child can generally be managed by prevention of itching, scratching, and secondary bacterial infection. However, varicella pneumonitis requires vigorous treatment with intravenous acyclovir (10 mg/kg 3 times daily for 7 to 10 days). For herpes zoster, valacyclovir (1 g 3 times daily) or famciclovir (750 mg tds) can be used to accelerate healing in severe cases of zoster provided it is commenced promptly, and this is particularly important when the eye is involved.

Vaccine

The Oka strain of VZV was isolated in Japan from a healthy child with natural varicella, and was attenuated through sequential passages in cultures of human embryonic lung cells, embryonic guinea-pig cells, and human diploid cell line WI-38. Routine vaccination with the Oka vaccine was introduced in the late 1990s in Japan, the United States, and a number of other developed countries. It is normally given by subcutaneous injection, the first dose at 12 to 15 months of age, the second dose at 4 to 6 years of age. For catch-up vaccination, the second dose is given at least 3 months after the first dose for children under the age of 13 years. The vaccine has led to a reduction by 80% to 90% in the incidence of reported varicella cases, hospitalizations, and deaths. The vaccine can induce fever and a few skin papules, occasionally in normal children but more frequently in children who are immunocompromised. This attenuated vaccine often establishes a latent infection in dorsal ganglia and may lead to zoster in later years, but such reactivations are mild and less frequent than those associated with a natural varicella infection.

In the United States shingles (zoster) vaccine is recommended for people aged 60 years and older (it is indicated for people older than 50, but not formally recommended so as to become standard practice). In large clinical trials the vaccine has been shown to be effective in preventing (51%) or markedly ameliorating (66%) the symptoms of shingles and postherpetic neuralgia (60%). In people who have had shingles the vaccine helps prevent future recurrences of the disease. The vaccine employs the same Oka strain attenuated virus, but at 14× the dose used in varicella vaccine for children. New vaccines employing a single varicella recombinant glycoprotein (E) and the complex adjuvant ASO1B appear promising in trials.

Since adults with latent VZV are now being re-exposed less often to naturally circulating virus, there have been fears that protective levels of immunity might wane. Indeed the incidence of zoster in older persons has been slowly increasing, but this increase began prior to the introduction of the vaccine. Vaccine wild-type recombinant strains of virus have been occasionally isolated from cases of herpes zoster in vaccinated children, a finding that complicates the ongoing surveillance of vaccine-related adverse effects.

CYTOMEGALOVIRUS INFECTION

In much of the world, cytomegalovirus (CMV) infection is acquired sub-clinically by the majority of people during childhood, but in some more affluent communities infection tends to be delayed until an age when it can result in clinical disease. Significant clinical impact occurs most often in the very young, during pregnancy or in immunosuppressed individuals, in whom both primary infection and reactivation can bring serious, sometimes life-threatening consequences. For example, primary infections during pregnancy can lead to severe congenital abnormalities in the fetus. Iatrogenic infections may follow blood transfusion or organ transplantation, and CMV can be a major cause of blindness or death in AIDS patients.

Clinical Features

Prenatal Infection and Disease

Cytomegalovirus infection during pregnancy is a major viral cause of congenital abnormalities in the newborn, and has been recognized as the leading priority in the control of congenital infections in the developed world. Less than two percent of babies are born with asymptomatic CMV infection, and only about 1 in 2000 has signs of cytomegalic inclusion disease (CID). The classic syndrome is not always seen in its entirety. The infant is usually small, with petechial hemorrhages, jaundice, hepatosplenomegaly, microcephaly, encephalitis, and sometimes chorioretinitis or inguinal hernia, or both. The abnormalities of the brain and eyes are associated with mental retardation, cerebral palsy, impairment of hearing, and rarely impairment of sight. Many of these infants require special care for life. More subtle syndromes result in socially and educationally important intellectual or perceptual deficits such as hearing loss, subnormal IQ, epilepsy, and behavioral problems which may not become apparent until as late as two to four years after birth.

Infectious Mononucleosis

Most infections acquired after birth, by whatever route, are subclinical. On occasion, however, a syndrome resembling EB virus mononucleosis is seen, particularly among young adults and in recipients of blood transfusions. This syndrome is usually less severe than that caused by EB virus and occurs on average a decade later, that is in early-mid adulthood. Typically, the patient presents with a prolonged fever and on examination is found to have splenomegaly, abnormal liver function, and lymphocytosis, often with "atypical lymphocytes" but usually at lower levels than those observed in EB virus infections (see below). In contrast to EB virus mononucleosis, however, pharyngitis and lymphadenopathy are uncommon, and heterophile antibodies are absent.

CMV Infection in the Immunocompromised Host

Cytomegalovirus infection has been one of the commonest and most difficult to manage of opportunistic infections among either immunocompromised patients or in premature infants receiving a blood transfusion. However this problem has been considerably reduced in developed countries over the past few years following routine use of pre-emptive therapy or prophylaxis with either ganciclovir or its prodrug equivalent, valganciclovir. If untreated, infection may be widely disseminated and almost any organ may be seriously affected. The spectrum and severity of disease and the timing of its development vary, depending on the underlying basis for the immunosuppression, the nature of the transplant, and whether infection is due to reactivation of endogenous virus or *de novo* introduction of virus into a seropositive or seronegative recipient. The most important presentations are interstitial pneumonia, hepatitis, chorioretinitis, arthritis, carditis, chronic gastrointestinal ulcerative lesions, such as gastritis or colitis, and various CNS diseases, especially encephalitis, Guillain-Barré syndrome, and transverse myelitis. Untreated CMV infection can also trigger graft rejection; there is also an association with the accelerated atherosclerosis that is seen in transplant patients, although there may not be a direct causative link.

Pathogenesis, Pathology, and Immunity

Once infected with CMV, an individual carries the virus for life and may shed it intermittently in saliva, urine, semen, cervical secretions, and/or breast milk. Up to 10% of people may be shedding virus at any one time, especially young children. The intermittent nature of CMV shedding and the fluctuations observed in antibody levels suggest that asymptomatic reactivation occurs regularly throughout life. Reactivation occurs more commonly during pregnancy, rising markedly as term approaches. Hormonal factors may be responsible for this, but immunosuppression is generally the most powerful trigger. CMV is a common cause of death in untreated AIDS patients as well as in transplant recipients, especially hemopoietic stem cell transplantation patients. The virus can be isolated from over 90% of patients profoundly immunosuppressed for organ and tissue transplantation. Such infections generally are the result of

reactivation of a latent, dormant infection in cells of either the donor or the graft recipient.

Relatively little is known about the pathogenesis of CMV infection and the mechanism of latency. It is likely that virus commonly enters via the epithelium of the upper alimentary, respiratory or genitourinary tracts, but infection can also bypass epithelial surfaces, as shown by transmission by blood transfusion or organ transplantation. During acute infection, whether primary or reactivated, cell-free virus is not found in the blood, but virus DNA can be recovered from monocytes, neutrophils, and less commonly from T lymphocytes. That these and other cells are potentially permissive has been confirmed by *in vitro* cultivation of CMV in monocytes, endothelial cells, vascular smooth muscle cells, and some CD8+ T cells, but not B cells. However, it is almost impossible to reactivate CMV by co-cultivation of leukocytes from healthy carriers with susceptible fibroblasts *in vitro*. PCR or *in situ* hybridization studies reveal that only about one percent of peripheral blood mononuclear cells from carriers contain the viral genome but in such instances only the major immediate early gene (IE1) is transcribed and translated. There is evidence to suggest that the viral genome persists principally in lineage-committed myeloid cells, including progenitors giving rise to granulocytes, macrophages, and dendritic cells, and also in endothelial cells, stromal cells, and/or ductal epithelial cells in salivary glands and renal tubules, from which virus is shed into the saliva and urine respectively. It is now believed that the state of cell differentiation is a factor controlling permissiveness; as primary peripheral blood monocytes differentiate to monocyte-derived macrophages, reactivation of latent CMV can be detected. There has been a debate whether persistence is maintained by a continuous low-level chronic productive infection, or by true latency in which episomal viral DNA is maintained, but expression of most genes is restricted until reactivation occurs. However, recent evidence suggests there is a distinct subset of genes that are expressed in latency but not in productive infection.

Cell-mediated immunity appears to be principally responsible for controlling CMV. Studies of murine CMV in mice and of human CMV in humans have shown that NK cells are important to host defense early in infection, and that CD8+ T lymphocytes directed at the major immediate early protein, IE1 and the tegument protein pp65, confer protection, although many other viral proteins are recognized by these cells in different subjects. Neutralizing antibodies directed mainly against the envelope glycoprotein gB, and to a lesser extent gH, may contribute to protection, but exogenous reinfection can occur. It is uncertain whether this is generally with a different strain; more than one strain has been isolated concurrently from a single individual.

Although CMV infection elicits a virus-specific cell-mediated immune response, a number of characteristics of infection may contribute to its persistence in the body. First, the virus multiplies extremely slowly and can spread contiguously from cell to cell by fusion, thereby escaping neutralization by antibody. Second, a number of CMV-coded proteins modulate and interfere with the immune response. For example, several viral proteins inhibit MHC class I presentation of viral antigens by multiple mechanisms during acute infection, thus perhaps protecting infected cells to some degree from recognition and lysis by cytotoxic lymphocytes. A CMV protein, UL18, which displays homology with the heavy chain of MHC class I proteins, can interact with β_2-microglobulin, thereby not only interfering with T cell recognition but also coating the free virion and protecting it from antibody. Further, in common with other herpesviruses, CMV codes for a protein with the functional characteristics of an Fc receptor, and this protects the plasma membrane of the infected cell against immune attack in a manner similar to that described above for HSV. It is possible that delaying immune clearance of infection helps to allow infection of myeloid progenitor cells and the establishment of latency. Finally, CMV infection can also be generally immunosuppressive, thereby also predisposing to secondary infection with bacterial or fungal agents.

Congenital CMV Infection and Infection in Infancy

Transplacental infection with CMV is now the commonest viral cause of prenatal damage to the fetus and newborn. Approximately one percent of all babies become infected *in utero*. A majority of these infections are derived from endogenous recurrences (reactivation) in the mother and are generally uneventful. In contrast, most cases of overt congenital disease result from primary infections occurring in the mother during the first 6 months of pregnancy. Hence this syndrome tends to be a disease of affluence, as over 50% of women in developed countries, compared with less than 10% in developing countries, are still seronegative as they enter child-bearing years. Primary infection during the first six months of pregnancy carries a 30% to 40% risk of prenatal infection and 10% to 15% risk of clinical abnormalities in the neonate. The risk of fetal infection following recurrent CMV infection during pregnancy is about one percent, with a low risk of fetal abnormality. Higher amounts of virus are transmitted to babies from mothers with primary than with recurrent infection. The pre-existing immunity present in the latter also confers considerable protection against disease in the fetus, and major deficits other than unilateral deafness are rare in babies infected prenatally as a result of reactivation of maternal CMV.

A minority of babies with clinical abnormalities at delivery are stillborn or die shortly after birth. Autopsy may reveal fibrosis and calcification in the brain and liver. Typical cytomegalic cells may be found in numerous organs; indeed,

they may be found in salivary glands or renal tubules of many normal children. Progressive damage may occur not only throughout pregnancy, but also after birth. The affected infant synthesizes specific IgM antibodies, and immune complexes are plentiful, but CMV-specific and non-specific cell-mediated immune responses are markedly depressed.

Postnatal or natal (intrapartum) infection is more common than prenatal infection and may occur via at least two different routes. Approximately 10% or more of women shed CMV from the cervix at the time of delivery; some of the babies who become infected sub-clinically may acquire the infection during delivery. More severe disease including pneumonitis and hepatosplenomegaly is common in premature infants, particularly in infants of seronegative mothers who acquire infection as the result of blood transfusion. Second, 10% to 20% of nursing mothers shed CMV in their milk, and their infants have a 50% chance of becoming infected via breastfeeding; such infections are subclinical and perhaps the most common mode of transmission of CMV in the neonatal period.

CMV Infection in Later Life

Oropharyngeal secretions are believed to constitute the principal vehicle of transmission, not only in childhood but again in adolescence, either via direct contact or contamination of hands, eating utensils, etc. Kissing and sexual contact no doubt account for the sudden increase in CMV seropositivity from 10% to 15% to 30% to 50% between the ages of 15 and 30 in countries such as the United States, though it is difficult to determine unequivocally the relative importance of each of the two routes. CMV is also shed intermittently in cervical secretions and in semen, hence sexual transmission may be significant, for example, CMV infection is almost universal among promiscuous male homosexuals.

The two remaining mechanisms of acquiring CMV, blood transfusion and organ transplantation, constitute special cases of iatrogenic infection. Almost all those who receive multiple transfusions of large volumes of blood develop CMV infection at some stage. Most such episodes are subclinical, but the mononucleosis syndrome is not uncommon. More serious manifestations of primary infection may occur following transfusion of seropositive (infected) blood into seronegative premature infants, pregnant women, or immunocompromised patients. CMV infection of recipients of kidney, heart, liver, or bone marrow transplants may also be of exogenous origin, introduced via the donated organ or via accompanying blood transfusions. Such an exogenous infection may be primary or may be a reinfection with a different CMV strain. On the other hand, the profound immunosuppression demanded for organ transplantation, or indeed for other purposes such as for cancer therapy, is sufficient for the reactivation of a previous

TABLE 17.4 Syndromes Caused by Cytomegalovirus Infections

Age or Immunocompetence	Route of Acquisition	Disease Caused by Primary Infection
Prenatal	Transplacental	Encephalitis, hepatitis, thrombocytopenia
		Long-term sequelae brain damage, nerve deafness, retinopathy
Perinatal	Cervical secretions, breast milk, saliva	Nil
Any age	Blood transfusion	Pneumonitis, disseminated disease
	Saliva or sexual intercourse	Mononucleosis, mild hepatitis
	Blood transfusion	Mononucleosis
Immunocompromised[a]	Saliva, sex, organ graft, blood transfusion	Pneumonia, hepatitis, retinitis, encephalitis, myelitis, gastrointestinal disease

[a]*Diseases shown occur less commonly after reactivation of a latent infection.*

infection. Primary CMV infections, in particular, are serious and often lethal in immunocompromised patients. Primary infections are also often associated with rejection of the transplanted organ, for example, glomerulopathy in a transplanted kidney. Reactivation of latent CMV is also one of the commonest opportunistic infections leading to death among untreated AIDS patients.

The relationship between the circumstances of transmission of CMV and the more common clinical outcomes is summarized in Table 17.4.

Laboratory Diagnosis

The strategies and techniques used for laboratory diagnosis of cytomegalovirus infection vary according to the clinical setting.

Congenital CMV Infection in the Newborn

Laboratory confirmation of CMV in a newborn infant is important as it affects the medical and educational management of the child and facilitates parental counseling. In order to confirm a congenital infection, samples must be taken within a few days of birth in order to distinguish it from the more common, clinically benign, perinatal infection. Classically, virus can be isolated in cultured human fibroblasts from urine or saliva samples. The virus replicates slowly, but because these infants often have high titers of virus, cultures may become positive within a few days instead of the one to three weeks required to isolate smaller amounts of virus. The sensitivity and speed of the method can be enhanced by use of culture-enhancement centrifugation systems such as the shell vial assay, with staining for early antigen. Detection of CMV DNA by PCR is now more widely used, and can be done on those dried blood spots collected for genetic screening. Serology is of less value, as a positive result for IgG antibodies will reflect passive transfer of maternal antibodies to the infant; the presence of CMV IgM antibodies indicates a congenital infection, as IgM antibodies do not cross the placenta, but this may be present in only 70% of infected babies.

Prenatal Diagnosis

Pre-existing maternal infection is readily identified during pregnancy by detection of CMV IgG antibodies. It is more difficult to reliably identify a maternal primary infection with its more serious prognostic implications, since paired sera with appropriate timing to demonstrate a rising titer are usually not available. CMV IgM antibody tests are not always reliable or consistent, and IgM antibodies can sometimes be detected during virus reactivations or persisting for more than a year after the acute episode. An alternative approach is the demonstration of low avidity IgG antibodies, because antibody avidity progressively increases after the initial infection as the immune response matures.

Prenatal diagnosis of fetal infection can be undertaken in women with primary or undefined CMV infection contracted during the first half of pregnancy. This includes ultrasound examination for fetal abnormalities, culture of amniotic fluid for virus, and the quantitation of CMV DNA in amniotic fluid by PCR; low viral DNA levels suggest an asymptomatic fetal infection, whereas viral loads $>10^3$ copies/mL are suggestive of disease.

Immunocompromised Hosts

CMV infections can be a life-threatening risk in immunocompromised persons, and early diagnosis is needed to ensure appropriate antiviral therapy is instigated. It is important not only to detect CMV but to discern whether the patient is undergoing an active acute infection rather than an asymptomatic chronic infection accompanied by virus shedding, and if so, whether that acute episode is a primary infection of exogenous origin or an endogenous reactivation of a previous persistent infection. Serological tests for IgG and IgM antibody and antibody avidity can be used to distinguish primary from pre-existing infection as described above. The most useful tests are the detection of the viral pp65 antigen by immunofluorescence in circulating polymorphs, and the detection of viral DNA in plasma and urine by qualitative and quantitative PCR. The detection of pp65 antigen is less sensitive than PCR and may not be easy to do in neutropenic patients, but both a positive antigen test and high DNA levels are valuable indicators of invasive disease, and serve as an indication to initiate antiviral therapy. Other specimens appropriate to particular clinical presentations include bronchoalveolar lavage and samples from various organs taken at biopsy or autopsy.

There is only a single serotype of human CMV, but different strains can be distinguished by kinetic neutralization, restriction endonuclease mapping, or sequencing. These techniques can also be employed in identifying the source of virus.

Epidemiology, Prevention, Control, and Therapeutics

Iatrogenic infection via blood transfusion or organ transplantation can be reduced by screening both donor and recipient for evidence of CMV infection through the detection of IgG antibodies. The presence of antibodies in the intended recipient indicates a significant degree of immunity, whereas antibodies in the donor provide a warning that virus, or reactivatable viral episomes, may be transmitted to the recipient with adverse consequences. This potentially dangerous combination is averted if seronegative recipients are given organs or blood taken only from seronegative donors. The amount of testing involved is not cost-effective for routine blood transfusions but CMV-negative blood should be used when the recipients are premature infants, pregnant women, or immunocompromised individuals. Removal of the leukocytes by filtration of donor blood effectively prevents transmission of CMV.

Some authorities have proposed routine screening in pregnancy to identify non-immune women. This would allow adoption of simple hygiene measures to reduce the risk of infection, including care with the handling of body fluids, particularly in nurseries and daycare centers where CMV excretion is common. For women who do seroconvert, CMV hyperimmune globulin may be of value in preventing perinatal transmission, and CMV hyperimmune globulin and/or antivirals may reduce the risk of sequelae in infants who do become infected, although objective evidence of benefit is still limited. Universal screening is not generally accepted at present, but screening may be clinically justified in particular risk situations.

Therapeutics

Ganciclovir and valganciclovir (see Chapter 12: Antiviral Chemotherapy) have dramatically reduced the impact of CMV disease in immunosuppressed patients, and new therapeutic and prophylactic regimens continue to be explored. Ganciclovir is active against herpes simplex, but is also 100 times more active than acyclovir against CMV. CMV does not possess a homolog of the HSV thymidine kinase gene, and the activity of ganciclovir in CMV-infected cells is via direct phosphorylation by a CMV-coded protein kinase, the product of the UL97 gene. Intravenous ganciclovir is the drug of choice for the treatment of severe CMV infections such as pneumonia, chorioretinitis, or colitis in AIDS patients and recipients of organ grafts. When administered intravenously for up to 3 months ganciclovir often leads to fever, rashes, diarrhea, and hematological toxicity: resistant viral mutants can also emerge.

As ganciclovir must be administered intravenously, it has been replaced in many indications by its valine ester valganciclovir which is delivered orally: valganciclovir is cleaved to ganciclovir in the gut and the liver. It is the drug of choice for routine CMV prophylaxis for bone marrow and solid organ transplant patients. It can be used to treat CMV chorioretinitis in AIDS patients, and has also been considered for prophylaxis against CMV in HIV-infected people who have severe immunosuppression.

Useful second-line drugs when CMV develops resistance to ganciclovir include foscarnet, cidofovir, maribavir, and high-dose valaciclovir or related compounds (see Chapter 12: Antiviral Chemotherapy).

Vaccines

The development of a vaccine against CMV has been recognized as a high priority and is likely to be cost-effective. The major aim would be a reduction in congenital CMV disease, the leading viral cause of sensorineural hearing loss and neurodevelopmental delay. CMV presents particular scientific challenges; natural immunity following infection is only partial, latent infection invariably persists despite an immune response, primary infection is usually asymptomatic and not recognized, and good animal models are not available. Nevertheless the successful introduction of the vaccine against VZV, a virus with some biological similarities to CMV, has provided some encouragement.

Live attenuated CMV vaccine strains, for example, the Towne strain, have been developed by serial passage in human fetal fibroblasts. Although these vaccines confer some protection against disease they do not protect these against infection, and therefore presumably latent infection can still develop. The question of the safety of live CMV vaccines must also be addressed, particularly in relation to adequacy of attenuation, establishment of persistence, subsequent reactivation, and possible oncogenicity. Many alternative approaches are at various stages of development; these include a recombinant gB envelope subunit vaccine, DNA vaccines, peptide vaccines, and vaccines based on either poxvirus or adenovirus vectors. Several of these have recently shown encouraging results in phase II trials.

HUMAN HERPESVIRUSES 6 AND 7 (HHV-6 AND HHV-7) INFECTION

HHV-6 was discovered in human lymphocytes in 1986. Since then it has been shown to be ubiquitous, infecting most children worldwide in the first year or two of life. The closely related HHV-7 was first isolated in 1990 from CD4+ T cells. Both viruses cause a generally harmless febrile illness sometimes associated with a rash which has been known since the early 20th century as exanthem subitum, roseola infantum, or sixth disease.

Properties of HHV-6 and HHV-7

The biological properties of HHV-6 resemble those of CMV, and sequencing of the genome has confirmed a taxonomic relationship. The genome occurs as a single isomer consisting of a unique 142-kb segment flanked by a direct repeat sequence of 10 to 13 kb as a single copy at each end. HHV-6 is classified as a species within the genus *Roseolovirus*, within the subfamily *Betaherpesvirinae*. There are two strains, HHV-6A and HHV-6B, with 95% sequence homology, but these are now viewed as separate species as the two viruses can be distinguished by restriction endonuclease mapping and reactivity with subsets of virus-specific monoclonal antibodies. HHV-6A strains have been isolated mainly from adults, whereas most exanthem subitum isolates have been characterized as due to HHV-6B.

Interestingly, HHV-6 is the only human herpesvirus in which latency involves integration of the viral genome into chromosomal DNA rather than persistence as episomal DNA. The direct repeat termini of the HHV-6 genome contain the hexanucleotide TTAGGG identical to a sequence present in the telomeres of mammalian chromosomes, and HHV-6 integrates into sub-telomeric regions of preferred chromosomes. A limited number of genes are transcribed during latency, and reactivation of replicating virus occurs intermittently. Up to one percent of the population has been found to have chromosomally integrated HHV-6 ("CIHHV-6"), in which every cell including the germ line contains integrated and complete HHV-6 genomes, with high viral loads (e.g., >5.5 log10 copies/mL) may be found in whole blood, but usually low levels in plasma. The infection is vertically transmitted in a Mendelian manner. Such infections are usually asymptomatic but may be misdiagnosed as an acute primary infection.

Clinical Features

Although the majority of infections are asymptomatic, HHV-6B causes the childhood illness roseola infantum, also known as exanthema subitum or sixth disease. The illness begins with the abrupt onset of a febrile phase lasting three to five days, sometimes accompanied by bilateral periorbital edema and/or febrile convulsions. This may be followed by an erythematous maculopapular rash starting on the trunk and then spreading centrifugally to the face and limbs, coinciding with the return of body temperature to normal. Commonly, not all of these aspects may be seen, and HHV-6 is a frequent cause of an acute non-specific febrile illness or febrile convulsions in children under two years of age. Meningoencephalitis may occasionally occur, and there have been isolated reports of other possible clinical associations, such as hepatitis. Adults have been reported with fever, lymphadenopathy, and a mononucleosis-like syndrome with negative test results for CMV and EBV.

Symptomatic infection with HHV-6A is largely seen in immunocompromised hosts after solid organ or bone marrow transplantation, where it can cause multiorgan involvement, accelerate graft rejection, and contribute to morbidity and mortality. HHV-7 is less well studied but appears similar in manifestations to HHV-6B although clinical roseola is less common.

HHV-6 has been investigated for a possible etiological or triggering role in multiple sclerosis, chronic fatigue syndrome, Hashimoto's thyroiditis, and other conditions. Needless to say, the widespread natural distribution of the virus, both within populations and within the body, makes confirming or disproving such proposals a demanding exercise.

Pathogenesis, Pathology, and Immunity

HHV-6 replicates most efficiently in dividing CD4+ T lymphocytes and continuous T cell lines. Infected T cells show ballooning, often with more than one nucleus and nuclear or cytoplasmic inclusions, or both, before eventually dying. Macrophages are persistently infected and may represent an important reservoir. Transformed B lymphocytes, NK cells, megakaryocytes, glial cells, fibroblasts, and epithelial cells have also been reported to support the productive replication of certain HHV-6 strains in culture. HHV-6A is readily demonstrable by *in situ* hybridization and immunofluorescence in salivary glands, and is regularly isolated from saliva, suggesting that salivary glands may represent a major reservoir, with saliva the main route of transmission. Reactivation is precipitated by immunosuppression, as in transplantation and in AIDS patients.

HHV-6 infection leads to multiple dysregulation of the immune system. There is impairment of immune function and a variety of mechanisms have been identified, including suppression or depletion of CD4+ T cells, cytotoxic effector cells, macrophages, dendritic cells, cytokine production, and innate antiviral responses. The virus codes for proteins analogous to chemokines and chemokine receptors, and this molecular mimicry may facilitate latency in a way yet to be defined. HHV-6 also codes for gene products that transactivate HIV-1 long terminal repeat (LTR)-directed gene expression. Co-infection with HHV-6A has been shown to accelerate disease progression in pigtailed macaques infected with simian immunodeficiency virus, and there is evidence of its presence accelerating the progression of HIV infection in humans.

Laboratory Diagnosis

Exanthem subitum may be mistaken clinically for measles or rubella. Laboratory diagnosis of primary infection can be made by demonstrating a four-fold rise in antibody titer measured by ELISA or immunofluorescence (IFA), but IgM antibody tests have not proved as reliable or specific. Detection of significant viral DNA levels by PCR in plasma can give presumptive evidence of primary infection, since viral loads are much lower or not detected in chronic infection. Reactivation can be suspected if a qualitative PCR detects viral DNA in serum, or if a quantitative PCR demonstrates a high titer of virus in a whole blood sample. Distinguishing between HHV-6A, -6B, and -7, and diagnosing primary HHV-7 infection in an individual persistently infected with HHV-6, requires more specific tests. Virus can be isolated from peripheral blood mononuclear cells of roseola patients in the early febrile stage of the illness, or from saliva of adults intermittently throughout life, by centrifugation-enhanced infection of phytohemagglutinin-activated peripheral blood leukocytes. Alternatively, virus can be detected using T cell lines: HHV-6 antigen is identified in infected cells by immunofluorescence using an appropriate monoclonal antibody.

Epidemiology, Prevention, Control, and Therapeutics

Serological surveys indicate that almost all children have been infected with HHV-6 by two to three years of age, strongly suggesting intrafamilial spread, probably from the mother shortly after maternal antibodies have declined in the infant. The high incidence of shedding of virus in saliva points to this as the likely, but not necessarily the only, natural route of transmission. HHV-7 tends to be acquired later, in the first five or six years of life; the reasons for this difference are not known.

Therapeutics

Although no international guidelines have yet been developed, life-threatening infections with HHV-6 (e.g.,

encephalitis, severe infections in immunosuppressed) have been treated with ganciclovir, foscarnet, or cidofovir, based on *in vitro* sensitivity data and limited experience in patients.

EPSTEIN-BARR VIRUS INFECTION

Clinical Features

Infectious Mononucleosis (Glandular Fever)

In young children EBV infections are asymptomatic or very mild. However, when infection is delayed until adolescence, as happens in developed countries, the result is often infectious mononucleosis. Following a long incubation period (four to seven weeks), the disease begins insidiously with headache, malaise, and fatigue. The clinical presentation can vary, but three regular features are fever, pharyngitis, and generalized lymphadenopathy, although some patients may present without pharyngitis. The fever is high and fluctuating, and the pharyngitis is characterized by a white or gray malodorous exudate covering the tonsils that may occasionally be so severe as to obstruct respiration. The spleen is often enlarged and liver function tests abnormal. In around 10% of patients a faint, widespread non-itchy maculopapular rash may last for about a week. This is different from the more intense, itchy, maculopapular rash on extensor surfaces and pressure points that commonly appears following misguided treatment with amoxicillin or ampicillin (but not penicillin), and is thought to represent some kind of unexplained hypersensitivity reaction. The clinical disease usually lasts for two to three weeks, but convalescence may be very protracted.

An extraordinary range of further complications can occasionally occur. Neurological complications include Guillain-Barré syndrome, Bell's palsy, meningoencephalitis, and transverse myelitis. Other complications include hemolytic anemia, thrombocytopenia, carditis, nephritis, and pneumonia. Splenic rupture, either spontaneous or following trauma, is a potentially serious complication seen in 0.1% to 0.5% of patients. As may occur after a number of acute systemic infections, a very small minority of patients can progress to develop chronic fatigue syndrome (see Chapter 39: Viral Syndromes). This sometimes disabling entity is diagnosed using a specific panel of distinct clinical criteria and is associated with a range of disturbances in cellular immunity that may represent a final common pathway following a number of different infections.

Infection in Immunocompromised Hosts

A variety of syndromes associated with uncontrolled progression of infection occur in individuals with congenital or an acquired inability to mount an adequate immune response to EBV. For example, a rare fatal polyclonal B cell proliferative syndrome caused by EBV infection occurs in families with an X-linked recessive immunodeficiency associated with a reduced ability to synthesize interferon γ (X-linked lymphoproliferative syndrome). There is an inexorable expansion of virus-infected B cells and concomitant suppression of normal bone marrow cells, and about half of boys die within a month from sepsis or hemorrhage; the remainder develop dysgammaglobulinemia or die from malignant B cell lymphomas.

Much more common is progressive lymphoproliferative disease, seen in immunodeficient children, AIDS patients, or transplant recipients (post-transplantation lymphoproliferative disorders, PTLDs). In these immunocompromised individuals, the absence of cell-mediated immunity permits unrestrained replication of EBV, acquired exogenously or reactivated from the latent state. Some of the infections present as mononucleosis, but others are atypical, presenting, for example, as pneumonitis or hepatitis. Infants with AIDS develop a lymphocytic interstitial pneumonitis; adults with AIDS may develop EBV-associated lymphoproliferative conditions of various kinds.

Burkitt's Lymphoma, Nasopharyngeal Carcinoma, and Other Malignancies

The reader is referred to Chapter 9: Mechanisms of Viral Oncogenesis for a discussion of the highly malignant neoplasms Burkitt's lymphoma and nasopharyngeal carcinoma, and their relationship to EBV. The virus is also associated with a proportion of Hodgkin and non-Hodgkin lymphomas, other NK/T cell lymphomas, lymphoepithelial-like carcinomas, and gastric carcinomas. A cerebral lymphoma is characteristic.

The oncogenesis of many EBV-related tumors is not fully understood. However, it is clear that a breakdown in the immune surveillance of EBV-infected cells contributes to the survival of the virus-bearing tumor cells. Tumor cells carrying EBV genomes usually show a "latency profile" of partial genome expression.

Treatments of EBV-related malignancies include conventional approaches (surgery, chemo-, and radiotherapy) as well as a reduction in immunosuppression and the transfer of EBV-specific T cells from genetically related donors.

Pathogenesis, Pathology, and Immunity

In primary infection, EBV replicates in both locally infiltrating B cells and epithelial cells of the nasopharynx and salivary glands, especially the parotid, and infectious virions are released into the saliva (Fig. 17.8A). Proliferating B cells resembling lymphoblastoid cell line (LCL)-like cells, with latency III form of expression (see below), arise in the tonsillar tissue and are found in the circulation. Latently infected B cells can exist in three different phenotypes (known as latency I, latency II, and latency III, depending on the respective restricted but distinctive pattern of viral gene expression). The genome persists as a circular, self-replicating episome in the B cell nucleus. At least one of the viral gene products, EBNA-2, immortalizes the B cell. Each resulting B cell clone secretes its own characteristic monoclonal antibody. The "heterophile" antibodies that result from this polyclonal B cell activation represent just one manifestation of the immunological chaos that characterizes the acute phase of infectious mononucleosis. By the end of the incubation period (30 to 50 days), up to 25% to 50% of circulating memory B cells are latently infected. There is also a general depression of cell-mediated immunity with an increase in regulatory T cells. Meanwhile, clones of CD8+ cytotoxic T lymphocytes that recognize class I MHC-bound peptides derived from EBV proteins EBNA-2 to EBNA-6 and latent membrane protein (LMP) become activated to proliferate and lyse B cells and oropharyngeal epithelial cells expressing LMP protein on the cell surface. It is these cytotoxic T lymphoblasts, not the infected B cells, that comprise the pathognomonic "atypical lymphocytes" so characteristic of infectious mononucleosis (Fig. 17.8B). They also are the major contributors to the lymphadenopathy. In individuals with congenital or acquired T cell immunodeficiencies, unrestrained virus replication continues, and B cell lymphomas or other lethal conditions may overwhelm the patient.

Most infected B cells are killed by the host immune defense, but approximately 1 in 100,000 persist as small, non-proliferating memory B cells, with either no EBV protein expression or latency I or II. These cells are important as a potential long term reservoir of virus: these cells begin to produce infectious virus when cultivated *in vitro*. After trafficking to adjacent epithelial cells in the oropharynx, the memory B cells can become stimulated to produce virus once more and deliver virus to permissive epithelial cells, from whence virus is shed in saliva to reach a new host. In the long term, a remarkable state of equilibrium is set up between active immune elimination of infected B cells recognized as bearing viral targets, and the selection of persistently infected B cells representing a more benign latent state.

Strains of EBV fall into two classes, based mainly on differences in sequence of the nuclear antigen genes EBNA-2 and EBNA-3. EBV-1 is 10 times more common than EBV-2 in Europe and the United States, whereas EBV-2 is more common in Africa, New Guinea, and among immunocompromised, particularly HIV-infected patients.

Laboratory Diagnosis

The clinical picture of infectious mononucleosis can be so variable that laboratory confirmation is normally required.

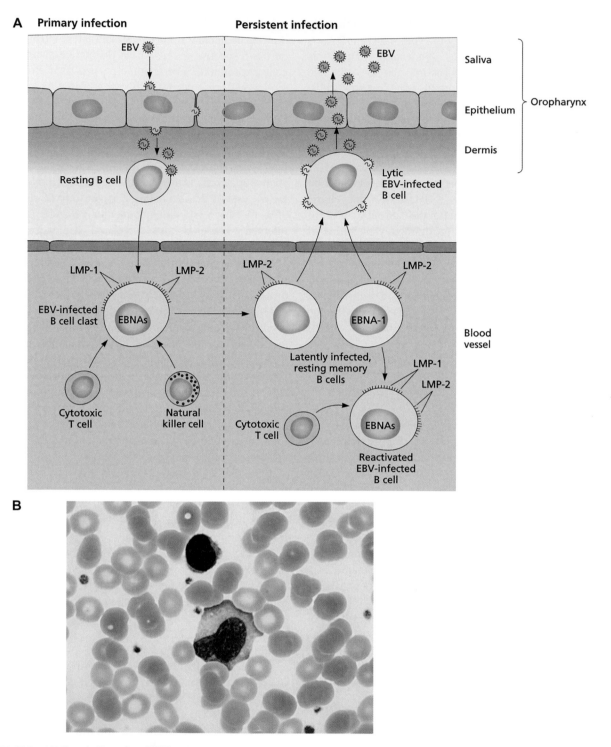

FIGURE 17.8 (A) Epstein-Barr virus (EBV) primary and persistent infection. (Left) Primary infection. Epstein-Barr virus infects epithelial cells in the oropharynx (e.g., the tonsils), and then infects resting B cells in the lymphoid tissue. Virus-infected B cells produce the full complement of latent viral proteins and RNAs (e.g., LMP-1 and LMP-2), and are stimulated to enter mitosis and proliferate. These cells may enter the circulation and produce antibody and function as B cells. During the acute phase, these latently infected B cells are attacked by natural killer cells and CTLs, and the clinical syndrome of infectious mononucleosis is seen. (Right) Persistent infection. Most infected B cells are killed as a result of innate and immune defenses, but a few (approximately 1 in 100,000) persist in the blood as small, non-proliferating memory B cells that synthesize only LMP-2A mRNA. These memory B cells are presumably the long-term reservoir of Epstein-Barr virus *in vivo* and the source of infectious virus when peripheral blood cells are removed and cultured. (B) Blood film of patient with infectious mononucleosis, showing an atypical lymphocyte (reactive T cells) with large pleomorphic lobulated nuclei. A normal lymphocyte is shown above for comparison. *(A): Reproduced from Flint, S.J., et al., Principles of Virology, Vol. II, 3rd edition, p. 157. ASM Press, with permission. (B): Reproduced from Cooke, R.A., Infectious Diseases, McGraw Hill, with permission.*

This rests on (1) differential white blood cell counts, (2) heterophile antibodies, and (3) EBV-specific antibodies.

By the second week of the illness, the white blood cell count is raised 10,000 to 20,000 per cubic millimeter or even higher. Lymphocytes plus monocytes account for 60% to 80% of this number. Of these, at least 10%, and generally more than 25%, are atypical lymphocytes, that is lymphocytes that are larger than normal with large pleomorphic lobulated nuclei and deeply basophilic vacuolated cytoplasm (Fig. 17.8B), that persist for two weeks to several months.

In 1932 John Paul and William Bunnell made the empirical observation that sera from glandular fever patients agglutinate sheep erythrocytes. These agglutinins are just one of the "non-Forssmann" heterophile antibodies elicited by EBV infection: heterophile antibodies are a group of IgM antibodies that react with red blood cell antigens from other species. Heterophile antibodies are often detected by the time of onset of symptoms, reach a peak between two and five weeks later, and can be measured for up to 1 year. The Paul–Bunnell test is still useful. Commercially available kits use red blood cells or sensitized latex particles as an indicator of agglutination, although test sera may require prior absorption to remove non-specific "Forssmann" antibody activity. In some patients, the test may remain negative for one to two weeks following the onset of symptoms, and in some cases it remains negative, therefore virus-specific antibody tests also play a key confirmatory role.

IgM antibodies against the EBV capsid antigen VCA develop to high titer early in the illness and then decline rapidly over the next three months or later; this therefore represents a good index of primary infection. VCA or other cloned EBV antigens can also be used to screen for IgG antibodies. Because antibodies to EBNA first appear a month or more after onset of disease and then persist, a rising titer is diagnostic. It should be noted, however, that antibody to EBNA-1 is specifically missing from many immunocompromised patients with severe chronic active infection. On the other hand, antibodies against the early antigen EA-D are diagnostic of acute, reactivated, or chronic active infection, as they decline rapidly and are not detected in asymptomatic carriers. IgG (or total) antibodies against VCA represent the most convenient measure of past infection and the immune status of the individual.

Isolation of virus is rarely used as a diagnostic procedure because no known cell line is fully permissive for EBV. The only method for isolating EBV *in vitro* is to inoculate infected oropharyngeal secretions or peripheral blood leukocytes onto umbilical cord lymphocytes and to demonstrate the immortalization of the latter into a lymphoblastoid cell line that can be stained successfully with fluoresceinated monoclonal antibody against EBNA. The other problem with virus isolation for routine diagnosis of EBV-induced

disease is that a high proportion of asymptomatic individuals carry the latent genome and/or shed virus for life.

Epidemiology, Prevention, Control, and Therapeutics

Following an attack of mononucleosis, virus is found in saliva for several months, and more sensitive assays reveal that most EBV-seropositive people secrete the virus at a lower titer thereafter, continuously or intermittently, probably for life. Chronic shedding is highest in young children, in early pregnancy, and in immunocompromised patients. In most developing countries EBV infection is ubiquitous. Almost all infants become infected in the first year or two of life, probably by salivary exchange, contamination of eating utensils, and other oropharyngeal contact. At this age almost all infections are subclinical; glandular fever is virtually unknown in developing countries. By way of contrast, in countries with higher standards of living many persons reach adolescence before first encountering the virus. When primary infection occurs in adolescence, there is a much higher likelihood of clinical disease developing. The intimate salivary contact involved in kissing is the principal means of transmission. Mononucleosis is a disease of 15- to 25-year-olds. More than 90% of people eventually acquire infection and are permanently immune (see Fig. 13.5).

As with CMV, conventional respiratory transmission of EBV by droplets does not appear to be significant. For example, casual roommates of mononucleosis patients are not at increased risk and much closer contact seems to be required. By analogy with CMV it may be reasonable to speculate that EBV can also be transmitted by sexual intercourse. In this context male homosexuals have a very high rate of seropositivity. Blood transfusions can also rarely transmit the virus.

Therapeutics

Although acyclovir inhibits EBV replication by inhibiting the DNA polymerase, oral or intravenous administration of acyclovir has not shown clinical benefit in acute infections, presumably because the symptomatic phase results from the immunopathology in response to the EBV-transformed B lymphocytes. Available antiviral drugs have not shown any benefit against chronic infection or EBV-related malignancies.

Vaccines

It is now clear that cytotoxic T cells play a major role in controlling both lytic and latent EBV infection, with immunodominant T cell responses demonstrable against both lytic and latent stage viral antigens. Possible vaccine strategies include a prophylactic vaccine to prevent primary infection, thereby reducing both glandular fever in developed countries and EBV-related malignancies, or a therapeutic vaccine to treat malignancies associated with latent infection. Approaches under current development include subunit and recombinant vaccines based on the major membrane glycoprotein complex gp350/gp220 to generate neutralizing antibodies; peptide epitope vaccines, either in adjuvant or loaded onto dendritic cells; and a poxvirus vector coding for antigenic viral fusion proteins. In a phase II human trial, a gp350 subunit vaccine reduced the incidence of clinical infectious mononucleosis but did not affect the rate of EBV infection. For therapeutic vaccines, the antigens selected include viral proteins LMP-1 and -2 and EBNA-1 as these antigens are displayed on EBV-associated nasopharyngeal and Hodgkin's lymphoma cells and thus offer targets for immune attack.

HUMAN HERPESVIRUS 8 (HHV-8: KAPOSI'S SARCOMA-ASSOCIATED HERPESVIRUS) INFECTION

Classical Kaposi's sarcoma is an uncommon multicentric tumor of endothelium occurring typically in older Mediterranean men. A more aggressive form of Kaposi's sarcoma in younger men emerged in nearly 40% of persons infected with HIV in the mid-1980s and became recognized as a marker for AIDS. There was speculation that, in addition to the HIV-related loss of tumor immune surveillance, pathogenesis of Kaposi's sarcoma might include a role for the enhanced acquisition of an exogenous oncogenic pathogen. The eventual discovery of a new herpesvirus HHV-8 resolved this debate. An endemic type of Kaposi's sarcoma in sub-Saharan African populations, and an iatrogenic type in immunosuppressed individuals, are also recognized. All four syndromes require HHV-8 infection.

HHV-8 is classified within the subfamily *Gammaherpesvirinae*, and while having some similarities to EBV it is considered as a separate species within the genus *Rhadinovirus*. A number of different subtypes and clades have been described, all of which tend to cluster by geographical region, and virus isolated from a single individual does not vary over time. HHV-8 can produce a latent infection with limited expression of genes from the viral episome, or a lytic infection with the ordered cascade of gene expression typical of herpesviruses. In infected patients, HHV-8 DNA has been found in circulating B lymphocytes as well as in prostatic epithelial cells and salivary epithelial cells, the latter possibly contributing to the shedding of virus in saliva and semen. It expresses many immune-evasion genes unlike EBV which gains the same effect by manipulating cellular genes. In Kaposi's sarcoma

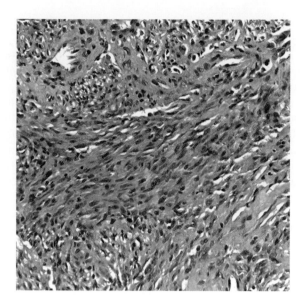

FIGURE 17.9 H&E stained section of Kaposi sarcoma, showing malignant spindle cells arranged in whorls forming vascular channels.

tumors, the majority of Kaposi's sarcoma-associated spindle cells (a tumor cell that expresses markers of both endothelial cells and macrophages) contain HHV-8 DNA, as a latent or non-productive infection in most cells (Fig. 17.9).

Serological surveys of blood donors have demonstrated that the prevalence of HHV-8 varies widely, ranging from >70% in parts of Africa and Mediterranean countries to 3% or less in North America and many other regions of the world. HHV-8 is frequently detected in the saliva of seropositive individuals, and transmission is thought to be by saliva within families and by sexual intercourse. Clinically, primary infection with HHV-8 has been linked to a febrile illness with a maculopapular rash in children, or with lymphadenopathy in homosexual men and in organ transplant recipients. A lifelong latent infection may be complicated by the late development of HHV-8-associated angioproliferative and lymphoproliferative disorders.

Kaposi's sarcoma presents as a group of scattered flat violaceous plaques or nodules of characteristic appearance. These are rarely painful or life-threatening except when there is extensive involvement of the lungs or gastrointestinal tract. Treatments used include surgery, local irradiation, liposomal anthracyclines, cytotoxic agents, and interferon. Ganciclovir treatment prior to the onset of Kaposi's sarcoma may reduce its incidence, and adequate anti-HIV chemotherapy can reduce the subsequent development of disease by up to 90%. HHV-8 is also associated with the late onset of two lymphoproliferative disorders. Multicentric Castleman's disease (MCD) involves lymphadenopathy, splenomegaly, recurrent fevers, and autoimmune phenomena, with different forms of histology in biopsies taken from infected lymph nodes. It frequently progresses to Kaposi's sarcoma or to lymphoma. Primary effusion lymphoma and related lymphomas arise usually in HIV-positive patients. Abnormal tumor cells of B cell origin are found in body cavity effusions; these tumor cells invariably carry HHV-8 genetic material and sometimes also EBV genomes. The disease is associated with a poor prognosis.

B VIRUS (MACACINE HERPESVIRUS 1) INFECTION

Macaques are often persistently infected with an alphaherpesvirus known variously as B virus, herpesvirus simiae, or cercopithecine herpesvirus 1. Its natural history is very like that of HSV-1 infection in humans. A number of fatal human cases of ascending paralysis and encephalitis in animal handlers have followed the bite of macaques, and B virus is a continuing risk to personnel working in zoos and laboratories. The level of risk following a macaque bite can be assessed based on the nature of the wound, the thoroughness and timeliness of the wound cleansing procedure, and the condition of the animal. A treatment regimen using valaciclovir, acyclovir, or ganciclovir should be adopted whenever there is any risk of infection. Because of the rarity and severity of this infection, in the United States a special reference diagnostic laboratory has been established and consensus guidelines for treatment have been prepared by a Centers for Disease Control and Prevention Working Group.

FURTHER READING

Ambinder, R.F., Cesarman, E., 2007. In: Arvin, A., Campadelli-Fiume, G., Mocarski, E. (Eds.). Clinical and pathological aspects of EBV and KSHV infection. In. Human Herpesviruses: Biology, Therapy, and Immunoprophylaxis, Cambridge University Press, Cambridge.

Cohen, J.I., Mocarski, E.S., Raab-Traub, N., Corey, L., Nabel, G.J., 2013. The need and challenges for development of an Epstein-Barr virus vaccine. Vaccine 31 (Suppl. 2), B194–6. Available from: http://dx.doi.org/10.1016/j.vaccine.2012.09.041.

Cohen, J.I., Davenport, D.S., Stewart, J.A., Deitchman, S., Hilliard, J.K., Chapman, L.E., B virus Working Group, 2002. Recommendations for prevention of and therapy for exposure to B virus (Cercopithecine herpesvirus 1). Clin. Infect. Dis. 35, 1191–1203. Epub 2002 Oct 17.

Griffiths, P., Plotkin, S., Mocarski, E., Pass, R., Schliess, M., Krause, P., Bialik, S., 2013. Desirability and feasibility of a vaccine against cytomegalovirus. Vaccine 31S, B197–B203.

Kimberlin, D.W., 2004. Neonatal herpes simplex infection. Clin. Microbiol. Rev. 17, 1–13.

Schmid, D.S., Jumaan, A.O., 2010. Impact of varicella vaccine on varicella-zoster virus dynamics. Clin. Microbiol. Rev. 23, 202–217.

Walker, S.P., Palma-Dias, R., Wood, E.M., Shekleton, P., Giles, M.L., 2013. Cytomegalovirus in pregnancy: To screen or not to screen. BMC Pregnancy and Childbirth 13, 96.

Ward, K.N., 2005. The natural history and laboratory diagnosis of human herpesviruses-6 and -7 infections in the immunocompetent. J. Clin. Virol. 32, 183–193.

Chapter 18

Adenoviruses

In 1953 Wallace Rowe and his colleagues observed that certain explant cultures of human adenoids degenerated spontaneously. On prolonged culture, they isolated a new virus they referred to as *adenovirus*. It quickly became evident not only that many adenoviruses persist for many years as latent infections of lymphoid tissues, but also many are a significant cause of respiratory and ocular disease. Later studies also implicated related viruses as the cause of genitourinary tract infections, and still later adenoviruses were found associated with gastroenteritis and with infections of immunocompromised patients. Following the discovery that certain human adenoviruses produce malignant tumors in neonatal rodents, molecular biologists turned their attention to the biochemistry and molecular biology of adenovirus replication and the events leading to oncogenesis. Although adenoviruses were eventually shown to play no role in human cancer, the spin-off from this research has had a major impact on our understanding of the expression of mammalian as well as viral genes. One landmark contribution was the discovery of RNA transcript splicing, for which Philip Sharp and Richard Roberts were awarded the Nobel Prize in Physiology or Medicine in 1993.

Adenoviruses cause 5% to 10% of all febrile illnesses in infants and young children. Most individuals have serological evidence of prior adenoviral infection by the age of 10. Adenovirus infections are especially prevalent in daycare centers and in households with young children. Many epidemics of adenoviral disease have been described, including pharyngoconjunctival fever in summer camps and public swimming pools, keratoconjunctivitis in medical facilities, and serious acute respiratory disease in military recruits.

Adenoviruses have been used as vectors for recombinant-DNA vaccines and for gene therapy due to the very high yield of adenoviruses in cell cultures coupled with a lack of oncogenic properties in humans.

CLASSIFICATION

The family *Adenoviridae* consists of five genera. Human adenoviruses together with other adenoviruses infecting mammals belong to the genus *Mastadenovirus*, whereas adenoviruses of birds, sheep, cattle, frogs, and fish are grouped into the separate genera *Aviadenovirus*, *Atadenovirus*, *Siadenovirus*, and *Ichtadenovirus*.

The classification of human adenoviruses is in transition, switching from a system based upon classical serological methods to genome sequencing methods. Serotypes had been distinguished by hemagglutination and neutralization assays which involve antigens on virion fibers and hexons. Human adenoviruses 1 to 52 were classified in this way. The discovery and classification of human adenoviruses types 52 to 68 have been based on genome sequencing and bioinformatic analysis. Biological properties, lack of cross-neutralization, sequence relatedness, pathology, and other properties have been used to group these human viruses into seven species (human adenoviruses A to G), each of which is associated with a distinct disease-association profile, including a varying capacity to induce tumors in experimental animals (Table 18.1). The species human adenovirus-D contains the most members, including a substantial number identified during the first two decades of the AIDS epidemic. Adenovirus epidemiology keeps evolving: in recent years, one virus, human adenovirus 14, has been associated with severe, even fatal outbreaks of pneumonia in residential facilities and

TABLE 18.1 Classification of Human Adenoviruses

Subgroup	Serotypes	Tropism	Production of Tumors in Animals
A	12, 18, 31	Intestinal	High
B1	3, 7, 14, 16, 21, 50	Respiratory	Moderate
B2	11, 14, 34, 35	Renal	Moderate
C	1, 2, 5, 6	Respiratory	Low or none
D	8 to 10, 13, 15, 17, 19, 20, 22 to 30, 32, 33, 36 to 39, 42 to 49, 51, 53, 54	Ocular and other	Low or none
E	4	Respiratory	Low or none
F	40, 41	Intestinal	Low or none

Fenner and White's Medical Virology. DOI: http://dx.doi.org/10.1016/B978-0-12-375156-0.00018-7

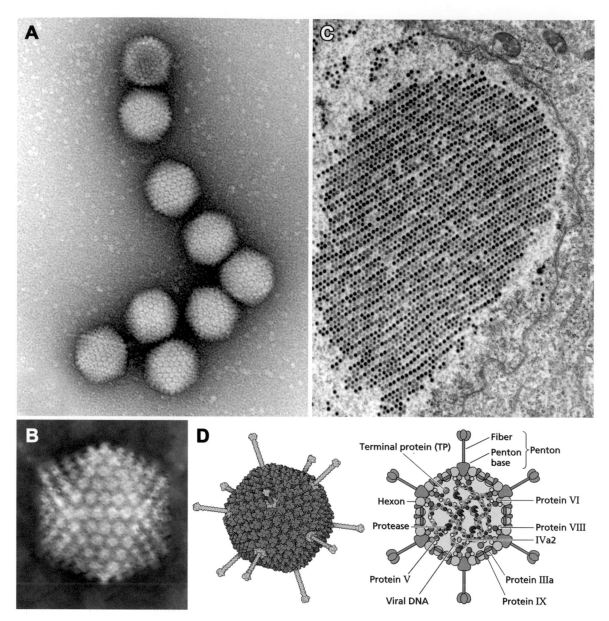

FIGURE 18.1 (A) Negative contrast electron microscopy of human adenovirus 5—clinical respiratory diagnostic specimen. (B) Single adenovirus virion showing its surface architecture. (C) Thin-section electron microscopy of human adenovirus 5 in human embryonic kidney cell culture, showing a typical paracrystalline array of virions in the nucleus of an infected cell. (D, left) Model of an adenovirus virion showing the virion surface composed of hexons and pentons—penton fibers are located at each vertex of the icosahedral virion. (D, right) Diagrammatic section of a virion showing the location of the major proteins. The structure of the nucleoprotein core has not been established so the location of core proteins is hypothetical. *(D, right) Reproduced from Flint S.J., et al., 2009. Principles of Virology: Pathogenesis and Control, third ed. ASM Press, Washington, DC, with permission.*

military bases. Homologous recombination in nature appears to play a major role in generating genome diversity.

PROPERTIES OF ADENOVIRUSES

Adenovirus virions are non-enveloped, with a 70 to 90 nm diameter capsid that is a perfect icosahedron. Structural studies of adenoviruses have focused on adenovirus serotypes 2 and 5. The capsid is composed of 252 capsomers: 240

hexons constitute the 20 equilateral triangular facets, while 12 pentons are located over the 12 vertices. A fiber protrudes from each penton to give adenoviruses one of the most distinctive morphological appearances among all viruses (Fig. 18.1). The distal ends of the fibers bear ligands for cellular receptors and determine the host species specificity of the virus. The genome, which is associated with an inner protein core, consists of a single linear molecule of dsDNA, 26 to 48 kbp in size, with inverted terminal repeats. The

DNA, in association with a 55 K protein which is covalently linked to each 5′ terminus, is infectious when transfected into susceptible cells.

There are about 11 viral structural proteins, among which protein II is the hexon, protein III the penton, protein IV the fiber, and a 23 kDa cysteine protease is responsible for processing other viral precursor proteins. Fibers of the majority of human adenoviruses comprise only one protein, but human adenoviruses 40 and 41 have two distinct fibers, of different lengths and primary sequence, present in the virion in equimolar amounts; this may extend the host ranges of these serotypes. The other minor proteins have been shown either to stabilize the hexons under different chemical environments, to participate in disrupting the endosomal membrane in viral invasion, or to assist in assembly of the virion. Polypeptide VII is the major core protein of the virus, while the minor core protein V functions to attach the core to the capsid.

Due to possessing a dsDNA genome, compact icosahedral capsid and the lack of a lipid envelope, adenoviruses are relatively resistant to the environment, consistent with recorded transmission in swimming pools. The virus is stable between pH 6.0 and 9.5, which aids the transmission via the fecal–oral route. Adenoviruses in simulated conjunctival samples can be shipped at ambient temperatures without loss of titer. Adenoviruses are resistant to chloroform, ether, and fluorocarbons. Recent studies recommend disinfecting ophthalmology equipment with 70% ethyl alcohol or approximately 5000 parts per million chlorine. When adenoviruses are heated to 56°C, the virus disintegrates, and the core is released (Table 18.2).

TABLE 18.2 Properties of Adenoviruses

Five genera; *Mastadenovirus* includes 68 human serotypes grouped into 7 genera designated A to G, and serotypes that infect other mammals. Genus and type-specific classification is mainly based on serological assays, supplemented by phylogenetic analysis

Non-enveloped virion 70 to 90 nm in diameter, hexagonal outline, icosahedral symmetry, with 240 hexons, 12 pentons with 12 fibers that mediate attachment to receptors

Contain a linear double-stranded DNA genome, 26 to 48 kb, with inverted terminal repeats and a protein primer at each 5′ end

Transcription, DNA replication, and assembly occur in the cell nucleus

Complex sequential program of early, intermediate, and late mRNA transcription (before and after DNA replication); extensive splicing of RNA transcripts

Some viruses are oncogenic in laboratory animals, but none have been associated with human cancer

---

VIRUS REPLICATION

Most adenoviruses are species-specific and generally will undergo a complete replication cycle only in cells derived from the native host species. In common with many DNA viruses, adenoviruses replicate within the cell nucleus where nascent virions are also assembled.

The virion enters cells by attachment of fiber "knobs" to a primary docking protein, for example, to the Coxsackie adenovirus receptor (CAR), or to CD46 or CD80, after which viral binding and entry are facilitated by the interaction between the RGD motif in the penton base and different cell surface integrins. Recent studies have shown that human adenovirus 5 (HAdV5) can attach to dendritic cells via a bridging mechanism involving the binding of lactoferrin to the C-type lectin receptor DC-SIGN. This may not only affect the host range of human adenovirus 5 in cells, but also has implications for using adenoviruses in gene delivery. Interestingly, human adenovirus 41, a cause of gastroenteritis, has two fibers, with only one attaching to CAR: the second is thought to attach to enterocytes. After attachment, internalization occurs through the interaction of the penton base and cellular integrins via clathrin-coated vesicles. The complex is then transferred into endosomes, where the viral capsid is partially degraded by a virus-coded protease, and viral DNA transported into the nucleus. Within the nucleus, viral DNA becomes attached to the nucleus matrix via its terminal protein, followed by the well-regulated transcription of early and late mRNAs.

Transcription of adenovirus mRNAs follows a sequential pattern, divisible into early and late stages, closely paralleling a similar pattern of expression among the well-characterized DNA bacteriophages. The viral genome contains five early transcription units (E1A, E1B, E2, E3, and E4), three delayed early transcription units (IX, IVa2, and E2 late), and one late transcription unit (major late), the latter processed to generate five families of late mRNAs (L1 to L5). All mRNAs are transcribed by cellular RNA polymerase II. Each of the early units has its own separate promoter, while the major late unit uses just a single promoter. Each of these units gives rise to multiple mRNAs that are created by alternative splicing, thereby making use of the limited genome sequences to translate a large number of proteins. Multiple splicing of the early transcripts leads to the production of around 30 mRNAs, coding for early proteins that are mainly involved in viral replication.

The E1A region of the viral genome encodes proteins that are essential for three main outcomes of early adenovirus transcription: (1) induction of cell cycle progression to provide an optimal environment for DNA synthesis and viral replication, (2) protection of infected cells from host antiviral immune defenses, such as from tumor necrosis factor (TNF) activity or from apoptosis, and (3) synthesis of viral proteins necessary for viral DNA replication. E1A and E1B gene products are also responsible for cell transformation and hence for the animal oncogenicity of some adenoviruses. Both proteins interact with the cellular tumor suppressor gene p53 to compromise its normal cellular function and thus allow cell cycle progression. Proteins translated from E2 are directly involved in viral DNA replication, and include a DNA polymerase, an ssDNA binding protein, and a precursor to the terminal protein. The E3 region is not essential for adenovirus replication in cell cultures and can be deleted or replaced without disrupting viral replication *in vitro*. It is therefore favored as an insertion site for foreign DNA when constructing adenovirus vectors. E3 proteins are known to interact with host immune defense mechanisms, thus modulating the host response to adenovirus infection. Inhibition of class I major histocompatibility antigen expression by infected cells and inhibition of tumor necrosis factor are two examples of immune evasion mediated by E3 encoded proteins. Studies of E4 mutants have shown that adenovirus inhibits the cellular DNA damage response.

Multiple splicing of the late genes gives rise to at least 18 mRNAs that code for the L1 to L5 families of late proteins which are involved in the assembly of progeny virions. The phenomenon of RNA splicing was discovered owing to the very high abundance of late mRNAs processed during adenovirus transcription. In addition to the above, adenoviruses contain two VA (viral associated) genes (VAI and VAII), which are transcribed by host RNA polymerase III to produce short VA RNA molecules that mimic the function of dsRNA due to the presence of two stem loop structures. These VA RNAs are not translated but are essential for lytic virus replication, and can inhibit both the host interferon system and host RNAi (Figs. 18.2 and 18.3).

Viral DNA synthesis is initiated at the end of inverted terminal repeats at each end of the genome, using the 5′-linked virus-coded precursor protein primer to supply the priming function normally provided by the 3′OH of an upstream DNA strand. DNA synthesis then proceeds from both ends of the genome by asymmetric strand synthesis that involves copying one strand and displacing the other. The elongation of viral DNA requires the participation of adenovirus DNA polymerase, ssDNA binding protein and a cellular topoisomerase. The displaced single-stranded DNA molecule forms a panhandle-like structure by annealing of its terminal repeat sequences, which then act as the site of origin for replication of the complementary DNA strand, finally giving rise to a full-length double-stranded progeny molecule.

Following DNA replication, late mRNAs are transcribed and translated into structural proteins. The shutdown of host cell macromolecular synthesis occurs progressively during the later stages of the replication cycle. The assembly of progeny virus occurs within the nucleus. During assembly, the viral protease cleaves at least four viral products in order to make the released virion more stable and thus enhance infectivity. Many thousands of assembled virions can be observed in the infected cell nucleus, arranged into a paracrystalline array. The release of progeny virions is aided by an increase in the permeability of the nuclear membrane mediated by a viral non-structural protein. The release of progeny virions is dependent upon cell lysis. In cell cultures, many adenoviruses induce severe condensation and margination of the host cell chromatin, making the nuclei appear abnormal, and appearing as the inclusion bodies so characteristically observed in adenovirus-infected cells.

Abortive adenovirus infection can occur with the expression of early genes, when the combined actions of E1A and E1B can induce cell transformation or tumor formation in animal models. There are also situations where adenoviruses persist *in vivo* and will remain as life-long infections that may lead to life-threatening disease in an immunocompromised host.

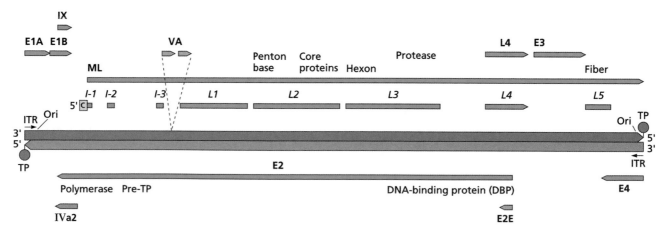

FIGURE 18.2 Adenovirus genome organization (human adenovirus 2). The genome of adenoviruses consists of a single linear molecule of double-stranded DNA, 26 to 48 kbp in size, with inverted terminal repeats. The genome encodes approximately 40 proteins that are transcribed by cellular RNA polymerase II according to a complex program involving both DNA strands and complex RNA splicing. Different virion proteins are transcribed in different directions (arrows). There are early and late transcriptional units, each under the control of different promoters. Viral DNA replication proceeds from both ends by a strand-displacement mechanism. *ITR*, inverted terminal repeat; *TP*, terminal protein; *ML*, major late tripartite leader. *Reproduced from Flint, S.J., et al., 2009. Principles of Virology: Pathogenesis and Control, third ed. ASM Press, Washington, DC, with permission.*

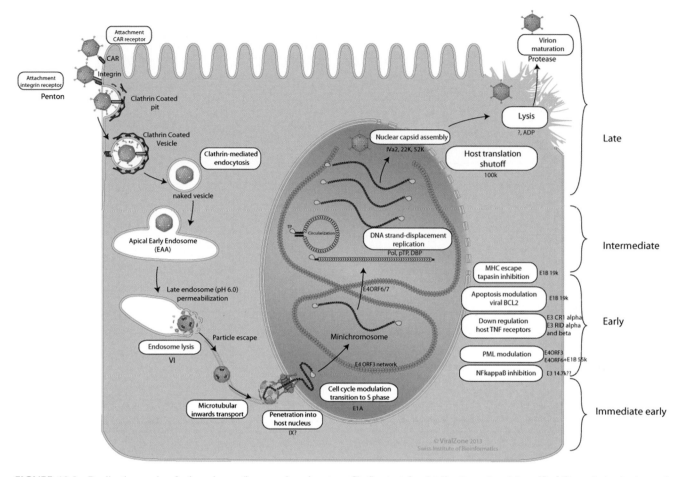

FIGURE 18.3 Replication cycle of adenoviruses (human adenovirus type C). See text for details. *Reproduced from Viral Zone, Swiss Institute of Bioinformatics, with permission.*

ADENOVIRUS DISEASES

Clinical Features

Only about half of the known human adenoviruses have been causally linked to disease (Table 18.3). Adenoviruses 1 to 8 are much the commonest species worldwide and are responsible for most cases of adenovirus-induced disease. Some 5–10% of acute respiratory illnesses in children under the age of five years (but less than 1% of those in adults) have been ascribed to adenoviruses. Enteric human adenoviruses, human adenoviruses 40 and 41, have been claimed to cause up to 10% of infantile gastroenteritis. Several common adenoviruses, including adenoviruses 8, 19, and 37 are major causes of eye infections and also cause genital infections.

Respiratory Infections

Acute respiratory infections are seen particularly in young children and are clinically similar to infections caused by other respiratory viruses. The infant presents with a cough, nasal congestion, and fever; the throat is inflamed

and there is often an exudative tonsillitis that resembles group A streptococcal infection. Adenoviruses 1 to 7 are usually responsible for these common sporadic infections that are relatively trivial except when otitis media or pneumonia supervene. Acute respiratory disease occurs in epidemic form when military recruits assemble in camps. Adenoviruses 4 and 7 are most often responsible.

Pneumonia, often severe and occasionally fatal, may develop in young children infected with any of the common adenoviruses, but particularly human adenoviruses 7 and 3. In some of the colder parts of the world, such as northern China and Canada, adenoviruses are an important cause of pneumonia in infants under the age of two. Adenovirus pneumonia may be associated with disseminated infection involving the heart, liver, kidney, pancreas and CNS with a fatality rate of 10% to 30%, and many of the survivors show permanent lung damage. Acute respiratory disease in military recruits also occasionally progresses to a pneumonitis. Severe infections, including pneumonia, seen in immunosuppressed patients are of increasing importance in this era of organ transplantation and AIDS.

TABLE 18.3 Diseases Caused by Human Adenoviruses

Disease	Age	Common Serotypes[a]	Major Subgenus	Major Source
Respiratory Infections				
Pharyngitis	Young children	**1, 2**, 3, 5, 6, 7	B, C	Throat
Acute respiratory disease	Military recruits	3, **4, 7**, 14, 21	B, E	Throat
Pneumonia	Young children	1, 2, **3**, 4, 5, 7, 21	B, C	Throat
	Military recruits	**4, 7**	B, E	Throat
Ocular Infections				
Pharyngoconjunctival fever	Children	1, 2, **3, 4**, 6, **7**	B, C, E	Throat, eye
Epidemic keratoconjunctivitis	Any age	**8**, 19, **37**	D	Eye
Genitourinary Infections				
Cervicitis, urethritis	Adults	19, **37**	D	Genital secretions
Hemorrhagic cystitis	Young children	**11**, 21	B	Urine
Enteric Infections				
Gastroenteritis	Young children	31, **40, 41**	A, F	Feces
Infections in Immunocompromised Individuals				
Encephalitis, pneumonia	Any age, including AIDS patients	7, **11, 34, 35**	B	Urine, lung
Gastroenteritis	AIDS patients	Many D including 43 to 47	D	Feces
Generalized	AIDS patients	2, 5	C	Blood

[a]Only the commonly occurring serotypes are listed; those most commonly associated with particular syndromes are in bold type.

Ocular Infections

Pharyngoconjunctival fever tends to occur in outbreaks, for example, at childrens' summer camps ("swimming pool conjunctivitis"), and is associated with types 3 and 7. Adenovirus 4 has caused a number of nosocomial outbreaks of conjunctivitis or pharyngoconjunctival fever among hospital staff.

Epidemic keratoconjunctivitis is a more severe eye infection, commencing as a follicular conjunctivitis and progressing to involve the cornea (keratitis). The disease once known as "shipyard eye," is highly contagious and often occurs as epidemics caused by adenoviruses 8, 19, and 37. Adenovirus 8 has been the principal cause, but in 1976 adenovirus 37 suddenly appeared, spread worldwide, and today is the predominant cause of epidemic keratoconjunctivitis.

Genitourinary Infections

Cervicitis and urethritis are common manifestations of venereal infection with adenovirus 37, first identified in prostitutes. Hemorrhagic cystitis, seen mainly in young boys, is caused by adenovirus 11 and more rarely adenovirus 21. Adenoviruses commonly establish asymptomatic persistent infection of the kidney and may be shed in the urine for months or years. This is observed particularly in immunocompromised individuals, such as renal transplant recipients.

Enteric Infections

Gastroenteritis in infants is commonly caused by adenoviruses 40 and 41. These enteric adenoviruses, previously visualized by electron microscopy in feces and long regarded as difficult to isolate, can now be grown in cultured cells. The recovery of these viruses from outbreaks of gastroenteritis, and recovery significantly more frequently from symptomatic patients than from controls, confirms that human adenoviruses do indeed cause gastrointestinal disease. However, many other adenoviruses that replicate in the intestine or in the throat are excreted asymptomatically in the feces for weeks or months, hence carefully controlled studies are required before assigning these viruses an etiological role in gastroenteritis.

Adenoviruses have also been suspected or suggested to be involved in other diseases such as myocarditis, cardiomyopathy, meningoencephalitis, and hepatitis, because of the recovery of virus (or detection by PCR) from series of patients with these diseases.

Infections in Immunocompromised Patients

Adenoviruses have emerged as an important cause of life-threatening infections in at least three groups of immunocompromised patients. In children with severe combined immune deficiency disease common adenoviruses can cause serious conditions such as pneumonia or meningoencephalitis. Transplant recipients, particularly children following stem cell transplantation, and patients with AIDS, often shed subgroup B adenoviruses (11, 34, and 35) in their urine for prolonged periods, and may develop high fever, pneumonia, hemorrhagic cystitis, encephalitis, hepatitis, or nephritis. AIDS patients also may excrete adenoviruses 43 to 47 in their feces.

PATHOGENESIS, PATHOLOGY, AND IMMUNITY

Adenoviruses are widespread in many animal species and can be readily isolated from healthy individuals, persisting life-long in lymphoid tissues. Adenoviruses multiply initially in the pharynx, conjunctiva, or small intestine, and generally do not spread beyond the draining cervical, pre-auricular, or mesenteric lymph nodes. Usually, the disease process remains relatively localized, and the incubation period is short (five to eight days). Most enteric infections and some respiratory infections are subclinical. Generalized infections are occasionally seen, especially in immunocompromised patients, and also in those undergoing transplantation. Deaths do occur, particularly from adenovirus 7, the most virulent human adenovirus. At autopsy, lungs, brain, kidney, liver, and other organs reveal the characteristic basophilic nuclear inclusions referred to above. One possible molecular mechanism that may enhance dissemination of virus is the release of excess viral penton fiber protein from infected cells at cell lysis. The fiber protein binds the epithelial junction protein desmoglein 2 (DSG2), thereby triggering intracellular signaling and transient opening of the junctions between epithelial cells. This may facilitate lateral spread and dissemination of virus in epithelial tissues.

Infection with the common endemic types 1, 2, and 5 can persist asymptomatically for years in the tonsils and adenoids of children, with virus continuously shed in the feces for many months after the initial infection, then intermittently for years thereafter. The mechanism of this persistence is uncertain; perhaps viral replication is held in check by the antibody synthesized by these lymphoid organs. Fluctuation in shedding indicates that latent adenovirus infections can be reactivated. For example, this can occur during infection with *Bordetella pertussis*, and measles can be followed by adenovirus pneumonia. Most adenovirus infections are localized in the eyes and pharynx, but in some cases there is contiguous extension into the lungs. Although most adenoviruses replicate harmlessly in the intestine, human adenoviruses 40 and 41 can cause gastroenteritis. Some adenoviruses are frequently shed in the urine of immunocompromised persons; adenoviruses 34 and 35 were originally isolated from renal transplant recipients and can be recovered commonly from AIDS patients. As there is no evidence of ascending infections, urinary bladder infection suggests that the virus probably is viremic at some stage in order to reach this organ. In addition, a wide variety of other adenoviruses have been recovered from the feces of AIDS patients. Many of these viruses are new or rare and some appear to be genetic recombinants. Presumably these novel viruses and "intermediate" strains arose in this cohort as a result of mixed infections and the greater levels of replication resulting from immunosuppression. Infections caused by other adenoviruses are characterized by prolonged latency in lymphoid tissue, but can also be reactivated in AIDS patients and frequently recovered from the blood.

Outbreaks of adenovirus infections can be caused by new adenovirus variants with a high virulence for patients without an immunodeficiency. For example, community based outbreaks of severe respiratory disease occurred at three different locations in USA in 2006–2007, caused by a human adenovirus 14 variant. Of 140 patients with acute Ad14 respiratory disease, 38% required hospitalization, 17% needed admission to an intensive care unit, and there were 9 deaths. Therefore the continuous monitoring of adenovirus infections is advised together with genetic analysis of viral strains over time in order to forecast the risk of an outbreak in the community.

In contrast to most other respiratory viral infections, adenovirus infections lead to lasting immunity to reinfection with the same serotype, perhaps because of the extent of involvement with lymphoid cells in the alimentary tract and the regional lymph nodes. Maternal antibody generally protects infants under the age of 6 months against severe lower respiratory disease. Relatively little is known of cell-mediated immune responses against adenoviruses in humans. CD4+ and CD8+ T cell epitopes exist in the conserved regions of the hexon protein and these epitopes are cross-reactive among some adenovirus species. However it is not known whether or not these T cell epitopes are protective against adenovirus. The virus can also modulate host T cell responses by expression of several early gene products. For example, one small protein encoded by a gene within the E3 transcription unit binds to the heavy chain of the class I MHC antigen, and prevents transport of class I MHC to the cell surface, thereby decreasing the presentation of adenovirus peptides to cytotoxic T lymphocytes. Another E3 gene-product protects infected cells against lysis by tumor necrosis factor, whereas other products stimulate clearance and degradation of the receptors for Fas ligand, TRAIL and epidermal growth factor from the cell surface, thereby interfering with intracellular signaling by these ligands. Due to the increasing use of adenoviruses for gene delivery, the nature of immune responses against adenoviruses is of particular interest in assessing the efficacy of heterologous gene expression from adenovirus vectors. For example, the longstanding solid immunity following infection is a problem in the use of certain serotypes as vectors in immune individuals.

LABORATORY DIAGNOSIS

Depending on clinical presentation, appropriate specimens include feces, pharyngeal swabs, nasopharyngeal aspirates, transtracheal aspirates, or bronchial lavage. Eye infections can be diagnosed by the taking of conjunctival swabs, corneal scraping, or tears. Other syndromes may involve sampling genital secretions, urine, biopsy (e.g., of liver or spleen) or autopsy (e.g., lung or brain) samples.

Enzyme immunoassay (EIA) is the method of choice for the detection of soluble viral antigen in feces or nasopharyngeal secretions. A monoclonal antibody to a hexon epitope common to all adenoviruses or polyclonal serum suffices to identify the family, then if desired, a type-specific monoclonal antibody can be used to identify the particular adenovirus concerned. Rapid point-of-care immunochromatographic tests have also shown good sensitivity and specificity. Immunofluorescence can also be employed to demonstrate adenoviral antigen in cells from the respiratory tract, eye, urine, or biopsy or autopsy material, after low-speed centrifugation of the specimen followed by fixation of the pelleted cells; however it tends to have lower sensitivity than EIA, particularly among adults.

The most sensitive method for virus detection is polymerase chain reaction (PCR) using genus-specific primer sets. This can be incorporated into a multiplex assay to detect a panel of common respiratory pathogens, and is also useful to quantitate the adenovirus DNA load, as a high load is more often associated with active disease. PCR can also be used for environmental detection of viruses, for example, in wastewater, surface water, and combined sewer overflows.

Virus isolation is still an approach used by some diagnostic and reference laboratories. Cell culture of adenoviruses is time-consuming because many of the viruses are very slow-growing. Human malignant cell lines such as HeLa, HEp-2, KB, or A-549, or diploid human embryonic fibroblasts are the substrates of choice. The fastidious enteric adenoviruses 40 and 41 have only recently been cultivated *in vitro* and require special cell lines such as Graham-293 which express the human adenovirus 5 E1A and E1B genes, or special conditions, for example, the use of a low-serum cell medium. The common adenoviruses (adenoviruses 1 to 7) generally produce cytopathology within one to two weeks; the cells become swollen, rounded, and refractile, cluster together like a bunch of grapes, and reveal characteristic basophilic intranuclear inclusions. Appropriate type-specific antisera, or monoclonal antibodies directed to type-specific epitopes on the fiber, can then be chosen to type the isolate by HI and/or neutralization.

EPIDEMIOLOGY, PREVENTION, CONTROL, TREATMENT

Adenoviruses are mainly associated with disease of the respiratory tract and eye. Virus is often transmitted by respiratory droplets or contact, particularly in outbreaks of pharyngoconjunctival fever in children (adenoviruses 3, 4, and 7), or acute respiratory disease in military recruits (adenoviruses 4 and 7) in late winter and spring. Types 1, 2, 5, and 6 are mostly associated with sporadic respiratory infections. A live-attenuated virus vaccine for human adenoviruses 4 and 7 has been used to prevent respiratory adenovirus infections in military recruitment camps in the United States, but such a vaccine has not been extended to young children.

Eye infections may be acquired by transfer of respiratory secretions on the fingers. Two important settings for direct entry to the eye are "swimming pool conjunctivitis" where the chlorination of water has been inadequate, and nosocomial infection—a number of major outbreaks of epidemic keratoconjunctivitis have been traced to the surgeries of particular ophthalmologists or hospitals where aseptic technique is inadequate.

Spread occurs also by the enteric (fecal–oral) route, with very large numbers of adenovirus particles shed into feces (10^{11} virions per gram) usually for 1 to 14 days. However, shedding can last for several months, thus representing a continuing source of infection. Adenoviruses 40 and 41 are endemic worldwide and often cause asymptomatic infection, but they have been clearly demonstrated to cause gastroenteritis in children the year around, with occasional outbreaks, for example, in schools or hospitals.

Adenovirus can be excreted in urine, while adenoviruses 19 and 37 can presumably be transmitted venereally as well as by contact in eye infections because they can cause genital ulcers and urethritis in both sexes.

A number of antiviral drugs including ribavirin, ganciclovir, and cidofovir have shown variable *in vitro* activity against adenoviruses, but in clinical use ribavirin has shown little efficacy. Cidofovir may be of some benefit, and it is sometimes used together with intravenous immunoglobulin for the treatment of severe infections.

Prevention of infection relies on careful attention to respiratory isolation particularly during outbreaks, handwashing, proper disinfection of ophthalmic equipment, adequate chlorination of swimming pools, and similar measures to prevent spread.

Beginning in 1971, military recruits in the United States received a highly effective, enteric-coated, oral live vaccine with adenovirus types 4 and 7; manufacture of this vaccine was discontinued in 1996. The incidence of adenovirus respiratory disease, including pneumonia, increased substantially, so the vaccine program was reinstituted. In 2011, a new live, oral adenovirus vaccine against adenovirus serotypes 4 and 7 was approved for use in the United States military personnel aged 17 to 50 years of age.

ADENOVIRUSES AS VECTORS FOR THE DELIVERY OF HETEROLOGOUS DNA

Adenoviruses offer considerable advantages for gene delivery based on the following points: (1) virus stocks can be reliably propagated to high levels, (2) the lack of integration into host chromosomes means that the chance of insertional oncogenesis is negligible, and the duration of the expression of the transgene will be limited, especially in cells with rapid turnover, (3) diseases caused by adenoviruses are mostly mild or subclinical, (4) adenovirus vaccines have been used in military populations, and there is experience regarding the immune responses induced by this virus. In addition, genetic manipulation of the virulence genes can further attenuate adenoviruses, while engineering of receptor-binding domains can be done to promote infection of specific cell targets, making the delivery of foreign DNA more effective.

Adenoviruses as vectors can be classified into two main categories, replication-defective and replication-competent. Replication-defective vectors serve as an inert vehicle to deliver the transgene into the target cell, whereas for replication-competent vectors virus replication in the target cell is part of the intended mechanism of action. In the first generation of replication-defective vectors, E1A and E1B regions were deleted and replaced by the insertion of the transgene. In addition the non-essential E3 region was also deleted to increase cloning capacity. After further refinements, replication-defective vectors are now being used for several therapeutic purposes. These include the delivery of genes to control cell growth and apoptosis, for the treatment of cancer through administration of cytotoxic genes. In separate studies, adenoviruses are being used to prevent the overgrowth of the arterial wall during the healing phase after angioplasty for the opening of blocked cardiac arteries.

A further use of adenovirus vectors is to deliver DNA expressing an epitope or antigen as a vaccine. Both the humoral and cell-mediated immune responses can be stimulated by these approaches. Antigens that have been inserted into adenoviruses include the hepatitis B surface or core antigens, HIV-1 env, gag, or p24 proteins, pseudorabies virus gD protein, Epstein-Barr virus glycoprotein 340/220, vesicular stomatitis virus structural glycoprotein, rotavirus VP4, rabies virus glycoprotein, bovine parainfluenza virus 3 glycoprotein, and feline immunodeficiency virus envelope glycoprotein.

One important obstacle that limits the *in vivo* use of adenovirus as a vector is the high prevalence of neutralizing antibodies against human adenoviruses 5, which has been most commonly used as gene therapy vector. Pre-existing antibodies or antibodies generated by repeated systemic administration of human adenovirus 5, will neutralize the vector, resulting in failure of the therapy. To circumvent this, researchers have been developing vectors based on other less common serotypes, for instance human adenovirus 48 or non-human adenoviruses that are mostly resistant to neutralizing antibodies against human adenovirus 5.

Replication-competent vectors may be more promising for the treatment of cancers. Oncolytic adenovirus vectors kill cancer cells as part of the natural adenovirus life cycle, so following replication the virions are released from the lysed tumor cell to infect surrounding cells within the tumor.

Despite the promise of adenoviruses for gene therapy, one significant failure was the HIV-1 vaccine STEP trial, which was halted after two years in 2007 because individuals seropositive for adenovirus 5 showed increased rates of HIV-1 acquisition after vaccination with a human adenovirus 5 vaccine. It was shown that using the adenovirus-based vaccine in individuals with pre-existing immunity against human adenovirus 5 resulted in the preferential expansion of HIV-susceptible activated CD4+ T cells homing to mucosal tissues, thereby increasing the number of virus targets. This led to a greater susceptibility to acquiring HIV. This setback demonstrates the complexity of using adenovirus for gene delivery and gene immunization in clinical trials. More work is needed to fully reveal the appropriate use of adenoviruses in gene therapy and gene immunization.

FURTHER READING

Hoeben, R.C., Uil, T.G., 2013. Adenovirus DNA replication. Cold Spring Harb. Perspect. Biol. 5, a013003.

Majhen, D., Calderon, H., Chandra, N., et al., 2014. Adenovirus-based vaccines for fighting infectious diseases and cancer: progress in the field. Human Gene. Ther. 25, 301–317.

Tebruegge, M., Curtis, N., 2012. Adenovirus: an overview for pediatric infectious disease specialists. Pediatr. Inf. Dis. J. 31, 626–627.

Chapter 19

Papillomaviruses

It has long been known that skin warts in humans and in many animal species are caused by different papillomaviruses. In the 1930s Richard Shope identified the cottontail rabbit papillomavirus as a cause of skin and mucosal warts in rabbits, and Peyton Rous demonstrated that these warts could in some cases undergo malignant transformation after treatment with chemical carcinogens. In recent decades the increased occurrence of sexually transmitted genital warts in humans has become recognized as a significant public health problem. However the full clinical impact of papillomaviruses only became evident with the demonstration of a causative role for papillomaviruses in common squamous cell carcinomas of the cervix in women, and of other genital and non-genital cancers in both sexes. The Nobel Prize in 2008 was shared by the German virologist Harald zur Hausen for his "discovery of human papillomaviruses (HPVs) causing cervical cancer." Most recently, the successful development and introduction of a papillomavirus vaccine is a landmark in the prevention of cancer by vaccination.

CLASSIFICATION

Host species–specific papillomaviruses have been found in many animals and birds. Most cause benign papillomas in the skin or mucous membranes. There are now more than 200 human papillomaviruses, and more than 80 other papillomaviruses of other host species, most displaying a predilection for a particular site in the body. Some have oncogenic potential.

The classification of papillomaviruses is currently in a state of transition. Until recently, newly discovered viruses were described as "types" and named after the host species from which the viruses were isolated ("human papillomavirus," "bovine papillomavirus") followed by a number indicating the order of discovery, for example, HPV16, BPV1. A new system now divides the family into 16 different *genera*, each designated by a letter of the Greek alphabet; members of each genus share more than 60% identity in the L1 gene (see below) and show less than 60% sequence identity with other genera. A *species* is designated for those members of a genus that share at least a 60 to 70% identity, and a *type* for those that have in common 71 to 90%. Using this latter system, papillomaviruses infecting

humans fall into five of the recognized genera: alpha, beta, gamma, mu, and nu. Confusingly, nearly all the literature on human papillomaviruses to date has used the historical nomenclature. Furthermore, human papillomaviruses can also be grouped biologically according to the usual site of infection, for example, cutaneous (β-clade) or mucosal (α-clade), and the type of pathology (see Table 19.1). The reader should be aware of both approaches.

Remarkably, the genetic relatedness between different papillomaviruses closely parallels the evolutionary

TABLE 19.1 Diseases Caused by Human Papillomaviruses

Site	Clinical Presentation	Types[a]
Genital tract	Subclinical infection	All genital types
	Condyloma acuminatum, anogenital warts	**6, 11**, 42, 43, 44, 55, and others
	Cervical cancer	**16, 18**, 31, 33, 35, 39, 45, 51, 52, 56
	Vulvar, vaginal, penile, anal cancers	16
Respiratory tract	Recurrent respiratory papillomas	**6, 11**
Eye	Conjunctival papillomas	**6, 11**
Mouth	Focal epithelial hyperplasia	**13, 32**
	Oral papillomas	**2, 6**, 7, **11**, 16, 32
	Oropharyngeal cancer	16
Skin	Plantar wart	**1**, 2, 4
	Common wart	**2, 4**, and others
	Flat wart	**3, 10**, 28, 41
	Butchers' warts	**7**
	Epidermodysplasia verruciformis[b]	**5, 8**, 9, 12, 14, 15, 17, 19–25, 36, 46, 47

[a]Common types in bold type.
[b]Types 5, 8, and less commonly 17, 20, and 47 are the principal types so far associated with malignant change in epidermodysplasia verruciformis.

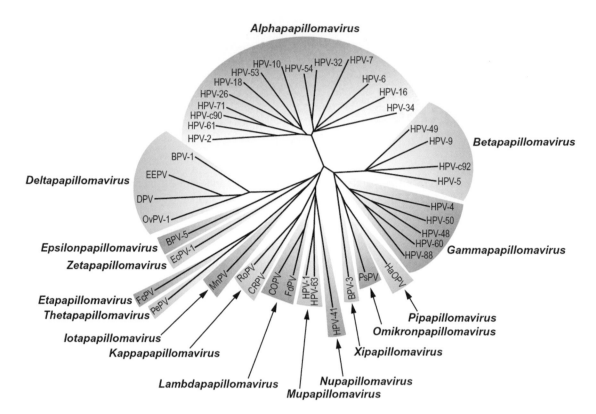

FIGURE 19.1 Phylogenetic tree representing the sequences of 118 papillomaviruses, based on a region of L1 ORF. *BPV*, bovine papillomaviruses; *COPV*, canine oral papillomavirus; *FcPV*, chaffinch papillomavirus; *CRPV*, cottontail rabbit papillomavirus (= Shope papillomavirus); *CRPV*, crowpox virus; *DPV*, deer papillomavirus; *EcPV*, equus caballus papillomaviruses (= Equine papillomaviruses); *EEPV*, european elk papillomavirus; *FdPV*, feline papillomavirus; *HaOPV*, hamster oral papillomavirus; *MnPV*, mastomys natalensis papillomavirus; *OvPV*, ovine papillomaviruses; *PsPV*, phocoena spinipinnis (Burmeister's Porpoise) papillomavirus; *PoPV*, possum papillomavirus; *PePV*, psittacus erithacus timneh (Timneh parrot) papillomavirus; *ROPV*, rabbit oral papillomavirus; *RhPV*, rhesus monkey papillomaviruses. *From Rodric Page, University of Glasgow. Reproduced from King, A.M.Q., et al., 2012. Virus taxonomy. In: Ninth Report of the International Committee on Taxonomy of Viruses. Academic Press, London, p. 247, with permission.*

relationships of each respective host (Fig. 19.1). Moreover, within the human virus types, for example, HPV16 and 18, the greatest sequence diversity is found among viruses from the most diverse racial and geographical host populations. This suggests that much of virus transmission is vertical, that is, from mother to infant, or between close family members.

PROPERTIES OF PAPILLOMAVIRUSES

The papillomavirus virion is a non-enveloped icosahedron (55 nm) with 72 capsomers (Fig. 19.2). The virion is stable to acid, ether, and heat. The genome consists of a covalently closed supercoiled circular dsDNA molecule of 8 kbp, coding eight to ten proteins with distinct functions.

The "late" L1 and L2 proteins make up the capsid, the "early" E1 and E2 proteins are involved in viral replication and its regulation, while E5, E6, and E7 induce the host cell DNA to replicate. Transcription of all mRNAs occurs in the same direction, that is, all ORFs are encoded in the same DNA strand. Between the 5′ region of E6 and the

3′ region of L1 is a ~1 kb region of the genome that does not contain an ORF, and is known alternatively as the long control region (LCR) or the upstream regulatory region: this contains regulatory elements and the origin of DNA replication (Fig. 19.3).

VIRAL REPLICATION

Experimental propagation of human papillomaviruses has only been achieved either in raft cultures of human keratinocytes treated with the phorbol ester TPA to increase cell differentiation, or using human skin or mucosal xenografts in immunocompromised rodents. Thus, our knowledge of virus replication and the mechanism of carcinogenesis has been acquired principally from either *in vitro* transfections with molecularly cloned HPV DNA, or from studies using athymic or transgenic mice.

HPV DNA is readily cloned and amplified in cloning systems, and recombinant L1 protein alone, or L1 and L2 together, can be induced to form virus-like particles (VLPs). These have proved exceedingly useful for both studies of

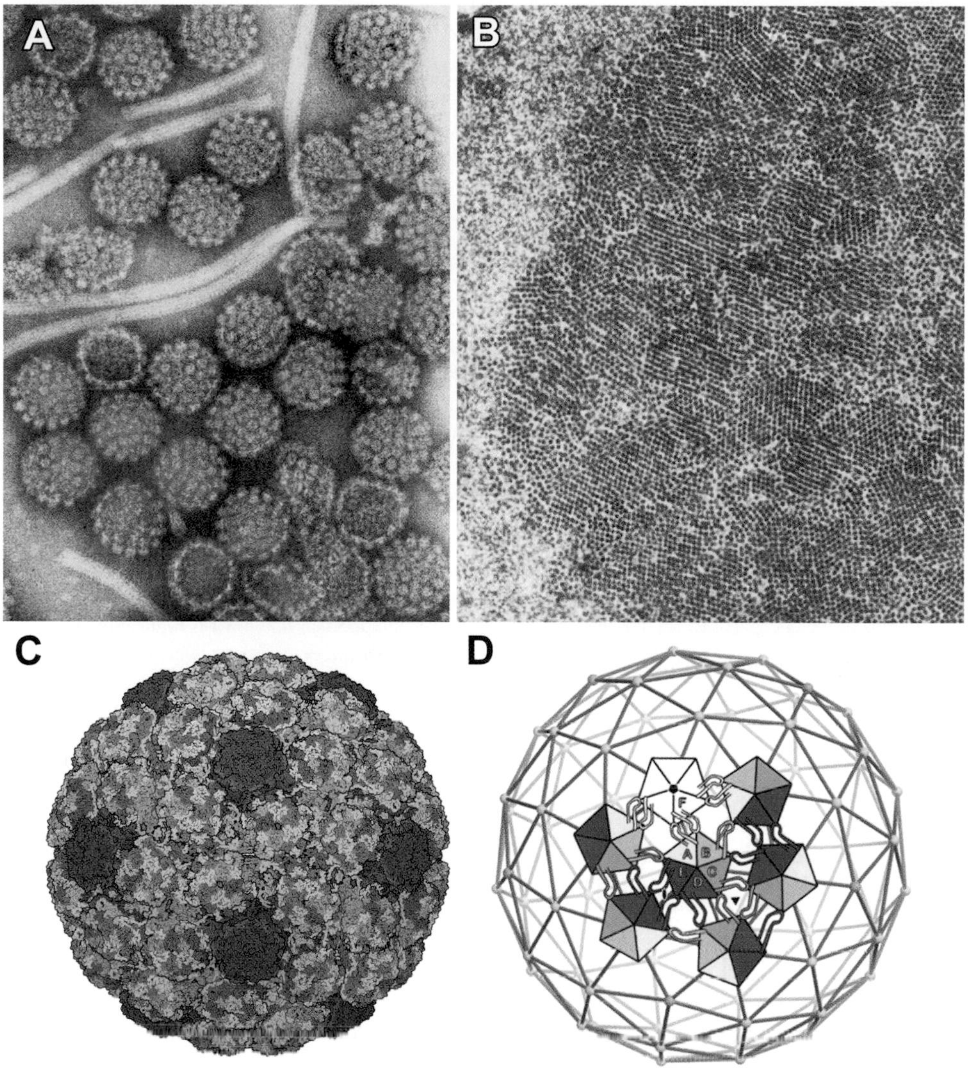

FIGURE 19.2 Papillomavirus morphology and structure. (A) Negative contrast electron microscopy. (B) Thin-section electron microscopy of the nucleus of a squamous epithelial cell containing a massive number of virions. (C) Model of capsid structure, surface view, view along axis of twofold symmetry, with vertex pentons (pentamers) shown in purple and hexons (hexamers) in multicolors—reconstruction from cryo-electron microscopy images. (D) Diagram of packing of structural units—there are 72 pentons composed of the major capsid protein (L1) arrayed in a **T = 7d** icosahedral lattice. Twelve of these (white) are centered on fivefold axes of symmetry and the 60 others (colored) are at coordinated positions between. *(C) From Jean-Yves Sgro, University of Wisconsin, with permission. (D) Reproduced from Wolf, M., et al., 2010. Subunit interactions in bovine papillomavirus. Proc. Natl. Acad. Sci. U.S.A. 107, 6298–6303, with permission.*

virus cell biology and as vaccines. The precise mode of virus attachment has not been clarified although virus can bind to both α6 integrins, heparin sulfate, and cell-surface glycosaminoglycans. Entry is clathrin-dependent and is followed by receptor-mediated endocytosis and uncoating. The viral genome migrates to the nucleus together with the L2 protein, a viral protein that has multiple roles including directing localization of the incoming genome. Transcription, DNA replication, and virion assembly then occur in the nucleus. The early genes are copied from a single promoter (P97) and the transcripts subjected to differential splicing to generate the mRNAs for the seven early proteins.

These include regulatory proteins, some with transactivating properties that derepress the host genes for certain cellular enzymes and stimulate cellular DNA synthesis. Replication of the viral genome occurs bidirectionally and is initiated by binding of non-structural proteins E1 and E2 to the unique origin of replication on the viral DNA.

Cells in the basal and outer layers of the epithelium show very different states of permissiveness for virus replication, a crucial point in understanding papillomavirus pathogenesis. Only early HPV genes are expressed in the replicating cells present in the basal layer of the epidermis but this suffices to augment cellular proliferation (hyperplasia); the viral

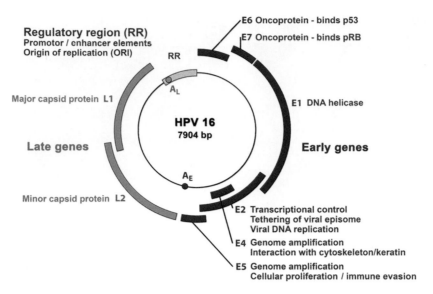

FIGURE 19.3 Organization of the HPV genome. The genome contains early and late genes, all of which are transcribed in a clockwise direction from the same DNA strand. Control of early gene transcription and genome replication is conferred by the upstream regulatory region (*URR*). E6 and E7 encode the major transforming proteins of the oncogenic HPVs, but these proteins also play an important role in virus replication. L1 encodes the major capsid protein and L2 the minor capsid protein. *Reproduced from Stanley, M.A., 2012. Epithelial cell responses to infection with human papillomavirus. Clin. Microbiol. Rev. 25, 215–222, with permission.*

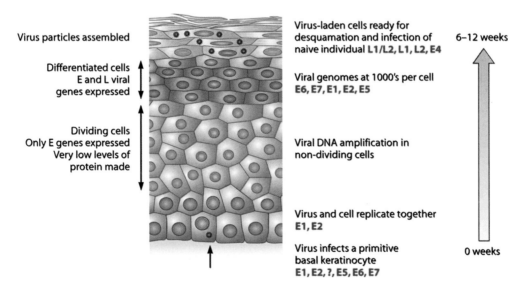

FIGURE 19.4 Expression of the papillomavirus genome is tightly regulated by the state of differentiation of the host cell. Infection first becomes established in the basal cell layer following abrasions. In the basal proliferating layers of the epithelium, the virus and cell replicate together and the viral DNA copy number is maintained at around 50 to 100 copies/cell. When the cell stops dividing and begins to differentiate into a mature keratinocyte, all the viral genes become activated and the viral genome copy number increases to thousands/cell. In the case of incipient malignancy, regulation of E6 and E7 expression is lost and cellular gene expression becomes deregulated. In the outermost layers large amounts of L1 and L2 proteins are expressed and many thousands of genomes are encapsidated to form infectious virions. The time taken between infection and the generation of infectious virus is at least three weeks. HPV thus has a very long replication cycle, has no blood-borne phase and does not cause cell death. *Reproduced from Stanley, M.A., 2012. Epithelial cell responses to infection with human papillomavirus. Clin. Microbiol. Rev. 25, 215–222, with permission.*

genome replicates only very slowly as an autonomous plasmid in the nuclei of these cells. However, in the terminally differentiated cells found in the outer layers of the epithelium, full expression of the viral genome occurs and viral DNA is amplified to high copy number. Late genes are transcribed from a second promoter (P742); these encode

the structural proteins, L1 and L2, which after various post-translational modifications, are then directed to the nucleus via nuclear localization signals, prior to assembly within the nucleus into nascent virions (Fig. 19.4).

The E5, E6, and E7 proteins play a role in transformation through interaction with a number of different host proteins and pathways involved in the regulation of cellular growth. For example, E5 binds to the epidermal growth factor receptor (EGFR), causing a prolonged activation of signaling cascades. E6 binds to the p53 tumor suppressor protein, targeting it for ubiquitination and proteasomal degradation, and the E7 protein binds to the Rb protein, leading to inappropriate entry of the cell into the S phase of cell division. Differences have been found between the high cancer risk and low-risk HPV types as to patterns of interaction with cellular signal transduction pathways.

CLINICAL FEATURES

Genital Infections

Many infections of the genital tract are subclinical, but there are several important clinical presentations. *Condyloma acuminatum, anogenital warts*, and *exophytic warts* are the names given to the large, moist pedunculated excrescences of soft papillomas found on the external genitalia, perineum, vaginal introitus, penis, or anus, caused commonly by HPV types 6 and 11 (Fig. 19.5). *Condyloma planum* (*flat wart*) is the more usual presentation in the cervix, with types 6

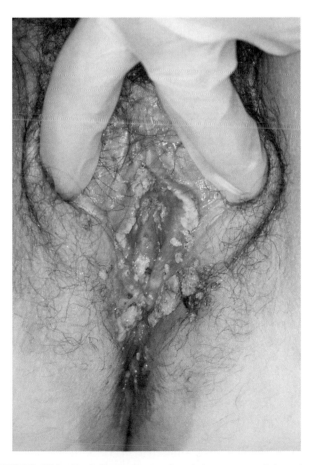

FIGURE 19.5 Typical appearance of genital warts around the vulva. *From Public Health Image Library, CDC.*

and 11 again being the most common. *Cervical carcinoma* may develop 20 to 50 years after infection with certain mucosal HPV types. There is a slow progression through three stages of lesions variously called *cervical dysplasia, cervical intraepithelial neoplasia* (CIN), or *squamous intraepithelial lesions* (SIL). The flat condyloma is sometimes classified as CIN-1 or low-grade SIL (LSIL); CIN-3 is often designated *carcinoma in situ* or high-grade SIL (HSIL). The fully malignant, invasive carcinoma is usually but not always of the squamous type. However, adenocarcinomas of the cervix are becoming more frequent, and have also been shown to be associated with HPV infection. Over 99% of all cancers of the cervix contain HPV DNA, most commonly of type 16 or 18. The same HPV types are associated with carcinomas of the vulva, vagina, penis, and anus (Fig. 19.5).

Cervical Cancer

In 1842 the Italian physician Domenico Antonio Rigoni-Stern noted that nuns in Verona had higher rates of breast cancer, but lower rates of "cancer of the womb," compared to married women. However, it took more than 120 years of controversy and false leads before the epidemiological relationship was clarified between sexual activity and cervical cancer. *Cervical cancer* is the second most common cancer among women worldwide, with approximately 500,000 new cases and 275,000 deaths each year. Rates vary greatly from country to country and with socioeconomic and ethnic factors. Unlike most forms of cancer which are more common in developed countries, cervical cancer is more frequently seen in developing countries, where it sometimes represents the commonest cancer of women. This may be attributed to the shorter life expectancy in some developing countries coupled with the fact that cervical cancer typically presents at a younger age than most forms of cancer. Other factors affecting these rates include the variable availability of Pap screening, treatment and vaccination programmes, and cofactors including HIV infection, smoking, and use of estrogen-progestogen oral contraceptives (Fig. 19.6).

HPV DNA from high-risk types is found in almost all cases of cervical cancer and HPV16 and HPV18 together account for around 70% of all cases, while less common HPV types are found in nearly all of the remainder. Viral DNA is integrated into cellular DNA in around 90% of cancers but in a lower proportion of the earlier grades of dysplasia. Only a restricted number of viral genes are expressed, with E6 and E7 being most commonly expressed and the likely drivers of progression to the transformed state (see Box 19.1).

Respiratory Papillomatosis

The genital HPV types infect the respiratory tract, but rarely, usually in children aged younger than 5 or in young

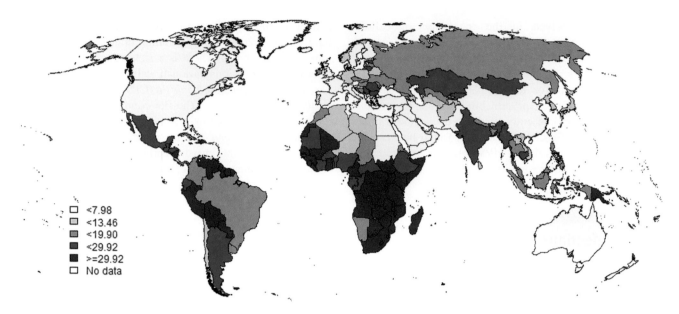

FIGURE 19.6 Incidence of cervical carcinoma/100,000 population in different regions of the world, 2014. Unlike many forms of cancer, this disease has higher prevalence in developing countries. *From International Agency for Research on Cancer, Lyons, with permission.*

BOX 19.1 Association Between Genital HPV Infection and Cervical Cancer

- 1842—Rigoni-Stern in Verona noted that "cancer of the womb" was more common in married women than in nuns, and the converse was true for breast cancer
- 1942—Papanicolaou described exfoliative cytology of the cervix to demonstrate premalignant cells and "warty" cells—"koilocytes"
- 1980s—Harald zur Hausen assembled epidemiological, histological and molecular evidence for a causative role
- 2010+—HPV immunization greatly reduces development of precancerous lesions
- "High-risk" HPV DNA is detected in ~99% cervical cancers—HPV16 in 60%, HPV18 in 10%, other virus types in the remainder
- Odds ratio for cervical carcinoma in high-risk HPV infection is 150:1
- Laboratory molecular evidence provides plausible mechanisms for HPV oncogenesis

adults. Classically, the lesions begin in the larynx and can grow copiously, obstructing the airway, with a tendency to spread also to other parts of the respiratory tract. Malignant change may develop in 3–5% of patients. Papillomas caused by the same genital types are also seen occasionally on the conjunctiva.

Oral Infections

Focal epithelial hyperplasia (Heck's disease), associated almost exclusively with HPV types 13 and 32, is a condition characterized by multiple nodular lesions in the mouth and is particularly prevalent in the Inuit people (indigenous peoples inhabiting the northern circumpolar region) and in indigenous South and Central American people. *Oral papillomas* of the more conventional kind can be caused by the sexually transmitted types 6, 11, and 16. Common warts on the lips are usually type 2. A subset of *oropharyngeal carcinomas* is associated with HPV. HPV-positive oropharyngeal cancers have increased significantly in several industrialized countries recently, for example, by a factor of four in the United States between 1984 to 1989 and 2000 to 2004; these have a better prognosis than HPV-negative cancers.

Skin Warts

Perhaps 10% of all schoolchildren and young adults experience a crop of warts on the skin, probably acquired in the course of close contact recreational activities. Generally, warts regress within two years. *Common warts*, seen on prominent anatomical sites subjected to abrasion such as hands and knees, are raised papillomas with a rough surface, often caused by type 2 or 4. Type 7 is the agent of "butchers' warts," an occupational disease of meat handlers. *Flat warts*, sometimes called *plane warts*, are smaller, flatter, smoother, and more numerous, seen especially on the arms, face, and knees of youngsters. Types 3 and 10 are often involved. *Plantar warts* largely due to type 1 are painful deep endophytic warts found on the weight-bearing regions of the heel and sole of the foot; palmar warts are similar.

The recently expanding list of HPV types and the application of PCR testing to other skin conditions have given tantalizing hints that HPVs may play some role in other common non-melanoma skin cancers, for example, squamous cell carcinomas. Although HPV, particularly the types associated with *Epidermodysplasia verruciformis* (see below), is found in other skin cancers in only around 30 to 40% of cases, the possibility that some form of "hit-and-run" mechanism may be involved cannot be excluded. The picture is confused by the fact that different HPV types, in particular types 5, 8, and 23, can be found in the healthy skin of many healthy individuals.

Epidermodysplasia verruciformis is a rare condition seen in people with particular forms of cell-mediated immunodeficiency. Most cases are due to an autosomal recessive hereditary mutation of the genes coding for the EVER1 and EVER2 integral membrane proteins found on the endoplasmic reticulum, but sporadic, sex-linked, and autosomal-dominant hereditary forms have also been described. The disease commonly presents in early childhood as multiple flat warts on the dorsal aspect of the hands, extremities, face, and neck, these usually persisting for life. The lesions are of two varieties: (1) flat warts, commonly caused by HPV types 3 and 10, as in normal children, and (2) reddish-brown macular scaly patches from which can be isolated any of nearly 20 rare HPV types found almost exclusively in these patients. *Squamous cell carcinoma* (SCC) arises in about one-third of all *epidermodysplasia verruciformis* patients after many years, in one or often several of the macular lesions situated on areas of the skin exposed to sunlight, the latter is therefore an important cofactor in the genesis of this malignancy. The tumors, usually carrying the genome of HPV type 5 or 8, are often slow-growing *in situ* carcinomas but may be metastasing and invasive squamous cell carcinomas. Renal transplant recipients and immunosuppressed patients are also at increased risk of developing *Epidermodysplasia verruciformis* lesions associated with HPV-5 or HPV-8 DNA. In addition, the disease is seen, albeit rarely, in some apparently immunocompetent individuals.

PATHOGENESIS AND IMMUNITY

Papillomaviruses are not only host species-specific but also restricted as to tissue tropism, multiplying only in epithelial cells of the skin or certain mucous membranes. Different HPV types display a preference for distinct sites of the body. Autoinoculation of virus to cause new superficial lesions is not uncommon, but systemic dissemination does not occur.

Broadly speaking, HPV types fall into two groups: cutaneous types (infecting skin) and mucosal types (infecting the genital tract and sometimes the respiratory tract, oral cavity, or conjunctiva) as outlined in Table 19.1. Furthermore, full replication and production of virus particles occur only in the differentiated superficial layers of squamous epithelium.

Warts have a long incubation period of up to two years. The virus penetrates the skin through an abrasion and infects the basal cell layer. In these relatively undifferentiated replicating cells, only early viral genes are expressed and limited viral DNA replication maintains the genome as a stable nuclear plasmid (episome). Expression of early viral genes stimulates proliferation of the basal cells; the resulting hyperplasia leads to acanthosis (thickening) and generally to a protruding papilloma after an incubation period of several months. Capsid proteins are synthesized and virions produced only in the terminally differentiated keratinocytes present in the outer layers of the epithelium; these cells produce keratin but are no longer dividing. Histologically, the differentiated layers of the epithelium contain *koilocytes*: large cells with perinuclear vacuoles and hyperchromatic, distorted nuclei. Hyperkeratosis is a prominent feature of skin warts.

Warts tend to disappear, usually synchronously, within a couple of years. This sudden regression is not related to the titer of antiviral antibodies, being generally ascribed to the T cell-mediated immune response. Indeed, where multiple warts are present the lesions normally regress simultaneously, and removal of one wart can be followed by spontaneous regression of those remaining, perhaps due to immune stimulation by released antigens following treatment. However, the nature of the mechanism whereby T cell immunity is induced, and conversely why this response is so delayed, is not clear. More widespread, persistent skin warts can be a problem in immunosuppressed individuals (Fig. 19.5). However, the later recurrence of warts after treatment, for example, in the larynx, and the appearance of multiple warts after immunosuppression suggest that longlasting latent infection of basal epithelial cells may be very common (Fig 19.7; Box 19.2).

Although the papillomas induced by HPV on the external genitalia are fundamentally similar to those described earlier, lesions around the cervix display some important differences. The virus enters during sexual intercourse, possibly via a minor abrasion in the vicinity of the squamocolumnar border, where cells are proliferating. After an incubation period of one or more (average three) months, a flat condyloma develops. Infections with certain HPV types may progress over a period of years through the various stages of cervical intraepithelial neoplasia to invasive squamous carcinoma. Papanicolaou smears of cervical dysplasia reveal the characteristic koilocytes. The HPV genome may persist for years as a non-integrated nuclear episome, not only within the lesion but also in histologically normal surrounding cells of the mucous membrane for up to a few centimeters. In contrast, cervical carcinomas harbor the genome of HPV, most commonly type 16, 18, 31, 33, or 35, in the form of incomplete copies of the viral DNA integrated at random sites within host cell chromosomes.

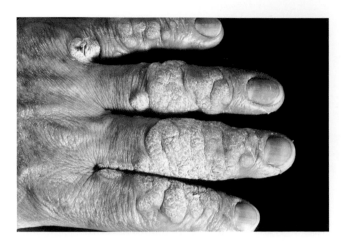

FIGURE 19.7 Multiple warts on the fingers of an immunocompromised patient. *Reproduced from Cooke, R.A. Infectious Diseases, McGraw Hill, with permission.*

BOX 19.2 Pathogenesis of HPV Infection

- Papillomaviruses are highly species and tissue specific
- Virus enters basal cells via abrasions, leading to expression of early genes, limited DNA replication, viral DNA persists as an episome
- Early proteins stimulate basal cell proliferation
- Incubation period of warts 6 to 20 weeks
- Significant viral DNA replication, production of capsid proteins, and virions, only occur in outer differentiated keratinocytes
- Infection does not spread beyond the basement membrane into deeper tissues
- Lesions usually disappear, usually synchronously, possibly after 9 to 18 months after the development of cell-mediated immunity

NATURAL HISTORY OF GENITAL HPV INFECTION

PCR testing has revealed that genital HPV infections are far more widespread than previously suspected. In various populations, from 2% to 44% of women after a single test were found to be HPV positive, depending on the age and the method used. Among women aged 15 to 19 years in the United Kingdom, the cumulative incidence over three years was 44% increasing to 60% over five years. However, the median duration of these infections was one year, and 80% had cleared the infection after 18 months. Recent data confirm that transient infection at any age is benign, and progression to malignancy only occurs in those with persistent infection. Similar data for males are not available, although it is clear that genital HPV infection in males must be similarly widespread; anogenital HPV-related tumors in males are also well described but are not as common as in females. Assembling these and similar data has allowed the outline of the natural history of genital HPV infection in females shown in Fig. 19.8.

In most populations, genital infection with one or more HPVs is acquired after commencing sexual activity in the majority of women, some of whom develop clinical warts or low-grade cervical cytological lesions (CIN1, LSIL) after an incubation period of 6 to 20 weeks. At least 80% of individuals then clear any detected infection within 18 months, whilst the remaining ~20% may remain persistently infected and may progress within two to five years to more high-grade cytological lesions (CIN2-3, HSIL). A proportion of the latter then progress to invasive cervical carcinoma after a further interval of 10 to 20 years. It has also been proposed that *de novo* detection of HPV infection in older women may reflect reactivation of virus from latently infected basal epithelial cells, as well as newly acquired infection. Such reactivation may be facilitated by a decline in immunological competence and/or local trauma, but the role of these factors in disease progression remains to be clarified.

Similar information about the natural history of papillomavirus infection at other sites is limited, but there is evidence of high-grade anal intraepithelial neoplasia in HIV-infected men leading to invasive anal cancer.

LABORATORY DIAGNOSIS

Clinically, the diagnosis of skin warts and condyloma acuminata on the external genitalia generally poses little in the way of significant problems, but a biopsy is required whenever malignancy is suspected. Cervical condylomas can be seen by colposcopy, but the lesions are often flat and indistinguishable from cervical intraepithelial neoplasia. Abnormal patches of mucosa are more visible after applying dilute acetic acid ("aceto-white" areas).

The Papanicolaou ("Pap") smear procedure allows exfoliated cells to be examined for HPV-related changes (koilocytes) and malignant cells. It is widely used for routine screening of women for early diagnosis of premalignant and malignant cell changes in the cervix, and is a very important public health measure.

Conventional approaches to laboratory diagnosis are not particularly rewarding in the case of human papillomaviruses. Virus isolation is impractical because these viruses can only be grown with difficulty in complex tissue culture systems. The detection of antibodies to the L1 proteins of common HPV types has been used to study cumulative past or present infection rates in epidemiological surveys, but the findings are sometimes difficult to interpret and are not reliable for diagnosing current infection. Antigen detection in exfoliated cells can also be unreliable because viral genomes can persist for years with very limited viral protein expression in basal cells, as episomal DNA in benign papillomas or premalignant cervical dysplasia, or integrated in cancer cells.

Thus, the only generally applicable diagnostic approach is detection of viral DNA (see Chapter 10: Laboratory Diagnosis of Virus Diseases, for a detailed

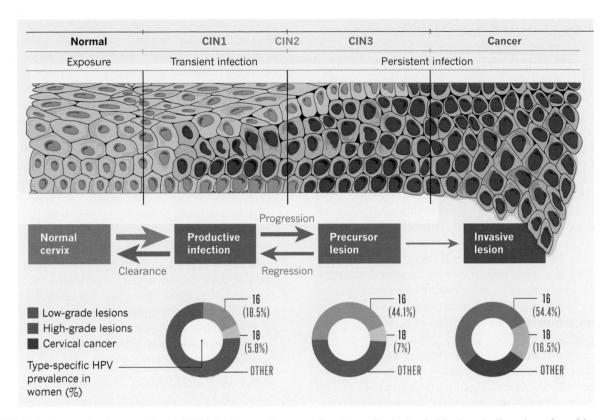

FIGURE 19.8   Progression from a productive HPV infection to malignancy. A few abnormally sized and oddly shaped cells on the surface of the cervix is classified as Cervical Intraepithelial Neoplasia I (CIN1), a low-grade lesion that typically develops within 1 to 3 years from infection and disappears within a few months without treatment. HPV DNA is present as an episome and the lesion extends less than one-third of the full thickness of the epithelium. In CIN3, a large number of precancerous cells are seen on the surface of the cervix that are distinctly different from normal cells and the lesion extends the full thickness of the epithelium. High-grade CIN3 lesions are still reversible spontaneously or through treatment. Progression to malignancy may take 20 years. Malignant cells show a loss of differentiation, no viral replication, high levels of E6 and E7 oncoprotein expression, and frequently have integrated HPV DNA. In invasive cervical carcinoma, the disease spreads beyond the basement membrane and invades connective tissue, and may show metastatic spread to lymph nodes and distant sites. HPV types 16 and 18 make up only 2.7% and 1.1% of initial cervical infections, respectively, but together account for 70% of cervical cancer cases. *Reproduced from Crow, J.M., 2012. HPV: the global burden, Nature 488, S2–S3. doi:10.1038/488S2a, with permission.*

discussion). Dot-blot hybridization, Southern blot, and *in situ* hybridization have largely been replaced by PCR techniques. This latter approach has been invaluable for defining the natural history of genital HPV infection, and can be used clinically to add information in the case of doubtful or atypical Pap smears: the use of PCR can also be valuable in the identification of higher-risk patients needing more frequent testing, for example, those patients infected with the oncogenic types such as HPV16 and 18.

EPIDEMIOLOGY

Transmission of skin warts occurs mainly among school-age children via direct contact through abrasions, with the possibility of subsequent spread by scratching (autoinoculation). Transmission is facilitated by the very high numbers of virus particles present in skin warts and the relatively long survival of virus in the environment. Plantar warts are readily picked up from the wet floors of public swimming pools and bathrooms. Genital warts, on the other hand, are spread by sexual contact; this may include close skin contact as well as contact with body fluids, and hence condoms may reduce but do not fully prevent transmission. Not surprisingly, the incidence of genital warts skyrocketed in parallel with the sexual revolution of the late 1960s and 1970s, and in many parts of the world, this infection has become the commonest single reason for STD clinic attendance. As might be expected, the prevalence of infection is correlated positively with age of first intercourse and the number of sexual partners. The age of onset of respiratory papillomatosis is bimodal, peaking in young children and again in young adults; infection may be sexually acquired, but most infections in young children are thought to be acquired during passage of the baby through an infected birth canal. Fortunately, the efficiency of transmission in this manner must be very low, considering the rarity of laryngeal papillomatosis vis-à-vis the high frequency of cervical infection.

The epidemiology and distribution of rarer HPV types found in particular settings, for example, in *Epidermodysplasia verruciformis* lesions or associated with butchers' warts, remain a mystery.

TREATMENT AND PREVENTION

Interruption of transmission may be considered in some situations. Condom usage reduces but does not eliminate transmission of genital warts, but safe sex practices when undertaken to reduce HIV and other STDs, do bring the additional benefit of reducing genital HPV infection. Cesarean section for infants carried by HPV-infected mothers has been considered for prevention of respiratory papillomatosis but not generally adopted as the low rate of respiratory infection does not warrant this intervention.

The fact that skin warts regress spontaneously encourages the perpetuation of mythical cures, including hypnosis and the fictional witchcraft-inspired remedies described by Mark Twain. A comparable rate of success can be assured by letting nature take its course. However, skin warts can be removed by cryotherapy or caustic chemicals, laryngeal papillomas by surgery or laser treatment, external genital warts by cryotherapy, podophyllin, imiquimod, laser, or diathermy, and the more severe grades of cervical dysplasia by laser or diathermy. Invasive carcinoma requires surgery, with or without radiotherapy and chemotherapy. Interferon α or β injected intramuscularly and/or into the lesion itself has been reported to cause genital papillomas to regress in a majority of cases, but its use for recurrent respiratory papillomatosis has not been so successful.

Two fundamental measures are proving critical and very effective in reducing the burden of cervical cancer.

1. Screening asymptomatic women by regular Papanicolaou (Pap) smears allows cytologists to identify HPV-affected cells (koilocytes) and also premalignant cells. Many countries encourage routine universal public health Pap screening programs. Many different screening regimens and frequencies are used, depending on the age group and population involved. Policies for subsequent management also vary, bearing in mind that most cases of early dysplasias (CIN1, LSIL) will spontaneously regress and can be managed by close observation, whereas more advanced dysplasias need prompt and adequate surgery. Moreover, the coexistence of a high-risk HPV type worsens the prognosis for any dysplasia and is an indication for more frequent screening.

2. In the last few years, the introduction of prophylactic immunization using non-replicating virus-like particles (VLPs) is beginning to revolutionize the impact of this disease. Vaccination with inactivated virions was previously shown to protect against infection in several animal papilloma models. The problem of producing HPV antigens in sufficient quantity and in an immunogenic, non-infectious form was overcome by Jian Zhou and Ian Frazer in Brisbane, Australia, who demonstrated that L1 proteins can self-assemble into highly immunogenic VLPs. In several landmark placebo-controlled trials, vaccines containing these particles and adjuvant were shown to confer a high level of type-specific protection against genital HPV infection and HPV-related CIN1. This success story has been translated into widescale application; more than 40 million courses of vaccine have now been given, mostly to girls, using one of two products—"Gardasil" (containing VLPs of HPV16, 18, 6, and 11) and "Cervarix" (containing HPV16 and 18 only)—and vaccines containing additional important HPV types are under development. The safety record has been excellent, with a very low reported rate of minor side effects. The vaccine is not currently recommended for use during pregnancy, but experience with inadvertent administration to pregnant women has not shown any adverse effects on the mother or fetus. Current debate centers on cost and access in developing countries, and whether boys should also be vaccinated. A study in one country reported that since introducing universal vaccination of girls, diagnosis of genital warts dropped by 59% in girls and young women aged 12 to 26, and by 39% in young men, indicating that vaccination of females helped protect males.

Similar vaccine constructs incorporating the HPV-16 E6 and E7 genes are undergoing human trials in the hope that they may be used therapeutically to boost the cytotoxic T cell response to E6/E7 peptides presented on the surface of cervical intraepithelial neoplasia or invasive carcinoma cells.

FURTHER READING

Aldabagh, B., Angeles, J.G.C., Cardones, A.R., Arron, S.T., 2013. Cutaneous squamous cell carcinoma and human papillomavirus: is there an association? Dermatol. Surg. 39, 1–23.

Bosch, F.X., Broker, T.R., Forman, D., et al., 2013. Comprehensive control of human papillomavirus infections and related diseases. Vaccine 31S, G1–31.

Frazer, I.H., 2014. Development and implementation of papillomavirus prophylactic vaccines. J. Immunol. 192, 4007–4011.

Gravitt, P.E., 2011. The known unknowns of HPV natural history. J. Clin. Invest. 121, 4593–4599.

McBride, A.A., 2008. Replication and partitioning of papillomavirus genomes. Adv. Vir. Res 72, 155–205.

McLaughlin-Drubin, M.E., Meyers, J., Munger, K., 2012. Cancer associated human papillomaviruses. Curr. Opin. Virol. 2, 459–466.

Moscicki, A.-B., Schiffman, M., Burchell, A., Albero, G., et al., 2012. Updating the natural history of human papillomavirus and anogenital cancers. Vaccine 30S, F24–F33.

Scotto, J., Bailar, J.C., 1969. Rigoni-Stern and medical statistics: a nineteenth century approach to cancer research. J. Hist. Med. Allied Sci. 24, 65–75.

Stanley, M.A., 2012. Epithelial cell responses to infection with human papillomavirus. Clin. Microbiol. Rev. 25, 215–222.

Zheng, Z.-M., Baker, C.C., 2006. Papillomavirus genome structure, expression and post-transcriptional regulation. Front. Biosci. 11, 2286–2302.

Chapter 20

Polyomaviruses

The prototype member of the family *Polyomaviridae* is the polyoma virus of mice. This virus induces many different types of malignant tumors when artificially injected into infant rodents, such as hamsters, although it causes only harmless inapparent infections in mice when spread by natural routes. Another polyomavirus, simian virus 40 (SV40), causes subclinical infection in monkeys but induces tumors after inoculation into newborn rodents. During the 1960s and 1970s these two viruses were important models for the biochemical investigation of virus-induced malignancy, in both cultured cells and experimental animals (see Chapter 9: Mechanisms of Viral Oncogenesis).

Subsequently two human polyomaviruses, BK and JC, were discovered in 1971. BK virus was recovered from the urine of a renal transplant recipient (with the initials "BK"), and JC virus from the brain of a patient with a rare demyelinating condition, progressive multifocal leukoencephalopathy (PML). As with SV40, BK and JC viruses are oncogenic for newborn hamsters and transform mammalian cells *in vitro*, but evidence that these viruses cause human cancer is lacking. Both BK and JC viruses are ubiquitous in humans, producing inapparent infections that persist for many years in the urinary tract, and undergo intermittent reactivation particularly following immunosuppression.

Since 2007, this field has been re-invigorated by the discovery of a further nine new polyomaviruses from humans and several from non-human primates, revealing that this family is in fact larger and more divergent than previously recognized. These findings raise the possibility of a plethora of new polyomaviruses that remain to be discovered, as well as raising new questions about the evolution, tropism, latency, reactivation, and pathogenicity of this virus family. With the exception of Merkel cell carcinoma-associated polyomavirus (MCPyV) and trichodysplasia spinulosa-associated polyomavirus (TSPyV), any association between these more recently described polyomaviruses and human disease remains to be clarified.

CLASSIFICATION, PROPERTIES, AND REPLICATION

Historically, papillomaviruses and polyomaviruses were classified within a single family *Papovaviridae* because both possess a double-stranded circular DNA genome and a similar virion capsid structure, as well as replicating and assembling in the host cell nucleus. However the seventh report of ICTV (2000) designated *Polyomaviridae* as a separate family from *Papillomaviridae*, on the basis of distinct molecular differences including a lack of major sequence homology, different genome organization, and the fact that polyoma transcription is bidirectional whereas papillomavirus transcription occurs in one direction only and from a single DNA strand.

At least 12 different polyomavirus species infecting mammals and one infecting birds are recognized within the single genus *Polyomavirus*. Additional unclassified human polyomaviruses have been identified (Table 20.1) using polymerase chain reaction (PCR) with random primers on concentrated patient samples, characterizing cDNAs made from cell RNA transcripts, or rolling circle amplification of skin swab material. A number of unclassified polyomaviruses have also been isolated from bats. A proposal to recognize three separate genera (*Orthopolyomavirus*, *Wukipolyomavirus*, and *Avipolyomavirus*) is currently being considered.

Polyomavirus virions are approximately 40 to 45 nm in diameter (cf. 55 nm for papillomaviruses), and do not have an envelope, the icosahedral capsid being composed of 72 capsomers (Fig. 20.1). The genome is a single molecule of closed circular, double-stranded DNA of approximately 5 kbp, which exists in the mature virion as a supercoiled, chromatin-like structure in association with host cell histone proteins. The genome includes a non-coding control region (NCCR), in which are located the origin of DNA replication (ORI) and also promoter and enhancer elements with binding sites for different DNA-binding proteins and transcription factors (Fig. 20.2).

Replication and transcription of the viral genomes as well as virion maturation, collectively take place in the nucleus,

Fenner and White's Medical Virology. DOI: http://dx.doi.org/10.1016/B978-0-12-375156-0.00020-5

TABLE 20.1 Polyomaviruses Isolated from Humans

Virus	Site of Isolation/Excretion	Year Described	Disease
BKV	Urine	1971	Nephropathy/graft rejection in renal transplant recipients; hemorrhagic cystitis in stem cell transplant recipients
JCV	Urine	1971	PML in HIV/AIDS, or in patients treated with humanized mcAbs, e.g., natalizumab
KIPyV	Nasopharyngeal aspirate	2007	None known
WUPyV	Nasopharyngeal aspirate	2007	None known
MCPyV	Merkel cell carcinoma tissue	2008	Merkel cell carcinoma, a rare neuroendocrine skin tumor
HPyV6	Normal skin swabs	2010	None known
HPyV7	Skin swabs	2010	None known
TSPyV	Trichodysplasia spinulosa (TS) skin spicules	2010	TS (skin papules, spicules, alopecia in immunosuppressed patients)
HPyV9	Serum, urine, skin	2011	None known
MWPyV	Stools	2012	None known
STLPyV	Stools	2012	None known

and, given the limited coding capacity of the polyoma virus genome, the virus makes extensive use of host cell functions. Early viral proteins are produced from a series of RNAs generated by differential splicing from a single main transcript originating near the non-coding control region. These proteins are known as T (tumor) antigens, for example large T, middle (m) T (not found in polyomaviruses of primates), small t antigens, and other intermediates; these proteins interact with cell cycle regulatory proteins including p53 and Rb, and the viral proteins can induce cell transformation, de-repression of some host enzymes, and stimulation of cellular DNA synthesis. After initiation of viral DNA synthesis, RNA transcription then begins from the NCCR in the opposite direction to produce late mRNAs for three structural proteins VP1, VP2, and VP3, and in SV40, VP4. The structural proteins eventually self-assemble in the nucleus to form full and empty capsids, the major component of the outer shell being VP-1.

Viral DNA synthesis is initiated by binding of the non-structural viral T antigen to the origin of viral DNA replication (ORI) site. DNA replication proceeds bidirectionally with involvement of the host DNA polymerase and host nuclear transcription factors, and terminates approximately 180° from this point.

PATHOGENESIS

Primary infection of mammals is normally asymptomatic, followed by a prolonged persistent infection. In humans, BK and JC viruses are known to remain latent in the tonsils, other lymphoid tissues, bone marrow, and urinary tract.

Using virus-like particles (VLPs), ELISA assays specific for the different human polyomaviruses have been used to map the prevalence of these infections. Seroprevalence rates for the various human polyomaviruses range from 35 to 99% of the population. However, despite the ubiquitous nature of these viruses, disease is a rare consequence of infection with only some of these viruses, and usually only in immunosuppressed individuals, with no evidence of human disease being associated with the remaining viruses (Table 20.1).

Different pathological mechanisms have been proposed in polyomavirus infections in different settings (see Box 20.1). These include (1) direct cytopathic effect accompanying high-level virus replication, as with the loss of oligodendrocytes infected with JC virus in the brains of patients with progressive multifocal leukoencephalopathy (PML), (2) immune reconstitution inflammatory syndrome (IRIS), an inflammatory response against polyoma virus antigens seen following a partial restoration cell-mediated immunity, (3) a combination of viral cytopathology and inflammation, (4) induction of autoantibodies, and (5) cell transformation and oncogenesis.

Mammalian polyomaviruses usually grow best *in vitro* in cells derived from a native host species. This restriction is thought to be determined at two levels: both by the presence or absence of appropriate receptors at the cell surface, and by the presence or absence of intracellular factors allowing full gene expression and virus replication. The finding that polyomavirus infection of cells of different species from the natural host can induce cell transformation without a full virus replication cycle has led to polyomaviruses

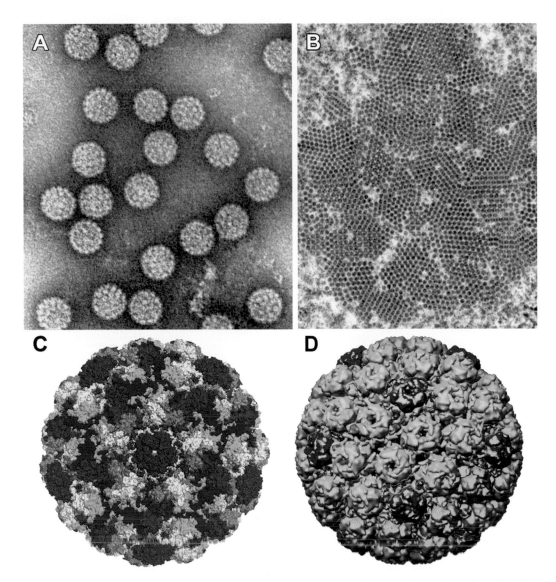

FIGURE 20.1   Polyomavirus morphology and structure. (A) Negative contrast electron microscopy of a polyomavirus. (B) Thin-section electron microscopy of masses of JC virions in the nucleus of an oligodendrocyte in brain of an immunosuppressed patient with progressive multifocal leukoencephalopathy (PML). (C) Graphic rendering of murine polyomavirus capsid assembly, composed of 72 pentamers of protein VP1—the capsid is colored according to distinct bonding environments of various VP1 proteins, showing how a single protein can be assembled into a very complex structure. (D) Reconstruction of an SV40 virion from cryoelectron microscopy images, showing vertex penton (pentomer) capsomers (blue) and hexon (hexomer) capsomers (green). *(A) From Erskine Palmer, U.S. Centers for Disease Control and Prevention; (B) from Shigeki Takeda and Hitoshi Takahashi, Niigata Neurosurgical Hospital, Japan; (C) from Research Collaboratory for Structural Bioinformatics (RCSB), Protein Data Bank (PDB); (D) from Jean-Yves Sgro, University of Wisconsin. All with permission.*

being extensively used as models for studying oncogenesis. Cells transformed by polyomaviruses express early viral proteins and usually contain integrated viral DNA. However, despite considerable research, very little evidence exists for the involvement of polyomaviruses in human cancers. In particular, many individuals were inadvertently exposed to SV40 virus as a contaminant of early batches of polio vaccine. Detailed follow-up studies of these populations have found no evidence for increased rates of cancer or a significant increase in mortality among the vaccine recipients.

BK DNA has been reported in some human tumors from a number of sites, but the significance of this is controversial because widespread asymptomatic carriage of virus occurs. Nevertheless, a WHO working group recently classified BK and JC viruses as "possibly carcinogenic for humans."

The one clear exception is MCPyV and Merkel cell carcinoma; integrated MCPyV DNA can be found within tumor cells, and RNA interference has shown an essential role for T-antigen expression, thereby suggesting at least a major etiological role for this virus in this form of cancer.

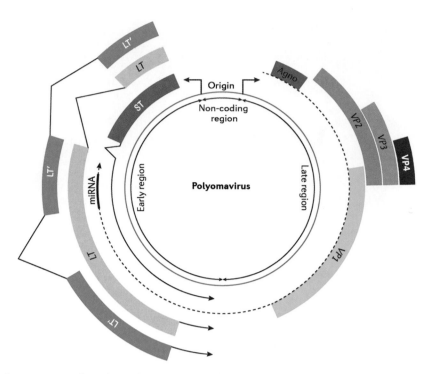

FIGURE 20.2 Scheme of a prototype murine polyomavirus genome showing closed circular, double-stranded DNA with three main regions: the non-coding control region (top) containing the early and late promoters, their transcription start sites and the origin of replication; an early region encoding large T antigen (LT) and small t antigen (ST), and an alternatively spliced LT (LT′); and a late region encoding the viral capsid proteins VP1, VP2, VP3, and VP4. VP2, VP3, and VP4 are translated in the same reading frame and they terminate at the same site, but translation starts at successive initiating AUG codons to generate the different proteins. VP4 has been confirmed only in the primate virus SV40. VP1 is read in a different reading frame. Agnoprotein (Agno) is encoded by a late transcript in JC and BK polyomaviruses, but has yet to be confirmed as present in the newly described polyomaviruses. Also, a microRNA present in the late transcripts, that overlaps and hence targets the large T-antigen transcript, is shown. *Reproduced from DeCaprio, J.A., Garcea, R.L., 2013. Nat. Rev. Microb. 11, 264–276, with permission.*

BOX 20.1 Proposed Pathological Mechanisms in Polyomavirus-Induced Disease

Mechanism	Example
1. Direct cytopathic effect associated with high-level virus replication	Oligodendrocytes infected with JC virus in PML
2. Immune-reconstitution inflammatory syndrome (IRIS)	(i) BK associated hemorrhagic cystitis after stem cell transplantation
	(ii) Worsening PML in AIDS patients after starting antiretroviral treatment, or after removal of natalizumab by plasmapheresis in multiple sclerosis
3. Cytopathic-inflammatory: high-level virus replication and inflammatory response	Polyomavirus-associated nephropathy in renal grafts
4. Auto-immune disease	T antigens complexed to DNA and nucleosomes inducing autoantibodies
5. Oncogenesis: Early gene expression leading to cell transformation	Merkel cell carcinoma

BK POLYOMAVIRUS

BK virus infects most children before the age of ten, often subclinically although sometimes it is associated with mild upper respiratory symptoms, suggesting that transmission may occur via the respiratory route. The viral genome persists for life in a number of organs, including the kidney and lower urinary tract, the tonsils, lymphoid tissue, and bone marrow, without any apparent ill effects. Reactivation occurs during the last trimester of about 3% of pregnancies, causing asymptomatic shedding of virus intermittently in the urine. Approximately 20% of renal transplant recipients show BK viremia within one year of transplantation, and BK virus nephritis is a significant and increasing problem leading to renal graft failure in a majority of cases. Among bone marrow transplant recipients BK virus infection is associated with hemorrhagic cystitis, and systemic infection leading to meningitis, retinitis, pneumonia, or vasculopathy has also been reported. However, the virus has not been associated with PML.

Diagnosis of BK virus infection is usually made by PCR for virus DNA in the blood and/or urine. However since asymptomatic urinary shedding and viremia are not uncommon in immunosuppressed patients, quantitative measurements of virus load are used to assess clinical significance. Other investigations are urine cytology to detect "decoy" cells containing viral inclusions, and renal biopsy. Isolation of BK virus from urine in cultured human diploid fibroblasts, or JC virus from urine or brain in human fetal glial cells, are not often done.

JC POLYOMAVIRUS

JC virus has a similar natural history to BK virus, although primary infections may occur somewhat later in childhood and only about 75% of older healthy adults have antibody. Again, lifelong persistence is established in the kidney, and virus is shed in the urine sporadically throughout life, more frequently during pregnancy or immunosuppression. JC virus causes progressive multifocal leukoencephalopathy (PML), an uncommon subacute demyelinating disease of the CNS that is invariably fatal. PML occurs as a complication of advanced disseminated malignant conditions such as Hodgkin's disease or chronic lymphocytic leukemia, and also in primary or secondary immunodeficiency syndromes, or following immunosuppression for organ transplantation. JC virus became more common in the early years of the AIDS pandemic, affecting around 5% of AIDS patients. Indeed, the HIV-1 Tat protein has been shown to transactivate transcription of the late JC viral genes. However, AIDS-related PML has since become rare with the advent of improved antiretroviral therapy. The target cell is the oligodendrocyte, in which the virus undergoes a lytic productive infection; neurons are unaffected. Histologically, the disease is characterized by multiple foci of demyelination in the brain, accompanied by proliferation of giant bizarre astrocytes. The surrounding oligodendrocytes are enlarged, with swollen nuclei occupied by a prominent inclusion body, which in fact contains a crystalline aggregate of many thousands of virions.

Clinically, PML presents as focal neurological defects, usually insidious in onset and slowly progressing as the regions of demyelination expand (Fig. 20.3). Initial diagnosis relies on clinical findings and neuro-imaging including MRI, and demonstration of JC virus DNA in the CSF by PCR provides strong confirmation. Occasionally a brain biopsy is required, and this may show characteristic tissue cytopathology, including oligodendrocytes with intranuclear inclusions, bizarre astrocytes, and lipid-laden macrophages. The presence of JC virus can be detected by immunohistochemistry, *in situ* nucleic acid hybridization, or electron microscopy. Treatments including antiviral, immunomodulatory, and psychotropic drugs have been investigated but are of no proven benefit. However PML in patients with AIDS can be arrested and in some cases show some improvement following initiation of appropriate antiretroviral therapy.

JC virus variants isolated from the brain of PML patients usually differ from the "archetype" found in the urine of asymptomatic carriers, showing extensive deletions and duplications in the nucleotide sequences within the promoter/enhancer region of the genome. These modifications may alter cell tropism and affect the switch between lytic and latent infection.

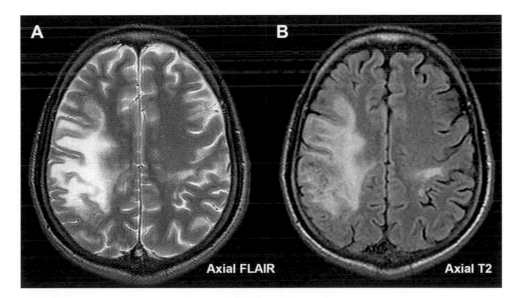

FIGURE 20.3 Progressive multifocal leukoencephalopathy (PML), CT images, Axial FLAIR (A), and Axial T2 (B) formats. Extensive right frontoparietal white matter abnormality extending to involve the subcortical white matter but sparing the cortex. There is also a focal site of infection in the left hemisphere. The patient was confirmed as having high viral loads of JC virus, consistent with the diagnosis of PML. *From Radiopedia.org, with permission.*

OTHER RECENTLY DESCRIBED HUMAN POLYOMAVIRUSES

KI and WU polyomaviruses (named after Karolinska Institute and Washington University respectively) were detected in 2007 in the nasopharyngeal aspirates from children with respiratory infections using PCR and random

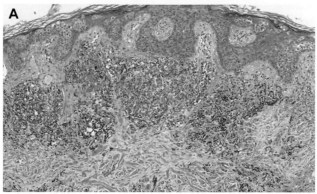

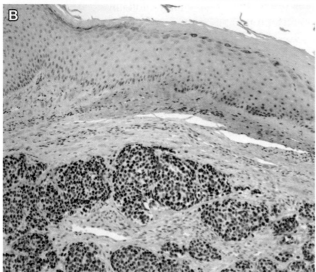

FIGURE 20.4 Merkel cell virus (MCV) was identified in Merkel cell carcinomas (MCCs) using digital transcriptome subtraction. Immunohistochemistry of lesions has clearly implicated the virus in the etiology of MCC, a rare but aggressive form of skin cancer that is increasing in incidence. (A) MCC infiltrating dermal layers of skin; H&E. (B) MCC in subepithelial layers of skin; immunohistochemistry using antibody to MCV large T protein. *Reproduced from Feng, H., et al., 2008. Clonal integration of a polyomavirus in human Merkel cell carcinoma [MC polyomavirus]. Science 319, 1096–1100, with permission.*

primers. Serological population surveys indicate that half to more than 90% of adults have been infected, with most infections occurring in childhood. At the present time no firm evidence has been found for a pathogenic role for these viruses in either human respiratory disease or cancers.

In 2008 a polyomavirus was identified using cDNAs of RNA transcripts isolated from cells from Merkel cell carcinoma, a rare aggressive neuroendocrine skin tumor occurring in the elderly or immunosuppressed (Fig. 20.4). Antibodies to MCPyV are also widespread in human populations, and around 80% of cases of Merkel cell carcinoma contain integrated MCPyV DNA. The integrated DNA is frequently truncated in the LT antigen and VP1 regions (Fig. 20.2), supporting a view that the integrated DNA may have lost its replicative ability whilst retaining the property of transformation. Extensive studies of other skin tumors and malignancies from many other sites have not revealed any association with other forms of cancer. The full mechanism of the oncogenic role of this common infection in this one specific rare form of cancer remains to be better understood. The finding of MCPyV stimulated further molecular searches for additional polyomaviruses in skin samples, with the result that HPyV6 and HPyV7 were discovered in 2010: neither appears to have any disease association at the present time.

Trichodysplasia spinulosa is a rare skin disease of immunosuppressed patients characterized by facial spines, papules, and alopecia, and patient samples had been reported in 1999 to contain intracellular virus particles. In 2010, extracts of patients' plucked facial spines were shown using rolling circle amplification to contain polyomavirus DNA fragments. TSPyV has not been found in a limited range of other tissues, and its role in human disease remains to be clarified. Similarly the significance of other polyomaviruses (HPyV9 and MWPyV) recently identified in human samples is unknown.

FURTHER READING

Dalianis, T., Hirsch, H.H., 2013. Human polyomaviruses in disease and cancer. Virology 437, 63–72.

DeCaprio, J.A., Garcea, R.L., 2013. A cornucopia of human polyomaviruses. Nature Rev. 11, 264–275.

White, M.K., Gordon, J., Khalili, K., 2013. The rapidly expanding family of human polyomaviruses: recent developments in understanding their life cycle and role in human pathology. PLoS Pathogens 9, e1003206.

Chapter 21

Parvoviruses

Members of the family *Parvoviridae* can only replicate either in dividing cells or in the presence of a helper virus, owing to the severe restriction in genetic coding capacity of the small single-stranded DNA genome. This requirement for cell division accounts for the tropism of parvoviruses for cells of the bone marrow, gut, and the developing fetus. Human parvovirus B19, originally discovered fortuitously in the serum of asymptomatic blood donors, has subsequently turned out to be associated not only with a very common exanthematous disease of children, but also linked to aplastic crises in patients with chronic hemolytic anemia, as well as with hydrops fetalis. The genus bocavirus includes several newly recognized human parvoviruses that are associated with respiratory infections and gastroenteritis. Parvoviruses of cats, dogs, and mink cause panleukopenia and enteritis, while a rat parvovirus causes congenital malformation of the fetus. The unique genomic structure and mechanism of replication of parvoviruses have attracted considerable interest for molecular biologists.

CLASSIFICATION

The family *Parvoviridae* contains two subfamilies *Parvovirinae* and *Densovirinae*. Members of the latter subfamily infect arthropods and will not be discussed further, but within the *Parvovirinae* there are three genera that include viruses of humans, *Erythrovirus*, *Dependovirus*, and *Bocavirus*, together with the two genera *Parvovirus* and *Amdovirus* containing important viruses of animals. The genus *Erythrovirus* contains the important human pathogen parvovirus B19 and similar viruses of macaques and other non-human primates. The genus *Dependovirus* was so named because, with a few exceptions, the replication of these viruses depends upon co-infection with helper adenoviruses or herpesviruses; there are a number of species infecting different animals, with the human representatives being the adeno-associated viruses (AAV) types 1 to 5. The genus *Bocavirus*, named originally because it included *bo*vine and *ca*nine representatives, now includes human bocaviruses 1 to 4. Other veterinary examples include canine parvovirus strain CPV-N, mink enteritis virus, the

quaintly named minute virus of mice (all species within the genus *Parvovirus*), and Aleutian mink disease, the type species of the genus *Amdovirus* (Table 21.1).

PROPERTIES OF PARVOVIRIDAE

The virions of parvoviruses are among the smallest viruses known, with diameters between 21 and 25 nm (Table 21.2). Virions consist of a simple non-enveloped icosahedral capsid surrounding a linear single-stranded DNA molecule with a restricted coding potential (5 kb for human parvovirus B19 and 4.7 kb for human dependoviruses). The capsid is composed of 60 protein subunits, mainly the VP2 protein, but around 5% comprises the larger VP1 protein whose sequence overlaps VP2 with an extension at the N-terminal end. Most of the outer surface consists of α-helices that are responsible for antigenic determinants and receptor binding

TABLE 21.1 Human Infections Associated with *Parvoviridae*

Genus	Virus	Disease
Erythrovirus	B19	Erythema infectiosum (fifth disease)
		Arthritis (especially in young women)
		Aplastic crisis in chronic hemolytic anemia
		Chronic anemia in immunodeficiency syndromes
		Hydrops fetalis
Dependovirus	AAV (1–5)	None
Bocavirus	HBoV1	Acute lower respiratory tract infection, wheezing[a]
	HBoV2–4	Acute gastroenteritis[a]

[a]Although HBoV1 and HBoV2–4 are regularly found in clinical samples from children with respiratory and gastrointestinal symptoms respectively, their pathogenic role is yet to be defined.

Fenner and White's Medical Virology. DOI: http://dx.doi.org/10.1016/B978-0-12-375156-0.00021-7

TABLE 21.2 Properties of *Parvoviridae*

Five mammalian genera: *Parvovirus, Erythrovirus, Dependovirus, Amdovirus,* and *Bocavirus*
Icosahedral virion, 20–25 nm, capsid composed of 60 protein subunits
Linear minus sense ssDNA genome, 5 kb, palindromic hairpins at each end
Unique "rolling hairpin" mechanism of DNA replication
Replicate in nucleus of dividing cells, using cellular enzymes
Relatively stable to heat (60°C) and pH (3–9)
Dependovirus usually requires helper virus; persists by integration

(Fig. 21.1). The virus is highly resustant to desiccation and disinfectants, resisting 60°C for some hours and a wide range of pH from pH 3 to 9. The virions are approximately 75% protein and 25% DNA, and do not contain lipid or carbohydrate.

The ssDNA parvovirus genome is of negative polarity, that is complementary to viral mRNA, but some virions (up to 50% in the case of the genus *Dependovirus* and the human B19 parvovirus) contain a DNA positive strand. Both ends of the genome contain palindromic sequences, 120 to 550 nucleotides in length according to the virus, enabling each end of the molecule to fold back on itself to form a hairpin structure; these structures are essential for replication. With some viruses, for example dependoviruses, the sequences at each end of the dependovirus genome may be related and may be viewed as terminal repeats. In contrast, the terminal hairpins at each end are unrelated in the case of members of the genus *Parvovirus* (Fig. 21.2).

VIRAL REPLICATION

In contrast to double-stranded DNA viruses that induce resting host cells to enter the S phase, for example polyomavirus (see Chapter 20: Polyomaviruses), the small ssDNA parvoviruses lack this capacity and hence can replicate only in dividing cells, either as a naturally occurring event (autonomous parvoviruses) or in concert with its induction by a helper virus (dependoviruses). Parvoviruses replicate in the nucleus, where both transcription and replication of the genome occur, the non-structural proteins accumulate, and subsequently nascent virions assemble. A cellular DNA polymerase is used to transcribe the viral ssDNA into dsDNA, which then serves as the template for transcription of mRNA by cellular DNA-dependent RNA polymerase II. There are no enzymes present in the virion nor coded by the viral genome itself to perform these functions. Non-structural proteins required for transcription and DNA

replication are coded by a gene cassette (known as the REP ORF) at the left side of the genome, and structural proteins of the capsid by a second cassette (the CP ORF) on the right. For most parvoviruses, both cassettes are read from the same DNA strand, although for some including densoviruses, the REP and CP functions are read from the 5′ ends of opposite sense strands (an ambisense strategy). Alternative splicing patterns give rise to a number of different mRNA species, and this process together with leaky scanning and post-translational protein cleavage leads to a greater number of different proteins than the limited coding potential of the short genome might suggest. The splicing program differs from one parvovirus to another. Certain non-structural proteins serve to transactivate the viral promoter(s) and to down-regulate transcription from certain cellular promoters.

The mechanism of replication of the genome, termed a "rolling hairpin" is extraordinary. The palindromic 3′-terminal hairpin serves as a self-primer for initiation of synthesis of a plus sense DNA strand to give a covalently closed duplex molecule. The viral protein NS1 then performs a "hairpin transfer" reaction by nicking the ligated molecule allowing formation of a "rabbit's ear" structure, after which progeny minus strands are in turn transcribed from the newly synthesized plus strand, again using the hairpin structure as a primer. Subsequent rounds of DNA duplication lead to the formation of double-stranded dimeric and tetrameric concatemers, from which individual genomic monomeric duplexes are then excised by endonuclease cleavage (Fig. 21.3).

In the case of adeno-associated viruses, infection may be productive (in the presence of helper virus) or it may be latent involving integration of tandem viral genome multimers into the host cell chromosome.

PARVOVIRUS B19

A parvovirus designated B19, now allocated to the genus *Erythrovirus*, was originally detected by Yvonne Cossart and colleagues in 1975 while screening sera from healthy blood donors for hepatitis B surface antigen. The virus has turned out to be the cause of a range of quite distinct clinical syndromes (Table 21.1). The commonest, erythema infectiosum, is a mild self-limited condition seen in normal children and adults. The rarer but more serious manifestations of infection occur only in patients with (1) an underlying congenital or acquired immunodeficiency, or (2) a requirement for accelerated erythropoiesis, for example, in chronic hemolytic anemia, or (3) pregnancy, in which both (1) and (2) also apply to some degree.

PATHOGENESIS AND IMMUNITY

Our understanding of the pathogenesis of parvovirus B19 infection was illuminated by a study in human volunteers—something possible only in the case of a relatively harmless

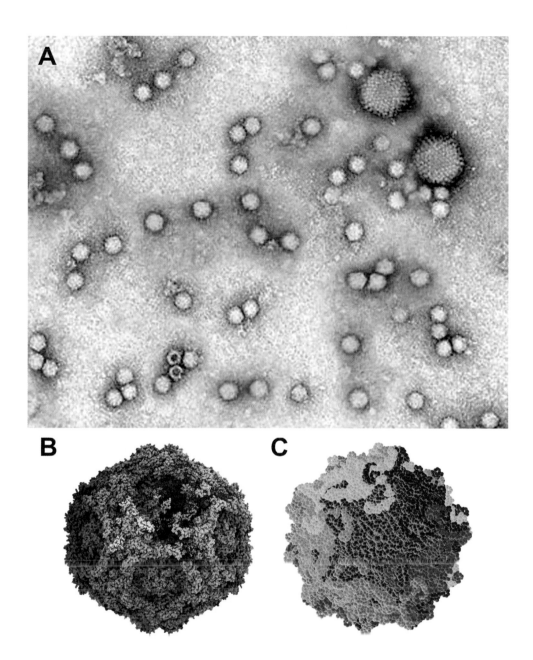

FIGURE 21.1 Parvovirus morphology and structure. Parvoviruses are extremely resistant to harsh physical conditions; this is based in their compact icosahedral structure and more so in particular details of their capsid construction. (A) Adeno-associated virus 2, now being used widely as a recombinant DNA vector for heterologous genes such as in vectored vaccines, along with two helper adenovirus virions. Negative contrast electron microscopy. (B and C) Human parvovirus B19 models constructed from X-ray crystallographic and cryoelectron microscopy image analyses. These models show that even though the parvovirus capsid is constructed from only one protein, complex interdigitation of each molecule results in sturdy virions. This is shown in (C), where some individual protein molecules are colored differently showing their complex interdigitation. *(B) From Jean-Yves Sgro, Institute for Molecular Virology, University of Wisconsin; (C) from Michael Rossmann, Division of Biological Sciences, Purdue University, with permission.*

virus such as this. Just over a week after intranasal inoculation of seronegative volunteers, a short-lived but high-titer viremia reached a peak, with virus being shed from the throat for a few days. By about the tenth day, neither erythroid precursors could be detected in bone marrow, nor reticulocytes in the blood; hemoglobin declined only very slightly, and there were no symptoms of anemia. Clinically, the volunteers displayed a biphasic illness. The first episode between day 8 and day 11 comprised fever, malaise, myalgia, and chills, coinciding with the peak levels of virus in the bloodstream and the destruction of erythroblasts in the bone marrow. Later, the rash and arthralgia occurred between days 17 and 24, that is after the viremia had disappeared but at a time when IgM antibodies had peaked and IgG had begun to rise. This timing is consistent with the hypothesis that the rash and arthritis are mediated by immune complexes, as in the case of Aleutian mink disease, a parvovirus infection of animals (Fig. 21.4).

family *Parvoviridae*
genus *Erythrovirus*, human parvovirus B19
genus *Dependovirus*, adeno-associated viruses

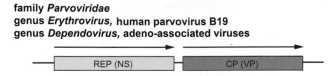

FIGURE 21.2 Gene organization of different members of the family *Parvoviridae*. Non-structural proteins are encoded by the REP ORF, and the capsid proteins by the CP ORF. All parvoviruses infecting humans belong to the subfamily *Parvovirinae*. *Reproduced from King, A.M.Q., et al., 2012. Virus Taxonomy. In: Ninth Report of the International Committee on Taxonomy of Viruses, Academic Press, p. 407, with permission.*

A transient aplastic crisis may ensue when chronic hemolytic anemia patients undergo a naturally occurring parvovirus B19 infection. Erythroblasts then reticulocytes vanish just a few days after viremia reaches a maximum. In these patients, however, the aplastic crisis is profound because the average life of circulating red cells may be only 15 to 20 days (compared with 120 days in a normal individual) and, despite the efforts of the bone marrow to compensate with increased erythrocyte production, the hemoglobin level is already low.

The remarkable selectivity of parvovirus B19 for erythrocyte progenitors was first demonstrated directly in cultures of bone marrow cells; efficient replication occurs in erythrocyte precursors, and has also been reported in cell lines of megakaryoblastoid or erythroleukemic origin. The virus receptor is the glycolipid globoside, also known as the blood group P antigen, which is present on cells of the erythroid series. P antigen is also expressed on fetal myocardial cells and these cells may become infected,

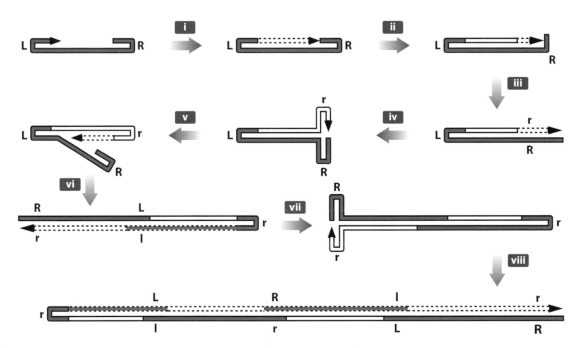

FIGURE 21.3 Parvovirus DNA replication, called rolling hairpin replication, is remarkably complex—and made more complex in that some parvoviruses have a positive-sense genome, some a negative-sense, and many encapsidate positive- and negative-sense DNA (e.g., human parvovirus B19, equal amounts of each). Here, minute virus of mice replication is illustrated: it has negative-sense genomic DNA. Parental genomic DNA is blue, newly synthesized DNA is represented by dashed lines, with its 3′-end capped with an arrowhead. Inverted palindromic terminal sequences are denoted as (R) and (r). In step [i] the left telomere of virion DNA folds back on itself (hairpin), allowing priming and synthesis of a complementary positive-sense strand. This creates a monomer-length duplex intermediate in which the two strands are joined at their left end. The 3′ end of the new DNA strand is ligated to the 5′ end of the right-hand hairpin by a host ligase to generate a covalently closed duplex copy. Further replication requires expression of NS1, which nicks the ligated strand, allowing copying of the hairpin sequence (steps [ii] and [iii]) and refolding into two hairpins (step [iv]). This creates a rabbit-ear structure that reverses the path of the fork and redirects synthesis back as shown (step [v]). The fork then progresses back along the duplex, displacing the original strand and replacing it with a covalently continuous new strand. During fork progression, the left hairpin is unfolded and copied, leading first to the synthesis of a duplex dimer (step [vi]), which is then similarly processed (step [vii]) to form a tetramer intermediate (step [viii]). Overall, the result is that duplex dimeric and tetrameric concatemers are generated, in which alternating unit-length viral genomes are fused. Unit-length genomes are enzymatically excised from these intermediates and their palindromic termini are then added. *Reproduced from Cotmore, S.F., Tattersall, P., 2014. Parvoviruses: small does not mean simple. Annu. Rev. Virol. 1, 517–537, with permission.*

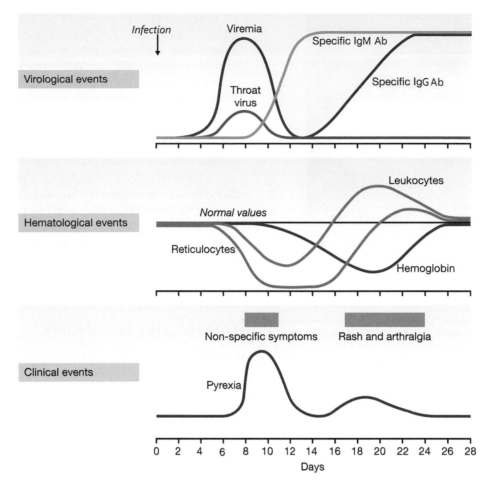

FIGURE 21.4 Schematic representation of the virologic, hematologic, and clinical events in parvovirus B19 infection. *Reproduced from Zuckerman, A.J., et al., eds., 2000. Principles and Practice of Clinical Virology, fourth ed., Wiley, Chichester, with permission.*

thereby contributing to the direct myocardial effects seen in fetal infections. It is also possible that other types of cells, for example neutrophils, are permissive or semi-permissive for parvovirus infection.

The dominant immune response to parvovirus B19 infection appears to be humoral, because antibody production coincides with the disappearance of viremia, IgG antibodies appear to determine immunity to reinfection, and patients with persistent infection show low or absent levels of antibody. The immunological aspects of disease progression in chronic arthritis, and in chronic anemia in the immunocompromised as well as in hydrops fetalis, merit further investigation.

CLINICAL FEATURES

Erythema Infectiosum

Erythema infectiosum, once known as "fifth disease," is an innocuous contagious exanthem of childhood which has been well-known to pediatricians for over a century although its cause remained a mystery until the 1980s. The classical rash evolves through three stages. First, there is a striking flushed cheek appearance (hence the familiar synonym "slapped cheek syndrome"). This is followed one to two days later by an erythematous maculo-papular rash on the limbs and trunk. During the third stage, the rash fades rapidly, sometimes developing the appearance of fine lace, and may fluctuate and occasionally be pruritic; this stage may last one to three weeks (Fig. 21.5). Though fleeting, the rash may reappear during the next few weeks or months following stimuli such as bathing or exposure to sunlight. There is wide variation in the form of the rash, which may be indistinguishable from rubella. As with rubella, arthralgia is seen occasionally in children and occurs in about 50% of adults, especially women, with a predilection for the peripheral joints of the hands, wrists, knees, and ankles. Indeed, polyarthralgia, sometimes without a rash, is often the dominant feature and may smolder on for weeks or months. The arthralgia usually lasts one to three weeks, but can on occasions persist for months, but joint destruction does not occur. Approximately 25 to 50% of infections are asymptomatic.

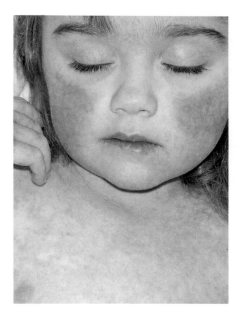

FIGURE 21.5 Fifth disease, human parvovirus B19. A 4-year-old girl with cyclic vomiting developed fever and malaise in association with a diffuse reticulated exanthem with slapped cheek erythema and purpura particularly on the hands and feet. PCR for parvovirus B 19 positive. *From Bernard Cohen, MD, Johns Hopkins Medical Center, with permission.*

TRANSIENT APLASTIC CRISIS

Transient aplastic crisis is a temporary but potentially life-threatening complication of various forms of chronic hemolytic anemia, such as sickle cell anemia, thalassemia, or hereditary spherocytosis. It was suspected from the 1960s that an infectious agent might play a role, but from the 1980s it became clear that the great majority of these episodes are directly attributable to infection with parvovirus B19. The patient presents with the pallor, weakness, and lethargy characteristic of severe anemia, often following a viral-like illness, and is found to have suffered a sudden drop in hemoglobin associated with almost total disappearance of erythrocyte precursors from the bone marrow and reticulocytes from the blood. There is usually no rash. Recovery generally occurs spontaneously within a week but can be expedited by blood transfusion, which is sometimes lifesaving.

OTHER COMPLICATIONS OF PARVOVIRUS B19 INFECTION

Chronic anemia in immunodeficient patients has been observed when parvovirus B19 infects patients with acute leukemia on chemotherapy, AIDS patients, bone marrow transplant recipients, or children with congenital immune deficiency. Anemia can be severe requiring transfusion, but clinical features of erythema infectiosum are absent because antigen–antibody complexes do not form due to the absence of an antiviral immune response.

Infection during pregnancy does not cause fetal malformations, but is associated with a 30% risk of transmission to the fetus and a 5 to 10% risk of fetal loss, more commonly in the first half of pregnancy. Fetuses lost in the second and third trimester show clinical signs of *hydrops fetalis*, a condition marked by generalized edema. This is presumed to result from severe anemia and perhaps also myocarditis, both contributing to congestive cardiac failure. The causes of this rather rare condition include immunological conditions, for example Rh incompatibility, and a range of infections including a primary parvovirus B19 infection occurring a few weeks beforehand.

Reports have associated a number of other clinical conditions with parvovirus B19 infection including rheumatoid arthritis, but the widespread nature of this infection and its persistence both in the environment and the host may mislead; careful controlled studies have usually failed to validate such reports.

LABORATORY DIAGNOSIS

Currently, the cornerstone for diagnosis of acute parvovirus infections in a child or adult is the demonstration of IgM antibodies (or a significant rise in IgG antibodies) by enzyme immunoassay (EIA) using recombinant capsid antigens. IgM antibodies are detected in almost all cases of erythema infectiosum at the time of presentation: they appear within a few days of the onset of transient aplastic crisis, and persist for two to three months after acute infection. Interpreting levels of IgG antibodies is less reliable, because of a high prevalence (50 to 60%) in the community and variability in level between patients. Detection of viral DNA by PCR is required to diagnose persistent infection, where antibody production is often very low or absent, and PCR can also be useful early in a transient aplastic crisis. An EIA for viral antigen in acute-phase serum has also been used, especially in aplastic crisis. Care is needed in interpreting these tests, as PCR may detect low levels of viremia that can persist for extended periods of time after an acute infection or even in healthy adults. Diagnosis of fetal infection is made by the demonstration of virus DNA in amniotic fluid and/or cord blood; maternal infection can be demonstrated by seroconversion, but maternal IgM antibodies may be negative at the onset of hydrops.

Parvovirus B19 has been cultured in cells from human bone marrow or fetal liver cells in the presence of erythropoietin and interleukin-3. The virus can also be grown in human erythroleukemic and megakaryoblastoid cell lines, for example MB-02, grown in the presence of the cytokine GM-CSF, following erythroid differentiation as a result of treatment with erythropoietin. However these systems are not practical for everyday use.

Histologically, giant pronormoblasts are seen in bone marrow. The infected erythroid precursor cells display characteristically large eosinophilic intranuclear inclusions with surrounding margination of the nuclear chromatin. Electron microscopy reveals crystalline arrays of virions in the nucleus. *In situ* hybridization for parvovirus DNA can also be used to demonstrate the presence of the genome in acute or chronic infections and hydrops fetalis.

EPIDEMIOLOGY

Parvovirus B19 is ubiquitous, common, and highly contagious. It is present year-round with a tendency to produce spring epidemics among schoolchildren four to ten years old. Because parvovirus B19 is readily transmitted by respiratory secretions and close contact, the attack rates among susceptible individuals during epidemics within day-care centers and schools are about 25%, and may reach 50% in household contacts. Over 50% of adults are seropositive and immune, and new infections continue at a low rate throughout life. Transmission probably occurs during the incubation period of erythema infectiosum, the patient being no longer infectious by the time the rash appears (Fig. 21.5). In contrast, chronic hemolytic anemia patients are infectious for up to one week after onset of an aplastic crisis, and immunocompromised patients with chronic B19 anemia may excrete virus for months or even years.

Transplacental transmission represents a second route by which the infection may be acquired. About 30% of primary maternal infections lead to infection of the fetus, but prospective studies suggest that less than 10% of primary maternal infections cause fetal death: congenital malformations rarely if ever occur. Transmission can also occur via blood transfusion, but infectious donations are so rare that the routine screening of blood is not justified. Factor VIII and IX administered to hemophiliacs is a greater problem, as the virus is sufficiently heat-stable to survive in clotting factor concentrates, and as a consequence since 2002 many blood product manufacturers have instituted screening for B19 DNA to reduce the risk of iatrogenic transmission.

TREATMENT AND CONTROL

Erythema infectiosum does not require treatment, but aplastic crises in chronic hemolytic anemia can be life-threatening, often requiring supportive care and occasionally blood transfusion. Intravenous administration of normal human immunoglobulin has been shown to be beneficial in the treatment of severe persistent anemia in immunocompromised patients, and infection usually responds well either to discontinuing immunosuppressive therapy, or the use of anti-retroviral therapy.

There is a substantial risk of transmission of parvovirus B19 to susceptible children and staff in schools and day-care centers, hospitals during nosocomial outbreaks, or when nursing chronic hemolytic anemia patients with an aplastic crisis. Hence, pregnant non-immune women, immuno-compromised individuals, or those with chronic hemolytic anemia, should be appraised of the potential risks.

DEPENDOVIRUSES

There are five antigenically distinct species of human adeno-associated viruses (AAV), and further AAV species have been isolated from animal hosts. These viruses were so named being first isolated from the throat or feces of humans concurrently infected with an adenovirus (Fig. 17.1A). These new viruses were shown as defective through their requirement of an adenovirus as a helper virus, hence the generic name *Dependovirus*. In the absence of a helper virus, a tandemly repeated double-stranded form of the AAV genome becomes integrated into the cellular genome where it persists indefinitely until rescued by subsequent superinfection with adenovirus; moreover, various adenovirus (or herpesvirus) "early" proteins transactivate AAV gene expression and provide other functions that facilitate the replication of AAV by modifying certain cellular activities. Very probably, this is how AAV, despite its genetic limitations, survives in nature. However, this requirement is not absolute, as under certain conditions, for example the synchronization of cell division with hydroxyurea, or treatment with mutagens, these viruses are also capable of replicating autonomously. Thus, the dependoviruses may not be as different from the "autonomous" parvoviruses as originally thought.

There is no evidence that adeno-associated viruses cause any disease. They have, however, a unique form of extreme parasitism. They have also been developed as a potential vector for gene therapy, because of the ability of virus to persist in the absence of any pathogenicity.

HUMAN BOCAVIRUSES

Parvoviruses of dogs (canine minute virus) and cattle (bovine parvovirus) have been recognized for decades, and assigned to a unique genus *Bocavirus*. From time to time small round viruses of uncertain significance have been seen by electron microscopy in human fecal samples, and limited sequencing of DNA isolated from such material showed a similarity to the B19 parvovirus. In 2005 a human bocavirus (HBoV1) was identified in respiratory secretions from children in Sweden, and in 2009 HBoV2 and HBoV3 were reported in stool samples from children in the United States and Australia. More recently a fourth member HBoV4 was also reported in stool samples. The human bocavirus genome encodes 4 proteins—non-structural protein 1 (NS1), a DNA-binding protein involved in DNA replication and gene transcription, the capsid proteins VP1 and VP2, and a nuclear phosphoprotein NP-1 of uncertain function.

HBoV1 appears to be primarily a respiratory agent and infection is widespread. Because it is commonly found in respiratory samples from sick children, with or without other recognized pathogens, it is not clear whether it can be a primary cause of disease, whether it contributes to more severe disease, or is usually a harmless passenger. HBoV2 and 3 are found in stool samples, particularly from cases of acute gastroenteritis, and in one study were the third most common agents found after rotavirus and astrovirus. However, repeated careful case control studies are required to fully define the true pathogenic role of these viruses.

FURTHER READING

Heegaard, E.D., Brown, K.E., 2002. Human parvovirus B19. Clin. Micro. Rev. 15, 485–505.

Jartti, T., et al., 2012. Human bocavirus—the first five years. Rev. Med. Virol. 22, 46–64.

Young, N.S., Brown, K.E., 2004. Parvovirus B19. NEJM 350, 586–597.

Chapter 22

Hepatitis B and Hepatitis Delta Viruses

HEPATITIS B VIRUS

Hepatitis B is one of the world's major unconquered diseases. Some 300 million people are chronic carriers of the virus, and a significant minority of these go on to develop cirrhosis or cancer of the liver. It is estimated that up to 2 billion people have been exposed to this virus at some time in their life. Reliable diagnostic procedures and vaccines have been available for several decades. Notwithstanding, worldwide there are still more than 600,000 deaths each year. The risk of post-transfusion hepatitis has been all but eliminated in most economically developed regions of the world, and childhood vaccination has been implemented in those countries where virus prevalence is high. Owing to the large number of chronically infected individuals, however, it may take several generations before such implementation has a significant impact on the incidence of chronic liver disease and related hepatocellular carcinoma (HCC).

In 1963 Baruch Blumberg, a geneticist investigating hereditary factors in the sera of isolated racial groups, discovered an antigen in the serum of an Aboriginal Australian that reacted with sera from multiply transfused American hemophiliacs. In due course the antigen was demonstrated to be present on the surface of particles with three different morphological forms known to be associated with serum hepatitis, now known as hepatitis B. The 22-nm particles of "Australia antigen," subsequently renamed as hepatitis B surface antigen (HBsAg), do not contain nucleic acid, and subsequently the discovery of the 42-nm particles by David Dane identified the true infectious virions capable of transmitting hepatitis. The unique characteristics of these viruses led to a new family, the *Hepadnaviridae*, reflecting the association with hepatitis and the presence of a DNA genome. These ground-breaking discoveries (for which Blumberg was awarded the Nobel prize) and the work that flowed from them completely rewrote the understanding of chronic liver disease and provided new concepts that led to our current management and prevention of blood-borne infections.

Classification

The family *Hepadnaviridae* contains two genera: hepatitis viruses specific for man, woodchucks, ground-squirrels, and other mammals are grouped in the genus *Orthohepadnavirus*, while those of birds (ducks, herons, snow geese, and others) are placed into the genus *Avihepadnavirus*. A number of hepadnaviruses of bats have been described as closely related to hepadnaviruses of rodents, calling into question the historical view that hepadnaviruses have evolved through contact with infected non-human primates. Studies of duck (DHBV) and woodchuck hepatitis virus (WHV) hepadnaviruses have been pivotal in our understanding the replication and pathogenesis of hepatitis B (Fig. 22.1).

Virion Properties

The virion consists of a 27-nm icosahedral nucleocapsid ("core") constructed from 180 capsomers, surrounded by a closely fitting "envelope." The virion is relatively heat-stable, but sensitive to acid and to lipid solvents. The genome consists of a 3.2 kbp molecule of circular double stranded DNA with an unusual structure (Fig. 22.2). The positive strand is incomplete, leaving 15% to 50% of the molecule single-stranded; the negative strand is complete but contains a discontinuity ("nick") at a unique site. The 5' termini of the positive and negative strands overlap by about 240 nucleotides, and include short direct repeats, DR1 and DR2, producing "cohesive" ends that base-pair to maintain the genome in a relaxed circular configuration. The "terminal protein" of the polymerase is covalently bound to the 5' end of the negative strand, whereas a 5'-capped oligoribonucleotide primer is attached to the 5' end of the positive strand. The negative strand contains four open reading frames: pre-S/S, pre-C/C, P (or POL), and X (Fig. 22.2). The P gene, which comprises 80% of the genome and overlaps all the other genes, encodes a polymerase with three distinct enzymatic functions: DNA polymerase, reverse transcriptase (RT), and RNase H, and also encodes the terminal protein primer. Gene X, spanning the cohesive ends of

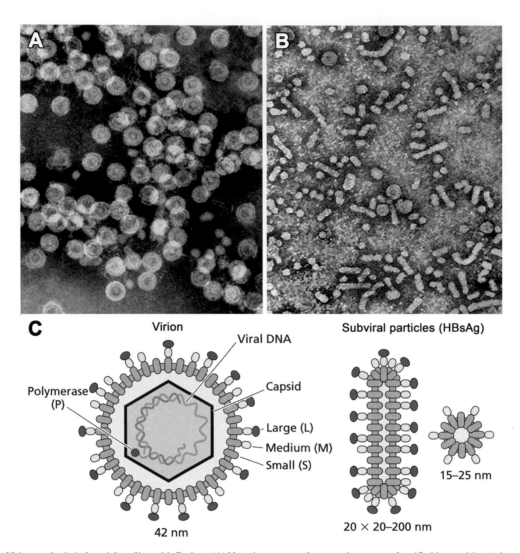

FIGURE 22.1 Virions and subviral particles of hepatitis B virus. (A) Negative contrast electron microscopy of purified intact virions (also called Dane particles), showing core and envelope of particles. (B) Negative contrast electron microscopy of subviral particles—hepatitis B surface protein (HBsAg) organized as 22 nm spheres, and as 22 to more than 200 nm rod-shaped particles. A few intact virions are also present. Note: in most patients' sera, the 22 nm spherical particles usually outnumber the other forms by 10- to 1000-fold. (C) Model of an intact virion and subviral particles showing constituents. *(C) Reproduced from Flint, S.J., et al., 2015. Principles of Virology: Pathogenesis and Control, fourth ed. Washington, DC, ASM Press, with permission.*

the genome, encodes a transactivating protein that upregulates transcription from all the viral and some cellular promoters. The *C* gene has two in-frame translation initiation sites that mark the 5′ ends of the *Pre-C* and a *C* region, respectively; translation of *Pre-C* + *C* produces the HBeAg protein, while translation of *C* alone produces HBcAg. The *Pre-S/S* gene encodes the envelope protein, S, which occurs in three forms: a large (L) protein, translated from the first of the three in-phase initiation codons, is a single polypeptide encoded by the *Pre-S1*, *Pre-S2* plus *S* regions of the genome and is found in the envelope of infectious virions; a middle-sized (M) protein comprises Pre-S2 plus S; and finally, the most abundant product is the S protein, the basic constituent of non-infectious HBsAg particles, comprising only the product of the S ORF. All three forms are glycosylated and Pre-S1 is myristoylated.

Hepadnaviruses code for three major antigens, designated surface (HBsAg), core (HBcAg), and *e* (HBeAg). The coding mechanisms for HBsAg and HBcAg are described above. HBeAg is formed from the polyprotein product of the *Pre-C* + *C* ORFs. *Pre-C* contains a hydrophobic signal sequence at its 5′-end that directs the HBeAg molecule to the endoplasmic reticulum where it is removed by proteolytic cleavage. Subsequently, after further cleavage from the C terminal end, it is secreted as an 18 K molecular weight monomeric protein HBeAg.

A number of different HBsAg subtypes of hepatitis B virus (HBV) were recognized early; these are defined serologically by various combinations of antigenic determinants present on the 22-nm HBsAg S protein. All have the same group-specific determinant, *a*, but there are four major subtype-specific determinants, certain pairs of which (*d* and *y*; *r* and *w*) tend to behave as alleles, that is, as mutually exclusive alternatives. Different subtypes tend to show characteristic geographical distributions, though they often overlap. There appears to be an incomplete relationship between genotypes and serotypes (Table 22.1).

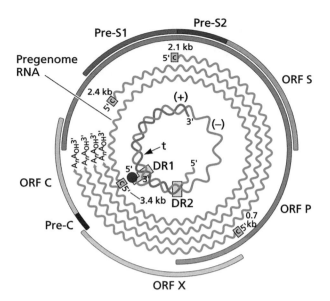

FIGURE 22.2 Genome organization and regulatory elements of hepatitis B virus (HBV). The inner blue lines show the circular DNA genome, which is double-stranded except for a short variable gap in the positive strand. The 5' end of the full-length, negative-sense strand is linked to the viral DNA polymerase. The partially double-stranded DNA is rendered fully double-stranded by completion of the (+) sense strand and removal of a protein molecule from the (−) sense strand and a short sequence of RNA from the (+) sense strand. Non-coding bases are removed from the ends of the (−) sense strand and the ends are rejoined. The outermost circles show the four known genes, called C, X, P, and S. The core protein and pre-core protein are encoded in gene C (HBcAg). HBeAg is produced by proteolytic processing of the pre-core protein. The DNA polymerase is encoded in open reading frame P. Open reading frame S encodes the surface antigen (HBsAg)—it has one long open reading frame but contains three in frame "start" (ATG) codons that divide the gene into three sections, pre-S1, pre-S2, and S. Because of the multiple start codons, polypeptides of three different sizes are translated: large (pre-S1 and 2 and S), middle (pre-S2, S), and small (S). The function of the protein coded for by gene X is not fully understood but it is associated with the development of hepatocellular carcinoma. The green circles show the various species of mRNA and the "pregenomic" RNA, which is greater than unit length and is the template for synthesis of the viral DNA genome by reverse transcription. *Reproduced from Flint, S.J., et al., 2016. Principles of Virology fourth ed., ASM Press Washington, DC, with permission.*

A comparison of complete HBV genomes has revealed 10 genotypes, A to J, as well as closely related genotypes of hepadnaviruses from chimpanzees, gibbons, and orangutans. The differences are based on intergroup divergence of at least 8%. The distribution of HBV genotypes varies according to region and can determine both disease outcome and response to antiviral treatment. Increasingly, isolates from the same genotype are being further classified into sub-genotypes on the basis of nucleotide sequence divergence of >4%. Importantly, currently available hepatitis B vaccines protect against all known HBV genotypes, although variability in the S protein is often recorded in samples from chronically infected individuals. The rapidity of HBeAg seroconversion (see below), the extent of mutational patterns in the Pre-C and C promoter regions of the genome, and disease severity all appear to be linked to HBV genotype. HBV genotype distribution is closely allied to the ethnic background of HBV-infected individuals.

Genotype A is common in Europe and North America, whereas genotypes B, C, and I predominate in Southeast Asia and the Far East. Genotype D is particularly prevalent around the Mediterranean basin and on the Indian subcontinent, whereas genotype E is found in African countries. The remaining genotypes occur less frequently, genotypes F and H being reported in Central and Latin America, and genotype G in France and the United States. Genotype I has been reported only from Southeast Asia and genotype J from a Japanese war veteran who was found in the Borneo jungle. Sub-genotypes are known: for example, the B1 genotype is found only in Japan, genotype B2 is prevalent in China and bordering countries and the dominant genotype in Polynesia is D4. Genotype C1 is prevalent in Southeast Asia, whereas C2 circulates in the more northerly latitudes of Korea, Japan, and northern China. At least some sub-genotypes are thought to be the result of recombination events in dual infections (Fig. 22.3).

TABLE 22.1 Properties of Hepatitis B Virus

Spherical, enveloped virion, 42 nm, enclosing inner icosahedral 27 nm nucleocapsid (core) composed of 180 capsomeres
Envelope contains the glycoprotein HBsAg of three different size species with common C-termini, L-, M-, and S-HBsAg
Core contains the phosphoprotein HBcAg, plus polymerase with three enzyme activities: reverse transcriptase, DNA polymerase, and RNase H
Relaxed circular dsDNA genome, 3.2 kb, cohesive 5' ends; minus strand nicked, 5' end covalently bound to polymerase; plus strand incomplete, 5' RNA primer
Four overlapping open reading frames: *pre-S1 + pre-S2 + S, C, P*, and *X*
Genome converted to supercoiled covalently closed circular (ccc) form and transcribed in nucleus to produce full-length pregenome RNA and subgenomic mRNAs
RNA pregenome in cytoplasmic core particles is reverse transcribed to dsDNA; some return to nucleus to augment pool of viral supercoiled DNA
Cores bud through endoplasmic reticulum, acquiring lipid membrane containing HBsAg; non-cytocidal
Serological subtypes of HBsAg differ in allelic pairs determinants (*d* or *y; r* or *w*); 10 genotypes of HBV (A–J)
Infected individuals secrete excess envelope lipoprotein, circulating in the plasma as 22-nm round particles and similar rod forms of variable length

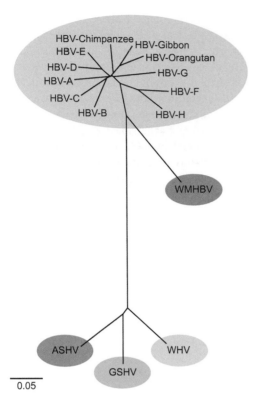

FIGURE 22.3 Phylogenetic relationships within the genus *Orthohepadnavirus*. Complete sequences from different HBV genotypes and isolates found in chimpanzee, orangutan, and gibbon were aligned using Clustal W with genomes from woolly monkey hepatitis B virus (WMHBV), woodchuck hepatitis virus (WHV), ground squirrel hepatitis virus (GSHV), and Arctic squirrel hepatitis virus (ASHV). The alignment was tested with the neighbor-joining method. Calibration bar: substitutions per site. *Reproduced from King, A.M.Q., et al., 2012, Virus taxonomy. In: Ninth Report of the International Committee on Taxonomy of Viruses. Academic Press, London, p. 453, with permission.*

Viral Replication

Prior to the development of DNA transfection of hepatoma cell lines, much of our current understanding of the replication of hepadnaviruses came from studies of animal hepadnaviruses.

The HBV genome is remarkably compact and makes use of overlapping reading frames to produce seven primary translation products from only four ORFs ("genes"): *S*, *C*, *P*, and *X*. Transcription and translation are tightly regulated via the four separate promoters and at least two enhancers plus a glucocorticoid-responsive element. Transcription occurs in the nucleus, whereas replication of the genome takes place in the cytoplasm within immature cores that represent intermediates in the morphogenesis of the virion. Replication of the double-stranded DNA genome occurs via a unique mechanism involving the reverse transcription of DNA from an RNA intermediate. Thus, hepadnaviruses are sometimes categorized as "retroid" viruses because of the similarity in replication strategy to the retroviruses,

although in a sense the two strategies are mirror images of one another. The key difference is that, in the case of the retroviruses, the positive-sense single-stranded RNA is packaged as the genome of the virion, whereas in the case of the hepadnaviruses, the single-stranded RNA is the intracellular intermediate in the replication of the double-stranded DNA genome present in the virion (Fig. 22.4).

The virion attaches to the hepatocyte predominantly via domains on Pre-S1 although other determinants on the S protein may also contribute to hepatocyte binding. The human sodium taurocholate cotransporting polypeptide (NTCP) has been identified recently as a high-affinity receptor for both HBV and HDV, although other cofactors may also be required.

Following removal of the envelope, the nucleocapsid is translocated to the nucleus and the viral genome released. The short (positive) strand of viral DNA is then completed to produce a *relaxed circular* double-stranded DNA molecule. This in turn is converted into a *covalently closed circular* ("ccc") form by removal of the protein primer from the negative strand and of the oligoribonucleotide primer from the positive-sense strand, elimination of the terminal redundancy from the negative strand, supercoiling, and ligation of the two ends of the DNA. This cccDNA form is the template for transcription by cellular RNA polymerase II. The "negative" strand only is transcribed to give mRNAs of 2.1 and 2.4 kb, plus a 3.4-kb RNA transcript known as the *pregenome*: this is actually longer than the genome itself because it contains terminally redundant sequences. Following transport to the cytoplasm, the 3.4-kb species is translated to yield the C (core) antigens and the polymerase, while the 2.1- and 2.4-kb transcripts are translated from three different initiation codons to produce the three forms of S (surface) antigens. A fourth mRNA species of 0.46 kb encodes the X protein.

Replication of the viral genome occurs via a mechanism distinct from that of any other DNA virus (Fig. 22.4). The newly synthesized RNA pregenome associates with the polymerase/reverse transcriptase and the core protein to form an immature core particle in the cytoplasm. Within this structure the reverse transcriptase domain, primed by the virus-coded terminal protein, transcribes a complementary (negative) strand of DNA. Meanwhile, the RNase H progressively degrades the RNA template from its 3′ end, leaving only a short 5′ oligoribonucleotide that serves as the primer for the DNA polymerase to transcribe a DNA positive-sense strand. Some of these core particles, containing newly synthesized viral DNA, are recycled back into the nucleus to amplify the pool of HBV genomes available for transcription. The remainder is assembled into virions before the positive-sense strand of the genome has been completed. Nascent nucleocapsids bud through those areas of endoplasmic reticulum into which the L, M, and S proteins have been inserted, thereby acquiring an HBsAg-containing lipid envelope. Vesicles transport the virions to the exterior without cell lysis.

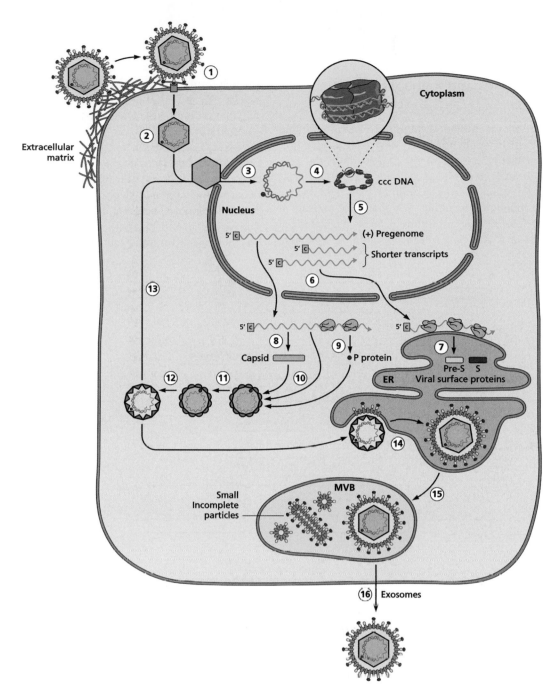

FIGURE 22.4 Replication cycle of hepatitis B virus. (1) The virion attaches to a susceptible hepatocyte, most likely via weak interaction with cell-associated heparin sulfate proteoglycans and then via the high-affinity receptor, sodium taurocholate-cotransporting polypeptide. (2) Details of entry and removal of the envelope are unknown, although host cell caveolin-1 has been implicated in entry. The viral core is transported to the nucleus, presumably along microtubules. (3) The core binds to proteins in the nuclear pore and the viral genome is released into the nucleus. (4) Repair of the gapped (+) DNA strand is likely accomplished by cellular enzymes. The product is a covalently closed circular form called cccDNA, which associates with histones to form a minichromosome. (5) The (–) strand of cccDNA is the template for transcription by cellular RNA polymerase II of a longer-than-genome-length RNA called the pregenome and shorter, subgenomic transcripts, all of which serve as mRNAs. (6) Viral mRNAs are transported from the nucleus. (7) Subgenomic viral mRNAs, which codes for the viral envelope protein, are translated by ribosomes bound to the endoplasmic reticulum (ER). Proteins destined to become anchored in the viral envelope, as well as in incomplete particles, enter the secretory pathway. (8) The pregenome RNA is translated to produce capsid protein. (9) The P protein (the viral reverse transcriptase) is also produced from pregenome RNA but at low efficiency; the ratio of capsid to P protein translation is 200 to 300 to 1. Following its synthesis, P binds to the packaging signal at the 5′-end of its own transcript, where viral DNA synthesis is eventually initiated. (10) Concurrently with capsid formation, and aided by the host heat shock protein chaperones Hsp90/70, the RNA-P protein complex is packaged and DNA replication is primed from a tyrosine residue in the polymerase. (11) Reverse transcription of the pregenome occurs within the capsid. (12) After completion of DNA synthesis, the newly assembled "cores" acquire the ability to interact with envelope proteins. (13) However, at early times after infection, some core particles are transported to the nucleus, where the viral genomes give rise to additional copies of cccDNA. Eventually, 10 to 30 molecules of cccDNA accumulate, thereby boosting the production of viral mRNA. (14) At later times, and possibly as a consequence of the accumulation of sufficient envelope proteins, the core particles acquire envelopes by budding into the ER, where viral surface proteins have been synthesized. (15) Viral assembly is believed to be completed in multivesicular bodies. (16) Progeny enveloped virus particles, and numerous small genome-lacking incomplete particles, are released from the cell by exocytosis. *Reproduced from Flint, S.J., et al., (2015). Principles of Virology, fourth ed. ASM Press, Washington, DC, with permission.*

Clinical Features and Pathogenesis of Hepatitis B

Acute infection in childhood is usually subclinical, but up to one-third of adult infections may lead to jaundice. After an incubation period of 40 to 180 days, the *pre-icteric* (prodromal) phase develops with malaise, lethargy, anorexia, and often nausea, vomiting and pain in the right upper abdominal quadrant. A minority of patients at this time develop a type of serum sickness characterized by urticarial rash and polyarthritis resembling a benign, fleeting form of rheumatoid arthritis. From two days to two weeks after the onset of the prodromal phase, a proportion of patients develop the *icteric* phase, heralded by dark urine (bilirubinuria) and closely followed by pale stools and jaundice. The *convalescent* phase can be long and drawn out, with malaise and fatigue lasting for weeks. Finally, an uncommon but devastating outcome is *fulminant* hepatitis occurring in approximately 1% of clinical infections, in which hepatocyte necrosis continues inexorably and the patient dies in 6 to 12 weeks from acute liver failure unless an emergency liver transplantation is performed (Fig. 22.5).

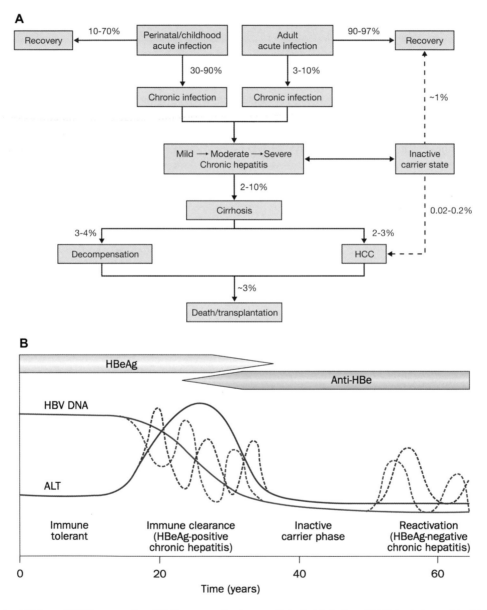

FIGURE 22.5 Natural history of HBV progression. (A) Sequence of outcomes. Note that infection at birth leads to persistent infection in >90% of infants, while in childhood this figure drops progressively, and in adults only around 3% to 10% remain infected. Development of chronic active hepatitis and cirrhosis can show a fluctuating time course that is affected by progression through the different immunological stages of disease. (B) Progression of immunological control of chronic HBV infection. Classically, four consecutive phases are identified: the immune tolerant phase; the immune clearance phase (HBeAg positive chronic hepatitis); the inactive carrier phase; and the reactivation phase (HBeAg-negative chronic hepatitis). Alanine aminotransferase (ALT) levels may be intermittently or persistently increased during the immune clearance and reactivation phases. HBV DNA levels are high during the immune tolerant phase. These decrease during the immune clearance phase and the inactive carrier phase, but may increase again during the reactivation stage. HBeAg levels decline over time, giving way to anti-HBe antibody seroconversion—this occurs during the immune clearance phase (intermittently or persistently). *(B) Reproduced from Kwon, H., Lok, A.S., 2011. Hepatitis B therapy. Nat. Rev. Gastrol. Hepatol. 8, 275–284, with permission.*

From 3% to 10% of adult patients with acute HBV infection progress to chronic infection, a condition defined as the continuing presence of HBsAg in the blood for six months or longer. This rate is higher at around 20% for those infected in childhood to over 90% for perinatal infection. Chronic HBV infection can evolve through five distinct phases. The duration of each phase can vary greatly from patient to patient, not all patients progress through all phases and some phases may be bypassed.

The first is an immune tolerance phase whereby high levels of serum HBV DNA and the presence of HBeAg indicate HBV replication. Among infants infected perinatally, this immune tolerance phase may last for several decades: during this period serum alanine aminotransferase (ALT) levels are normal, there is little sign of an inflammatory response, and the risk of cirrhosis and the subsequent development of HCC is low. In older children or adults this immune tolerance phase may be short-lived or even absent.

The second phase is one of immune-mediated clearance, characterized by high levels of HBV DNA in the blood, the presence of HBeAg, and elevated levels of serum ALT. Liver biopsy may reveal a significant inflammatory response, particularly around the portal tracts. A feature of this phase is the regular occurrence of spikes in the level of serum ALT, indicating an ongoing immune-mediated cytolysis of infected hepatocytes. The frequency and duration of these flares correlates with the risk of cirrhosis and the development of HCC. Recurrent flares are seen more frequently among males. Seroconversion to an anti-HBe status marks the end of this phase. This seroconversion event occurs earlier among Asian patients infected with genotype B as opposed to genotype C.

The third phase is the inactive, HBsAg positive, carrier state where serum ALT levels are normal and serum HBV DNA is at low titer or undetectable. Thus the presence of serum HBsAg is the principal marker of infectivity. This inactive state, equivalent to the older concept of chronic persistent hepatitis, may last for many years, and indeed among those carriers infected at an early age who reach this phase there is no difference in survival rates compared to healthy individuals. Liver biopsy shows a mild hepatitis and minimal fibrosis. Reactivation of HBV replication may occur as a result of immunosuppression or spontaneously due to an unrecognized cause.

The fourth phase is one of reactivated HBV replication in the absence of serum HBeAg but with measurable levels of circulating HBV DNA and elevated serum ALT. Liver biopsies from such cases show an ongoing necroinflammatory reaction. Some patients progress almost at once to this stage from HBeAg positive, especially older individuals. A particular feature of this phase is the marked fluctuation in biochemical markers of liver injury, and thus serial serum samples should be taken to differentiate these particular patients from inactive carriers. Importantly, the prevalence of this phase of HBeAg-negative chronic hepatitis

B has increased over the last two decades and has important implications for antiviral treatment. The higher rate of cirrhosis in this cohort of patients (up to 10%) is related to age and the more advanced liver disease at presentation. Additional factors may accelerate the progression to cirrhosis in this group, for example, concurrent HCV or HIV infection and excessive alcohol intake. The risk of cirrhosis is also much higher for individuals infected with genotype C.

Many patients in this fourth phase show viral mutations in the pre-core region and/or in the basal core promoter (BCP). The common pre-core mutation (G1896A) introduces a stop codon that prevents the translation of the pre-core protein and completely abolishes the production of HBeAg. Some BCP mutations are linked to enhanced viral DNA replication *in vivo* and in cell culture, and to suppression of HBeAg synthesis. Although still not fully clarified, there is evidence that these mutations may add to the worsening outcome in this cohort of patients, and also may be linked to fulminant hepatitis. Genotype variation may determine which of these two sets of mutations are favored, and conversely the relative ease of acquiring these mutations may contribute to different outcomes with different genotypes. For example, the G1896A mutation is more likely to occur when the nucleotide on the opposite site of the stem-loop structure is T rather than C: as genotype C viruses almost always have a C at this position (nt1858), PreC variants occur less often among patients in north America and Europe where genotype A is predominant. In contrast, pre C variants are most common among genotype D patients. The net result in both instances, however, is a loss of HBeAg production and the stimulation of an anti-HBe antibody response.

The continuing presence of HBV DNA in liver tissue in the absence of HBV DNA or HBsAg in the blood is a feature of the fifth phase, occult infection. Such individuals are either positive for anti-HBc/HBe antibodies or completely negative for serum markers of previous or ongoing HBV infection. Using both serology and sensitive assays for quantifying HBV DNA, it has become clear that HBV can persist, albeit at very low levels, in the liver and sometimes in other tissues and the blood of patients with serological evidence of recovery. Some cases may be due to extensive mutations in the HBV envelope proteins resulting in a failure of licensed immunoassays to detect the presence of HBsAg particles; however, in most cases this is likely to result from prior wild-type virus infection where virus replication is strongly suppressed. There is now evidence that this may contribute to the development of cirrhosis and even to carcinogenesis. Reactivation of HBV replication has also been reported among individuals positive for anti-HBs antibodies when undergoing immunosuppression.

Hepatocyte injury is thought to be a consequence of the host immune response, both humoral and cellular, directed against viral antigens expressed by infected cells.

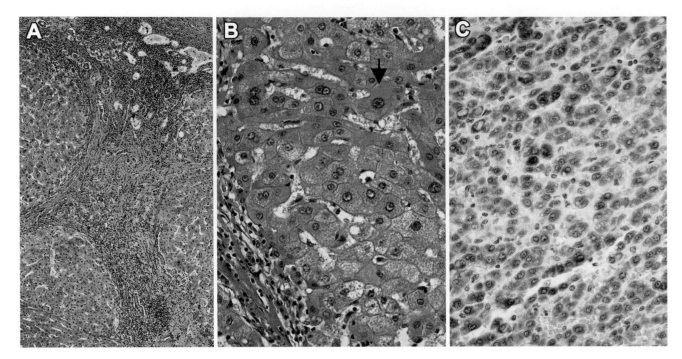

FIGURE 22.6 Chronic hepatitis B virus (HBV) infection. (A) Moderately severe chronic hepatitis is characterized by extensive portal and periportal mononuclear inflammation and extensive lobular hepatocellular necrosis and portal fibrosis, affecting almost all portal tracts. (B) "Ground glass" hepatocytes (arrow) with swollen, homogeneous flat hazy and uniformly dull cytoplasm. This is caused by the accumulation of HBsAg in these cells. (C) HBsAg in the cytoplasm of hepatocytes. (A) and (B), H&E. (C) Immunohistochemistry, avidin-biotin-peroxidase (ABC) technique. *Reproduced from Cooke, R.A., Infectious Diseases, McGraw-Hill, with permission.*

In addition, inflammatory cytokines may contribute via non-cytolytic intracellular inactivation of HBV replication. Further complexity in the pathogenesis of HBV results from the effect of genotypic variation of the virus and the accumulation of mutations with time.

The course of infection is likely influenced also by the degree by which the host innate response is modulated. There is evidence that HBeAg expression suppresses innate immunity through interactions such as cell surface Toll-like receptors and intracellular suppression of cascades regulated by NF-κB, leading to a reduced interferon response. In addition, the severity and outcome of infection is likely to depend on the rate and strength of the adaptive immune response, as shown by the different outcomes in the very young and in immunosuppressed individuals (Fig. 22.5).

Laboratory Diagnosis

Routine biochemical tests of liver function distinguish viral hepatitis from the many non-viral causes of jaundice, for example, obstructive jaundice or liver damage due to chemicals. Characteristically, levels of serum transaminases (aminotransferases) are elevated markedly (5-fold to 100-fold) in acute symptomatic viral hepatitis regardless of causative virus. ALT and aspartate aminotransferase (AST), rise together late in the incubation period to peak about

the time jaundice appears; the levels of both enzymes gradually revert to normal over the ensuing two months in an uncomplicated case. Serum bilirubin may rise anything up to 25-fold, depending on the severity of the disease, but may of course be close to normal in anicteric viral hepatitis.

There are reports of successful HBV replication in primary cultures of human hepatocytes, but virus culture is impracticable for routine diagnostic use. Thus serology forms the basis of the diagnosis of hepatitis B, its differentiation from other causes of viral hepatitis, and in distinguishing one clinical stage of hepatitis B from another. While many types of immunoassay have been successfully applied to HBV, the most widely used and most sensitive have been radioimmunoassay and enzyme immunoassay. Six markers, all found in serum, are of particular diagnostic importance: HBsAg, HBV DNA, HBeAg, antibody to HBsAg (anti-HBs), anti-HBe, and anti-HBc. Understanding the evolution of these markers during an uncomplicated transient HBV infection, and during an infection that progresses to chronicity, provides the basis for interpreting serological test results (Fig. 22.7).

Table 22.2 summarizes the patterns of serological markers characteristic of various outcomes of hepatitis B infection. Note that the key markers are HBsAg, anti-HBs, and anti-HBc antibodies, and HBV DNA; the profile of each can distinguish most of the important clinical outcomes of

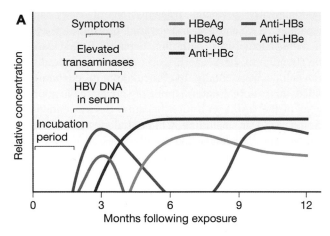

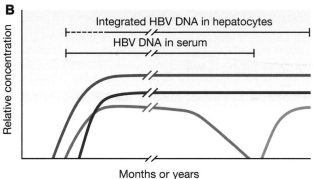

FIGURE 22.7 Typical serological profiles of (A) acute resolving hepatitis B infection, and (B) acute infection leading to persistence. *Modified from Zuckerman, A.J., et al., 2000. Clinical Virology, fourth ed. John Wiley & Sons, UK.*

HBV infection. Note also that (1) the single most reliable marker of past or present HBV infection is the presence of anti-HBc antibodies, (2) persistence of HBV DNA in chronic active hepatitis portends an unfavorable prognosis, and (3) anti-HBs antibodies, indicative of a neutralizing antibody response, appears only after HBsAg is no longer detected, hence are a reliable indicator of recovery and therefore of immunity to reinfection.

Hepatocellular Carcinoma

HCC is the fifth most common cancer and the third leading cause of cancer deaths worldwide, with more than 80% of cases occurring in Southeast Asia and sub-Saharan Africa. HCC is one of the most heterogeneous and complex of human cancers. Surgical resection or liver transplantation remain the only effective interventions. There is approximately a 100-fold to 200-fold greater risk for HCC in HBV-infected individuals as compared to HBV-negative persons. In comparison, smoking increases the risk of lung cancer by only about 10-fold to 25-fold compared to those who have never smoked. Over a 50-year period, the risk of developing HCC is about 15% among HBV carriers, higher if cirrhosis develops at an earlier age.

Patients over the age of 40 years with HBV DNA levels above 10^5 copies/mL are at a significant risk of developing cirrhosis or HCC, regardless of the level of serum aminotransferases. Additional risk factors include genotype C and the presence of BCP mutations. Some experts recommend six-month surveillance for HCC in males over the age of 40 years and females over the age of 50 years with chronic HBV infection. Liver ultrasound is the primary scanning method, supplemented by measurement of serum α-fetoprotein levels.

TABLE 22.2 Interpretation of Serological Markers of HBV Infection

HBsAg	HBeAg	Total Anti-HBc	IgM Anti-HBc	Anti-HBs	Interpretation
−	−	−	−	−	Never infected
+	+ or −	−	−	−	Early acute infection
+	+	+	+	−	Acute infection
−		+	+	+ or −	Resolving acute infection
−	−	+	−	+	Recovered from past infection; immune
+	+	+	−	− (+)	Persistent infection—early immunotolerant phase
+	−	+	−	− (+)	Persistent infection; inactive or immune escape phase
−	−	+	−	−	Past infection with undetectable anti-HBs; "low-level" persistent infection; false-positive result; (in infant) passive transfer of maternal antibody
−	−	−	−	+	Immune, if level is >10 mIU/mL after vaccination; passive transfer after HBIg administration

There is overwhelming evidence for a causal relationship between HCC and HBV infection. HCC and active HBV infection share a common geographical distribution and most HCC cases arise in HBV carriers. A prospective study of 22,000 Taiwanese men showed that HBsAg-positive males had over 200 times the rate of HCC, compared to those negative for HBsAg. This effect was specific for cancer of the liver, and was seen only in individuals with current infection. Most HCC tumors contain HBV DNA. Integrated HBV DNA is found in 85% of tumors. There is often little detectable HBV protein expression, and the HBV genome is often deleted and rearranged. Importantly, tumors are usually clonal with respect to the viral integrated sequences—which indicates that viral integration occurred prior to the clonal expansion of tumor cells. Furthermore, 90% of newborn woodchucks that are experimentally infected with WHV develop HCC within 2 to 3 years.

One of two mechanisms may explain a direct role of HBV in HCC. First a viral protein may directly contribute to the deregulated growth of infected cells. One candidate is the X gene product, which is a potent transactivator of gene expression and which can, in some settings, transform cells. It has been noted that the X protein may be required for early stages of oncogenesis, but not for later stages, since HBV gene expression is close to zero in HCC tumors. Around 40% of integrated HBV genomes are restricted to the region where the X and core genes are located. Alternatively, HBV integration may cause insertional mutagenesis and the deregulation of cellular genes controlling cell growth. For example, WHV-associated tumors are often associated with rearrangement of *N-myc* and activation of *N-myc* expression. However, no common rearrangements have so far been detected in human HCC tumors. New technologies have given insight into the complex interactions of HBV proteins: over 140 host cell proteins have been identified, many of which interact with HBcAg and many of which are associated with oncogenesis.

In addition, HBV infection may lead indirectly to hepatoma development following chronic liver injury, due to immune attack on virally infected hepatocytes. Chronic liver regeneration would then predispose to tumor development. This concept is supported by the fact that alcoholic cirrhosis predisposes to HCC.

Cofactors may also contribute to the development of HCC. Dietary aflatoxin is one such cofactor in some areas, for example, West Africa. In addition, mutation of cellular genes such as the p53 tumor suppressor gene may occur (a mutational hot-spot at codon 249 of p53 has been detected in several hepatomas).

Epidemiology

Hepadnaviruses of humans and primates are transmitted both perinatally and sexually. This contrasts with the predominantly perinatal transmission of other animal hepadnaviruses and the transovarial transmission typical of avihepadnaviruses. Although hepatitis B first came to the attention of the Western world as an iatrogenic disease transmitted accidentally by inoculation of contaminated blood, this is not the natural mode of spread. In the populous areas of the world with high hepatitis B virus endemicity (Southeast Asia including China, Indonesia, Philippines and the Pacific islands, the Middle East, Africa, and the Amazon basin), where the majority of people are antibody positive and 8% to 15% are chronic carriers, most become infected at birth or in early childhood. In these regions some 5% to 12% of pregnant women are HBsAg positive, almost half of whom may be positive for HBeAg and hepatitis B virus DNA positive; in this group, perinatal transmission occurs in 70% to 90% with a high probability of the infant becoming a carrier. As the infant generally becomes HBsAg positive only one to three months after birth, it is considered that most perinatal infections result from contamination of the baby with blood during parturition, rather than transplacentally; breast milk or maternal saliva are thought to be occasionally responsible. However, up to half of all children in medium to high prevalence communities who become carriers acquire the infection between 1 and 5 years of age from intrafamilial contact with chronically infected siblings or parents secreting virus in oozing skin sores, blood, or saliva. The probability of becoming a chronic carrier following infection in infancy is of the order of 25%. In adolescents and adults transmission is principally by sexual intercourse and blood contact; in contrast, less than 5% of primary infections acquired in adulthood progress to chronicity (Fig. 22.8).

The picture is substantially different in the economically developed world where the carrier rate is generally less than 1% except among ethnic minorities (e.g., Asian immigrants) and in intravenous drug users. Perinatal spread is correspondingly less common, and sexual (including homosexual) transmission among adolescents and adults is a significant risk. Percutaneous transmission by medical skin penetration procedures or by intravenous drug use represent the most commonly identified mode of spread, with injecting drug users constituting the largest cohort of carriers. Post-transfusion hepatitis B and infection of hemophiliacs by contaminated factor VIII, once a common complication, has now almost disappeared as a result of routine screening of blood and organ donors, but hepatitis B remains a significant occupational risk for unvaccinated laboratory and healthcare workers who are vulnerable to accidental infection from spilt blood or needle-stick injury.

Professionals occupationally at risk include dentists, surgeons, pathologists, mortuary attendants, technicians; and scientists working in serology, hematology, biochemistry, and microbiology laboratories in hospitals or public health institutions, blood banks, or hemodialysis units. However, the

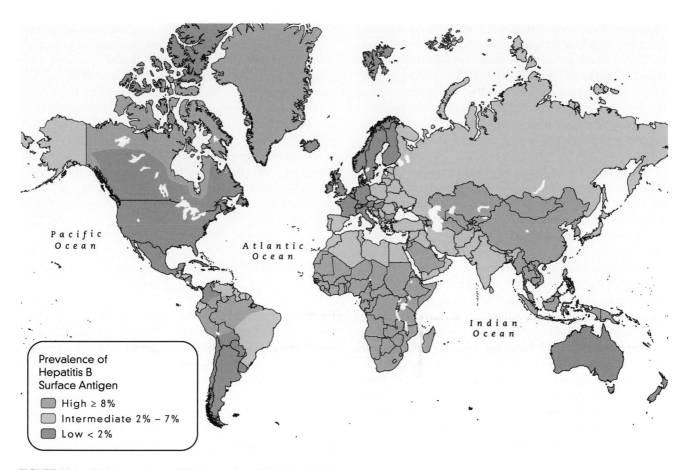

FIGURE 22.8 Global prevalence of HBsAg carriage, 2014. *From CDC.*

availability of an effective vaccine has greatly reduced this occupational risk. Tattooing, acupuncture, and ear piercing without rigorous sterilization of equipment constitute other potential routes of transmission, as do certain body contact sports such as wrestling and rugby football.

Occult HBV infection is increasingly recognized as a challenge for blood transfusion and organ transplantation. Such patients may act as a source of infection when acting as organ donors (rarely as blood donors) and may undergo acute reactivation when immunosuppressed.

Management of Hepatitis B Patients

After many years of disappointing results, antiviral therapy for chronic HBV infection has recently become an exciting and rapidly advancing area. The aim is to improve survival by preventing progression to cirrhosis and HCC. The most widely used marker of a successful response is the sustained reduction of serum HBV DNA to undetectable levels; however, seroconversion from HBeAg positive to anti-HBe positive, or less often from HBsAg positive to anti-HBs positive, are also markers of benefit. Six antivirals are licensed for the treatment of hepatitis B in many countries.

These drugs are either immunomodulators (interferon-alpha and pegylated-interferon-alpha), nucleoside analogs (lamivudine, telbivudine, and entecavir), or nucleotides (adefovir, tenofovir). Of these, entecavir, tenofovir, and pegylated-IFN-alpha are the most effective and thus in widespread use. International therapeutic guidelines are available and are regularly updated. Indications for treatment in chronic infection are based on serum HBV DNA levels (e.g., $>2 \times 10^3$ IU/mL), elevated serum ALT levels, and the severity of liver disease as judged by biopsy or a non-invasive test such as fibroscan or the APRI-test (AST to platelet ratio index). Treatment is not generally indicated in the third inactive phase of chronic infection. Antiviral drugs may also be used for acute fulminant hepatitis B, and during late pregnancy to reduce the rate of vertical transmission.

Interferon (see Chapter 12: Antiviral Chemotherapy) has had an established place in the treatment of chronic hepatitis B for over 20 years but the long-term outcomes have been generally disappointing. Pegylated interferon-alpha is given by injection once a week for 6 months to a year, and brings about a response in up to 50% of cases but there is often a relapse after cessation of therapy. The drug frequently causes side effects including influenza-like symptoms, myelosuppression, autoimmune thyroid disease, and depression (see Chapter 12: Antiviral Chemotherapy). Its efficacy seems to be largely limited to a particular cohort of patients with low levels of HBV DNA ($<10^7$ copies/mL), high serum aminotransferase levels, and a high activity histology score prior to treatment. Seroconversion to an anti-HBe antibody-positive status is more likely in patients infected with genotypes A or B as opposed to genotypes C and D.

Antiviral compounds can be regarded as being of two types depending upon the likelihood of resistance developing over time. Lamividine, a chain-terminating nucleoside analog that interferes with the reverse transcriptase activity of both HBV and HIV-1, is safe and only minimally toxic. Unfortunately within 5 years of monotherapy around 70% of patients develop resistance due to a single mutation (M204S/I/V) within the highly conserved YMDD motif at the catalytic center of the HBV polymerase reverse transcriptase (RT) function. Adefovir also has a low barrier to resistance, and resistance mutations eventually develop at A181V/T and N236T. Telbivudine, another potent inhibitor of HBV replication, also leads to a high incidence of resistance. In contrast, entecavir and tenofovir are potent HBV inhibitors and show a high barrier to resistance; both these drugs are recommended for first-line monotherapy. However, entecavir alone is not recommended for the treatment of HBV/HIV co-infected patients as it can promote the emergence of the M184V mutation in the HIV reverse transcriptase.

Ongoing monitoring is essential, and will identify patients showing either a primary non-response, partial virological response, or virological breakthrough. Breakthrough resistance is a major issue in the treatment of chronic hepatitis B; it signals an indication to change to the most effective agent that does not share cross resistance with the previously used drug. With all antiviral therapies used to date, none has the ability to completely eliminate the HBV cccDNA from within hepatocytes: thus there is always the risk of further flares and recrudescence many years after the disappearance of markers of ongoing replication. There is recent evidence that long-term maintenance treatment with entecavir or tenofovir reduces the incidence of HCC in patients with pre-existing cirrhosis. While combination therapy has been the norm for the treatment of HIV for nearly two decades, a combination strategy for the treatment of HBV has not been universally adopted, although this is likely to change over the next few years.

Management of patients with persistent infection must include appropriate education as to the likely course of disease; and in particular, clear information about how and where they may or may not pose a risk of infection to others, thereby avoiding either a mindset where the patient is inappropriately fearful of infecting others, or actions that put others unnecessarily at risk.

Despite the above advances, liver transplantation may still be the only remaining option for patients with end-stage HBV disease. Earlier results of transplantation were disappointing, with graft reinfection rates of 80% to 100% being reported. Since then, management includes prophylaxis with a combination of hepatitis B immune globulin (HBIG) and antiviral drugs. In one such study, reinfection rates were 50%, 7.5%, and 0% in patients whose serum HBV DNA at the time of transplant was $>10^5$, 2×10^2–10^5, and <200 copies/mL. Overall, survival rates for patients transplanted for HBV-related cirrhosis have now been reported as 85% at 1 year and 75% at 5 years. Newer antiviral drugs may further increase survival after transplantation.

Prevention and Control

Prevention of Transmission

Routine screening of blood donations for HBsAg using sensitive immunoassays has almost eliminated post-transfusion hepatitis B in economically developed regions. Intravenous drug users are frequently the targets of educational campaigns in order to reduce the high risk of transmission accompanying the sharing of contaminated syringe needles. The prevention of sexual transmission is addressed by general education advocating safe sex practices, monogamy, and screening of sexual partners for HBsAg. Partners of known carriers should be vaccinated and encouraged to avoid contact with the carrier's blood or other secretions, for example, by using condoms, covering skin sores or abrasions, and avoiding the sharing of toothbrushes, razors, eating utensils, etc. Perinatal transmission to newborn infants of carrier mothers can be minimized by inoculation at birth with the combined use of hepatitis B vaccine and hepatitis B immune globulin.

Prevention of infection in healthcare workers and their patients is based upon vaccination and universal precautions in the ward, theater, and laboratory founded on the presumption that any patient may be infectious. Barrier techniques include wearing of gloves, gowns, masks, and eye-shields to prevent exposure to blood in high-risk situations, such as during invasive procedures, avoidance of mouth-pipetting and of eating or smoking while working, meticulous handwashing routines, careful attention to the disposal of blood and body fluids and to the cleaning-up of blood spills with appropriate chemical disinfectants (such as 0.5% sodium hypochlorite [5000 ppm available chlorine]—glutaraldehyde and formalin are no longer used in most countries because of the cancer risk to staff), special precautions in disposal of used needles, the use of disposable equipment wherever possible, and appropriate procedures for the sterilization of reusable equipment. These approaches to aseptic technique and sterilization of equipment should apply equally to dentists, acupuncturists, tattooists, etc.

Post-Exposure Immunoprophylaxis

Hepatitis B immune globulin (HBIG), obtained by plasmapheresis of subjects with a high titer of anti-HBs antibodies, is effective for the post-exposure prophylaxis of hepatitis B when given in the first 48 hours after exposure, for example, in unvaccinated individuals who have been recently exposed to infection with HBsAg-positive blood through needle-stick accidents. Vertical transmission from infected mothers is reduced by more than 90% by the routine administration of HBIG to neonates within 48 hours of birth, combined with commencing a course of vaccination.

Vaccination

The first hepatitis B vaccines were produced by the purification of HBsAg particles from human plasma followed by chemical treatment to inactivate any accompanying HBV or other contaminating virus. These "plasma-derived" vaccines were widely used for many years, particularly in developing countries where the need is greatest, but have been progressively replaced by genetically engineered products.

In the early 1980s the HBsAg gene, cloned into a plasmid, was used to transfect the yeast *Saccharomyces cerevisiae*, and the non-glycosylated form of HBsAg particle produced was extracted and purified for use as a vaccine. As with the plasma-derived vaccine, the recombinant vaccine is adsorbed with the adjuvant aluminum hydroxide, stored cold but not frozen, and administered by intramuscular injection into the deltoid muscle in a course of three doses, separated by one month then five months, respectively. There are no side effects other than an occasional (5% to 20%) sore arm. Protective levels of neutralizing antibodies are elicited in >90% of recipients and as high as 95% in neonates. Importantly, all licensed hepatitis B vaccines confer immunity against all genotypes and serotypes of HBV.

Among non-responders, only half seroconvert following a second full course, suggesting the absence of an appropriate immune response gene in those consistently non-responsive. Renal dialysis patients, and immunodeficient or elderly recipients also respond sub-optimally. Protection studies demonstrate that immunity in immunocompetent vaccine recipients remains solid for a decade or so, and that those who do become infected during this period generally develop an anamnestic anti-HBs response as well as anti-HBc antibodies, and do not progress to chronicity. Nevertheless, longer-term studies have confirmed that a booster dose of vaccine is desirable after about 10 years, particularly in those individuals with a high risk of exposure.

Recombinant HBsAg vaccines are continually being refined. First, inclusion of the nucleotide sequence encoding pre-S1 and pre-S2 in the cloned gene construct enhances immunogenicity of the resultant protein particles.

Second, HBsAg produced by mammalian cell lines such as Chinese hamster ovary (CHO) cells are glycosylated normally, thus more closely resemble the natural human product. Third, the HBsAg gene has been incorporated into live vaccinia virus or adenovirus recombinants and demonstrated to confer protection, but such vaccines are not generally available.

Vaccination strategies differ from country to country. In those populous areas of the world where the HBsAg carrier rate is high, the top priority is to interrupt the chain of perinatal and horizontal transmission among the young by vaccinating all infants at birth. For convenience, the course of injections should be synchronized with those for diphtheria/pertussis/tetanus and/or with oral polio vaccine (OPV), as recommended by the World Health Organization's Expanded Program of Immunization.

The strategy of universal immunization of newborn infants should be the aim of all countries, but current policy in most of the developed world where the HBsAg carrier rate is less than 1% is to target only high-risk groups in the community. Several of these cohorts are notoriously difficult to access and the experience in the United States has been that the incidence of hepatitis B virus has actually increased in spite of strenuous efforts to implement this policy. Moreover, the cost of locating and testing for HBsAg carriage approaches that of the vaccine. As a result, many health authorities have advocated a policy of universal vaccination of all children.

In many countries vaccination against HBV is mandatory for healthcare workers and all those working with body fluids, including blood. Vaccination is also widely recommended for transplant recipients, the partners of persons positive for HBV infection and other individuals at high risk of exposure. The immunization of women of childbearing age is especially important for the interruption of transmission from mother to offspring.

HEPATITIS DELTA VIRUS

In 1977 an Italian physician, Mario Rizzetto, detected a novel antigen, which he called δ (delta), in the nuclei of hepatocytes from particularly severe cases of hepatitis B. Delta antigen was also found inside 36-nm virus-like particles, the "delta agent," the outer coat of which was serologically indistinguishable from HBsAg. It transpired that the δ agent, now known as hepatitis delta virus (HDV), is a defective *satellite* virus, found only in association with its helper virus, hepatitis B virus. The tiny RNA genome of hepatitis delta virus, smaller than that of all known animal viruses, encodes its own nucleoprotein (the delta antigen) but the outer envelope of the hepatitis delta virus is composed of HBsAg, encoded by the genome of hepatitis B virus co-infecting the same cell (Table 22.3).

TABLE 22.3 Properties of Hepatitis Delta Virus

Roughly spherical, enveloped virion, 36 to 43 nm in diameter. The envelope contains all three envelope proteins of HBV and encloses the genome and 60 to 70 copies of delta antigen HDAg (both large and small forms)

The HDV genome is a covalently closed, single-stranded RNA molecule of negative-sense, 1679 nucleotides in length; extensive intramolecular base-pairing creates an unbranched rod-like structure

Co-infection with HBV is required for complete replication and production of HDV virions; however, HDV RNA replication alone can take place in a variety of cell types without HBV co-infection

Genome replication involves a rolling circle mechanism to create linear oligomeric molecules which self-cleave to monomers; these are ligated to yield closed circular monomers, that in turn act as templates to produce complementary sense RNA monomers by the same mechanism

A subgenomic polyadenylated linear positive-sense RNA transcript is produced and acts as mRNA. During replication, two forms of HDAg (p24 and p27) are derived from the same ORF. This occurs through RNA editing of the anti-genomic (plus) strand, whereby the amber (UAG) termination codon at the end of the ORF for p24, is converted by the host enzyme double-stranded RNA adenine deaminase (ADAR) to a tryptophan codon (UGG), allowing read through and production of large HDAg (P27)

HDAg-p24 promotes HDV replication, while HDAg-p27 inhibits replication and is required for virion assembly

Classification

Hepatitis delta virus is unique among human viruses. Some regard it as a subviral agent falling outside the definition of a virus. It displays features characteristic of several different classes of plant pathogens known variously as viroids, virusoids, satellite RNAs, and satellite viruses, all of which have RNA genomes resembling that of hepatitis delta virus in certain respects despite not coding for a coat protein or indeed any protein at all. The hepatitis delta virus genome, similar to that of several of these subviral plant pathogens, is a covalently closed circle of single-stranded RNA with self-cleaving (*ribozyme*) and self-ligating activities. Although hepatitis delta virus is currently the only known mammalian example of this class of agents, it is likely that other comparable infectious agents of humans await discovery.

Virion Properties

The hepatitis delta virus virion is roughly spherical with a mean diameter of 36 nm. The outer coat is derived from HBV through a process of co-infection and includes all three envelope proteins of HBV (see above). The genome is a covalently closed circle of negative-sense RNA of only 1679 nucleotides; extensive base-pairing creates a secondary structure enabling it to fold into an unbranched rod-like structure with 76% intramolecular base-pairing. There is no sequence homology with either hepatitis B virus DNA or cellular DNA. Each RNA genome is coated with around 200 copies of the hepatitis delta antigen (HDAg). The latter has at least three functional domains: (1) an RNA-binding domain, which accounts for its intimate association with the viral genome, (2) a nuclear localization signal, which directs the

infecting HDV genome and newly synthesized HDAg to the site of viral transcription and replication in the nucleus, and (3) a leucine zipper, which is thought to promote interaction between HDAg and HBsAg in hepatitis delta virion assembly (Fig. 22.9).

Viral Replication

The hepatitis delta virus replication cycle has been elucidated by transfection experiments using either genetically cloned DNA copies of the hepatitis delta virus RNA genome or isolated recombinant genomic RNA delivered in liposomes to cells that have been engineered to express HDAg. A wide range of mammalian cells will support hepatitis delta virus RNA replication in the absence of hepatitis B virus, but the latter is necessary for the production of hepatitis delta virus particles.

Following entry and uncoating, the genome coated with HDAg is transported to the nucleus where the host RNA polymerase I and II transcribes complementary RNA of two distinct forms: (1) full-length circular positive-sense RNA, the "anti-genome," which serves as the intermediate in replication of the genome, and (2) a shorter polyadenylated linear transcript that is exported to the cytoplasm to serve as mRNA for translation into HDAg. Replication of the genome occurs in the nucleus by a double rolling circle mechanism similar to that of plant viroids. The genomic template rolls as the anti-genome is synthesized (RNA polymerase II), to produce excessively long linear copies (Fig. 22.9). This multimeric transcript then self-cleaves to yield a unit-length anti-genome, one specific nucleotide sequence serving as the "substrate" (cleavage site) and another as a magnesium-dependent "enzyme." A ligase activity then joins the ends of

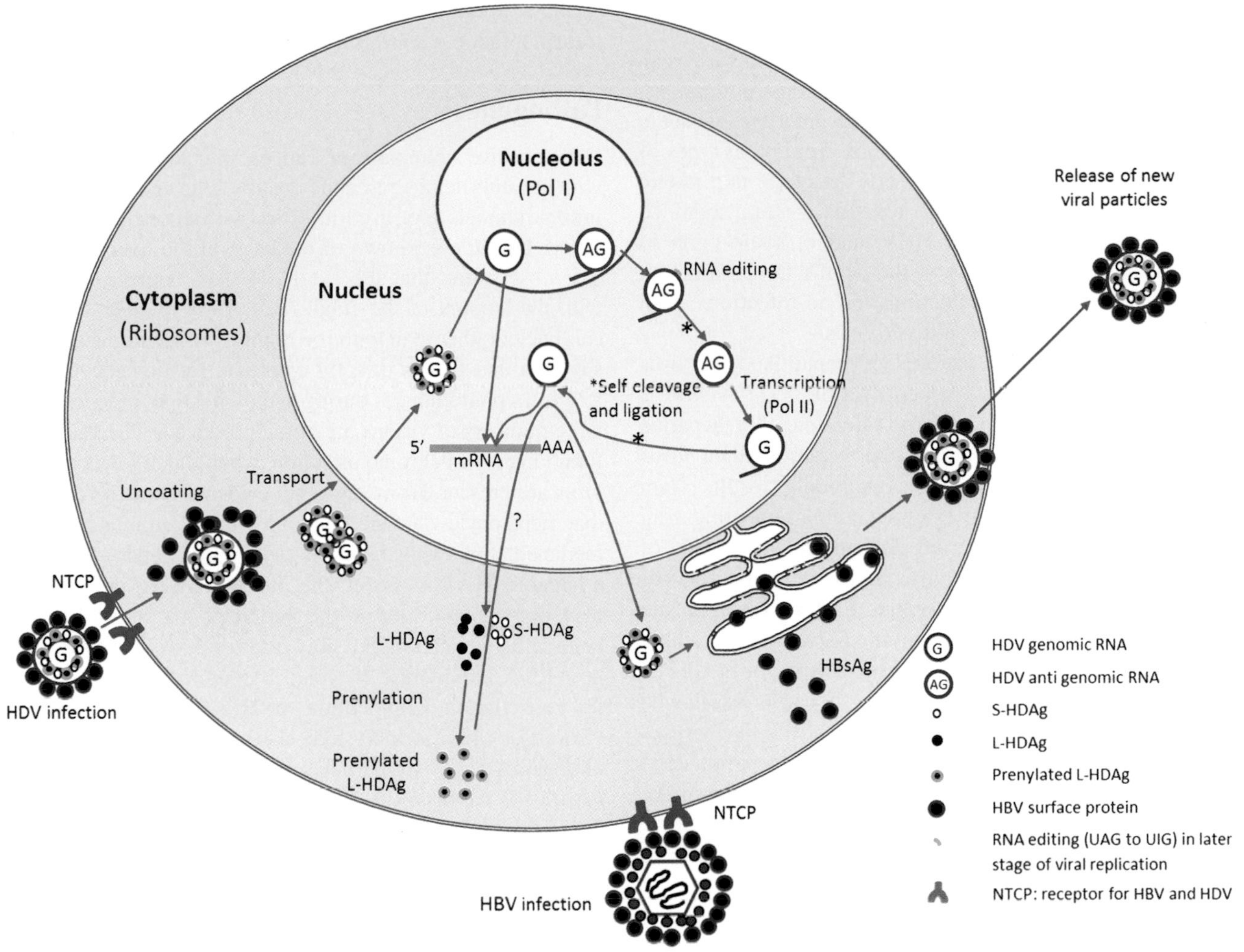

FIGURE 22.9 Hepatitis delta virus (HDV) replication. This defective satellite virus, which requires co-infection with hepatitis B virus as a helper for its replication, contains a viroid-like circular RNA that replicates via a double rolling circle mechanism mediated by cellular RNA polymerases. Following virus entry and uncoating, the viral genome is translocated into the nucleus. The genomic RNA is then replicated in the nucleolus by a rolling circle mechanism by RNA polymerase I to generate a greater-than-unit length anti-genomic RNA that is cleaved to the correct length by the self-encoded ribozyme, and ligated by a host cellular enzyme to form a circular RNA anti-genome. A functional genome is then produced from the anti-genome via this same rolling circle process except that in this case it is catalyzed in the nucleoplasm by host RNA polymerase II. Unit-length genomes are released by the viral ribozyme and self-ligated to form (−) strand circular genomic RNAs. Functional subgenomic mRNAs are produced from the genomic RNA by RNA polymerase II and exported to the cytoplasm for translation. A unique facet of HDAg expression is the production of two different forms of this protein from the same coding region; S-HDAg is translated from the anti-genomic sequence derived from the input virus, while L-HDAg (essential for assembly of new virions) is produced after an RNA editing step that mutates the stop codon of the shorter form. Late in the replication cycle the HDV core ribonucleoprotein (RNP) is surrounded by ~100 molecules of HBsAg, which in association with host membrane-derived lipids becomes the viral envelope. Thus, HBsAg is essential for entry, assembly and infectivity of the virus. *Reproduced from Guo, Z., King, T., 2015. Therapeutic strategies and new intervention points in chronic hepatitis delta virus infection. Int. J. Mol. Sci. 16:19537–19552, with permission. See also Macnaughton, T.B., et al., 2002. Rolling circle replication of hepatitis delta virus RNA is carried out by two different cellular RNA polymerases. J. Virol. 76:3920–3927.*

the linear transcript to yield the monomeric closed circular positive-sense RNA anti-genome. A similar sequence of events henceforth generates new copies of the genome, using the anti-genome as a template. During RNA replication an extraordinary example of RNA editing is observed; a specific mutation (UAG to UGG) occurs in the termination codon at the end of the ORF for the truncated (small) form of HDAg (P24), enabling read-through translation of the

mRNA to yield the large form of HDAg (P27). P24 is required for hepatitis delta virus RNA replication whereas P27 inhibits it, hence the switch from production of P24 to P27 suppresses further genome replication and promotes its packaging. Assembly of the genome–HDAg complex with HBsAg produced by the co-infecting HBV genome then produces the complete hepatitis delta virus particle in the cytoplasm prior to release from the cell.

Clinical Features

Similar to hepatitis B infection, hepatitis delta virus can establish both acute and chronic infections in humans. Clinically and histologically the presentation of acute hepatitis D does not differ significantly from other types of viral hepatitis, except in its severity. Anicteric and icteric infections are seen. Fulminant hepatitis is seen about 10 times more often than with HBV, and in some parts of the world, a high proportion of the deaths from fulminant hepatitis represent superinfections or co-infections with hepatitis delta virus and hepatitis B virus.

There are two main patterns of hepatitis delta virus infection, co-infection and superinfection. *Co-infection* is defined as simultaneous primary infection of a hepatitis B virus-susceptible individual with hepatitis delta virus and hepatitis B virus. It most commonly results from parenteral transmission, for example, among injecting drug abusers. The incubation period depends upon the size of the inoculum of hepatitis B virus and is similar to that of hepatitis B (six weeks to six months). Both viruses replicate simultaneously or sequentially, giving rise either to a single episode of clinical hepatitis B plus delta virus, or in 10% to 20% of cases to two discrete episodes—hepatitis B followed by hepatitis D—depending on the ratio of the two viruses in the original inoculum. Co-infections with hepatitis delta virus and hepatitis B virus can be more severe than the disease caused by hepatitis B virus alone, with the incidence of fulminant hepatitis being higher, but the proportion of acute infections that progress to chronicity, 1% to 3%, is no greater than following infection only.

Superinfection by hepatitis delta virus of an individual chronically infected with HBV is the commoner occurrence and the more serious. Because large numbers of hepatocytes are already producing HBsAg, hepatitis delta virus is able to replicate without delay and, after a relatively short incubation period (3 weeks), a viremia of up to 10^{11} virions/mL and a severe clinical attack of hepatitis delta-induced hepatitis ensues. Case fatality rates of up to 20% from fulminant hepatitis have been observed in some studies. Amongst others, acute self-limited hepatitis delta virus infection may occur, but most go on to develop persistent hepatitis delta virus infection with chronic active hepatitis, and progression to cirrhosis can be rapid. About 60% to 70% of individuals with chronic hepatitis delta virus infection will develop cirrhosis—a rate that is about three times higher than the rate for hepatitis B or hepatitis C virus infection (Fig. 22.10).

Following liver transplantation a third form of hepatitis delta virus infection may be seen, latent hepatitis delta virus, whereby the hepatitis delta-virus re-infects the graft but HBV does not due to the administration of HBIG during and after transplantation. Over the next two to three months, intra-hepatic hepatitis delta virus genome expression declines in the absence of HBV reinfection, and thus liver transplantation can be curative for hepatitis delta.

Pathogenesis

HBV-positive chimpanzees can be infected experimentally with hepatitis delta virus, and hepatitis delta virus is also found in some animals (e.g., in co-infection with woodchuck hepatitis virus), but little is known of the basis of the liver pathology. The observation that the extent of liver damage correlates with the amount of the small (P24) form of HDAg in liver cell nuclei rather than with the number of inflammatory cells suggests that P24 is directly cytotoxic for hepatocytes. The P24 form predominates during acute infection, a period when large numbers of virions are being produced, whereas P27 is found mainly in chronic infection, when much lower titers of virus are present. However, other evidence supports the view that hepatocyte damage may be mainly immunologically mediated. Following hepatitis delta virus superinfection of a hepatitis B virus carrier, the titer of HBsAg in the serum may be depressed during the period of maximum hepatitis delta virus replication, possibly as a result of competition for host RNA polymerase II which is required by both viruses for transcription. Competition for HBsAg, required by both viruses for virus assembly, may also be a factor but the number of hepatitis delta virus particles produced often exceeds that of hepatitis B virions by 1000-fold.

Laboratory Diagnosis

The method of choice for diagnosis of ongoing hepatitis delta virus infection is reverse transcriptase polymerase chain reaction (RT-PCR). Fig. 22.10 depicts the time course of the several diagnostic serologic markers throughout the progression of hepatitis delta virus/hepatitis B virus co-infection (upper panel), or hepatitis delta virus superinfection of an HBsAg carrier (lower panel). In simultaneous co-infection towards the end of the incubation period, HBsAg appears, followed by the detection of HDAg in serum by enzyme immunoassay or immunoblotting, and hepatitis delta virus RNA by RT-PCR. HDV RNA levels can reach 1×10^{12} genomes/mL. A rise in transaminases, anti-HBc IgM and anti-HDV IgM antibodies then occurs. In most instances hepatitis delta virus is cleared rapidly and the only measured marker in the later stages of the acute disease may be IgM anti-HD antibodies identified by use of an enzyme immunoassay. Total antibody to hepatitis delta virus is demonstrable for longer periods although even the IgG antibody response tends to be short-lived following resolution of the infection, thus screening for anti-HDV antibodies is not a reliable indicator of past hepatitis D virus infection.

Acute superinfection is distinguished from acute co-infection by the absence of IgM anti-HBc. In chronic hepatitis delta infection, ALT levels fluctuate but remain

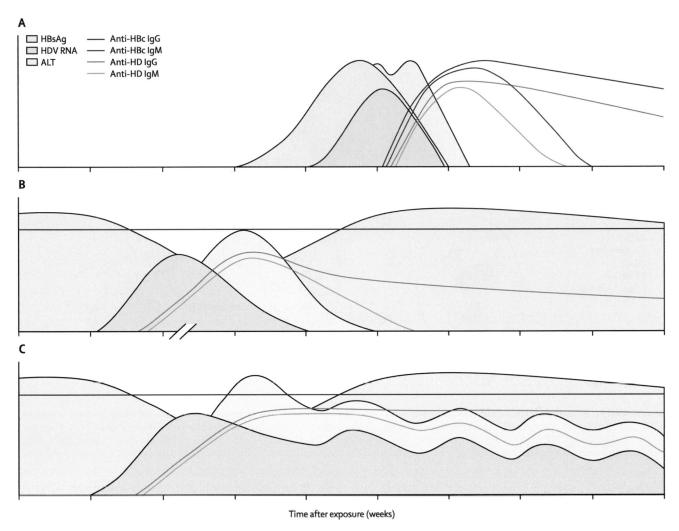

FIGURE 22.10 Typical evolution of serological and virological markers in hepatitis delta virus (HDV) infection. (A) Simultaneous co-infection with HBV and hepatitis delta virus, resulting in clearance of both viruses in almost all patients. (B) Hepatitis delta virus superinfection of an HBV carrier with self-limited outcome. The spontaneous clearance of HDV RNA might take years to occur (indicated by break on x-axis) and, in few cases, can herald the loss of HBsAg (not depicted). (C) Hepatitis delta virus superinfection of a HBV carrier with chronic persistent viral replication; the more common outcome after superinfection. *ALT*, alanine aminotransferase; *HBsAg*, hepatitis B surface antigen. *Reproduced from Hughes, S.A., et al., 2011. Hepatitis delta virus, Lancet 378;73–85, with permission.*

elevated, while anti-HD IgM and IgG antibodies together with hepatitis delta virus RNA remain demonstrable in the serum for months or years.

Epidemiology

Transmission of hepatitis delta virus occurs by exactly the same routes as for hepatitis B virus, namely parenteral, perinatal, sexual, and close contact. In developing countries where there is high hepatitis delta virus prevalence it is likely that most spread occurs horizontally among children by contact with open skin lesions, etc., and among adolescents and young adults by sexual intercourse. Perinatal transmission is less common than with hepatitis

B virus. In developed countries where prevalence is low, such as North America, temperate South America, Australia, and western and northern Europe, the majority of hepatitis delta virus infections are acquired parenterally, particularly by intravenous drug users sharing needles. Prolonged epidemics occurred in the latter cohort during the 1970s, with about half of all HBsAg-positive injecting drug users in some Western countries becoming infected. The overall trend more recently has been a decline in the prevalence of hepatitis delta virus in this high-risk group. Sexual transmission to partners occurs frequently, but is less efficient than sexual transmission of hepatitis B virus. Hepatitis delta virus outbreaks have also been reported in hemophiliacs receiving contaminated clotting factors and

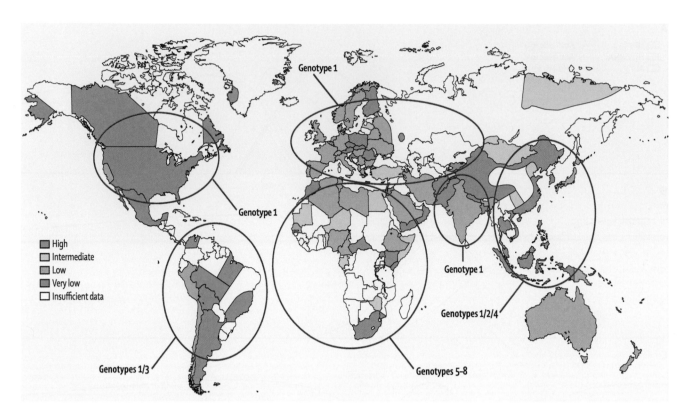

FIGURE 22.11   Worldwide prevalence of hepatitis delta virus (HDV) and the geographic distribution of its genotypes. Two subtypes have been identified in HDV genotype 1, designated 1A (predominant in Asia) and 1B (in the United States). Both are common in the Mediterranean. HDV genotype 2 occurs in the Far East. HDV genotype 3 occurs exclusively in the northern part of South America and is linked to HBV genotype F. Genotypes 5–8 have been identified primarily in patients from Africa. *Reproduced from Hughes, S.A., et al., 2011. Hepatitis delta virus. Lancet 378:73–85, with permission.*

in hemodialysis units, as well as among hepatitis B virus carriers in institutions for the developmentally disabled.

Hepatitis delta virus is transmitted mainly by blood and blood products similarly to hepatitis B virus. Hepatitis delta virus is much less common than HIV-1, suggesting that it is not spread easily by a sexual route. In the United States, about 70,000 people have chronic hepatitis delta virus infection, and hepatitis delta virus causes about 1000 deaths per year.

The world distribution of hepatitis delta virus roughly parallels that of its helper hepatitis B virus, but its prevalence in HBV carriers varies widely; for example, hepatitis delta virus is rare in the high hepatitis B virus prevalence countries of east Asia, including most parts of China. High hepatitis delta virus prevalence regions include the western Amazon basin, parts of central Africa, Eastern Europe, and isolated provinces of China and parts of Micronesia. In these areas over 20% of HBsAg carriers and over 60% of chronic hepatitis cases have markers of hepatitis delta virus infection. For example, a remote tribe of indigenous Venezuelan Indians has been found to suffer a 10% annual hepatitis delta virus infection rate among hepatitis B virus carriers, with a high case fatality rate from fulminant hepatitis or from rapidly progressive chronic hepatitis. Prevalence is moderately high in Southern and Eastern Europe, the Middle East, the central Asian republics,

the Indian subcontinent, and Central and South America. Hepatitis delta virus prevalence is reported to be increasing in parts of the Middle East and Asia, and decreasing significantly in some Western countries in recent years (Fig. 22.11).

Eight genotypes of hepatitis delta virus are currently described: genotype 1 is widely distributed, while genotype 2 is mainly found in Japan; genotype 3, from South America, is associated with a severe form of hepatitis delta characterized by high mortality and a characteristic histological lesion in the liver called a morula cell. Genotype 4 has been reported from Taiwan and genotypes 5 to 8 from Africa.

Treatment and Prevention

There is no reliable established treatment for hepatitis delta virus infection. Interferon-alpha, 5 megaunits daily, produces some amelioration of the disease in about half of all chronic hepatitis delta patients and should be continued as long as possible while the patient is still hepatitis delta virus RNA positive. Most other antivirals, including those that inhibit HBV, are of no benefit. However, recent studies with the lipopeptide Myrcludex-B, an entry inhibitor of HBV, have shown excellent control of hepatitis delta virus when used in combination with pegylated-IFN-alpha. Moreover, the prenylation inhibitor lonafarnib, which interferes with viral assembly, has also recently been shown to be active against hepatitis delta virus with or without combination with pegylated-IFN-alpha. Further trials are underway, but liver transplantation currently offers the only therapeutic option in the management of end-stage liver failure and may be curative for hepatitis delta virus (see earlier).

Immunization against HBV provides complete protection against hepatitis delta virus, since hepatitis delta virus requires hepatitis B virus in order to replicate. Control of hepatitis delta virus focuses on the prevention of co-infection with hepatitis B virus or of superinfection of hepatitis B virus carriers, and hence requires all the measures that apply to the prevention of hepatitis B virus infection, including vaccination against hepatitis B virus. Individuals with persistent HBV infection are at particular risk and should avoid possible exposure to hepatitis delta virus.

FURTHER READING

Kremsdorf, D., et al., 2006. Hepatitis B virus-related hepatocellular carcinoma: paradigms for viral-related human carcinogenesis. Oncogene 25, 3823–3833.

Liang, T.J., Block, T.M., McMahon, B.J., et al., 2015. Present and future therapies of hepatitis B: from discovery to cure. Hepatology 62, 1893–1908.

Pascarella, S., Negro, F., 2011. Hepatitis D virus: an update. Liver International 31, 7–21.

Squadrito, G., Spinella, R., Raimondo, G., 2014. The clinical significance of occult HBV infection. Ann. Gastroenterol. 27, 15–19.

Taylor, J.M., 2015. Hepatitis D virus replication. Cold Spring Harb. Perspect. Med. 5, a021568.

Trepo, C., Chan, H.L.Y., Lok, A., 2014. Hepatitis B infection. Lancet 384, 2053–2063.

Chapter 23

Retroviruses

Retroviruses are fascinating biological agents possessing a complex and unexpected replication strategy. Retroviruses are likely to have had a unique role in the evolutionary history of higher animals, and have provided reagents that are now an essential part of the molecular biologist's toolkit. The family *Retroviridae* includes examples that cause diseases of major importance to humans and domestic animals. The importance of retroviruses has been underlined by the award of Nobel Prizes in this area on no less than four separate occasions (see Box 23.1).

The threat posed by AIDS triggered an unprecedented effort, by research scientists and government agencies alike, to understand and conquer this disease. Indeed, HIV has set the pace in virus research for many years. New concepts and techniques were pioneered by HIV virologists in every area of the discipline, including regulation of viral replication, molecular pathogenesis, laboratory diagnostic methods, novel approaches to antiviral therapy and vaccinology, epidemiological monitoring, and ethical and patient privacy issues. These advances have since become the cutting edge serving as a yardstick for researchers in other areas.

PROPERTIES OF *RETROVIRIDAE*

Retroviruses are widely distributed amongst vertebrates. Many exist as horizontally transmitted infectious agents, with or without producing disease. Others are represented by endogenous genetic material (proviruses) that have resulted from incorporation of retroviral genetic material into the germ line. These endogenous retroviruses are usually incapable of producing infectious virus because of acquired deletions or mutations, and are transmitted vertically as Mendelian genes. Genetic material of this nature can constitute as much as 10% of the genome of some species, and there is debate about the evolutionary processes leading to the incorporation and maintenance of these genetic footprints.

Retrovirus virions are spherical, 80 to 100 nm diameter, and have a characteristic three-layered structure. Innermost is the genome–nucleoprotein complex, closely associated with several molecules of the three viral enzymes—reverse transcriptase, integrase, and protease. This structure is enclosed within a capsid, which appears icosahedral and is centrally located in the case of human T cell lymphotropic virus (HTLV), spumaviruses, and the so-called C-type retroviruses of animals and birds but which appears as a cone or rod in the case of the lentiviruses. This in turn is surrounded by a matrix protein layer, and the whole is enclosed by a lipid envelope from which glycoprotein peplomers project.

The retroviral genome is unique among viral genomes in several respects: (1) it is the only genome existing as two copies within the virion, (2) it is the only viral RNA genome that is synthesized and processed by the mRNA-processing

---

BOX 23.1 Nobel Prizes Awarded for Work Associated with Retroviruses

1. In 1911 Peyton Rous discovered that a malignant sarcoma of chickens could be transmitted by a cell-free filtrate containing an infectious virus. The immense significance of this discovery was belatedly recognized by the award of a Nobel prize in 1966, more than half a century later.

2. Equally controversial initially, Howard Temin proposed that genetic information could flow "against the tide," from RNA to DNA. This iconoclastic proposal was confirmed unequivocally by the subsequent discovery in 1970, independently by Temin and David Baltimore, of the enzyme reverse transcriptase in retroviruses, for which they received the Prize in 1975. This enzyme has underpinned many of the subsequent spectacular advances in recombinant DNA technology and genetic engineering.

3. Michael Bishop and Harold Varmus received the Prize in 1989 for their earlier discovery of oncogenes and their role in oncogenic viruses and in cancer generally.

4. In 1984 Françoise Barré-Sinoussi and Luc Montagnier isolated what we now designate human immunodeficiency virus (HIV), the causal agent of the devastating pandemic of acquired immunodeficiency syndrome, AIDS, and shared the 2008 Nobel Prize for this work. This discovery was confirmed and extended by Bob Gallo and his colleagues, who in 1980 had described HTLV-1, the first human retrovirus and cause of adult T cell leukemia.

Fenner and White's Medical Virology. DOI: http://dx.doi.org/10.1016/B978-0-12-375156-0.00023-0

machinery of the cell, (3) it is the only genome associated with a specific tRNA whose sole function is to prime replication, and (4) it is the only plus sense ssRNA genome that does not serve as mRNA soon after initiation of infection. Each copy of the genome is a linear, single-stranded, plus sense molecule of 7 to 10 kb, with a 3′ polyadenylated tail and a 5′ cap. The specific cellular tRNA, which serves as the primer for transcription, is base-paired to a site near the 5′ end of each viral RNA monomer.

The genome of all non-defective retroviruses contains three essential major genes, each coding for two or more polypeptides. Reading from the 5′ end, (1) the *gag* gene (standing for group-specific antigen) codes the virion core (capsid) proteins, (2) the *pol* gene codes for the enzymes reverse transcriptase (polymerase), integrase and protease, and (3) the *env* gene codes the viral envelope proteins. Many exogenous oncogenic retroviruses of animals and birds also contain an oncogene of cellular origin at the 3′ end that is transmitted as a fourth gene within the infecting virus. Other retroviruses contain a variety of additional regulatory genes coding for proteins involved in virus RNA synthesis and export, or in combatting different host restriction factors. For example, the only known human oncogenic retrovirus, HTLV-1, although without an oncogene, does however contain two regulatory genes, one of which, *tax*, transactivates expression of all viral genes as well as certain cellular genes. The non-oncogenic lentivirus genome usually contains six such regulatory genes. Importantly, each end of the retrovirus genome contains an identical copy of a distinctive component known as the *long terminal repeat* (LTR), which plays an essential role in transcription, integration, and regulation of expression of the integrated cDNA provirus.

Retroviruses are currently classified into seven genera, only three of which include pathogens of humans (Table 23.1 and Fig. 23.1). Genera differ in virion morphology, regulatory genes, and replication strategy, with no antigens in common. Within genera, however, there are some shared "group-specific" antigenic determinants. Antibodies to type-specific determinants on envelope glycoproteins neutralize viral infectivity.

The phylogenetic relationships within the genus *Lentivirus* are discussed in more detail at the end of this chapter.

VIRAL REPLICATION

Overview: The replication of retroviruses proceeds through several unique biological steps that in turn affect pathogenesis. Very soon after virus entry into the cell, the genetic information encoded in the viral RNA is copied to DNA by the action of the reverse transcriptase carried in the virion. This DNA migrates to the nucleus, becomes integrated into chromosomal DNA of the cell, and is then transcribed into new viral RNA by a similar mechanism to the transcription of cellular genes. The newly made virus RNA, with or without subsequent splicing, is exported

TABLE 23.1 Members of the Family *Retroviridae* Subfamily Orthoretrovirinae

Genus	Examples of Species
Alpharetrovirus	Avian leukosis virus; Rous sarcoma virus
Betaretrovirus	Mouse mammary tumor virus
Gammaretrovirus	Murine leukemia virus; feline leukemia virus; Moloney murine sarcoma virus
Deltaretrovirus	Bovine leukemia virus; human T-lymphotropic virus 1 and 2; simian T-lymphotropic virus 1 and 2
Epsilonretrovirus	Walleye dermal sarcoma virus
Lentivirus	Human immunodeficiency virus 1 and 2; simian immunodeficiency virus; feline immunodeficiency virus; equine infectious anemia; bovine immunodeficiency virus; caprine arthritis encephalitis virus; visna virus
Subfamily Spumaretrovirinae	
Spumavirus	Simian foamy virus

to the cytoplasm where it provides both mRNA for viral protein synthesis and new RNA genomes for progeny virions. Finally, new virions are assembled in the cytoplasm. The information flow in retrovirus replication represents a mirror image of that employed by hepadnaviruses, in which the form of the genome present in extracellular virus is DNA not RNA, and where reverse transcription takes place as the viral RNA is leaving rather than entering the cell (Table 23.2; see Chapter 22: Hepadnaviruses and Hepatitis Delta).

Steps in replication: (see Fig. 23.2 for the replication of HIV as an example). Retrovirus virions adsorb to specific cell receptors via one of the two viral envelope glycoproteins. Among different retroviruses entry occurs either by receptor-mediated endocytosis or by direct fusion with the plasma membrane. Uncoating then releases the nucleoprotein complex into the cytoplasm. Using the genome-associated tRNA as a primer, the reverse transcriptase begins to synthesize a minus sense cDNA from the viral genome. Concomitantly, the viral positive sense RNA template is digested by a second domain of the reverse transcriptase molecule, which carries RNase H enzymatic activity. The ongoing minus strand DNA synthesis then relocates from the 5′ end to the 3′ end of the RNA template and continues eventually to yield a full-length minus viral DNA strand. Oligonucleotides remaining from the viral RNA template, due to incomplete hydrolysis of polypurine tracts (PPTs) of RNA, serve as primers for the synthesis of plus sense cDNA using the newly made minus sense cDNA as template. Synthesis of positive sense cDNA similarly jumps from the 5′ end of the minus strand DNA template to continue from the opposite 3′ end. The 5′ and 3′ ends of the viral RNA

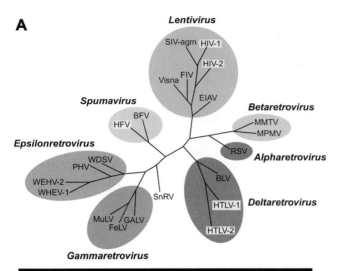

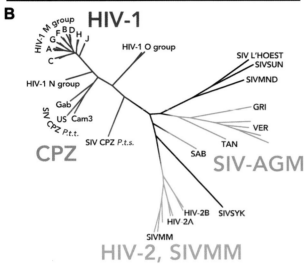

FIGURE 23.1 Phylogeny of the retroviruses. (A) Overview of the relationship of the genera in the family *Retroviridae*, based on conserved regions of the polymerase gene of the various viruses, including human immunodeficiency viruses 1 and 2 (HIV-1, HIV-2), human T-lymphotropic viruses 1 and 2 (HTLV 1, HTLV 2) and human foamy virus (HFV) (B) Relationship of member viruses of the genus *Lentivirus* that are related to HIV, based upon meta-analysis of studies of several genes—these include human and non-human primate viruses. Human immunodeficiency virus (HIV) includes HIV-1 M (Main) group with subtypes A–J (responsible for the pandemic); HIV-1 O (Outlier) group, most commonly found in West Africa; and HIV-1 N group, found in a very small number of individuals in West Africa. HIV-2 subtypes A and B. Simian immunodeficiency viruses (SIV) include: SIV-CPZ-P.t.t., from chimpanzee *Pan troglodytes troglodytes* (with subtypes GAB, Cam3 and US); SIV-CPZ-P.t.s., from chimpanzee *Pan troglodytes shweinfuthii*; SIV-AGM, African green monkey; SIV-AGM TAN, tantalus monkey; SIV-AGM-VERTYO, vervet monkey; SIV-AGM-GRI, grivet monkey; SIV-AGM-SAB1C, sabaeus monkey; SIVMM, sooty mangaby and also found in captive macaques; SIVLHOEST, SIV L'hoest; SIV-MND, mandrill; Others: *ALV*, avian leukosis virus. *BFV*, bovine foamy virus; *EIAV*, equine infectious anemia virus; *FeLV*, feline leukemia virus; *FIV*, feline immunodeficiency virus; *HFV*, human foamy virus; *MMTV*, mouse mammary tumor virus; *MPTV*, Mason-Pfizer tumor virus; *MuLV*, murine leukemia virus; *PHV*, perch hyperplasia virus; *RSV*, Rous sarcoma virus; *BLV*, bovine leukemia virus; *SnRV*, snakehead retrovirus; *Visna*, visna virus; *WDSV*, walleye dermal sarcoma virus; *WEHV*, walleye epidermal hyperplasia virus. *(B) Based on work by the Theoretical Biology and Biophysics Group, Los Alamos National Laboratory, with permission.*

TABLE 23.2 Replication of Retroviruses and Hepadnaviruses Are Mirror-Images of the Same Essential Processes

	Retroviruses	Hepadnaviruses
Genetic material in the virion	RNA (positive sense, two copies per virion)	DNA[a] (double-stranded with a single-stranded gap)
Stage in replication where reverse transcription occurs	"Afferent" arm, that is, between RNA entry and its reaching the nucleus	"Efferent" arm, i.e., after RNA leaves the nucleus but before it exits from the cell[a]
Requirement for integration of newly made proviral DNA into chromosomal DNA	Essential	Does occur but infrequently
Template within the nucleus for RNA transcription	Integrated proviral DNA	"Episomal" covalently closed circular (ccc) DNA molecules

[a]*Spumaviruses are similar to hepadnaviruses in these two properties.*

contain sequences identified as R/U5 (5′ end) and U3/R (3′ end), and because of these jumps, the linear double-stranded DNA resulting from reverse transcription contains identical long terminal repeats (LTRs) at each end composed of sequences derived from each end of the viral RNA, namely U3, R, and U5 in that order. The LTRs at each end of the proviral DNA are important because they contain critical regulatory sequences, including an enhancer/promoter region with binding sites for viral and/or cellular regulatory proteins. The dsDNA, within a reverse transcription complex, migrates to the nucleus along microtubules, and several such molecules become integrated as provirus at random sites into cellular DNA. Integration requires the viral integrase and involves removal of two nucleotides from the ends of the viral DNA and generation of a short duplication of cell sequences at the integration site, enabling joining of the ends of the viral DNA to cellular DNA.

Integration is a prerequisite for virus replication in order that the integrated provirus can be transcribed by cellular RNA polymerase II. The complete RNA transcript is identical to the original genomic RNA (that is, with R/U5 at the 5′ end, and U3′/R at the 3′ end) and serves as mRNA for translation into the Gag (core) polyprotein and, after frameshifts, the Gag-Pol precursor polyprotein. A second, shorter mRNA, spliced from the full-length RNA, is translated, again in a different reading frame, to yield the Env precursor. In the case of HTLV and HIV, a range of additional differently spliced mRNAs predominate early in replication and codes for various regulatory proteins. All

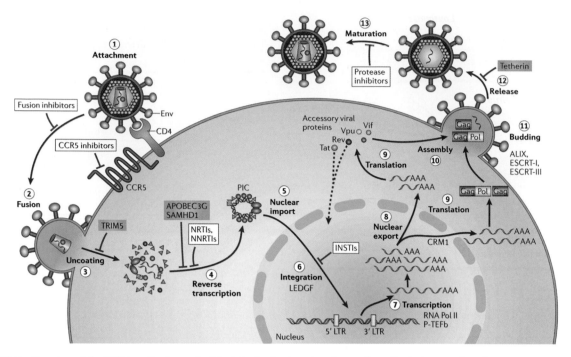

FIGURE 23.2   Overview of the HIV-1 replication cycle. The infection begins when the envelope (Env) glycoprotein spikes engage the receptor CD4 and the membrane-spanning co-receptor CC-chemokine receptor 5 (CCR5) (step 1), leading to fusion of the viral and cellular membranes and entry of the viral core particle into the cell (step 2). Partial core shell uncoating (step 3) facilitates reverse transcription (step 4), which in turn yields the pre-integration complex (PIC). Following import into the cell nucleus (step 5), PIC-associated integrase orchestrates the formation of the integrated provirus, aided by the host chromatin-binding protein lens epithelium-derived growth factor (LEDGF) (step 6). Proviral transcription (step 7), mediated by host RNA polymerase II (RNA Pol II) and positive transcription elongation factor b (P-TEFb), yields viral mRNAs of different sizes, the larger of which require energy-dependent export to leave the nucleus via host protein CRM1 (step 8). Some mRNA molecules serve as templates for protein production and other RNA molecules provide full-length progeny viral genomes (step 9), which are incorporated into nascent budding viral particles along with the several viral proteins (step 10). Viral-particle budding (step 11) and release (step 12) from the cell are mediated by complex network of host proteins, ESCRT (endosomal sorting complex required for transport) and ALIX complexes and is accompanied or soon followed by protease-mediated maturation (step 13) to create the infectious mature virion. Each step in the HIV-1 life cycle is a potential target for antiviral drug intervention; the sites of action of clinical inhibitors (white boxes) and cellular restriction factors (blue boxes) are indicated: *INSTI*, integrase strand transfer inhibitor; *NNRTI*, non-nucleoside reverse transcriptase inhibitor; *NRTI*, nucleoside reverse transcriptase inhibitor. *Reproduced from Engelman, A., Cherepanov, P., 2012. The structural biology of HIV-1: mechanistic and therapeutic insights. Nat. Rev. Microbiol. 10:279–290, with permission.*

the mRNAs share a common sequence at their 5′ end. The viral protease is responsible for post-translational cleavage of the Gag polyprotein to yield the matrix, capsid, and nucleocapsid proteins, and also for cleavage of the Gag-Pol polyprotein to yield the reverse transcriptase/RNase H and the integrase. On the other hand, the Env polyprotein is cleaved by a cellular protease to yield two envelope glycoproteins: a transmembrane protein and a receptor-binding protein. Identification and discussion of mature retroviral proteins have been facilitated by adoption of standard nomenclature (Tables 23.3A and B).

The details of assembly of the virion are not fully understood and differ from genus to genus. Two covalently linked molecules of the full-length plus sense genomic RNA assemble into a core structure by interaction of a packaging signal in the leader sequence of the RNA with a zinc-finger motif in the nucleocapsid protein. Capsids generally assemble at the cell surface. Myristylated and glycosylated envelope proteins enter the plasma membrane and form the

peplomers of the envelope acquired during exocytosis. The final proteolytic cleavage steps occur during, and possibly, after budding (Table 23.4).

HUMAN T CELL LYMPHOTROPIC VIRUSES

HTLV-1, the first human retrovirus discovered, was described in 1980. As with most of the previously known retroviruses of other animals and birds, HTLV-1 was found to be oncogenic—in this case responsible for an unusual form of T cell leukemia in adults—but the mechanism of oncogenesis is unique. Six different subtypes are endemic in different parts of the world, for example, subtype A in Japan, subtypes B, D, and F in Central Africa, subtype C in Melanesia, and subtype E in South and Central Africa. A second species, HTLV-2, is prevalent in various Native American populations and among injecting drug users in the United States, but its association with disease has not been consistently confirmed.

TABLE 23.3 (A) Standard Nomenclature of Retroviral Proteins

Gene	Standard Protein Designation	Common Name (HIV)
env (envelope)	SU (surface)	gp120
	TM (transmembrane)	gp41
gag (group-specific antigen)	MA (matrix)	p17
	CA (capsid)	p24
	NC (nucleocapsid)	p7
		p6
pol (polymerase)	PRO (protease)	
	INT (integrase)	
	RT (reverse transcriptase)	

(B) Accessory and Regulatory Proteins of HIV

Protein	Size (kDa)	Function
Regulatory		
Tat	14	Enhances efficiency of transcription from HIV promoter
Rev	19	Assists nuclear export of unspliced or singly spliced (intron-containing) mRNA
Accessory		
Nef	27	Down-regulates MHC-I and -II, CD4 to enhance T cell activation; required for development of clinical AIDS
Vif	23	Enhances yield of infectious virus by inhibiting APOBEC deaminase enzymes that inactivate HIV proviral DNA
Vpr	15	Transport of pre-integration complex to the nucleus; activates transcription from HIV LTR; arrests cell cycle in G2
Vpu (HIV-1 only)	16	Promotes degradation of CD4; enhances virus release by budding
Vpx (HIV-2 only)	15	Transport of pre-integration complex to the nucleus; induces degradation of SAMHD1, a host restriction factor that depletes dNTPs

TABLE 23.4 Properties of Retroviruses

Spherical enveloped virion, 80–100 nm diameter

Ribonucleoprotein within central capsid (icosahedral in type C viruses; truncated cone in lentiviruses) surrounded by matrix protein and envelope with glycoprotein peplomers

Two copies of linear plus sense ssRNA genome, each 7 to 10 kb, non-covalently linked; *gag, pol, env* genes; some also contain regulatory genes; some carry an oncogene

Reverse transcriptase transcribes DNA from virion RNA; following formation of long terminal repeats, dsDNA is integrated into cellular chromosomes as provirus

Full-length RNA transcripts encode core and capsid proteins, protease, reverse transcriptase/RNase H, integrase; shorter spliced transcripts encode envelope glycoproteins and regulatory proteins

In productive infections, virions assemble at and bud from plasma membrane

Infection may be cytocidal, non-cytocidal, or transforming; oncogenic retroviruses may be replication-competent or defective, and induce malignancy by transduction, *cis*-activation, or *trans*-activation

Viral Replication

HTLV-1 is a replication competent retrovirus that can induce cancer in spite of the fact that its genome does not carry a viral oncogene, and nor is there *cis*-activation of a nearby cellular oncogene following integration of proviral DNA. In addition to the usual *gag-pol-env* gene order of retroviruses, HTLV-1 contains a unique fourth region, PX, coding for regulatory proteins Tax, Rex, p12, p13, and p30 (Fig. 23.3). Tax acts in *trans* both to up-regulate transcription of all the viral genes in the integrated HTLV proviral DNA and also to initiate the leukemogenic process (see Chapter 9: Mechanisms of Viral Oncogenesis). A second HTLV regulatory gene, *rex*, encodes a protein Rex that increases the proportion of unspliced RNA transcripts by promoting the production and transport to the cytoplasm of unspliced viral RNA and the incompletely spliced *gag/pol* and *env* mRNAs, so allowing the production of new virions. Rex has a negative effect on its own production and that of Tax, because multiple splicing is required to produce the mRNAs for these two regulatory proteins (Fig. 23.3). This negative regulation of Tax and Rex expression by Rex would then lead to decreased expression of all the viral genes, and might enhance the establishment of latency. Thus it has been postulated that Rex may help to orchestrate a state of latency alternating with periods of virus production.

A further regulatory gene product, HTLV-1 basic leucine zipper factor (HBZ), encoded by the minus strand of the provirus, inhibits tax-mediated viral transcription

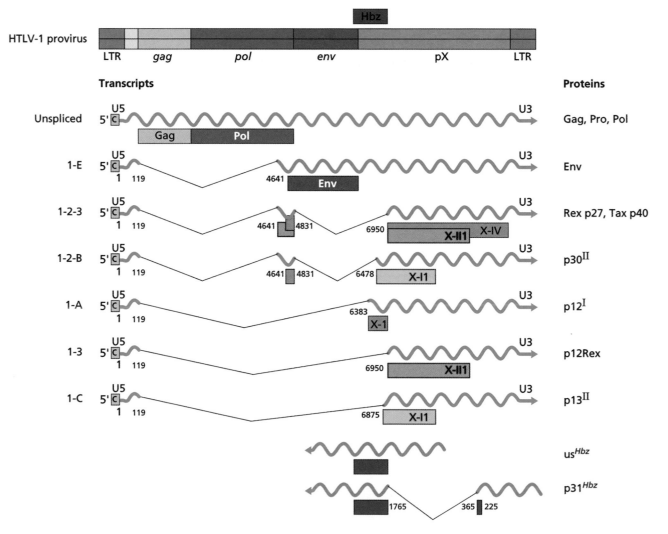

FIGURE 23.3 Transcription map of human T-lymphotropic virus type 1 (HTLV-1) proviral DNA showing major gene-coding regions and their functions, and intron splicing patterns. Transcription of all but the Hbz gene is initiated between the unique 3′ (U3) and R regions in the 5′ LTR. The Hbz gene is transcribed either from multiple Sp1-promoted initiation sites in the unique 5′ (U5) and R regions of the 3′ LTR, which produce spliced mRNA transcripts (sHbz), or from an initiation site within the Tax gene to form the unspliced transcript (usHbz). *Adapted from P. Kannian and P. L. Green, Viruses 2:2037–2077, 2010.*

and promotes cell proliferation, a high proviral load, and persistence *in vivo*. Both HBZ and Tax are thought to play key roles in leukemogenesis.

Pathogenesis

HTLV probably enters the body principally inside infected CD4+ lymphocytes in semen or blood, as well as vertically from mother to infant via breast milk and possibly across the placenta. It establishes a lifelong persistent infection which generally remains subclinical but occasionally induces disease after an incubation period of months to 40 years. With HTLV-1, the pathogenesis of one of the two major clinical manifestations, adult T cell leukemia/lymphoma, was discussed in detail in Chapter 9: Mechanisms of Viral Oncogenesis. Briefly, the proteins HBZ and Tax are both thought to play important roles in oncogenesis. HBZ inhibits tax-mediated 5′ LTR transcription and modulates cell signaling pathways involved in cell growth, T cell differentiation and immune responses. Tax transactivates transcription not only from the proviral LTR but also from certain cellular transcription factors and cell cycle regulators, as well as from the cellular gene encoding the interleukin-2 (IL-2) receptor, thereby inducing autocrine stimulation of T cell proliferation. In ATL cell lines and patient T cells, HBZ is often the only viral protein expressed, and continued expression of Tax is not necessary for maintenance of malignancy. As will be seen later in this chapter, expression of the HTLV *tax* gene also induces the production of the pleiotropic cellular DNA-binding transcription factor NF-κB and may thereby serve as a cofactor in the progression of AIDS by inducing transcription of the integrated HIV provirus.

In contrast, HTLV-2 has not been implicated in leukemia or other forms of malignancies. HTLV-2 has a homologue of HBZ, known as APH-2, which has different effects from HBZ on cellular functions such as transforming growth factor β (TGF-β) signaling and interferon regulatory factor 1 (IRF-1) trans-activation, and these functional differences may be involved in differences in oncogenic potential.

Clinical Features

Although most HTLV-1 infections remain asymptomatic, there is a 1% to 4% risk of disease following an incubation period of 10 to 40 years, greatest usually between the ages of 30 and 50. There are two distinct clinical manifestations but only rarely are both encountered in the same patient.

Adult T cell leukemia/lymphoma (ATLL or ATL) generally takes the form of an acute aggressive leukemia of mature CD4+ T lymphocytes seen in middle-aged adults. The patient presents with lymphadenopathy, hepatosplenomegaly, hypercalcemia, lytic bone lesions, and sometimes leukemic cell infiltrates in the skin. The malignant T cells are pleomorphic with large convoluted nuclei. The consequent immunosuppression may lead to various opportunistic infections. Death generally occurs within one to two years. Less commonly, the disease may follow a more chronic course with fewer leukemic cells in the blood but resembling a non-Hodgkin's lymphoma, with or without skin deposits. There also appears to be an association with polymyositis.

Tropical spastic paraparesis (TSP), otherwise known as HTLV-1 associated myelopathy (HAM), is a progressive demyelination of the long motor neuron tracts in the spinal cord. Patients show high HTLV-I viral loads and high levels of Tax-specific cytotoxic lymphocytes, and molecular mimicry between viral and cellular antigens has been proposed as a mechanism.

There is an incubation period varying from months to decades. Seen more frequently in women than in men, the condition typically starts with lumbar back pain radiating down the legs and progresses to weakness and spastic paralysis of both lower limbs, accompanied by urinary frequency or retention. Patients may also present with, or develop, visual changes due to uveitis, arthritis, polymyositis, or keratoconjunctivitis sicca. The natural course of the disease is variable; most progression occurs in the first few years after diagnosis, and thereafter the condition remains relatively stable or deteriorates slowly over many years, although up to half of patients eventually need a wheelchair.

Laboratory Diagnosis

HTLV-1 infection, which is usually subclinical, is diagnosed by demonstrating the presence of antibodies by enzyme immunoassay (EIA). Because of the ever-present possibility of false-positive results between different EIAs, use of a confirmatory procedure is essential for all initially reactive samples. Different confirmatory algorithms include the use of multiple EIAs from more than one manufacturer, Western blotting, and polymerase chain reaction (PCR). Western blotting is also necessary to distinguish between HTLV-1 and HTLV-2 because of the shared epitopes between these viruses.

PCR is used to demonstrate HTLV DNA in lymphocytes and can also discriminate between HTLV-2 and HTLV-1 provided suitable pairs of primers are chosen. PCR can also be used to quantify viral load, and higher loads may reflect the likelihood of ATL or HAM/TSP developing in HTLV-1 carriers. Viral load is expressed as the number of viral genome copies per fixed number of peripheral blood mononuclear cells.

Virus can be propagated only with difficulty, classically by *in vitro* culture of the patient's peripheral blood leukocytes, with either a mitogen or an HTLV-transformed T cell line. Because virus is almost exclusively cell-associated, infection occurs by fusion of infected and uninfected CD4+ lymphocytes, leading to transformation and immortalization of the latter. Cells other than T lymphocytes can also be infected, albeit with difficulty, by co-cultivation with infected CD4+ T cells.

Hematological findings in ATL include pleomorphic lobular leukemic cells with mature CD4+ T cell markers.

Epidemiology

The geographic distribution of HTLV-1 is very patchy, with a tendency to cluster in certain countries, mainly in the tropics. The highest prevalence (>1% of the population) has been observed in southern Japan, the Caribbean, equatorial Africa, parts of South America, eastern Siberia, Pacific islands such as Papua New Guinea, and certain Inuit populations. This suggests that the virus is of ancient origin and has been maintained by vertical transmission in isolated racial groups, dating back to the days before the major waves of intercontinental migration 40,000 to 100,000 years ago, a time when the average human lifespan was shorter than the incubation period of the diseases seen in HTLV-positive patients today. Spread to the Caribbean may have coincided with the much more recent slave trade from Africa.

The observed clustering of infection within particular families indicates that vertical transmission or close contact is required. Mother-to-child transmission occurs during breast-feeding, with an efficiency of up to 20% with a longer duration of breast-feeding. Transplacental transmission occurs but is less common at around 5%. The second important route of transmission is sexual intercourse, with the receptive partner being most at risk; this accounts for the somewhat higher incidence in females. Third, parenteral transmission can be significant; transfusion of infected blood has a 40% to 60% transmission rate: intravenous drug use is an important route of HTLV-2 transmission in North America.

Prevalence in blood donors varies from less than 0.1% in the United States to 5% or more in countries of high endemicity. Transmission by blood transfusion has now become a rarity in those countries that have adopted routine screening of blood donors. Hemophiliacs appear not to be at risk, presumably because the virus is cell-associated and is effectively lost during the extraction of factor VIII from plasma. On the other hand, intravenous drug abuse is an increasing hazard, so much so that around 5% of intravenous drug users are already infected with HTLV; fortunately, most infections appear to be due to HTLV-2, with which no disease has yet been linked. Little is yet known of the natural history of HTLV-2, found most frequently among post-pubertal Guaymi Indians in Panama, most probably spread by sexual intercourse. Recently, viruses designated HTLV-3 and HTLV-4 have been discovered in humans in Central Africa, and the search continues for additional HTLV types and for possible additional disease associations (Fig. 23.4).

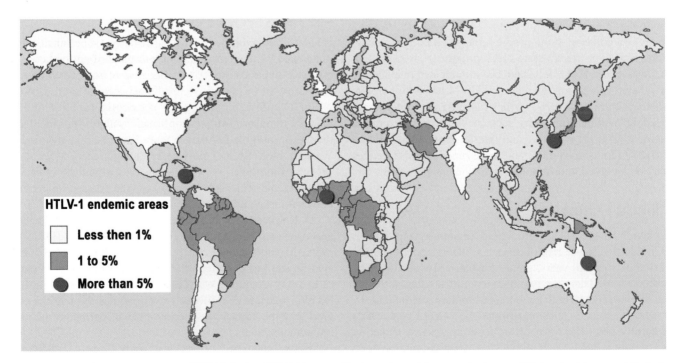

HTLV-1 endemic areas

☐ Less then 1%

▨ 1 to 5%

● More than 5%

FIGURE 23.4 Global distribution of HTLV-1 infection. HTLV-1 is not a ubiquitous virus—it is present in clusters of high endemicity, especially in the southwestern part of Japan, sub-Saharan Africa, areas of South America, the Caribbean basin, the Middle East and Melanesia. *Adapted from Goncalves et al., 2010. Clin. Microb. Rev. 23, 577–589.*

Control

The screening of blood and organ donors for HTLV antibody has all but abolished transmission whenever this policy is implemented. However screening is not universally adopted in high-prevalence countries due to cost. Preventing mother-to-child transmission in endemic areas would ultimately have a major impact on disease. Prenatal screening is recommended in high-risk areas, and seropositive mothers are advised not to breast-feed. Policies to prevent sexually transmitted infections should be emphasized, including condom use, avoiding multiple and unknown sexual partners, and commercial sex. Intravenous drug users must be discouraged from sharing needles and syringes.

Experimental recombinant and peptide vaccines have been developed, but the major difficulties include the lack of suitable laboratory animal models, and the problem of interrupting cell-mediated transmission. Antiviral regimens used for treating HIV infection, including zidovudine with or without lamivudine, raltegravir, and histone deacetylase inhibitors, have been evaluated for the treatment of HTLV-1 infections, for example, for an effect on viral load or clinical benefit, but to date have not been shown to be of value.

HUMAN IMMUNODEFICIENCY VIRUSES

HIV/AIDS represents a classical example of the introduction of a new viral infection into the human population. Within a decade of its introduction the infection became firmly established in most parts of the world. The first human case of AIDS was recognized in 1981, and by 2013 the World Health Organization (WHO) estimated that over 35 million people were currently infected with HIV, including 3.3 million children under 15 years of age. In that year 1.6 million deaths and 2.3 million new infections occurred. In the United States AIDS became the commonest cause of death in young males during the 1990s. However, the impact in many African countries has been particularly devastating, with 15% to 25% of the adult (15 to 49 years) population in a number of sub-Saharan countries being HIV-positive. In Southeast Asia, India, and China, infection rates continue to climb, particularly among intravenous drug users and those at high risk for sexual acquisition including men who have sex with men (MSM), sex workers, and transgender women. AIDS can threaten the very fabric of society, not least because it usually targets productive age groups in the community and because of the socially disrupting effects of the stigma often associated with the disease. Moreover, large numbers of infected children, many of whom have lost both their parents to AIDS, represent an enormous social problem. However, the last decade has also seen the widespread introduction of combination antiviral therapy (ART), leading to very successful suppression of virus and disease progression. If ART is initiated at the right time, life expectancy for someone living with HIV is now similar to a person without HIV infection. Furthermore, antiviral therapy substantially reduces sexual transmission, and when given either in pregnancy or to neonates can greatly reduce vertical transmission, thereby offering a practical measure to reduce the future burden of infection. Thus, for those with access to ART, a diagnosis of HIV infection is now compatible with many years of active and productive life. However, despite our enormous store of knowledge about this infection, development of a successful vaccine and a cure is still proving an enormously difficult challenge.

Properties of Human Immunodeficiency Viruses

Two species of HIV are currently recognized within the *Lentivirus* genus. HIV-1 is the predominant species causing the current pandemic. HIV-2 is at present largely confined to Africa, and appears to be less virulent than HIV-1. There is limited serological cross-reactivity in ELISA tests between HIV-1 and HIV-2 Gag proteins but no cross-reactivity between Env proteins. HIV-2 displays only about 50% nucleotide sequence similarity (homology) with HIV-1 and contains a unique gene, *vpx*, in lieu of the *vpu* of HIV-1.

Among HIV-1 isolates four distinct groups can be distinguished (M, N, O, and P), and group M ("main") includes nine discrete clades (A, B, C, D, F, G, H, J, and K) differing by 30% to 35% in *env* and *gag* sequences respectively. Overall, within any one clade HIV-1 strains can differ by up to 20% by nucleotide sequence. The extent of genetic drift in an individual patient is much less, usually less than 3%, hence PCR and sequencing can be used to identify the source of infection (e.g., from a particular batch of factor VIII, from a mother to a child, or from a single blood donor).

At least 15 circulating recombinant forms have been described. CRFs are each identified by a number and the different parental clades. CRF12_BF, for example, is a recombinant between subtypes D and F.

The diploid (2×9.2 kb) plus sense ssRNA genome of HIV contains, in addition to the standard *gag, pol*, and *env* genes, six additional genes *vif, vpr, vpu, tat, rev*, and *nef*, whose products regulate the synthesis and processing of virus RNA and also help combat host restriction factors (Fig. 23.5).

Unlike other retrovirus genera where the virion internal cores are roughly spherical, HIV virions contain a cone-shaped core largely composed of the *gag* cleavage product CA. Indeed, purified HIV CA-NC protein can assemble *in vitro* to form cylinders and cones, a process that is facilitated by the presence of viral RNA. Projecting from the envelope are 72 peplomers, trimers of the *env* gene product (gp160) after it has been subsequently cleaved into two non-covalently linked components, SU and TM. The receptor-binding ligand and the most important antigenic domains, notably the V3 loop, are present on the SU molecule; SU is

HIV-1 genome
Proviral DNA

Genome expression
Genomic RNA, Gag-Pol mRNA, pre-mRNA

5'Ⓒ〜〜〜〜〜〜〜→ A_nA_{OH}3'

9.1 kb

Singly spliced mRNAs: Vif, Vpr, Vpu, Env

5'Ⓒ〜〜〜〜→ A_nA_{OH}3'

5'Ⓒ〜〜〜→ A_nA_{OH}3'

~4.3 kb

Multiply spliced mRNAs: Tat, Rev, Nef

5'Ⓒ〜〜〜→ A_nA_{OH}3'

(plus others)   ~1.8 kb

FIGURE 23.5 The genome and transcription scheme of human immunodeficiency virus 1 (HIV-1). Proviral genes are located in all three reading frames, as indicated by the overlaps. HIV-1 mRNAs fall into three classes. The first is an unspliced transcript of 9.1 kb. The second class comprises singly spliced mRNAs (average length 4.3 kb) that result from splicing from a 5′ splice site upstream of the gag gene to any one of a number of 3′ splice sites near the center of the genome. One of these mRNAs specifies the Env polyprotein precursor; others specify the accessory proteins. The third class comprises mRNAs (average length 1.8 kb) derived by multiple splicing from 5′ and 3′ splice sites throughout the genome. They include mRNAs that specify the regulatory proteins Tat, Rev, and Nef and are the first to accumulate after infection. *Reproduced from Flint, S.J., et al., 2009. Principles of Virology: Pathogenesis and Control, third ed. ASM Press, Washington, DC, with permission.*

the most extensively glycosylated viral protein known, and it is presumed that this "sugar coating" is a protective device to impede access of neutralizing antibodies. The hydrophobic membrane anchor is provided by TM, which is also responsible for viral entry into the host cell through a process of membrane fusion. The viral envelope also contains some cellular proteins, notably class I and II MHC antigens.

The inner surface of the envelope is lined by a myristylated matrix protein MA, a cleavage product of the 55 kDa *gag* gene product. The most abundant virion protein is another *gag* gene product, the phosphoprotein CA, of which the regular lattice of the core is constructed; the other two are NC (p7) and p6, also closely associated with the genome in the core of the virion. In addition associated with

the genome in the core are the three viral enzymes: reverse transcriptase (also carrying RNase H activity), integrase (endonuclease), and protease (Figs. 23.6A and B).

Replication of HIV

The replication of HIV is notable in a number of respects. Since CD4+ T lymphocytes support HIV replication only when dividing, yet most T cells in the body are resting, many virus–cell interactions result in a long-standing latent infection that may later be converted to a productive infection following activation of the T cell. Both cellular and viral proteins are involved in complex pathways regulating almost every step in the viral replication cycle. The synthesis, processing, and ultimate function, of the newly synthesized viral RNAs are regulated by feedback loops involving the viral gene products Tat and Rev. Although the HIV genome contains only nine genes, three coding for structural proteins and six for regulatory proteins, multiple splicing options yield over 30 distinct RNA species, the functions of many have yet to be understood. HIV spreads much more efficiently from cell to cell by cell–cell fusion to form syncytia, but most of our knowledge of the replication cycle has come from cell culture studies involving conventional infection by free virions.

The virion attaches via the SU envelope protein to the CD4 differentiation antigen, which is the primary HIV receptor on CD4+ T lymphocytes and cells of the macrophage lineage. This binding event exposes a high-affinity chemokine receptor-binding site on SU. Different strains of HIV then interact specifically with different chemokine co-receptors, the commonest being CCR5 and CXCR4. The SU–chemokine interaction results in conformational changes in TM to expose the fusion peptide, allowing pH-independent fusion between the viral envelope and cell membrane. However, recent evidence suggests that entry of the virion core depends predominantly on endocytosis rather than entry at the cell surface, as entry requires the action of dynamin, a GTPase that promotes the scission of endocytotic vesicles. Reverse transcription takes place within partially disassembled cores in the cytoplasm, producing dsDNA copies of the genome, and these are then transported to the nucleus assisted by nuclear localization signals present on one or more of the viral core proteins including integrase. This use of cellular nuclear transport processes allows infection to occur without first requiring mitosis, in contrast to many other retroviruses, and HIV is therefore able to establish infection in a wider range of differentiated non-dividing cells, for example, macrophages (Fig. 23.7).

At this point in the replication cycle the integrase protein then facilitates integration of some of the linear dsDNA molecules into cellular DNA. However, various unintegrated forms of HIV DNA are also found in abundance within the nucleus during infection, including one-LTR circles, two-LTR circles, and linear forms. There is increasing

A

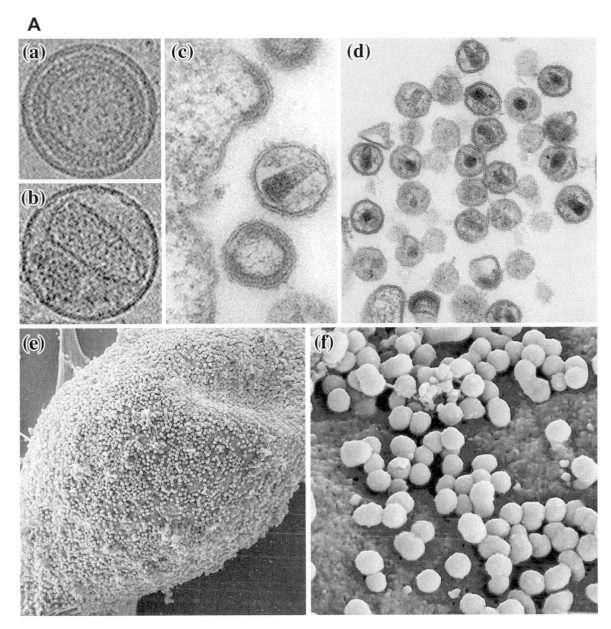

FIGURE 23.6-A   HIV virion model and image of virus. (a) Cryoelectron microscopy of HIV-1, immature virion. (b) Cryoelectron microscopy of HIV-1, mature virion. (c) Thin-section electron microscopy of HIV-1 immature and mature virions and an early stage of virus budding. (d) Thin-section electron microscopy of HIV-1 virions showing the usual variety of mature virion profiles. (e) Scanning electron microscopy of the surface of an infected T-lymphocyte covered with budding/budded virions. (f) Scanning electron microscopy, high magnification of the image in (e). *(a) and (b) from Wikipedia; (c) and (d) from Public Health Image Library, U.S. Centers for Disease Control and Prevention; (e) and (f) reproduced from Roingeard, P., Brand, D., 1998. Images in clinical medicine. Budding of human immunodeficiency virus. N. Engl. J. Med. 339:32, with permission.*

recognition that these unintegrated forms may play a role in replication as a template for the early expression of some genes, and possibly as another potential reservoir for HIV genetic material available for rescue later during the course of infection. This unintegrated material may be increased as a result of anti-integrase therapy.

The integrated provirus may remain latent indefinitely. Transcription by cellular RNA polymerase II can only occur in activated cells. Activation of the T cell can be brought about by mitogens, cytokines, or trans-activating proteins encoded by certain other viruses such as HTLV-1, herpes simplex virus (HSV), Epstein-Barr virus (EBV), cytomegalovirus (CMV), human herpesvirus 6 (HHV-6), or hepatitis B virus. Many of these agents activate the cell by activating expression of members of the NF-κB family of cellular proteins, which are usually sequestered in the cytoplasm complexed with an inhibitor; activation promotes disassembly of the complex and translocation of NF-κB to

B

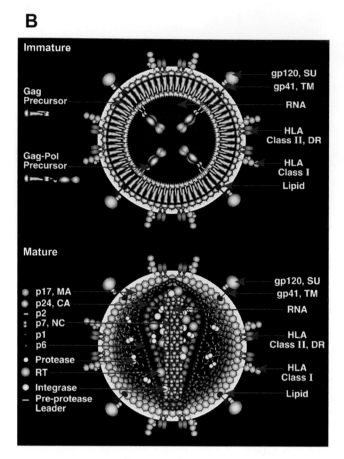

Immature

Gag Precursor

Gag-Pol Precursor

gp120, SU
gp41, TM
RNA
HLA Class II, DR
HLA Class I
Lipid

Mature

• p17, MA
• p24, CA
− p2
: p7, NC
− p1
. p6
• Protease
• RT
• Integrase
− Pre-protease Leader

gp120, SU
gp41, TM
RNA
HLA Class II, DR
HLA Class I
Lipid

FIGURE 23.6-B Diagrams of immature and mature forms of HIV-1, showing components of the virion. The Gag and Gag-Pol polyproteins are shown with different domains in different colors to indicate the mature proteins formed from these precursors. The SU and TM components of the gp120 envelope protein are shown jutting out from the lipid envelope, as are some host proteins that may be incorporated into the viral envelope. Maturation of the virion includes proteolysis of Gag and Gag-Pol to produce individual virion proteins, and condensation of these products to form the conical internal core that is characteristic of lentiviruses. *Originally from Louis, E., Henderson and Larry Arthur, 1994. National Institute of Allergy and Infectious Diseases, NIH, public domain.*

the nucleus. The normal function of these host transcription factors is to bind to discrete enhancer elements in host cell DNA which control expression of certain cellular genes, but these factors also recognize and bind to two κB enhancer elements in the U3 region of the HIV proviral LTR. Binding of NF-κB, together with several additional host transcriptional factors that are expressed constitutively, enables the proviral genome to be transcribed, albeit at a relatively low level. HIV RNA production in monocytes is very limited unless stimulated; after the differentiation of monocytes into macrophages, the permissiveness to viral replication increases dramatically.

There are three major classes of HIV mRNA, the result of different splicing mechanisms: an unspliced full-length (9.2 kb) RNA transcript, corresponding to the viral genome, is used as mRNA for translation of the Gag and Gag-Pol polyproteins; singly spliced 4.5 kb RNAs encode the Env, Vif, Vpr, and Vpu proteins; and several multiply spliced 2 kb RNA transcripts encode the Tat, Rev, and Nef proteins. In the initial round of transcription, the only RNA transcripts exported to the cytoplasm are the latter multiply spliced 2 kb mRNAs (Fig. 23.5). The *gag-pol* transcript is translated into a Gag polyprotein at a rate of approximately once in every 10 to 20 times, translational readthrough following a ribosomal frameshift occurs, producing a Gag-Pol polyprotein.

The viral genome assembles in the cytoplasm with the Gag and Gag-Pol precursors of the core proteins and migrates to associate with areas of the plasma membrane containing the Env glycoprotein gp160. The Pol polyprotein is cleaved to yield the three enzymes contained within the core of the virion, whereas the Gag polyprotein is cleaved to yield the four structural proteins p24, p17, p9, and p7. The viral protease is responsible for these processing events, which occur during and continue to occur after budding of the virion. The host cell protease furin cleaves gp160 into the two *env* products gp120 and gp41.

Function of the Accessory Proteins

The six accessory proteins play multiple roles in enhancing and modulating virus replication within the cell, inhibiting various host cell antiviral defense mechanisms and enhancing infection in the body. Tat is a potent transactivator which binds to a responsive element known as TAR (trans-activating response), which is present in the LTR of both proviral DNA and the common 5′ end of all the viral RNA transcripts. In association with cellular factors, Tat enhances the efficiency of transcription by cellular RNA polymerase from the HIV promoter up to 100-fold, both by preventing premature termination of transcription and by facilitating remodeling of nucleosomes in the chromosomal strand downstream of the transcription initiation site. All the viral genes are transcribed, but transcripts encoding all proteins other than Tat, Rev, and Nef are retained in the nucleus until significant levels of Rev have accumulated.

Rev is the second essential regulatory protein encoded by HIV. It is a phosphoprotein that binds to a sequence known as RRE (Rev responsive element) in viral RNA. Because RRE is located within the *env* gene, it is present in the 9.2 kb and 4.5 kb HIV RNA transcripts but not in the 2 kb transcripts encoding Tat, Rev, or Nef. In the absence of Rev, the 9.2 kb and 4.5 kb transcripts are retained unspliced in the nucleus. The RRE contains several stem-loops, one of which binds an arginine-rich RNA-binding domain of Rev. A short nuclear export signal in the C-terminal region of Rev then directs the export of the Rev–RNA complex. After dissociation of this complex in the cytoplasm, Rev molecules can recycle to the nucleus guided by a nuclear localization signal. Thus, the

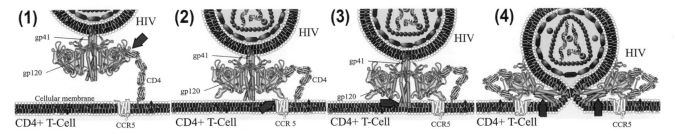

FIGURE 23.7 Mechanism of HIV entry/membrane fusion. (1) Initial interaction between gp120 and CD4. (2) Conformational change in gp120 allows for secondary interaction with CCR5. (3)The distal tips of gp41 are inserted into the cellular membrane. (4) gp41 undergoes significant conformational change; folding in half and forming coiled-coils. This process pulls the viral and cellular membranes together, fusing them. *Reproduced from Wikipedia, with permission.*

virus-coded protein Rev can overcome the normal inability of cells to export intron-containing RNAs from the nucleus, and as Rev accumulates in the nucleolus, the full range of HIV mRNAs becomes exported and available for translation.

Nef is an important HIV accessory protein. It is myristylated, synthesized in large amounts after integration of the provirus, and becomes localized at the inner surface of the plasma membrane. Its actions include the down-regulation of MHC-I and -II on APCs and target cells, together with down-regulating expression of CD4 and CD28 expression on CD4+ T cells. These effects lower the threshold for T cell activation by exogenous stimuli, thereby leading to the generation of dividing cells susceptible to HIV. Nef was shown to be required for the development of simian AIDS in the simian immunodeficiency virus (SIV)–rhesus monkey model. A similar role in promoting disease was suggested by a cohort of patients infected with a nef deletion mutant in whom a much slower progression to clinical AIDS was seen.

Vif (viral infectivity factor) enhances the yield of infectious virus by inhibiting the action of APOBEC3G and similar host cytosine deaminases. In the absence of Vif, these enzymes deaminate C to U residues in newly synthesized HIV minus strand DNA, resulting in a G-A mutation in the HIV proviral DNA. Vpu is an integral membrane phosphoprotein resembling the ion channel M2 protein of influenza virus. Vpu promotes degradation of CD4 in the endoplasmic reticulum by linking it to an E3 ubiquitin ligase, and also enhances release of the virion by budding. Vpr, which is found in the virion itself, is critical for efficient viral replication; it assists transport of the pre-integration complex into the nucleus and is also a weak transcriptional activator of the HIV LTR and of various cellular promoters.

Pathogenesis

The clinical hallmark of AIDS is depletion of CD4+ T lymphocytes, and these cells are the principal target of HIV infection in lymph nodes and blood. CD4+ T cells support the replication of the virus only when activated, but integrated viral DNA can persist in resting lymphocytes as a latent infection. In addition, cells of the monocyte lineage are susceptible, and macrophages constitute a reservoir of persistent virus in tissues, including the brain (Fig. 23.8).

HIV is most commonly transmitted from person to person via genital secretions, notably semen, and less commonly via direct blood contact. Knowledge of the early pathogenic events is sketchy because of the difficulty in identifying infected individuals soon after exposure. Simian immunodeficiency virus (SIV) infection of non-human primates has provided useful information. The first cells to become infected may be resident tissue macrophages, Langerhans cells, or submucosal CD4+ lymphocytes in the genital tract or rectum. The virus is then transported as either free virus, virus attached to dendritic cells, or virus-infected cells, to the draining lymph nodes, where it replicates extensively. Significant viremia is quickly developed, and the infection becomes established in the gut-associated lymphoid tissue (GALT) where it induces major depletion of CCR5+ memory CD4+ cells in the lamina propria of the intestine.

One to three weeks after infection many patients exhibit a brief glandular fever-like illness, associated with a high titer of virus in the blood (up to 10^7 copies/mL), and a dramatic decline in circulating CD4+ T cells. A vigorous cellular and humoral immune response ensues, and within a month or so the viremia declines to a lower steady state, the absolute level of which varies between individuals. CD8+ cytotoxic T cells, natural killer (NK) cells, and antibody-dependent cell-mediated cytotoxicity (ADCC) may all contribute to this decline by lysing infected cells. During this phase, long-term reservoirs of virus are established, in particular within lymphoid tissue (GALT, and within the follicular dendritic cell network of lymph nodes), as well as within latently infected resting CD4+ cells. Persistence of virus in these reservoirs is a major stumbling block to eradication of virus from the body by antiviral treatment (Fig. 23.8).

There then follows a long asymptomatic period lasting from one to 15 years or longer (average about 10 years),

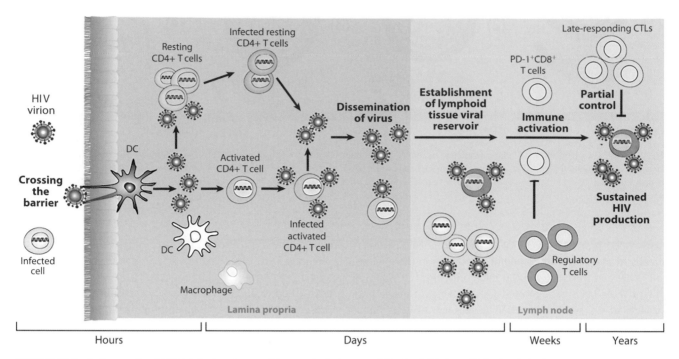

FIGURE 23.8　Pathogenesis of HIV-1 infection: Proposed immunological events following initiation of infection. Infection begins with transmission across a mucosal barrier, either by a cell-free virus, infected cell, or virion attached to dendritic cells (DCs) or Langerhans cells (LCs). Early low-level propagation probably occurs in partially activated CD4+ T cells, followed by massive propagation in activated CD4+ T cells of the gut-associated lymphoid tissue lamina propria. Dissemination of HIV to other secondary lymphoid tissues and establishment of stable tissue viral reservoirs ensue. Immune response lags behind the burst of viremia and provides only partial control of viral replication. *CTL*, cytotoxic T-lymphocyte; *PD-1*, programmed death 1. *Adapted from Moir, S., et al., 2011. Pathogenic mechanisms of HIV disease. Annu. Rev. Pathol. Mech. Dis. 6:223–248.*

before any further clinical evidence of disease becomes apparent. During this period, virus and viral antibodies are demonstrable in the blood, but only a tiny minority of circulating CD4+ T cells produce virus. The actual level of circulating virus during this phase is a predictor of the rate of progression of the disease in different individuals. However, beneath this relatively benign clinical appearance, virus replication, T cell destruction and T cell regeneration actively continue. During this stage, there is very rapid turnover of both circulating virus and infected cells, with estimated half-lives of 6 hours and one day respectively; there is a daily release of at least 10^{10} virions per day. When this is combined with the very high error rate of the reverse transcriptase, it is easy to understand how a swarm of related genetic variants of the original viral sequence are continually being produced and selected for survival fitness. In the lymphoid organs, follicular dendritic cells entrap free virus and present it to lymphocytes moving through the organ, leading to lymph node hyperplasia. As time passes there is a steady decline in the number of circulating CD4+ T cells, owing to involution of lymphoid follicles. Although immunological dysfunction exists throughout infection, it is particularly so when CD4+ cell counts fall below about 200 to 400 per μL that opportunistic infections with various microorganisms and specific malignancies may occur, and

eventually the depleted immune system is unable to cope. Activation of CD4+ T cells renders these cells permissive for HIV replication, and other viruses such as HHV-6, CMV, and HTLV-1 also transactivate HIV expression. Large numbers of virions spill over from the degenerating lymph nodes into the blood. The increasing burden of virus and the decreasing numbers of immunocompetent helper T cells collude to convert a low-level chronic infection to one that is more rapidly progressive. Death follows as a result of secondary infections, malignancy, or a cachexia-like state.

The dramatic loss of CD4+ T cells is likely to be due to a number of mechanisms. Most obviously, HIV infection can be directly cytolytic for activated CD4+ T cells. There is recent evidence that 95% of cells that die are undergoing abortive or non-productive infection, and that cell death may be triggered by incomplete DNA products. In contrast, other recent studies suggest that cell killing is triggered by viral integration in productively infected cells. Alternatively, infection can cause cells to fuse via the HIV envelope glycoprotein with uninfected CD4+ cells to form a labile syncytium. Infected T cells are also subjected to immune cytolysis by cytotoxic T lymphocytes, ADCC, or antibody/complement-mediated lysis. It has been proposed that, as a result of cross-linking of CD4 by HIV gp120, or by other mechanisms, T cells may become activated

TABLE 23.5 Interactions between HIV and CD4+ T Cells Leading to Immune Dysfunction

Viral cytolysis of CD4+ T cells, especially memory cells
Fusion of infected with uninfected cells to form syncytia
Inhibition of cytokine expression, CD4 expression, and biological functions of CD4+ T cells
Lysis of infected CD4+ T cells by cytotoxic T cells, NK cells, ADCC, antibody/complement
Induction of apoptosis of infected CD4+ T cells
Infection of stem cells
Dysregulation of helper T cell differentiation (Th$_1$ decrease, Th$_2$ increase)
Autoimmune destruction of infected CD4+ T cells

to commit suicide by apoptosis. Indeed, the SU and TM glycoproteins are toxic for cultured cells, but whether this mechanism is significant *in vivo* is not known. To explain the fact that dying CD4+ T cells are not totally replenished, it has been postulated that HIV might also infect stem cells, or other cells that normally secrete cytokines required for such replacement (Table 23.5).

Macrophages are the principal target cells for certain lentiviruses of other animals which are not notably immunosuppressive but, similarly to HIV, attack the brain. HIV replicates slowly in differentiated non-replicating human tissue macrophages, and less well in monocytes. Chronic infection of these phagocytes can continue for weeks. Although functions such as cytokine production may be affected, there is little cytopathology or syncytium production. In macrophages, HIV was thought to bud into cytoplasmic vacuoles rather than through the plasma membrane, which could insulate these cells from immune cytolysis. However, more recent work has questioned this interpretation. Infection of macrophages may predispose the patient to opportunistic infections such as tuberculosis and toxoplasmosis.

There is evidence from cell culture studies that HIV can infect other types of cells, including neurons, endothelial cells, oligodendrocytes, and astrocytes in the brain, and M cells and enterochromaffin cells in intestinal mucosa, thus accounting, respectively, for the diverse CNS manifestations of AIDS and the regular occurrence of chronic diarrhea and malabsorption. Cell culture studies have identified alternative virus entry mechanisms in addition to the classical CD4+/chemokine receptor route. These include direct entry via cell contact, and endocytosis after binding to galactosylcerebroside or Fc receptors as antibody–virus complexes. There is a debate about whether the protean effects of HIV on the CNS are attributable solely to infection

of the principal target cells, the resident brain macrophages known as microglia, with a neurotropic variant of the virus, or whether infection of perivascular macrophages, glial cells, astrocytes, and brain capillary endothelial cells is also involved. Furthermore, it is not clear whether the damage is attributable to direct lysis of cells, to the destruction of these cells by cytotoxic T cells, or to release of neurotoxic viral proteins and/or cellular products, for example, cytokines, proteases, or to more subtle effects such as interference with neurotransmitters. Certainly, it is clear that antiviral therapy can lead to clinical improvement in many cases of AIDS dementia, but HIV-associated neurological disease can also progress despite control of virus replication.

Genetic Variation among HIV Isolates

Such is the infidelity of reverse transcription that those viral genomes recovered from any given individual diverge steadily with time from the original infecting strain, giving rise to a constellation of countless distinct mutants collectively known as "quasi-species." This variation is most marked in the "variable" regions of the *env* gene. Such variants are continually being selected for replicative fitness in the face of evolving immune responses, the changing antiviral drug milieu, permissiveness and non-permissiveness of different cell types, and other selection pressures.

Two major classes of variants often appear sequentially during the course of infection. Early in infection, particularly after sexual transmission, macrophage-tropic variants tend to predominate; these generally use the chemokine receptor CCR5 as a co-receptor (R5 viruses). Later in the course of infection, such variants are usually replaced by T cell tropic variants; these variants generally use CXCR4 as a co-receptor (X4 viruses). Since X4 viruses appear in the later stages of disease, there has been debate whether these viruses are inherently more virulent and drive the disease, or merely arise by selection in patients with a more deranged immune system.

Neutralization-escape mutants can be selected readily by growing HIV in the presence of monoclonal antibodies *in vitro*, or even in the presence of monoclonal cytotoxic T cells. Such escape mutants often display amino acid substitutions in the hypervariable regions of the envelope glycoprotein(s), as is found in natural human isolates. It is, however, difficult to prove whether they are selected by the immune response, generated randomly, or associated with favored changes in tropism or virulence, for example those that map to the hypervariable V3 loop of gp120. Nevertheless, the increasing heterogeneity of virus sequences that evolve during the asymptomatic phase is likely to be due, at least in part, to antigen escape mutants. In contrast, specific point mutations associated with resistance to particular antiviral drugs have been well documented: these mutations develop more rapidly during long-term therapy when only single or two drugs are used, hence the use of three drugs (triple therapy) is now routine practice.

Thus, as the long period of asymptomatic infection progressively gives way to advanced disease, the viral mutants that emerge tend to be those displaying the properties of extended cellular tropism, enhanced cytopathogenicity for CD4+ T cells, more rapid kinetics of replication, and escape from neutralizing antibodies.

Latency and Virus Reservoirs

It is clear that high rates of virus replication are found throughout the asymptomatic phase so there is never a period of true virological latency. However, at the cellular level, latently infected CD4+ T cells are established early following infection; these are lost slowly during antiviral treatment and go unrecognized by the immune system, thereby representing a long-term reservoir capable of regenerating and disseminating virus if therapy is interrupted. The latent reservoir is established very early after acute infection. At the cellular level, evidence suggests that one important reservoir comprises resting memory CD4 T cells, most likely having become infected while activated before undergoing de-activation, or through direct infection in the setting of chemokines or antigen presenting cells in tissue. Other T cell subsets including transitional memory cells, memory stem cells, and naïve T cells, macrophages, dendritic cells, and hematopoietic stem cells have also been implicated. The heterogeneity of reservoir compartments is demonstrated by the observation that on commencement of antiviral treatment, circulating virus levels decay with at least two kinetically distinct components. Various mechanisms for the biochemical basis of latency have been investigated including (1) transcriptional interference mediated by nearby host gene promoters, (2) limited availability of host transcription and elongation factors, (3) transcriptional repressors, (4) chromatin restriction, that is epigenetic silencing of integrated proviral DNA by modified histone tail domains, and (5) inappropriate HIV accessory factor effects, for example, insufficient Tat activity. Latently infected cells also undergo homeostatic proliferation. It is also possible that other reservoirs of latently infected cells exist, for example, in more privileged sites such as infection of T follicular helper cells in the B cell follicle in lymphoid tissue, and, in the genital tract and CNS. These reservoirs of quiescent, latently infected cells are considered the single biggest obstacle to the control of virus following cessation of ART.

Clinical Features

Over the past 12 to 15 years the clinical features and course of established HIV infection seen in many countries have dramatically changed due to the widespread use of antiretroviral therapy. The "classical" course of disease described below is now not commonly seen if therapy is instituted early in infection. However, both the US Centers for Disease Control and Prevention (CDC) and WHO have developed and maintained criteria for defining HIV infection and classifying HIV-infected individuals based on the classical progression of disease; these criteria remain of particular use for epidemiological and clinical reporting purposes.

HIV Seroconversion Illness

Two to three weeks after infection there is often a brief influenza-like illness that may resemble infectious mononucleosis. The features include acute-onset fever with or without night sweats, myalgia, arthralgia, lethargy, malaise, diarrhea, depression, lymphadenopathy, sore throat, maculopapular skin rash, and mucocutaneous ulceration. Sometimes neurological manifestations often present clinically as headache, photophobia, and retroorbital pain. This illness is often mild and disregarded or misdiagnosed; a high level of clinical suspicion should be triggered if there are relevant lifestyle considerations. Examination of the blood reveals a temporary reduction in CD4+ (and CD8+) T cell count, followed by a predominantly CD8+ lymphocytosis and an inverted CD4/CD8 ratio. Virus, viral nucleic acid, or viral p24 antigen may be detected at this stage of the infection. The development of antibodies coincides with resolution of the illness or follows shortly thereafter. Circulating virus RNA levels are usually very high in this phase (10^5–10^6 copies/mL), indicating that patients in this period are likely to be highly infectious.

Clinical Latent Period

With a few exceptions, the patient recovers from the primary (seroconversion) illness within two to three weeks, and the majority go on to a phase of relatively good health. The length of this asymptomatic phase may range from less than one year to 20 years or more in different patients, while in the absence of treatment the median time to development of AIDS is 10 to 11 years. Between 1% and 5% of individuals maintain CD4+ cell counts ≥500 cells/μL for more than 10 years without any antiretroviral therapy; these individuals, previously referred to as long-term non-progressors (LTNPs), have more recently been shown to eventually develop HIV disease. The term "elite controllers" is used for a small group (approximately 0.6%) who maintain undetectable viral loads in the absence of therapy, many of whom have been shown to have stronger HIV-specific immune responses than those who do not control viral replication. While some such cases may be infected with a strain of HIV thought to be less virulent, many such individuals have been found to carry genetic traits conferring some degree of resistance; the most important of these appear to be mutations in HLA-I genes (responsible for antigen presentation to CD8+ T cells) and to a lesser extent in HLA-II genes (responsible for antigen presentation to CD4+ T cells). Other possible factors include mutations in the fucosyl transferase 2 (FUT2) gene which determines secretor status, in mitochondrial DNA, in co-receptors (e.g., the δ32 variant of the CCR5 gene impairs the ability of cells to be infected by HIV), or in the host cytosine deaminase APOBEC3G.

Early Symptomatic Infection

Symptomatic HIV disease may develop for some years before it reaches the criteria that satisfy the definition of AIDS. Persistent generalized lymphadenopathy (PGL), defined as involving two or more non-contiguous sites, may continue from the seroconversion illness or may become apparent much later, but does not have any clear prognostic relevance. However, other conditions listed in Table 23.6 are markers of clinical immune dysfunction and do predict progression to AIDS. Reflecting the polyclonal activation of the immune system, autoimmune conditions may occur during this period.

When the CD4+ T cell count falls below about 500 per μL the patient may develop a constellation of constitutional symptoms (fever, night sweats, oral candidiasis, diarrhea, and weight loss), all collectively known previously as the AIDS-related complex (ARC). Early opportunistic infections begin to be seen. At this intermediate stage of immune depletion these infections are generally not life-threatening. Dermatological, oral, and constitutional symptoms may also develop during this time.

Mycobacterial infections frequently occur in these patients, and this has led to an alarming resurgence of tuberculosis in both developing and developed countries. Reactivation of latent herpesviruses, particularly herpes simplex and zoster, also occurs (see Chapter 17: Herpesviruses). Gastrointestinal infections, caused by any of a wide variety of organisms, including the yeast *Candida albicans* and parasites such as Cryptosporidia, are common.

AIDS

When CD4+ T cells drop below 200 per μL the numbers of these cells generally begin to decline at an accelerated rate, the titer of virus in the blood increases markedly, and the tempo of progression of the illness increases. For epidemiological purposes, AIDS is defined to be present when there is confirmed serological evidence of HIV infection plus either the CD4+ cell count is less than 200, or when any of the conditions listed in Table 23.7 are present.

TABLE 23.6 Clinical Conditions that May Develop Later During the Early Symptomatic Stage

Dermatological	Herpes zoster, warts (both skin and ano-genital), molluscum contagiosum, bacterial folliculitis, eosinophilic folliculitis, seborrheic dermatitis, tinea
Oral	Aphthous ulceration, oral and esophageal candidiasis, oral hairy leukoplakia, linear gingival erythema, acute necrotizing ulcerative gingivitis
Constitutional	Episodes of fever, weight loss, headache, fatigue, diarrhea, myalgia, arthralgia
Autoimmune	Guillain-Barré syndrome, chronic demyelinating neuropathy, Reiter's syndrome, polymyositis, cranial nerve palsy, idiopathic thrombocytopenic purpura, Sjögren's syndrome, psoriasis

---

TABLE 23.7 Stage-3-(AIDS)-Defining Illnesses

- Bacterial infections, multiple or recurrent
- Candidiasis of bronchi, trachea, or lungs
- Candidiasis of esophagus
- Cervical cancer, invasive
- Coccidioidomycosis, disseminated or extrapulmonary
- Cryptococcosis, extrapulmonary
- Cryptosporidiosis, chronic intestinal (>1 month's duration)
- Cytomegalovirus disease (other than liver, spleen, or nodes), onset at age >1 month
- Cytomegalovirus retinitis (with loss of vision)
- Encephalopathy, HIV related
- Herpes simplex: chronic ulcers (>1 month's duration) or bronchitis, pneumonitis, or esophagitis (onset at age >1 month)
- Histoplasmosis, disseminated or extrapulmonary
- Isosporiasis, chronic intestinal (>1 month's duration)
- Kaposi's sarcoma
- Lymphoid interstitial pneumonia or pulmonary lymphoid hyperplasia complex
- Lymphoma, Burkitt (or equivalent term)
- Lymphoma, immunoblastic (or equivalent term)
- Lymphoma, primary, of brain
- *Mycobacterium avium* complex or *Mycobacterium kansasii*, disseminated or extrapulmonary
- *Mycobacterium tuberculosis* of any site, pulmonary, disseminated, or extrapulmonary
- *Mycobacterium*, other species or unidentified species, disseminated or extrapulmonary
- *Pneumocystis jirovecii* (previously *Pneumocystis carinii*) pneumonia
- Pneumonia, recurrent
- Progressive multifocal leukoencephalopathy
- *Salmonella* septicemia, recurrent
- Toxoplasmosis of brain, onset at age >1 month
- Wasting syndrome attributed to HIV

Source: Based on MMWR, December 5, 2008, 57, 1–14, Appendix A.

These conditions can be usefully considered predominantly under the headings (1) opportunistic infections, (2) malignancy, (3) neurological disorders, or (4) constitutional syndromes. Many of these are no longer commonly seen in communities with good access to treatment, except in those who present late with untreated infection and severe immunodeficiency (Table 23.8).

This stage of HIV infection is associated with more severe opportunistic infections. Particular infections tend to appear at a relatively predictable level of CD4+ T cells (Fig. 23.9). The commonest of these opportunistic infections (in the absence of chemoprophylaxis) is pneumonia caused by *Pneumocystis jirovecii* (*carinii*), a fungus that is ubiquitous but rarely causes disease in immunocompetent people. Others include EBV-associated interstitial pneumonitis (particularly in pediatric AIDS), esophageal candidiasis, cryptosporidial and microsporidial enteritis, infections with members of the *Mycobacterium avium* complex, cytomegalovirus retinitis, enteritis, or encephalitis, and infections of the brain with *Toxoplasma gondii* or *Cryptococcus neoformans*.

Malignant tumors may also appear as CD4+ T cell counts decline. The most common of these is Kaposi's sarcoma, seen mainly in male homosexuals and associated with human herpesvirus 8 (Kaposi's sarcoma-associated herpesvirus—see Chapter 17: Herpesviruses). Previously known as a relatively harmless collection of indolent skin tumors in old people of Mediterranean descent, Kaposi's sarcoma presents in young homosexual HIV-infected males in a more invasive but slowly progressive form, often early in the onset of AIDS. Other types of malignancy, notably aggressive B cell lymphoma or non-Hodgkin's lymphoma, may develop in patients with very low CD4 cell counts; genital cancers are also frequently recorded.

Neurological disease is an extremely common and sometimes the first observed manifestation of HIV infection. The spectrum of HIV-associated neurological disease includes dementia and its early forms, a severe encephalopathy (especially in children), myelopathy, and motor dysfunction. The patients may notice diminished concentration and memory, together with motor disturbances such as action tremor and loss of balance similar to Parkinson's disease. These patients often also display signs of co-existent myelopathy and peripheral neuropathy (e.g., ataxia and parasthesia). Other CNS manifestations include cerebral toxoplasmosis, cryptococcal meningitis, primary CNS lymphoma, CMV-associated encephalomyelitis, and progressive multifocal leukoencephalopathy (see Chapter 20: Polyomaviruses).

The Effect of Antiretroviral Treatment on HIV Infection

The widespread use of antiretroviral therapy has led to major changes in the above patterns of disease in those countries where it is available. Notably, the dramatic suppression of viral replication leads to immune restoration as shown by increasing CD4+ T counts, more so in younger individuals. AIDS-related deaths and disease rates have declined and in most countries life expectancy is now normal, a significant tribute to the efficacy of current therapy. Moreover, the proportion of deaths in HIV-infected individuals attributable to non-AIDS causes, for example, cardiovascular, hepatic disease, or malignancy, has increased. Some of the conditions listed in Table 23.7, particularly Kaposi's sarcoma and *Pneumocystis jirovecii* pneumonia, are now seen infrequently in treated cohorts.

TABLE 23.8 Criteria for Staging of HIV Infection

Laboratory criteria for defining HIV infection	Positive result for HIV antibody or combination antigen/antibody test (e.g., by EIA) *PLUS*—positive confirmatory test (e.g., Western blot, immunofluorescence)
	OR—Positive virological test (e.g., HIV RNA or DNA, HIV p24 antigen, virus isolation)
Stages of infection	Meets laboratory criteria for defining HIV infection (above) *PLUS*
Stage 1	[a]No AIDS-defining condition and either CD4+ count ≥500 cells/μL or CD4+ cells ≥26% total lymphocytes
Stage 2	[a]No AIDS-defining condition and either CD4+ count 200–499 cells/μL or CD4+ cells 14–25% total lymphocytes
Stage 3 (AIDS)	[a]Either CD4+ count <200 cells/μL or CD4+ cells <14% total lymphocytes or documentation of a Stage-3-defining condition (Table 23.7)

Source: Based on MMWR, April 11, 2014/63(RR03); 1–10.
[a]*Slightly different values are applied for children <6 years of age. See source.*

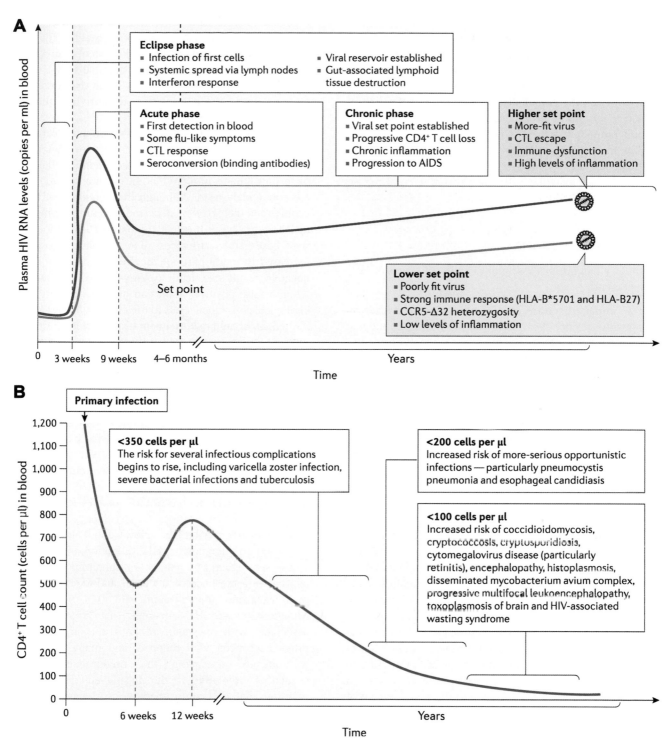

FIGURE 23.9 Progression of HIV infection. (A) HIV plasma RNA levels first become detectable after several days and peak a few weeks later, after which the adaptive immune response results in partial control. A steady state of viremia (set point) eventually ensues, reflecting a complex balance between virus production and removal, and target cell proliferation and killing. A higher set point level correlates with more rapid disease progression. (B) As the CD4+ T cell count declines, the risk of various disease complications increases. The clinical condition AIDS is defined (CDC) as the presence of HIV infection together with either a CD4+ T cell count of <200 cells/μL or an AIDS-defining complication. *Adapted from Deeks, S.G. et al., 2015. HIV infection. Nat Rev Disease Primers 1, 1–22. Article number: 15035 doi: 10.1038/nrdp.2015.35.*

Treatment has become dramatically simplified, and most individuals take medications co-formulated into a single tablet with few adverse effects. In the majority of patients who can adhere to contemporary treatment regimens, HIV replication can be fully suppressed and drug resistance is rare. However, the potentially devastating complication of immune restoration inflammatory syndrome (IRIS) can occur in individuals who initiate ART with a low CD4 count, leading to a rapid activation of a dysregulated immune response against a secondary pathogen that is already present (e.g., tuberculosis, herpes zoster, *Cryptococcus*, viral hepatitis B and C).

Laboratory Diagnosis

Laboratory diagnosis is essential to guide all stages of HIV management.

1. The primary diagnosis of HIV infection can be made by a positive result either for HIV antibody, for p24 antigen, and/or for HIV RNA. For many years the mainstay has been antibody detection, usually by EIA, followed by a confirmatory test (usually Western blotting) to identify the now small percentage of initially reactive results that turn out to be false positive. Recently, fourth-generation combination EIA assays have become available that detect p24 antigen in addition to HIV-1 and HIV-2 antibodies. Use of these newer assays allows detection of HIV sooner after infection than using assays for antibody alone, and helps overcome problems with the accuracy of Western blot assays early in infection or with HIV-2. Recent guidelines issued by the Centers for Disease Control and Prevention in the USA recommend that samples reactive in such a combination assay should then be tested in an immunoassay that differentiates HIV-1 and HIV-2 antibodies; where this differentiation assay gives a non-reactive or indeterminate result, a nucleic acid test (e.g., PCR) is used for final confirmation.

2. Because virtually all HIV infections persist for life, the detection of antibody can be taken as an indication of current infection rather than of recovery. A confirmed positive or negative result for HIV antibody is critical for identification, counseling, and the management of the infected individual, for reassurance of uninfected individuals, and for population surveillance. Patient consent for a test, and confidential pre- and post-test discussion with the patient of the implications of an HIV test result, are important aspects of the test process, fundamental to the well-being of the patient, and to the community's confidence in a testing policy. The confidentiality of results, and policies about screening of groups such as immigrants, prisoners, life insurance candidates, etc., have been subject to vigorous debate and thoughtful policy formulation. Donated blood or organs are screened for infectivity prior to use, either by antibody testing as above or often by PCR for viral RNA. Furthermore, by careful selection of appropriate primers PCR can be used to distinguish HIV-2 from HIV-1, as well as to distinguish some drug-resistant mutants from those sensitive to antiviral therapy (Fig. 23.10).

Because access to reliable HIV antibody testing is essential in many communities that do not have ready access to modern laboratories and to facilitate the ease of testing, low-technology "point of care" (POC) tests are now widely used. Although the frequency of false-positive and -negative results is more common compared to laboratory-based testing, the ease of use associated with these tests means access to testing has dramatically increased in many parts of the world. Clearly a false-negative result can lead to unwarranted reassurance while a false-positive result can have devastating psychological consequences as to life choices. Therefore for individuals at high risk, frequent testing is recommended as well as confirmation with a laboratory-based assay.

3. Monitoring of progress. The progressive decline in immune competence is monitored by CD4+ cell counts (which roughly correspond to the *extent of accumulated* damage resulting from the infection) which are generally performed every 6 months, and the level of virus replication by quantitative PCR for plasma RNA ("viral load" expressed in copies/mL—corresponding to the *rate of ongoing* damage). Current recommendations now recommend treatment for all infected individuals irrespective of CD4 counts or viral load. PCR testing for viral RNA is also sometimes used to identify individuals in the "window" period (see below), and it also provides a method for diagnosing HIV infection in neonates, whose serological test results are unhelpful due to the confounding presence of maternal antibodies acquired by transplacental transmission.

4. Mutations in the HIV genome that are commonly associated with resistance to particular antiretroviral drugs have been well mapped, and many guidelines recommend sequencing the predominating viral sequences circulating in the plasma and planning or adjusting antiviral drug therapy as dictated by the predominant sequence, prior to the initiation of ART.

5. Diagnosis of opportunistic infections often involves microbiological investigations that may be rarely needed for non-HIV patients.

6. Other manifestations of late-stage disease, and investigation of possible complications of therapy, all may require separate investigation.

The "window" period of two to three weeks or occasionally longer between acquiring HIV infection and full seroconversion as defined by both fourth-generation

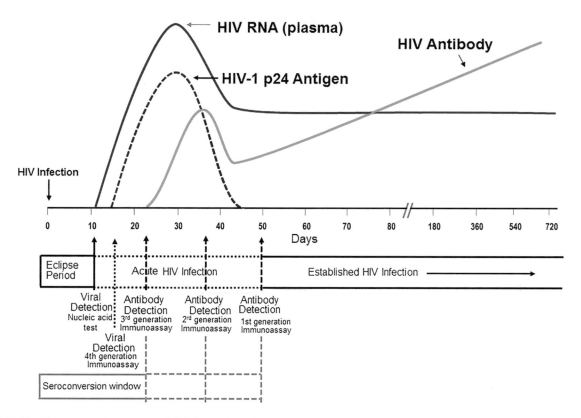

FIGURE 23.10 Time course of appearance of HIV diagnostic markers. Note that the window period between acquisition of infection and the appearance of an antibody marker has become progressively shorter with later generation antibody tests. *Adapted from Laboratory Testing for the Diagnosis of HIV Infection: Updated Recommendations, 2014. U.S. Centers for Disease Control and Prevention, public domain.*

EIA testing and Western blotting, can pose a dilemma. Prospective blood donors in some countries are required to sign a legally binding declaration that they have not engaged in any of certain defined risky behaviors during the preceding six months. On the other hand, where HIV infection is suspected clinically, there are two general solutions. The first is simply to repeat the antibody assay after a suitable interval. The second is to select a more sensitive test, that is, p24 antigen EIA or PCR, capable of establishing a diagnosis earlier after infection. However, frequent PCR testing of antibody-negative individuals at risk is not advised, because of cost and risks of false-positive results. Antibody testing at 3- to 6-monthly intervals is recommended for individuals at higher risk of HIV including men who have sex with men.

HIV can be isolated early in infection from peripheral blood leukocytes (PBL), or less readily from plasma, cerebrospinal fluid (CSF), genital secretions, or various organs such as the brain or bone marrow. The procedure is slow, often difficult, and of course must be conducted under conditions of strict biocontainment, PC3 level, or equivalent in most countries. The patient's PBL are co-cultured with mitogen-activated PBL from a seronegative donor; after

several days or weeks, reverse transcriptase or p24 antigen can be detected in the culture medium. In general, virus is present in PBL in high titer only early and late during the prolonged course of HIV infection (Fig. 23.10). Virus isolation is almost never performed now for diagnostic purposes, following the advent of PCR diagnostics.

The core antigen, p24, is one of the most abundant proteins made by HIV-infected cells and is detected in the plasma by EIA a few weeks before seroconversion. It disappears after antibodies appear, remains undetected for years, and rises again at the time clinical manifestations of AIDS develop. The reappearance of antigenemia is associated with a poor prognosis. Conversely, decline in the titer of p24 antigen can be used to monitor the success of antiviral therapy.

Epidemiology of HIV Infection

The three principal routes of transmission of HIV were elucidated within a few years after the first cases of AIDS in Los Angeles were recognized: sexual intercourse, exchange of blood, and perinatal transmission. However, the clinical and social settings where these routes operate in practice are very different in different parts of the world, and in different subcultures. For example, in China three separate expanding foci of infection occurred at the same time in different regions, fueled by intravenous drug use, inadvertent medical (transfusion)-related transmission, and sexual transmission respectively. HIV dynamics can be heterogeneous, akin to many small fires advancing on different fronts at different rates (Fig. 23.11):

1. Sexual intercourse is the major route of transmission in most settings. The risk of transmission is higher to the passive (receptive) partner, generally the female, and higher following anal intercourse (estimated at 1% per episode) than vaginal intercourse (estimated about 0.1% per episode). Another concurrent sexually transmitted disease enhances the risk by an order of magnitude, especially if genital ulcers are present, as in syphilis or chancroid. Obviously the risk is also greatly enhanced the greater the number of sexual partners. In some countries, for example, the United States and Australia, sexual transmission of HIV is greatest among men who have sex with men, while in developing countries, particularly Africa, heterosexual transmission is more common.

2. Perinatal transmission occurs in 15% to 25% of infants born to infected mothers, and the rate nearly doubles with breast-feeding (35% to 40%). The mechanism probably includes prenatal transmission across the placenta, intrapartum transmission via blood and/or genital secretions during parturition, and postnatal transmission via breast milk. Risk factors have been shown to include advanced maternal disease (maternal virus titers being highest late in the mother's infection), low maternal titers of neutralizing antibodies, prematurity at delivery, and breast-feeding; obstetric procedures may also be significant. A major advance has been the demonstration that antiviral therapy can significantly reduce transmission. Triple drug therapy during pregnancy, additional antivirals during labor, and short-term ART prophylaxis to the exposed neonate, can reduce the transmission rate to less than 1%. This intervention has the potential to have a major impact on the rate of new infections, and governments in different countries and international agencies have gone to considerable lengths to make antenatal screening and drug access available to all women and children.

3. Transmission by blood–blood contact occurs wherever the opportunity exists for infected blood to cross the barrier of an intact epithelium. The efficiency of transmission of HIV infection by transfusion of infected blood is greater than 90%. Blood transfusion and the

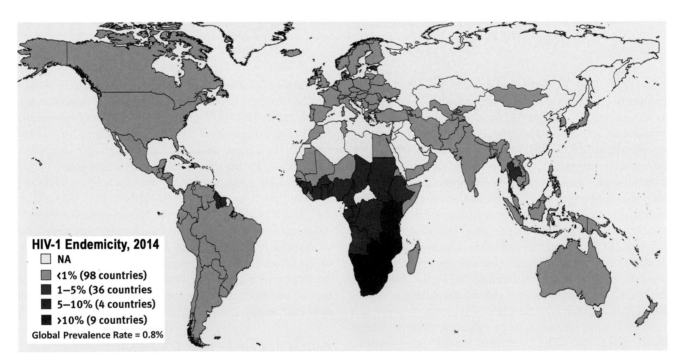

HIV-1 Endemicity, 2014
- NA
- <1% (98 countries)
- 1–5% (36 countries)
- 5–10% (4 countries)
- >10% (9 countries)

Global Prevalence Rate = 0.8%

FIGURE 23.11 The global prevalence of HIV-1. Worldwide, there are approximately 35 million people currently infected with HIV and many millions of people have died of AIDS-related causes since the beginning of the epidemic. While new cases continue to be reported from all regions of the world, 95% of new infections occur in individuals in low- and middle-income countries, particularly in sub-Saharan Africa. About half of new infections are among those under age 25. While most people infected with or at risk for HIV infection do not have access to adequate prevention, care and treatment, massive strides have been made in a number of countries. The number of people receiving antiviral therapy in resource–poor countries has increased (from 400,000 in 2003 to 9.7 million in 2012) and continues to rise dramatically. *Data from WHO.*

administration of blood products such as factor VIII to hemophiliacs now account for exceedingly rare transmission in most countries where blood donations are routinely screened. Transmission between injecting drug users (IVDUs) is usually through the sharing of needles, although contact with other blood-contaminated equipment may sometimes be involved. In some regions, for example, parts of Scotland, Italy, and some Southeast Asian countries, the spread of HIV between IVDUs has been a major health problem, while elsewhere IVDU-related infections represent only a small fraction of the total infection burden. Accidental transmission can also occur in the medical setting by needlestick injury to a staff member (in approximately 0.3% of such events that involve a proven infected source), or where the same used syringe and/or needle is shared between patients, and in other rare scenarios usually involving surgical procedures where infection has been documented to have spread between consecutive patients or from an infected staff member to patients. Notwithstanding such occasional documented instances of unexpected blood–blood transmission, there is no risk of transmission in the absence of clear blood–blood exposure. Trace amounts of virus can be detected from time to time in a range of bodily secretions including saliva, but there is no evidence that infection can be transmitted from person to person by kissing or oral sex.

The epidemiological patterns are extremely diverse. For example, in North America, Western Europe, and Australasia, prevalence is highest in men who have sex with men. In countries where access to clean needles is available, the prevalence in people who inject drugs is low. In such countries, groups at highest risk also include ethnic and social minorities, immigrant groups, and the disadvantaged who may not access health prevention messages.

In sub-Saharan Africa, heterosexual transmission predominates, with infection being roughly equal between males and females: perinatal transmission to infants occurs but this is declining. In some countries HIV is highly prevalent, reaching 30% to 40% amongst young women who are at greatest risk. Many other factors contribute to the high rate, including high rates of genital ulceration, poor access to medical services (including antibody testing, drug treatment, preventive education), poverty, and gender inequality. Likewise the impact of HIV on these communities has been devastating; the infection has struck down large numbers of people in their child-raising and economically productive phase of life. This has now dramatically changed, however, with the widespread availability of ART.

In Asia, infection has spread following a variety of patterns. According to UNAIDS data, overall HIV prevalence is relatively low but foci of higher rates have been found in IVDUs, sex workers, and military populations.

Interactions between these groups have enhanced the spread. Countries with highest rates of new infections include Indonesia, Pakistan, and the Philippines. Epidemics are also expanding in India and China. In Latin America, the majority of infections are in or around networks of men who have sex with men. However there is also a significant burden in IVDUs, sex workers, and their clients.

In many parts of the world, the widescale adoption of ART has led to a reduction in AIDS deaths and in rates of new HIV infections since 2000.

Management of the HIV-Infected Patient

The management of HIV has been transformed by the development of many potent antiviral drugs and the understanding of how to use them; by the application of sophisticated diagnostic tests for monitoring and adjusting therapeutic regimens; and by an increased understanding of the social impact of HIV infection together with the ethical, legal, and humanitarian responsibilities in dealing with infected individuals. The successful management of infected individuals, and the success or otherwise of community education programs to limit spread of infection, are now understood to be entwined with strong partnerships between the scientific and affected communities. The goals of therapy are to reduce circulating virus levels (reflecting virus replication) to undetectable levels, for example, less than 20 copies of virus RNA per µL, which has been shown to be associated with recovery of normal life expectancy. This needs to be combined with management of the clinical side-effects of drug treatment, combined with active prevention of co-infections. It has long been clear that successful suppression of virus replication with antiviral drugs is not followed by eradication of virus from reservoirs of latency within the body, and therefore antiviral treatment needs to be maintained for life.

The development of antiviral drugs has targeted virus-specific steps within the virus' replication cycle—reverse transcription, cleavage of newly synthesized viral protein by the viral protease, integration of the HIV proviral DNA into cellular DNA, and the elaborate mechanism of virus entry into the cell. The first reverse transcription inhibitor, azidothymidine (AZT or zidovudine), was introduced in 1987 and led to immediate but limited clinical benefit when used as monotherapy. This was followed by later nucleoside/nucleotide analogue reverse transcriptase inhibitors (NRTIs), each accompanied by distinct adverse effects, prescribing characteristics, and dosage regimens. Non-nucleoside reverse transcriptase inhibitors (NNRTIs) bind to the reverse transcriptase enzyme leading to disruption of the active site, but do not compete directly with template or substrate. Inhibitors of the viral protease (first licensed in 1995) prevent cleavage of the *gag-pol* polyprotein that occurs during maturation of the HIV virion. This results in the production of immature, non-infectious virus. Inhibitors of entry that are currently available either interfere with the binding of gp120 (env) to the chemokine receptor CCR5, or bind to gp41 thereby interfering with its fusion function. Integrase inhibitors inhibit the strand transfer step in integration of proviral DNA into cellular DNA. Further drug targets, for example, the accessory proteins tat and rev, and further drug vehicles, for example, antisense oligonucleotides, are likely to be exploited in the search for further antiviral agents (Table 23.9).

TABLE 23.9 Major Antiretroviral Drugs Used in the Management of HIV

Nucleos(t)ide reverse transcription inhibitor (NRTI)	Zidovudine (AZT)
	Didanosine (ddI)
	Stavudine (d4T)
	Lamivudine (3TC)
	Emtricitabine (FTC)
	Abacavir (ABC)
	Tenofovir disoproxil fumarate (TDF)
	Tenofovir alanfenamide fumerate (TAF)
Non-nucleoside reverse transcription inhibitor (NNRTI)	Nevirapine (NVP)
	Efavirenz (EFV)
	Delavirdine (DLV)
	Etravirine
	Rilpivirine
Protease inhibitor (PI)	Atazanavir (ATV)
	Fosamprenavir (FOS)
	Indinavir (IDV)
	Lopinavir/ritonavir (LOP/r)
	Ritonavir (RTV)
	Saquinavir (SQV)
	Darunavir (DRV)
	Tipranavir (TPV)
	Nelfinavir (NFV)
Integrase inhibitor	Raltegravir
	Dolutegravir
	Elvitegravir
Binding inhibitor (CCR5 antagonist)	Maraviroc
Fusion inhibitor	Enfuvirtide (T-20)

An initial assessment should include a full appraisal of general, sexual, and psycho-social health; education about HIV infection, its natural history, how infection is transmitted, and the effects of therapy; and assessment of virological (viral load) and immunological (CD4+ T cell count) status and relevant co-morbidities including co-infection with HBV or HCV, and risk of cardiovascular, bone, and malignant disease. Treatment is now recommended for all infected individuals on the basis of multiple cohort and observational studies as well as the recent results from the Strategic Timing of Antiretroviral Therapy (START) study. This study recruited 4685 subjects with HIV infection and followed them over a mean of three years—randomized to commence antiretroviral therapy at a CD4 count of greater than 500 cells/µL or defer therapy until CD4 was below 350 or clinical progression occurred. The deferred group developed significantly greater non-AIDS events.

Individuals should be started on a combination of three or four drugs from the outset. This principle of combination therapy has been practiced to reduce drug resistance against tuberculosis since the 1960s, and is fundamental for the management of HIV infection.

From the time of diagnosis, the patient will require continuous assessment for response to therapy, side-effects and interactions between drugs, issues of adherence to drug regimens, and prevention of co-infections. In individuals who present with advanced immunodeficiency (which is now increasingly rare), careful observation and prophylaxis against opportunistic infections are recommended. This may include cotrimoxazole (for *Pneumocystis* and *Toxoplasma* infections), azithromycin and/or isoniazid (mycobacterial infections), acyclovir or ganciclovir (herpesviruses and CMV), and ensuring appropriate vaccinations (pneumococcal disease, influenza, and hepatitis A and B). Drug combination regimens usually need to be adjusted from time to time to counteract the emergence of resistant virus or the appearance of serious side-effects of the drugs. Overall, life expectancy has increased considerably since the introduction of these regimens, but patients now show increasing late problems of HIV-associated neurological disease, cardiovascular disease, and malignancies.

Control

Control of the AIDS epidemic is a global problem requiring a concerted, multifaceted, international approach. Although the detailed implementation of national programs is greatly influenced by such variables as the extent of the epidemic within any one region, the general level of health, education, and affluence, local social customs and superstitions, political realities, and available funding, the major objectives are universal.

Reducing HIV Transmission

1. Reducing sexual transmission is one mainstay of HIV control campaigns. The underlying principles include:

   a. sexual health campaigns, use of condoms, and avoidance of unprotected sex, particularly with new partners

   b. education to encourage limiting the number of sexual partners

   c. voluntary HIV testing particularly of individuals with high-risk behavior and appropriate counseling about the implications of a positive or a negative result

   d. prompt treatment of other sexually transmitted diseases. Not only does prompt treatment decrease the chance of contracting HIV, but patients attending an STD clinic constitute a captive audience ready to receive and implement advice about AIDS prevention.

   The use of antiretroviral agents to prevent transmission can take three forms. Treatment of those infected, to achieve undetectable viral loads, has been shown to reduce transmission to very low levels (HPTN 052 study), which has led to the "treatment as prevention (TasP)" strategy. Post-exposure prophylaxis with antiretroviral agents has been in use for many years in both the occupational and non-occupational settings in which exposure to HIV-containing body fluids may have occurred. The third use of ART in prevention is as pre-exposure prophylaxis. A number of clinical trials have recently confirmed the effectiveness of this approach in which a non-infected individual with frequent ongoing exposure to HIV can reduce significantly the risk of infection. Pre-exposure prophylaxis can involve the use of systemic antiviral agents or topical agents at the mucosal surface potentially in contact with HIV. The former can be given as daily prophylaxis (e.g., tenofovir/FTC, marketed commercially as Truvada), or as a long acting drug, for example, the integrase inhibitor cabotegravir. There is also a recognized role for an effective vaginal microbicidal agent; this would empower women to control their own destinies with a greater degree of certainty. Different approaches are being explored, including a vaginally inserted ring that releases the antiretroviral drug dapirivine and is active for up to one month, thereby reducing compliance/adherence issues.

   In different communities and cultures, achieving these goals has required different approaches, has encountered different political and social obstacles, and has met different degrees of success. Much effort has gone into educational research and targeted education campaigns, including ways to empower women in sexual transactions, dealing with discrimination against homosexuality, and discrimination against HIV-infected individuals. At the political level a balance has to be struck between coercing people into responsible behavior by legislation (e.g., compulsory testing and disclosure of results) versus protection of individual rights and privacy. The whole community needs to be educated about the danger and routes of transmission of HIV, what constitutes high-risk behavior, and practical approaches to minimize those risks. The message needs to be targeted in different ways to different audiences, but particularly to the young and sexually active, through mass media, in schools, and utilizing relevant, non-threatening peer-group leaders speaking the same language. It is crucial not to alienate high-risk groups at the social margins by judgmental or discriminatory attitudes.

2. The second target group includes people who inject drugs. Many countries have introduced ready access to new disposable syringes and needles free of charge, and some countries have set up "safe injecting rooms" where administration of drugs is backed up by access to professional staff. Access to clean needles has had a major impact on reduction in HIV infection among people who inject drugs in Australia and many other countries. Provision of needles and syringes in prisons has also demonstrated significant success with reduction of new HIV infections. Needless to say, such an approach also can meet with significant opposition from conservative elements within a community.

3. In the health care and laboratory environment, the crucial principle is to treat all biological material of human origin, particularly visible blood, as potentially infectious. Strict standards of aseptic technique are required in handling such materials in the ward or laboratory, regardless of the results of screening tests that may or may not have been done. The details are beyond the scope of this text, but some of the procedures are described in Chapter 14. Control, Prevention, and Eradication.

4. Almost all countries have implemented effective procedures for the routine screening of blood, sperm, and organ donors for HIV antibody. Furthermore, blood products such as factor VIII used for the management of hemophilia are now routinely heat-treated to destroy the labile HIV virion. In most countries compulsory screening for HIV infection (by EIA for anti-HIV antibodies) is confined to blood, semen, and organ donors; universal population screening is not considered to be cost-effective in low-prevalence countries. However, voluntary screening of those at high risk of infection is to be strongly encouraged. High-risk categories include the following: (1) men who have sex with men, (2) individuals attending STD clinics, (3) travelers returning after unprotected sex in countries of high HIV endemicity, (4) injecting drug

use, (5) anyone who received blood or a blood product between 1980 and 1985, (6) health workers accidentally exposed (e.g., via needle stick or blood spill), and (7) anyone with a clinical presentation consistent with immunodeficiency.

Counseling before and after testing is important. Management of an individual found to be seropositive for the first time includes skilled counseling about the implications for the patient's lifestyle, family, social and sexual contacts, staging of disease, and future medical management strategy. Confidentiality must be respected; however, sexual partners should be made aware preferably by the patient him/herself, and counseled.

Vaccines

Despite the recent significant achievements of chemotherapy and measures to reduce virus transmission, the ultimate control of AIDS will be achieved only by the use of an effective vaccine. In countries where HIV is largely restricted to identifiable groups, the spread of HIV could be restricted by vaccinating the major risk groups, namely, sexually promiscuous male homosexuals, prostitutes, injecting drug users, and newborn children of infected mothers. However, where AIDS is spread heterosexually and perinatally, control could only be achieved by universal infant immunization.

HIV may enter the body not only as free virions, neutralizable by antibody, but as infected leukocytes in semen or blood, in which the virus or provirus would be protected against antibody or T cell-mediated cytolysis. Unless the infecting cells are rapidly destroyed by allograft rejection, infection could spread directly from cell to cell by fusion. Also, latently infected cells carrying an integrated cDNA copy of the HIV genome, which do not express viral proteins, will not be recognized by antibody or cell-mediated immunity. Moreover, infected cells may be sequestered from the immune response in such sites as the CNS. Thus, unless all latently infected cells can be eliminated, immune surveillance will need to be maintained for the rest of the individual's life to contain any future reactivation of virus replication.

As the normal mode of transmission of HIV is via the mucosal route, the first line of defense in a vaccinated host would preferably be mucosal IgA antibodies and submucosal lymphocytes. This would argue for mucosal delivery, preferably of a live or recombinant vaccine, perhaps vaginally, orally, or intra-nasally, to take advantage of the common mucosal pathway. However, what constitutes protective immunity against HIV is yet to be established: there has been continuing debate whether neutralizing antibody, cell-mediated immunity, or both modalities combined, represent the magic key to success. Whereas the ideal vaccine would prevent infection, clinicians may be obliged to lower their sights somewhat in the case of HIV and settle for reduction in the virus load such that latent infection is established in only a limited scale, in the hope that the immune response might reduce the rate of spread and delay the onset of disease. A strategy of "therapeutic" vaccination given to those already infected, in order to retard disease progression, is an alternative approach.

Although numerous problems stand in the way of the development of an effective HIV vaccine (Table 23.10), passive infusion of neutralizing antibodies has been shown to protect non-human primates against challenge with a chimeric simian-human immunodeficiency virus (SHIV). Recently, it has been recognized that broadly reacting neutralizing antibodies do develop in a significant proportion of infected individuals after some length of time, and broadly reacting monoclonal antibodies can be generated against quaternary epitopes on the envelope glycoprotein gp120 or against gp120 trimers; this is one area of active investigation.

In the face of this background, enormous research effort has been applied for many years with in most cases disappointing or frustrating results. Because of the lethality of the virus, the certainty of establishment of latent infection,

---

TABLE 23.10 Obstacles to the Development of a Vaccine Against HIV

1. No clear population exists with natural immunity. For smallpox, milk-maids were recognized as immune following prior cowpox infection. For HIV, some groups of sex workers, for example, in Nairobi, appear to have partial immunity to repeated sexual exposure, and attempts are being made to understand and exploit this

2. No correlates or surrogate laboratory markers of immunity are known

3. The animal models for vaccine research are difficult to access and limited in value

4. Antigenically, HIV shows great diversity between clades and within strains, and continues to mutate during the course of an individual's infection

5. Viral envelope glycoproteins are heavily glycosylated leading to physical shielding of epitopes by extensive N-linked glycans

6. Potential conserved epitopes, for example, receptor-binding sites, are disguised by the conformational flexibility of the envelope glycoprotein structure

7. Virus DNA integrates into various reservoir cell populations. Hence a successful vaccine must maintain long-term surveillance to suppress virus re-emergence

8. Commercial equation. Because of its difficult scientific feasibility and low financial market attractiveness, commercial funding of HIV vaccine development scores poorly on both sides of the commercial balance sheet. Most vaccine development is funded by non-profit organizations, for example, National Institutes of Health, Gates Foundation, etc.

and the likelihood of subsequent mutation, it is improbable that any live attenuated HIV vaccine would be licensed for human use, with the possible exception of a deletion mutant analogous to the *nef*-mutant of the simian immunodeficiency virus, which is avirulent for rhesus monkeys. The main approaches have been (1) conventional inactivated whole-virus vaccines or purified envelope glycoproteins, (2) recombinant live vectors carrying at least the gene for the HIV envelope protein(s) (gp120 ± gp41), (3) DNA vaccines, or (4) various combinations of the above in prime-boost protocols. Experimental vaccines of all these kinds have been produced and tested in chimpanzees and/or humans, and details of several recent major human efficacy trials are shown in Table 23.11. The modest but clear protection reported in the RV144 trial gave the whole field significant encouragement, but the disappointing outcome of the more recent HTVN505 trial underlined how difficult this goal is proving to be and the uncertainty about the optimal approach.

Simian Immunodeficiency Virus and the Origins of HIV/AIDS

The primate lentivirus group of the genus *Lentivirus* also contains the simian immunodeficiency virus (SIV), and different unique strains of SIV have been recovered from a range of African monkey species in nature (Fig. 23.1B). It is likely that each of these strains has evolved in adaptation with individual hosts, a view that is supported by the fact that in most cases these viruses do not produce overt disease in the original host. However, when some of these infect other species of monkeys, severe disease can result. For example SIV_{agm} naturally infects a high proportion of African green monkeys in the wild causing little harm, but if introduced into the Asian rhesus macaques (*Macaca mulatta*) it leads to a disease not unlike AIDS in humans. Another strain, SIV_{mac}, has not been found in rhesus macaques in the wild in Asia, but can spread within colonies of captive macaques where it causes an AIDS-like disease. SIV_{mac} is now known to have originated from a strain SIV_{smn} that was originally an infection of sooty mangabeys, an African monkey.

These observations are interesting and important for a number of reasons. First, they have been used in many studies to examine possible mechanisms affecting virulence. When SIV_{mac} infection in rhesus monkeys and in sooty mangabeys was compared, both species showed early massive activation of immune response genes including interferon response genes, but only the sooty mangabeys could bring this under control, while in the rhesus monkeys that developed disease this activation response was sustained. Second, a major bottleneck in HIV vaccine development is access to appropriate systems to test the plethora of possible strategies that have been proposed. SIV models are thus valuable in testing alternative vaccination approaches.

Third, this work has improved our understanding of the likely origins of the human HIV pandemic. It is thought that lentiviruses that were naturally present in chimpanzees have been transmitted across species to humans on multiple occasions, in most cases with very limited impact or spread. One strain designated SIVcpzPtt found in the subspecies *Pan troglodytes troglodytes* in southeastern Cameroon, is very similar to HIV-1 group M (the main cause of AIDS globally); it is thought that on one, or a number of occasions, cross-species transmission of this or a closely related strain led to persistence and dissemination through the human population, giving rise to the AIDS pandemic. Interestingly, the SIVcpzPtt

TABLE 23.11 Recently Completed Major HIV Vaccine Trials

Trial Name	Site and Date of Completion	Vaccine Design	Efficacy
Vax 004	N America and Europe, 2005	Recombinant clade B gp120	No prevention of acquisition, or modification, of infection
Vax 003	Thailand, 2006	Recombinant clade B and E gp120	
Step (HTVN502)	International, 2008	Replication-defective adenovirus 5 vector/clade B gag, pol, nef	Trial halted; no efficacy; suggestion of increased rate of infection in subgroups
RV144	Thailand, 2009	*Prime*: replication-defective recombinant canarypox vector (ALVAC)/gp120	31% reduction in acquisition of infection, but duration of protection short-lived
		Boost: Clade B/E recombinant gp120 protein	
HTVN 505	USA, 2009–13	*Prime*: DNA plasmid encoding EnvA, EnvB, EnvC, gagB, polB, nefB	Halted in 2013; no reduction in acquisition of infection, or viral load
		Boost: Recombinant adenovirus 5/EnvA, EnvB, EnvC, gag/polB	

strain may have only been present in chimpanzees for a relatively short time, because subspecies of chimpanzees from West Africa do not harbor this virus. The other three main lineages of HIV-1 in humans, designated O, N, and P, are likely to represent three similar independent cross-species events. A similar mechanism involving SIVmac is thought to have given rise to HIV-2 in humans, probably as a result of at least eight separate cross-species transmissions from sooty mangabeys in West Africa (see also Chapter 15: Emerging Virus Diseases and Fig. 15.2).

Following such cross-species transmission events, the potential for epidemic spread could then be determined by many factors including adaptation of the virus to the new species (involving replication competence, ability to counteract host restriction factors, production and release of infectious virus), and the characteristics of the new host species favoring or hindering virus spread (including host behavior and population density).

SUBFAMILY *SPUMAVIRINAE*

This subfamily contains only one genus, *Spumavirus*. This comprises a number of species of "foamy viruses," so named owing to possessing a characteristic cytopathology, namely, vacuolation in cultured cells. The proviral genome contains the typical retrovirus elements *gag, pol,* and *env* and the usual LTR sequences. In addition the genome codes for two non-structural proteins Tas (transcriptional activator of *Spumavirus*) and Bet (whose exact regulatory function is unclear). However none has been reported to contain an oncogene. Spumaviruses are remarkable in that reverse transcription occurs during budding and assembly on the efferent or exit pathway rather than between virus entry and integration as with other members of the *Retroviridae*. As a result of this strategy, in common with hepadnaviruses the extracellular virion form of the genome exists as DNA rather than RNA.

Species-specific spumaviruses are widespread in many mammalian species, including cats, cattle, horses, and most primate species other than humans, and cause inapparent persistent infections. Spumaviruses are frequently found as contaminants in primary cell cultures from non-human primates, causing large multinucleated syncytia and vacuolation and rendering the cell cultures unsuitable for virus cultivation. Rare human infections have been documented as a result of zoonotic transmission from non-human primates, but no human to human spread has been reported. The original "human" foamy virus was isolated in 1971 from a primary nasopharyngeal carcinoma tissue culture from a Kenyan patient, and designated human foamy virus (HFV). However subsequent work suggests this patient may have acquired a chimpanzee foamy virus via zoonotic contact. Later studies of large human populations have not found any convincing cases of human foamy virus infection supported by PCR and serological evidence, except for individuals suffering from a zoonotic infection usually following severe monkey bites. Thus current evidence suggests that HFV is not naturally prevalent in the human population.

FURTHER READING

Centers for Disease Control, USA. Laboratory testing for the diagnosis of HIV infection: updated recommendations 2014.

Deeks, S.G., Overbaugh, J., Phillips, A., Buchbinder, S., 2015. HIV infection. Nat Rev Disease Primers 1, 1–22. Article number: 15035 Available from: http://dx.doi.org/10.1038/nrdp.2015.35.

Goncalves, D.U., Proietti, F.A., Ribas, J.G.R., et al., 2010. Epidemiology, treatment, and prevention of human T cell leukaemia virus type 1-associated diseases. Clin. Microb. Rev. 23, 577–589.

Hoy, J., Lewin, S., Post, J.J., Street, A. (Eds.), 2009. HIV Management in Australasia, Pub. Australasian Society for HIV Medicine, Darlinghurst, NSW, Australia.

Kwong, P.D., Mascola, J.R., Nabel, G.J., 2012. The changing face of HIV research. J. Int. AIDS Soc. 15, 17407.

Kwong, P.D., Mascola, J.R., Nabel, G.J., 2011. Rational design of vaccines to elicit broadly neutralizing antibodies to HIV-1. Cold Spring Harb. Perspect Med., A007278.

Margolis, D.M., 2010. Mechanisms of HIV Latency: An Emerging Picture of Complexity. Curr HIV/AIDS Rep 7, 37–43. Available from: http://dx.doi.org/10.1007/s11904-009-0033-9.

Moir, S., Chun, T.-W., Fauci, A.S., 2011. Pathogenic mechanisms of HIV disease. Annu. Rev. Pathol. Mech. Dis. 6, 223–248.

Sharp, P.M., Hahn, B.H., 2011. Origins of HIV and the AIDS pandemic. Cold Spring Harb Perspect Med. 1, a006841.

Wilen, C.B., Tilton, J.C., Doms, R.W., 2012. Molecular Mechanisms of HIV Entry. In: Rossmann, M.G., Rao, V.B. (Eds.) Viral Molecular Machines, 223 Advances in Experimental Medicine and Biology, pp. 726. Available from: http://dx.doi.org/10.1007/978-1-4614-0980-9_10.

Chapter 24

Reoviruses

The name reovirus is an acronym, short for *r*espiratory *e*nteric *o*rphan because the first members of this family to be discovered, now classified as the genus *Orthoreovirus* of the family *Reoviridae*, were found to inhabit both the respiratory and the enteric tract of humans and animals, but to be "orphans" in the sense that they are not associated with disease. The discovery of human rotaviruses in 1973 changed all that, for members of the genus *Rotavirus* are recognized to be the most important cause of infantile gastroenteritis throughout the world, as well as causing economically important diarrheal disease in many livestock species. In addition, dozens of arboviruses, at least one of them causing disease in humans, have been allocated to the genera *Orbivirus* and *Coltivirus*. Yet other genera contain pathogens that infect both plants and insects, raising the question of whether these fascinating viruses that cross kingdoms so readily might have evolved in insects.

Furthermore, reoviruses have attracted much attention from molecular biologists because of the unusual nature of the genome. Composed of double-stranded RNA, the genome is segmented into 10 to 12 separate molecules, each representing a different, generally monocistronic gene. The ease with which these viruses undergo genetic reassortment has been exploited to exchange genes from temperature-sensitive (*ts*) mutants and thus determine the role of individual genes in pathogenesis and virulence.

More recently, at least 10 further unrelated virus families with double-stranded RNA genomes have been described, that infect plants, insects, bacteria, and fungi.

PROPERTIES OF REOVIRUSES

Members of the family *Reoviridae* have striking icosahedral or quasi-spherical virions (Fig. 24.1), with one, two, or three concentric layers of capsid proteins surrounding an inner core. Two subfamilies are distinguished. The *Spinoreovirinae* contain nine different genera, including *Orthoreoviruses* and *Coltiviruses* that infect humans, and have virions with a single complete capsid layer to which large spikes or turrets are attached at the 12 icosahedral vertices; an outer incomplete capsid layer is usually also present, giving a double-shelled appearance. The *Sedoreovirinae* (*sedo*, Latin, smooth) comprise six genera, including *Rotaviruses* of humans and animals, and *Orbiviruses* which include the important animal pathogens bluetongue viruses and African horse sickness viruses; virions of the *Sedoreovirinae* have a double-shelled capsid and in some cases a third outer core layer, and appear more spherical or smooth, The genome comprises 9, 10, 11, or 12 unique molecules of dsRNA, total size 18–27 kilobase pairs, depending on the genus (Table 24.1), and in most cases each RNA species codes for a single protein. The positive strands of each duplex are 5′ capped and the negative strands 5′ phosphorylated; the 3′ termini are not polyadenylated. In addition to the variable number of structural capsid proteins, virions contain at least three internal proteins with enzyme functions: these include an RNA-dependent RNA polymerase, involved in the synthesis of both new positive strand RNA from a dsRNA template and new negative strand RNA from a positive strand RNA template; helicase; NTPase; guanylyl transferase; and transmethylase activities. Genera share no antigens, but within genera there are genus-specific and species-specific antigens (Table 24.2).

VIRAL REPLICATION

The replication cycle has been studied in most detail with reovirus 3, a member of the genus *Orthoreovirus* (Fig. 24.2). The intact virion may enter the cell by receptor-mediated endocytosis, or, alternatively, intermediate subviral particles that are generated by digestion with chymotrypsin in the intestine, may pass into the cytoplasm either via the endosomal pathway or directly. Both types of particles are then degraded to become core particles; within these, the virion-associated transcriptase and capping enzymes repetitively transcribe 5′ capped (but not 3′ polyadenylated) mRNA molecules, which are extruded through the hollow apices of the icosahedral core particles. Only certain genes are transcribed initially; the others are derepressed following the synthesis of an early viral protein. The efficiency of translation also varies between the different mRNAs.

The replication of genomic RNA occurs later, after early mRNA synthesis, and is complex. Various non-structural

Fenner and White's Medical Virology. DOI: http://dx.doi.org/10.1016/B978-0-12-375156-0.00024-2

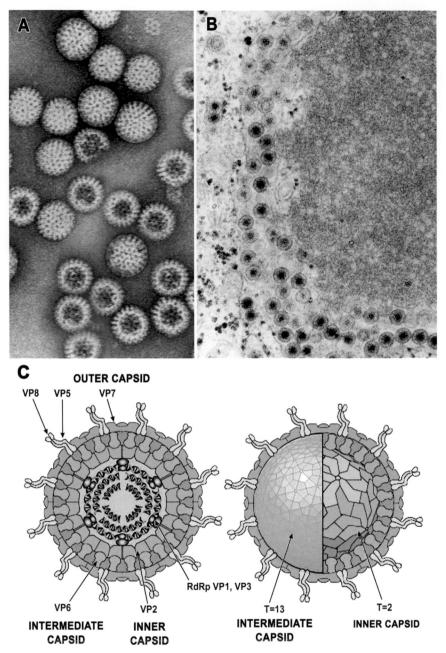

FIGURE 24.1 Morphology and structure of member viruses of the family *Reoviridae*. (A) Negative contrast electron microscopy of human rotavirus, showing the multiple capsid layers—some particles are complete with outer capsid layer clear (the largest particles), whereas some are intermediate capsids (the smaller particles). (B) Thin-section electron microscopy of human rotavirus in cell culture, showing the relationship of virion maturation to an intracytoplasmic inclusion body (viral factory). (C) Model of rotavirus virion, showing its triple capsid layer construction, about 80 nm in total diameter. The inner capsid has $T = 2$ icosahedral symmetry; the intermediate capsid has a $T = 13$ symmetry and the outer capsid has viral protein projections that serve as cell binding ligands. *C, Reproduced from ViralZone, Swiss Institute of Bioinformatics, with permission.*

TABLE 24.1 Properties of *Reoviridae*

Subfamily *Spinoreovirinae* contains 9 genera, including *Orthoreovirus* and *Coltivirus* that infect humans; subfamily *Sedoreovirinae* contains 6 genera, including *Rotavirus* and *Orbivirus* that infect humans

Non-enveloped spherical or icosahedral virion (*Coltivirus and Orthoreovirus*, 60 to 85 nm; *Rotavirus*, 100 nm; *Orbivirus*, 90 nm)

Virion composed of two or three outer capsid shells surrounding inner core; core contains RNA polymerase and associated enzymes involved in mRNA synthesis (RNA elongation, capping, methylation, helicase)

Double-stranded RNA segmented genome; *Rotavirus*, 11 segments, total size 18 kbp; *Coltivirus*, 12 segments, 27 kbp; *Orbivirus*, 10 segments, 19 kbp; *Orthoreovirus*, 10 segments, 24 kbp

Cytoplasmic replication; entry may require cleavage of capsid protein; transcriptase in "core particles" transcribes early mRNAs; genome replication occurs by transcription from mRNA molecules to form dsRNA in nascent progeny subviral particles; transcription of late genes for structural proteins; virions undergo self-assembly, released by cell lysis

Genetic reassortment occurs between species

TABLE 24.2 Members of the Family *Reoviridae* of Medical Significance

Reovirus Genera	Virus Species of Medical Significance	Transmission	Host Species	Clinical Disease
Spinoreovirinae				
Orthoreovirus	Mammalian orthoreoviruses	Respiratory and fecal–oral	Humans	No clinical disease
Coltivirus	Colorado tick fever virus (CTFV)	*Dermacentor andersoni* wood tick	Rodents, humans	Fever, meningo-encephalitis, hemorrhagic fever
Sedoreovirinae				
Rotavirus	Rotavirus	Fecal–oral	Humans domestic animals, birds	Vomiting, fever, diarrhea
Orbivirus	Kemerova virus Tribeč virus	Ticks	Rodents, humans	Fever, encephalitis

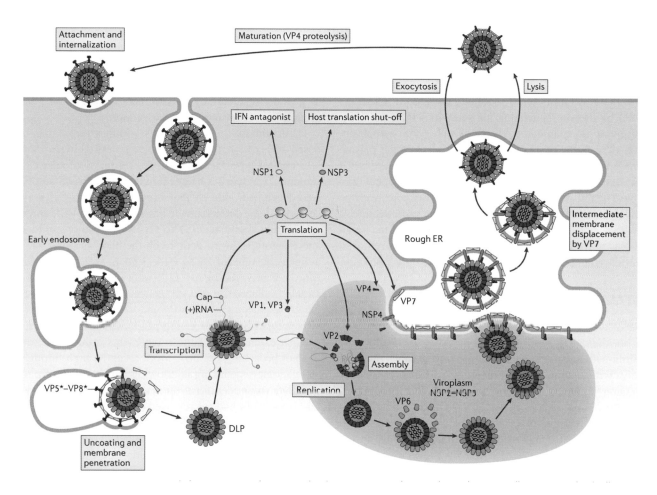

FIGURE 24.2 Rotavirus replication cycle. Rotaviruses first bind to target cell surface sialic acids via VP8* which lies at the tips of the virion spikes; they then enter the cell via receptor-mediated endocytosis. Uncoating of the triple-layered infectious particles (TLPs) occurs within early endosomes; loss of the outer capsid layer and release of the double-layered particle (DLP) into the cytosol activates the virion polymerase complex (VP1 and VP3) within the incoming virus, initiating the transcription of capped positive-sense RNA ((+) RNAs) from each of the 11 double-stranded RNA (dsRNA) genome segments. These mRNAs are translated and become associated with newly synthesized viral proteins to form RNase-sensitive subviral particles. (+) RNAs serve either as mRNAs for synthesis of viral proteins or as templates for synthesis of negative-sense RNA ((−) RNA) replicative intermediates. Non-structural protein 2 (NSP2) and NSP5 interact to form large intracytoplasmic inclusion bodies (viroplasms) that sequester components required for genome replication and virion assembly. Genome packaging involves the formation of "assortment complexes," wherein single copies of each dsRNA segment are bound together in the nascent virion core structure. This triggers more dsRNA synthesis by VP1. The intermediate capsid protein, VP6, then assembles onto the nascent core to form the DLP. Assembly of the outer capsid is not well understood; one model proposes that NSP4 recruits DLPs and outer capsid protein VP4 to the face of the endoplasmic reticulum (ER) membrane, from where DLP–VP4–NSP4 complexes bud into the ER. Subsequent removal of the ER membrane and NSP4 permits assembly of TLPs via VP7 crosslinking. Particles released from infected cells are exposed to trypsin-like proteases of the gastrointestinal tract, resulting in the specific cleavage of VP4 into VP5* and VP8*, yielding fully infectious virions. *Modified from ViralZone, Swiss Institute of Bioinformatics, by Trask, S.D., et al., 2012. Structural insights into the coupling of virion assembly and rotavirus replication. Nat. Rev. Microbiol. 10, 165–177, with permission.*

and structural viral proteins associate with a complete set of mRNA molecules to form nascent progeny subviral particles, within which a single round of complementary minus sense RNA strand synthesis occurs to produce dsRNA molecules. These in turn serve as templates for the transcription of more mRNAs, which this time are uncapped and are translated preferentially to yield a large pool of viral structural proteins. New virions self-assemble and accumulate in cytoplasmic inclusions termed virus factories or viroplasms, before being released by cell lysis.

These general principles are similar for rotavirus replication. However, certain distinctive features mark the early and late stages of the rotavirus replication cycle. The full triple-layered virus particle containing the outer capsid proteins VP4 and VP7 is required for attachment. Attachment and entry is a complex, multistep process involving binding of the outer capsid hemagglutinating protein VP4 to glycans including sialic acid and histo-blood group antigens, or to $\alpha2\beta1$ integrin, followed by interaction with other members of the integrin family and heat shock protein 70. Cleavage of VP4 by trypsin, generating the subunits VP5* and VP8*, is required to enable the virion to enter the cytoplasm, which may occur by both direct penetration of the plasma membrane and receptor-mediated endocytosis. After removal of the outer shell, RNA transcription then takes place within double-layered particles (DLPs) in the cytoplasm. At virus assembly, newly synthesized progeny particles acquire a temporary envelope as virions bud into cisternae of the rough endoplasmic reticulum (RER). This envelope is then removed and VP7 remains to complete the outer capsid glycoprotein.

ROTAVIRUS GASTROENTERITIS

In 1973 a new human virus was discovered in Australia, first in duodenal biopsies then in feces of Melbourne children with gastroenteritis. Electron micrographs revealed virions with a unique appearance (*rota*, wheel). Quickly it became apparent that rotaviruses are the commonest cause of gastroenteritis in infants, frequently leading to hospitalization; rotaviruses are also responsible for up to 500,000 diarrheal deaths each year, predominantly in the developing world.

Properties of Rotaviruses

The rotavirus virion is particularly photogenic (Fig. 24.1). Smooth and round in outline, the infectious virion is an icosahedral particle with three concentric protein layers. The 50 nm core of the virion, composed mainly of the structural protein VP2 (but also the transcriptase/replicase VP1 and the guanylyltransferase VP3), is surrounded by a 60 nm intermediate icosahedral capsid composed of the major group-specific protein VP6; this in turn is enclosed within a 75 nm outer icosahedral capsid composed mainly of the glycoprotein VP7. Dimers of the hemagglutinin/cell attachment protein VP4 are attached to VP6 and project from the surface as 60 short spikes; 132 channels pass through both the outer and inner capsid to communicate with the core. The genome consists of 11 molecules of dsRNA, total size 18 kbp.

Rotaviruses are widely distributed but have only been found in vertebrates. Essentially every species of domestic animal or bird that has been thoroughly searched has at least one indigenous rotavirus, causing diarrhea in the newborn. Rotaviruses are classified into five species or groups (*Rotavirus A* to *Rotavirus E*) and three additional unclassified viruses, on the basis of (1) differences between the highly conserved major group-specific capsid antigen VP6, (2) sequence homology in conserved genome segments, and (3) host range. Most human rotaviruses (and the well-studied simian rotavirus SA11) belong to the species *Rotavirus A*, but isolates of *Rotavirus B* have caused sporadic outbreaks of human disease in adults, and self-limiting outbreaks of human gastroenteritis due to *Rotavirus C* isolates also have been described. Exchange of genome segments by reassortment occurs readily within groups, but not between different groups.

Differentiation into serotypes within group A has been based on cross-neutralization assays using hyperimmune sera against type-specific epitopes on the two outer capsid proteins. A binary system of classification of serotypes was developed, akin to that used for influenza viruses. G serotypes were defined on the basis of different VP7 antigenic specificities designated G1 to G14 (VP7 is a *g*lycoprotein), and P serotypes were defined on the basis of different VP4 antigens designated P1 to P14 (VP4 is *p*rotease sensitive). More recently, the binary G/P serotyping system has been largely replaced by a G/P genotyping system based on cDNA sequencing of the VP4 and VP7 genes. Currently, 27 G genotypes and 35 P genotypes have been described, but to what extent each of these represents a distinct serotype has not been fully defined. More recently, high-throughput full genome sequencing is allowing development of a more complete strain identification system that takes into account all 11 segments, and has revealed the favored clustering of all 11 dsRNA segments into different linked gene constellations.

Large numbers of electropherotypes have been differentiated by polyacrylamide gel electrophoresis (PAGE) of the viral RNA. Because these patterns reflect differences in the migration of any of the 11 RNA segments, there is no direct relationship between electropherotypes and serotypes, but they have been a valuable tool for tracking the epidemiology of individual strains.

Pathogenesis and Immunity

Following oral ingestion, VP4 is cleaved by trypsin-like proteases in the intestine to generate two subunits VP5* and VP8*; this is important to activate infectivity and allow entry into the cell. Differentiated enterocytes at the tips of intestinal villi are infected and destroyed, leading to a succession of pathological consequences described in Chapter 8: Patterns of Infection. At least three factors play important roles leading to massive fluid and electrolyte loss; the direct toxic effect on the mucosa of the viral protein NSP4, the first viral enterotoxin to be described; the decrease of brush border enzymes (e.g., maltase, sucrose, lactase) leading to malabsorption of sugars and osmotic diarrhea; and the activation of the enteric nervous system. Different products of rotavirus replication that are not yet fully identified, are recognized as non-self and act as pathogen-associated molecular patterns (PAMPS) to stimulate interferon and early antiviral gene products of the host innate defense. Rotaviruses in turn have evolved mechanisms to avoid or inactivate this defense. These include: (1) sequestration of viral RNAs in cytoplasmic viroplasms, and dsRNA in virus particles, (2) degradation of the interferon regulatory factors (transcription factors) and NF-κB by the rotavirus non-structural protein NSP1, (3) selective inhibition of cellular but not viral mRNA translation through actions of NSP3. There is also evidence from animal rotavirus experiments that the tropism of different virus strains for different host species may be governed by the effectiveness of these various viral strategies to avoid host defenses.

Maternal antibodies of the IgG class transmitted across the placenta do not protect the newborn, but IgA antibodies in the colostrum do. Breast-fed infants are significantly less susceptible than bottle-fed infants during the first few days of life. Thereafter, the antibody titer in milk drops off rapidly, but infections acquired during the first few months still tend to be inapparent, except in the poorest communities. Neonatal infections themselves elicit a transient neutralizing secretory IgA ("coproantibody") response in the gut, as well as a neutralizing IgG response in serum that persists for a few months, but they do not confer significant immunity against reinfection. Reinfections do, however, tend to be clinically less severe, presumably as a result of priming for an anamnestic response. Immunity following clinical disease occurring at a later age may be more durable, and there is some evidence for low levels of cross-reactive (heterotypic), as well as homotypic, immunity. Severe disease is less common on reinfection, and in older children. However, longitudinal studies are required to sort out whether increasing resistance is simply age-related or attributable to homotypic or heterotypic immunity, whether secretory IgA, serum IgG, or possibly T cell-mediated immunity is the most important, and whether neutralizing antibodies to VP4 are more relevant than those to VP7.

Clinical Features

Asymptomatic infection is the rule in neonates and is quite common in older children and adults, contributing to virus shedding and a source of virus dissemination. Clinical illness is seen principally between the ages of 6 and 24 months, with the peak around 12 months, but the peak is earlier in developing countries. After an incubation period of one to three days, vomiting and fever generally precede non-bloody diarrhea, which can last for three to eight days and can lead to severe dehydration. Respiratory signs and symptoms have been reported in 30 to 50% of children, but it is not clear to what extent this is due to rotavirus or to other co-existing respiratory virus infection. Death is rare in well-nourished children, but large numbers die in the poorer tropical countries. Rotavirus infection has been associated with complications including necrotizing enterocolitis, intussusception, biliary atresia, and central nervous system involvement, but a definite relationship with rotavirus infection and any of these conditions has not been confirmed. Infection in adults, if symptomatic, is typically milder.

Laboratory Diagnosis

Rotaviruses were discovered by electron microscopy, which remains one approach to rapid diagnosis (see Chapter 10: Laboratory Diagnosis); the virions are plentiful in feces and are so distinctive that they cannot be mistaken for anything else. However, EM is cumbersome for large numbers of samples, and enzyme immunoassay (EIA) is a more practicable and more sensitive method for most laboratories. Virus can be detected one to two days before the onset of symptoms and remains detectable in three-quarters of cases after four to eight days of illness. Latex particle agglutination and reverse passive hemagglutination are also sensitive, specific, and simple (see Chapter 10: Laboratory Diagnosis). The specificity of any of these tests can be manipulated at will by selecting either type-specific or broadly cross-reactive monoclonal antibodies, for example in immune-electron microscopy, or as capture and/ or indicator antibodies in antigen-capture EIAs. For example, group B or C rotaviruses can be distinguished by using antibodies against the group-specific internal antigen VP6.

RNA detection methods can be used to identify viral genomes directly from feces. Reference has already been made to the use of polyacrylamide gel electrophoresis to resolve the 11 gene segments; for instance, rotavirus groups A, B, and C can be clearly distinguished by RNA electrophoretic pattern alone. RT-PCR detection of viral dsRNA extracted from feces is the most sensitive method but such sensitivity is not usually required; primer pairs are chosen appropriate for the degree of specificity desired (e.g., primers for VP6 to distinguish rotavirus groups, or VP7 and VP4 to distinguish G and P genotypes respectively). These approaches are usually only used for research or epidemiological purposes.

Human group A rotaviruses are not particularly difficult to culture *in vitro* provided that trypsin is incorporated in the (serum-free) medium, to cleave the relevant outer capsid protein VP4 and thus facilitate entry and uncoating of the virus. However, group B and C rotaviruses are grown with difficulty only, and virus isolation is not widely practiced. Primary rhesus monkey kidney cells, the monkey kidney cell line MA104, and the human colon carcinoma line CaCo-2 are the most commonly used. The cytopathic effect is not striking; immunofluorescence is used to identify rotaviral antigen in infected cells. Neutralization tests using appropriate polyclonal antisera or MAbs (plaque reduction or fluorescent focus reduction) can be used to determine the serotype of the isolate. Serum antibodies can be measured by EIA or neutralization tests.

Epidemiology

Rotaviruses occur in all parts of the world, with some geographical clustering of different strains, and most children develop anti-rotavirus antibodies by the age of 2. Transmission of animal rotaviruses to humans is rare and probably does not cause clinical illness, although reports do exist of disease in humans due to animal-like strains, possibly natural reassortants. Rotaviruses are shed in large numbers, up to 10^{10} particles per gram of feces, usually for 3 to 7 days; this can be prolonged in the case of immunocompromised children, but no true carrier state has been described. Fecal–oral spread occurs mainly by close personal contact and by fomites, and can be decreased by rigorous attention to hygiene, including hand washing, disinfection, and proper disposal of contaminated diapers. Nevertheless, nosocomial outbreaks in hospitals, nurseries, and day-care centers are a common problem. Waterborne epidemics involving adults also occur. Like enteroviruses, rotaviruses can remain viable in water supplies for a considerable time and are relatively resistant to chlorination. Descriptions of respiratory symptoms in some children and the winter prevalence of rotavirus in temperate climates raise the question of whether transmission can also occur via the respiratory route. The antigenic similarity between certain serotypes of animal and human rotavirus and the facility with which human rotaviruses can be grown in piglets, calves, and dogs raise the possibility of zoonotic infections.

Most human rotavirus infections are caused by group A serotypes, principally in infants 6 to 24 months of age. The viruses are endemic year-round in the tropics, but display an autumn, winter, or spring peak in temperate countries. Group B human rotaviruses are much less common but have been responsible for some extensive waterborne outbreaks in China, involving adults as well as children. Group C rotaviruses occur mainly in pigs and affect humans occasionally.

Of the group A rotaviruses associated with diarrheal disease, genotypes G1P[8], G2P[4], G3P[8], G4P[8], and G9P[8] are the most common world wide. Many may co-circulate in the same season, creating conditions for reassortment between genotypes. The strains that predominate in the one location may change from year to year, raising challenges for vaccine design. The "nursery" strains of group A isolated from nosocomial infections in newborn babies may be of G type 1, 2, 3, or 4, and are generally of P type 3. They appear to be relatively avirulent, although secretory IgA and trypsin inhibitors in breast milk, and the delayed appearance of proteolytic enzymes in the neonatal gut, may help explain the mild clinical outcome.

Long-term prospective studies indicate that in some hospital nurseries and pediatric wards a high proportion of all newborn infants may become infected. Usually no disease occurs at this early age, but a degree of immunity develops. Disease is commoner in the 6-month to 2-year-old group. A 7-year prospective investigation of the molecular epidemiology of rotaviruses in children admitted to a Melbourne hospital revealed that, while one particular strain tends to predominate for months or years before being replaced by another, several electropherotypes are often found to be circulating simultaneously. The possible impact that widescale vaccination against rotavirus may have on the selection and circulation of virus strains has not yet been fully assessed.

Treatment and Control

Raising the standard of nutrition and hygiene in the developing world is the long-term answer to the horrific mortality from infantile gastroenteritis. In the meantime many lives can be saved by fluid and electrolyte replacement. A suitable mixture of glucose and electrolytes for administration by mouth is approved by the World Health Organization (WHO) for universal use. Indeed, such has been the success of oral therapy that intravenous therapy may be necessary only for those infants with shock or unusually severe vomiting.

Passive immunization using orally administered rotavirus antibody has been explored. Cows immunized with human or bovine rotaviruses produce colostrum and milk containing high-titer rotavirus antibodies, sold as a nutritional supplement (Gastrogard). Such preparations have been used either prophylactically, for control of cross-infection in hospitals, or therapeutically, to treat immunocompromised infants who are chronic rotavirus excretors. Anti-rotavirus antibody from bovine colostrum has also been shown to reduce symptoms and shorten virus excretion in established infection, but wide use of this approach is logistically difficult.

Early live rotavirus vaccines were derived from bovine or rhesus rotavirus strains using a "Jennerian" approach (using a strain of animal origin that showed attenuated properties in humans). In 1998, the tetravalent (G serotypes 1 to 4) rhesus rotavirus-based vaccine Rotashield (RRV-TV) was licensed in the United States and introduced for routine immunization of infants. However this was withdrawn within a year because of an increased risk of intussusception (approximately one case per 10,000 vaccine recipients). In the mid-2000s, the live vaccines RV5 (RotaTeq) and RV1 (Rotarix) were licensed for use in many countries of the world. RV5 contains five human-bovine re-assortant strains that were generated by crossing the naturally attenuated bovine strain WC3 with five unique human strains each contributing different VP7 or VP4 genes. On the other hand, RV1 contains a single human G1P1A strain that was attenuated by serial passage in cell culture. There is conflicting evidence on the extent that one vaccine strain can generate heterotypic protection, that is against other G-types, but several studies have concluded that Rotarix can prevent severe diarrhea caused by G2P viruses and there is no convincing evidence to date for differences between the vaccines in their efficacy against different serotypes. The vaccines are given as two (RV1) or three (RV5) oral doses beginning at two months of age.

In middle- and high-income countries, a course of either vaccine has been shown to prevent rotavirus gastroenteritis of any severity in 70% of recipients, and to prevent severe disease in 85 to 100% of recipients. Consistent with the above, introduction of routine vaccination has led to reductions in rotavirus-associated hospitalizations by 60 to 75% in a number of countries, with large savings in medical treatment costs (Fig. 24.3). Efficacy is more variable in resource-poor settings, for reasons that are not fully understood. Vaccine viruses are shed in the feces, particularly after the first dose, and some transmission to unvaccinated twins has been reported, but the overall significance of possible natural transmission of vaccine strains has not been studied. Clinical trials have not demonstrated an increased risk of intussusception with either vaccine, although one post-licensing study suggested a very slight increase in the seven days after the first dose of either vaccine. No other important adverse effects, including vomiting, diarrhea or fever, have been consistently shown, although vaccine-associated gastroenteritis is very occasionally seen. The impact of mass vaccination on the natural circulation of rotaviruses, and on new strain selection and virus diversity, remains to be assessed after these vaccines have been in use longer; however, the existence of many related animal virus strains, and their potential for reassortment, raises parallels with the situation in the case of influenza viruses. In any case, what has been a major pediatric viral disease, especially in poorer regions of the world is now being progressively dealt with (Fig. 24.4).

ORBIVIRUSES

The genus *Orbivirus* gets its name from the large doughnut-shaped or ring-shaped "capsomers" (*orbis*, ring or circle) that make up the inner capsid. The diameter of the virion is 80 nm, and the genome comprises 10 molecules of dsRNA. There are dozens of species, all being arboviruses. The genus includes important animal pathogens causing bluetongue of sheep and African horse sickness.

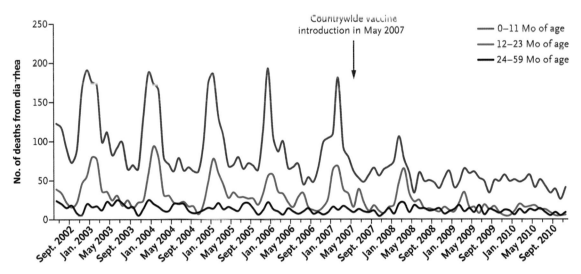

FIGURE 24.3 Diarrhea-related deaths among children 59 months of age or younger from July 2002 through December 2010 in Mexico, by age group. Because of year-to-year variations in diarrhea rates, cautious interpretation of early findings was warranted, but with longer term studies it was found that compared with baseline, diarrhea mortality fell by 56% (95% CI: 49 to 63%) during rotavirus seasons after vaccination began. Reductions were primarily among children under 1 year of age in the 2007–2008 season and extended to older ages in subsequent seasons. These results translate to an annual reduction of approximately 880 deaths related to childhood diarrhea. *Reproduced from Richardson, V., et al., 2011. Childhood diarrhea deaths after rotavirus vaccination in Mexico. N. Engl. J. Med. 365, 772–773, with permission.*

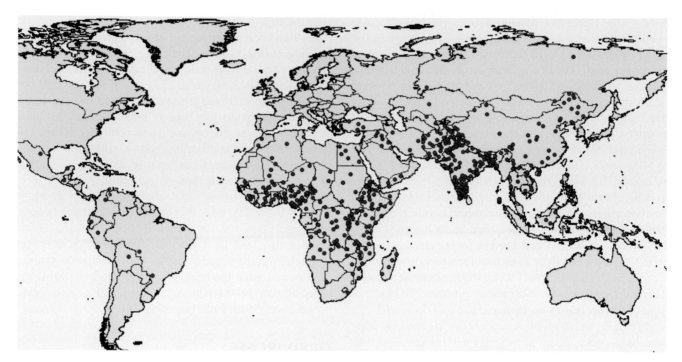

FIGURE 24.4 Global distribution of rotavirus mortality. Each dot = 1000 deaths. As of 2014 it was estimated that worldwide there were about 527,000 rotavirus-related deaths per year in children under 5 years of age. This number is expected to plummet as rotavirus vaccine becomes part of childhood vaccination programs in most countries. *Reproduced from CDC, with permission.*

The only orbiviruses believed to infect humans are the tick-borne Kemerovo viruses of Siberia and the related Tribeč virus found in Slovakia. The primary reservoirs are rodents and birds, and infection of humans following a tick bite can lead to a febrile meningoencephalitis but without paralysis, similar to the better-known tick-borne encephalitis (TBEV) a flavivirus.

COLTIVIRUSES

Coltiviruses represent a genus within the *Spinovirinae* subfamily and possess 12 segments of dsRNA. Colorado tick fever is contracted from ticks by campers, hikers, hunters, and forest workers in the Rocky Mountains of North America. Colorado tick fever virus (CTFV) is maintained by the wood tick, *Dermacentor andersoni*, being transmitted transstadially and overwintering in hibernating nymphs and adults. Nymphs feed on small mammals such as squirrels and other rodents, which serve as a reservoir for the virus. Adult ticks feed on larger mammals including humans during the spring and early summer.

After an incubation period of three to six days, the onset of illness is sudden. A characteristic "saddle-back" fever, headache, retroorbital pain, severe myalgia in the back and legs, and leukopenia are the cardinal features; convalescence can be protracted, particularly in adults. More serious forms of the disease, notably meningoencephalitis and hemorrhagic fever, occur in perhaps 5% of cases,

mainly in children. Virus can be isolated from red blood cells or detected inside them by immunofluorescence, even several weeks after symptoms have disappeared. This is a remarkable situation, as erythrocytes have no ribosomes and cannot support viral replication. It seems that the virus multiplies in erythrocyte precursors in bone marrow, then persists in the mature red cells throughout its life span, protected from antibody during a prolonged viremia.

Eyach virus is an antigenically related species in the coltivirus genus, isolated from ticks in France and Germany. Antibodies have been reported in some patients with meningoencephalitis and polyneuritis, but no causal relationship has been established.

ORTHOREOVIRUSES

The first reovirus was isolated in 1951 from the feces of an Australian Aboriginal child, by inoculation of infant mice. This is now recognized as one of three serotypes of the species mammalian orthoreoviruses (MRV) within the genus *Orthoreovirus*, which are not only ubiquitous in humans but also widespread in virtually every species of mammal that has been carefully studied. The genus includes four additional species (avian, baboon, Nelson Bay, and reptilian orthoreoviruses), which represent evolutionarily distinct lineages and are able to reassort 10 dsRNA genome segments within, but not between, species. Orthoreoviruses infect only vertebrates and are spread by respiratory or fecal–oral routes. These viruses replicate well in many types of cell culture. A cytopathic effect (CPE) is slow to develop and not particularly striking; however, on staining, characteristic crescentic perinuclear acidophilic inclusions are seen in the cytoplasm. The three serotypes of MRV share common antigens but can be distinguished by neutralizing antibodies against the outer capsid protein σ1.

Orthoreoviruses have been intensively studied by molecular biologists because the segmented genome lends itself to detailed biochemical and genetic analysis. The position and role of each type of viral protein in the virion, the function of individual gene products in replication, and the molecular pathogenesis of infection in mice have been studied in detail.

Infection of humans is almost invariably inapparent, yet must be very common because the majority of people acquire antibodies to all three serotypes by adulthood. However, a possible link with mild upper respiratory tract symptoms was suggested by some studies; furthermore at least one orthoreovirus (Melaka virus) belonging to the Nelson Bay species (which circulates in fruit bats) has been associated with human respiratory disease. Orthoreoviruses in human infections are shed in feces for up to several weeks, are regularly found in sewage and polluted water, and are assumed to spread mainly by the fecal–oral route but also by the respiratory route. Different orthoreovirus infections of monkeys, mice, and birds may range from inapparent to causing severe disease, depending on the virus strain.

FURTHER READING

Arnold, M.M., Sen, A., Greenberg, H.B., Patton, J.T., 2013. The battle between rotavirus and its host for control of the interferon signaling pathway. PLOS Pathog. 9, e1003064.

Hu, L., Crawford, S.E., Hyser, J.M., Estes, M.K., Prasad, B.V.V., 2012. Rotavirus non-structural proteins: structure and function. Curr. Opin. Virol. 2, 380–388.

Mamelli, C., Fabiano, V., Zucotti, G.V., 2012. New insights into rotavirus vaccines. Hum. Vaccin. Immunother. 8, 1022.

Chapter 25

Orthomyxoviruses

Few viral diseases have played a more central role in the historical development of virology than influenza. The pandemic that swept the world in 1918, just as the First World War ended, was the deadliest single event in recorded human history, killing an estimated 40 to 50 million people—more than the war itself. The eventual isolation of the virus in ferrets in 1933 was a milestone in the development of virology as a laboratory science. Other seminal advances involving influenza include understanding of virus evolution through antigenic shift and drift, and the structure–function correlations of virus envelope peplomers.

Influenza A viruses are arguably the most important of the viral zoonoses. Aquatic birds, especially ducks, shorebirds, and gulls are natural reservoirs; the viruses replicate in avian intestinal epithelium and are excreted in high concentrations in feces, resulting in an efficient fecal–oral transmission pattern. Migrating aquatic birds carry viruses between the continents and thereby play a key role in the continuing process of virus evolution. There are periodic exchanges of viral genes or whole viruses between these reservoirs and other species—domestic fowl, especially ducks, mammals, pigs, and humans. Domestic pigs are important intermediate hosts and China is the epicenter for viral movement from reservoir hosts, passing through intermediate hosts and into humans. There are human epidemics almost every year and pandemics arise whenever a major antigenic variant virus emerges. Much of the excess mortality that occurs each winter, particularly among the elderly, has been shown to be directly or indirectly due to influenza.

Phylogenetic analyses have indicated that influenza viruses have evolved into five host-specific lineages: lineages in pigs and humans are closely related, derived from a common avian ancestor. The dates when some lineages diverged from an ancestral virus can be estimated: the ancestor of the human virus that caused the 1918 pandemic diverged from a classic swine virus between 1905 and 1914. Avian viruses, unlike mammalian viruses, show low evolutionary rates; in fact, viruses in aquatic birds appear to be in evolutionary stasis. Paradoxically, nucleotide changes have continued to occur at a similar rate in avian and mammalian influenza viruses, but these changes no longer result in amino acid changes among the avian viruses. This suggests that avian viruses are approaching or have reached a state of adaptation where further changes provide no selective advantage. Thus the source of genes for future epidemics in humans already exists in the aquatic bird reservoir.

Transmission is by aerosol, droplets, and fomites in humans, and is water-borne among ducks. Thogoto and Dhori viruses are transmitted by ticks and replicate in both ticks and mammals.

PROPERTIES OF ORTHOMYXOVIRUSES

Classification

The family *Orthomyxoviridae* comprises the genera *Influenzavirus A*, *Influenzavirus B*, *Influenzavirus C*, *Thogotovirus*, and *Isavirus*. Influenza A viruses are pathogens of humans, horses, pigs, mink, seals, whales, and fowl. Influenza B viruses are pathogens of humans only, and although influenza C viruses infect humans and pigs, these viruses rarely cause serious disease. The member viruses of the genus *Thogotovirus* are tick-borne arboviruses infecting humans and livestock in Africa, Europe, and Asia. The genus *Isavirus* is named for its type species, infectious salmon anemia virus.

The genus *Influenzavirus A* contains one species *influenza A virus*, which consists of a cluster of strains behaving as a continuous lineage. Classification of influenza strains is influenced considerably by the practical need to assess the risk represented by the emergence of new variant viruses and the question of whether herd immunity against previously circulating strains will dampen spread, and whether existing vaccines will need to be reformulated. The emergence of variant viruses not only depends upon antigenic drift, due to genetic point mutations (nucleotide substitutions, insertions, deletions), but also upon antigenic shift, that is genome segment reassortment. Genetic drift and shift of two genes, the viral hemagglutinin and the neuraminidase genes, are most important. This has prompted a classification system for influenza viruses whereby isolates are placed into genera (and species) A, B, or C, then into 18 hemagglutinin and 11 neuraminidase types, for example, H3N2. Viruses are further categorized into subtypes by host (swine, horses, birds, etc.), geographical origin, strain

Fenner and White's Medical Virology. DOI: http://dx.doi.org/10.1016/B978-0-12-375156-0.00025-4

number, and year of isolation. Recently, further division of highly pathogenic influenza viruses into hemagglutinin clades has been formalized as part of global surveillance for emerging highly pathogenic pandemic variants—the purpose of such surveillance is to allow development of vaccine candidates in preparation for future pandemics.

Thus, the full identification of an influenza virus is codified to be both precise and informative, for example:

1. Influenza virus A/swine/Iowa/15/30 (H1N1), the prototypic strain of swine influenza virus, first isolated in 1930 and considered to be the closest relative to the virus that caused the pandemic of 1918.
2. Influenza virus A/New Jersey/8/76 (H1N1), a prototypic H1N1 influenza virus.
3. Influenza virus A/Japan/305/57 (H2N2), the virus that caused the pandemic of 1957.
4. Influenza virus A/Hong Kong/1/68 (H3N2), the virus that caused the pandemic of 1968.
5. Influenza virus A/Hong Kong/156/97 (H5N1) and influenza virus A/chicken/Hong Kong/258/97 (H5N1), virtually identical viruses isolated from a human and a chicken, causing severe, in some cases fatal, infection in humans in Hong Kong in 1997.
6. Influenza virus B/Lee/40, the prototypic influenza B virus.

Virion Properties

Orthomyxovirus virions are pleomorphic, often ovoid or spherical but predominantly filamentous in fresh isolates, 80 to 120 nm in the smallest dimension (Fig. 25.1). Virions consist of an envelope with large peplomers surrounding eight (genera *Influenzavirus A, Influenzavirus B,* and *Isavirus*) or seven (genus *Influenzavirus C*), or six (genus *Thogotovirus*) helically symmetrical nucleocapsid segments of different sizes. There are two kinds of glycoprotein peplomers: one consists of homotrimers of the hemagglutinin protein, the other homotetramers of the neuraminidase protein (Fig. 25.1). Influenza C viruses have only one type of glycoprotein peplomer, consisting of multifunctional hemagglutinin-esterase molecules (HE). Nucleocapsid segments have a loop at one end and consist of a molecule of viral RNA enclosed within a capsid composed of helically arranged nucleoprotein (NP) molecules. Associated with the RNA are three proteins that make up the viral RNA polymerase (PB1, PB2, and PA). Virion envelopes are lined internally by the matrix protein (M1) and spanned by a small number of ion channels composed of tetramers of a second matrix protein, M2. The genome in eight, seven, or six segments consists of linear negative-sense single-stranded RNA, 10 to 13.6 kb in overall size. The genome segments have terminal repeats at both ends, those on the 3′-ends being identical on all segments.

Influenza viruses are sensitive to heat (56°C, 30 minutes), acid (pH 3), and lipid solvents, and are thus very labile under ambient environmental conditions.

Viral Replication

The viral hemagglutinin is activated in permissive tissues by cleavage into two constituents, HA1 and HA2, linked to each other by disulfide bonds. Virions attach to cells via the binding of activated hemagglutinin to sialic acid receptors on the plasma membrane, with different orthomyxoviruses using sialic acid molecules with different carbohydrate side chains as receptors. Entry is via receptor-mediated endocytosis; transcription complexes (nucleocapsids with associated RNA polymerase) are released into the cytoplasm after fusion between the viral envelope and the endosomal membrane (Fig. 25.2). This fusion is triggered by low pH within the endosome, the result being a further conformational change in hemagglutinin structure. Transcriptional complexes are transported into the nucleus where transcription and RNA replication take place.

As with all other viruses with negative-sense RNA genomes, the genome segments of orthomyxoviruses serve two functions, first as template for the synthesis of mRNAs, second as template for the synthesis of positive-sense replicative intermediate RNA that in turn serves as template for nascent RNA genome synthesis. mRNA transcription and genome replication both take place in the nucleus. Primary transcription involves an unusual phenomenon known as cap-snatching: the viral endonuclease (PB2) cleaves the 5′-methyl-guanosine cap plus about 10 to 13 nucleotides from heterogeneous cellular RNAs. These cap structures are then used by the virus as primers for transcription of viral genes by the viral RNA polymerase (transcriptase; PB1). Among the primary RNA transcripts produced from the eight gene segments of influenza A and B viruses, five are monocistronic and are translated directly. The other three undergo splicing, each yielding two mRNAs which are translated in different reading frames, each producing two proteins; this nuclear-dependent control of viral mRNA synthesis results in the expression of M1 and M2 from segment 7, NS1 and NS2 from segment 8, and PB1 and PB1-F2 from segment 2. Not expressed by all influenza A viruses, PB2-F2 migrates into the cytoplasm and becomes associated with mitochondria, possibly accelerating apoptosis of infected cells.

Influenza B virus uses a different strategy, involving alternative translation start sites and reading frames. Viral mRNAs are 3′-polyadenylated and lack the 5′-terminal 16 nucleotides of the corresponding RNA genome segment.

Viral protein synthesis occurs in the cytoplasm using the cellular translation machinery. Interestingly, the orthomyxoviruses have evolved several mechanisms to increase genome coding capacity: splicing of mRNAs, coupled stop-start translation of tandem genes, and frame shifting. The proteins associated with the virion RNA are transported to the nucleus during the first hours after infection, migrating later to the cytoplasm.

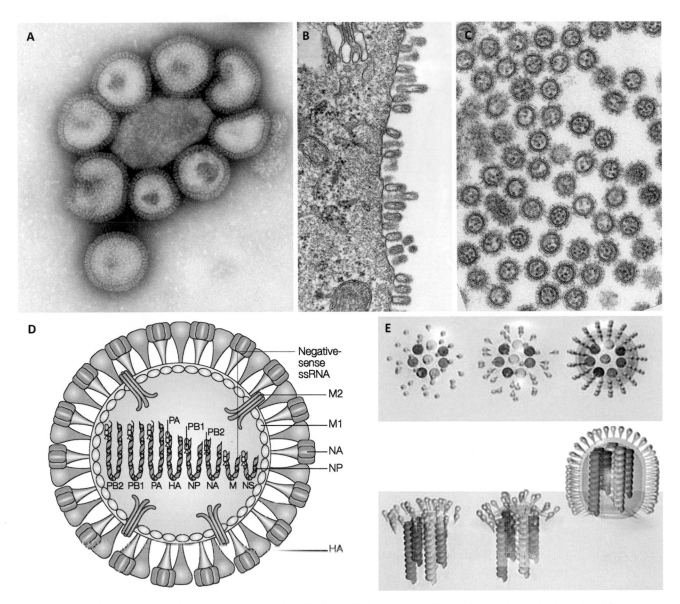

FIGURE 25.1 Influenza virus morphology and structure. In initial isolates of influenza viruses from patient respiratory tract samples virions are most often filamentous, but upon even a single passage in cell culture virions become uniformly "pill-shaped" (short round-ended cylinder shaped) or spherical (A) Negative contrast electron microscopy (A/Hong Kong/1/1968 [H3N2] virus), as would be seen in diagnostic settings. (B and C) Thin-section electron microscopy of influenza virus budding from the plasma membrane of a cell in culture—cross-sectional view and tangential view. (D) Diagram of influenza virus showing envelope and hemagglutinin (HA), neuraminidase (NA) and matrix (M2) projections, as well as the eight nucleocapsid segments (RNA and associated proteins) of influenza A viruses. (E) The highly organized and coordinated influenza virus budding process viewed by a "tomography" diagram (top, cell surface view; bottom, cross-sectional view)—left, the eight nucleocapsid segments are brought together along with associated proteins beneath the plasma membrane of the infected cell; center, viral projection (spikes, peplomers) rafts lead the nascent budding process; right, the budding process is completed by "pinching off" mediated by NA. *(B–E) Reproduced from Yoshihiro Kawaoka: Noda, T., et al., 2006. Architecture of ribonucleoprotein complexes in influenza A virus particles. Nature 439, 490–492, with permission.*

Replication of RNA genome segments requires the synthesis of full-length positive-sense RNA intermediates, which unlike the corresponding mRNA transcripts, must lack 5'-caps and 3'-poly(A) tails. Newly synthesized nucleoprotein binds to these full-length positive-sense RNA templates, facilitating the synthesis of nascent RNA genome segments. Late in infection, the matrix protein M1 enters the nucleus and binds to nascent RNA molecules, thereby down-regulating transcription and permitting export from the nucleus and subsequent assembly of nucleocapsids into virions.

Virions are formed by budding, incorporating M protein and nucleocapsids aligned below patches on the plasma membrane in which hemagglutinin and neuraminidase peplomers have already been inserted. As virions bud, the

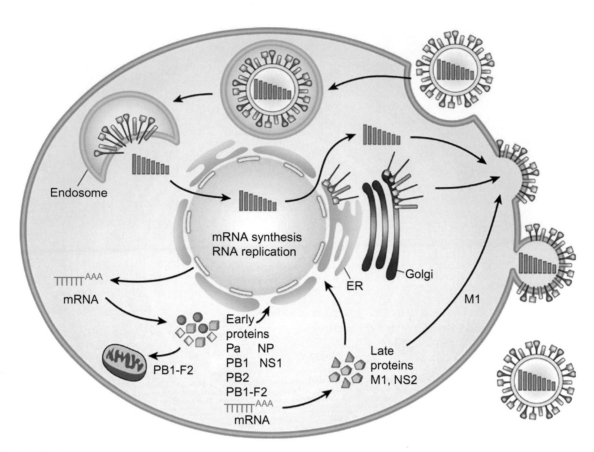

FIGURE 25.2 Replication cycle of influenza virus. ER, endoplasmic reticulum; M1, M2, matrix proteins; NP, nucleoprotein; NS1, NS2, non-structural proteins; Pa, PB1, PB2, proteins of the viral RNA polymerase complex; PB1-F2, non-essential strain-dependent protein linked to virulence. *Adapted from MacLachlan, N.J., Dubrovi, E.J., 2011. Veterinary Virology, Academic Press, p. 358.*

neuraminidase peplomers facilitate the "pinching-off" and release of virions by destroying receptors on the plasma membrane that would otherwise recapture virions and retain progeny virions at the cell surface.

At least one copy of each of the genomic RNA segments must be packaged for a single virion to initiate a productive infection—there is evidence that the vast majority of virions contain no more than eight segments. Until recently, the process by which this was achieved was poorly understood, but a clearer picture has begun to emerge of a mechanism, mediated by *cis*-acting packaging signals in the RNA segments. Packaging is thought to involve a series of protein–protein interactions between the cytoplasmic tails of the viral proteins (Table 25.1).

INFLUENZA

Clinical Features

There is a tendency for patients, and regrettably even some physicians, to label all respiratory ailments as "flu." In reality the syndrome "influenza" is a distinct clinical entity characterized by abrupt onset of fever, sore throat, non-productive cough, myalgia, headache, and malaise. The uncomplicated syndrome resolves over a period of three to seven days although a cough and weakness may sometimes persist for another week or more. This syndrome, however, can be caused by a number of other respiratory viruses, and infection with influenza viruses can cause a spectrum of disease ranging from mild upper respiratory symptoms to pneumonia.

Complications depend upon the age of the patient. Young children may develop croup, pneumonia, or middle ear infection. However, most deaths occur in the elderly and are most frequently attributable to secondary bacterial pneumonia (usually due to *Staphylococcus aureus*, *Streptococcus pneumoniae*, or *Haemophilus influenzae*), and/or to exacerbation of a pre-existing chronic condition, such as obstructive pulmonary disease or congestive cardiac failure.

In most influenza seasons, a majority of influenza-related deaths occur in people over the age of 65, especially those over 75, and those with chronic conditions affecting the pulmonary, cardiac, renal, hepatic, or endocrine systems. Further, the success of modern medicine in keeping alive many children with congenital diseases such as cystic fibrosis and immunodeficiencies, and patients of any age

TABLE 25.1 Properties of Orthomyxoviruses

Five genera: *Influenzavirus A, Influenzavirus B, Influenzavirus C, Thogotovirus,* and *Isavirus*
Virions are pleomorphic, spherical, or filamentous, 80 to 120 nm in diameter
Envelope peplomers consist of rod-shaped hemagglutinin (HA) homotrimers and mushroom-shaped neuraminidase (NA) homotetramers; the envelope is lined by matrix protein (M1) on the inner surface and contains a small number of pores composed of protein M2
The genome consists of linear negative-sense, single-stranded RNA, divided into eight (genera *Influenzavirus A* and *Influenzavirus B*), or seven (genus *Influenzavirus C*), or six (genus *Influenzavirus C*) segments, 10 to 13.6 kb in overall size
Virions contain 8, 7, or 6 separate helical nucleocapsids corresponding to each gene; for influenza A and B, each contains nucleoprotein (NP) and RNA polymerase complex (PA, PB1, PB2) in association with genome
Transcription and RNA replication occur in the nucleus; capped 5'-termini of cellular RNAs are cannibalized as primers for mRNA transcription; budding takes place upon the plasma membrane
Defective interfering particles and genetic reassortment frequently occur

with organ transplants or AIDS, has increased the number of younger people at risk of death during an influenza epidemic. In contrast, the peak age group for influenza-related deaths in the 1918/19 pandemic was in the 20 to 40-year-old age group; many theories for this have been proposed but it remains essentially unexplained (Fig. 25.3).

Pathogenesis

Influenza viruses replicate in epithelial cells of the upper and lower respiratory tract. Infection causes inflammation, and this leads to a serous nasal discharge. The most important changes occur in the lower respiratory tract, however, and include laryngitis, tracheitis, bronchitis, bronchiolitis, and interstitial pneumonia accompanied by congestion and alveolar edema. Myocarditis has been reported rarely. Secondary infections may result in conjunctivitis, pharyngitis, and interstitial pneumonia. Viremia has been reported but is rare, and systemic spread does not occur. The correlation between clinical symptoms and the progressive pathological changes is discussed in Chapter 7: Pathogenesis of Virus Infections.

Factors contributing to innate resistance include: (1) the mucous blanket protecting the underlying epithelium with its continuous beating of cilia that clears virus from the respiratory tract ("the mucociliary escalator"), (2) soluble lectins, lung surfactants, and sialoglycoproteins present in mucus and transudates that bind virions, and (3) alveolar macrophages. If the individual has been infected previously, anti-hemagglutinin antibodies may intercept and neutralize invading virions. Secretory IgA is generally believed to be the most relevant antibody, in the upper respiratory tract at least, but serum IgG may provide protection in the lung especially after infection and the inflammatory response have progressed.

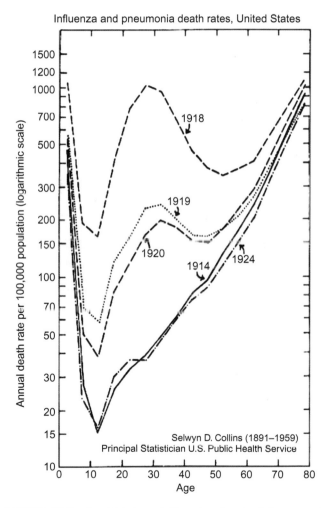

FIGURE 25.3 Age-related mortality rates due to influenza and pneumonia for each of three waves of infection associated with the 1918 influenza pandemic (1918, 1919, and 1920), compared with mortality before (1914) and after (1924) the pandemic. *Reproduced from the US Public Health Service, with permission.*

The study of infection in experimental animals has shown that activated macrophages and natural killer (NK) cells, plus two important cytokines, interferon-γ and interleukin-2, as well as specific CD4 and CD8 T lymphocytes are collectively important in clearing virus from the lower respiratory tract. Class I-restricted CD8 cytotoxic T cells are most effective in ensuring cross-protection when long-lived memory T cells are reactivated by a new infection caused by a different strain of virus. Class II-restricted CD4 T cells are important in secreting cytokines that attract and activate macrophages, NK cells, and other T cells.

Laboratory Diagnosis

Formerly, laboratory diagnosis of influenza relied on virus culture, or on direct antigen detection in exfoliated respiratory cells and nasal secretions. Increasingly nowadays, clinical diagnosis is based upon virus antigen detection using rapid immunochromatographic lateral flow tests delivered at the point of care. A number of these rapid tests have been approved for use in the United States and several European countries. The sensitivity of these tests is less than that of confirmatory tests and false-positive results are also occasionally seen, but these tests are useful for rapid patient triage and informing decisions as to antiviral treatment. Such presumptive tests are backed up by reverse transcriptase polymerase chain reaction (RT-PCR) and/or virus culture for confirmation.

Point-of-care tests are most often done on throat swabs, but are much more reliable using throat washings or nasopharyngeal aspirates. RT-PCR is sufficiently sensitive that any one of these clinical specimens is suitable.

Virus isolation by inoculation of chick embryos has now been largely replaced by virus isolation in the MDCK cell line at 33 to 34°C in the presence of trypsin to cleave the HA of progeny virions and to enable virus spread to adjacent cells. Cytopathology is not usually conspicuous but growth of virus can be recognized after one to ten (typically three to seven) days by hemadsorption, and the isolate identified by immunofluorescence on the fixed monolayer. More rapid confirmatory diagnosis can be achieved by 24- to 48-hour cultivation in MDCK cells, followed by enzyme immunoassay to identify viral antigens in detergent-disrupted cells.

Direct detection of antigen by immunofluorescence or enzyme immunoassay has also been used, with enzyme immunoassay being somewhat more sensitive, quicker, and easier. Monoclonal antibodies enhance both sensitivity and specificity.

Serological diagnosis, by demonstrating a rising hemagglutination inhibition (HI) antibody titer, is less than satisfactory. All but very young patients have previous experience of, and therefore antibody against, one or more strains of influenza virus. Upon subsequent infection with a later strain of the same subtype such individuals develop a memory antibody response directed against earlier strains to which their B cells are primed ("original antigenic sin"), thus complicating the interpretation of HI results.

Epidemiology

Influenza viruses are highly contagious and are spread rapidly by infectious secretions and aerosols generated by frequent sneezing and coughing. Virus is excreted during the incubation period and individuals remain infectious for at least five days after the onset of clinical disease. Close contact seems to be necessary for rapid transmission; however, fomites may also contribute to virus distribution. The rapid international spread of influenza is caused by the year-round movement of people, especially on aircraft.

Since the re-emergence in humans of the H1N1 subtype in 1977, strains of two subtypes of influenza A, H1N1 and H3N2, have co-circulated along with two separate lineages of influenza B. One, or sometimes two strains tend to dominate at any given time in a particular region of the world. In the tropics influenza can occur throughout the year, but in the temperate and colder regions of both hemispheres the disease occurs as winter epidemics. These epidemics can escalate with alarming speed because the incubation period is exceptionally short (one to four days) and very large numbers of virions are shed in droplets. Many symptomatic people together with those who are subclinically infected, remain ambulant in the population, further amplifying spread. Transmission is enhanced if the community lacks immunity to a novel subtype arising by antigenic shift, or displays inadequate immunity to strains arising by antigenic drift.

Schools and day-care centers are principal sites for transmission and children bring the virus home to the family; parents then play a part in disseminating the infection in the workplace. The most reliable indicator of the scale of an epidemic is a sudden leap in the incidence of absenteeism from schools and large places of employment; this is followed by an increase in hospital admissions and deaths, particularly among the elderly. Epidemiologists use an indicator known as "excess deaths" to indicate the increase in mortality during an influenza epidemic. This is measured by comparison with the average number during comparable winters without an influenza epidemic. In each of the 20 influenza epidemics recorded in the United States between 1957 and 1987, there were at least 10,000 (and occasionally up to 50,000) excess deaths as a direct or indirect consequence of influenza. Indeed, influenza constantly ranks among the top ten causes of death in the United States and Europe, both as a direct cause and as a trigger to pneumonia or cardiac deaths.

Only influenza A viruses undergo antigenic shift, roughly every 10 to 40 years, in a manner that is unpredictable. In addition, influenza A and B both undergo antigenic drift. Considerable effort has been expended in trying to devise ways of anticipating influenza epidemics, but this remains a very inexact science. The nature of the next strain of

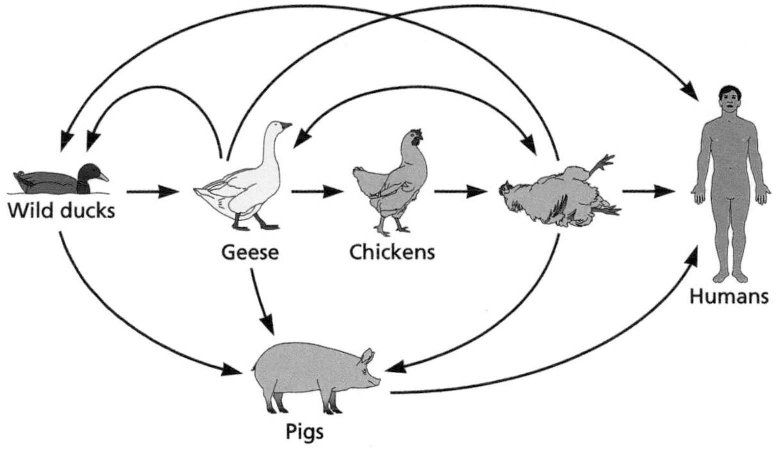

FIGURE 25.4 Emergence and transmission of H5N1 influenza virus. H5N1 has its origins in wild waterfowl, where it was relatively non-pathogenic. Infection was then thought to have spread to domestic ducks and chickens, where it evolved to be highly pathogenic in chickens. It then spread back to domestic ducks and geese, where it re-assorted its genome with those of other influenza viruses of aquatic birds. The resulting new re-assortants spread directly to domestic chickens, humans, and swine. These infections were facilitated by mutations in their PB2, HA, NA, and NS genes that made them more pathogenic to domestic and wild waterfowl and humans. However, to date H5N1 viruses in humans have not acquired properties allowing ready transmission between humans. *Reproduced from Flint, S.J., et al., 2009. Principles of Virology, Vol. II, p. 348. ASM Press, with permission.*

virus cannot be predicted with any certainty, nor when it will emerge, how widely it will spread, and how virulent it will be. In general, however, a novel subtype of influenza A arising by antigenic shift causes a pandemic with high morbidity and significant mortality. Between pandemics a succession of strains arise by antigenic drift and, as time goes on, infect people with partial immunity because of prior infection with another related strain of that subtype. Since the reappearance of the H1 subtype of influenza A viruses in 1977, most people old enough to have experienced that subtype before it disappeared in the 1950s were found to display a degree of immunity to contemporary H1 strains. Influenza B infections can be just as severe as influenza A infections, but because they do not undergo antigenic shift these are not associated with pandemics on the same scale.

The natural reservoir of influenza A viruses is a diverse genetic pool distributed through numerous species of aquatic birds (Fig. 25.4). The highly pathogenic avian H5N1 influenza virus which appeared in humans in 1996 illustrates the role of virus receptor usage in affecting tropism. Asymptomatically infected birds are thought to have transmitted the virus to domestic poultry from whence humans became infected through close contact. Avian intestinal mucosal epithelial cells have an abundance of α2-3-linked sialic acids on their apical surfaces, and avian strains have a preference for this specific receptor. In contrast, influenza isolates from humans show enhanced binding to α2-6-linked sialic acids. In the human respiratory tract, α2-6-linked sialic acids predominate in the upper respiratory tract whereas α2-3-linked sialic acids are found mainly in the lower tract. These observations provide a rational explanation for the lack of significant human-to-human

transmission of avian H5N1 virus, but may account for the severe disease and mortality associated with this virus in individual cases where the lower respiratory tract becomes infected. It is thought that the latter may happen on repeated exposure to H5N1 virus. By 2013 H5N1 virus had accounted for over 620 reported human cases, including 374 deaths. The constant interchange of influenza strains between wild birds and poultry provides an almost limitless opportunity for genetic exchange and mutations, increasing further the risk of cross-transmission to humans as a result.

In the spring of 2009 the number of human cases of influenza began to escalate in the northern hemisphere at a time when normally the influenza season ends. Beginning first in Mexico, this pandemic spread rapidly to all corners of the world within a few weeks. The causative virus was rapidly identified as an H1N1 virus antigenically related to viruses normally responsible for swine influenza. This swine-origin virus (now known as 2009 pandemic influenza A(H1N1)) represented an unusual triple reassortment containing RNA genome segments from swine, human, and avian sources (Fig. 25.5). This pandemic caused notable concern because the transmissibility of the virus seemed to be greater than that of typical seasonal influenza virus strains; however, in contrast to initial reports it turned out to cause only moderately severe illness in the majority of all population groups.

Treatment

Several antivirals are available for treating influenza virus infections. These fall into two broad categories as to mode of action: first, compounds that interfere with the function of the M2 ion channel through the viral envelope (amantadine,

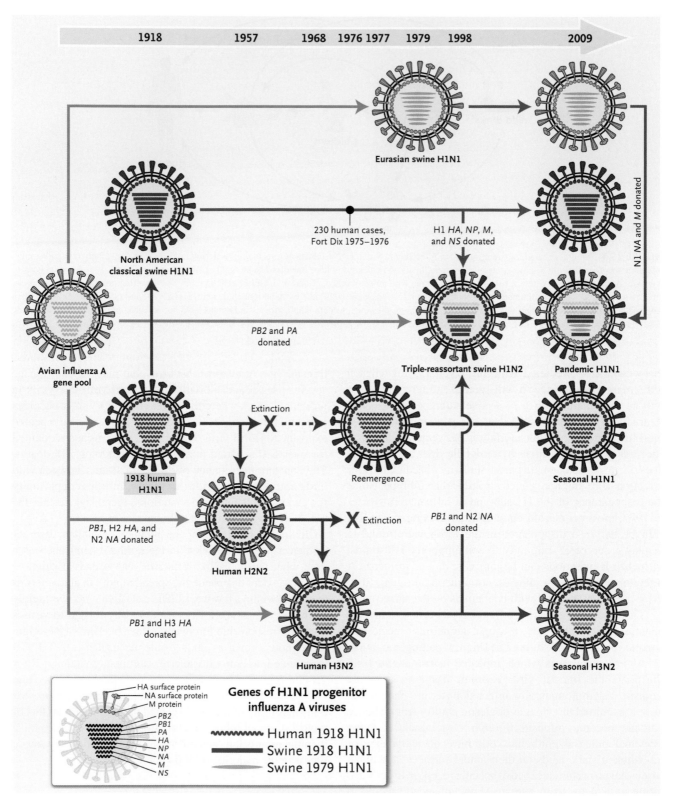

1918   1957   1968   1976 1977   1979   1998   2009

Eurasian swine H1N1

North American classical swine H1N1

230 human cases, Fort Dix 1975–1976

H1 *HA*, *NP*, *M*, and *NS* donated

Avian influenza A gene pool

PB2 and *PA* donated

N1 *NA* and *M* donated

Triple-reassortant swine H1N2   **Pandemic H1N1**

1918 human H1N1

Extinction

Reemergence

PB1, H2 *HA*, and N2 *NA* donated

Seasonal H1N1

Human H2N2

X Extinction

PB1 and N2 *NA* donated

PB1 and H3 *HA* donated

Human H3N2

Seasonal H3N2

HA surface protein
NA surface protein
M protein
PB2
PB1
PA
HA
NP
NA
M
NS

Genes of H1N1 progenitor influenza A viruses

〜〜〜〜 Human 1918 H1N1
━━━━ Swine 1918 H1N1
━━━━ Swine 1979 H1N1

FIGURE 25.5   Genetic relationships and paths of emergence of human and relevant swine influenza viruses, 1918–2009, from the ever-present global gene pool in wild birds, to the 1918 H1N1 pandemic strain, to the 1957 H2N2 "Asian" pandemic strain, to the 1968 H3N2 "Hong Kong" pandemic strain, to the 2009 H1N1 "swine influenza" strain. Green arrows reflect donation of one or more genes from the avian influenza gene pool. The dashed red arrow indicates a period without known circulation. Solid red arrows indicate the evolutionary paths of human influenza virus lineages; solid blue arrows, of swine influenza virus lineages; and the blue-to-red arrow, of a swine-origin human influenza virus. The genes of the 1918 human and swine H1N1 and the 1979 H1N1 influenza A viruses were all descended from avian influenza A genes shortly before each virus emerged – some of these viruses then donated genes to the 2009 H1N1 pandemic strain. *Reproduced rom David Morens, Jeffrey Taubenberger and Anthony Fauci, U.S. National Institute of Allergy and Infectious Diseases, with permission.*

rimantadine), and second, inhibitors of neuraminidase function (oseltamivir and zanamivir).

Amantadine and rimantidine are synthetic primary amines with a long history of use for the treatment of influenza. Both bind to amino acid side chains lining the M2 ion channel through the viral membrane, thereby increasing the pH within endosomes and reducing the efficiency of virus uncoating. Unfortunately resistance frequently develops to the extent that viruses such as the highly pathogenic H5N1 virus that has been active in Asia in recent years are almost totally resistant, and the drugs are of no value against influenza B viruses. Amantadine is readily secreted in the urine whereas rimantadine is metabolized in the liver, making the latter preferable for use, especially in elderly patients. Both consistently reduce the severity of symptoms as well as the duration of illness, but they have a narrow dosage window between efficacy and doses leading to side effects.

Oseltamivir (Tamiflu) and zanamivir (Relenza) are licenced in many countries for both treatment and prophylaxis of influenza A and B in adults and children, and are now the drugs of choice. Both drugs inhibit the release of newly made virus particles from the infected cell. Clinical trials with either antiviral have shown an approximate 70% efficacy in reducing the severity of symptoms, provided they are administered within 48 hours of the onset of illness. The duration of the acute infection is reduced by two to three days. Although this is an apparently modest reduction in the duration of illness, some evidence indicates that the risk of later stage pneumonia is also reduced. An important observation is that the use of neuraminidase inhibitors does not reduce the effectiveness of any immune response induced by influenza vaccines.

Oseltamivir is delivered orally, whereas zanamivir is administered as a nasal spray. The occasional side effects of nausea and vomiting associated with oseltamivir are much reduced through the use of zanamivir.

Immunization

No disease illustrates better than influenza the difficulties of control by immunization and the ingenuity required to elevate the level of herd immunity. Existing vaccines are continually being rendered obsolete by antigenic shift and drift, and the composition of the vaccine needs to be modified almost every year to match the current strains of circulating virus. The World Health Organization, through its extensive network of laboratories around the globe, is constantly on the alert for such changes, and supplies the vaccine manufacturers with seed stocks of the latest strain(s) of influenza so that their product may be updated if necessary prior to each influenza season.

In the usual process, the current strains of influenza A(H1N1), A(H3N2) and B are grown separately in the allantois of chick embryos, inactivated with an appropriate chemical such as β-propiolactone, then purified by zonal ultracentrifugation, disrupted with detergent, and pooled. The resulting polyvalent inactivated vaccine is administered each autumn. Recently, vaccines consisting of virus grown in cell cultures (MDCK cells) have also been approved in the United States and Europe; this approach is expected to avoid problems in recipients who are allergic to eggs, and also to speed up vaccine production. In some countries it is considered cost-effective to recommend routine annual vaccination to all persons aged over six months who do not have contraindications, with a view to limiting the circulation of virus in the community thereby protecting the whole community. An alternative strategy is to immunize only the most vulnerable cohorts, namely the elderly above the age of 65 years, residents of nursing homes and other chronic care facilities, and those of any age with chronic debilitating disease of the pulmonary, cardiovascular, renal, or endocrine systems (e.g., asthma, emphysema, chronic bronchitis, cystic fibrosis, diabetes, etc.). Immunization is also recommended for those with compromised immune function. In the worst epidemic years, immunization may be extended to medical personnel and others providing vital community services, as well as close relatives and home-care personnel attending high-risk invalids.

The commonest side effect is a mild local reaction: some tenderness, redness, and swelling occur around the injection site in about 15% of recipients. Less frequently fever, malaise, and myalgia may develop within hours and disappear a day later. The only major contraindication for influenza vaccine is a known allergy for eggs; though rare, inoculation of such persons can produce an immediate allergic reaction.

A particularly serious problem, Guillain-Barré syndrome, was encountered in one in every hundred thousand Americans vaccinated against influenza A/New Jersey/76 (H1N1), during a mass campaign in 1976–77 to protect the population against an outbreak of "swine flu" which in the event did not spread widely, but such an association has not been reported since (see Chapter 11: Vaccines and Vaccination).

In 2010, a higher rate of febrile convulsions (up to 1%) was seen in children under five years of age who received an inactivated seasonal influenza vaccine obtained from one particular manufacturer. The reasons for this are not clear, and thus caution is advised in selecting inactivated vaccines for use in children less than ten years of age. This experience reinforces the need for continuing monitoring of all vaccination programs.

Efficacy is greatest amongst the young and least in the old. There are probably two reasons for this difference. First, immune responsiveness declines with age; second, "original antigenic sin" tends to divert the response to influenza vaccines in the elderly, which is unfortunate as this cohort constitutes the principal target group for annual vaccination campaigns. However, even though antibody titer rises following vaccination in the elderly are often disappointing and there is generally only a 30 to 70% reduction in the incidence of influenza, there is a 60 to 90% reduction in pneumonia, hospitalization, and mortality, providing that there is a good match between the vaccine strain and the circulating strains of virus. Higher dose vaccines are recommended for those aged over 65 years. Current inactivated influenza vaccines, while less than perfect, offer a worthwhile degree of protection.

Live attenuated vaccines administered intranasally are an alternative approach. These have theoretical advantages of producing a mucosal IgA, IgG, and T cell memory response of broad cross-reactivity. Such vaccines have been used in Russia for many years, and current preparations are now gaining more widespread use. For example, a quadrivalent live attenuated influenza vaccine (LAIV4, FluMist) uses a parent influenza virus strain that is attenuated, cold-adapted (for optimal growth at 25°C), and temperature-sensitive (unable to grow at 37°C); this is tailored by genetic reassortment to also contain HA and NA genes corresponding to the current circulating strains. Administered intranasally, the vaccine strains replicate in the nasopharynx for up to 28 days; although virus shedding can occur, no evidence for virus reversion or illness in others has been found. Because this strategy has been shown to give less protection in older persons, it is currently recommended for those aged 2 to 49 years. There is evidence that live attenuated influenza vaccines are more efficacious than inactivated vaccines against laboratory-confirmed influenza in children aged from 6 to 72 months. However, recent concerns that current formulations were less effective have led the US Centers for Disease Control and Prevention to withdraw their endorsement of a live attenuated vaccine for 2016–2017. The commonest side effects of these vaccines are mild upper respiratory tract symptoms.

Other studies are seeking to improve non-replicating vaccines. Some of these are directed at increasing the immunogenicity of inactivated virions, or of solubilized or genetically cloned hemagglutinin (with or without neuraminidase), for example by addition of adjuvants, coupling to carriers, or incorporation into "liposomes," "virosomes," or "iscoms" (see Chapter 11: Vaccines and Vaccination).

Prevention and Control

Surveillance of influenza activity occurs at many levels, including regular reporting by "sentinel" GP practices of "influenza-like" illnesses; monitoring of absenteeism by large employers; monitoring of hospital admission rates and attendance at casualty departments; laboratory diagnosis rates; and mortality data. In addition, routine virus isolates are regularly submitted by local laboratories to a network of WHO Influenza Collaborating Centres for characterization and comparison with previous strains.

When a new outbreak or epidemic is detected in a local area, its pandemic potential needs to be assessed, based on

BOX 25.1 Lessons from the 1918/19 Pandemic

- The great influenza pandemic spread progressively to almost all parts of the world within 12 months, aided by the troop movements and the social disruption due to the First World War.
- The mortality ranged from 5% to 40% of clinical cases, with a major peak in the 20- to 40-year-old age group, unlike other pandemics before or since. Reasons for this are not fully understood.
- Death occurred through one of three syndromes—a very rapid viral pneumonia similar to respiratory distress syndrome, a necrotizing/hemorrhagic viral bronchopneumonia, or a later bacterial bronchopneumonia as a result of superinfection.
- At least three successive waves of infection swept parts of the world between July 1918 and late 1919.
- The measures adopted to try to contain the spread of the virus, and the community response to the pandemic, were based on contemporary knowledge, but contain lessons about the implementation of control measures and the management of community responses that are relevant today (Fig. 25.6).
- The 1918 virus has recently been recovered from RNA fragments still present in frozen cadavers and pathology museum specimens dating from that time. Viable virus has been assembled from eight different RNAs copied from cDNAs.

This virus, handled under very high containment, has been shown to be highly virulent for mice, ferrets, and macaques; this property is derived from the combined effect of a number of genes, but the HA and polymerase genes are particularly important determinants.

evidence about its ease of human-to-human spread, case fatality, and the molecular and antigenic relationship of the virus to previous strains. Control measures that may need to be instituted include quarantining of cases and contacts coupled with contact tracing; chemoprophylaxis for high-risk individuals using Tamiflu or Relenza; and sometimes rapid initiation of the production of a new vaccine against the new strain. Because the possibility of another pandemic similar to that in 1918/19 is ever-present (Box 25.1), these decisions can be agonizing. For example, the first reports of the new H1N1 swine-related strain in 2009 in Mexico suggested that it had high virulence, leading to national stockpiling of antivirals and fast-tracking of large supplies of new vaccine. These measures subsequently turned out to be largely unnecessary in the light of further experience of the low virulence of the pandemic, yet if they had not been instituted to counter a true lethal pandemic, the world would have paid a high price.

Similarly, after an extensive influenza epidemic in chickens in China in 1997, a number of human cases occurred in Hong Kong of infection with H5N1 influenza, a subtype previously known to occur only in birds. Many of the human cases were severe and the mortality was around 30%.

FIGURE 25.6 Measures to manage the 1918 influenza pandemic. (A, B) Posters containing public warnings. (C) Mass graves in Philadelphia. *Reproduced from Barry, J.M., 2004. The Great Influenza, Penguin Books.*

Investigations also showed that infection was widespread among different species of birds in the region, that there was some human-to-human transmission, but most human infections occurred through contact with infected chickens. Surveillance in the area was increased, all 1.2 million chickens in Hong Kong were slaughtered, premises were disinfected, and strategies were developed for production of a human vaccine. The hemagglutinin of the virus was found to contain multiple basic amino acids adjacent to its cleavage site and an insert at the same site, a feature characteristic of highly pathogenic avian influenza A viruses; experimentally, the virus was found to be highly pathogenic for chickens. The fact that the virus remained lethal to chickens even after passage through a human raised the possibility that a few infected people traveling beyond Hong Kong could spread this virus to chickens in other countries—intense surveillance of poultry flocks at possible risk was quickly initiated.

FURTHER READING

Barry, J.M., 2004. The Great Influenza. Penguin Books..

Belshe, R.B., 2005. The origins of pandemic influenza: lessons from the 1918 virus. N. Engl. J. Med. 353, 2009–2011.

Centers for Disease Control, 2014. Prevention and control of seasonal influenza with vaccines: recommendations of the Advisory Committee on Immunization Practices. Morb. Mortal. Wkly. Rep. 63, 691–697.

Garten, R.J., Davis, C.T., Russell, C.A., et al., 2009. Antigenic and genetic characteristics of swine origin 2009 (H1N1) influenza viruses circulating in humans. Science 325, 197–201.

Hulse-Post, D.J., Sturm-Ramirez, K.M., Humberd, J., et al., 2005. Role of domestic ducks in the propagation and biological evolution of highly pathogenic H5N1 influenza viruses in Asia. Proc. Natl. Acad. Sci. U.S.A. 102, 10682–10687.

Knossow, M., Skehel, J.J., 2006. Variation and infectivity neutralization in influenza. Immunology 119, 1–7.

Kobasa, D., Takada, A., Shinya, K., et al., 2004. Enhanced virulence of influenza A viruses with the haemagglutinin of the 1918 pandemic virus. Nature 431, 703–707.

Olsen, B., Munster, V.J., Walenstein, A., et al., 2006. Global patterns of influenza A virus in wild birds. Science 312, 384–388.

Subbarao, K., Shaw, M.W., 2000. Molecular aspects of avian influenza (H5N1) viruses isolated from humans. Rev. Med. Virol. 10, 387–348.

Taubenberger, J.K., Baltimore, D., Doherty, P.C., Markel, H., Morens, D.M., Webster, R.G., et al., 2012. Reconstruction of the 1918 influenza virus: unexpected rewards from the past. mBio 3/5, e00201–e00212.

Trifonov, V., Khiabanian, H., Rabadan, R., 2009. Geographic dependence, surveillance and origins of the 2009 influenza A (H1N1) virus. N. Engl. J. Med. 361, 115–119.

Chapter 26

Paramyxoviruses

The family *Paramyxoviridae* is included in the order *Mononegavirales*, along with the families *Rhabdoviridae*, *Filoviridae*, and *Bornaviridae*. This order was established to bring together viruses with distant, ancient phylogenetic relationships that are also reflected in similarities in both gene order and strategies for gene expression and replication. All these viruses are enveloped and, other than bornaviruses, have prominent envelope glycoprotein spikes. All viruses included in the order have genomes consisting of a single molecule of negative-sense, single-stranded RNA. The features that differentiate the individual families of the order include genome size, nucleocapsid structure, site of genome replication and transcription, manner and extent of messenger RNA (mRNA) processing, virion size and morphology, tissue specificity, host range, and the nature of the disease(s) caused in the respective hosts.

Several of the most important viral diseases of humans (and animals) are caused by members of the family *Paramyxoviridae* (Table 26.1). In particular, the viruses causing measles, mumps, respiratory syncytial virus infection, together with parainfluenza and metapneumovirus infection have arguably caused more morbidity than any other single group of related viruses in history. Today the lethal impact of most of these pathogens has been dramatically reduced through the use of vaccines. Yet acute respiratory infection due to respiratory syncytial virus (RSV) remains the leading cause of mortality in young children under 5 years of age, accounting for nearly one-fifth of childhood deaths worldwide. Human parainfluenza viruses (HPIVs) types 1, 2, 3, and 4, human metapneumovirus, and RSV cause about half of childhood lower respiratory tract disease (croup, bronchiolitis, pneumonia) in developed countries.

The animal pathogens in the family *Paramyxoviridae* also represent some of the most devastating causes of human suffering. Rinderpest virus, eradicated globally in 2008, affected human development across Africa and Eurasia. Respiratory pathogens affected and still affect livestock (cattle, sheep, goats, horses) industries and avian viruses substantially affect poultry industries globally. One school of thought is that measles and other human paramyxoviruses share a common ancestor to paramyxoviruses of livestock, human paramyxoviruses evolving closely with the development of agriculture some 10,000 years ago.

Of recent concern are the viruses of the genus *Henipavirus*, Hendra and Nipah viruses; both naturally infect various species of fruit bats, but cause high mortality rates in infected humans and some domestic animals. As wildlife species come more in contact with humans and domesticated animals through changes in habitat resulting from human incursion, the opportunities increase for cross-species infections by other, as yet unidentified, zoonotic paramyxoviruses.

PROPERTIES OF PARAMYXOVIRUSES

Classification

Within the order *Mononegavirales*, family *Paramyxoviridae* is subdivided into two subfamilies, the *Paramyxovirinae* and *Pneumovirinae*. The subfamily *Paramyxovirinae* is divided into five genera: *Respirovirus*, *Rubulavirus*, *Avulavirus*, *Morbillivirus*, and *Henipavirus*, and the subfamily *Pneumovirinae* into two genera, *Pneumovirus* and *Metapneumovirus*. Which human pathogen is placed in which taxon is best described by the table (Table 26.1). How the family and genera fit into the order *Monanegavirales* is best described by the graphic phylogenetic tree (Fig. 26.1).

Virion Properties

Member viruses of the family *Paramyxoviridae* are enveloped, 150 to 300 nm in diameter, ranging in shape from spherical to almost filamentous. The envelope contains two major glycoproteins present as 8 to 14 nm spike-like projections: the hemagglutinin-neuraminidase (HN) and fusion (F) proteins, respectively. The envelope encloses a helical nucleocapsid structure containing the viral RNA genome and associated proteins, including the nucleocapsid (N), phosphoprotein (P), and large (L; RNA polymerase) proteins. The matrix (M) protein, the most abundant protein in the virion, assembles between the viral envelope and the

Fenner and White's Medical Virology. DOI: http://dx.doi.org/10.1016/B978-0-12-375156-0.00026-6

TABLE 26.1 Human Paramyxoviruses

Subfamily	Genus	Virus
Paramyxovirinae	*Rubulavirus*	Human parainfluenza viruses 2 and 4, mumps virus
	Respirovirus	Human parainfluenza viruses 1 and 3
	Morbillivirus	Measles virus
	Henipavirus	Hendra virus, Nipah virus
Pneumovirinae	*Pneumovirus*	Human respiratory syncytial virus
	Metapneumovirus	Human metapneumovirus

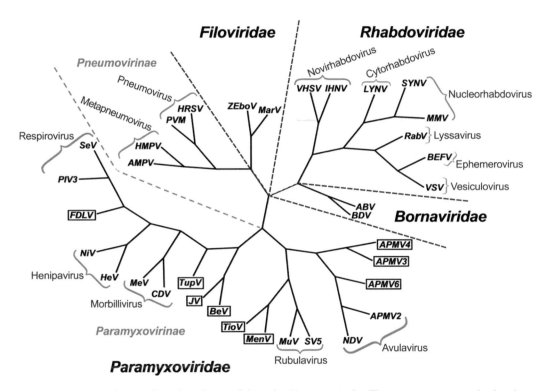

FIGURE 26.1 Unrooted phylogenetic tree of member viruses of the order *Mononegavirales*. The tree was constructed using the sequences of the conserved domain of the polymerase protein. Human pathogens are underlined; nine paramyxoviruses, not yet assigned to genera, have been included (boxed). Genera are shown in red and sub-families in green. Abbreviations: ABV, avian bornavirus; AMPV, avian metapneumovirus; APMV2, avian parainfluenza virus types 2; BDV, Borna disease virus; BEFV, bovine ephemeral fever virus; CDV, canine distemper virus; HeV, hendra virus; HMPV, human metapneumovirus; HRSV, human respiratory syncytial virus; IHNV, infectious hemorrhaghic necrosis virus; LNYV, lettuce necrotic yellows virus; MarV, Marburg virus; MeV, measles virus; MMV, maize mosaic virus; MuV, mumps virus; NDV, Newcastle disease virus; NiV, Nipah virus; PIV3, parainfluenza virus type 3; PVM, pneumonia virus of mice; RabV, rabies virus; SeV, Sendai virus; SV5, simian virus 5; SYNV, Sonchus yellow net virus; VSV, vesicular stomatitis Indiana virus; VHSV, viral hemorrhagic septicemia virus; ZeboV, Zaire ebolavirus. *Reproduced from King, A.M.Q., et al., 2012. Virus Taxonomy. In: Ninth Report of the International Committee on Taxonomy of Viruses, Academic Press, p. 657, with permission.*

nucleocapsid; it organizes and maintains virion structure and is involved in virus entry and budding (Table 26.2; Fig. 26.2).

The genome consists of a single linear molecule of negative-sense single-stranded RNA 15 to 19 kb in size. There are six to ten genes encoding the viral proteins,

these separated by conserved non-coding sequences with termination, polyadenylation, and initiation signals. The gene order is generally consistent within the family: $3' \leftarrow$ N-P-M-F-HN-L$\rightarrow 5'$. Some members, for example, RSV, encode a number of additional proteins (Fig. 26.3). Members of the subfamily *Paramyxovirinae* possess a P gene of remarkable

TABLE 26.2 Properties of the Family *Paramyxoviridae*

Two subfamilies: *Paramyxovirinae*, containing the genera *Respirovirus*, *Rubulavirus*, *Henipavirus*, and *Morbillivirus*: and *Pneumovirinae*, containing the genera *Pneumovirus* and *Metapneumovirus*

Virions are enveloped, 150 to 300 nm in diameter, pleomorphic in shape with spherical and filamentous forms

Virion envelope contains two viral glycoproteins 8 to 14 nm in length, the hemagglutinin-neuraminidase (HN) and the fusion (F) proteins and one or two non-glycosylated proteins

Matrix protein (M) lines the inner aspect of the viral envelope

Virions contain a helically symmetrical nucleocapsid, 600 to 800 nm in length, consisting of the genome and proteins N (nucleocapsid), P (phosphoprotein), and L (RNA polymerase)

Genome consists of a single linear molecule of negative sense, single-stranded RNA, 13 to 19 kb in size, with seven to eight open-reading frames coding for 10 to 12 proteins, including N, P, M, F, HN (or H or G), and L, common to all genera

Cytoplasmic replication, commences with mRNA transcription by the virion-associated L polymerase, using the -ve sense virion RNA as template

Progeny viruses bud from the plasma membrane

flexibility in terms of maximizing coding potential, expressing from two to five functionally distinct proteins. This is achieved through "RNA editing," a process whereby the transcription of the P gene allows insertion of one or two additional G residues at specific points along the transcript in a fraction of the RNA molecules; this alters the translational reading frame, and thus allows the same genomic RNA sequence to code for more than one distinct proteins, e.g., P and V proteins possessing identical N-terminal domains but different C-terminal domains due to the frame-shift.

A secondary mechanism involves ribosome initiation at alternative translation codons in order to produce distinct proteins from the same transcript. P gene products also control the expression and replication of the negative-sense viral RNA.

The hemagglutinin-neuraminidase (HN) and fusion (F) proteins play a key role in the pathogenesis of all paramyxovirus infections. Cell attachment is mediated via the HN proteins. This glycoprotein elicits neutralizing antibodies that inhibit adsorption of virus to cellular receptors. The fusion proteins are present on newly formed virions in an inactive precursor form and this is subsequently activated following proteolytic cleavage by a cell surface protease. Once activated, the newly generated amino terminal sequence of the protein has a hydrophobic domain and it is postulated as directly involved in fusion. Cleavage of the fusion protein is essential for viral infectivity: when a host cell does not contain the appropriate proteases, nascent virus is not infectious. Furthermore, viral virulence has been correlated with the presence or absence of specific amino acid motifs recognized by host proteases. Since the fusion protein is essential for viral infectivity and for direct cell-to-cell spread via fusion, it plays a pivotal role in disease development, including persistent infections (Table 26.3).

Paramyxoviruses have narrow host ranges and have been identified mostly in vertebrates, primarily mammals, birds, and reptiles. All are sensitive to heat or desiccation and are readily inactivated by lipid solvents and common disinfectants.

Viral Replication

Paramyxoviruses replicate within the cytoplasm of infected cells. Virions attach via the HN protein to cellular sialoglycoproteins or glycolipid receptors. The F protein then mediates fusion of the viral envelope with the plasma membranes at physiological pH. The liberated nucleocapsid remains intact, and all three associated proteins (N, P, and L) are required for the transcription of the incoming viral genome. This nucleocapsid structure thus acts as a template for the initiation of infection through a process of transcription of viral mRNA from the negative-sense viral RNA.

During the early stages of the replication cycle, the viral genome directs the synthesis of a (+) sense leader and six to ten discrete, unprocessed mRNA species. This transcription is initiated from a single promoter site at the 3′ end of the RNA genome, and proceeds by an interrupted termination-reinitiation process. The amount of each mRNA transcribed decreases progressively with increasing distance from the 3′ end of the template, in the order N>P>M>F>HN>L. The mRNAs are capped and polyadenylated. When the concentration of N protein reaches a critical level it binds to the nascent (+) sense antigenome intermediates, with the result that the polyadenylation, termination, and reinitiation of full-length RNA molecules is prevented and transcription then switches to the replication of full-length genome intermediates; these in turn serve as templates for the production of nascent full-length (−) sense viral genomes.

Newly synthesized RNA associates with the N proteins and L protein to form helical nucleocapsids. Virus maturation involves four stages: (1) the incorporation of viral glycoproteins into patches (also called rafts) on the infected cell plasma membrane, (2) the aligning of matrix protein (M) and other non-glycosylated proteins under these patches, (3) the alignment of nucleocapsids with these patches, and (4) the formation and release of virus particles via budding through the plasma membrane.

To control the complexity of paramyxovirus genome expression and replication, a number of viral accessory proteins are produced but these are not essential for virus replication in cell culture.

Several different cell types are used to grow different paramyxoviruses. Cell cultures derived from homologous

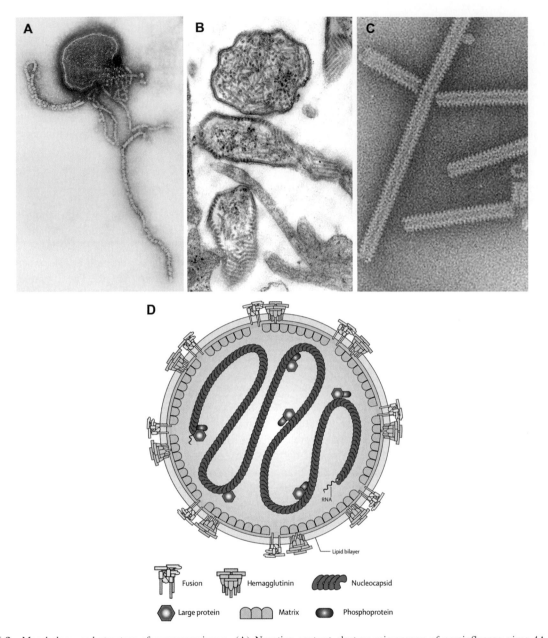

FIGURE 26.2 Morphology and structure of paramyxoviruses. (A) Negative contrast electron microscopy of parainfluenza virus 4A, showing a ruptured virion with spilling nucleocapsid. Projection (spike, peplomer) glycoproteins are densely packed on the virion surface. (B) Thin-section electron microscopy of mumps virus in Vero cell culture, showing massive amounts of nucleocapsid in virions of various sizes and shapes. Paramyxoviruses are pleomorphic (ranging in size from 110 to 540 nm) and virions encapsidate multiple genome copies – studies have shown that virions may contain up to six genome copies. In some instances, there is no discernible order to nucleocapsid packaging (virion at top), but in others nucleocapsids appear lined up just under the viral envelope (virion at center). (C) At high magnification the characteristic herringbone arrangement of N protein molecules on the genomic RNA of measles virus is clear. (D) Schematic diagram of measles virus, showing the hemagglutinin-neuraminidase, fusion protein, nucleocapsid protein, large protein (the viral polymerase), matrix protein and a phosphoprotein. *(C) Reproduced from Desfosses, A., et al., 2011. Nucleoprotein-RNA orientation in the measles virus nucleocapsid by three-dimensional electron microscopy. J. Virol. 85, 1391–1395, with permission. (D) Reproduced from Moss, W.J., Griffin, D.E., 2012. Measles. Lancet 379, 153–164, with permission.*

species are usually used for morbilliviruses and pneumoviruses, but these viruses are not readily cultivated and adaptation by passage is usually required. Viral replication in cell cultures is usually lytic, although persistent infections in cell culture can be generated. Formation of syncytial cells in culture is frequently observed, and when seen in vivo with acidophilic cytoplasmic inclusions, are characteristic of human paramyxovirus infections.

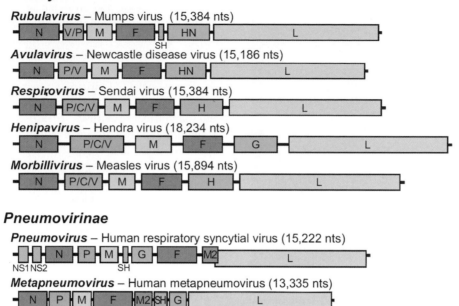

FIGURE 26.3 Maps of genomic RNAs (3′–5′) of viruses in the family *Paramyxoviridae*. Each box represents a separately encoded mRNA. Multiple distinct ORFs within a single mRNA are indicated by slashes. The lengths of the different boxes are approximately to scale, but the intervening or preceding spaces are not to scale. For further details see text and ICTV report. *Reproduced from King, A.M.Q., et al., 2012. Virus Taxonomy. In: Ninth Report of the International Committee on Taxonomy of Viruses, Academic Press, p. 675, with permission.*

TABLE 26.3 Viral Proteins—Functions and Terminology

Genus *Rubulavirus*	Genus *Morbillivirus and Respirovirus*	Genus *Pneumovirus*	Protein Name: Function
HN	H	Gª	H: Hemagglutinin, attachment protein, induction of neutralizing antibodies / N: Neuraminidase, virus release from cell, destruction of mucin inhibitors
F	F	I	Fusion protein: Cell fusion, virus entry, cell–cell spread, protective immunity
N	N	N	Nucleoprotein: Protection of RNA genome, regulation of RNA synthesis
L and P/C/V	L and P/C/V	L and P	Transcriptase: RNA-dependent RNA transcription
M	M	M	Matrix protein: Virion stability
(SH)		SH, M2	

ªNo hemagglutinating activity.

RESPIRATORY SYNCYTIAL VIRUS INFECTION

Respiratory syncytial virus (RSV) is the most important respiratory pathogen of childhood, accounting for a quarter of all cases of pneumonia during the first few months of life. Immunity following infection is not robust, and reinfections occur regularly during life, usually accompanied by less severe disease. Despite its large health impact and economic burden, and considerable research, neither antiviral treatment for severe cases, nor a successful licensed vaccine to prevent infection, is available.

Clinical Features

The commonest presentation of respiratory syncytial virus infection in all age groups is a febrile rhinitis or pharyngitis, or both, with limited involvement of the bronchi. However, the distinctive more severe manifestation of respiratory syncytial virus is bronchiolitis, which can progress to pneumonia particularly in infants experiencing primary infection (Fig. 26.4). Almost 1% of all babies develop a respiratory syncytial virus infection severe enough to require admission to hospital and of these about 1% die, particularly those with congenital heart defects, bronchopulmonary

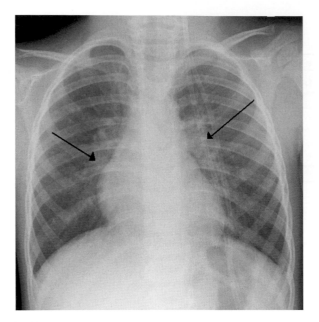

FIGURE 26.4 Lung X-ray of a child with RSV infection, showing mild perihilar infiltrates (arrows). The lung is hyperinflated due to partial obstruction of the bronchioles. *From James Heilman.*

dysplasia, very low birth weight, or immunodeficiency. Otitis media is a common complication among older children. Reinfections are common, usually decreasing in severity and presenting as rhinorrhea, sore throat, hoarseness, sinusitis, and exacerbations of asthma. In the elderly, the presentation is similar to influenza, but severe pneumonia can occur during winter epidemics of respiratory syncytial virus, particularly in immunosuppressed transplant patients.

While all strains of respiratory syncytial virus are considered to comprise a single species, strains may be divided into two major groups by genome sequencing and antigenic analysis. The F protein is serologically cross-reactive but the G protein is group-specific. Group A strains are somewhat commoner than group B and may more often be associated with severe disease.

Pathogenesis, Pathology, and Immunity

The virus multiplies in the mucous membrane of the nose and throat; in the very young and very old it may involve the trachea, bronchi, bronchioles, and alveoli. Fatal cases usually show extensive bronchiolitis and pneumonitis.

The immune responses after respiratory syncytial virus infection are notoriously weak, partially explicable by the fact that the immune response of infants to the protective F and G glycoproteins is very poor. Antibody levels decline to low or undetectable levels within a year of primary infection, but respond briskly again after reinfection. Maternal antibody does not fully protect infants, but higher levels of maternal antibody in cord blood do correlate with less severe disease and an older age of primary infection. A role for cell-mediated immunity, presumably cytotoxic T cells, is indicated by the finding that immunization with a vaccinia recombinant bearing the respiratory syncytial virus internal N nucleoprotein expedites recovery. Furthermore, respiratory syncytial virus is a much less effective inducer of interferon synthesis in normal infants than are influenza and parainfluenza viruses (PIV). Surfactant protein A, a lung C-type lectin involved in innate host defense, opsonizes RSV and enhances phagocytosis. Surfactant A appears to act by binding specifically to the RSV F protein, and infants with severe RSV disease have a reduction in surfactant concentration and function.

In the 1960s, an experimental formalin-inactivated respiratory syncytial virus vaccine was tested in children. However, when the immunized children encountered RSV during a subsequent epidemic, they actually suffered significantly more serious lower respiratory disease compared to non-immunized controls. It was subsequently demonstrated that, although the formalin-inactivated virus induced a good antibody response, the antibodies lacked neutralizing activity. The exact immune mechanism for enhancement of disease has not been fully clarified. Anti-respiratory syncytial virus IgE antibodies as well as histamine and leukotrienes have been reported in respiratory secretions of infants with RSV bronchiolitis. There is also evidence in rodent models that cytotoxic T cells, while expediting recovery by eliminating infected epithelial cells, may nevertheless exacerbate clinical signs by augmenting the inflammatory response.

Laboratory Diagnosis

Nucleic acid detection assays based on reverse transcriptase polymerase chain reaction (RT-PCR) are now widely available for RSV, often in a multiplexed panel for detection of multiple respiratory virus pathogens. These tests are typically more sensitive than any of the virus isolation or protein-based detection assays used in the past. This enhanced sensitivity is especially helpful when testing adults, who often shed virus in very low titers. Positive RT-PCR tests need to be interpreted in the context of the clinical scenario, since the tests can remain positive for prolonged periods of time after infection, well beyond the period during which infectious virus can be isolated. Since children may experience symptomatic respiratory infections every few weeks during the winter, caution must be used in interpretation of positive PCR tests, especially when multiple viruses are detected simultaneously in a sample. In the past nasopharyngeal aspirates or nasal washes were preferred to nasal or throat swabs but the increased sensitivity of RT-PCR-based assays has increased the usefulness of some kinds of swabs.

Three other diagnostic methods are still in use:

Enzyme immunoassay for the detection of viral antigen. Using appropriate monoclonal antibodies as capture and detector antibodies, enzyme immunoassay can detect as little as 10 ng of RSV antigen, and EIA methods are readily automated in large diagnostic or reference laboratories. Rapid antigen-detection kits are commercially available, reliable and sensitive, taking 15 to 20 minutes from start to finish.

Immunofluorescence (IF) on exfoliated cells. This method had been a mainstay of direct diagnosis of respiratory pathogens for many years, but requires laboratory equipment and expertise and thus is unsuitable for large numbers of samples.

Isolation of the virus in cell culture. The extreme lability of respiratory syncytial virus makes it essential that the specimen is taken early in the illness and added to cultured cells without delay and certainly without preliminary freezing. Human heteroploid cell lines such as HeLa or HEp-2 are the most sensitive, and a CPE generally becomes visible in three to five days. Fixation and staining generally reveal extensive syncytia containing acidophilic cytoplasmic inclusions, but some strains produce only rounded cells. An absence of hemadsorption distinguishes RSV from all the other paramyxoviruses. Isolates can be allocated into groups A and B, further subdivisible into various subgroups, by enzyme immunoassay using monoclonal antibodies to the G protein.

Epidemiological surveys can be undertaken by measuring serum antibodies using enzyme immunoassay with recombinant-derived antigens, but serology is not customarily used for diagnosis.

Epidemiology, Chemotherapy, Control, and Prevention

Respiratory syncytial virus is highly contagious, being shed in respiratory secretions for several days, sometimes weeks. Nosocomial infections are frequent, and outbreaks occur in neonatal wards of maternity hospitals, sometimes inflicting high mortality. Careful studies of transmission have shown that both aerosol inhalation, and contact with contaminated objects, for example, hands, paper towels, and other ward surroundings, are significant in RSV transmission. Strict measures are necessary to reduce the spread of infection between hospitalized children and healthcare workers both as sources and targets of infection.

Improvements in intensive care facilities and practices have led to a marked reduction in the mortality from respiratory syncytial virus pneumonia. Ribavirin administered as a small particle aerosol is licensed in some countries but its efficacy is controversial. The success of neutralizing anti-G or anti-F antibodies for the prevention or treatment of experimental respiratory syncytial virus infections in cotton rats has encouraged clinical trials of purified, high-titer human IgG antibodies. These preparations have been administered intravenously for the treatment of RSV bronchiolitis/pneumonia in infants or for prophylaxis in high-risk infants during winter epidemics. A humanized monoclonal antibody preparation (Palivizumab) directed against the fusion protein has been shown to reduce hospitalization rates by 45 to 55% and is recommended for prevention of infection in high-risk infants.

Because of the potential for disease enhancement, traditional viral vaccines are presently considered to be inappropriate for pediatric use. The present focus is in designing pediatric RSV vaccine centers on live-attenuated virus vaccines and several subunit (F protein)-based vaccines—several of these are in advanced clinical trials.

PARAINFLUENZA VIRUS INFECTIONS

Parainfluenza viruses generally produce relatively mild upper respiratory tract infections (URTI), but they are also the commonest cause of a more severe condition in young children known as "croup" and, occasionally, pneumonia. Human parainfluenza viruses 1 and 3 belong to the genus *Respirovirus*, whereas human parainfluenza viruses 2, 4a, and 4b are now classified with mumps virus in the genus *Rubulavirus*.

Epidemiology and Clinical Profile

Parainfluenza viruses (PIV) are highly transmissible, infecting most children by the age of five. PIV 3 infects the majority of infants within the first year of life and can spread within hospitals and nurseries causing outbreaks of pneumonia and bronchiolitis. PIV 1, 2, and 3 are second only to RSV as a cause of serious respiratory tract disease in infants and children, whereas disease caused by PIV4a and 4b occurs less frequently and is less severe.

Croup (laryngotracheobronchitis) is a distinctive syndrome of children usually under five years of age, often preceded by a cough and runny nose for one to two days, and is marked by severe inspiratory stridor and respiratory distress, caused by narrowing of the subglottic region of the trachea. PIV1 is the major cause, but other PIVs can also be responsible.

Antibodies against most or all of the major proteins can be detected in sera from infected individuals. The HN and F proteins, however, are the only antigens known to induce antibodies that neutralize infectivity. Neutralizing antibodies mediate long-term resistance, especially in the lower respiratory tract, while protection mediated by CTLs and secretory antibodies is relatively short-lived.

Laboratory Diagnosis

Nucleic acid detection assays based on RT-PCR are now widely available for PIV3, and somewhat less often available for the other parainfluenza viruses; testing is often in a multiplexed panel for detection of a large array of respiratory pathogens. In some settings immunofluorescence (IF) and enzyme immunoassays are employed, using nasopharyngeal aspirates. IF is employed to demonstrate antigen in exfoliated cells, and enzyme immunoassay is sufficiently sensitive to detect free antigen in mucus suitably solubilized to liberate intracellular as well as extracellular antigens.

Treatment and Prevention

Several experimental parainfluenza 3 vaccines have shown some degree of protection in rodents or primate animal models. However, formidable problems mitigate against parainfluenza vaccines achieving wide acceptability. First, the immunity that follows natural infection with parainfluenza viruses is poor. Second, a vaccine needs to be administered shortly after birth, when immune responses are weakest and maternal antibodies are present. Third, in light of the worrying experience with earlier respiratory syncytial virus and measles vaccine candidates that actually potentiated the disease occurring upon subsequent challenge, subunit vaccines would also need to undergo very careful clinical trials before licensing. As with RSV vaccine research and development, the present focus in designing pediatric parainfluenza virus vaccines centers on live-attenuated virus vaccine candidates and several subunit (F protein)-based vaccines, all employing intranasal instillation—several of these are in early-stage clinical trials.

HUMAN METAPNEUMOVIRUS INFECTION

A new paramyxovirus was described in the Netherlands in 2001 and soon shown to be a significant cause of lower respiratory tract disease in children worldwide. It is now classified in a separate genus *Metapneumovirus* in the *Pneumovirinae* subfamily.

hMPV infections occur at a slightly older age than RSV, but virtually all children have been infected by the age of 5 and it accounts for 5 to 10% of all hospitalizations of children for acute respiratory virus infection. Clinically, it is very similar to RSV infection. Primary infection usually causes upper respiratory symptoms (fever, cough, rhinorrhea), but bronchiolitis, croup, pneumonia, and exacerbation of asthma also can occur. Repeat infections are common, usually associated with milder upper respiratory tract symptoms and/or otitis media. Diagnosis is usually made by RT-PCR in multiplex configurations.

MEASLES

Until recently measles virus was perhaps the most important of all the common childhood diseases. Due to the development of an effective live-virus vaccine and vigorous implementation of a policy for the immunization of children, the incidence of the disease has dramatically declined in many countries. Globally, between 2000 and 2012 more than 1 billion children in high-risk countries were vaccinated, leading to a 78% drop in measles deaths. However, it is still one of the leading causes of mortality among young children, being responsible for approximately 120,000 deaths in 2012.

Clinical Features

Following a prodromal period marked by fever, cough, coryza, and conjunctivitis, an exanthem appears on the head and spreads progressively over the chest, trunk, and limbs. The rash consists of flat macules that fuse to form blotches rather larger than those of other viral exanthems. These patches are slow to fade and often leave the skin temporarily stained (Fig. 26.5).

Common complications include otitis media, croup, bronchitis, and bronchopneumonia. Bacterial pneumonia is the usual cause of death among malnourished children. Immunologically deficient children can die from giant cell pneumonia or from progressive infectious encephalitis (measles inclusion body encephalitis) with no sign of a rash. However, the most dangerous complication of measles is acute postinfectious encephalitis, which occurs in about one in every thousand cases, and has a mortality rate of around 15% and is the cause of permanent neurological sequelae in many survivors. Subacute sclerosing encephalitis (SSPE) is a very rare complication, developing in only about one in every 300,000 cases, often some years after apparent recovery from the original infection.

Other complications include nausea, vomiting, diarrhea, and diffuse abdominal pain, seen in 30 to 60% of cases; in developed countries these symptoms usually resolve with the rash, but in developing country settings these can persist and aggravate preexisting malnutrition. Similarly, ocular complications (conjunctivitis and punctate keratitis) usually resolve, but in malnourished children can progress to corneal ulceration, bacterial superinfection, and blindness.

Pathogenesis, Pathology, and Immunity

Measles is a systemic infection that begins with infection of the respiratory tract. During the incubation period, the virus replicates initially in epithelial cells in the upper airways and the virus spreads in lymphatics, mostly in cell-associated form, to local draining lymph nodes and other lymphoid tissues. Virus entry into tissues from the vascular bed is thought to occur mostly by infected leukocyte transmigration. Replication in local lymph nodes is followed about 6 days postinfection by a secondary viremia and the dissemination of virus to many organs, including lymph nodes, spleen, thymus, orophyrynx,

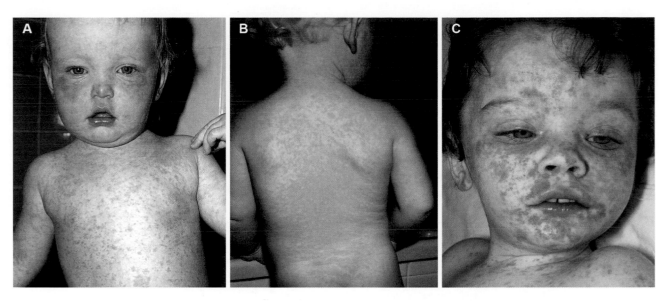

FIGURE 26.5 (A and B) Measles in a young child. Note blotchy red maculopapular rash all over the body. (C) Severe measles rash, accompanied by conjunctivitis. *Reproduced from Cooke, R.A., Infectious Diseases, McGraw Hill, with permission.*

conjunctiva, skin, kidney, bladder, gastrointestinal tract, and liver. Because the epithelia of the conjunctiva and the respiratory tract are only 1 to 2 cells deep, there is often focal necrosis, usually seen some 9 to 10 days postinfection. Mucosal foci can ulcerate on about the 11th day, to produce characteristic Koplik's spots in the mouth. By the 14th day as circulating antibodies are first detected, the characteristic maculopapular rash appears and the fever falls. This skin rash is due in large part to the virus-specific immune response with activation of virus-specific CD4+ and CD8+ T cells. Cytokines produced are consistent with activation of Th1 CD4+ and CD8+ T cells, followed by Th2 CD4+ T cells and regulatory cells. The histopathology of the measles rash suggests that the initial event is infection of dermal endothelial cells and the consequent cytokine/inflammatory response, followed by spread of infection into the overlying epidermis with infection of keratinocytes leading to focal keratosis and edema. Epithelial giant cells form and mononuclear cells accumulate around vessels. Clearance of infectious virus is approximately coincident with fading of the rash and the production of IgM and IgG antibodies, but clearance of RNA is slower. Concurrently, immune suppression and increased susceptibility to other infections often become evident. Production of antibody and cellular immune responses to new antigens are impaired and reactivation of tuberculosis has been reported. Measles also provides a classic example of the increased severity of a disease in association with malnutrition (Fig. 26.6).

Much of what is known about measles pathogenesis comes from the rhesus macaque model; the infection in macaques closely resembles the human infection. In the macaque model, the predominant cell types infected are lymphocytes/monocytes/macrophages and dendritic cells. Two cellular receptors have been identified: CD46 and SLAM (CD150). Circulating T cells are decreased in number during the acute phase of measles and the normal CD4:CD8 ratio is lowered.

Occasionally there is central nervous system involvement, usually after the infection in visceral organs has come to an end. Three syndromes have been recognized. Acute postinfectious measles encephalitis is rare in children less than 2 years of age, but occurs in about one in 1000 cases of measles in older children, with a case-fatality rate of 15%. There is little or no production of virus in the brain, but myelin basic protein is found in the cerebrospinal fluid, and patients' T lymphocytes are often reactive to myelin. The pathogenesis appears to involve autoimmune demyelination. In contrast, measles inclusion body encephalitis occurs only in immunocompromised children, usually within six months of the rash; it may be rapidly progressive and is attributable to failure to eliminate virus-infected cells because of the lack of cytotoxic T cells. Finally, subacute sclerosing panencephalitis (SSPE) occurs years after the acute disease; it too is always fatal, and is characterized by very slow replication and spread of measles virus in the brain. Although it is difficult to isolate measles virus from the SSPE brain, some neurons contain very large accumulations of measles virus nucleocapsids; sequencing of the viral genome reveals numerous mutations in the M gene and to a lesser extent in other genes. The replication of measles virus usually causes death of cells in culture. The elements determining persistent non-cytopathic versus lytic infection are not completely understood, but properties

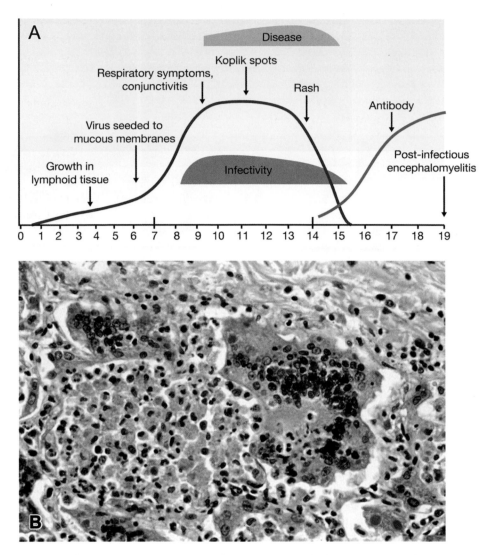

FIGURE 26.6 (A) Sequence of clinical and pathogenetic events during acute measles infection. Note that virus in the bloodstream is nearly all cell-associated, mainly involving monocytes. Around day 14, circulating antibodies become detectable, the maculopapular rash develops and fever falls. Infectivity is mainly due to shedding from the respiratory tract. (B) H&E stained section of lung in measles infection, showing multinucleate giant cells.

of both the infected cell and the infecting virus seem to contribute (see Chapter 8: Patterns of Infection).

In less-developed countries of Africa and South America the case-fatality rate of measles is of the order of 3 to 6%, sometimes higher; this is several hundred times more than that seen in developed nations before universal vaccination programs began. Factors associated with the increased severity of measles in developing countries include young age at the time of infection, lower socioeconomic status, crowding, concomitant diarrhea, malnutrition (including vitamin A deficiency), lack of access to health care, and underlying immunodeficiency from a variety of causes. In addition to the direct mortality caused by measles, increased overall mortality rates are found in children who have had measles during the previous 9 months compared with children who have not.

Essentially all primary measles infections give rise to clinical disease. The resulting immunity is effectively lifelong; second attacks of measles are virtually unknown. There is only one serotype of measles virus. While adoptive transfer of neutralizing antibodies against either the H or F glycoprotein confers excellent solid short-term protection against challenge, T cells raised to the internal nucleoprotein (N) are also protective, presumably by expediting recovery.

Laboratory Diagnosis

The diagnosis of measles on clinical grounds used to be sufficient. However laboratory diagnosis is now desirable in those countries where natural circulation of virus has been eliminated, since other causes such as parvovirus B19, enteroviruses, and HHV-6 (roseola) are usually more common. Moreover, most younger healthcare workers may have never seen the disease, and a new imported case requires a significant public health response. EIA testing for serum IgG and IgM have been the mainstay for diagnosis. IgM antibodies are detected in 75% of patients three days after the onset of the rash, rising to 100% by day 7. RT-PCR of a respiratory sample, where available, gives a rapid result and can be positive for up to three weeks after the onset of rash.

Cultivation of the virus is difficult and slow, and is not widely used. Measles antigen can be identified by immunofluorescence in cultured cells or, more simply and quickly, in cells aspirated directly from the nasopharynx. Serology is also employed to screen populations for their immune status. HI was traditionally used for this purpose but enzyme immunoassay is both more convenient and more sensitive.

Epidemiology, Prevention, and Control

Measles is highly contagious, and is shed in respiratory secretions from the latter part of the incubation period to the first two or three days of the rash. In countries from which the infection has been eliminated, recognition of a new imported case is important to allow identification, quarantining, and passive immunization where indicated.

A live measles vaccine was developed originally by John Franklin Enders. This was then further attenuated to produce the Schwarz vaccine used today in the Western world, and a number of other effective live attenuated vaccines have been developed thereafter. In the developed world vaccination is recommended after maternal antibody has completely disappeared. Seroconversion occurs in 95 to 98% of recipient infants at 15 months of age, compared with as few as 50% if vaccinated at six months. The antibody titers that result from vaccination are 10 fold lower than following natural infection, but they do generally persist for many years at protective levels. Trivial side effects are not uncommon, particularly mild fever (in about 10%) and/or transient rash (5%).

Globally, endemic measles transmission was eliminated from the Americas in 2002, and only sporadic imported cases are now seen. Since 2001 there has been a 93% reduction in cases in the African region and large reductions in other World Health Organization (WHO) regions; however a significant increase has occurred in Western Europe in individuals who are not vaccinated due to religious or philosophical beliefs, lack of access to health care, or antivaccination movements focused on the unfounded belief of MMR vaccines being the cause of autism. Thus, it is important to redouble the effort to vaccinate all children, concentrating particularly on unimmunized immigrants. Several outbreaks have occurred in college students and office workers, most of whom were vaccinated in infancy.

Serological surveys indicate that the percentage of people with protective levels of immunity falls progressively from near 98% over the years following vaccination. Moreover, with relatively little virus circulating in the community to boost such waning immunity, there is a danger that non-immune or unvaccinated adults might first encounter the virus at an age when complications are more serious. It is now recommended that all children receive two doses of measles vaccine: the first at 12 to 15 months (as combined measles/mumps/rubella vaccine), plus a booster just prior to entering either kindergarten/first grade (4 to 6 years of age) or junior high/secondary school (11 to 12 years).

In the developing world, where there is high mortality from measles in infants in the first year of life, immunization is a top priority of the World Health Organization where it is part of the Expanded Immunization Programme. In these areas maternal immunity declines more rapidly than in developed countries, and infants become susceptible to measles earlier. The World Health Organization recommends measles vaccination at 9 months or as soon as possible thereafter. Ideally, the optimum age for vaccination should be determined for each individual country and a second dose should be given later in the case of a primary vaccination failure. Maintenance of the "cold-chain" is also vital in the tropics. Today, vaccines are less heat-labile and are supplied freeze-dried, permitting storage at 4°C until reconstitution immediately before use.

Passive immunization still has a place in protecting unvaccinated children following exposure to measles, particularly those children who are immunocompromised. If administered promptly, pooled normal human immunoglobulin will abort the disease; if given a few days later, the disease may still be modified. No antiviral agent is effective against measles but bacterial superinfections, such as pneumonia or otitis media, require vigorous antibacterial therapy.

MUMPS

The predominant clinical feature is a painful edematous enlargement of the parotid and other salivary glands, with the patient unable to talk or eat without undue discomfort (Fig. 26.7). Among postpubertal males, the infection in 25% of cases is complicated by epididymo-orchitis and may lead to atrophy of the affected testicle. In 10% to 30% of cases, the orchitis may be bilateral. A wide variety of other glands may be involved, including commonly the pancreas, ovary, thyroid, and, more rarely, the breast. Benign meningeal signs are detectable in at least 10% of all cases of mumps, and clinical meningitis, sometimes presenting without parotid involvement, is quite common. Fortunately, the prognosis is consistently better than with bacterial meningitis and adverse sequelae are rare. Mumps encephalitis on the other hand, though much less frequent,

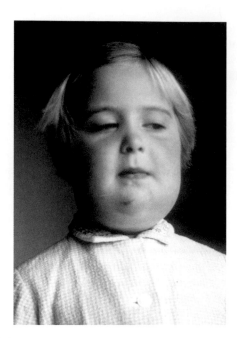

FIGURE 26.7 Mumps infection in an 8-year-old girl. Note bilateral swelling of the parotid and submandibular glands. *From Cooke, R.A., Infectious Diseases, McGraw Hill, with permission.*

is a more serious development. Unilateral nerve deafness may occur and can be an important long-term consequence. Among infants, mumps infection is often symptomless or present only as a mild respiratory disease.

The virus is spread via droplets and is one of the most transmissible human virus infections. During the incubation period, the virus first multiplies in the upper respiratory tract, spreading to mucosal lymph nodes and disseminating further via a transient viremia to other sites. Mumps virus can infect virtually all tissues and organs, causing acute inflammation. Termination of viral excretion in saliva correlates with the local appearance of virus-specific secretory IgA and plasma viremia disappears with the development of a mumps-specific humoral response. Lifelong immunity follows natural infection.

Although humans are the natural host of mumps, experimental infection of animals has been used to elucidate its pathogenesis, particularly neurotropism. Virus invasion of the CNS occurs across the choroid plexus, with infected blood-borne mononuclear cells crossing the endothelium of the chorid plexus stroma and subsequently leading to infection of the choroidal epithelium. In addition, virus can penetrate brain parenchyma and infect neurons by contiguous spread from infected cells lining the ventricular cavities of the brain. Once within neurons, the virus most likely spreads along neuronal axons. Cases of postinfectious encephalitis, an autoimmune attack on CNS myelin sheaths triggered by the neural breakdown products of the primary infection, have been reported.

Although the classic case of mumps (Fig. 26.7) can be identified by the clinician without help from the laboratory, atypical cases and meningo-encephalitis present a diagnostic challenge. Mumps is more commonly diagnosed by serology: IgM capture enzyme immunoassay for rapid diagnosis of mumps meningo-encephalitis, while enzyme immunoassay can be used to screen for IgG antibodies to monitor immune status among vaccine recipients. RT-PCR is being used increasingly, particularly for mumps virus CNS infection. Mumps virus can be isolated from the saliva with swabs taken from the orifice of Stensen's duct, from urine (mumps virus is one of the few human viruses that can be successfully isolated from this source), or from cerebrospinal fluid in patients with meningitis. Cell cultures combined with hemadsorption, and immunofluorescence are used to confirm the diagnosis.

A dramatic decline in the incidence of mumps followed the introduction of vaccination (Fig. 26.8). A trivalent, live attenuated MMR vaccine includes measles and rubella viruses as well as mumps virus. However, the reduction in the incidence of mumps since the introduction of the vaccine is not being consistently maintained, with cases often occurring among unvaccinated adolescents and young adults in whom the risk of complications is much higher. In an outbreak in Iowa, the United States, nearly half the patients had received two or more doses of MMR vaccine, suggesting that many individuals had either mounted an insufficient initial response, or had waning immunity. As with measles virus, attaining universal coverage is essential whenever a policy of widespread childhood immunization is introduced, and reinforces the desirability of a booster dose of MMR vaccine for children entering high school or college.

HENIPAVIRUS INFECTIONS

In the last few years a number of new paramyxoviruses have been discovered of public health importance. In 1994, a new virus was isolated from an outbreak of fatal respiratory disease among racehorses and their trainer in Hendra, a suburb of Brisbane, Australia. Sporadic cases have since occurred of what is now termed Hendra virus at various places in the state of Queensland and northern New South Wales. Although the fatality rate is high, the total number of cases has been relatively few.

Four years later, a new virus was isolated from seriously ill pig farmers in Malaysia. First isolated in the Malaysian states of Perak and Negri Sembilan, this virus is now known to be widespread throughout Southeast Asia and the northeastern Indian subcontinent. This virus, named Nipah virus, is now considered to be a significant zoonotic pathogen in the region, especially in Bangladesh and the Indian state of West Bengal.

During epidemiological investigations of pig farms in Australia stimulated by these findings, a further

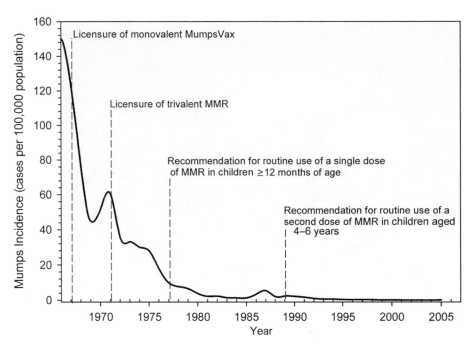

FIGURE 26.8 Decline in the incidence of reported cases of mumps in the United States since 1965. *From CDC.*

paramyxovirus was discovered: Menangle virus is, however, a member of the genus Rubulavirus and does not apparently cause significant illness in humans.

Epidemiology of Henipavirus Infections

Both Hendra and Nipah viruses share similar epidemiological properties, notably that fruit-eating bats of the genus *Pteropus* act as animal reservoirs. Transmission to livestock and horses occurs through exposure to contaminated bat feces, urine, and possibly by aerosol transmission. Infected animals then act as an intermediate host with humans being infected as a result of continuing close contact with either horses (Hendra virus) or pigs (Nipah virus). In Bangladesh there is circumstantial evidence to suggest direct transmission of Nipah virus from bats to humans. It should be noted that a number of domesticated animals, including dogs and cats, are susceptible to Nipah virus and thus act as alternative sources of infection.

Hendra virus. The incubation period is between five and eight days. In 1994 the virus first came to light when a racing horse trainer and a member of his staff presented with an acute and severe respiratory illness. The trainer subsequently died from the infection and the stable hand only recovered after a prolonged six months of illness. The infection quickly spread to other horses in the yard, with a mortality of around two-thirds. Sporadic cases have occurred since, including cases among veterinarians and their assistants undertaking necropsy investigations on horses that have succumbed to undiagnosed illnesses.

At present, there is no evidence of either human-to-human transmission or asymptomatic infection in humans. Secondary cases among horses have not been reported, although prolonged exposure to horses remains the major risk factor for humans.

Nipah virus first became apparent when clusters of encephalitis cases began to emerge in Malaysia in September 1998. The largest number of these occurred in the village of Sungai Nipah, the community that subsequently gave its name to the virus. Most cases were among males with a history of handling pigs. The mortality rate was high at 40% and authorities first believed the cases were due to Japanese encephalitis virus. The outbreak was only halted after a ban on the transport of pigs was introduced and followed by extensive culling.

In April 2001 a cluster of encephalitis cases was reported in the Meherpur district of Bangladesh. First mistaken for measles, serology quickly revealed these were further cases of Nipah virus infection. In contrast to previous experiences in Malaysia, however, there seemed to be little history of contact with farm animals among the Bangladeshi cases. There is anecdotal evidence of probable direct exposure to infected fruit bats as a result of the infected individuals having previously collected fruit or been involved in the processing of date palm oil. Human-to-human spread is also regularly reported.

More recently, henipa- or henipa-like viruses have been found in fruit bats of several genera, insectivorous bats and microbats in Central and South America, Africa, China, and many SE Asian countries (Fig. 26.9). Phylogenetic analyses of isolates from northeast India to Cambodia suggest that Nipah virus is rapidly evolving within geographical

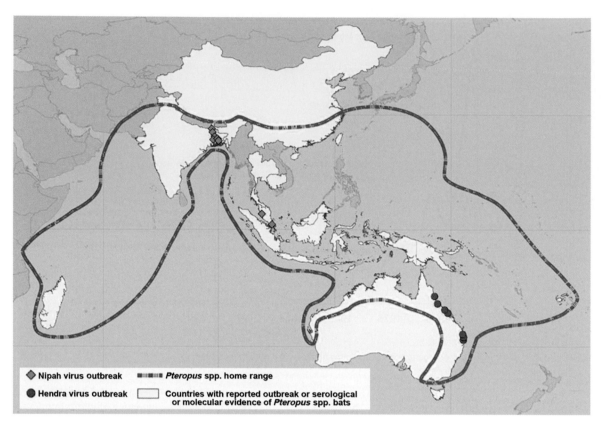

FIGURE 26.9 Global distribution of reported human cases of Hendra and Nipah virus infection, together with the geographical range of *Pteropus* bat species. *From CDC, 2014.*

localities. There is good evidence that both Hendra and Nipah viruses are natural pathogens of *Pteropus* bats and that further outbreaks are to be expected as a result of ecological disturbances due to either climate change or manmade changes to the environment.

Virological Properties

The genomes of Hendra and Nipah viruses closely resemble those of other paramyxoviruses. However, the envelope glycoprotein (G) lacks hemagglutination and neuraminidase activities. The genome is also far larger than other paramyxoviruses, due principally to an extended P gene and longer tracts of non-coding sequences between most of the viral genes. Fig. 26.3 shows a map of the henipavirus genome relative to other paramyxoviruses. Henipaviruses attach to host cells by recognizing either ephrin-B2 and ephrin-B3: these molecules are widespread in the endothelial cells and neurons of many different vertebrate species, which may account for the neurological involvement of henipaviruses, and the associated wide host range. The process of virus maturation requires the cleavage of F0 to functionally active subunits F1 and F2 by the host protease cathepsin L, a process distinct from the cleavage by furin as is the case for other paramyxoviruses.

The complexity of P gene expression may also facilitate a wide range of susceptible species compared to other members of the *Paramyxoviridae* family. No less than four translation products are produced from the P gene through the use of alternative reading frames. Interferon signaling is inhibited once the products of the P gene (P, V and W proteins) form complexes with the STAT protein, thereby blocking the Jak-STAT pathway. and modulating the host innate immune response.

Clinical Features and Pathology

Patients with Hendra virus present with both respiratory and neurological symptoms. Although only seven human cases have been recorded, the syndrome appears similar to Nipah virus infection. Cases have an acute influenza-like illness that may progress to pneumonia or encephalitis. Autopsy has confirmed both lung and brain involvement. Pneumonitis was present, characterized by the presence of syncytia, giant cell formation, and inclusion bodies. Necrotizing alveolitis with extensive cellular infiltration

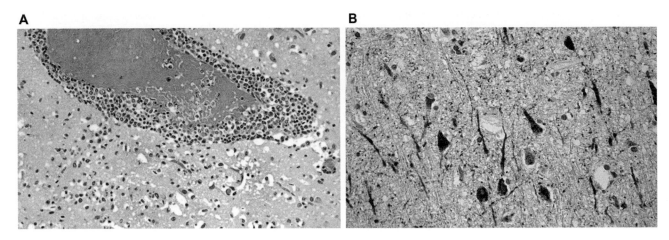

FIGURE 26.10 Nipah encephalitis. (A) H&E stained section showing intense perivascular cuffing with inflammatory cells. (B) Immunohistochemistry, showing virus antigen-staining neurons. *From Zaki, S., CDC, with permission.*

was evident. The close resemblance to cases of measles virus led initial reports to refer to Hendra virus as equine morbillivirus.

More detail is available for Nipah virus, where the incubation period ranges from a few days up to 14 days. In contrast to Hendra virus, there is clear evidence that mild or asymptomatic infections occur. Patients first develop an influenza-like illness followed quickly by early signs of encephalitis accompanied by myalgia and drowsiness. The symptoms may progress to confusion, dizziness, abnormal reflexes, hypertension, and tachycardia (Fig. 26.10). A decline in mental capacity is a notable feature. Serious cases progress within two weeks to seizures, respiratory difficulties, and coma. This presentation may be accompanied by pulmonary disease, with cough and dyspnea that may progress to pneumonia and acute respiratory distress syndrome, especially among cases in Bangladesh. It is possible that there are differences in clinical presentation between localities, possibly reflecting the rapid evolution of Nipah virus.

Marked thrombocytopenia is seen in a third of cases, with occasional leukopenia, and the CSF contains lymphocytes, elevated protein levels, and anti-henipavirus antibodies. Disseminated vasculitis with endothelial syncytia, leading to thrombosis and perivascular hemorrhage, is a hallmark of the pathological changes with both Hendra and Nipah viruses. Radiography shows interstitial pneumonia and alveolar consolidation. There is long-term neurological impairment among those patients who recover, including personality changes, speech impairment, and cognitive dysfunction.

Treatment and Prevention

Treatment of henipavirus infections is essentially supportive. Ribavirin has been administered to human cases of Nipah virus infection with some possible benefit. A recombinant

human monoclonal antibody against the G glycoprotein is under investigation, but there is as yet no vaccine available for use in humans against either henipavirus. However, since November 2012 a commercial subunit Hendra vaccine for use in horses has been available, which protects horses from lethal disease: as it eliminates viral shedding in horses, it should also play a valuable role in breaking the chain of transmission to humans.

Prevention of Hendra involves protecting horses, for example by ensuring horses are not sheltered under trees with bat roosts, and vaccination of horses at potential risk; and educating veterinarians and animal handlers about the recognition of, and safe handling of, infected horses. Prevention of Nipah infection of humans is targeted at three main routes for human acquisition ingestion of fresh date palm sap (which is frequently contaminated by bats), contact with infected domestic animals, and direct contact with bat secretions (e.g., by people climbing trees). Fruit collected within the endemic area from trees close to bat roosts for Nipah virus should be thoroughly washed prior to consumption. Human-to-human spread can be reduced by implementing standard infection control precautions, particularly involving saliva. Early identification of infected domestic animals is important, and requires education of farmers and access to laboratory testing. During established outbreaks, additional measures may be required such as restricting movement of animals between farms, culling of animals, and temporary closure of abattoirs.

FURTHER READING

Croser, E.L., Marsh, G.A., 2013. The changing face of the henipaviruses. Vet. Microbiol. 167, 151–158.
El Najjar, F., Schmitt, A.P., Dutch, R.E., 2014. Paramyxovirus glycoprotein incorporation, assembly and budding: A three way dance for

infectious particle production. Viruses 6, 3019–3054. Available from: http://dx.doi.org/10.3390/v6083019.

Giodson, J.L., Seward, J.F., 2015. Measles 50 years after use of measles vaccine. Infect. Dis. Clin. N Am. 29, 725–743.

Meng, J., Stobart, C.C., Hotard, A.L., Moore, M.L., 2014. An overview of respiratory syncytial virus. PLOS Pathogens 10 (issue 4), e1004016.

Pernet, O., Lee, B., 2012. Henipavirus receptor usage and tropism. Curr Top Microbiol. Immunol. 359, 59–78. Available from: http://dx.doi.org/10.1007/82_2012_222.

Principi, N., Esposito, S., 2014. Paediatric human metapneumovirus infection: Epidemiology, prevention and therapy. J. Clin. Virol. 59, 141–147.

Chapter 27

Rhabdoviruses

The family *Rhabdoviridae* encompasses more than 175 viruses of vertebrates, invertebrates (mostly arthropods), and plants. Virions have a distinctive bullet shape. The only important human pathogen is rabies virus with vesicular stomatitis being an occasional, zoonotic disease. In 2005, the estimated global expenditure for rabies prevention exceeded US\$ 1 billion—moreover, the resources and costs of rabies control programs, including post-exposure prophylaxis, are expected to rise dramatically in all countries where rabies is endemic, particularly in regions where old nervous-tissue-based vaccines are replaced by cell culture-based vaccines.

Rabies is one of the oldest and most feared diseases of humans—The Babylon Eshnuna code (23rd century BCE) contains the first known mention of rabies. The Greek philosopher Democritus provided the first clear description of animal rabies in about 500 BCE. Perhaps the most lethal of all infectious diseases, rabies also has the distinction of having stimulated one of the great early discoveries in biomedical science. In 1885, before the nature of viruses was comprehended, Louis Pasteur and his colleagues developed, tested, and applied a rabies vaccine, thereby opening the modern era of infectious disease prevention by vaccination.

Rabies is also a disease where much is known with regard to transmission, and it provides an illustration of how a virus can evolve in terms of temporo-spatial epidemiology and the changing risk such evolution presents to human health. It is also one of the few examples whereby the control of a human disease has been attempted successfully by controlling the infection among animals, including wildlife.

PROPERTIES OF RHABDOVIRUSES

Classification

The family *Rhabdoviridae* is a member of the order *Mononegavirales*, along with the families *Bornaviridae*, *Filoviridae*, and *Paramyxoviridae*—the order was created to unify all viruses with genomes comprising a single molecule of linear, negative sense, single-stranded RNA.

Based on virion properties, serological relationships and genome sequencing, six genera have been recognized in the family *Rhabdoviridae*, four of which contain animal viruses (*Lyssavirus, Vesiculovirus, Ephemerovirus,* and *Novirhabdovirus*) (Table 27.1). Animal rhabdovirus species are distinguished genetically by genome sequencing and serologically, by the use of neutralization tests.

The genus *Lyssavirus* comprises rabies virus and closely related viruses, including Mokola virus, Lagos bat virus, and Duvenhage virus from Africa, as well as European bat viruses 1 and 2 and Australian bat lyssavirus. In addition, there are several newly discovered lyssaviruses, mostly isolated from bats in Eurasia. Each of these viruses is considered capable of causing rabies-like disease in both animals and humans. The genus *Vesiculovirus* contains about 35 serologically distinct viruses, most importantly vesicular stomatitis-Indiana virus and vesicular stomatitis-New Jersey virus, as well as six other viruses that are known to cause vesicular disease in horses, cattle, swine, and occasionally humans. The genera *Ephemerovirus* and *Novirhabdovirus* do not contain any known human pathogens, having members of importance to livestock industries and aquaculture. Genera *Cytorhabdovirus* and *Nucleorhabdovirus* contain exclusively plant viruses,

Virion Properties

Rhabdovirus virions are 70 nm in diameter and 170 nm long (although some are longer, some shorter) and consist of an envelope with large peplomers within which is a helically coiled cylindrical nucleocapsid. The precise cylindrical form of the nucleocapsid gives rise to the distinctive bullet or conical morphology of virus particles as seen by electron microscopy (Fig. 27.1). The genome is a single molecule of linear, negative-sense, single-stranded RNA, 11 to 15 kb in size. The genome of the Pasteur strain (CVS) of rabies virus consists of 11,932 nucleotides. The genome encodes five genes in the order 3'-N-NS-M-G-L-5'; some viruses have additional genes, or pseudogenes, interposed (Fig. 27.2). Rhabdoviruses generally have five proteins: L, the RNA-dependent RNA polymerase which functions in transcription and RNA replication; G, the glycoprotein which forms trimers that make up the peplomers (spikes); N, the nucleoprotein, the major component of the viral nucleocapsid; P, a component of

Fenner and White's Medical Virology. DOI: http://dx.doi.org/10.1016/B978-0-12-375156-0.00027-8

TABLE 27.1 Rabies and the Rabies-Like Viruses (Member Viruses of the Genus *Lyssavirus*)[a]

Virus	Distribution	Reservoir
Rabies virus	Worldwide with the exception of Antarctica, Australia[b], New Zealand[b], United Kingdom[b], and some islands. Some regions traditionally affected (e.g., western Europe) have recently eradicated the disease and attained rabies-free status	*Canivora*: many species of domestic and wild canids, foxes, mongooses, skunks, and the raccoon; *Chiroptera*: (Americas only) several species of insectivorous, vampire, and frugivorous bats
Lagos bat virus	Several African countries including Central African Republic, Ethiopia, Nigeria, Senegal, South Africa, Zimbabwe	*Megachiroptera*: Fruit bats including *Eidolon helvum*, *Rousettus aegyptiacus*, and *Epomophorus wahlbergi*
Mokola virus	Several African countries including Cameroon, Central African Republic, Ethiopia, Nigeria, South Africa, Zimbabwe	Reservoir unknown—possibly shrews (*Crocidura* spp.) and rodents (*Lopyhromys sikapusi*); most reported cases in domestic cats and dogs
Duvenhage virus	African nations including Guinea, South Africa, Zimbabwe	*Microchiroptera*: *Miniopterus schreibersii*, *Nycteris gambiensis*, and *N. thebaica*
European bat lyssavirus 1	Europe including Denmark, France, Germany, The Netherlands, Poland, Russia, Spain, Ukraine	*Microchiroptera*: *Eptesicus serotinus*
European bat lyssavirus 2	Several countries of western Europe particularly The Netherlands, Switzerland, Finland and the United Kingdom	*Microchiroptera*: *Myotis* spp., especially *M. dasycneme* and *M. daubentonii*
Australian bat lyssavirus	Australia and possibly areas of SE Asia	*Megachiroptera*: *Pteropid* spp.; *Microchiroptera*: *Saccolaimus flavicentris*
Aravan virus	Kyrgyzstan	*Microchiroptera*: *Myotis blythi*
Irkut virus	Irkutsk province, Russia; a closely related virus (*Ozernoe* virus) recovered in Primorye Territory, Russia	*Microchiroptera*: *Murina leucogaster*; *Ozernoe* virus from a human case after exposure to an unidentified bat
Khujand virus	Tajikistan	*Microchiroptera*: *Myotis daubentonii*
Shimoni bat virus	Kenya	*Hipposideros commersoni* (Commerson's leaf-nosed bat)
West Caucasian bat virus	Krasnodar region, Russia	*Microchiroptera*: *Miniopteris schreibersi*
Bokeloh bat lyssavirus	Germany	*Myotis nattererii* (Natterer's bat)
Ikoma virus	Tanzania	Isolated from an African civet (*Civettictis civetta*); bat reservoir likely
Lleida bat lyssavirus	Spain	*Miniopterus schreibersii*

[a]*Even though evidence is lacking with some of these viruses, it must be presumed that all of them may cause rabies-like disease in humans—laboratory biosafety concerns must be set accordingly.*
[b]*Lyssaviruses closely related to rabies are endemic in bats, but not terrestrial animals, in these countries (see below).*

the viral polymerase; and M, a matrix protein that facilitates virion budding by binding to the nucleocapsid and to the cytoplasmic domain of the glycoprotein. Three proteins (N, P, L), in association with viral RNA, constitute the nucleocapsid. The glycoprotein contains neutralizing epitopes that are targets of vaccine-induced immunity; it and the nucleoprotein have epitopes involved in cell-mediated immunity. Virions also contain lipids, the composition of which mimics the composition of host cell membranes, as well as carbohydrate as side-chains of the envelope glycoprotein. Rhabdovirus infectivity is relatively stable in the environment especially when the pH is alkaline—vesicular stomatitis viruses can survive in water troughs for many days—but the viruses are thermo-labile and sensitive to UV-irradiation in sunlight. Rabies and vesicular stomatitis viruses are easily inactivated by detergent-based disinfectants.

Viral Replication

Viral entry into the host cell occurs by attachment of the G protein to cell surface receptors, followed by endocytosis via coated pits and low pH fusion of viral and endosomal membranes; all replication occurs in the cytoplasm. Replication first involves transcription of five monocistronic mRNAs from the (−) sense virion RNA by the virion polymerase (L) (Fig. 27.2). There is but a single promoter

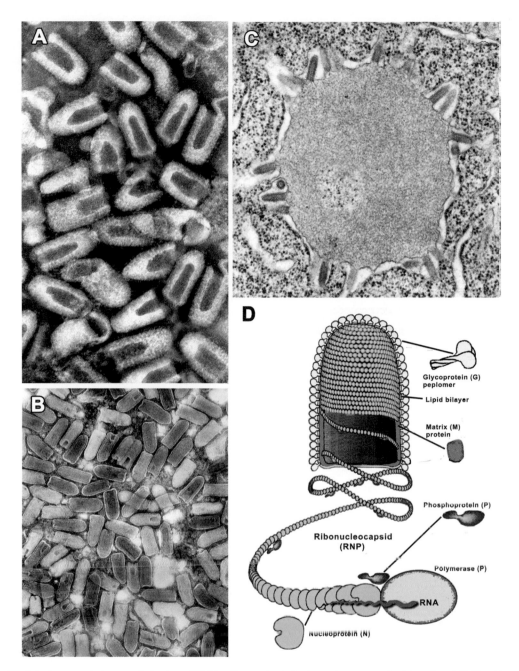

FIGURE 27.1   Rhabdovirus morphology and structure. (A) Negative contrast electron microscopy of rabies virus. (B) Negative contrast electron microscopy of vesicular stomatitis virus. (C) Thin-section electron microscopy of a street rabies virus isolate in hamster brain. Neuronal cytoplasm with virus budding from endoplasmic reticulum membrane in association with an inclusion body (Negri body of light microscopy), which is composed of massed viral nucleoprotein. (D) Diagram of a rhabdovirus virion with its coiled nucleocapsid structure. *(D) Reproduced from King, A.M.Q., et al., 2012. Virus Taxonomy, Ninth Report of the International Committee on Taxonomy of Viruses, Academic Press, London, with permission.*

site, located at the 3′-end of the viral genome; the polymerase attaches to the genome at this site and as it moves along the RNA template it encounters stop/start signals at the boundaries of each viral gene. Only a proportion of the polymerase molecules moves past each junction to continue the transcription process. This mechanism of interrupted transcription results in relatively greater quantities of mRNA being made from genes that are located at the 3′-end of the genome, generating a gradient of descending numbers of transcripts from downstream genes, namely N > P > M > G > L. This allows for the production of larger amounts of the structural proteins such as the N (nucleocapsid) protein relative to the amount of the L (RNA polymerase) protein, the latter being needed in lesser amounts.

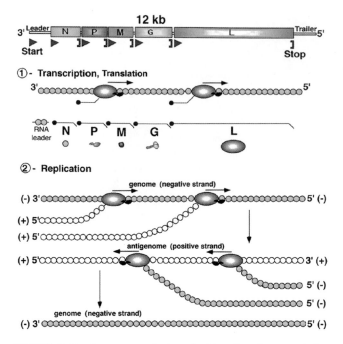

FIGURE 27.2 Genome organization of rabies virus, its transcription, translation, and genome replication strategy. Top: Using virion RNA as a template, the viral transcriptase transcribes five subgenomic mRNA species. There is only a single promoter site, located at the 3′ end of the viral genome; the polymerase attaches to the genomic RNA at this site and as it moves along the viral RNA it encounters stop-start signals at the boundaries of each of the viral genes. Only a fraction of the polymerase molecules move past each junction and continue the transcription process. This mechanism, called stop-start or stuttering transcription, results in more mRNA being made from genes that are located at the 3′ end of the genome and a gradient of progressively less mRNA from downstream genes. Bottom: Replication of the negative sense genome via a positive sense antigenome intermediate. The switch from transcription to replication appears to be regulated by the N protein. G, glycoprotein; N, nucleocapsid; P, phosphoprotein, M, matrix protein; L, RNA polymerase. Rabies virus has an additional transcription unit (−psi) between the G and L genes that is missing from other rhabdoviruses. It does not appear capable of encoding a protein, and has been called a pseudogene. *Reproduced from MacLachlan, N.J., Dubovi, E.J., 2011. Veterinary Virology, fourth ed. Academic Press, London (Figure 4.1), with permission.*

Later, using the protein products of this early phase of transcription, there is production of full-length (+)-stranded templates that in turn are used for the synthesis of the RNA genome. This switch from transcription to genome replication appears to be regulated by the nucleoprotein. Attachment of N protein to newly formed full-length (−) sense genome RNA molecules induces the self-assembly of helically wound nucleocapsids. Through the action of M protein, nucleocapsids are in turn bound to cell membranes at sites where viral peplomers are inserted. The budding of nucleocapsids through cell membranes forms virions. Rabies virus buds mainly through intracytoplasmic membranes of infected neurons, but almost exclusively through plasma membranes of salivary gland epithelial cells, a characteristic

pivotal in facilitating transmission to humans via an animal bite. Vesicular stomatitis viruses bud from basal plasma membranes. Vesicular stomatitis viruses usually cause rapid cytopathology, perhaps the most rapid of any virus. The replication of wild-type rabies virus is slower and usually non-cytopathic (called "street virus" strains first by Pasteur) because these viruses do not shut down host cell protein and nucleic acid synthesis. Rabies virus produces prominent cytoplasmic inclusion bodies (Negri bodies) within infected cells.

Laboratory-adapted ("fixed") rabies virus and vesicular stomatitis viruses replicate well in many kinds of cell cultures: Vero (African green monkey kidney) and BHK-21 (baby hamster kidney) cells being the most common substrates for animal rabies vaccines. Rabies and vesicular stomatitis viruses replicate to high titer in suckling mouse and suckling hamster brain.

During rhabdovirus replication, defective interfering (DI) virus particles are commonly formed. These are complex deletion mutants, possessing greatly truncated RNA genomes that interfere with the normal viral replication processes (see Chapter 4: Virus Replication) (Table 27.2).

RABIES

A known animal exposure followed several weeks later with the appearance of classic signs and symptoms should alert the astute infectious disease physician or public health official to the possibility of rabies as the cause. However, rabies should also be considered in any suspected viral encephalitis of unknown origin in rabies endemic areas, regardless of a history of animal bite—the exposure may have been unrecognized, forgotten, or discounted. There is no practical method for detecting virus in humans until late in the course of the central nervous system infection. Use of appropriate laboratory techniques is the sole method for confirming rabies virus infection.

Clinical Features

Following the bite of a rabid animal the incubation period is usually between 14 and 90 days, but may be considerably longer. A few human cases have been observed in which the last opportunity for exposure occurred from two to seven years before the onset of clinical disease. In such cases the virus has always been found to have a genotype originating from a reservoir animal species (usually the domestic dog) in the country where the patient has been resident or has traveled.

The risk of acquiring infection and the incubation period are affected by the type and extent of exposure. A bite on exposed skin is more likely to produce rabies than a bite through thick clothing, which may remove much of the salivary contamination. Multiple bites are more likely to

TABLE 27.2 Properties of Rhabdoviruses

Virions are enveloped, bullet-shaped, 70 nm in diameter, and 170 nm long, and comprise an envelope with large spikes, within which is a helically coiled cylindrical nucleocapsid
The genome is a single molecule of linear, negative-sense, single-stranded RNA, 11 to 15 kb in size
Rhabdoviruses generally encode 5 proteins: L, RNA-dependent RNA polymerase involved in transcription and RNA replication; G, glycoprotein which forms trimers that make up the peplomers (spikes); N, nucleoprotein, the major component of the nucleocapsid; P, component of the viral polymerase; M, matrix, a protein that facilitates virion budding by binding to the nucleocapsid and to the cytoplasmic domain of the glycoprotein. Gene order is 3′-N-P-M-G-L-5′
Viral entry into the cell occurs by attachment of the G protein to cell surface receptors, followed by endocytosis via coated pits and low pH fusion of viral and endosomal membranes; replication occurs in the cytoplasm
Virion polymerase transcribes five monocistronic mRNAs from incoming viral genome, in decreasing amounts from the 3′ end. Later, a switch from transcription to replication generates full-length + ve strand RNA that is used to synthesize −ve sense genomic RNA
Attachment of N protein to newly formed full length (−) sense genome RNA molecules induces the self-assembly of helically wound nucleocapsids. Virions are formed by budding through intracytoplasmic or plasma membranes
Some viruses, such as vesicular stomatitis viruses, are rapidly cytopathic, while others such as street rabies virus, are non-cytopathic

transmit than a single bite, and bites on the face or neck are more likely to transmit, and to lead to a shorter incubation period, than bites on the extremities.

There is a prodromal phase prior to overt clinical disease, consisting of fever, malaise, and often paresthesia around the site of the bite. Later, muscles become hypertonic and the patient becomes anxious, with episodes of hyperactivity, aggression, and convulsions. The patient often cannot swallow water because of paralysis of pharyngeal muscles, giving rise to the old name for the disease, "hydrophobia." There is often excessive salivation, exaggerated responses to light and sound, and hyperesthesia. Terminally, there are often delirium, convulsive seizures, coma, and respiratory arrest, with death occurring 2 to 14 days after the onset of clinical signs.

Infected animals may show a prodromal phase with subtle behavioral changes, followed by one of two clinical forms: "furious" rabies, in which the animal becomes restless, aggressive, and may lose its fear and attack humans without provocation; or "dumb" rabies, in which the animal develops paralysis, may develop a characteristic "dropped jaw" due to paralysis of the masseter muscle, and may even be misdiagnosed as having a foreign body in the throat, a dangerous situation if it leads to attempted removal.

Pathogenesis, Pathology, and Immunity

The risk of developing rabies after exposure depends upon the location and severity of the bite or scratch and the species of animals involved (foxes can carry up to 10^6 infectious units of virus per mL of saliva). The bite of a rabid animal usually delivers virus deep into striated muscles and connective tissue, but infection can also occur, albeit with less certainty, after superficial abrasion of the skin. From its entry site, virus must gain entry into peripheral nerves; this may happen directly, but in many instances virus is amplified by first replicating in local muscle cells. The virus invades the peripheral nervous system through sensory or motor nerve endings. Several receptors are implicated in virus binding, including neuronal cell adhesion molecule (NCAM) at neuromuscular junctions and the nicotinic acid acetylcholine receptor (AChR), facilitating entry into nerve endings. Additionally the virus is thought to bind to the neuronal growth factor receptor p75NTR.

Neuronal infection and centripetal passive movement of the viral genome within axons deliver virus to the central nervous system, usually via the spinal cord. An ascending wave of neuronal infection and neuronal dysfunction then occurs. The virus reaches the limbic system of the brain, where it replicates extensively, and causes the release of cortical control of behavior—this leads to the fury seen clinically. Spread within the central nervous system continues, and when replication occurs in the neocortex the clinical picture changes to the dumb or paralytic form of the disease. Depression, coma, and death from respiratory arrest quickly follow.

In animal species that serve as reservoir hosts of rabies virus, late in infection virus moves centrifugally from the central nervous system through peripheral nerves to a variety of organs: the adrenal cortex, pancreas, and most importantly, the salivary glands. Most virus produced in nervous tissue buds from intracytoplasmic membranes; however, in the salivary glands virions bud upon plasma membranes at the apical (lumenal) surface of mucous cells and are released in high concentrations into the saliva. Thus, at the time when viral replication within the central nervous system causes the infected animal to become furious and bite indiscriminately, the saliva is highly infectious.

With street virus (wild-type) strains, histopathological examination shows only a modest level of neuronal damage compared to the lethality of the infection. Electron microscopy, immunofluorescence, or immunohistochemistry prove that many neurons are infected, but there is little frank cytopathology or inflammatory cell infiltration. The apparent minimal target damage accompanied by lethal neurological dysfunction has always been considered a great paradox, although there is evidence that this feature may reflect neuronal dysfunction arising from virus-induced inhibition of protein synthesis. In dumb or paralytic rabies, levels of viral RNA and antigen in the CNS tend to be lower and cellular infiltrates are more marked.

Although rabies proteins are highly immunogenic, neither humoral nor cell-mediated responses can be detected during the stage of movement of virus from the site of the bite to the central nervous system, probably because very little antigen is delivered to the immune system—most is sequestered in muscle cells or within nerve axons. However, in this early stage of infection the virus is accessible to antibody; hence the efficacy in exposed humans of post-exposure prophylaxis (consists of rabies immune globulin and vaccine, see below). Immunological intervention is effective for some time during the long incubation period because of the delay between the initial viral replication in muscle cells and the entry of virus into the protected environment of the nervous system.

The host innate immune system is inhibited by the viral phosphoprotein P, which is expressed as both full-length product and also as up to four truncated versions. Interferon synthesis is inhibited at several points in the induction pathway.

Laboratory Diagnosis

In most countries where rabies is endemic, laboratory diagnosis is done only in approved laboratories by qualified, experienced personnel. Most testing is done on animals suspected of having transmitted virus to a human via bite, scratch, or otherwise. However, the same methods are used in the infrequent circumstances of testing for human infection.

Historically, brain tissue from the subject was examined for Negri bodies, which are specific intracytoplasmic inclusion bodies (mostly concentrations of viral nucleocapsids) observable under the light microscope after staining sections or impressions with Seller's or Giemsa stain. The largest Negri bodies occur in neurons in Ammon's horn of the hippocampus, Purkinje cells of the cerebellum, and pyramidal neurons of the cerebral cortex; however, Negri bodies may be absent in up to a third of rabies cases, while artifacts and inclusions produced by other agents may be confounding. Also historically, intracerebral inoculation of brain material into mice was used widely to confirm rabies diagnosis, even though the time required, up to 21 days of observation, made it less immediately useful.

Today, the direct immunofluorescence method is the test of choice for fresh or frozen tissue, owing to its sensitivity, specificity, and the regular availability of high-quality reagents. It has been the test recommended by WHO for many years. Touch impressions on glass slides are made from brain tissue (medulla, cerebellum, and hippocampus); these are then fixed in acetone, reacted with commercially produced and certified fluorescein isothiocyanate-conjugated hyperimmune rabies antibodies (IgG, polyclonal, or monoclonal), before examination under UV-light. Negri bodies and smaller aggregates of viral antigen appear as apple-green forms against a dark (or counterstained) background (Fig. 10.3A).

In some laboratories, those with appropriate reagents and facilities, post-mortem diagnosis can also be performed using RT-PCR; this is done with primers that amplify both genome RNA and viral mRNA sequences. The method is 100- to 1000-fold more sensitive than standard methods and is a great benefit when specimens are unsuitable for other testing, for example when the suspect animal has been buried for some time.

Ante-mortem diagnosis, using immunofluorescence or RT-PCR assays, is only done in suspected human rabies cases, and then only when there is clinical evidence of encephalitis. For this, a skin biopsy taken from the highly innervated area of the nape of the neck is snap-frozen and then processed as thin sections suitable for either immunofluorescence or fixed for immunohistochemistry (Fig. 27.3A and B). In past years corneal impressions or saliva specimens were tested, but the value of this approach was proven marginal. In any event, only positive results are of diagnostic value, since sites of infection may be missed.

Pre-exposure rabies vaccination is recommended for all occupationally exposed individuals. Since rabies diagnosis often involves opening the skull of suspect animals, appropriate methods and personal protection equipment, including dedicated laboratory clothing; heavy protective gloves and a face shield should be employed. Decontamination of surfaces should be done with standard disinfectants and suspect carcasses should be incinerated or buried after diagnostic specimens have been obtained.

Epidemiology

The number of human deaths caused each year by rabies is estimated to be 40,000 to 100,000, worldwide. An estimated 10 million people receive post-exposure treatments each year after being exposed to animals suspected of being infected with rabies virus.

In usual circumstances the only risk of rabies virus transmission to humans is by the bite or scratch of a rabid

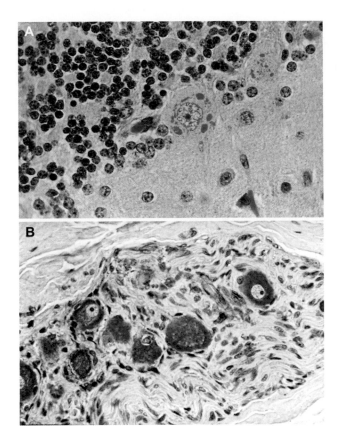

FIGURE 27.3 (A) Brain, human, cerebellum, Purkinje cells containing magenta intracytoplasmic inclusion bodies (Negri bodies). Giemsa stain. In areas of the world where ante-mortem diagnosis is still not available, post-mortem confirmation of the clinical diagnosis of rabies is commonly performed by histology. (B) Spinal dorsal root ganglion showing massive amounts of rabies viral antigen in the cytoplasm of sensory nerve neuronal bodies. Formalin-fixed section, immunohistochemistry. In recent years this methodology has been added in reference diagnostic centers for post-mortem confirmation of rabies, in some instances when the clinical diagnostic rule out did not include rabies. Perhaps because of its rarity in many parts of the world, human rabies is too often not included in differential diagnosis algorithms.

animal, although in bat caves, where the amount of virus may be very high and the extremely high humidity may stabilize the virus, transmission may occur via aerosol. Human-to-human transmission is only a problem in organ transplantation: a significant number of human rabies cases have resulted from the surgical implantation of infected corneas from donors with fatal but undiagnosed rabies. In addition, rabies cases have occurred in Europe and the United States in recipients of kidneys, liver, and arterial segments, again from donors with fatal undiagnosed rabies.

In recent years, understanding of the epidemiology and natural history of rabies virus in its various reservoir host animal species has followed the genome sequencing of large numbers of isolates. It was found that reservoirs among different animal species are perpetuated largely as independent cycles of infection—"fox-bites-fox-bites-fox"—favoring genotype- and pathotype-specific drift; variants then evolve with characteristics that are beneficial to the perpetuation of the virus in the niche. For example, in North America, there are six rabies virus genotypes in terrestrial animals (and many more in various species of bats): (1) a skunk genotype in north-central states of the United States and south-central Canada, (2) a second skunk genotype in south-central states and a third in California, (3) an Arctic fox and red fox genotype in Alaska and Canada, (4) a gray fox genotype in Arizona and western Texas, (5) a dog/coyote genotype in southern Texas and northern Mexico, and (6) a raccoon genotype in eastern states extending into eastern Canada. These genotypes reflect the evolutionary consequence of host preference. After an unknown number of passages the virus becomes a distinct genotype, still able to kill other species but most efficiently transmitted by its own reservoir host (Fig. 27.4).

Separately, there are a number of bat genotypes, each associated with a different bat species, again each filling an independent ecological niche. Such variant viruses are fully virulent for other mammals within a particular region but are best transmitted among the particular reservoir species and tend to die out when infecting other species. There is also spillover of variants into dead-end hosts, for example bat variants into cats and important spillover into hosts that initiate new infection chains and eventually evolve into new variants, for example bat variant into skunks in Arizona, United States.

Dog rabies (the cosmopolitan rabies virus genotype and variants) is still of great importance in many parts of the world and the cause of most human rabies cases, especially in India, China, the Philippines, and many countries in Africa—incidence data are largely inadequate, especially as the numbers of cases of human disease are being reduced greatly in some countries by vaccination and dog rabies control programs. The disease occurs throughout the world, with certain exceptions: rabies is absent in Japan, Australia, New Zealand, United Kingdom, Antarctica, and many smaller islands such as Hawaii and most of the islands of the Caribbean and the southwest Pacific. The rabies-free status of the United Kingdom and Australia is only confounded by the presence of the related lyssaviruses of bats. Rabies is also absent from Switzerland, France, Germany, Belgium, and several other European countries as a result of wildlife vaccination programs; in these countries there was only fox rabies, so enabling bait-delivered vaccination programs to be highly focused on specific geographical areas (Fig. 27.5).

Differences in rabies virus dynamics in different populations involve everything from varying intrinsic resistance to infection, varying immune responsiveness, varying pathogenic factors, to varying animal behavioral characteristics, such as movement over long distances, typical group behavior, seasonal behavior, etc., and a

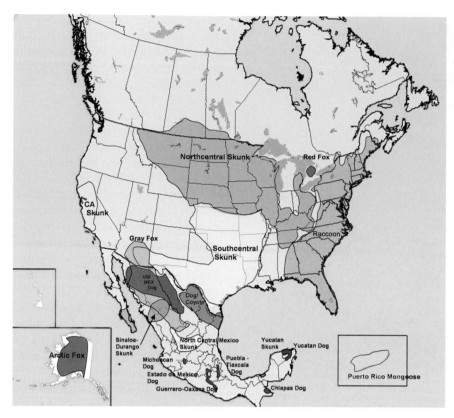

FIGURE 27.4   Geographic occurrence of terrestrial wildlife rabies virus variants in North America. The distribution of terrestrial rabies virus variants in North America is delineated by the primary reservoir host species or the common name of the group of terrestrial carnivore species serving as reservoir hosts. Numerous other rabies virus variants occur in a number of insectivorous bat species throughout the main part of the continent from Alaska to Mexico and South America. *From Andrés Velasco-Villa, U.S. Centers for Disease Control and Prevention, with permission.*

varying ability to bite. Because of their biting capabilities, dogs and other carnivores are the most significant.

Post-Exposure Prophylaxis

The first step in dealing with a possible human rabies exposure is thorough cleansing of the wound: the immediate and vigorous washing and flushing with soap and water is crucial. Next, an appraisal must be made of the likely risk of infection, based on the type of exposure, the epidemiology of rabies in that region, and the animal species involved, together with the circumstances of the exposure incident, for example whether provoked or unprovoked (Table 27.3). Prompt consultation with local public health officials is essential in order to obtain advice about the current incidence of rabies in the area.

Often an algorithm is used to help guide decisions stemming from clinical evaluation, the known distribution of the virus in reservoir animal populations, etc. The specifics vary from country to country, but in general are similar to those of the WHO and the U.S. Advisory Committee on Immunization Practices (Table 27.3). Circumstances of possible rabies exposure and clinical issues can be complex;

where rabies is rare, clinicians may seek the help of public health authorities for advice regarding post-exposure prophylaxis.

Bat rabies poses particular issues: exposure may occur even when a bite wound is not evident. Bat rabies may occur in areas where there is no terrestrial animal rabies and the level of suspicion is low. Rabies surveillance in bats is not commonly done in many areas of the world.

As mentioned above, post-exposure prophylaxis employs hyperimmune globulin and a course of vaccine. Human rabies immune globulin may be used at 20 IU/kg of body weight, ensuring it is thoroughly infiltrated into and around the wound, with any remaining volume injected intramuscularly. There are two licensed human rabies vaccines available in many developed countries: human diploid cell-culture vaccine and a purified chick embryo cell vaccine. If the exposed person has been vaccinated previously, human rabies immune globulin is not used and only two doses of vaccine are given, one each on days 0 and 3. Slightly different regimens are recommended by authorities in various countries, often in response to the very high cost of HRIG and the human diploid cell or chick embryo cell vaccines used in developed countries.

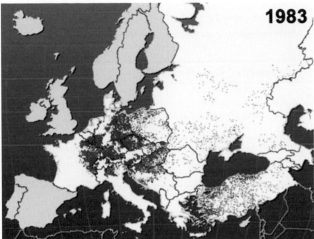

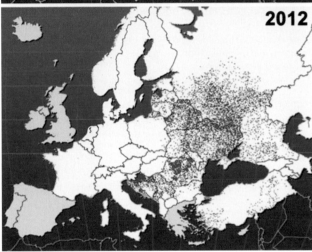

FIGURE 27.5 (top) Rabies, animals, Europe, 1983 – after 30 years of conventional rabies control (trapping, poisoning, shooting foxes) there was little evidence of control of the public health problem. (bottom) Rabies, animals, Europe, 2012 – after 29 years of bait-delivered oral vaccination of foxes there was a remarkable effect. Since the fox was the only reservoir host in Western Europe the program eliminated the threat entirely. As the wildlife vaccination program continues, reaching into more areas of Eastern Europe, the apparent increase in cases in Eastern Europe may be an artefact of increasing surveillance. Additionally, as the program reaches into Russia and surrounding countries, another reservoir host, the raccoon dog, must be dealt with. *From the European Centre for Disease Prevention and Control, with permission.*

Remarkably, these regimens can reduce the mortality from this frightening disease virtually to zero, although the issue of cost often remains an unresolved issue.

Pre-Exposure Vaccination

Pre-exposure vaccination of veterinarians and other individuals occupationally or otherwise at risk, for example laboratory personnel working with rabies virus, animal control, and wildlife workers in rabies-endemic areas, and certain international travelers has become standard practice in developed countries. Because of several variables in the level and nature of risk and the likelihood of immediate availability of post-exposure prophylaxis, an algorithm is used to guide recommendations for primary vaccination and boosters (Table 27.4).

Treatment

There is no specific treatment once symptoms appear. Many antiviral agents have been tried and found to be of no benefit, and in one series all patients succumbed within 17 days despite aggressive symptomatic management, intensive care, and candidate anti-rabies treatments. In recent decades, four exceptional cases have been reported where patients have survived, in some cases with apparently full recovery. Each of these had received some form of post-exposure treatment after onset of symptoms, but when further cases have been treated with the same regimens there have been disappointing results. The most recent example being what has been called the "Milwaukee (or Wisconsin) Protocol" in which patients have been placed in a drug-induced coma and given a combination of antiviral medications that included ribavirin, ketamine, and amantadine. The index patient survived with few sequelae, but several subsequently treated patients did not and hence the survival rate has been questioned.

Control

The control of rabies in different regions of the world poses very different problems, depending upon which reservoir hosts are present and the prevalence of infection among such hosts:

Rabies-free countries Rigidly enforced quarantine of dogs and cats for various periods before entry has been effectively used to exclude rabies from Japan, the United Kingdom, Australia, New Zealand, Hawaii, and many islands. Rabies had never become endemic in wildlife in the United Kingdom and was eradicated from dogs in that country in 1902, and again in 1922 after its re-establishment in the dog population in 1918. Since then, there had been no rabies in the United Kingdom until recently when bats were found infected with European bat lyssavirus type 1, a close variant of rabies virus. This virus was the cause of death in one person in Ukraine and the related virus, European bat lyssavirus type 2, has caused two fatal infections, one in Finland, the other in Scotland. Similarly, rabies is not present in Australia, although the closely related Australian bat lyssavirus (8% sequence difference in its N gene from the cosmopolitan rabies virus genotype) is endemic in fruit bats in several areas of Australia and there have been three recognized human cases and two equine cases, all fatal. The distinction between rabies virus versus these rabies-like viruses has been maintained in Europe and Australia,

TABLE 27.3 WHO Guidelines: Categories of Rabies Contact and Recommended Post-Exposure Prophylaxis (PEP). Many Countries Also Offer Additional Guidelines Based on the Local Prevalence of Rabies Infection in Various Animal Species

Categories of Contact with Suspect Rabid Animal	Post-Exposure Prophylaxis Measures
Category I—touching or feeding animals, licks on intact skin	None
Category II—nibbling of uncovered skin, minor scratches, or abrasions without bleeding	Immediate vaccination and local treatment of the wound
Category III—single or multiple transdermal bites or scratches, licks on broken skin; contamination of mucous membrane with saliva from licks, contacts with bats	Immediate vaccination and administration of rabies immunoglobulin; local treatment of the wound

All category II and III exposures assessed as carrying a risk of developing rabies require PEP. This risk is increased if:
- the biting mammal is a known rabies reservoir or vector species;
- the animal looks sick or has an abnormal behavior;
- a wound or mucous membrane was contaminated by the animal's saliva;
- the bite was unprovoked; and
- the animal has not been vaccinated.

In developing countries, the vaccination status of the suspected animal alone should not be considered when deciding whether to initiate prophylaxis or not. PEP for category II exposures includes wound disinfection and administration of vaccine only. PEP for category III exposures includes in addition rabies immunoglobulin. PEP for category II and III exposures:
- IM administration requires 1 mL or 0.5 mL (depending on the type of vaccine) into the deltoid muscle (or anterolateral thigh in children aged < 2 years)
- (i) the 5-dose regimen prescribes 1 dose on each of days 0, 3, 7, 14, and 28
- (ii) the 4-dose regimen prescribes 2 doses on day 0 (1 in each of the 2 deltoid or thigh sites) followed by 1 dose on each of days 7 and 21
- An alternative regimen for healthy, fully immunocompetent people who receive wound care plus high-quality rabies immunoglobulin (RIG) plus WHO prequalified rabies vaccines is 4 doses administered IM on days 0, 3, 7, and 14
- ID administration
- The 2-site regimen prescribes ID injection of 0.1 mL at 2 sites (deltoid and thigh) on days 0, 3, 7, and 28
- This regimen may be used for people with category II and III exposures in countries where the ID route has been endorsed by national health authorities

TABLE 27.4 Rabies Immunization: Criteria for Pre-Exposure Immunization, United States

Exposure Category	Nature of Risk	Typical Populations	Pre-Exposure Regimen
Continuous	Virus present continuously, often in high concentrations. Specific exposures may go unrecognized. Bite, non-bite, or aerosol exposures	Rabies research lab workers. Rabies biologics production workers	Primary course. Serologic testing every 6 months. Booster immunization if antibody titer falls below acceptable level
Frequent	Exposure usually episodic with source recognized but exposure also might be unrecognized. Bite, non-bite, or aerosol exposures	Rabies diagnostic lab workers, cavers, veterinarians and staff, and animal control and wildlife workers in areas where rabies is enzootic. All persons who frequently handle bats	Primary course. Serologic testing every 2 years. Booster vaccination if antibody titer is below acceptable level
Infrequent (greater than population at large)	Exposure nearly always episodic with source recognized. Bite or non-bite exposures	Veterinarians and animal control staff working with terrestrial animals in areas where rabies is uncommon to rare. Veterinary students. Travelers visiting areas where rabies is enzootic and immediate access to appropriate medical care including biologics is limited	Primary course. No serologic testing or booster vaccination
Rare (population at large)	Exposure always episodic. Bite or non-bite exposure	Population at large, including individuals in rabies-epizootic areas	No vaccination necessary

Adapted from guidelines from the U.S. Centers for Disease Control and Prevention: *Pre-exposure immunization:* Pre-exposure immunization consists of three doses of human diploid cell vaccine (HDCV) or primary chick embryo cell vaccine (PCEC) vaccine, 1.0 mL, IM (i.e., deltoid area), one each on days 0, 7, and 21 or 28. Administration of routine booster doses of vaccine depends on exposure to risk category as noted. Routine pre-exposure booster immunization consists of one dose of HDCV or PCEC, 1.0 mL/dose, IM (deltoid area), or HDCV, 0.1 mL ID (deltoid). *Post-exposure immunization:* All PEP should begin with immediate thorough cleansing of all wounds with soap and water. *Persons not previously immunized:* HRIG, 20 IU/kg body weight, as much as possible infiltrated at the bite site (if feasible), with the remainder administered IM; 4 doses of HDCV or PCEC, 1.0 mL IM (i.e., deltoid area), one each on days 0, 3, 7, and 14. *Persons previously immunized:* Two doses of HDCV or PCEC, 1.0 mL, IM (i.e., deltoid area), one each on days 0 and 3. HRIG should not be administered.

despite the diseases in humans appearing similar, to ensure maintenance of strict quarantine for all imported animals, hopefully thereby preventing classical rabies from becoming endemic among terrestrial wild and domestic animals.

Developing countries. In most countries of Asia, Latin America, and Africa, endemic dog rabies is a serious problem, marked by significant domestic animal and human mortality. In these countries, very large numbers of doses of human vaccines are used and there is a continuing need for comprehensive, professionally organized, and publicly supported rabies control agencies. That such agencies are not in place in many developing countries is a reflection of their high cost; nevertheless, great progress is being made. Substantial decreases in rabies incidence have been reported in recent years in India, China, Thailand, and Sri Lanka, following implementation of programs for vaccination of dogs and improved post-exposure prophylaxis of humans. The decline in rabies cases in Latin America has been remarkable—the Pan American Health Organization (PAHO) has reported a 95% reduction in human rabies between 1980 and 2012, all due to proven control strategies: mass vaccination of dogs, promotion of responsible pet ownership, timely pre- and post-exposure prophylaxis, epidemiological surveillance and laboratory diagnosis, and community health education. Remaining problems are centered in Bolivia, Brazil, the Dominican Republic, Guatemala, Haiti, and Peru.

Developed countries. In most developed countries, even those with a modest disease burden, publicly supported rabies control agencies operate in the following areas: (1) stray dog and cat removal and control of the movement of pets (quarantine is used in epidemic circumstances, but rarely), (2) immunization of dogs and cats, so as to break the chain of virus transmission, (3) laboratory diagnosis, to confirm clinical observations and obtain accurate incidence data, (4) surveillance, to measure the effectiveness of all control measures, and (5) public education programs to assure cooperation.

European countries. Historically, in developed countries, rabies control in wildlife was based upon animal population reduction by trapping and poisoning, but in recent years the immunization of wild animal reservoir host species, especially foxes, by the aerial distribution of baits containing attenuated virus vaccine, has become the method of choice. Since 1990, fox rabies, the only endemic rabies in much of Europe, has been virtually eliminated (Fig. 27.4). The success of wildlife vaccination programs, whether reaching the level of virus elimination from a region or just greatly reduced prevalence, has in turn reinvigorated other rabies control strategies and more conventional animal control programs, for example the elimination of dog rabies as a problem in KwaZulu-Natal, South Africa.

North American countries. In recent years, there has been a significant increase in the numbers of animal rabies cases reported in the United States, mostly because of the epidemicity of raccoon rabies, but also because of increases in coyote rabies in southern Texas and Arctic fox rabies in Alaska. Rabies in Arctic foxes is an endemic, refractory problem extending across most areas of the northern polar regions of the world; it periodically extends into regions of Canada and the United States inhabited by the red fox resulting in wave front epidemics. Skunk rabies in central North America, from Texas to Saskatchewan, is the principal cause of rabies in cattle. The epidemic of raccoon rabies in the eastern United States has been traced to the translocation of raccoons from an old rabies endemic area in Florida to West Virginia in 1977. Its continuing spread, by 1997 as far as eastern Canada and across the Appalachian mountains into Ohio, is still the cause of massive prevention and control efforts.

Fox rabies had been a problem in the eastern United States and Canada, but in the past few years, especially in Ontario, aerial distribution of vaccine-containing baits has minimized the problem. The great merit of this approach over animal population reduction is that the econiche remains occupied, in this case by an immune population, and is not subject to the sort of reproductive "population boom" that the fox is capable of when faced with an empty niche. The same approach has also been very successful in intercepting the northern movement toward large population centers of a coyote-dog rabies virus variant in southern Texas—this large focus has been completely eliminated. Attenuated virus rabies vaccine is immunogenic in foxes and coyotes when administered *per os* in baits, but less so in raccoons and skunks. Further, the habits of the latter two species require use of much higher baiting densities, thereby greatly increasing the cost of control programs. Nevertheless, raccoon vaccination programs are in place across the entire area where raccoon rabies has become endemic.

Rabies in Bats

Bat rabies represents a unique problem, and maintenance of the virus in bats is not well understood. There are over 1200 species of bats, representing 25% of all mammal species, with major differences in habitats and life styles. While infection in bats can cause clinical disease and death, there is evidence that non-lethal immunizing infections and continuing circulation of viruses between different bat colonies are important. Bat rabies occurs in areas where there is no other reported host involvement. Moreover, bats have been an increasing source of human rabies cases in some countries, and in many of these cases there has been no history of a bite—in the United States over 70% of human rabies cases over the past 30 years have been caused by bat genotype viruses. It has been suggested that the viral genotype in some bat species might have enhanced invasiveness, causing infection even after the most trivial, unrecognized bite as bat incisors may leave a wound as small as that made by a 30-gauge syringe needle.

Between 1977 and 2014, more than 1000 cases of bat rabies (that is bat lyssavirus infection) were detected in Europe; the majority of positive bats originated from Denmark, the Netherlands, Germany, Poland, France, Spain, Switzerland, Great Britain, Czech Republic, Slovakia, Hungary, Belarus, Ukraine, and Russia. European bat lyssavirus type 1 has a specific association with the Serotine bat (*Eptesicus serotinus*, in Spain *E. isabellinus*), while European bat lyssavirus type 2 has been isolated from Daubenton's bats (*Myotis daubentonii*) in the United Kingdom, Switzerland, Finland, and Germany, and from Pond bats (*M. dasycneme*) in the Netherlands. European bat lyssavirus types 1 and 2 have been responsible for several confirmed human fatalities.

In several countries of Latin America, rabies in vampire bats is a problem both for the livestock industries and humans. There are three species of vampire bats, the most important being *Desmodus rotundus*. Here, control efforts have depended upon the use of bovine vaccines and upon the use of anti-coagulants such as diphenadione and warfarin. These anti-coagulants are either fed to cattle as slow-release boluses (cattle are very insensitive to their anti-coagulant effect) or mixed with grease and spread on the backs of cattle. When vampire bats feed upon the blood of treated cattle or preen themselves on each other to remove the grease, they suffer fatal hemorrhages in the animals' wing capillaries.

Australian bat lyssavirus, first discovered in a black flying fox (fruit bat, *Pteropus alecto*), is now known to extend along the entire east coast of Australia and to infect all four of the pteropid bat species of Australia and insectivorous bats as well. Fatal human infections have followed encounters with both fruit and insectivorous bats, such as the yellow-bellied sheath-tailed bat (*Saccolaimus flaviventris*).

VESICULAR STOMATITIS AND OTHER RHABDOVIRUSES

In the Western Hemisphere, vesicular stomatitis viruses and some other vesiculoviruses are zoonotic, being transmissible to humans (typically, farmers and veterinarians) from vesicular fluids and tissues of infected animals. The disease in humans resembles influenza, presenting with an acute onset of fever, chills, and muscle pain. The infection resolves without complications within 7 to 10 days. Human cases are not uncommon during epidemics in cattle and horses, but because of lack of awareness few cases are reported. Human cases can be diagnosed retrospectively using serological methods. There are no practical measures for preventing occupational exposure.

Besides the well-known viruses, vesicular stomatitis Indiana virus and vesicular stomatitis New Jersey virus, found in cattle, pigs, and horses in North, Central, and South America, there are other vesiculoviruses that are capable of causing disease in humans, often seeming to be somewhat more incapacitating than the better known viruses: these are vesicular stomatitis Alagoas, Cocal, and Piry viruses in South America, Isfahan virus in Iran, and Chandipura virus in India and Africa. Chandipura virus infection has been recognized in India for some years as causing acute influenza-like illness with neurologic dysfunction mostly in children. However, in the past decade, explosive outbreaks involving hundreds of children have been seen in rural areas of India. The illness has been further characterized by acute onset of fever, altered sensorium, seizures, diarrhea, vomiting, and coma and in many instances death (reported case fatality rate is 55 to 75% in subjects 2 to 16 years of age). Death or recovery occurs rapidly, within two to three days; there are no sequelae among survivors.

FURTHER READING

Advisory Committee on Immunization Practices, 2008. Human Rabies Prevention—United States, current version. MMWR Vol. 57 #RR-03.

Delmas, O., Holmes, E.C., Talbi, C., et al., 2008. Genome diversity and evolution of lyssaviruses. PLOS One 3, 6.

Hemachudha, T., Ugolini, G., Wacharapluesadee, S., et al., 2013. Human rabies: neuropathogenesis, diagnosis and management. Lancet Neurol 12, 498–513.

Mackenzie, J.S., Childs, J.E., Field, H.E., Wang, L.F., Breed, A.C., 2008. The role of bats as reservoir hosts of emerging neurological viruses. In: Reiss, C.S. (Ed.), Neurotropic Viral Infections, Cambridge University Press, pp. 382–406. Chapter 21.

Streiker, D.G., Turmelle, A.S., Vonhof, M.J., 2010. Host phylogeny constrains cross-species emergence and establishment of rabies virus in bats. Science 329, 676–679.

WHO. Rabies—Guide for post-exposure prophylaxis. www.who.int/rabies/human/postexp/en (undated).

Chapter 28

Filoviruses

Until the 1990s, all filovirus disease outbreaks were self-limiting. However, this has changed—especially with the recent prolonged epidemic of Ebola virus disease (Ebola hemorrhagic fever) in the West African countries of Sierra Leone, Liberia, and Guinea. This epidemic, involving 28,616 cases and 11,310 deaths (WHO final data) has stimulated a watershed in filovirus research, the numbers of cases giving added urgency to the development of new treatment and prevention strategies.

The documented story of the filoviruses began in 1967, when 31 cases of hemorrhagic fever, with seven deaths, occurred among laboratory workers in Germany and Yugoslavia who were processing kidneys from African green monkeys (*Chlorocebus aethiops*) recently imported from Uganda. A new virus was isolated from patients and monkeys and subsequently named Marburg virus, now the prototype member of the family *Filoviridae*. Nine years later two further extraordinary epidemics of hemorrhagic fever occurred, one in villages in the rain forest of Zaire (now Democratic Republic of Congo), the other in southern Sudan, 700 km away. Altogether there were more than 550 cases and 430 deaths. A virus morphologically identical to but antigenically distinct from Marburg virus was isolated from Zaire by scientists at the United States Centers for Disease Control and named Ebola virus. A similar relationship was found by scientists in the United Kingdom between Marburg virus and the Ebola virus from the Sudan leading to worldwide speculation, since disproven, that the 1976 outbreaks shared a common source.

Since 1976, there have been many sporadic outbreaks of Ebola virus disease in Africa, caused by four genetically distinct viruses (Zaire, Sudan, Taï Forest, and Bundibugyo viruses), representing an increasing geographical range within Africa. Ebola virus was recognized outside Africa for the first time in 1989 and 1990 when monkeys imported from the Philippines into an import quarantine facility in Reston, Virginia, the United States, were found infected with another new virus, now called Ebola virus, subtype Reston. Infected monkeys at the facility became ill and many died. Four animal caretakers were infected but, in sharp contrast to African filoviruses, there was no clinically apparent disease. Filoviruses related to Ebola virus isolates were later discovered among swine in the Philippines and also in cave-dwelling bats in Africa and Europe. Antibodies to Reston virus have been found among workers in the swine industry of the Philippines.

Viruses of the family *Filoviridae* are intriguing for several reasons: (1) the viruses, although similar in genomic organization and mode of replication to other members of the order *Mononegavirales*, are morphologically the most bizarre of all viruses, (2) the viruses have caused large outbreaks among humans and monkey populations, and therefore are recognized as having substantial epidemic potential, (3) with the exception of Reston virus, the viruses cause a devastating clinical disease in humans, with extremely rapid and florid tissue damage and a very high mortality rate, and (4) the viruses cause a silent infection in insectivorous and fruit-eating bats, a possible exception being Lloviu virus, isolated in Spain, which causes a symptomatic disease in bats.

Biohazard: Marburg and Ebola viruses are *Restricted Pathogens* and their importation or possession is regulated by most national governments. The viruses are restricted to certain national laboratories for all research procedures; the viruses are also Biosafety Level 4 pathogens; all specimens must be handled in the laboratory under maximum containment conditions to prevent human exposure. Patients known or reasonably suspected to be infected with Ebola virus are subject to strict regulatory actions.

PROPERTIES OF BUNYAVIRUSES

Classification

The family *Filoviridae* belongs to the order *Mononegavirales*, comprising negative-sense single-stranded RNA viruses with similar genome organization and replication strategies that include the families *Paramyxoviridae, Rhabdoviridae, Bornaviridae*, and *Filoviridae*. These characteristics and conserved ancient domains found in their genome sequences support the notion of a common ancestry. Conserved domains, in nucleoprotein and polymerase genes, suggest that viruses of the family *Filoviridae* are most closely related to those of the genus *Pneumovirus* in the family *Paramyxoviridae*, rather than to viruses of the family *Rhabdoviridae* as might be expected from the similarly helically wound nucleocapsid structures seen in both taxa.

Fenner and White's Medical Virology. DOI: http://dx.doi.org/10.1016/B978-0-12-375156-0.00028-X

TABLE 28.1 Members of the Family *Filoviridae*

Genus	Species	Region	Approximate Mortality in Humans
Marburgvirus	Marburg	East Africa	>90%
Ebolavirus	Sudan	Africa	~50%
	Zaire	Africa	>90%
		West Africa	~50%
	Taï Forest	Cote D'Ivoire	Only one fatal case reported
	Reston	Philippines	Does not cause clinical disease
	Bundibugyo	Uganda	~50%
Cuervavirus	Lloviu	Spain	Not known

All the non-segmented negative-sense RNA viruses share several characteristics: (1) a similar gene order flanked by 5′- and 3′-UTR regions and devoid of a 3′-polyA tail, (2) a virion-associated RNA polymerase (transcriptase), (3) a helical nucleocapsid, (4) transcription of mRNAs by sequential interrupted synthesis from a single promoter, and (5) virion maturation via budding of preassembled nucleocapsids from plasma membrane sites containing patches of viral glycoprotein peplomers.

The family *Filoviridae* contains three genera, *Marburgvirus*, *Ebolavirus*, and *Cuervavirus* (Table 28.1). The degree of genetic stability of the filoviruses overall, and the absence of genetic variability among Ebola virus isolates obtained within an outbreak, match similar characteristics of other member viruses of the order *Mononegavirales* (Fig. 28.1).

Two lineages (>90% identical) are recognized within the genus *Marburgvirus*: one lineage comprises Marburg virus, the other Ravn virus. Five species have been recognized in the genus *Ebolavirus*: *Zaire ebolavirus*, *Sudan ebolavirus*, *Reston ebolavirus*, *Tai Forest ebolavirus* (formerly *Ivory Coast or*

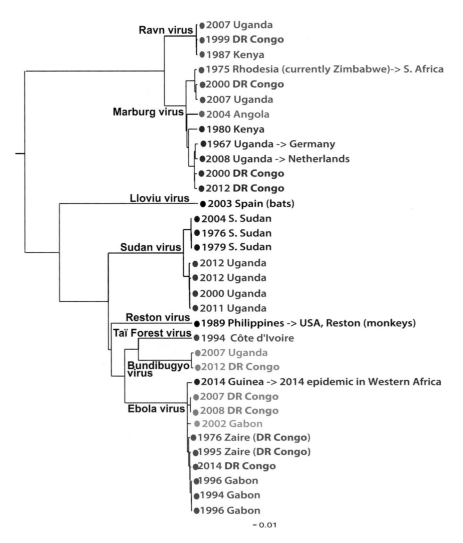

FIGURE 28.1 Phylogenetic tree showing the relationship between the Ebola virus subtypes and the two Marburg virus variants. The various isolates of each subtype, in some cases made over long intervals and from distant geographic locales, are remarkably similar and yet different enough to indicate that each has evolved separately in its own locale or econiche. The major epidemic that started in February 2014 in Guinea, West Africa, and then spread into Liberia and Sierra Leone was caused by an Ebola virus (prototype isolate, Makona) that is rather closely related to the clade that goes back to the original isolate from Zaire in 1976. The entire coding region of the genome of the viruses was used to construct this tree. The color coding is used to highlight closely related isolates. *Reproduced from Yusim, K., et al., 2016. Integrated sequence and immunology filovirus database at Los Alamos. Database (Oxford). pii: baw047. doi: 10.1093/database/baw047, with permission.*

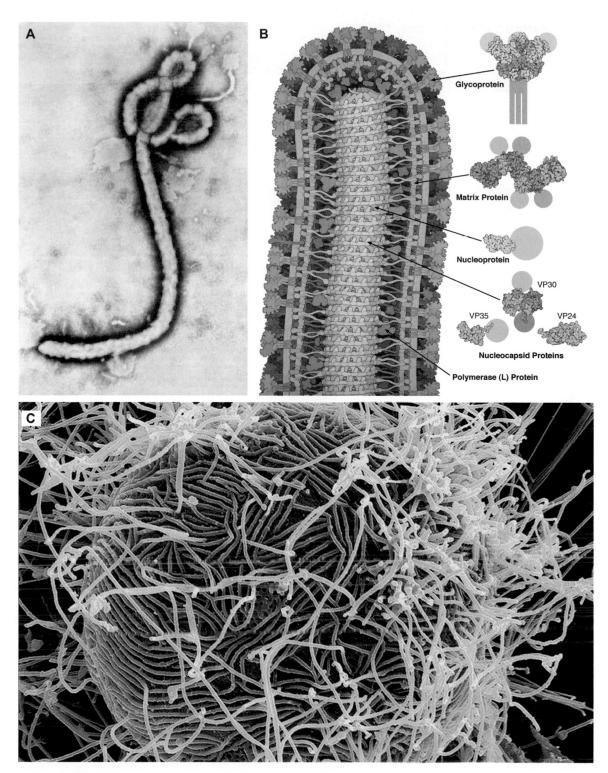

FIGURE 28.2 Filovirus morphology and structure. (A) Negative contrast electron microscopy of Ebola Zaire virus, from the original outbreak in 1976 in Yambuku, Zaire (now Democratic Republic of Congo). (B) Section through part of an Ebola virion, showing proteins in blue, green and magenta, the RNA genome in yellow, and the membrane-virus envelope in light purple. Components are shown on the right, with those for which structure has not been determined shown with schematic circles. (C) Scanning electron microscopy of a Vero cell infected with Ebola Zaire virus—exceptional amount of virus still associated with the cell surface after budding through the plasma membrane. *(B) From David Goodsell, The Scripps Research Institute and the Research Collaboratory for Structural Bioinformatics (RCSB), Protein Data Bank (PDB), with permission; (C) from the National Institute of Allergy and Infectious Diseases, NIH, with permission.*

Marburg virus genome, 19.1 kb

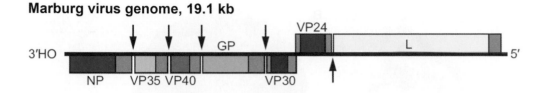

Ebola Zaire, Ebola Sudan, Ebola Taï Forest virus genomes, 18.9 kb

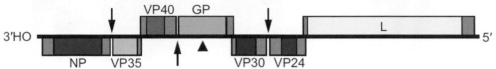

FIGURE 28.3 Genome organization of Marburg and Ebola viruses. Genes for the structural proteins associated with the virion RNA include the genes coding for the nucleoprotein (NP), VP30, VP35, and RNA polymerase (L) proteins. Genes for the envelope-associated proteins include the genes for the matrix (VP40), VP24, and glycoprotein (GP). Genes begin with a conserved transcriptional start site and end with a transcriptional stop (polyadenylation) signal; adjoining genes are either separated from one another by intergenic regions (arrows) or overlap one another. The primary gene product of the GP gene undergoes proteolytic cleavage, and in the case of Ebola virus there is also transcriptional editing (black triangle) to produce either the full-length spike glycoprotein GP1,2, or two non-structural secreted glycoproteins, sGP and ssGP. At the extreme 3′- and 5′-ends of the genomes are complementary leader and trailer sequences. *Reproduced from King, A.M.Q., et al., 2012. Virus Taxonomy, Ninth Report of the International Committee on Taxonomy of Viruses, Academic Press, with permission.*

Cote d'Ivoire ebolavirus), and *Bundibugyo ebolavirus*. A distinct complete filovirus sequence has been obtained from bats in Spain but the virus has not been propagated in cell culture or animals; this new filovirus, designated Lloviu virus, represents the single species *Lloviu cuevavirus* in a new genus *Cuevavirus*.

Virion Properties

Filovirus virions are pleomorphic, appearing as long filamentous, sometimes branched forms, or as "U"-shaped, "6"-shaped or circular forms. Virions have a uniform diameter of 80 nm and vary greatly in length (particles may be up to 14,000 nm long, but have unit nucleocapsid lengths of about 800 nm for Marburg and 1000 nm for Ebola virus). In any one preparation, nearly 50% of virions contain more than one genome copy, some so extensive in length to have over 20 copies. This may aid virus dissemination through epithelial layers of body tissues.

Virions are composed of a lipid envelope covered with peplomers, surrounding a helically wound nucleocapsid (50 nm in diameter with a periodicity of 5 nm) (Fig. 28.2). The genome is composed of a single molecule of negative-sense single-stranded RNA, 19.1 kb in size, the largest of all negative-sense RNA viruses. The filovirus genome encodes two unique proteins: VP30 that may have a homologue in the paramyxovirus genome, and VP24, which is unique. The gene order is: 3′-untranslated region—NP (the nucleoprotein)—VP35 (part of the nucleocapsid)—VP40 (membrane-associated matrix protein)—GP (the glycoprotein)—VP30 (part of nucleocapsid)—VP24 (membrane-associated protein)—L (the RNA polymerase or transcriptase)—5′-untranslated region. Genes are separated

either by intergenic sequences or by overlaps, that is, short (17 to 20 bases) regions where the transcription start signal of the downstream gene overlaps the transcription stop signal of the upstream gene. Ebola virus has three overlaps that alternate with intergenic sequences, Reston virus has two, and Marburg virus has a single overlap (Fig. 28.3).

The seven filovirus structural proteins can be subdivided into two categories, those that form the nucleocapsid and those that are associated with the envelope. The nucleocapsid-associated proteins are involved in the transcription and replication of the genome, whereas the envelope-associated proteins have a role in either the assembly of virions or virus entry (Table 28.2).

Marburg virus glycoprotein is coded by a single open reading frame, whereas two reading frames code for the Ebola virus glycoprotein. Expression of Ebola virus glycoprotein involves site-specific RNA transcriptional editing and translational frame shifting that joins the two open reading frames.

The full-length Ebola virion glycoprotein (GP$_{1,2}$, Mr 120,000–170,000) forms the trimeric surface peplomers, whereas a second glycoprotein (sGP, Mr 60,000), with a common N-terminal end but formed by proteolytic cleavage, is made in large amounts and secreted extracellularly. The nature of the participation of this soluble glycoprotein in the pathogenesis of Ebola virus disease is unknown, but it may serve as some sort of immune decoy, minimizing the immune response to the virus, and/or it may be immunosuppressive, affecting the host response to infection. Each surface trimer has a "glycan cap" which may play a role in immune evasion.

Of the other filovirus structural proteins (Table 28.2), it is remarkable how many are involved in evasion of the host inflammatory, innate and adaptive immune responses.

TABLE 28.2 Filovirus Genes and the Functions of Their Gene Products

Gene Order	Gene	Protein Function
1	Nucleoprotein (NP)	Major nucleoprotein; RNA encapsidation
2	Virion protein (VP) 35	Polymerase complex cofactor; interferon antagonist
3	VP40	Matrix protein; virion assembly and budding; interferon antagonist
4	Glycoprotein (GP)	Virus entry (surface peplomer); receptor binding and membrane fusion
	Soluble glycoprotein (sGP)	Unknown
	Secondary soluble glycoprotein (ssGP)	Unknown
5	VP30	Minor nucleoprotein; RNA encapsidation and transcription activation
6	VP24	Minor matrix protein; virion assembly; interferon antagonist
7	Polymerase (L)	RNA-dependent RNA polymerase; enzymatic component of polymerase complex

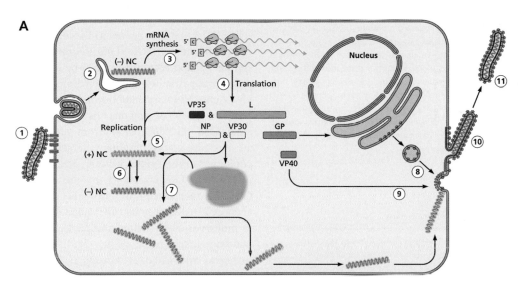

FIGURE 28.4A Replication cycle of Ebola virus. (1) Virus particles bind to a cell surface receptor, followed by uptake via macropinocytosis and trafficking to late endosomes. (2) Viral glycoprotein (GP) is cleaved by endosomal cysteine proteases and then binds to the membrane protein, NPC1, which is exposed in the endosomal lumen. This interaction leads to fusion of the viral and endosomal membranes and delivery of the viral nucleocapsid into the cytoplasm. (3) The nucleocapsid, which is composed of the viral RNA and L (polymerase), NP, and VP30 proteins, is the template for the synthesis of seven viral mRNAs in the cytoplasm. The mRNAs are synthesized from the (−) strand RNA template as the polymerase complex recognizes conserved start and stop sequences on the template. (4) These mRNAs are translated. (5) The concentrations of viral proteins, especially NP, regulate the switch from mRNA synthesis to genome replication, which begins with synthesis of full-length (+) strand copies of the viral RNA. (6) These are encapsidated by NP and, in turn, serve as templates for the synthesis of full length (−) strand RNAs. (7) Inclusion bodies are the sites of RNA synthesis and nucleocapsid assembly. (8) Assembly begins with the synthesis of GP in the ER and transport to lipid raft domains at the plasma membrane. (9) Octamers of VP40 are produced and transported to the GP-containing rafts. VP40 interacts with the C terminus of NP and directs viral (−) strand nucleocapsids to sites of virus budding. (10) Nucleocapsids form parallel to the plasma membrane, and (11) virus particles are released by budding. *Reproduced from Flint, S.J., et al., 2009. Principles of Virology, third ed. ASM Press, Washington DC, with permission.*

Virions also contain lipids, the composition of which reflects the composition of host cell membranes, and large amounts of carbohydrates as side-chains on the glycoproteins. Viral infectivity is relatively stable at room temperature, but sensitive to UV- and γ-irradiation, detergents, and common disinfectants.

Viral Replication

Filoviruses replicate to high titers in cell cultures, such as Vero (African green monkey kidney) cells (Fig. 28.4); infection is characterized by a rapid cytopathology and large intracytoplasmic inclusion bodies (composed of masses of viral nucleocapsids). Virions enter cells by receptor-mediated

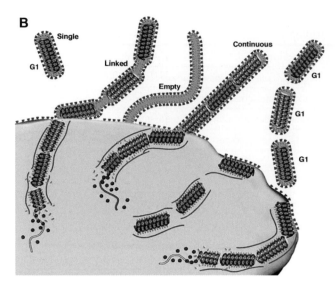

FIGURE 28.4B Schematic model of Ebola virus genome packaging. The virus appears pleomorphic in regard to particle length, but an underlying structural organization is maintained. There are three basic morphological forms of the particles; empty, linked, and continuous. Single genome (G1) virus and multigenome particles are shown budding from the cell. Genomes are assembled in the host cell and transported to the surface where end-to-end apposition takes place during (or prior to) budding and envelopment at the plasma membrane. Color-coding: nucleocapsids, red, yellow, and orange helices; nucleocapsid protein, purple spheres; VP40, green ovals, VP24/VP35 bridges, blue oval; GP spikes, red; microtubules, brown. *Reproduced from Beniac, D.R., et al., 2012. The organisation of ebola virus reveals a capacity for extensive, modular polyploidy. PLoS One 7(1): e29608, with permission.*

TABLE 28.3 Properties of Filoviruses

Virions are pleomorphic, appearing as long filamentous forms and other shapes; virions have a uniform diameter of 80 nm and vary greatly in length (unit nucleocapsid lengths of about 800 nm for Marburg and 1000 nm for Ebola virus)

Virions are composed of a lipid envelope covered with peplomers surrounding a helically wound nucleocapsid

The genome is composed of a single molecule of negative-sense single-stranded RNA, 19.1 kb in size, the largest of all negative-sense RNA viruses

Cytoplasmic replication; initially, interrupted (start-stop) transcription gives rise to individual monocistronic mRNAs, in decreasing copy number commencing from the 3′ end (structural proteins). Later, full-length (+) strand RNA is transcribed to generate a template for the synthesis of progeny RNA genomes

Assembly involves endosomal membrane invagination and formation of multi-vesicular bodies. Preassembled nucleocapsids bud from the plasma membrane at sites already containing patches of viral glycoprotein

Infection is extremely cytopathic in cultured cells and in host target organs

macropinocytosis and replication occurs in the cytoplasm. The host proteinase cathepsin B is required and entry is blocked by mutations in the Niemann-Pick C1 protein that controls endosomal-lysosomal cholesterol transport.

Transcription is initiated at a single promoter site, located at the 3′-end of the viral genome. Transcription yields monocistronic mRNAs, that is, separate mRNAs for each protein. This is accomplished by conserved transcriptional stop and start signals located at the boundaries of each viral gene. As the viral polymerase moves along the RNA genome these signals cause the enzyme to pause and sometimes to fall off the template and terminate transcription (called stuttering or stop/start transcription). The result is that more mRNA is made from genes that are located close to the promoter, and less from downstream genes. This regulates the expression of genes, producing large amounts of structural proteins such as the nucleoprotein and smaller amounts of proteins such as the RNA polymerase. Replication of the filovirus genome is mediated by the synthesis of full-length complementary-sense RNA that then serves as the template for the synthesis of virion-sense RNA. This requires that the stop/start signals needed for transcription be overridden by the viral polymerase—the immediate envelopment of newly formed viral-sense RNA by nucleoprotein appears to mediate this.

Maturation of virions occurs via budding of pre-assembled nucleocapsids from the plasma membrane at sites already containing patches of viral glycoprotein peplomers. The glycoprotein precursor GP is cleaved in the trans-Golgi network by the enzyme furin to yield GP1 (the receptor binding domain) and GP2 (a membrane anchored fusion protein). GP1 and GP2 are then transported to multi-vesicular bodies and to the cell membrane where budding takes place (Fig. 28.4B). The formation of 6-shaped and irregular virions is likely due to dynamic membrane invagination prior to complete envelopment of the nucleocapsid (Table 28.3).

EBOLA AND MARBURG VIRUS DISEASE (EBOLA AND MARBURG HEMORRHAGIC FEVER)

Clinical Features

Marburg virus and Ebola Zaire, Sudan, Taï Forest, and Bundibugyo viruses cause severe hemorrhagic fever in humans, and sporadic cases cannot be distinguished clinically from other causes of hemorrhagic fever including Lassa fever, Crimean-Congo hemorrhagic fever, yellow fever, or even malaria. Following an incubation period of usually five to ten days (extreme range 2 to 21 days for infection by the Zaire subtype of Ebola virus), there is an abrupt onset of illness with initial non-specific signs and symptoms, including fever, severe frontal headache, malaise, and myalgia, fever, weakness, and often muscle pains, and abdominal discomfort. Rash is also a common feature, but in dark-skinned patients, it is not always apparent. Bleeding, manifested as petechiae, ecchymoses, uncontrolled bleeding from venipuncture sites, and melena, is not the common manifestation that many believe is the hallmark of the disease—it occurs in less than 50% of laboratory-confirmed cases. There is a profound leukopenia, bradycardia, and conjunctivitis and there may be pharyngitis. Deterioration over the following two to three days is marked by nausea and vomiting, prostration, anuria, and a fall in body temperature. Abortion is a common consequence of infection. Infected pregnant woman and their infants invariably die. Death due to multiple organ failure and shock usually occurs six to nine days after the onset of clinical disease (range one to 21 days); at the time of death patients typically have high viremia, thus presenting a continuing potential source of contagion. The mortality rate has been very high: 25% with Marburg virus, 60% with the Sudan subtype of Ebola virus, 88 to 90% with the Zaire subtype of Ebola virus, and about 35% with the Bundibugyo subtype. Convalescence is slow and marked by weakness, prostration, weight loss, myalgias and arthralgias, and hair loss and desquamation of areas initially affected by rash. Also amnesia is common among survivors for the period of acute illness.

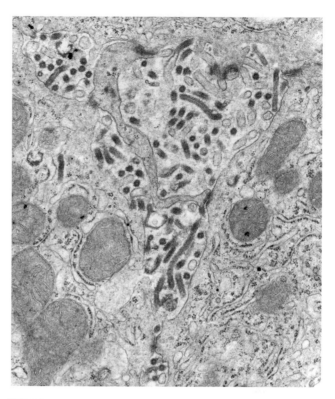

FIGURE 28.5 Liver from a rhesus monkey infected with Marburg virus. This image depicts an area where hepatocytes are still intact; at this site, virions fill the intercellular space as a result of budding from plasma membranes. Thin-section electron microscopy; magnification ×39,000.

Pathogenesis, Pathology, and Immunity

Marburg and Ebola viruses cause similar pathological changes in humans. Macrophages, monocytes, and dendritic cells are the primary sites of infection where the virus replicates to high titer. The most striking lesions are found in the liver, spleen, and lymph nodes. These include focal hepatic necrosis with little inflammatory response, and follicular necrosis found in lymph nodes and spleen. In experimentally infected rhesus, cynomolgus and African green monkeys, filoviruses replicate to a high titer in the reticuloendothelial system, endothelium, liver, and lungs; there is severe necrosis of these target organs, which is most evident in liver, and there is interstitial hemorrhage, most evident in the gastrointestinal tract (Figs. 28.5 and 28.6). The interaction between virus and dendritic cells is likely pivotal in determining disease outcome.

Virus shedding from infected primates occurs from all body surfaces and orifices, including the skin and mucous membranes, and especially from sites of hemorrhage. Of all the viruses causing a hemorrhagic disease, the filoviruses cause the most severe hemorrhagic manifestations and the most pronounced liver necrosis, not unlike that seen in livestock infected with Rift Valley fever virus. There is an early and profound leukopenia, followed by a dramatic neutrophilia, and very little inflammatory infiltration in sites of parenchymal necrosis in the liver. There is no evidence for long-term latency or persistence, although viral persistence in sequestered sites may continue for months.

Abnormalities in coagulation parameters include the appearance of fibrin split products in the blood and prolonged prothrombin and partial thromboplastin times—all indicative of disseminated intravascular coagulopathy (DIC) as a common terminal event.

Filovirus infections overwhelm specific host defense mechanisms of experimental animals by both the speed in which the disease progresses—animals often die before it might be expected that an effective primary specific inflammatory/immune response would be elicited; and viral tropism—early tropism for reticuloendothelial and lymphoid cells likely minimize the response that might be elicited otherwise.

Pro-inflammatory responses appear to play a pivotal role in determining disease outcome. Ebola virus effectively inhibits the host interferon type I response, mainly due to the inhibitory properties of VP35. Macrophage infection induces pro-inflammatory cytokines, thereby promoting endothelial leakage and bystander apoptosis of lymphocytes, although the latter does not abrogate the development of a specific CD8+ cytotoxic T cell response. Circulating cytokines also induce expression of procoagulant and surface adhesion molecules

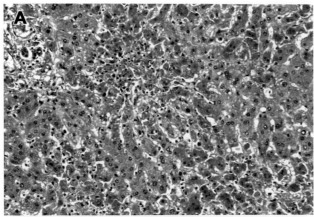

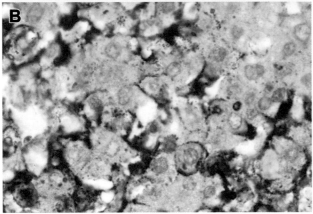

FIGURE 28.6 Ebola virus infection, human, Zaire (now Democratic Republic of Congo), 1976. (A) Massive, but focal hepatocellular necrosis marked by zones of cells which appear intact but are undergoing eosinophilic necrosis (brighter red than cells that appear normal), apoptotic pyknosis and dissolution of nuclei. H&E, ×400. (B) The same liver tissue block as in (A), but processed for immunohistochemistry by Sherif Zaki, U.S. Centers for Disease Control and Prevention. Massive amounts of Ebola virus antigens (bright red).

by endothelial cells, and infected macrophages express cell surface tissue factor, all of which promote disseminated intravascular coagulopathy (DIC). Platelet dysfunction and increasing hepatic impairment may also contribute to bleeding.

Domains on VP35 block the function of the helicase RIG-I, which in concert with MDA-5 are cellular sensors of virus infection and trigger a type I interferon response. Ebola virus VP24 also contributes to the inhibition of innate immunity by preventing the accumulation of antiviral STAT-1 in the nucleus of infected cells. There are some differences between how interferon signaling is suppressed between Marburg virus and Ebola viruses: Marburg virus blocks both STAT-1 and STAT-2 activation through the effects of VP40 rather than through VP35 as is the case with Ebola virus.

The combined effects of tissue destruction, cytokine effects, fluid shifts, interstitial hemorrhage, and tissue ischemia as a result of microvascular thrombi, all contribute towards a high rate of mortality.

Laboratory Diagnosis

It is essential that laboratory samples be handled under appropriate safety precautions whenever a diagnosis of Marburg or Ebola hemorrhagic fever is possible.

Today filovirus diagnosis is primarily based upon optimized reverse transcriptase polymerase chain reaction (RT-PCR), various iterations of which are deployed in the field, in mobile field laboratories, local hospitals, and regional laboratories, along with IgM and IgG capture ELISA serological methods. At the onset of outbreaks, clinical specimens are also sent to international reference centers for PCR-sequencing-based programs to assess viral genotype. The minor genotypic drift (but absence of any major genotypic or pathotypic shift), observed during the progress of the 2014 to 2015 epidemic of Ebola virus disease in West Africa has been observed due to this extended international laboratory diagnostic system.

Initially the diagnosis of filovirus infections was based upon virus isolation from blood or tissues in cell culture, such as Vero (African green monkey kidney) cells or MA-104 (fetal rhesus monkey kidney) cells. As wild-type filoviruses do not induce a florid cytopathic effect in cell cultures, the presence of virus was detected by immunofluorescence. Some isolates, particularly those of the Ebola, Sudan subtype, have been more difficult to isolate in cell culture, but are facilitated by blind passage.

Diagnosis was also made by the direct detection of viral antigen in tissues by immunofluorescence or antigen-capture enzyme immunoassay. Laboratory biosafety and security was a major factor in the use of these methods on unfixed human tissue/blood specimens. Virus antigens are also found in the skin, a feature which has been exploited to allow for diagnosis using immunohistochemistry on skin biopsies fixed in formalin—this approach has been of more use in surveillance than in clinical care.

Serological diagnosis of filovirus infections, originally based upon indirect immunofluorescence tests, has a long history of problems—there were many false-positives, which undermined confidence in serological surveys of humans and animals thought to be reservoir hosts. Many unsubstantiated results confounded understanding of the natural history and epidemiology of the viruses. Enzyme-linked immunosorbent capture assays were developed for the detection of both IgM antibodies in acute Ebola infections and IgG antibodies in individuals surviving infection—these have become the standard for human serological diagnosis and for serosurveys of animals, including wild and captive non-human primates. These assays also have had the advantage of being readily adapted for use in the field during outbreaks or outbreak investigations, providing appropriate personal protective equipment is used and clinical specimens are inactivated using heat and detergents. This has allowed testing patients quickly either onsite at local hospitals or at local/regional laboratories.

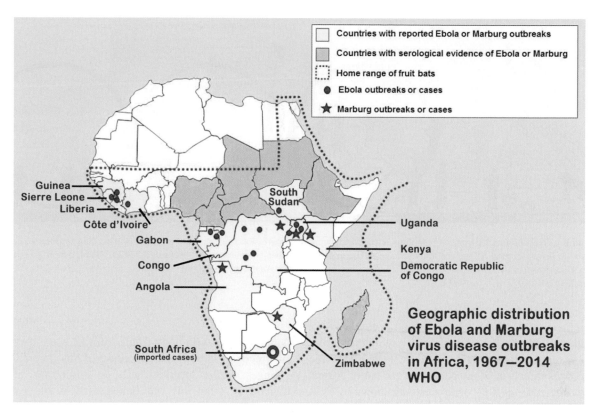

FIGURE 28.7   Map showing location of Marburg and Ebolavirus outbreaks from 1967 to 2014, those countries with serological evidence of Marburg or Ebola virus infection, and the geographical distribution of fruit bats. *From WHO.*

Epidemiology, Prevention, Control, and Treatment

As indicated by the detection of human infections, Marburg virus is endemic in a natural reservoir in Central and West Africa, particularly in woodland areas; Ebola virus is found in the humid rainforests of Central and West Africa. There has been an increasing incidence of Ebola hemorrhagic fever in West Africa over the past few years, evident as large and small outbreaks, involving an ever-increasing geographical range (Fig. 28.7). The fact that Ebola viruses causing human disease episodes have been genetically distinct from each other makes it clear that a common source transmission chain extending across sub-Saharan Africa is not the case—rather, different virus lineages lodged at or near each site of human disease episodes have been responsible.

The recent epidemic in West Africa has provided more accurate information as to the epidemiology of Ebola virus. The reproduction rate, R_0, of around 1.7 approximates that of pandemic influenza, although there is little evidence of aerosol transmission, the major route being close contact with body fluids or possibly fomites and droplets. The incubation period is usually five to ten days (extreme range 2 to 21 days for infection by the Zaire subtype of Ebola virus). The risk of transmission increases rapidly at the time clinical disease becomes apparent.

Human index cases have often occurred in or near the end of the tropical rainy season and have been associated with the tropical forest or the marginal zone between forest and savannah. Several index case investigations have centered on mines and caves, places where the primal ecology has been disturbed. In any case, the reservoir host(s) of the filoviruses inhabit what seem to be a diverse range of ecological niches.

It must be recognized that within the ecosystems under consideration, there is considerable isolation of individual micro-niches together with ecological insularity; that is, there are many sites within larger geographical areas in which different filoviruses may coexist with animal reservoirs but not be evident. Candidate reservoir host species have over the years included all mammals, particularly monkeys and rodents, but also birds, reptiles, amphibians, and arthropods.

Particular attention is now being given to bats as the primary reservoir hosts of the filoviruses; this follows upon several lines of evidence: (1) experimental virus inoculation wherein several species of insectivorous and fruit bats have been found to support the growth of Ebola

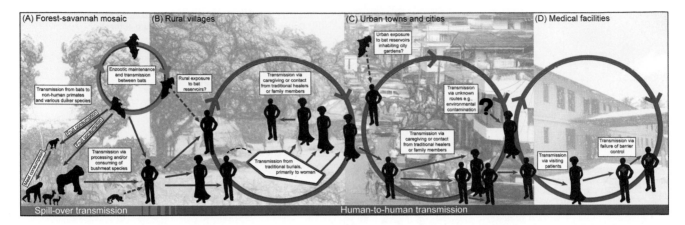

FIGURE 28.8 Transmission patterns of Ebola virus infection, showing the progression from enzootic infection of bats, then spillover to non-human primates and to people in rural human communities, and finally to people in urban communities and medical facilities. *Adapted from Alexander, K.A., Sanderson, C.E., Marathe, M., Lewis, B.L., Rivers, C.M., Shaman, J., et al., 2015. What factors might have led to the emergence of ebola in West Africa? PLoS Negl. Trop. Dis. 2015;9, 1-26.e0003652.*

and Marburg viruses, sustaining virus asymptomatically in their tissues and blood for weeks, (2) asymptomatic Marburg virus infection of the common fruit bat of Africa, *Rousettus aegyptiacus*, has been repeatedly shown by PCR and virus isolation, as well as serology—PCR and serology have implicated other bat species as well, (3) asymptomatic Ebola virus infection has been shown in several species of insectivorous bats trapped in Central Africa, by PCR and serology, and in some instances by virus isolation, (4) the genome of Lloviu virus was sequenced in the absence of virus isolation from Schreiber's bats (*Miniopterus schreibersii*) after a rapid increase in bat mortality was observed in animals frequenting the Cuerva del Lloviu in northeast Spain. Thus, overall, the evidence for a role of bats as reservoir hosts of Ebola and especially Marburg virus is strong, but whether or not there are other reservoirs remains unknown and an understanding of the full natural history of these viruses in nature is far from being answered.

The largest outbreak of Ebola virus disease recorded to date has been the West African outbreak beginning in 2014 affecting more than 28,000 individuals in Sierra Leone, Liberia, and Guinea, with a few cases occurring in Mali, Senegal, and Nigeria. A number of cases were also exported to Europe, Asia, and North America. This also represents the first large self-sustaining epidemic in crowded urban settings, which highlights a new future risk scenario for this disease. The causative virus varied very little from isolates of the Zaire subtype of Ebola virus of earlier years. Phylogenetic analysis supports the view that Ebola virus has most likely evolved independently in West Africa. Thus filoviruses have a much wider distribution throughout Africa than was previously considered.

In all Ebola hemorrhagic fever outbreaks, secondary spread between humans appears to have been principally due to contact with body fluids from an acute case. In the first recognized outbreaks the reuse of blood-contaminated syringes and/or needles was an important mode of transmission, but this was halted early in the outbreaks. In later outbreaks improper contact with patients without gloves, mask, or protective clothing or training in how to use personal protective equipment was a major factor in nosocomial spread. Cultural practices, especially burial traditions, that favored direct contact between patients and others were also of major importance (Fig. 28.8).

One of the major lessons of the recent West African Ebola virus disease epidemic has been the escalation in the level of personal protective equipment (PPE) recommended for healthcare providers and the systems used for patient isolation and decontamination. The system developed from the experiences of Médecins Sans Frontières is appropriate in Africa and in countries where exported cases are treated. This has evolved into a tiered approach, involving hospital and staff preparedness and training, oversight, logistics, and understanding of the great value of supportive patient care, especially hydration and nutrition. The system may have at one time resembled the "strict barrier nursing" schemes used in infectious disease wards, but today includes wearing of personal protective equipment that fully covers the body, with no skin exposed. Similarly, in today's system decontamination, using hypochlorite and detergents, is much more rigorous than in the past.

The West African epidemic spanning 2014 to 2015 has led to a rapid escalation in the development of novel therapeutic and immunization strategies. A cocktail of humanized monoclonal antibodies produced using transgenic tobacco plants (*Nicotiana benthamiana*) was used to treat several infected healthcare workers, several of whom recovered—further experience has led to less optimism. Another strategy, still in a pre-clinical research stage, involved the use of small interfering RNAs targeting the expression of Ebola viral L, VP24, and VP35 proteins. Up to seven successive injections of such siRNAs protected macaques against a lethal virus challenge.

Attempts to produce filovirus vaccines have focused on a number of antigen presentation systems. Encouraging results initially having been obtained in macaques injected first with DNA encoding Ebola virus glycoprotein and then boosted with a recombinant adenovirus 5 vector expressing the homologous viral protein. More recently, recombinant attenuated vesicular stomatitis virus encoding the glycoprotein of Marburg or Ebola virus has been shown to induce solid immunity to challenge in non-human primates. There are a number of other vaccine candidates, based upon several different recombinant DNA technologies, in preclinical and early clinical stages of development. Taken together, these developments show that there is optimism that protection can be conferred by immunization. Funding to bring the most promising vaccine candidates to licensure and use seems assured.

FURTHER READING

Bradfute, S.R., Bavari, S., 2011. Correlates of immunity to filovirus infection. Viruses 3, 982–1000. Available from: http://dx.doi.org/10.3390/v3070982.

Feldman, H., 2014. Ebola—a growing threat? N. Engl. J. Med. Available from: http://dx.doi.org/10.1056/NEJMp1405314.

Kuhn, J.H., Becker, S., Ebihara, H., et al., 2010. Proposal for a revised taxonomy of the family Filoviridae: classification, names of taxa and viruses, and virus. Arch. Virol. 155, 2083–2103.

Negredo, A., Palacios, G., Vazquez-Morón, S., Gonzalez, F., Dopazo, H., et al., 2011. Discovery of an Ebolavirus-like filovirus in Euorpe. PLoS Pathog. 7, e1002304. Available from: http://dx.doi.org/10.1371/jopurnal.ppat.1002304.

Olinger, G.G., Pettitt, J., Kim, D., et al., 2012. Delayed treatment of Ebola virus infection with plant-derived monoclonal antibodies provides protection in rhesus macaques. Proc. Natl. Acad. Sci. U.S.A. 109 (44), 18030–180305. Available from: http://dx.doi.org/10.1073/pnas.1213709109.

Response Team, WHO., 2014. Ebola virus disease in West Africa—the first 9 months of the epidemic and forward projections. N. Engl. J. Med. 371, 1481–1495.

Valmas, C., Grosch, M.N., Schümann, M., Olejnik, J., Martinez, O., et al., 2010. Marburg virus evades interferon responses by a mechanism distinct from Ebola virus. PLoS Pathog. 6 (1), e1000721. Available from: http://dx.doi.org/10.1371/journal.ppat.1000721.

Chapter 29

Bunyaviruses

The family *Bunyaviridae* is the largest family of RNA viruses with more than 350 named member viruses. It represents one of the most geographically diverse families of viruses, existing under virtually all climatic conditions and maintained in nature through a wide variety of reservoir hosts. Nearly all of the viruses are arboviruses, maintained in specific arthropod–vertebrate–arthropod cycles. This specificity is the basis for the usually narrow geographical and ecological niches occupied by each virus and is seen as ancient relationships between specific viruses and specific reservoir hosts. Also because of this host–vector specificity and diversity, bunyaviruses are often seen as the cause of new or emerging diseases, encountered as humans occupy previously uninhabited regions or are the cause of ecologic/environmental changes that favor new exposures to remote virus–host transmission cycles.

Hantaviruses are unique among the bunyaviruses in not being transmitted by arthropods—the hantaviruses are transmitted in vertebrate–vertebrate cycles without arthropod vectors; however, the hantaviruses share with other bunyaviruses exquisite specificity as to vertebrate reservoir hosts and narrow ecological niches. Another characteristic of hantavirus infection patterns is persistent, in some instances lifelong, silent infection in animal reservoir hosts.

Most bunyaviruses are not associated with human disease, but the exceptions are important. Diseases range from encephalitis to hepatitis, nephritis, acute respiratory distress syndrome and undifferentiated multiorgan syndromes. Two syndromes are particularly characteristic: acute hantavirus pulmonary syndrome (HPS) caused by pathogenic hantaviruses in the western hemisphere, and hemorrhagic fever with renal syndrome (HFRS) caused by several viruses found in the eastern hemisphere.

The most important bunyavirus pathogens are Rift Valley fever virus, Crimean-Congo hemorrhagic fever virus, La Cross virus, Oropouche virus, and the hantaviruses causing HPS and HFRS. Rift Valley fever virus is of particular concern to national and international disease control agencies. Recently several new bunyavirus pathogens of humans have been identified as threats: severe fever with thrombocytopenia syndrome virus (SFTSV) in China, Heartland virus in the United States, and others.

Some of the viruses described in this chapter are subject to importation or possession restrictions by law or regulation by many national governments. These viruses are restricted to specified laboratories for all research and diagnostic procedures. Many of the human pathogens are categorized as Biosafety Level 3 or 4 pathogens; these must be handled in the laboratory under high or maximum containment conditions to prevent accidental release and human exposure.

PROPERTIES OF BUNYAVIRUSES

Classification

The very large number and diversity of the bunyaviruses offer a taxonomic challenge. Genome properties are used to define each genus, particularly the organization of each RNA genome segment and the sequences of conserved nucleotides at the 5′ and 3′ termini. Classical serological methods have been used to define serological groupings within each genus. In general, antigenic determinants on the nucleocapsid protein are conserved and so serve to define broad groupings among the viruses, whereas shared epitopes on the envelope glycoproteins that are the targets of neutralization and hemagglutination inhibition assays define narrow groupings. Unique epitopes on envelope glycoproteins, also determined by neutralization assays, define individual viruses. With a few exceptions, viruses within a given genus are antigenically related to each other but not to viruses in other genera. These antigenic groupings are at present being transformed into genotypic groupings, which will be similar, but more practical given the high-throughput capacity of sequencing technology.

Genetic reassortment occurs when either cultured cells or arthropods or vertebrate hosts are co-infected with closely related bunyaviruses, and such events have almost certainly played some part in the evolution of the family. Within its particular ecological niche each bunyavirus evolves by genetic drift and Darwinian selective forces; for example, isolates of La Crosse virus from different regions in the United States differ considerably due to cumulative point

Fenner and White's Medical Virology. DOI: http://dx.doi.org/10.1016/B978-0-12-375156-0.00029-1

TABLE 29.1 Family Bunyaviridae

Genus	Virus	Geographical Distribution	Arthropod Vector	Target Hosts (Reservoirs, Amplifiers)	Human Disease
Orthobunyavirus					
	La Crosse and other California encephalitis group viruses	North America	Mosquitoes	Humans, chipmunks and other small mammals	Encephalitis
	Oropouche virus	Brazil	Mosquitoes, *Culicoides*	Humans, sloths, monkeys	Fever, arthralgia, myalgia
Phlebovirus					
	Rift Valley fever virus	Africa	Mosquitoes	Humans, sheep, cattle, buffalo	Flu-like illness, hepatitis, hemorrhagic fever, retinitis
	Sandfly fever viruses	Mediterranean, South America	Sandflies	Gerbil, forest rodents	Fever, myalgia, conjunctivitis
	Severe fever with thrombocytopenia syndrome virus	China	?Ticks	Goats, cattle, pigs	Fever, thrombocytopenia, multiorgan failure
Nairovirus					
	Crimean-Congo hemorrhagic fever virus	Africa, Asia	Ticks	Humans, sheep, cattle, goats	Hemorrhagic fever
Hantavirus (see Table 29.3)					

mutations and nucleotide deletions and duplications—such mutations are stabilized by occupying isolated ecological niches. The evolution of La Crosse virus has also involved genome segment reassortment—reassortant viruses have been isolated from mosquitoes in the field.

The more than 350 distinct bunyaviruses fall into four genera infecting vertebrates (all of which include human pathogens), and a fifth genus, *Tospovirus*, which contains plant viruses. Additionally, more than 80 viruses have not yet been assigned to a genus or subgroup (Table 29.1).

Genus *Orthobunyavirus* contains 18 subgroups and at least 160 viruses, most of which are mosquito-borne, but some are transmitted by sandflies or *Culicoides* spp (midges). The genus includes more than 30 pathogens of humans and domestic animals, including La Crosse, Bunyamwera, Oropouche, and Cache Valley viruses.

Genus *Phlebovirus* contains two subgroups and at least nine distinct viruses with over thirty other candidate members not yet fully characterized. Sandflies or mosquitoes transmit all. The genus contains important pathogens, including Rift Valley fever virus, the sandfly fever viruses, and Toscana virus. Uukiniemi virus also belongs to this genus, although notably it is tick-borne and does not code for a non-structural protein (NSm, see below) at the N-terminus of the envelope precursor glycoprotein. Although non-pathogenic for humans, Uukiniemi virus is mentioned here as this virus has been intensively studied as an example of the phleboviruses.

Genus *Nairovirus* contains seven distinct virus species divisible into many subgroups. Nearly all are tick-borne, including the pathogens Crimean-Congo hemorrhagic fever and Dugbe viruses.

Genus *Hantavirus* contains at least 23 viruses, many of which have been discovered in the past few years. All are transmitted by persistently infected reservoir animal hosts via urine, feces, and saliva; the same transmission pattern has occurred among rats in laboratory colonies. In humans, several of these viruses cause hemorrhagic fever with renal syndrome (HFRS); others cause acute respiratory distress syndrome [hantavirus pulmonary syndrome (HPS)].

Virion Properties

Bunyavirus virions are spherical, approximately 80 to 120 nm in diameter, and are composed of an envelope with glycoprotein peplomers, while internally there are three circular, helical nucleocapsid segments (Fig. 29.1). Nucleocapsid circles are formed by panhandles, that is non-covalent bonds between palindromic sequences on the 3′- and 5′-ends of each RNA genome segment. The terminal sequences are identical for all three RNA segments for all members of each genus, but these differ between the genera.

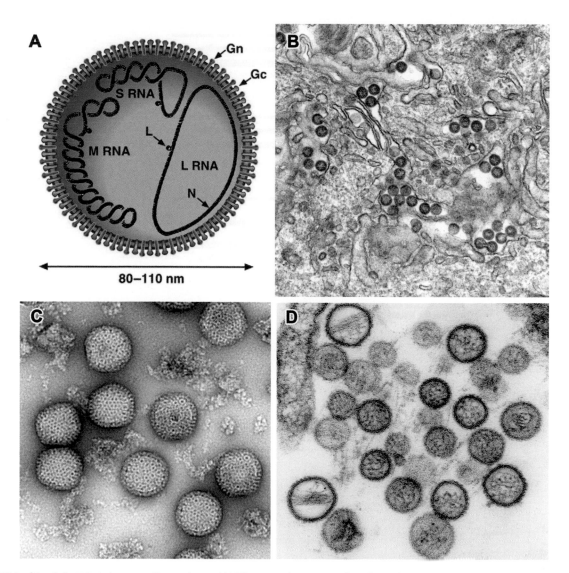

FIGURE 29.1 Morphology and structure of bunyaviruses. (A) Diagrammatic representation of an orthobunyavirus virion in cross-section. The surface spikes comprise two glycoproteins termed Gn and Gc. The three helical nucleocapsids are circular and comprise one each of the unique ssRNA segments (L, large; M, medium; S, small). (B) Thin-section electron microscopy of a hepatocyte of a rat infected with Rift Valley fever virus, showing virions budding into Golgi vesicles. (C) Negatively stained Rift Valley fever virus virions, showing tightly packed peplomers (spikes) on the surface of the 100 nm virions. (D) Thin-section electron microscopy of Sin Nombre virus (etiologic agent of hantavirus pulmonary syndrome) in Vero E6 cell culture. *(A) Reproduced from King, A.M.Q., et al., (Eds.), 2012. Virus Taxonomy: Ninth Report of the International Committee on Taxonomy of Viruses. Academic Press, with permission. (B) From an unpublished study by Joel Dalrymple and Frederick Murphy. (C) From Alexander Freiberg, with permission. (D) From Cynthia Goldsmith, with permission.*

The viral genome consists of three segments of negative-sense (or ambisense) single-stranded RNA, designated large (L), medium (M), and small (S). The RNA segments differ in size among the genera: the L RNA segment ranges in size from 6.3 to 12 kb, the M RNA segment from 3.5 to 6 kb, and the S RNA segment from 1 to 2.2 kb. Complete nucleotide sequences have been determined for many of the viruses, including representative viruses of each genus.

The L RNA encodes a single large (L) protein, the RNA-dependent RNA polymerase (transcriptase). The M RNA encodes a polyprotein that is processed to form two glycoproteins (Gn and Gc) and a non-structural protein (NSm). The S RNA encodes the nucleocapsid (N) protein and a non-structural (NSs) protein. The translation strategy of the S RNA differs between the genera. In some members of the genera *Orthobunyavirus* and *Hantavirus*, the S RNA encodes two overlapping reading frames for the N and NSs proteins, while in the genus *Nairovirus* the S RNA contains only one open reading frame encoding the N protein which is somewhat larger than for the other genera (Fig. 29.2).

An exception to the usual coding strategy of negative-sense RNA viruses occurs in the S RNA of the member viruses of the

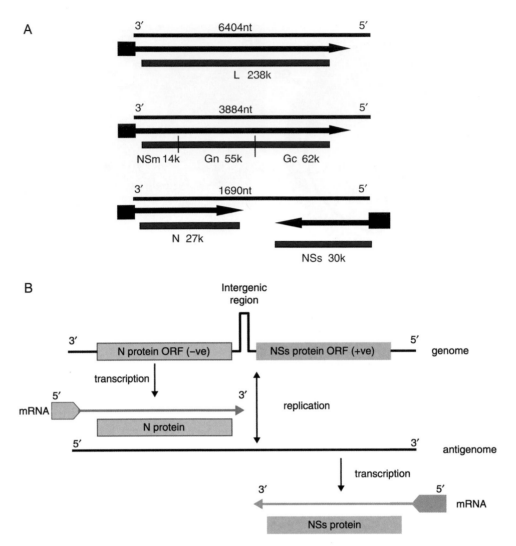

FIGURE 29.2 Most of the steps in the transcription, translation and replication of the member viruses of the family *Bunyaviridae* are typical of other negative-sense RNA viruses, but a unique sequence of events is necessary to produce viral proteins from the ambisense S segment of the phleboviruses such as Rift Valley fever virus. These viruses have two negative-sense (L and M segments) and one ambisense (S segment) RNAs that code for seven proteins. (A) The S segment encodes the nucleocapsid (N) protein in the negative-sense and a non-structural (NSs) protein in the positive-sense, though NSs cannot be translated directly from the S segment but rather from a specific subgenomic mRNA. These two domains are separated by an intergenic region that forms a hairpin structure. The three segments (L, M and S) are shown in descending order of size; RNAs are represented by thin lines and the mRNAs are shown as arrows with cellular RNA sequences at the 5′ end. Gene products are represented by colored boxes. (B) Transcription and replication scheme of the ambisense-sense phlebovirus S genome segment. (1) The open reading frame at the 3′-end of the genomic S segment is transcribed (from its 3′-end) by the viral RNA-dependent RNA polymerase (transcriptase) to produce a subgenomic mRNA of complementary-sense, which is capped at its 5′-terminus by a primer obtained from cellular mRNAs (viral cap-snatching); (2) this mRNA is translated into the nucleoprotein (N); (3) genomic replication produces full-length complementary-sense RNA (cRNA); (4) the open reading frame in the complementary-sense RNA then serves as template for transcription, from its 3′-end, of mRNA which is translated to produce the non-structural protein (NSs). *Reproduced from Elliott, R.M., Brennan, B., 2014. Emerging phleboviruses. Curr. Opin. Virol. 5, 50–57, with permission.*

genus *Phlebovirus*. With these viruses, S RNA encodes the N and NSs proteins, each translated from a separate subgenomic mRNA. The N protein is coded in the 3′-half of the S RNA, and as usual its mRNA is transcribed using the genome RNA as template; however, the NSs protein, occupying the 5′-half of the same S RNA, is encoded in the complementary-sense RNA species and is only translated after synthesis of full-length RNA intermediates—the S RNA segment is therefore said to exhibit an *ambisense* coding strategy. The NSs proteins

of orthobunyaviruses, phleboviruses, and hantaviruses function as interferon antagonists.

The four major virion proteins differ in size among the genera: the L protein [the RNA-dependent RNA polymerase (transcriptase)] has an Mr of 150,000 to 200,000, the N protein (nucleoprotein) has an Mr of 25,000 to 50,000 and G1 and G2, the two glycoproteins have Mr 40,000 to 120,000. Virions also contain lipids, the composition of which reflects the composition of host cell membrane glycoproteins.

The viruses are moderately sensitive to heat and acid conditions and are readily inactivated by detergents, lipid solvents, and common disinfectants.

Viral Replication

Bunyaviruses replicate to a high titer in many kinds of cells: Vero E6 (African green monkey kidney) cells, BHK-21 (baby hamster kidney) cells, and mosquito (*Aedes albopictus*) cells. Hantaviruses do not replicate in mosquito cells. The viruses are cytolytic for mammalian cells (except for hantaviruses and some nairoviruses), but non-cytolytic for invertebrate cells. Most of the viruses also replicate to a high titer in suckling mouse brain.

Viral entry into its host cell is by receptor-mediated endocytosis; all subsequent steps take place in the cytoplasm. Because the genome of the single-stranded negative-sense RNA viruses cannot be translated directly, the first step after penetration of the host cell and uncoating is the activation of the virion RNA polymerase (transcriptase) and its transcription of viral mRNAs from each of the three virion RNAs (the exception, as noted above, is that in the genus *Phlebovirus*, the 5′-half of the S RNA is not transcribed directly; instead, the mRNA for the NSs protein is transcribed following synthesis of full-length RNA complementary to genomic RNA). The RNA polymerase also has endonuclease activity, cleaving 5′-methylated caps from host mRNAs and adding these to viral mRNAs to prime transcription (another example of cap-snatching). After primary viral mRNA transcription and translation, replication of the three segments of the virion RNA occurs and a second round of transcription begins, amplifying in particular the structural proteins for virion synthesis.

Virions mature by budding through intracytoplasmic vesicles associated with the Golgi complex and are released by transport of vesicles through the cytoplasm and release by exocytosis from the basolateral plasma membrane (Table 29.2).

DISEASES CAUSED BY MEMBERS OF THE GENUS *ORTHOBUNYAVIRUS*

La Crosse Virus Encephalitis and Other California Serogroup Infections

The California serogroup in the genus *Orthobunyavirus* includes 14 viruses, each of which is transmitted by mosquitoes and exhibits a narrow range of vertebrate hosts and a limited geographical distribution. There is no evidence that there is any clinical disease associated with these viruses other than in humans; however, it is the infection in reservoir host animals and mosquito hosts that are key to dealing with the human diseases. The most important zoonotic pathogen in the California subgroup is La Crosse virus.

TABLE 29.2 Properties of Bunyaviruses

Four genera infect vertebrates: *Orthobunyavirus*, *Phlebovirus*, and *Nairovirus* are arthropod-borne; *Hantavirus*, non-arthropod-borne

Virions are spherical, enveloped, 80 to 100 nm in diameter

Virions have glycoprotein peplomers but no matrix protein

Three circular, helical nucleocapsid segments; circles are formed by panhandles

Negative-sense single-stranded RNA genome in three segments. L, 6.3 to 12 kb (coding for RdRp); M, 3.5 to 6 kb (coding for glycoproteins Gn and Gc and non-structural protein NSm); S, 1 to 2.2 kb in size (coding for nucleocapsid protein N and non-structural protein NSs)

The S segment of the RNA genome of the member viruses of the genus *Phlebovirus* has an *ambisense* coding strategy

Capped 5′-termini of cellular RNAs are captured as primers for mRNA transcription

Cytoplasmic replication; budding into Golgi vesicles

Generally cytocidal for vertebrate cells but non-cytocidal persistent infection in invertebrate cells can occur

Genetic reassortment occurs between closely related viruses

Clinical Features

La Crosse virus encephalitis is a classic example of a disease that was unrecognized until the development of laboratory methods permitting a specific etiological diagnosis. The virus was first isolated in 1964 from a fatal case of encephalitis in a four-year-old child. Using this isolate as a source of antigen, retrospective serological surveys revealed that the virus was an important cause of disease previously listed under the heading "viral meningitis of undetermined etiology." It is now estimated that there are well over 100,000 human infections with at least 100 cases of encephalitis annually in the United States and Canada—this disease had been occurring regularly for many decades, long before the time when the virus was discovered.

La Crosse virus causes acute encephalitis preceded by a non-specific febrile illness. The symptoms of the central nervous system disease include stiff neck, lethargy, nausea, headache, and vomiting in milder cases and seizures, coma, paralysis, and permanent brain damage in severe cases. The protein level in the cerebrospinal fluid is elevated in about 20% of cases; unlike most viral encephalitides, many of the cells found in the cerebrospinal fluid are polymorphonuclear neutrophils. Severe disease occurs most commonly in children under the age of 16. About 10% of children develop seizures during the acute disease, a few develop persistent paresis and learning disabilities, and the mortality rate is about 0.3%.

Pathogenesis, Pathology, and Immunity

Based on experimental studies it is estimated that the incubation period ranges between 6 and 15 days. After virus entry in a mosquito bite, virus multiplication probably occurs in vascular endothelial and reticuloendothelial cells; dissemination occurs in the blood and lymph. In the central nervous system the virus replicates to high titer causing neuronal necrosis/apoptosis. In experimental animals the peripheral site where the replicates to highest titer is the nasal turbinates—presumably the virus can enter the brain via the olfactory neuronal pathway.

Histological changes in the few human cases that have been studied have been those typical of viral encephalitis—perivascular and perineuronal mononuclear inflammatory foci in the cerebral cortex, basal ganglia, and pons.

Laboratory Diagnosis

In areas where La Crosse virus encephalitis is common, laboratory testing is available in state or local public health laboratories and in some referral hospitals. Serum and/or cerebrospinal fluid are assayed by a virus-specific IgM antibody capture ELISA. Virus may also be detected by RT-PCR and in fatal cases by histopathology and virus-specific immunohistochemistry. Virus isolation is rarely employed.

Epidemiology, Prevention, Control, and Treatment

La Crosse virus is maintained by transovarial transmission in *Aedes triseriatus*, a tree-hole breeding woodland mosquito, and is amplified by a mosquito–vertebrate–mosquito cycle involving silent infection of woodland rodents, such as squirrels and chipmunks. The virus occurs throughout the eastern and mid-western United States—recently, more cases have been reported from mid-Atlantic and southeastern states. A closely related virus, snowshoe hare virus, occupies a similar niche in Canada. Most human cases occur during the summer months in children and young adults who are exposed to vector mosquitoes in wooded areas. Humans are dead-end hosts, and there is no human-to-human transmission (Fig. 29.3).

There is no vaccine against La Cross virus infection, nor is there any specific antiviral therapeutic. Treatment is supportive. Prevention is based on avoidance of vector mosquitoes through use of insect repellants, wearing protective clothing, screening, and in some instances, vector control.

Jamestown Canyon virus (including the closely related Jerry Slough virus) is also an occasional cause of arboviral encephalitis in the United States. The virus is transmitted by *Culiseta inornata* mosquitoes and several *Aedes* species mosquitoes as vectors, and is broadly distributed across much of North America. Vertical transmission of the virus has been demonstrated in several *Aedes* species mosquitoes. White-tailed deer are the most likely vertebrate reservoir host. Unlike La Crosse virus, which mainly causes encephalitis in children, Jamestown Canyon virus appears to cause disease predominantly in adults.

Oropouche Fever

Oropouche virus was first isolated in 1955 in Trinidad; since then the virus has been found to have caused more than half a million cases of febrile illness. It was at first recognized mainly in the Amazon basin regions of northern Brazil and Peru, usually in the rainy season, but now it is known to occur widely in northern South America, Panama, and Trinidad and Tobago. The human disease is characterized by fever, headache, myalgia, arthralgia, and prostration, but there is no recorded mortality. Occasionally, patients also exhibit rash and/or meningitis. Laboratory diagnosis is best done with an IgM capture ELISA using rDNA-expressed N protein; RT-PCR-based assays are also available.

The major urban vector is the midge *Culicoides paraensis* and the virus may be maintained in a midge–human cycle during epidemics. The sylvatic cycle involves sloths, monkeys, and probably jungle mosquitoes, such as *Coquillettidia venezuelensis*. Control involves avoiding the buildup of rotting organic debris such as banana tree stalks or cacao husks in agricultural areas, that curtails the population buildup of *C. paraensis* and reduces the risk of seasonal epidemics. Advice about avoiding exposure to midges during their early evening feeding hours (treated netting, DEET repellents) has been a staple of public health messaging in epidemic areas. At least three genotypes are recognized: sequencing suggests each of the three RNA segments has an independent evolutionary history.

Bunyamwera Virus Infection

Bunyamwera virus is present throughout much of sub-Saharan Africa and is an important cause of acute febrile illness in humans. The virus has been isolated from humans in Uganda, Nigeria, and South Africa and antibodies have been detected in humans in most of sub-Saharan Africa, with a high prevalence (up to 82%) recorded in some locations. Isolation of the virus from several *Aedes* species mosquitoes has implicated them as the primary vector. Antibodies have been detected in domestic animals, non-human primates, rodents, and birds, and viremias capable of supporting mosquito transmission have been recorded in rodents, bats, and non-human primates. However, the full natural history of the virus remains unresolved.

Ngari Virus Infection

Ngari virus was first isolated from human hemorrhagic fever cases during a large disease outbreak in 1997 and 1998 in East Africa (Kenya and Somalia)—it was estimated that

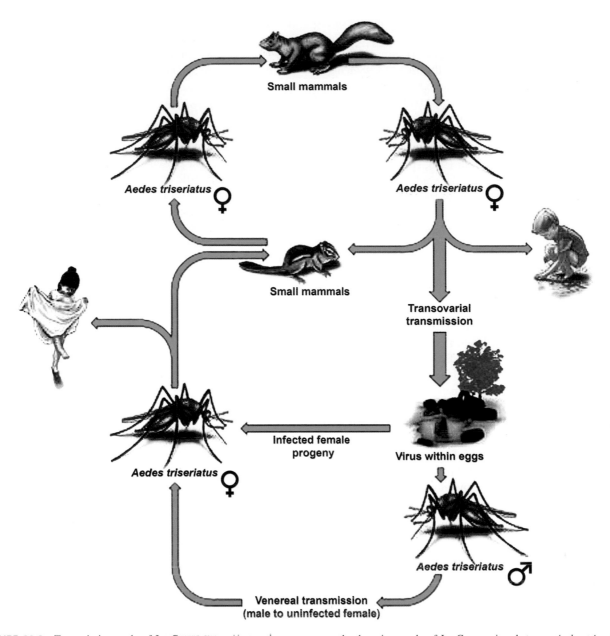

FIGURE 29.3 Transmission cycle of La Crosse virus. During the warmer months there is a cycle of La Crosse virus between *Aedes triseriatus*, a woodland mosquito, and chipmunks and tree squirrels. The virus is also maintained indefinitely in mosquitoes by transovarial transmission and is amplified by venereal transmission between male and uninfected female mosquitoes, which can in turn transmit either to a vertebrate by biting or to the next generation of mosquito by transovarial transmission. *From Barry Beaty, with permission.*

in this outbreak there were at least 89,000 cases and over 250 deaths. The disease was characterized by acute onset of fever and headache, followed by hemorrhage (with gastrointestinal and/or mucosal bleeding). A previously unidentified orthobunyavirus was isolated from cases—it was characterized as a naturally occurring reassortant between Bunyamwera virus (S and L segment donor) and Batai virus (M segment donor). Interestingly, neither of these two viruses separately causes hemorrhagic febrile illness in humans.

DISEASES CAUSED BY MEMBERS OF THE GENUS *PHLEBOVIRUS*

Rift Valley Fever

Epizootics of Rift Valley fever in sheep, goats, cattle, and humans have occurred at regular intervals in southern and eastern African countries from the time when intensive livestock husbandry was introduced at the beginning of the 20th century. An exceptionally devastating epizootic occurred in Egypt in 1977, resembling the biblical description of one of the plagues of ancient Egypt. In addition to many thousands of cases in sheep and cattle, there were an estimated 200,000 human cases, with 600 reported deaths. In 1997–98, a major epizootic spread from Somalia through Kenya into Tanzania, causing the death of many thousands of sheep, goats, and camels, and more than 90,000 human cases, with some 500 deaths. This epizootic, considered the largest ever seen in eastern Africa, was blamed on exceptional rainfall as a result of an El Nino weather pattern. In 1993, epidemic Rift Valley fever returned to Egypt—molecular epidemiological studies have since suggested that the virus was reintroduced from an unknown habitat in the Sudan. An extensive epizootic in Yemen and Saudi Arabia in 2000 was the first recognition of the disease outside Africa.

Clinical Features

Rift Valley fever virus infections vary from asymptomatic (30 to 60%), to an undifferentiated febrile illness, and in about 2% of cases severe hemorrhagic fever, encephalitis, retinitis, and death. The clinical disease begins after a very short incubation period (2 to 6 days) with chills, fever, severe headache, retro-orbital pain, photophobia, and a generalized "back-breaking" myalgia, diarrhea, vomiting, and hemorrhages. Usually the clinical disease lasts four to six days, followed by a prolonged convalescence and complete recovery. In a small percentage of patients, the development of one or more of the three distinct syndromes can occur: ocular disease (0.5 to 2%), meningoencephalitis (nuchal rigidity, confusion, hypersalivation, fatigue, malaise, stupor, and coma) (less than 1%), or hemorrhagic fever (less than 1%). Elevated aspartate aminotransferase (AST), lactate dehydrogenase (LDH), and alanine aminotransferase (ALT) and reduction of platelet count and hemoglobin level are typical in severe cases, indicative of hepatocellular damage. Ophthalmologic complications remain very important sequelae of the disease and can result in lifelong loss of central vision. The case-fatality rate among severely affected patients is about 5 to 10%.

Pathogenesis, Pathology, and Immunity

Rift Valley fever virus is one of the most pathogenic viruses known—it replicates rapidly in target tissue and to a high titer. After entry by mosquito bite or through the oropharynx, there is an incubation period of 30 to 72 hours, during which time the virus invades the parenchyma of the liver and reticuloendothelial organs, leading to widespread severe cytopathology. In fatal cases it is not uncommon at autopsy to find nearly total hepatocellular destruction. The spleen is enlarged and there are gastrointestinal and subserosal hemorrhages. Encephalitis, as evidenced by neuronal necrosis and perivascular inflammatory infiltration, is a late event, seen

in a small proportion of survivors. Experimentally, the virus infects a wide variety of laboratory and domestic animals, and is often lethal. In experimentally infected animals the two most frequent syndromes are hepatitis and encephalitis.

Laboratory Diagnosis

Because of its broad geographical distribution and its explosive potential for invading new areas, the laboratory confirmation of the presence of Rift Valley fever virus is treated as a diagnostic emergency. Owing to the need for a rapid response, various RT-PCR methods are of particular value—various isothermal amplification methods have also been developed for use in field laboratories. Serological diagnosis is done by IgM capture ELISA on single acute sera, or by standard enzyme immunoassays, or by neutralization or hemagglutination-inhibition assays on paired sera from convalescing patients. The virus replicates in a variety of cell cultures such as Vero E6 (African green monkey kidney) and BHK-21 (baby hamster kidney) cells—the virus is rapidly cytopathic and causes plaques. Laboratory workers need to exercise extreme caution in order to avoid becoming infected during processing of diagnostic materials. In some countries, there is an inactivated vaccine available for laboratory and field workers at high risk of infection.

Epidemiology, Prevention, Control, and Treatment

In eastern, western, and southern Africa, Rift Valley fever virus survives in a minimally evident endemic cycle for years; then, when there is a period of exceptionally heavy rainfall, the virus explodes in epidemics of considerable magnitude. Although such epidemics had been studied for many years, it was not until the late 1980s that the mechanism of this phenomenon was discovered. It was found that the virus is transovarially transmitted among floodwater *Aedes* spp. mosquitoes; the virus survives for extensive periods in mosquito eggs laid at the edges of usually dry depressions, called "dambos," which are common throughout grassy plateau regions. When the rains arrive and flood these dambos, the eggs hatch and infected mosquitoes emerge and infect nearby wild and domestic animals. This discovery involved one of the first successful applications of satellite remote sensing and geographical information system (GIS) technology (Fig. 29.4).

In an epidemic, the virus is amplified in wild and domestic animal populations by many species of *Culex* and *Aedes* mosquitoes. These mosquitoes become exceedingly numerous after heavy rains or when improper irrigation techniques are used; they feed indiscriminately on viremic sheep and cattle (and humans). A very high level of viremia is maintained for three to five days in infected sheep and cattle, allowing many mosquitoes to become infected. This

FIGURE 29.4   Irrigation canals, Egypt, breeding sites for *Aedes* spp. *Photograph by Colin Howard.*

amplification, together with mechanical transmission by biting flies, results in infection and disease in a very high proportion of animals (and humans) at risk. In its epidemic cycles Rift Valley fever virus is also spread mechanically by fomites and by blood and tissues of infected animals. Infected sheep have a very high level of viremia and transmission at the time of abortion via contaminated placentae and fetal and maternal blood is a particular problem. Abattoir workers and veterinarians (especially those performing necropsies) can become infected by contact with infected tissues and body fluids (Fig. 29.5).

The capacity of Rift Valley fever virus to be transmitted without the involvement of an arthropod vector raises concerns over the possibility of its importation into non-endemic areas via contaminated materials, animal products, viremic humans, or non-livestock animal species. The virus poses a considerable threat to the Tigris-Euphrates basin, other areas of the Middle East, and anywhere else where livestock are raised—the epizootic in Yemen and Saudi Arabia in 2000 has been taken as a warning of this risk.

As was the case in Egypt in 1977 and 1978 and more recently, many mosquito species capable of efficient virus transmission are present in most of the livestock-producing areas of the world; for example, experimental mosquito transmission studies have shown that more than 30 common mosquitoes in the United States could serve as efficient vectors.

Prevention of disease in humans at particular risk, such as veterinarians and livestock and abattoir workers and laboratory personnel, can be achieved by vaccination although vaccines are not generally available. An improved inactivated virus vaccine was produced by the US Department of Defense, but it has not been commercialized.

Control is based primarily upon livestock vaccination, but vector control by environmental management plays a role too: agricultural development projects in Africa must take into account the danger of creating new larval habitats, such as water impoundments and artificial dambos.

Mosquito larvicide and insecticide use is virtually impossible in the huge areas of Africa that are at risk—the involvement in epidemics of a wide range of vector species with different habits and econiche preferences, the usual wide geographical distribution of epidemics, and the need to intervene throughout long vector breeding seasons and years of little activity contribute to this reality. However, vector control would be a major element in control programs were the virus to be introduced outside Africa.

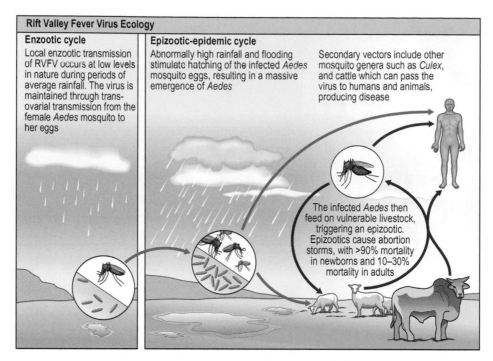

Rift Valley Fever Virus Ecology

Enzootic cycle	Epizootic-epidemic cycle
Local enzootic transmission of RVFV occurs at low levels in nature during periods of average rainfall. The virus is maintained through trans-ovarial transmission from the female *Aedes* mosquito to her eggs	Abnormally high rainfall and flooding stimulate hatching of the infected *Aedes* mosquito eggs, resulting in a massive emergence of *Aedes*

Secondary vectors include other mosquito genera such as *Culex*, and cattle which can pass the virus to humans and animals, producing disease

The infected *Aedes* then feed on vulnerable livestock, triggering an epizootic. Epizootics cause abortion storms, with >90% mortality in newborns and 10–30% mortality in adults

FIGURE 29.5 Transmission cycles of Rift Valley fever virus. *From Thomas Ksiazek.*

Sandfly Fever

Sandfly (phlebotomus) fever is a common but non-lethal disease caused by a group of related phleboviruses that are transmitted to humans by peridomestic sandflies (*Phlebotomus papatasii, P. perniciosus, P. perfiliewi*) in countries around the Mediterranean Sea and eastwards to central Asia and India. Human infections with Naples and Sicilian sandfly fever viruses are incapacitating but self-limiting (so-called "3-day fever")—patients present with influenza-like symptoms, including fever, myalgia, malaise, and retroorbital pain. Usually, these patients recover fully within seven days. However, infection with Toscana virus is the major cause of aseptic meningitis (95%) and meningoencephalitis (4.5%) and influenza-like illness during the summer season in central Italy. Toscana virus has also been found in France, Spain, Portugal, Slovenia, Greece, Cyprus, and Turkey. Recent findings of a much higher prevalence of infection, especially subclinical infection, in endemic countries has raised the question of a potential risk of transmitting the virus to naïve subjects through blood transfusions. The indigenous people in endemic regions are usually immune as a result of childhood infection, but travelers are at risk and epidemics have occurred in armies throughout history. A second focus occurs in Central and South America where forest-dwelling phlebotomines of the genus *Lutzomyia* are the vectors. No vertebrate host other than humans has been definitely incriminated but gerbils are the principal suspects in Europe and Asia and forest rodents in South America. The virus persists in sandflies by transovarial and transstadial transmission. Genetic reassortment between closely related viruses has been shown to occur in nature.

Severe Fever with Thrombocytopenia Syndrome Virus (SFTSV) and Other Phlebovirus Infections of Humans

From 2009 onwards a new phlebovirus has been reported from the central and southern provinces of the People's Republic of China. Patients have presented with fever, joint pains, and other symptoms including gastroenteritis, thrombocytopenia, leukopenia and multiorgan dysfunction.

The clinical syndrome progresses in three stages: the first stage or fever stage (days 1 to 7 after onset of illness) is characterized by marked thrombocytopenia and leukopenia (lymphopenia) and low platelet and white blood cell counts. Cardiac and liver enzymes are elevated. The second stage (the multiorgan dysfunction stage), occurs between days 7 and 13 after disease onset and is characterized by the development of hepatic, renal and CNS involvement as well as hemorrhagic fever. In fatal cases, platelet counts continue to decline and tissue enzymes increase until death. The third stage (the convalescent stage) in surviving patients occurs after day 13. The case-fatality rate has been 12 to 30%. There has been some evidence of person-to-person transmission by close contact.

The majority of affected patients have been farmers living in wooded and hilly areas who were exposed to ticks of the family *Ixodidae* (species *Haemaphysalis longicornis*).

High seroprevalence of SFTSV has been detected in goats, cattle and pigs in the endemic region. SFTSV is cultivated in DH82 and Vero E6 cells, but with great difficulty.

Another new phlebovirus with a clinical presentation not dissimilar to SFTSV has been reported since 2009—Heartlands virus, from Missouri, United States. Patients have presented with a history of tick bite, and with fever, diarrhea, headache, anorexia, nausea, elevated hepatic aminotransferase levels, thrombocytopenia, moderate neutropenia, and leukopenia. Recovering patients reported fatigue and short-term memory lapses. Heartland virus is related to SFTSV from China (but with less than 40% homology between the nucleocapsid genes of the viruses). A related virus has also been recovered from ticks in Australia, suggesting that related viruses may be geographically widespread, as yet unrecognized because of difficulty in laboratory detection. In common with Uukiniemi virus, another tick-borne phlebovirus, the SFTSV genome lacks the coding capacity for NSm.

The discovery of SFTSV and Heartlands virus within recent years has led to a re-examination of other phleboviruses isolated many years previously and hitherto not thought to be significant human pathogens. It appears that Bhahja and Lone Star viruses, for example, are part of the same phylogenetic complex of phleboviruses as Heartlands virus and SFTSV. Thus more discoveries are likely of phleboviruses with potential for causing significant human disease.

DISEASES CAUSED BY MEMBERS OF THE GENUS *NAIROVIRUS*

Crimean-Congo Hemorrhagic Fever

Crimean-Congo hemorrhagic fever is an emerging problem, with an increasing number of cases being reported each year from many parts of the world, paralleled by an increase in antibody prevalence among livestock populations. For example, more than 8% of cattle in several regions of Africa have been shown as seropositive

The disease was first described in the Crimean Peninsula in 1944, and independently in 1969 in the Congo, hence the name. Crimean-Congo hemorrhagic fever is the most dramatic of all human hemorrhagic fevers in the amount of hemorrhaging and the extent of subcutaneous and mucosal ecchymoses. The disease primarily affects farmers, veterinarians, slaughter-house workers and butchers and others coming into contact with livestock, and woodcutters, and anyone else coming into contact with infected ticks are also at risk within endemic areas.

Clinical Features

Crimean-Congo hemorrhagic fever presents in four phases: incubation, prehemorrhagic, hemorrhagic, and convalescence phases. The incubation period following a tick bite is usually short, ranging between three and seven days. The incubation period following contact with infected animal blood or tissues is usually five to six days, with a maximum of 13 days. Clinical manifestations in the prehemorrhagic phase include fever, malaise, myalgia, dizziness, and, in some patients, diarrhea, nausea, and vomiting, lasting for about 3 days. The hemorrhagic phase usually begins on days three to five after the onset of illness and is usually short, lasting on average two to three days. The first evidence of hemorrhagic disease is usually a flushing of the face and the pharynx and a skin rash that progresses to petechiae and ecchymoses, hemorrhage of the mucous membranes and conjunctiva, hematemesis, melena, epistaxis, hematuria, and hemoptysis. Cerebral hemorrhage, gingival bleeding, and bleeding from the nose, vagina, uterus, or urinary tract may occur, as well as internal bleeding into abdominal muscles. Cerebral hemorrhage and massive liver necrosis indicate a poor prognosis. Hepatomegaly and splenomegaly may occur in up to 40% of the patients. Death generally occurs on days 5 to 14 of illness and is attributed to hemorrhages, hemorrhagic pneumonia, or cardiovascular disturbances. Fatal cases typically do not develop an antibody response. The mortality rate is about 30% (range 5% to more than 80%); this wide variation is believed to reflect differences in levels of patient care as well as genetic variation among virus strains. The convalescence period is quite long and recovery can take up to a year, during which time patients may experience generalized weakness, tachycardia, hair loss, poor appetite, difficulty in breathing, polyneuritis, loss of hearing, and memory and impaired vision (Fig. 29.6).

Pathogenesis, Pathology, and Immunity

Clinical characterization of the pathogenesis of Crimean-Congo hemorrhagic fever has been described fairly extensively; however the lack of a good experimental animal model has impeded manipulative *in vivo* experimental pathology research. Thrombocytopenia is characteristic and most patients also have leukopenia and elevated levels of aspartate aminotransferase (AST), alanine aminotransferase (ALT), lactate dehydrogenase, and creatine phosphokinase. Serum AST levels are typically higher than ALT levels. Decreased platelet count and fibrinogen levels along with increased activated partial thromboplastin time are also indicators of hepatic and/or multiorgan disease, and indicators of a poor outcome (Fig. 29.7).

Laboratory Diagnosis

Useful diagnostics include IgM- and IgG-specific antigen-capture ELISAs and RT-PCR-based assays. Antibodies can be detected four to five days after the onset of clinical

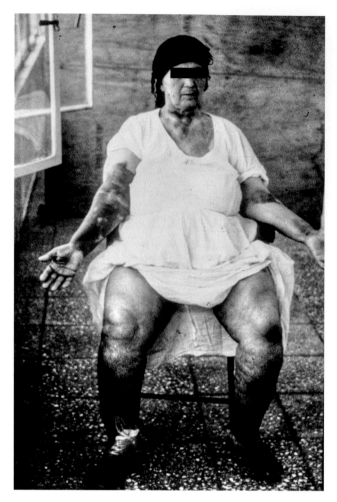

FIGURE 29.6 Patient with Crimean-Congo hemorrhagic fever in Iraq, showing hemorrhagic skin lesions. Petechiae progress to extensive ecchymoses. *Photograph by courtesy of the late David Simpson.*

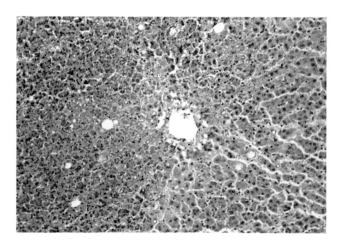

FIGURE 29.7 H&E section of human liver infected with Crimean-Congo hemorrhagic fever. Note large area of hemorrhagic necrosis. *Reproduced from Congo-Crimean Hemorrhagic Fever in Dubai: Histopathological Studies. Baskerville, A., et al., 1981. J. Clin. Pathol. 34, 871–874, with permission.*

disease, making ELISA a useful diagnostic tool for hospital use. The most definitive diagnostic tools include detection of antigen in tissues (usually by immunofluorescence), plaque reduction neutralization tests, and virus isolation and characterization. However, virus isolation has proven difficult: the virus is very labile, shipping of diagnostic specimens from usual locations of disease is often less than satisfactory, and all laboratory work must be done under maximum containment conditions (see Chapter 10: Laboratory Diagnosis of Virus Diseases).

Epidemiology, Prevention, Control, and Treatment

The virus is maintained by a cycle involving transovarial and transstadial transmission in *Hyalomma* spp. and many related ticks. Larval and nymphal ticks become infected when feeding on small mammals and ground-dwelling birds, and adult ticks when feeding on wild and domestic ruminants (sheep, goats and cattle). Infection in wild and domestic ruminants is very productive, resulting in viremia levels high enough that ticks become infected when feeding.

Crimean-Congo hemorrhagic fever has been recognized for many years in central Asia and Eastern Europe. However, this recognition was just the tip of the iceberg—today, the disease is known to extend from China through central Asia to India, Pakistan, Afghanistan, Iran, Iraq, other countries of the Middle East, Eastern Europe, and most of Saharan and sub-Saharan Africa. In recent years, there have been repeated outbreaks in the countries of the Persian Gulf, especially in connection with traditional sheep slaughtering and butchering practices (Fig. 29.8).

The virus is transmitted to humans by direct contact with subclinically infected viremic animals, or from a bite by an infected tick. Human-to-human transmission is also possible, especially in hospitals. Not uncommonly, nosocomial outbreaks are traced to surgical intervention when a patient presents with extensive gastric hemorrhaging—a lack of precautions leaves the surgical team at considerable risk of infection.

Crimean-Congo hemorrhagic fever virus infection is somewhat sensitive to ribavirin, provided it is administered sufficiently early. The use of immunoglobulin has been investigated although extensive trials of its use are yet to be reported.

A vaccine is not available, and prevention based on vector control is difficult because of the large areas of wooded and brushy tick habitat involved. One important prevention approach would be the enforcement in endemic areas of Asia, the Middle East, and Africa of occupational safety standards on farms and in livestock markets, abattoirs, and other workplaces where there is routine contact with sheep, goats, and cattle.

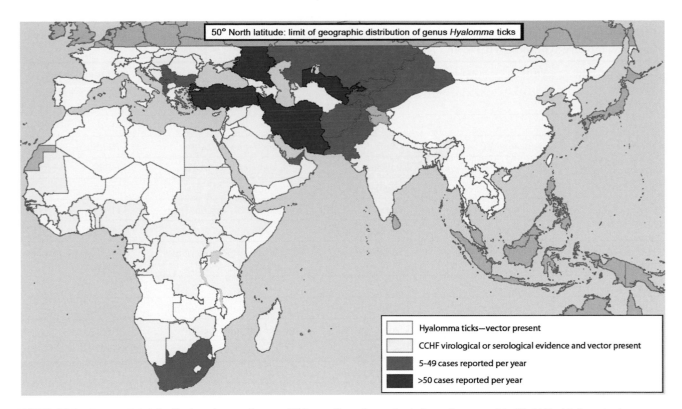

FIGURE 29.8 Current global distribution of reported cases of Crimean-Congo hemorrhagic fever. *Courtesy of the World Health Organization.*

DISEASES CAUSED BY MEMBERS OF THE GENUS *HANTAVIRUS*

The 23 confirmed member viruses of the genus *Hantavirus* comprise the only viruses in the family *Bunyaviridae* that are not arthropod-borne—these viruses are transmitted between small mammals by long-term shedding in saliva, urine, and feces. Several hantaviruses are zoonotic, and can cause severe human disease. Six Old World hantaviruses cause hemorrhagic fever with renal syndrome (HFRS); fourteen New World hantaviruses cause hantavirus pulmonary syndrome (HPS), clinically previously known as acute respiratory distress syndrome. The pathogenicity of many hantaviruses in terms of potential to cause infections among humans has not yet been determined (Table 29.3).

Hemorrhagic Fever with Renal Syndrome

The discovery of hantaviruses involves an intriguing history. During the Korean War of 1950–52, many thousands of UN troops developed a disease marked by fever, headache, hemorrhagic manifestations, and acute renal failure with shock; the mortality rate following infection was 5 to 10%. Despite intense research, the etiological agent of this disease remained a mystery for 28 years until the prototype hantavirus, Hantaan virus, was isolated from the striped field mouse *Apodemus agrarius*. Other, similar but often milder syndromes had long been recognized in the Balkans, in Finland (e.g., "nephropathia epidemica") and other countries, but the common link of hantaviruses as their origin was not made until the causative agents were identified. The hantaviruses discovered since then have also been difficult to isolate in cell culture or experimental animals, thus reverse-transcriptase polymerase chain reaction assays have become a key tool for obtaining diagnostic sequences from clinical materials (including acute-phase blood) and rodent tissues.

Clinical Features

The clinical course of severe hemorrhagic fever with renal syndrome (HFRS) in humans involves five overlapping but distinct stages: febrile, hypotensive, oliguria, diuretic, and convalescent stages, not all of which are seen in every case. The onset of the disease is sudden with intense headache, backache, fever, and chills. Hemorrhage, if it occurs, is manifested during the febrile stage, with injection of the conjunctiva and mucous membranes. A petechial rash may also appear. Sudden and extreme hypotension, albuminuria, nausea, and vomiting around day four, are features of severe disease. One-third of deaths occur during this stage owing to vascular leakage and acute shock, while up to 50% of deaths

TABLE 29.3 Old and New World Hantaviruses Causing Hemorrhagic Fever With Renal Syndrome (HFRS) or Hantavirus Cardiopulmonary Syndrome (HCPS) and Their Known Distribution and Hosts

Distribution and Subfamily of Rodent Host	Virus	Distribution	Primary Host Species	Human Disease
Old World				
Murinae (Old World rats and mice)	Hantaan virus	China, Korea, Russia	*Apodemus agrarius*	HFRS
	Dobrava (Belgrade) virus	Balkans	*Apodemus flavicollis*	HFRS
	Seoul virus	Global	*Rattus* sp.	HFRS
	Saaremaa virus	Europe	*Apodemus agrarius*	HFRS
	Amur virus	Far East Russia	*Apodemus peninsulae*	HFRS
Arvicolinae (voles and lemmings)	Puumala virus	Europe, Russia	*Myodes (Clethrionomys) glareolus*	HFRS (nephropathia epidemica)
New World				
Sigmodontinae (New World rats and mice)	Sin Nombre virus	North America	*Peromyscus maniculatus*	HPS
	Monongahela virus	North America	*Peromyscus leucopus*	HPS
	New York virus	North America	*Peromyscus leucopus*	HPS
	Black Creek Canal virus	North America	*Sigmodon hispidus*	HPS
	Bayou virus	North America	*Oryzomys palustris*	HPS
	Choclo virus	Panama	*Oligoryzomys fulvescens*	HPS
	Andes virus	Argentina, Chile	*Oligoryzomys longicaudatus*	HPS
	Bermejo virus	Argentina	*Oligoryzomys chocoensis*	HPS
	Lechiguanas virus	Argentina	*Oligoryzomys flavescens*	HPS
	Maciel virus	Argentina	*Bolomys obscures*	HPS
	Oran virus	Argentina	*Oligoryzomys longicaudatus*	HPS
	Laguna Negra virus	Argentina, Bolivia, Paraguay	*Calomys laucha*	HPS
	Araraquara virus	Brazil	*Bolomys lasiurus*	HPS
	Juquitiba virus	Brazil	*Oligoryzomys nigripes*	HPS

occur during the subsequent (oliguric) stage due to renal failure. Patients who survive and progress to the diuretic stage show improved renal function but may still die of shock or pulmonary complications. The convalescent stage can last weeks to months before recovery is complete. The disease is believed to be immunopathologically mediated, since antibodies are present usually from the first day the patient presents for medical care. Supportive therapy, with particular constraint in the use of fluids and in some cases the use of hemodialysis, has resulted in a decline of the mortality rate from 15% to 5% or less. The drug ribavirin shows antiviral activity against hantaviruses in cell culture, and has been used in clinical trials of HFRS with reports of reduced mortality.

Pathogenesis, Pathology, and Immunity

The pathogenesis of hantavirus infection in the rodent hosts that act as reservoirs is not well understood. The hallmark of infection is persistent (usually lifelong) inapparent infection with shedding of virus in the saliva, urine, and feces. In a landmark pathogenesis study, Ho-Wang Lee and his colleagues in Korea inoculated the reservoir rodent, *Apodemus agrarius*, with Hantaan virus and followed the course of infection by virus titration of organs, serology, and immunofluorescence. Viremia was found to be brief and disappeared as neutralizing antibodies appeared. However, virus persisted in several organs, including lungs and kidneys. Virus titers in urine and throat swabs were about 100 to 1000 times higher during the first weeks after inoculation than later, and animals were much more infectious for cage-mates or nearby mice during this period.

Human infection results from contact with contaminated rodent excreta, usually in winter when human–rodent contact is greatest, and human-to-human transmission is rare. The incubation period ranges from 7 to 21 days. Since the virus is not cytopathic in human cells, and virus-specific antibodies and T cells are detectable at the time that symptoms develop, immunological mechanisms may be involved in disease in humans. Hantaviruses infect endothelial cells and platelets, with cell surface integrins serving as virus receptors. Those hantavirus pathogenic for man predominantly bind β3 integrin whereas the non-pathogenic hantaviruses recognize β1. Integrins, especially β3, play an essential role in regulating platelet function and capillary permeability, thus, perturbance of integrin function may induce a severe disease. Hantaviruses also use the Decay Accelerating Factor (DAF) as a co-factor: binding to DAF leads to the opening of tight junctions between polarized cells, allowing the virus access to integrins restricted to the basal cell surface. Dysregulation of endothelial cell function results in vascular hyperpermeability and leakage through disruption of vascular endothelial growth factor (VEGF) and inhibition of VE-caderin function in maintaining tight junctions.

Laboratory Diagnosis

Diagnosis is made by the detection of antigen in tissues (usually by immunofluorescence) or serologically by detection of virus-specific IgG and IgM by capture enzyme immunoassays. There is considerable antibody cross-reactivity between different hantaviruses. The IgM capture enzyme immunoassay is the primary diagnostic tool in reference diagnostics centers, but RT-PCR assays are now routine. Virus isolation has proven very difficult: samples must be passaged several times in cell culture (most commonly Vero E6 cells) and detected by serological or molecular means, and results are usually negative. The shipping of diagnostic specimens from sites of disease is often less than satisfactory, and all laboratory work on Hantaan, Dobrava, Sin Nombre, and other highly pathogenic viruses must be done under maximum containment conditions (see Chapter 10: Laboratory Diagnosis of Virus Diseases).

Epidemiology, Prevention, Control, and Treatment

More than 200,000 cases of hemorrhagic fever with renal syndrome are reported each year throughout the world, with more than half in China. Russia and Korea report hundreds to thousands of cases; fewer are reported from Japan, Finland, Sweden, Bulgaria, Greece, Hungary, France, and the Balkan countries.

A matrix of four characters is needed to define the global context of the disease hemorrhagic fever with renal syndrome (Table 29.3): (1) the viruses, *per se*, (2) their reservoir rodent hosts, (3) the locale of human cases, and (4) the severity of human cases. Four viruses are involved: Hantaan, Dobrava, Seoul, and Puumala viruses. Hantaan and Dobrava viruses cause severe disease with mortality rates of 5 to 15%; Seoul virus causes less severe disease and Puumala virus causes the least severe form of the disease (mortality rate less than 1%) that is known in Scandinavia as nephropathia epidemica. There are three disease locale patterns: rural, urban, and laboratory-acquired. Hantaan virus causes rural disease, which is widespread in China, Eastern Russia, and Korea, while Dobrava virus causes severe HFRS in the Balkans and Greece. Puumala virus also causes rural disease, in northern Europe, especially in Scandinavia and Russia. Seoul virus causes urban disease; it occurs in Japan, Korea, China, and South and North America. Each virus has a specific reservoir rodent host: Hantaan virus, the striped field mouse, *Apodemus agrarius*; Dobrava virus, the yellow-neck mouse, *Apodemus flavicollis;* Seoul virus, the Norway rat, *Rattus norvegicus*; and Puumala virus, the bank vole, *Clethrionomys glareolus*. The relationship between each virus and its host defines the geographical limits of disease and confirms ancient virus–host relationships. For example, the relationship between Seoul virus and the Norway rat explains its worldwide distribution and presence in seaports. A major scientific challenge is to understand the Darwinian forces that drive the long-standing maintenance of these relationships.

The recovery of hantaviruses from moles and shrews indicates that the diversity and complexity of hantaviruses are even more complex than originally thought. Thottapalayan virus, isolated in India from a shrew (*Suncus murinus*) over 40 years ago, was considered an exception to the finding of hantaviruses predominantly in rodents. However, phylogenetic analyses now show that ancestral shrews or moles as opposed to murines are more likely to have been the original reservoirs for the progenitor virus of the hantavirus genus. The close association between hantaviruses and specific families of small mammals indicates a parallel evolution between virus and host species over the millennia.

Human infection has been recorded following exposure to infected laboratory rats and wild rodents brought into laboratories, involving transmission to animal caretakers and research personnel. There have been episodes of human disease in Belgium, Korea, the United Kingdom, and Japan, totaling more than 100 cases, with one death. Prevention of introduction of virus into laboratory rat colonies requires quarantined entry of new stock (or entry only of known virus-free stock), prevention of access by wild rodents, and regular serological testing. Cesarean derivation and barrier rearing of valuable rat strains can eliminate the virus from infected colonies. Prevention of viral spread should also involve the testing of all rat-origin cell lines before release from cell culture collections or other laboratories.

Since control of the wide-ranging rodent reservoirs of Hantaan, Seoul, and Puumala viruses is not possible in most settings where the disease occurs, the mainstay of prevention is reducing human exposure to rodents. This includes rodent-proofing of dwellings, disinfection of carcasses and droppings, together with the correct storage of food. Vaccine development has been hampered by the lack of animal models of disease. Korean and Chinese investigators have developed inactivated whole virus vaccines that are licensed for local use.

Hantavirus Cardiopulmonary Syndrome

In 1993 a new zoonotic hantavirus disease was recognized in the south-western region of the United States. The disease was manifest not as hemorrhagic fever with renal syndrome, but rather as an acute respiratory distress syndrome. The etiologic agent was found to be a new hantavirus, Sin Nombre virus, but in the following years quite a few other, related hantaviruses of the Western Hemisphere were also found to be etiologically associated with the same syndrome. As of 2015, a total of 690 cases had been reported in 34 states of the United States with a mortality rate of 36%. Cases have also been reported from the western provinces of Canada.

Clinical Features

Hantavirus pulmonary syndrome (HPS) typically starts with fever, myalgia, headache, nausea, vomiting, a non-productive cough, and shortness of breath. As the disease progresses, there is radiological evidence of bilateral interstitial pulmonary edema and pleural effusions. There is thrombocytopenia, which is a critical marker of disease progression, as mortality is correlated with the extent of platelet depletion. There is also a left shift in the myeloid series and large immunoblasts are found in the circulation. The course of the disease progresses rapidly, death often following in hours to days.

The majority of hantavirus pulmonary syndrome patients require intensive cardiopulmonary support, and many of these patients decompensate rapidly. Thus, early recognition of infection and close monitoring in a hospital setting for hypoxemia and shock are crucial to survival. Supportive therapy must be initiated before irreversible decompensation makes resuscitation difficult. Hypoxemia can be successfully managed by administration of high concentrations of inspired oxygen through a non-rebreathing or positive pressure mask. Some patients require intubation within a few hours of the beginning of the cardiopulmonary phase of the disease.

Recovery can be as rapid as the development of life-threatening clinical signs. Functional impairment of vascular endothelium and shock are central to the pathogenesis of the disease. Histopathological lesions include interstitial pneumonitis with congestion, edema, and mononuclear cell infiltration. Current therapy includes aggressive management of cardiovascular abnormalities and the pulmonary edema. Although no cases of human-to-human transmission have been identified in the United States, investigation of an epidemic in Argentina in 1995 provided strong evidence for person-to-person transmission; strict barrier nursing techniques are now recommended for the management of suspected cases.

Pathogenesis, Pathology, and Immunity

Much has been learned about the pathogenesis and pathology of New World hantavirus infections of reservoir rodent hosts. The viruses are minimally deleterious to the survival or reproduction of their reservoir rodent hosts, but virus is shed in saliva, urine, and feces of these animals for at least many weeks and perhaps for the lifetime of the animal. Transmission from rodent to rodent is believed to occur after weaning, by contact and perhaps also by biting and scratching. Transmission (and human infection) likely also occurs by inhalation of aerosols or dust containing infected dried rodent saliva or excreta. In fatal human cases, large amounts of viral antigen are present in pulmonary capillary endothelial cells and abundant inflammatory cells and macrophages are seen in the lungs (but not the kidney as is seen with HFRS), suggesting that local cytokine production may be involved in the mechanism of disease. Viral antigens are also found in myocardial capillaries, and impaired myocardial function may contribute to the pulmonary edema.

Laboratory Diagnosis

During 1993, before the virus causing the human disease outbreak in south-western United States had been isolated, serological tests, using surrogate antigens from related hantaviruses, and molecular procedures were developed and used to prove that the etiological agent was a previously unknown hantavirus. Viral RNA was amplified from human autopsy tissues and peripheral blood mononuclear cells using RT-PCR, followed by partial sequencing of amplified products. Where specimens were unsuitable for RT-PCR, immunohistochemical staining was used to detect viral antigens in fixed lung tissues. Later, PCR-amplified products were extended—that is, overlapping amplified products were sequenced and aligned until the entire genome of the new virus had been determined. Sequences encoding the G1 glycoprotein of Sin Nombre virus were then used in expression systems to produce homologous antigens and then in turn to produce reference antisera for further diagnostic tests and investigations. The same methods were applied to specimens from large numbers of rodents collected in the areas where patients lived. This proved

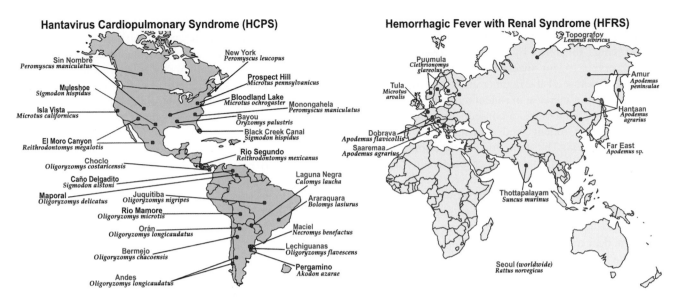

FIGURE 29.9   Geographical distribution of the pathogenic hantaviruses (in red), paired to each reservoir host (black, italics). Presently, there are at least seven hantaviruses associated with hemorrhagic fever with renal syndrome (HFRS), all with ancient niches in the Eastern Hemisphere. The most severe form of HFRS is caused by Hantaan virus in Asia and Dobrava virus in Europe—mortality rate from 3 to 12%. Saaremaa virus, related to Dobrava virus, is recognized as distinct and associated with a lower mortality rate. Puumula virus causes the mild form of HFRS, called nephropathia epidemica, in central and northern Europe, Russia and the Balkans, with a mortality of 0.1 to 0.4%. Seoul virus, distributed worldwide in *Rattus norvegicus* and *R. rattus*, causes a mild form of HFRS, mortality rate 1 to 2%. Amur and Far East viruses have been identified in severe HFRS cases in the far eastern region of Russia. Tula virus is widely distributed in rodents from central to eastern Europe—it has been linked to human disease in the Czech Republic, Switzerland and Germany. Topografov virus has only been linked to HFRS by historic association. Thottapalayam virus, isolated from an insectivore *Suncus murinus*, has been linked to human disease in India. Since the first recognition of hantavirus cardiopulmonary syndrome (HCPS) in the southwestern United States in 1993 and the isolation of its causative agent, Sin Nombre virus, more than 30 new hantaviruses associated with various rodent species throughout the Americas have been identified, many in connection with more than 2000 cases of HCPS. The most important pathogenic hantaviruses in the United States and Canada are: Monongahela virus, harbored by *Peromyscus maniculatus*; New York virus, harbored by *Peromyscus leucopus*; Bayou virus, harbored by *Oryzomys palustris*; and Black Creek Canal virus, harbored by *Sigmodon hispidus*. In Latin America, the most important pathogens are: Andes virus, harbored by *Oligoryzomys longicaudatus* in Argentina and Chile; Araraquara virus, harbored by *Necromys lasiurus* in Brazil; Leguna Negra virus, harbored by *Calomys laucha* in Paraguay; and Choclo virus, harbored by *Oligoryzomys fulvescens* in Panama. *Adapted and compiled from multiple sources.*

that at least eight species of rodents were involved, the primary reservoir host in the southwest being *Peromyscus maniculatus*, the deer mouse (10 to 30 and even up to 50% of these animals were found to harbor Sin Nombre virus in areas where there was human disease). The comparison of sequences obtained from specimens from different areas indicated that several different variant viruses, all previously unknown, were active in the United States (Table 29.3).

The serological diagnosis of hantavirus infection is typically done by IgM- and IgG-specific capture ELISAs, often using recombinant virion N protein as antigen (this is the most abundant viral protein and induces a strong humoral response). Virtually all patients are IgM positive and RT-PCR positive by the onset of acute illness. Lateral flow immunoelectrophoresis ("dipstick") assays are also widely used in the clinical setting. Virus antigen and RNA detection in lung tissue, blood clots and circulating mononuclear cells are done using immunohistochemistry and RT-PCR in specialty laboratories in the endemic region of western United States and in national laboratories in Western Hemisphere countries. Direct isolation of hantaviruses from acute human specimens or from tissues of suspect small mammalian hosts is difficult and time-consuming—the viruses are slow to adapt to growth in cell culture and the presence of virus must be established secondarily, typically by direct immunofluorescence or RT-PCR. Virus isolation and utilization of laboratory animals should only be carried out under appropriate biocontainment conditions.

Epidemiology, Prevention, Control, and Treatment

At least ten viruses have been etiologically associated with hantavirus pulmonary syndrome, with cases reported from Argentina, Bolivia, Brazil, Canada, Chile, Paraguay, Peru, and Uruguay as well as the United States (Fig. 29.9). Cases usually occur individually or in small clusters, and large outbreaks are rare and usually associated with large environmental dislocation; for example change in land use or heavy rainfall. Other New World hantaviruses have been discovered recently; their pathogenicity has not yet been determined.

Sin Nombre virus and other New World hantaviruses have been present for eons in the large area of the western United States inhabited by *Peromyscus maniculatus* and other reservoir rodent species; these were recognized in 1993 only because of the number and clustering of human cases. A massive increase in rodent numbers following two especially wet winters and a consequent increase in piñon seeds and other rodent food contributed to the number of human cases. The temporal distribution of human disease reflects a spring–summer seasonality (although cases have occurred throughout the year), again matching rodent reservoir host behavior. Just as with the Old World hantaviruses, each New World virus pathogenic for humans has a specific reservoir rodent host: for example, Sin Nombre virus, the deer mouse, *Peromyscus maniculatus;* New York virus, the white-footed mouse, *Peromyscus leucopus*, and Andes virus, the long-tailed pygmy rice rat, *Oligoryzomys longicaudatus* (Table 29.3).

Extensive public education programs have been developed to advise people about reducing the risk of infection, mostly by reducing rodent habitats and unsecured food supplies in and near homes, together with precautions when cleaning rodent-contaminated areas. The latter involves rodent-proofing food and pet food containers, trapping and poisoning rodents in and around dwellings, eliminating rodent habitats near dwellings, use of respirators as well as the wetting down of areas with detergent, disinfectant, or hypochlorite solution before cleaning areas that may contain rodent excreta. Recommendations have also been developed for specific equipment and practices to reduce risks when working with wild-caught rodents, especially when this involves obtaining tissue or blood specimens: these include use of live-capture traps, protective clothing and gloves, suitable disinfectants, and safe transport packaging.

FURTHER READING

Elliott, R.M., Brennan, B., 2014. Emerging phleboviruses. Curr. Opinion Virol. 5, 50–57.

Ergonul, O., 2012. Crimean-Congo hemorrhagic fever virus: New outbreaks, new discoveries. Curr. Opinion Virol. 2, 215–220.

Fulhorst, C.F., Koster, F.T., Enria, D.A.,, Peters, C.J., 2011. Hantavirus infection. In: Guerrant, R.L., Walker, D.H., Weller, P.F. (Eds.) Tropical Infectious Diseases: Principals, Pathogens and Practice (third ed.,), Elsevier, Philadelphia, pp. 470–480.

Vaheri, A., Strandin, T., Hepojoki, J., et al., 2013. Uncovering the mysteries of hantavirus infections. Nat. Rev. Microbiol. 11, 539–550.

Yanagihara, R., Gu, S.H., Arai, S., Kang, H.J., Song, J.W., 2014. Hantaviruses: Rediscovery and new beginnings. Virus Res. 187, 6–14.

Yu, X.J., Liang, M.F., Zhang, S.Y., Li, J.D., Sun, Y.L., et al., 2011. Fever with thrombocytopenia associated with a novel bunyavirus in China. N. Engl. J. Med. 364, 1523–1532. Available from: http://dx.doi.org/10.1056/NEJMoa1010095.

Chapter 30

Arenaviruses

Arenaviruses are important zoonotic pathogens having evolved closely with a number of rodent and other animal hosts that act as reservoirs of infection. The risk of transmission of each virus to humans relates to the nature of the infection in each rodent host: rodents usually undergo a persistent, asymptomatic infection accompanied by the shedding of virus throughout the life of the animal. The risk of human infection through contact with virus-laden rodent excreta closely corresponds to rodent population dynamics, animal behavior, and changes in agricultural practice. The consequences of human exposure include some of the most lethal hemorrhagic fevers known—Lassa, Lujo, Argentine, Bolivian, Venezuelan, and Brazilian hemorrhagic fevers.

The prototype arenavirus, lymphocytic choriomeningitis virus, has over many years played two disparate roles in comparative virology—wild-type strains are zoonotic pathogens and the subject of public health surveillance, whereas laboratory strains have provided much of the conceptual basis for our understanding of viral immunology and pathogenesis (Table 30.1).

Conceptual advances continue to flow from research with the lymphocytic choriomeningitis virus model, including vaccine development based on antiviral T-lymphocyte epitopes, and cytokine and lymphokine modulation of viral infections. Recent studies suggest that arenaviruses may also have evolved in snakes of the *Boidae* family, although there is no evidence of spread from snakes to their human keepers.

The increasing numbers of arenavirus infections indicate the need for much greater vigilance by public health workers. Arenavirus infection should always be considered as a cause of febrile illnesses among persons living within endemic areas, particularly those who are likely to have come into contact with rodents. There is increasing awareness of arenavirus activity in North America, spurred on by investigation of hantavirus epidemiology. Abnormal climate changes, together with woodland and forest clearance linked to changes in agricultural practice, are significant risk factors.

Biohazard: Human arenaviruses are *Restricted Pathogens* and importation or possession of these viruses is regulated by most national governments. Being classified as a Biosafety Level 4 pathogen, the handling of human arenaviruses is restricted to certain national laboratories for all research and diagnostic procedures; these viruses must be handled in the laboratory under maximum containment conditions to prevent human exposure (see Chapter 10: Laboratory Diagnosis of Virus Diseases). The exception is certain strains of LCM virus that may be handled in Biosafety Level 3 facilities. Classical laboratory-adapted strains, such as the Armstrong strain, frequently used for immunology and pathogenesis research, usually require Biosafety Level 2 containment.

PROPERTIES OF ARENAVIRUSES

Classification

The family *Arenaviridae* contains a single genus, *Arenavirus*, divided into two evolutionary subgroups based on the genetic and serological characteristics of each virus species. One subgroup includes lymphocytic choriomeningitis virus, which is associated with the common house mouse, *Mus musculus*, in North and South America, Europe, and perhaps elsewhere, and the Old World arenaviruses, associated with *Mastomys* spp. and *Paromys* spp. in Africa. This subgroup contains Lassa and Lujo viruses, both of which produce severe disease in humans, together with a few other viruses that infect humans but are not thought to cause clinical disease. The second subgroup includes the New World arenaviruses (also called the Tacaribe complex) that are associated with particular cricetids (many different rodents, muskrats, and gerbils) in North, Central, and South America. This subgroup contains the important human pathogens, Junin, Machupo, Guanarito, and Sabiá viruses. Several other viruses, for example, Whitewater Arroyo virus, are not as yet confirmed as significant pathogens for humans.

Phylogenetic analyses show a distant relationship between Old World and New World arenaviruses that is broadly consistent with the serological relationships between these viruses—these relationships suggest long evolutionary adaptation between an arenavirus and its reservoir animal hosts.

Fenner and White's Medical Virology. DOI: http://dx.doi.org/10.1016/B978-0-12-375156-0.00030-8

TABLE 30.1 Contributions of LCM Virus Research to Viral Immunology

Feature
Immunological tolerance
Virus-induced immunopathology
Immune complex disease
Virus-specific CTLs as mediators of viral clearance and immunopathology
Major histocompatibility complex (MHC) restriction in T-lymphocyte recognition
Natural killer cell activation and proliferation; and
Infection-induced impairment of specialized cell functions

Apart from those mentioned earlier, most known arenaviruses (Table 30.2) are found in rodents but do not normally infect other mammals or humans, and with a geographical range that is typically more restricted than that of any cognate rodent host (Fig. 30.1).

Virion Properties

Arenavirus virions are pleomorphic in shape, ranging in diameter from 50 to 300 nm, although most virions have a diameter of 110 to 130 nm (Fig. 30.2). Virions are composed of an envelope covered with club-shaped glycoprotein peplomers 8 to 10 nm in length, within which are two circular helical nucleocapsid segments, each with the appearance of a string of beads. The nucleocapsids are circular as a consequence of forming panhandles through the formation of non-covalent bonds between conserved complementary nucleotide sequences at the 3'- and 5'-ends of each RNA genome segment. The family derives its name from the presence within virions of cellular ribosomes resembling grains of sand (Latin: *arena*, sand) when observed by thin-section electron microscopy. The genome consists of two segments of single-stranded RNA, designated large (L), and small (S), 7.2 and 3.4 kb in size, respectively.

Most of the genome is of negative-sense polarity, but the 5'-half of the S segment and the 5'-end of the L segment are of positive sense; the term *ambisense* is used to describe this unusual genome arrangement which is also found in some members of the family *Bunyaviridae* (see Chapter 29: Bunyaviruses; Fig. 30.3). Specifically, the nucleoprotein mRNA is copied from the 3'-half of the genomic-sense S RNA, whereas the viral glycoprotein precursor mRNA is copied from the 3'-half of the antigenomic-sense S RNA. Similarly, the L protein mRNA is copied from the 3'-end of the genomic-sense L RNA and a zinc-binding protein (Z) is copied from the 3'-end of the antigenomic-sense L RNA. The Z protein is a matrix protein that may also modulate host innate immunity. Both L and S RNA contain hairpin configurations between the genes that function to terminate transcription from the viral and viral-complementary RNAs. The mRNAs are capped—5'-methylated caps are cleaved from host mRNAs by the viral RNA-dependent RNA polymerase (transcriptase), which also has endonuclease activity, and the caps added to viral mRNAs in order to prime transcription (referred to as cap-snatching).

Arenaviruses are sensitive to heat and acid conditions and are readily inactivated by detergents, lipid solvents, and common disinfectants.

Viral Replication

Arenaviruses replicate in the cytoplasm to high titer in many kinds of cells, for example, Vero E6 (African green monkey) cells and BHK-21 (baby hamster kidney) cells. The widely conserved host protein α-dystroglycan has been identified as the cellular receptor for Old World arenaviruses and some New World arenaviruses, although cofactors may also be involved. As the genome of all single-stranded negative-sense RNA viruses cannot be translated directly, the first step after attachment, penetration of the host cell, and uncoating is the activation of the virion RNA polymerase (transcriptase). By adopting an ambisense coding strategy, only nucleoprotein (NP) and polymerase (L) mRNAs are transcribed directly from virion RNAs prior to translation. The newly synthesized polymerase and nucleocapsid proteins then facilitate the synthesis of full-length virion–complementary-sense RNA that serves either as a template for the transcription of glycoprotein and zinc-binding protein mRNAs or as a template for the synthesis of more full-length virion-sense RNA. Both NP and Z proteins have additional functions in suppressing the host innate immune response: studies have shown that the NP of many arenaviruses inhibits the nuclear translocation and transcription of NF-κB, required for the activation of many proinflammatory cytokines in addition to interferon.

The envelope glycoproteins are derived by cleavage of a precursor, GPC, this translation product being first cleaved by a cellular signal peptidase to yield a stable signal peptide, and the whole is then processed further to generate the glycoproteins GP1 and GP2. Both envelope glycoproteins remain non-covalently attached to the signal peptide and form trimeric spikes consisting of three identical protein peplomers. Budding of virions occurs from the plasma membrane.

Arenaviruses have a limited lytic capacity for infected cells, readily producing a carrier state in which defective-interfering

TABLE 30.2 Members of the *Arenaviridae* Family (human pathogens are shown in bold)

Virus	Natural Host	Distribution
Lassa–LCM Complex (Old World)		
Ippy	*Arvicanthus* spp.	Central African Republic
Kodoko	*Mus (Nannomys)* spp.	West Africa
Lassa	*Mastomys natalensis*	West Africa
Lujo	(Not known)	Southern Africa
Lymphocytic choriomeningitis	*Mus musculus, Mus domesticus*	Worldwide except Australasia
Merino Walk	*Myotomus unisulcatus*	South Africa
Mobala	*Praomys jacksoni*	Central African Republic
Mopeia	*M. natalensis*	Mozambique
Tacaribe Serocomplex (New World)		
Clade A		
Allpahuayo	*Oecomys bicolor*	Peru
Bear Canyon	*Peromyscus californicus*	California, United States
Big Bushy Tank	*Neotoma* spp.	United States
Catarina	*Neotoma micropus*	Texas, United States
Flexal	*Neocomys* spp.	Brazil
Paraná	*Oryzomys buccinatus*	Paraguay
Pichinde	*Oryzomys albigularis*	Columbia
Pirital	*Sigmodon alstoni*	Venezuela
Skinner Tank	*Neotoma* spp.	United States
Tamiami	*Sigmodon hispidus*	Florida, United States
Tonto Creek	*Neotoma* spp.	United States
Whitewater Arroyo	*Neotoma albigula*	New Mexico, United States
Clade B		
Amapari	*Oryzomys gaedi, Neocomys guianae*	Brazil
Chapare	Not known	Bolivia
Cupixi	*Oryzomys capito*	Brazil
Guanarito	*Zygodontomys brevicuda*	Venezuela
Junín	*Calomys musculinus, Calomys laucha, Akadon azarae*	Argentina
Machupo	*Calomys callosus*	Bolivia
Sabiá	Not known	Brazil
Tacaribe	*Artibeus literatus* (bat)	Trinidad
Clade C		
Latino	*C. callosus*	Bolivia
Oliveros	*Bolomys obscurus*	Argentina
Pampa	*Bolomys* spp.	Argentina

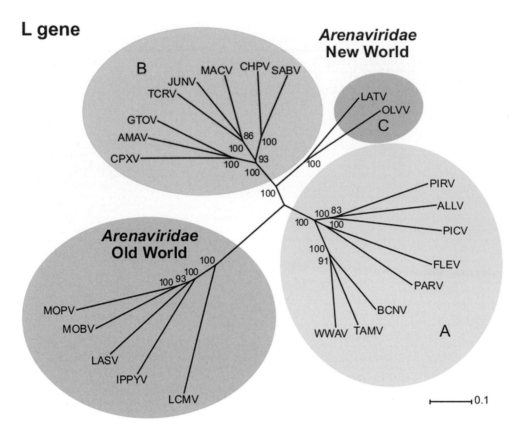

FIGURE 30.1 Phylogenetic relationships between the arenaviruses. (Skinner Tank, Tonto Creek, Big Bushy Tank, Pampa, and Merino Walk viruses omitted for clarity). Among the Old World viruses, Lassa (LASV), Mopeia (MOPV), and Mobala (MOBV) viruses are monophyletic, while Ippy virus (IPPYV) and lymphocytic choriomeningitis virus (LCMV) are more distantly related. LuJo virus, found in South Africa, is most closely related to Old World viruses but contains elements of New World sequence in its glycoprotein gene. The New World viruses can be divided into three groups on the basis of the sequence data. In group A are Pirital (PIRV), Pichinde (PICV), Paraná (PARV), Flexal (FLEV), and Allpahuayo (ALLV; Peru) viruses from South America, together with Tamiami (TAMV), Whitewater Arroyo (WWAV), and Bear Canyon (BCNV) viruses from North America. Group B contains the human pathogenic viruses Machupo (MACV), Junín (JUNV), Guanarito (GTOV), Sabiá SABV), and Chapare (CHPV) viruses as well as the non-pathogenic Tacaribe (TACV), Amapari (AMAV), and Cupixi (CPXV; Brazil) virus. Latino virus (LATV) and Oliveros (OLVV) viruses form a small separate group (group C). The division of the arenaviruses into Old World and New World groups, as well as the subdivision of New World arenaviruses into three groups, is strongly supported by all phylogenetic analyses (L (polymerase) gene bootstrap values are shown). The property of human pathogenicity appears to have arisen on at least two independent occasions during arenavirus evolution. *Reproduced from Howard, C.R., 2012. Lecture Notes on Emerging Viruses and Human Health: A Guide to Zoonotic Viruses and Their Impact. World Scientific Press: Singapore (Amazon) and King, A.M.Q., et al., 2012. Virus Taxonomy, Ninth Report of the International Committee on Taxonomy of Viruses. Academic Press, London, with permission.*

particles are produced. After an initial period of active viral transcription, translation, genome replication, and production of progeny virions, viral gene expression is down-regulated and cells enter a state of persistent infection wherein virion production continues for an indefinite period but at a greatly reduced rate.

Arenavirus virions are unusual in having internalized host cell ribosomes. The presence of ribosomes is not essential for replication and what function the presence of these structures plays in virus replication is as yet unknown (Table 30.3).

LYMPHOCYTIC CHORIOMENINGITIS VIRUS INFECTION

Lymphocytic choriomeningitis virus presents two zoonotic disease problems. First, the virus causes human disease after infected wild mice invade dwellings and farm buildings where dried virus-laden excreta may be transmitted by aerosols or fomites. Second, the virus causes major problems if established in laboratory mouse and hamster colonies, the virus not only posing a zoonotic threat to animal handlers

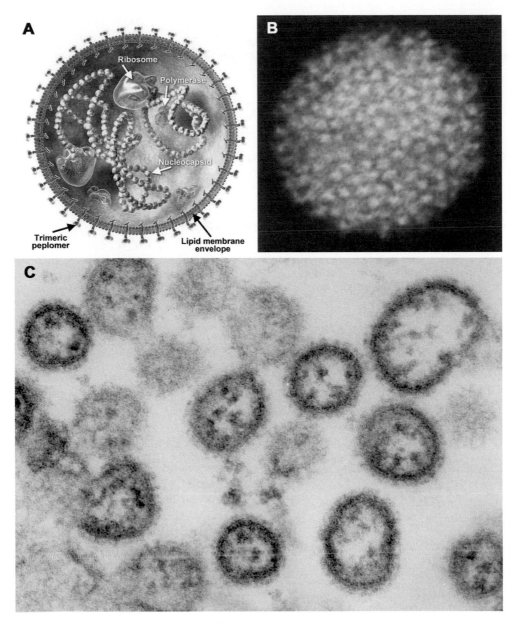

FIGURE 30.2 Arenavirus virion morphology and structure. (A) Diagram of arenavirus virion structure. (B) Negative-contrast electron microscopy of Tacaribe virus, showing the surface of a virus particle with closely apposed surface projections (peplomers). (C) Thin-section electron microscopy of Lassa virus in Vero cell culture, showing typical presence of ribosomes within the virus particles. *(A) Reproduced from King, A.M.Q., et al., 2012. Virus Taxonomy, Ninth Report of the International Committee on Taxonomy of Viruses. Academic Press, London, with permission.*

but also confounding research dependent upon virus-free animals, or cell cultures derived from these animals. For example, infected mouse tumors have been implicated in infections of laboratory workers. Mice and hamsters are the only species in which long-term, asymptomatic infection is known to exist, but guinea pigs, rabbits, rats, dogs, swine, and primates may also become infected in laboratory settings.

Clinical Features

Subclinical infection of humans is common, and some surveys have found 2 to 10% of human populations positive for LCMV antibodies. Infection may present as an influenza-like illness and may include fever, fatigue, malaise, anorexia, headache, sore throat, myalgia, photophobia,

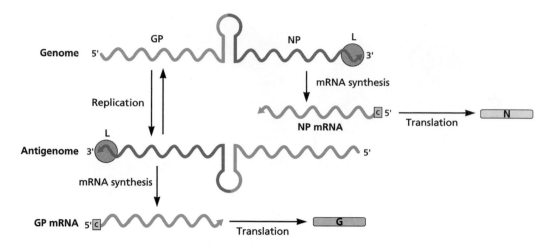

FIGURE 30.3 Genome organization and replication of arenaviruses. The viral genome comprises two ambisense RNA segments: large (L; 7.2 kb) and small (S; 3.5 kb). The L segment encodes the RNA-dependent RNA polymerase (L) and an accessory protein (Z) that functions in genome packaging, particle assembly, and budding. The S segment encodes a surface glycoprotein (GP), which binds to the viral receptor and mediates target cell recognition and entry, and a histone-like nucleocapsid protein (NP) that, with the viral RNAs, forms the nucleocapsid. For simplicity, only expression of genes on the S segment is shown, but the same process occurs for the L segment. Upon entry of the viral RNA into the host cell cytoplasm, the viral L protein, which enters the cell with the infecting particle, binds to the 3′-end of the RNA (shown as an orange ball) and synthesizes the (+) strand NP mRNA, which is then translated: copying of the rest of the genome may be temporarily blocked by the stem loop structure. Replication of the genomic RNA into a complementary full-length antigenome then allows synthesis of GP mRNA. This mechanism of gene expression results in temporal control of viral gene expression, which is also a common feature of the reproductive cycles of many viruses with DNA genomes. *Reproduced from Flint, S.J., et al., 2015. Principles of Virology, vol. 1, fourth ed. ASM Press, Washington, DC, with permission.*

TABLE 30.3 Properties of Arenaviruses

Property

One genus, *Arenavirus*, with two subgroups, one each for Old World and New World (Tacaribe complex) viruses

Virions are pleomorphic, enveloped, 50 to 300 (generally 110–130) nm in size

Virions contain non-functional host cell ribosomes

Virions contain two circular helical nucleocapsids with associated RNA-dependent RNA polymerase (transcriptase)

Genome consists of two segments, large (L, 7.2 kb) and small (S, 3.4 kb), of single-stranded RNA, both ambisense

Viral proteins: nucleoprotein (N), RNA-dependent RNA polymerase (L), two glycoproteins (Gp1, Gp2), zinc-binding protein (Z), plus minor proteins

RNA-dependent RNA polymerase (L) carried in the virion transcribes mRNAs for L and N proteins; newly synthesized L protein then transcribes full-length virion-complementary strand, which acts as a template for G and Z mRNAs and for progeny full-length virion-sense genome
Replication occurs in the cytoplasm, generally non-cytocidal, persistent infections

Maturation occurs by budding from the plasma membrane

Genetic reassortment occurs between closely related viruses

and gastrointestinal signs such as nausea and vomiting. Coughing, rashes, joint aches, and chest pain are sometimes seen. In most cases, the symptoms resolve without treatment within a few days. Less often patients present with aseptic meningitis; and rarely as a severe encephalomyelitis. Meningoencephalitis is characterized by profound neurological signs such as confusion, drowsiness, sensory abnormalities, and motor signs. Intrauterine infection has resulted in fetal and neonatal death, as well as hydrocephalus and chorioretinitis among infants.

In recent years increasing numbers of patients have been infected after solid organ transplantation. In the cases reported to date the initial clinical signs starting about one week after surgery included fever, lethargy, anorexia, and leukopenia, and quickly progressed to multisystem organ failure, hepatic insufficiency or severe hepatitis, dysfunction of the transplanted organ, coagulopathy, hypoxia, multiple bacteremias, and shock. Localized rash and diarrhea were also seen in some patients. Transplant-acquired lymphocytic choriomeningitis has a very high morbidity and mortality rate. In three clusters reported in the United States between 2005 and 2010, nine of the ten infected recipients died. In some instances the virus has been traced to the organ donor's pet hamster. The Dandenong isolate was obtained from an Australian transplant patient who received an organ from a deceased donor, the latter most likely infected while in the Balkans. Similarly, there have been increasing numbers of immunosuppressed patients, especially children, who have developed severe, often fatal lymphocytic choriomeningitis after being infected by pet hamsters derived from endemically infected colonies. In the United States, advice has been issued concerning this risk to immunosuppressed people.

Pathogenesis, Pathology, and Immunity

The outcome of lymphocytic choriomeningitis virus infection of mice depends on age, route of infection, and immunological status at the time of infection. Most strains of laboratory mouse infected *in utero* or during the first 48 hours after birth develop a persistent, tolerant infection. This infection may be asymptomatic, or over the course of several weeks to a year may become evident by weight loss, runting, blepharitis, and impaired reproductive performance. Terminal immune complex glomerulonephritis is a common result of the breakdown of immune tolerance. Animals infected peripherally after the first few days of life may overcome the infection or may show over a period of weeks decreased growth, rough hair coat, hunched posture, blepharitis, weakness, photophobia, tremors, and convulsions. The visceral organs of these animals, including the liver, kidneys, lungs, pancreas, and major blood vessels, become infiltrated by large numbers of lymphocytes. Animals inoculated intracerebrally (such as when the virus contaminates materials being used in research) usually develop an acutely fatal choriomeningoencephalitis, marked by florid accumulations of lymphocytes at all brain surfaces (meninges, choroid plexus, ependyma, and ventricles). In hamsters clinical disease is rare, and again the outcome depends upon age at the time of infection. As in mice, infection of young hamsters may lead to growth retardation, unthriftiness, weakness, conjunctivitis, dehydration, occasional tremors, and prostration. Lesions include vasculitis, glomerulonephritis, choriomeningoencephalitis, and lymphocytic infiltrations in visceral organs.

Because lymphocytic choriomeningitis virus provides such an important animal model for the mammalian immune response, a great deal is known about the pathogenesis, pathology, and immunology of the infection in mice. Sir Macfarlane Burnet and Frank Fenner first postulated the concept of immunological tolerance after studying LCM virus infection in mice: they concluded that exposure to viral antigens before the maturation of the immune system results in mice becoming tolerant and thus were unable to clear the infection. Subsequent work showed that persistently infected mice do make antiviral antibodies; however, but most of these antibodies are complexed with virus and complement and contribute to the progressive degenerative disease, including the glomerulonephritis, arteritis, and chronic inflammatory lesions.

Persistently infected mice were eventually shown to have *split-tolerance* to the virus; although they mount an antibody response, they do not generate those specific cytotoxic T-lymphocytes needed to clear the infection. In contrast, in acutely infected mice both the clearance of virus and the lethal choriomeningoencephalitis are due to a cytotoxic T-lymphocyte response. The accumulation of T-lymphocytes results in neuronal dysfunction and osmotic disturbances, often ending with fatal convulsions. In fact, antiviral cytotoxic T-lymphocytes were first demonstrated in the lymphocytic choriomeningitis virus model; Peter Doherty and Rolf Zinkernagel earned a Nobel Prize for work using this model to demonstrate the concept of major histocompatibility complex (MHC) restriction in antigen recognition by T-lymphocytes (see Chapter 6: Adaptive Immune Responses to Infection). The difference in disease outcome is thought most likely due to the extent of the innate immune response in the early stages of virus infection. An inadequate response among neonatal mice is insufficient to allow the CD8+ T-lymphocytes to clear the virus and persistence results.

One additional pathogenetic property of lymphocytic choriomeningitis virus is its ability to cause the loss of other specialized cellular functions beyond those required for cell survival and basic functioning. Michael Oldstone and his colleagues have shown that chronic infection can affect cellular homeostasis in subtle ways. For example, persistent infection in mice results in reduced neurotransmitter activity and reduced levels of growth and thyroid hormones. Reduced growth hormone synthesis is associated with runting in young mice.

Laboratory Diagnosis

For many years serological diagnosis of lymphocytic choriomeningitis was done by indirect immunofluorescence, using inactivated cell culture "spot-slides" as an antigen substrate. Such tests were also set up to measure IgM antibody indicating recent infection. In recent years IgM capture enzyme immunoassays have become the standard in national reference laboratories for the detection of recent infections, with IgG capture enzyme immunoassays used to detect the presence of the virus in mouse and hamster colonies. Virus is easily cultivated in a wide variety of mammalian cells, particularly Vero E6 (African green monkey kidney) cells and BHK-21 (baby hamster kidney) cells. Primary isolates often do not produce a cytopathic effect so cultures are assayed for antigen by immunofluorescence or enzyme immunoassay. With proper sampling in animal colonies, virus isolation may also be used to confirm the presence of virus.

Epidemiology, Prevention, Control, and Treatment

Lymphocytic choriomeningitis virus is maintained in nature by persistent infection of mice with long-term, even lifelong virus shedding in urine, saliva, and feces. Vertical transmission occurs by transovarial, transuterine, and various post-partum routes, including milk, saliva, and urine. Human infections follow contact with contaminated food and dust, the handling of dead mice, and mouse bites. The distribution of human cases is focal and seasonal, probably because mice move into houses and barns during winter months. Feral mice may also introduce the virus into laboratory and commercial mouse, rat, hamster, guinea pig, and primate colonies. In the United States the virus has been a particular problem in hamster colonies and immunocompromised (nude, SCID, etc.) mouse colonies, resulting in contaminated diagnostic reagents, failed research protocols, and clinical disease in laboratory and animal care personnel. Infection can be eradicated by both cesarean derivation of mouse and hamster breeding stock and the modification of facilities to prevent entry of wild mice. Animals, cell cultures, and tumors and tumor cell lines derived from laboratory mice must be subjected to regular testing so as to recognize any reintroduction of the virus. The increasing popularity of hamsters as pets has also resulted in many human disease episodes, some involving hundreds of cases. Again, similar attention to breeding stock, facilities, and testing is necessary to contain this zoonotic problem.

Lymphocytic choriomeningitis virus poses a particular threat to the golden lion tamarin (*Leontopithecus rosalia*), an endangered primate currently found only in a small area of Brazil. In 1995 there were at most 400 of these animals left in the wild and because of habitat destruction the number keeps declining. A global captive-breeding, habituation, and reintroduction program had been developed, involving zoos in several countries, but for several years episodic mortality due to a disease called callitrichid (marmoset) hepatitis threatened its success. It was found that this mortality was due to the exquisite susceptibility of these tamarins (and other tamarins and marmosets) to LCM virus previously introduced into primate facilities by wild mice. Rodent proofing facilities has succeeded in stopping this mortality; breeding and re-introduction programs in Brazil are continuing, and the number of golden lion tamarins in protected reserves is now in the thousands.

LASSA FEVER

In 1969 the death of a nurse from a missionary hospital in Lassa, a town in the northeast of Nigeria, was the first alert to the existence of this virus. A second nurse who attended her also died; another nurse who had assisted at the autopsy of the index case became desperately ill but recovered after evacuation to the United States. A new virus was isolated from her blood by virologists at Yale University, one of whom died and another of whom became severely ill but survived following transfusion with immune plasma from the surviving nurse. In succeeding years, Lassa fever, caused by Lassa virus, was found to be a common zoonotic disease in West Africa. The principal reservoir of the virus is the multimammate rodent, *Mastomys natalensis*, one of the most commonly occurring rodents in Africa.

Lassa fever is now recognized as endemic in rural West Africa, especially from Nigeria to Guinea. Serological surveys have shown that many people living in countries of West Africa have antiviral antibodies; there are over 100,000 new human infections a year, with as many as 3000 deaths. The disease to infection ratio is estimated at 20% and a case-fatality rate of 5 to 15%. Changing social circumstances have led to several major outbreaks. For example, in eastern Sierra Leone, where surface diamond mines brought together nearly 100,000 people from all over West Africa, "instant" villages and a currency-based economy resulted in a considerable increase in the *Mastomys* population with a corresponding increase in the risk of human contact with infected rodents. Nearly half the febrile patients admitted to two hospitals in the region had Lassa fever and the 16% case-fatality rate accounted for 30% of the deaths in hospital wards. Urban outbreaks have also broken out in Nigerian cities, displaying a similarly alarming mortality, with nosocomial spread in hospitals reminiscent of the original 1969 occurrence.

Clinical Features

Lassa fever typically presents after an incubation period of about 10 days (range: three to 21 days) with an insidious onset of fever, headache, generalized weakness, and malaise. Within a few days, these clinical signs may be followed by pharyngitis, cough, retrosternal chest pain, conjunctivitis, abdominal pain, and general discomfort. The minority of patients with florid disease may then manifest facial and neck swelling, subconjunctival hemorrhage, and mild oozing or bleeding from the nose, mouth, or genitourinary or gastrointestinal tract. Hypotension and tachycardia may be present, especially in more severe cases. Pleural effusions are common. A variety of encephalopathic and other neurological manifestations of unknown etiology have been noted in severely ill patients. Severely ill patients are more likely to be thrombocytopenic, are usually lymphopenic, and may have an elevated white blood cell (WBC) count with neutrophilia. Moderate hemoconcentration and proteinuria are usually present with an elevated blood urea nitrogen. Aspartate aminotransferase (AST) is often elevated, and higher values are predictive of a poor prognosis. The disproportionate elevation in AST compared to the alanine transaminase (ALT) suggests that its source is not solely the liver but rather may result from diffuse ischemic end-organ damage. Patients with severe illness may deteriorate rapidly, progressing to shock, delirium, respiratory distress, coma, seizures, and death. Surviving patients generally begin to defervesce within approximately 10 days of onset. With the exception of sensorineural deafness, which may affect 25% of survivors, recovery is usually complete. The case-fatality rate is about 15% of patients seeking medical treatment. Lassa fever in pregnant women carries a significantly elevated risk for both the mother and the fetus. Maternal mortality is particularly elevated during the third trimester, when fetal death approaches 100%. Pediatric Lassa fever is less well described but produces a spectrum of disease ranging from undifferentiated febrile illness to severe Lassa fever.

Pathogenesis, Pathology, and Immunity

Arenaviruses are thought to enter the body via skin abrasions or inhalation. Viremia follows and these pantropic viruses replicate in a variety of organs during the one to two week incubation period.

Common gross findings at post-mortem examination include ecchymoses and petechiae involving skin, conjunctivae, mucous membranes, and internal organs; these may be minimal in degree. Microscopically, congestion and variable degrees of necrosis are usually observed in multiple organ systems. Multifocal hepatocellular necrosis with cytoplasmic eosinophilia, Councilman body formation, nuclear pyknosis, cytolysis, and fatty change is the most prominent lesion (Fig. 30.4); inflammatory cell infiltrates in necrotic areas are usually minimal. Necrotizing lesions are also seen in the spleen, adrenal gland, kidney, and gastrointestinal mucosa. The pathophysiology of the hemorrhagic fever/shock syndrome is not fully understood. One major point of difference from hemorrhagic fevers caused by viruses of other families is that disseminated intravascular coagulation does not appear to play a significant role until perhaps the terminal phase of the illness.

In its reservoir host, *Mastomys natalensis*, virus persistence leads to chronic viral shedding in urine, saliva, and feces. In experimentally infected rhesus monkeys there is anorexia, progressive wasting, vascular collapse, and shock, with death occurring at 10 to 15 days after infection. The pathophysiological basis for the disease is not well understood although there is some hepatocellular necrosis, platelet dysfunction, and endothelial damage. Antibodies are found in recovered humans and experimental animals, but they usually do not neutralize the virus in standard assays until quite late after infection—it is presumed that cell-mediated immunity is the key to recovery and protection against reinfection.

Laboratory Diagnosis

Diagnosis is now based on the demonstration of IgM antibodies using an IgM capture enzyme immunoassay; a rapid lateral flow immunoelectrophoretic assay has also been used in clinical settings. However, IgM antibodies may persist for months or years, and direct detection of viremia by antigen-capture immunoassay, RT-PCR, or virus isolation is more definitive evidence for acute infection. Virus can also be detected in throat washings, urine, and pleural fluid. Importantly, virus can be excreted in the urine for over two weeks after the onset of illness. RT-PCR testing

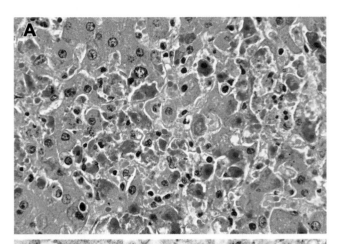

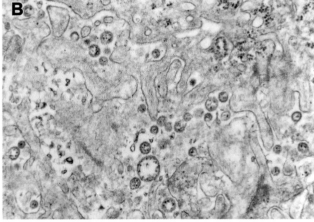

FIGURE 30.4 (A) Lassa fever, human, Sierre Leone, 1975. The liver is the main target organ in human Lassa virus infection; there is focal eosinophilic necrosis of hepatocytes with pyknosis and apparent apoptotic changes, and little evidence of an inflammatory response. H&E, ~ × 400. (B) The presence of virions in the extracellular space at the site of hepatocellular damage, without the presence of inflammatory cells, suggests that the damage is a direct effect of infection rather than an effect of an immunopathological process as is seen in LCM virus infection in experimental animals. Uncontrolled cytokine expression, and virus-induced immunosuppression, may also be involved. Thin-section electron microscopy. *(A) Reproduced from Winn Jr., W.C., Walker, D.H., 1975. The pathology of human Lassa fever. Bull. World Health Organ. 52(4–6):535–545, with permission. (B) Reproduced from Winn Jr, W.C., et al., 1975. Lassa virus hepatitis. Observations on a fatal case from the 1972 Sierra Leone epidemic. Arch. Pathol. 99(11):599–604, with permission.*

of blood samples has been shown to be useful and feasible in local laboratories in endemic regions in Africa. Virus antigen may also be detected in liver tissue in fatal cases by immunofluorescence or immunohistochemistry and virus may be isolated from blood or lymphoreticular tissues using Vero E6 cells aided by indirect detection of viral growth. A high level of biocontainment is required for any laboratory work where there is possible exposure to virus.

Infection in the early stages can be readily confused with other infectious diseases, particularly malignant malaria. The two most reliable prognostic markers are the titers of circulating virus and of AST, a biochemical marker of liver damage. Patients in whom virus titers exceed 10^4 pfu/mL and have an AST value above 150 IU have a poor prognosis.

Epidemiology, Prevention, Control, and Treatment

Lassa virus is enzootic in the West African multimammate mouse, *M. natalensis*, a peridomestic rodent that lives in or near human dwellings, breeds year-round and transmits the virus vertically to its offspring, and horizontally to humans by contaminating the house with urine. Uniquely among the arenaviruses, person-to-person spread of Lassa virus is also common, although not as frequent as infection from a rodent source; the fact that Lassa virus does not appear to be spreading outside of West Africa suggests that the virus is not capable of sustained person-to-person transmission.

Risk factors for human infection include contact with rodents (practices such as catching, cooking, and eating rodents), the presence of rodents in dwellings, direct contact with patients and the reuse of unsterilized needles and syringes.

Ecological changes account for much of the increasing occurrence of the disease; for example, in Sierra Leone diamond mining has led to the building of primitive villages in which *M. natalensis* populations flourish. In several areas of West Africa demonstration projects have shown the value of rodent elimination in villages; however, these programs have been difficult to sustain.

Treatment with high-dose intravenous ribavirin has been shown to reduce mortality, but unfortunately the drug is too often not available for immediate use in endemic areas. Specific guidelines on ribavirin use are available from WHO. Finally, a vaccinia virus-vectored vaccine carrying the genes for the glycoproteins and nucleoprotein of Lassa virus has been shown to be protective in experimental animals; unfortunately, there are no plans for its commercialization and use in West Africa.

LUJO VIRUS INFECTION

The emergence of this arenavirus in 2008 illustrates how a severe hemorrhagic fever can go unrecognized even in an epidemiological setting familiar to those experienced in identifying such infections. Although cases were few in number, it is worthwhile to consider the presentation and course of the disease in the index and secondary cases.

The index case, a travel agent living in Lusaka and with a smallholding on the outskirts of the city, first suffered a febrile illness accompanied by diarrhea, an extensive rash, facial edema, and severe arthralgia. Initially she was treated for food poisoning with antibiotics but on worsening she was evacuated by air to Johannesburg for further treatment. Once there, the patient developed cerebral edema, severe respiratory distress, and impaired renal function, succumbing to the infection 12 days after onset. A marked thrombocytopenia and elevated serum transaminases suggestive of a viral hemorrhagic fever were not noticed as such until a paramedic who had accompanied the index case to South Africa also became ill. Notably the index case was not subjected to strict barrier nursing techniques to limit potential nosocomial spread. Of the total of five cases, four died as a result of what was subsequently confirmed as a new arenavirus phylogenetically related to LCM and other Old World arenaviruses.

SOUTH AMERICAN HEMORRHAGIC FEVERS

Each of the five South American arenavirus hemorrhagic fevers occupies a separate geographical range: each is associated with a different reservoir host, and each represents an expanding zoonotic disease threat. The five viruses have similar natural histories, causing persistent, lifelong infections in individual reservoir rodent hosts, accompanied by long-term shedding of large amounts of virus in urine, saliva, and feces. The natural history of each human disease is determined by the pathogenicity of the virus, its geographical distribution, the habitat and habits of the rodent reservoir host, and the nature of the human–rodent contact. Human infections are usually rural and often occupational, reflecting the relative risk of exposure to virus-contaminated dust and fomites. Changes in ecology and farming practices throughout the region have increased concerns over the potential public health threat posed by these viruses.

Clinical Features

Of the South American hemorrhagic fevers, most is known about Argentine hemorrhagic fever; Bolivian and Venezuelan hemorrhagic fevers have similar clinical characteristics. The incubation period is 6 to 14 days (limits 4 to 21 days). The onset of illness is insidious, with fever, chills, malaise, anorexia, headache, and myalgia. After several days, there is flushing of the face, neck, and upper chest, conjunctival and periorbital congestion, edema and petechial bleeding and in some cases backache, retroorbital pain, epigastric pain, and nausea or vomiting. A morbilliform, maculopapular, or petechial skin rash is common in South American hemorrhagic fevers. The clinical signs are associated with thrombocytopenia, leukopenia, hemoconcentration, proteinuria, in some cases pulmonary edema, and in severe cases death from hypotension and hypovolemic shock. Neurological signs are common by the end of the first week; patients may be disoriented, irritable, and lethargic, with fine tremors, ataxia, cutaneous hyperesthesia, and a decrease in deep tendon and muscle reflexes—these signs may proceed to delirium, generalized convulsions, and coma. During the second week of illness, 70 to 80% of the patients begin to improve; in the remaining 20 to 30%, severe hemorrhagic or neurologic manifestations, shock, and superimposed bacterial infections appear along with hematemesis, melena, hemoptysis, epistaxis, hematomas, metrorrhagia, and hematuria. Convalescence may be protracted and patients may suffer a late neurologic syndrome. The case-fatality rate was about 15 to 20%, but prevention methods in the field, widespread use of vaccine, and modern clinical care have reduced this to less than 1%. Maternal mortality is particularly elevated during the third trimester, when fetal death approaches 100%.

During the acute phase, there is progressive leukopenia and thrombocytopenia, with cell counts falling to 1000–2000 white blood cells (WBCs) and 50,000–100,000 platelets/mL. Aspartate transaminase (AST), creatine phosphokinase (CPK), and lactate dehydrogenase (LDH) become moderately elevated.

Pathogenesis, Pathology, and Immunity

Argentine hemorrhagic fever has pathology typical of hemorrhagic fevers. Changes seen at postmortem include widespread petechial hemorrhages and mucosal bleeding, consistent with the thrombocytopenia; there is relatively little inflammation or necrosis, although some focal necrosis may be seen in the liver, kidney, and adrenal cortex; secondary changes associated with shock are common, and there may be interstitial pneumonitis and pulmonary edema.

In contrast to LCM and the Old World arenaviruses, Junin and Machupo viruses cause disease in their respective reservoir rodent hosts. Junin causes up to 50% mortality among infected suckling *Calomys musculinus* and *Calomys laucha*, and stunted growth in others. Machupo virus induces a hemolytic anemia, splenomegaly and fetal death in its rodent host, *Calomys callosus*. Nothing is known about the nature of the infection caused by Guanarito or Sabiá viruses in reservoir hosts. As with other arenavirus pathogens, the pattern of infection in wild rodents caused by the South American viruses differs with age, host genetic determinants, route of exposure and viral entry, together with the dose and genetic character of the virus. Natural transmission from rodent to rodent is horizontal and occurs

through contaminated saliva, urine, and feces. Junin and Machupo viruses also induce sterility in neonatally infected females, thereby minimizing the numbers of offspring destined to become persistently infected shedders of virus. Complex cyclic fluctuations in infection rates and population densities are thought to be a consequence of this.

Laboratory Diagnosis

Diagnosis of Junin and Machupo virus infections is now based on the demonstration of IgM antibodies using an IgM capture enzyme immunoassay. RT-PCR testing of blood samples has also been shown to be useful and feasible in local laboratories in the Junin virus endemic region of Argentina. Virus antigen may also be detected in liver tissue in fatal human cases and virus isolation may be done using Vero E6 cells and immunological detection of viral growth. Methods for the diagnosis of Guanarito or Sabiá virus infections are the same, although specific reagents are not widely available. A high level of biocontainment is required for any laboratory work where there is possible exposure to virus.

Epidemiology, Prevention, Control, and Treatment

Argentine Hemorrhagic Fever (Junin)

Argentine hemorrhagic fever, caused by Junin virus, was first recognized in the 1950s in Las Pampas region of Argentina. Most commonly affected are farm workers, explicable by the behavior of the rodent hosts, *C. musculinus* and *C. laucha*. These rodents are not peridomestic, but rather occupy hedgerows surrounding cornfields, exposing agricultural workers through contact with virus-infected dust and grain products. Virus is acquired through cuts and abrasions or through airborne dust generated primarily when rodents are caught up in harvesting machinery. Since the 1950s the disease has spread from an area of 16,000–120,000 km² containing a population of over 1 million people. There is a three to five year cyclic trend in the incidence of human cases, which exactly parallels cyclic changes in the density of *Calomys* spp.

Strategies for the prevention and control of Argentine hemorrhagic fever through rodent control are unrealistic; human exposure during harvesting will increase as more and more rodents thrive in the fields and hedgerows. However, much progress has been made using a vaccine; an attenuated virus vaccine is now in use throughout the endemic area in Argentina—the vaccine called "Candidate #1" has been used since 1991, with greater and greater production capacity evolving over the years, and with overall great success. Its use has resulted in a steady decline in the incidence of disease. The vaccine may also protect against Bolivian hemorrhagic fever, although it does not appear to cross-protect against other arenaviruses. Ribavirin, administered intravenously in high dosage, is used as a backup to this vaccine. Administration of immune plasma has also proved beneficial.

Bolivian Hemorrhagic Fever (Machupo and Chapare viruses)

Machupo virus emerged in Bolivia in 1952 when political instability forced those in rural areas to attempt subsistence agriculture at the borders of tropical grassland and forest; by 1962 more than 1000 cases of Bolivian hemorrhagic fever had been identified, with a 22% case-fatality rate. Abnormally low rainfall combined with the liberal use of DDT for insect control led to a decline in cat numbers and as a consequence a dramatic increase in the numbers of rodents. *Calomys callosus*, a forest rodent and the reservoir host of Machupo virus, adapts well to human contact—invasion of villages resulted in clusters of cases in particular houses in which substantial numbers of infected rodents were subsequently trapped. Peridomestic control of *C. callosus* in the endemic area by trapping resulted in the disappearance of the disease for many years, but in the 1990s cases reappeared, again starting on farms and then moving into villages. In one instance there was secondary spread to six of eight family members, all of whom died. As in the past, disease prevention could have been facilitated by reducing rodent numbers. However, such programs are difficult to sustain except for short periods in villages with exceptional transmission rates.

In 2004 a small cluster of acute febrile cases was reported from Chapare province in the department of Cochabamba, Bolivia, outside the known endemic region for Machupo virus. The index case, a tailor who also grew coca on a smallholding, suffered an acute febrile illness accompanied by severe myalgia and vomiting. There is limited information regarding this outbreak and the extent of infection with Chapare virus, the causative agent, although phylogenetic analyses indicate Chapare virus is most related to Sabiá virus. This is the only known outbreak of human Chapare infection to date.

Venezuelan Hemorrhagic Fever (Guanarito Virus)

Venezuelan hemorrhagic fever was first recognized in rural areas of Venezuela in 1989; it appeared as a result of forest clearance and subsequent preparation of land for agriculture. Through 2006, there were 618 cases; the case-fatality rate was 23%. An arenavirus, Guanarito virus, was isolated from cases and traced to a reservoir rodent host, *Zygodontomys brevicauda*. In the same area another new arenavirus, Pirital virus, was isolated from the rodent *Sigmodon alstoni*; however, there is no evidence that Pirital virus causes significant human infection. Given the similarities between the natural history of Guanarito and Machupo viruses, disease prevention in Venezuela might be facilitated by rodent trapping, but to date this has not been attempted.

Brazilian Hemorrhagic Fever (Sabiá Virus)

Sabiá virus was isolated from a fatal case of hemorrhagic fever in São Paulo, Brazil, in 1990; subsequently there have been two additional cases, one a naturally acquired fatal infection in an agricultural engineer, the second a non-fatal infection in a virologist working with the virus. The latter suffered a severe illness characterized by 15 days of fever, headache, chills, myalgia, sore throat, conjunctivitis, nausea, vomiting, diarrhea, bleeding gums, and leukopenia. Given the lack of knowledge of the natural history and epidemiology of Sabiá virus, little is known in regard to disease prevention and control.

Other Arenavirus Infections

A number of new North American arenaviruses have been identified in recent years (Table 30.2). Among these, Whitewater Arroyo virus has received particular attention with some evidence of seroconversion in a few cases of non-specific febrile illness and acute respiratory distress. However, at the time of writing doubt remains as to whether Whitewater Arroyo virus or any of the other North American arenaviruses are human pathogens.

FURTHER READING

Albarino, C.G., Palcios, G., Khristova, M.L., et al., 2010. High diversity and ancient common ancestry of lymphocytic choriomeningitis virus. Emerg. Infect. Dis. 16, 1093–1100.

Enria, D.A., Mills, J.N., Bausch, D., Shieh, W.-J., Peters, C.J., 2011. Arenavirus infections. In: Guerrant, R.L., Walker, D.H., Weller, P.F. (Eds.) Tropical Infectious Diseases. Principles, Pathogens and Practice (third ed.), Elsevier, Philadelphia, PA, pp. 449–461.

Howard, C.R., 2012. Lecture Notes on Emerging Viruses and Human Health: A Guide to Zoonotic Viruses and Their Impact. World Scientific Press, Singapore, ISBN 10-981-4366-90-0.

Yun, N.E., Walker, D.H., 2012. Pathogenesis of Lassa Fever. Viruses 2, 2031–2048.

Chapter 31

Coronaviruses

The family *Coronaviridae* encompasses a broad spectrum of animal and human viruses, all characterized by a distinctive morphology. Virions are enveloped and spherical (coronaviruses), or disc, kidney, or rod shaped (toroviruses). Each particle is surrounded by a fringe or "corona" representing the bulbous distal ends of embedded envelope glycoproteins. Prior to 2003 members of this family were believed to cause only mild respiratory illness in humans, other coronaviruses then known being largely of importance only to the livestock industry. But the emergence of severe acute respiratory virus (SARS-CoV) that year stimulated major research into these viruses, to the effect that many new coronaviruses have since been discovered, some with zoonotic potential of causing serious outbreaks of disease in humans. The more recent emergence of MERS-CoV is exemplary.

Coronaviruses are also noted for having the largest positive-sense RNA genome: coronavirus genes are mostly expressed by a complex procedure whereby nested mRNA transcripts are produced, the regulation of which governs the progression of the replication cycle.

PROPERTIES OF CORONAVIRUSES

Classification

The family *Coronaviridae* is one of three RNA virus families within the order *Nidovirales*, the other being the *Arteriviridae* and the *Roniviridae* containing pathogens of birds and insects, respectively. The family consists of two subfamilies, *Coronavirinae* and *Torovirinae*, the latter containing viruses causing mainly enteric infections of horses, cattle, pigs, cats, and goats. Although of economic importance, members of the *Torovirinae* subfamily are not as yet known to cause human infection, and thus are not dealt with further. All coronaviruses share a common morphology and possess a single-stranded RNA genome of up to 30 kb in length.

Members of the subfamily *Coronavirinae* are subdivided into four genera. The genus *Alphacoronavirus* contains the human virus HCoV-229E, one other human coronavirus (HCoV-NL63), and many animal viruses. The genus *Betacoronavirus* includes the prototype mouse hepatitis virus (MHV), the three human viruses HCoV-OC43, SARS-HCoV, and HCoV-HKU1, and the

SARS-related coronavirus, Middle Eastern respiratory syndrome (MERS) coronavirus, together with a number of animal coronaviruses. The genus *Gammacoronavirus* contains viruses of cetaceans (whales) and birds, and the genus *Deltacoronavirus* contains viruses isolated from pigs and birds.

Since 2005, dozens of new coronaviruses have been isolated from bats, and there is evidence that human respiratory coronaviruses, SARS coronavirus, and MERS coronavirus, may each have originally emerged from ancestral bat viruses (Table 31.1, Fig. 31.1).

Structure and Genome

Coronaviruses virions contain three major structural proteins: the very large (200 K) glycoprotein S (for spike) that forms the bulky (15 to 20 nm) peplomers found in the viral envelope, an unusual transmembrane glycoprotein (M) and the internal phosphorylated nucleocapsid protein (N). In addition, there is a minor transmembrane protein E, and some

---

TABLE 31.1 Properties of Coronaviruses

- Virion is pleomorphic spherical 80 to 220 nm (coronaviruses); or disc, kidney, or rod shaped 120 to 140 nm (toroviruses)
- Envelope with large, widely spaced club-shaped peplomers
- Tubular nucleocapsid with helical symmetry
- Linear, plus sense ssRNA genome 27 to 33 kb, capped, polyadenylated, infectious; untranslated sequences at each end
- Three or four structural proteins: nucleoprotein (N), peplomer glycoprotein (S), transmembrane glycoprotein (M), sometimes hemagglutinin-esterase (HE)
- Genome encodes 3 to 10 further non-structural proteins, including the RNA-dependent RNA polymerase made up of subunits cleaved from two polyproteins translated from the 5'-end
- Replicates in cytoplasm; genome is transcribed to full-length negative sense RNA, from which is transcribed a 3'-coterminal nested set of mRNAs, only the unique 5' sequences of which are translated
- Virions are assembled and bud into the endoplasmic reticulum and Golgi cisternae; release is by exocytosis
- Variant viruses arise readily, by mutation and recombination, and the use of different receptors influences the host range exhibited

---

Fenner and White's Medical Virology. DOI: http://dx.doi.org/10.1016/B978-0-12-375156-0.00031-X

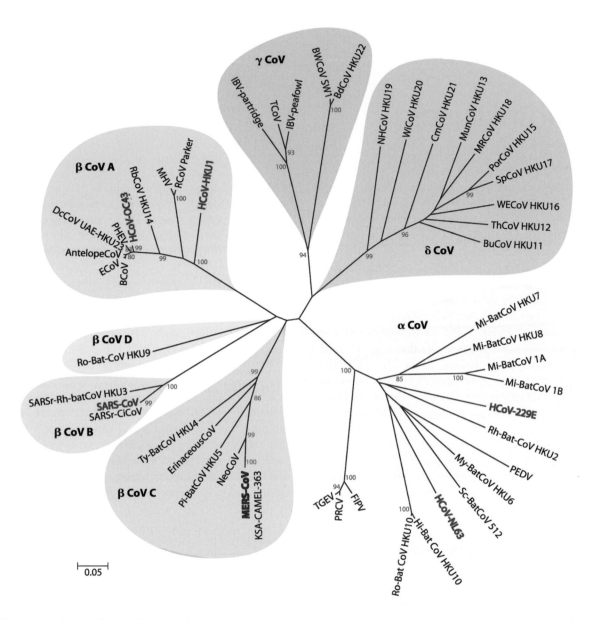

FIGURE 31.1 Phylogeny of coronaviruses. Phylogenetic tree of 50 coronaviruses constructed by the neighbor-joining method using MEGA 5.0 using partial nucleotide sequences of RNA-dependent RNA polymerase. The scale bar indicates the estimated number of substitutions per 20 nucleotides. Space does not permit providing full virus names, except for the human viruses, which are scattered among the viruses isolated from many other species (major pathogens shown in red): HCoV-229E, human coronavirus 229E; HCoV-HKU1, human coronavirus HKU1; HCoV-NL63, human coronavirus NL63; HCoV-OC43, human coronavirus OC43; KSA-CAMEL-363, KSA-CAMEL-363 isolate of Middle East respiratory syndrome coronavirus; MERS-CoV, Middle East respiratory syndrome coronavirus; MHV, murine hepatitis virus, the prototypic virus of the family; SARS-CoV, SARS coronavirus; SARSr-CiCoV, SARS-related palm civet coronavirus. A remarkable number of the viruses represented here are from bats, many different species of bats, and quite a few of these are rather closely related to SARS-CoV. *Modified from Chan, J.F., Lau, S.K., To, K.K., Cheng, V.C., Woo, P.C., Yuen, K.Y., 2015. Middle East respiratory syndrome coronavirus: another zoonotic betacoronavirus causing SARS-like disease. Clin. Microbiol. Rev. 28, 465–522, with permission.*

coronaviruses contain a further envelope protein with both hemagglutination and esterase functions (HE) (Fig. 31.2).

The 30 kb positive sense, single-stranded RNA genome is the largest RNA viral genome known. It is capped at the 5′-end and polyadenylated at the 3′-terminus, and is infectious. Due to its size the expression of individual genes occurs through a complex process whereby sets of nested mRNAs are produced, all sharing the same 5′-end sequence. Extensive

rearrangements may occur as a result of heterologous RNA recombination. At the 5′-end of the genome is an untranslated (UTR) sequence of 65 to 98 nucleotides, termed the leader RNA, which is also present at the 5′-ends of all subgenomic mRNAs. At the 3′-end of the RNA genome is another untranslated sequence of 200 to 500 nucleotides, followed by a poly(A) tail. Both untranslated regions are important for regulating RNA replication and transcription.

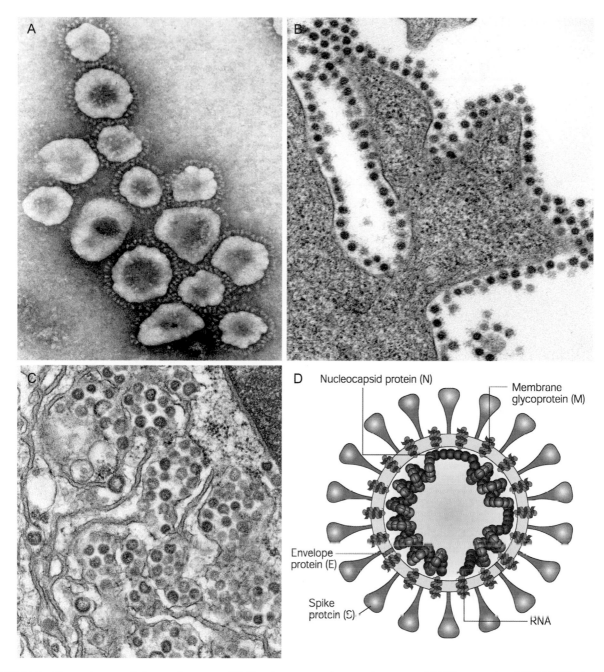

FIGURE 31.2   Coronavirus morphology and structure. (A) Negative contrast electron microscopy of SARS coronavirus (SARS-CoV), showing the large petal-shaped surface projections (spikes, peplomers). (B) Thin-section electron microscopy of SARS-CoV in cell culture, showing typical adherence of virions to the plasma membrane of a cell—virions adhere to infected and uninfected cells. (C) Thin-section electron microscopy of Middle Eastern respiratory syndrome virus (MERS-CoV) in cell culture, showing typical virion assembly in the lumen of the Golgi membrane system. (D) Model of coronavirus virion structure, showing the supercoiling of the viral nucleocapsid under the envelope. *(B) From Sandra Crameri, CSIRO, Geelong, Australia. (C) From Public Health Image Library, CDC. (D) Reproduced from Stadler, K., et al., 2003. SARS—beginning to understand a new virus. Nat. Rev. Microbiol. 1, 209–218. All with permission.*

The coronavirus genome contains 7 to 14 open reading frames (ORFs). Starting from the 5′-end, Gene 1, which comprises two-thirds of the genome, is about 20 to 22 kb in length. It consists of two overlapping ORFs (1a and 1b), collectively functioning as the viral RNA polymerase (Pol).

The order of the other four genes of structural proteins are 5′-S (spike)–E (envelope)–M (membrane)–N (nucleocapsid) -3′. These genes are interspersed with several ORFs encoding non-structural proteins and the HE glycoprotein, when present. Each gene differs markedly among coronaviruses

Mouse hepatitis virus, MHV (31,526 nts)

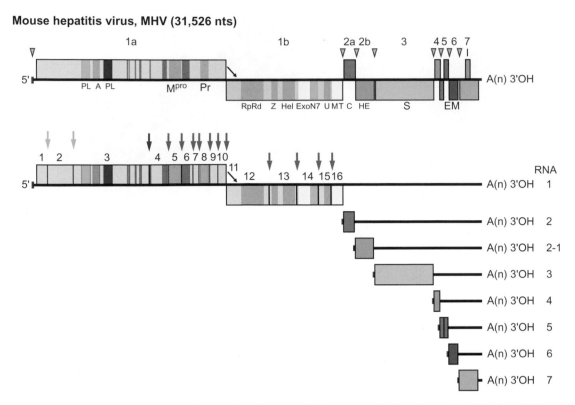

FIGURE 31.3 Coronavirus genome organization and expression. (Upper panel) Genome organization of mouse hepatitis virus. ORFs are represented by boxes, with a number above and the protein acronym below. The diagonal arrow between ORF 1a and 1b represents the ribosomal frameshift site. Red arrowheads indicate the locations of transcriptional-regulating sequences (TRSs). (Lower panel) Map of the viral mRNAs, which are overlapping and coterminal at the 3′ end. The different sub-genomic mRNAs are numbered by convention from large to small, beginning with the genome as RNA1. The two huge replicase polyproteins (1a and 1ab) are cleaved by virus-coded proteases as shown by colored ORFs and colored arrows; green, papain-like protease 1(PL1^{pro}); red, papain-like protease 2 (PL2^{pro}); blue, 3C-like main protease (M^{pro}). *Reproduced from King, A.M.Q., et al., 2012. Virus taxonomy. In: Ninth Report of the International Committee on Taxonomy of Viruses, Academic Press, London, p. 808, with permission.*

in number, nucleotide sequence, gene order, and method of expression, although these are conserved within the same serogroup. The SARS-CoV genome encodes several smaller ORFs located in the 3′ region of the genome not present in other coronaviruses. These ORFs are predicted to express eight novel proteins termed accessory proteins. Antibodies reactive against all of the SARS-CoV proteins have been detected in sera isolated from SARS patients, indicating that these proteins are expressed by the virus *in vivo* (Fig. 31.3).

ORFs 1a and 1b are first translated into two polyproteins, both identical at the N terminus but one of which has a C-terminal extension due to frame-shifting. These are precursors of proteins in the transcription–replication complex. All coronaviruses encode a chymotrypsin-like protease, termed M^{pro} (main protease) or 3CL^{pro} since it shares some similarities with the 3C proteases of picornaviruses. This protease is responsible for processing the remainder of the polyprotein, producing as many as 16 non-structural proteins. SARS-CoV contains the largest number of these non-structural proteins. For example, nsp3 is a multifunctional protein with protease and ADP-ribose 1″phosphatase activity. Two proteins (nsp 7 and nsp

8) form a cylinder-like structure that may be important in coronavirus RNA synthesis, and a single-strand RNA-binding protein (nsp 9). ORF1b encodes the viral RNA-dependent RNA polymerase and a multifunctional helicase protein. In addition to helicase activity, this protein has NTPase and dNTPase activities as well as 5′ triphosphatase activity.

These non-structural protein gene products are not essential for virus replication, but deletion of one or more often causes viral attenuation. At least one, the product of ORF3a, is now recognized to be a structural protein. The ORF3a product is an O-glycosylated, triple-membrane-spanning protein capable of binding to N, M, and S proteins, suggesting a role in viral biogenesis.

Viral Replication

The N-terminal half of the S protein contains the receptor binding domain, and the C-terminal half is membrane anchored and has fusogenic activity. The specificity of binding is important for the host spectrum of individual coronaviruses. Soon after the virus was isolated, the receptor for SARS-CoV was identified as angiotensin-converting enzyme 2 (ACE2), a cellular protein expressed on the surface of cells of the lungs, heart, kidney, and small intestine as well as other tissues. Other proteins, such as CD209L (L-SIGN), DC-SIGN, and LSECtin, can support the entry of SARS-CoV into cells, but cannot by themselves confer susceptibility to a cell lacking ACE2. Receptors of other coronaviruses have also been identified. For instance, the receptor for murine hepatitis virus (MHV) is a murine biliary glycoprotein belonging to the carcinoembryonic antigen (CEA) family in the Ig superfamily (CEACAM1). Individual strains of MHV may exhibit different preferences for viral entry. For example, only the MHV-3 strain infects T and B cells, with subsequent lymphopenia. Although most strains of coronavirus exhibit strict species specificity, like other RNA viruses, coronaviruses can readily mutate under selective pressure during passage *in vitro* or *in vivo* in response to environmental conditions, and an S protein with extended host range is thus selected. These variants can efficiently use human CEA as receptors for infection of human cells. Such a mechanism may account for the development of cross-species infection, exemplified in the outbreak of the SARS epidemic.

After binding to the cellular receptor, the viral S protein acts as a class I fusion protein, undergoing conformational changes that lead to fusion between the viral envelope with either the plasma or endosomal membranes. SARS-CoV requires an acidic pH for entry, but not for S glycoprotein-mediated cell-to-cell fusion. Subsequently, the viral nucleocapsid is released into the cytoplasm and the RNA is uncoated prior to transcription.

During the first stage of virus replication, the positive strand viral RNA serves as mRNA for translation of the two large open reading frames (ORF 1a and 1b), each encoding units of the RNA-dependent RNA polymerase. After cleavage, these proteins assemble to form the active RNA polymerase which transcribes full-length complementary (negative-sense) RNA. This, in turn, is transcribed not only into full-length genomic RNA, but also a nested set of 3′-coterminal overlapping subgenomic mRNAs described above. Each species of viral mRNA differs in length but shares a common 3′-end and a 70 nucleotide 5′-leader sequence. Only the unique sequence not shared with the next smallest mRNA in the nested set is translated to yield viral protein. Due to the large size of the genome, genetic recombination occurs at high frequency between the genomes of different but related coronaviruses. This mechanism may be important for generation of the genetic diversity of coronaviruses in nature.

During maturation and assembly, the S protein is cotranslationally inserted into the rough endoplasmic reticulum (RER) and glycosylated with N-linked glycans. Glycosylation is essential for the proper folding and transport of the S protein. The S protein forms trimers before it is exported out of the endoplasmic reticulum (ER), and then interacts with M and E proteins to migrate to the site of virus assembly. In infected cells, the M protein is localized mostly in the virus-budding compartment, while at a later stage of viral replication, N protein is transported to the site of virus assembly. SARS-CoV expresses another structural protein, 3a, that is not only associated with intracellular and plasma membranes, but is also secreted and induces apoptosis. S protein is crucial for virus entry, but not necessarily required for assembly, as spikeless virus-like particles can form in the absence of S protein.

Pathogenesis and Immunity

Most coronaviruses first replicate in epithelial cells of the respiratory or enteric tracts. Because coronaviruses are enveloped, virions are less stable in the environment and in clinical specimens compared to most non-enveloped viruses. Although transmission is usually associated with close contact, SARS-CoV is surprisingly stable on environmental surfaces. It is not clear how these enveloped viruses retain infectivity in the presence of bile and proteolytic enzymes present in the enteric tract. Perhaps the virions may be more resistant to proteolytic degradation because coronavirus envelope glycoproteins are extensively glycosylated.

The pathogenesis and immune responses of coronaviruses have been most studied in animal coronavirus infections. For example, mouse hepatitis virus (MHV), the prototype of betacoronaviruses, includes a spectrum of strains with different tropism, causing enteric, hepatic, respiratory, or CNS infections. Since neurological disease caused by MHV simulates multiple sclerosis (MS) in humans, the pathogenesis and immune responses have been studied in detail. Highly neurovirulent, moderately neurovirulent, and attenuated strains cause different clinical manifestations involving different patterns of infection of neurons, oligodendrocytes, microglia, and astrocytes. Demyelination can occur, which is mostly immune-mediated; irradiated or congenitally immunodeficient mice do not develop demyelination after infection, but when these mice, which lack T and B cells, are reconstituted with virus-specific T cells, demyelination rapidly develops. Demyelination is accompanied by infiltration of macrophages and activated microglia into the white matter of the spinal cord. Both CD4 and CD8 T cells are also required for virus clearance from the central nervous system (CNS), with CD8 T cells as the most important in this process.

On the other hand, the efficacy of the innate immune response also determines the extent of initial virus replication and the levels of virus seen. Coronaviruses have developed strategies to counter the innate immune responses. For example, N protein and the SARS-CoV accessory proteins ORF6 and ORF3b prevent IFN induction, and the N protein prevents nuclear translocation of proteins containing classical nuclear import signals, including STAT1, a crucial component of IFNα, IFNβ, and IFNγ signaling pathways.

HUMAN CORONAVIRUS INFECTIONS

Respiratory Coronavirus Infections

Prior to 2003 the interest in human coronaviruses was primarily driven by having a role in relatively mild upper respiratory tract infections. Viral strains HCoV-229E and HCoV-OC43 were isolated from patients with upper respiratory tract infections in the 1960s. Then, shortly after the discovery of SARS-CoV in 2003, two new human coronaviruses were isolated, HCoV-NL63 in the Netherlands and HKU1 in Hong Kong. The host cell receptor for HCoV-229E aminopeptidase N, while that for HCoV-NL63 is ACE2. Interest in the possible role of human coronaviruses in multiple sclerosis has initiated a search for human coronaviruses in human brain tissues. One study found HCoV-OC43 sequences in 36% of brains from MS patients and 14% in those of controls, but the significance of this remains unclear.

Clinical Features

The typical coronavirus "common cold" is mild and the virus remains localized to the epithelium of the upper respiratory tract and elicits a poor immune response, hence the high rate of reinfection. There is no cross-immunity between human coronavirus-229E and human coronavirus-OC43, and it is likely that new strains are continually arising by mutation selection.

Human volunteer studies have shown that these viruses cause an illness of some seven days, with typical symptoms of a sore throat, rhinorrhea, fever, cough, and headache, indistinguishable from the common colds caused by rhinoviruses. Asymptomatic infections are frequent as measured by the detection of virus in the upper respiratory tract.

Occasionally human coronaviruses HCoV-229E and HCoV-OC43 cause lower respiratory tract infections and otitis media. There is no evidence of either of these viruses causing enteric disease in humans, despite the finding of coronavirus-like particles in the stools of such patients.

Laboratory Diagnosis

Laboratory diagnosis of the "common" human respiratory coronaviruses is not often called for, but is sometimes included in the panel of respiratory pathogens incorporated in multiplex RT-PCR assay systems. These viruses are difficult to grow in cultured cells, hence are rarely recovered from clinical systems. The exception is HCoV-NL63 which, like SARS-CoV, grows readily in a number of cell lines; this may be linked to the finding that HCoV-NL63 and SARS-CoV are unique in using ACE2 as a receptor. Human coronavirus-OC43 and related strains were originally isolated in organ cultures of human embryonic trachea or nasal epithelium. Organ culture is too intricate a technique for a diagnostic laboratory, but some strains can be isolated directly in diploid fibroblastic lines from human embryonic lung or intestine. Foci of "granular" cells become evident after a week and may progress to vacuolation before disintegrating; syncytia may form in some cell types. Hemadsorption and hemagglutination are demonstrable with OC43 only.

Epidemiology

As with many respiratory infections, spread is by direct contact between infected individuals or via fomites. HCoV-229E and HCoV-OC43 viruses are most often detected between November and May in the northern hemisphere, although the incidence of infection varies from year to year. It is thought these two viruses together account for 5 to 30% of all common colds. Little is known with regard to the epidemiology of HC-V-NL63 and HCoV-HKU1.

SEVERE ACUTE RESPIRATORY VIRUS (SARS-CoV)

SARS-CoV infection of humans is a serious lower respiratory tract illness that emerged with dramatic suddenness in China in 2002. Up to 20% of infections needed intensive care; the overall fatality rate initially was around 10% but approached 50% in elderly patients and those with underlying illness. The epidemic was halted in 2003 by a highly effective national and global health response, and the virus is no longer circulating in humans, although it is endemic in horseshoe bats.

Clinical Features and Pathogenesis

The disease began initially with an influenza-like prodrome starting two to seven days after exposure. This was followed after a further three or more days by the lower respiratory tract phase, comprising dry cough, dyspnea, and increasing respiratory distress sometimes requiring mechanical ventilation. Most patients showed lymphopenia (70 to 95%), with a substantial drop in both CD4 and CD8 T cells. Asymptomatic or mild illness was uncommon, as illustrated by studies of exposed health-care workers in which less than 1% of those without a SARS-like illness had serological evidence of infection. However, in the year following the epidemic of 2003, only a few SARS patients were found and these had mild disease. SARS-CoV infected both upper

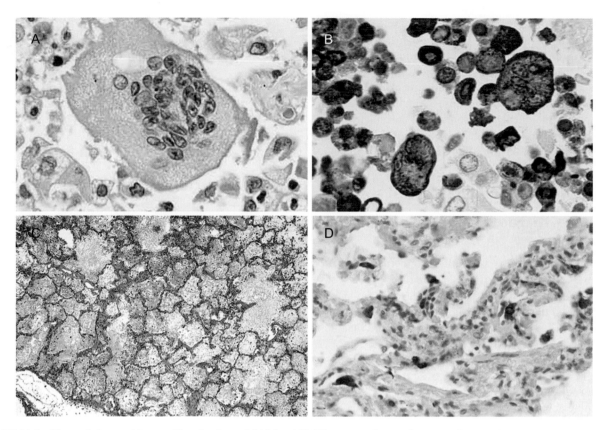

FIGURE 31.4   Histopathology and immunohistochemistry of SARS and MERS acute respiratory distress syndrome (ARDS). (A) SARS: lesions have consisted of diffuse alveolar damage at various levels of progression and severity—changes have included interstitial mononuclear inflammatory infiltration, hyaline-membrane formation, desquamation of pneumocytes and necrotic inflammatory debris into small airways, focal intra-alveolar hemorrhage, and as shown multinucleated syncytial cells. H&E. (B) SARS: immunohistochemical staining of SARS-CoV-infected cells free in interstitial space in small airway. Viral antigens in the cytoplasm of cells, including syncytial cells, with most intense immunostaining near margins of cells. Immunoalkaline phosphatase system, napthol–fast red substrate, hematoxylin counterstain. (C) MERS: pulmonary edema with mononuclear inflammatory cells in alveoli. H&E. (D) MERS: immunohistochemical staining of MERS-CoV-infected pneumocytes within alveoli at a site where normal lung architecture is still intact. Commercial avidin–alkaline phosphatase complex system with naphthol–fast red substrate. *From Thomas Ksiazek, University of Texas Medical Branch, Galveton and Sheruf Zaki, U.S. Centers for Disease Control and Prevention, and their many colleagues in these studies; Ksiazek, T.G., et al., 2003. A novel coronavirus associated with severe acute respiratory syndrome. N. Engl. J. Med. 348, 1953–1966; and Ng, D.L., et al. 2016. Clinicopathologic, immunohistochemical, and ultrastructural findings of a fatal case of middle east respiratory syndrome coronavirus infection in the United Arab Emirates, April 2014. Am. J. Pathol. 186, 652–658.*

airway and alveolar epithelial cells, resulting in lung injury. Virus or viral products were also detected in other organs, such as the kidney, liver, and small intestine, as well as in stools. Although the lung is recognized as the organ most severely affected by SARS-CoV, the exact mechanism of lung injury is controversial. Levels of infectious virus diminished as the clinical disease worsened, suggesting that lung injury was due to an immunopathological mechanism. Similar to mice with MHV-mediated demyelination, large numbers of macrophages were detected in infected lung. However, this conclusion derived from samples of live patients—from nasopharyngeal aspirates, not from the lungs or other organs (Fig. 31.4). In common with other coronaviruses, SARS-CoV infected macrophages and dendritic cells, but the replication cycle is not completed in these cells. Several proinflammatory cytokines and chemokines, such as IP-10, MCP-1, MIP-1, RANTES and MCP-2, TNF-α, and IL-6 are expressed by infected dendritic cells; many of these molecules were also expressed at elevated levels in the serum of infected patients.

Infection with SARS-CoV triggers a series of humoral and cellular immune responses in patients. Specific IgG and IgM antibodies against SARS-CoV were detected approximately two weeks postinfection, reaching a peak 60 days post-infection and remained at high levels for at least 180 days post-infection. High titers of neutralizing antibodies and SARS-CoV-specific cytotoxic T lymphocytes were detected in patients who had recovered from SARS, with high levels correlating well with a favorable outcome. This suggests that both humoral and cellular immune responses are crucial for the clearance of infection by SARS-CoV.

An important unresolved issue is how SARS-CoV caused such severe disease in humans. SARS-CoV infects several species of animals, including mice, ferrets, hamsters, cats, and cynomolgus macaques, but these animals develop either mild or subclinical disease. This may be due to the fact that these animals have already experienced subclinical infection with a coronavirus from the same group as SARS-CoV. Consistent with this, the clinical course of SARS-CoV in patients in Guangzhou, China, was mild in the year following the outbreak of severe cases of SARS.

Laboratory Diagnosis

Frequently a combination of serological and RT-PCR assays was used to detect and confirm infection. With very sensitive PCR assays (e.g., a nested or real-time PCR assay) and RNA extraction procedures that increased the amount of specimen used in the assay, the positivity rate in respiratory specimens increased from less than 40% to more than 80% during the second or third day of illness. Results by EIA showed the presence of SARS-CoV antibodies against the N protein in 50% to more than 80% of sera collected during the first week of illness, and in more than 50% of respiratory and stool specimens collected during the second and third weeks of illness.

SARS antibodies were detected as early as six days after onset of illness, but nearly always by 14 days after the onset of the illness, (in rare instances antibodies did not appear until four weeks after the onset of illness). Because sera from persons not infected with SARS during the 2002 to 2003 outbreak rarely tested positive for SARS-CoV antibodies, a single specimen positive for SARS-CoV antibodies was usually considered diagnostic for infection, and a negative test on a serum specimen collected late in the illness (28 days or later after onset of illness) could be used to rule out SARS-CoV infection. Because diagnosis of a re-emergence of SARS-CoV would have substantial public health, social, and economic impact, a future case putatively diagnosed as SARS would have to be confirmed by a national reference laboratory before international response actions could be initiated–false positives do occur.

During the SARS-CoV outbreak, rapid diagnosis was best made by RT-PCR using defined primers, usually derived from the viral N sequence. For early diagnosis, specimens from throat or nasal pharyngeal swabs can be used for RT-PCR, and serum samples were also used to detect viral RNA during the first week of illness. Both stool and respiratory specimens were assayed for viral RNA during the second week of illness. In contrast, serological assays usually provide the best way to confirm or rule out infections *ex post facto*. EIAs for viral N and S antigens were also developed to screen for suspected patients in rural areas.

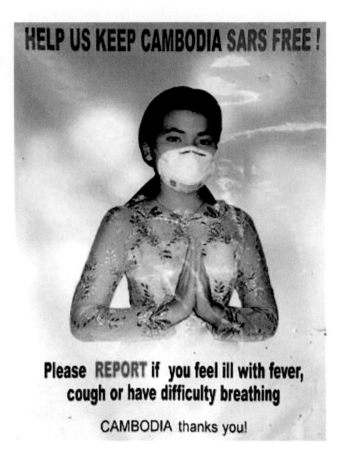

FIGURE 31.5 The sudden emergence of human cases of SARS, with its high mortality and rapid intercontinental dissemination, caused significant disruption to many international activities, and led to heightened awareness and respect for the possibility of new epidemics in the 21st century. *From a wall poster, Department of Immigration, Cambodia.*

Epidemiology

The high mortality and rapid intercontinental spread of clinical cases of SARS led to intensive epidemiological and virological investigations. It caused significant disruption to affected cities and many activities that involved travel. Anxiety and warnings impacted on many parts of the world (Fig. 31.5). Molecular analysis and field epidemiological studies for identifying the source of SARS-CoV revealed that horseshoe bats were most likely the natural reservoir of this virus. Sequences of several distinct SARS-like coronaviruses have been amplified from horseshoe bats from Hong Kong and several provinces in China, and 30% to 85% of this species of bat had antibodies to a SARS-like coronavirus. Initially, SARS-CoV was thought to originate in civet cats, but these were later found to act as intermediary hosts providing a source of transmission to humans. Multiple serological studies demonstrated that SARS-CoV had not circulated to any significant extent in humans prior to the outbreak in 2002 and 2003. Some persons working in wild animal markets in China had serological evidence of a SARS-CoV-like infection acquired before the 2003 outbreak, but reported no SARS-like respiratory illness.

Although animals were the original source of SARS, its global spread occurred by human-to-human transmission. Transmission appeared to occur through close contact or infectious droplets and probably aerosols in some instances. There was also substantial patient-to-patient variation in efficiency of transmission, which in part was associated with degree of severity of illness and possibly associated with the virulence of viral strains. Since the SARS outbreak was controlled in June 2003, only 17 cases of SARS have been confirmed and none of these occurred after June 2004.

The underlying basis for the mysterious outbreak and sudden disappearance of SARS has not yet been fully explained.

Treatment and Prevention

Prevention has been based on careful identification and isolation of cases and contacts until ten days after symptoms had cleared, combined with investigation of particular environmental circumstances responsible for clusters of cases. This approach successfully stopped the outbreak within 4 months of the start of its global spread.

For treatment of SARS, several drugs were tried clinically with no clear benefit, and since then a number of drugs including protease inhibitors used in HIV have been investigated for efficacy *in vitro*. The viral S protein has been suggested as a candidate for developing a preventative vaccine against SARS or other coronaviruses. However, the genetic variability of these viruses, and poor immunity after natural infection, indicate the challenges involved.

MIDDLE EAST RESPIRATORY VIRUS (MERS-CoV)

The increasing numbers of coronaviruses isolated from animals and birds illustrate that it is likely that more members of this family will emerge in the years to come, posing further threats to public health in affected areas. The first case of MERS-CoV occurred in Jeddah, Saudi Arabia in June 2012 with a patient presenting with acute pneumonia and renal failure. As of May 2016, 1733 laboratory-confirmed cases of MERS had been reported to WHO, including at least 628 deaths (case fatality rate 36%). Evidence of the presence of MERS coronavirus has been reported from 27 countries, but still mostly from Saudi Arabia and Korea.

Clinical Features and Pathogenesis

The clinical presentation ranges from asymptomatic infection, to a mild flu-like illness or severe pneumonia accompanied by acute respiratory distress syndrome (ARDS), septic shock, and multi-organ failure preceding death. The course of infection is more severe among immunocompromised individuals and those with underlying medical conditions, especially chronic renal failure, heart disease, and diabetes.

The disease starts with fever and cough, sore throat, chills, arthralgia, and myalgia, soon followed by dyspnea. The infection rapidly progresses to severe pneumonia (Fig. 31.4). Around one-third of cases also show gastrointestinal symptoms such as diarrhea and vomiting.

The course of the illness is relatively short, with the disease progressing rapidly through a number of stages by 7 to 11 days after presentation. The lower respiratory tract is involved early in acute infection. Chest radiographs are consistent with viral pneumonitis, with bilateral infiltrates, segmented or lobular opacities and pleural effusions.

In common with other human coronaviruses, MERS-CoV evades the host innate immune response and causes a rapid dysregulation of cellular transcription pathways. The result is extensive apoptosis of bystander cells, to a far greater extent than is the case in SARS.

The cellular receptor for MERS-CoV has been identified as CD26, a dipeptyl peptidase, in contrast to the angiotensin-converting enzyme 2 (ACE2) used by SARS-CoV. This receptor is involved in the regulation of cytokine responses as well as glucose metabolism. Antibodies to the S binding domain of MERS-CoV efficiently neutralize infectivity.

Laboratory Diagnosis

MERS-CoV RNA can be detected in blood, urine, and stool as well as in respiratory aspirates by RT-PCR. In investigating and controlling a new outbreak such as this, it is essential that a consistent policy for testing and interpreting test results is applied to the diagnosis, management, quarantine, and reporting of cases. WHO provide such recommendations on their website www.who.int.csr.

Epidemiology

Dromedary camels have been implicated as the primary source of infection for humans, with a high percentage possessing viral antibodies. However, the exact route of transmission from camels is not clear. A recent report of MERS-CoV sequences in bats trapped in Saudi Arabia adds to the puzzle. Although the majority of hospitalized cases are thought to be secondary cases resulting from human-to-human transmission in health-care settings, MERS-CoV is not considered to be highly transmissible with an R_0 at <1—there is no evidence for ongoing spread within the community (see Chapter 13: Epidemiology of Viral Infections for further details of virus transmissibility).

Phylogenetic analysis of related sequences suggests MERS-CoV has diverged from a common ancestor as recently as 2007 to 2010. The discovery of another coronavirus in camels closely related to bovine coronavirus suggests that camels could be acting as intermediate hosts. Closely related viruses have been recovered from bats, and two amino acid changes in the spike protein of one of these viruses allows it to be activated by human proteases and to infect human cells. This suggests a mechanism for this virus to jump species from bats to humans.

Treatment and Prevention

To date, treatment has been focused on supportive therapy in the absence of any specific intervention measures. The use of antimicrobials to minimize the risk of opportunistic infection has been employed. Attempts to reverse the progression of respiratory distress and fibrosis through the use of corticosteroids have been unsuccessful.

Prevention of infection involves avoiding exposure to camels including consuming raw camel milk and inadequately cooked meat, particularly for those with diabetes, chronic lung disease, renal impairment, the immunocompromised, or the elderly. Confirmed cases should be isolated to avoid nosocomial spread.

FURTHER READING

Coleman, C.M., Frieman, M.B., 2014. Coronaviruses: important emerging human pathogens. J. Virol. 88, 5209–5212.

Graham, R.L., Donaldson, E.F., Baric, R.S., 2013. A decade after SARS: strategies for controlling emerging coronaviruses. Nat. Rev. Microb. 11, 836–848.

Pyrc, K., Berkhout, B., van der Hoek, L., 2007. The novel human coronaviruses NL63 and HKU1. J. Virol. 81, 3051–3057.

The WHO MERS-CoV Research Group, 2013 Nov 12. State of Knowledge and Data Gaps of Middle East Respiratory Syndrome Coronavirus (MERS-CoV) in Humans, Edition 1. PLOS Currents Outbreaks. Available from: http://dx.doi.org/10.1371/currents.outbreaks.0bf719e352e7478f8ad85fa30127ddb8.

Chapter 32

Picornaviruses

The *Picornaviridae* comprises one of the largest and most important families of viruses. At least 68 enteroviruses including the polioviruses, and 16 parechoviruses, may infect the human enteric tract; hepatitis A virus is a major human pathogen, and over 100 rhinoviruses cause common colds.

Picornaviruses have featured in many of the milestones of virology. In 1897, foot-and-mouth disease was the first infection of any animal to be shown to be caused by an agent that could pass through a filter that held back all bacteria—thus the concept of "filterable viruses" was born. In 1949 John Franklin Enders and his colleagues first propagated poliovirus in monolayers of mammalian cells cultivated *in vitro*—a landmark achievement. Based on this technologic breakthrough, inactivated and live attenuated polio vaccines were developed by Jonas Salk and Albert Sabin in the United States in 1954 and 1961 to 1962 respectively. The use of both vaccines quickly led to the demise of infantile paralysis throughout the developed world. In 1981, poliovirus became the first RNA animal viral genome to be molecularly cloned and sequenced. Poliovirus type 1 and rhinovirus type 14 were the first human viruses for which the three-dimensional structure was solved by X-ray crystallography in 1985, opening up new approaches to antiviral chemotherapy and vaccinology. In 1991 authentic infectious virions were synthesized *in vitro* from poliovirus RNA in a cell-free cytoplasmic extract from uninfected cells—yet another first for poliovirus. Finally, in 2002 poliovirus cDNA chemically synthesized in the absence of a natural template was used to generate infectious poliovirus, showing for the first time that it is possible to synthesize an infectious agent solely by introducing into cells the correct genetic information.

PROPERTIES OF PICORNAVIRUSES

Classification

The nomenclature of picornaviruses infecting humans has been arbitrary and confusing. Different serotypes have been named on the basis of a combination of properties including the type of disease produced in man, the disease produced in suckling mice, and the place and historical order of discovery, for example, Coxsackievirus A16, Echovirus 11, and Enterovirus 68. The family *Picornaviridae* also includes important pathogens of animals including foot-and-mouth disease virus (FMDV) and other viruses infecting livestock and rodents (Table 32.1).

Virion Properties

Picornavirus virions are non-enveloped, 30 nm in diameter, with icosahedral symmetry and appear smooth and round in outline by electron microscopy (Fig. 32.1). The capsid is constructed from 60 copies each of four polypeptides, VP1, 2, 3, 4, derived by cleavage of a single polyprotein (see Fig. 32.2). The detailed structure of the virion was discussed in Chapter 3: Virion Structure and Composition (see Figs. 3.1 and 3.2).

The genome is a single linear molecule of single-stranded RNA of positive polarity, 7 to 8 kb in length, polyadenylated at the 3′ end, together with a protein, VPg, covalently linked to its 5′ end. Both the 5′ and 3′ ends of the RNA contain untranslated regulatory sequences (UTRs); the 5′ UTR varies in length from approximately 500 to 1200 nt and contains an internal ribosome entry site (IRES). Being a single molecule of plus sense, the RNA genome can act directly as a mRNA and is thus infectious for susceptible cells.

Virus Replication

For a number of decades poliovirus has been a major model for studying RNA viruses, processes such as the mechanism of viral RNA replication, post-translational proteolytic cleavage of polyproteins, and the assembly of a simple icosahedral virion. Indeed it has been suggested from time to time that poliovirus has been so well studied that there is little remaining to be learned ("is poliovirus dead?"), usually to be followed by a wave of new unsuspected discoveries.

The cellular receptors used by different picornaviruses are diverse and do not always follow the current taxonomy for picornaviruses. Polioviruses, coxsackie B viruses, and some rhinoviruses use members of the immunoglobulin

Fenner and White's Medical Virology. DOI: http://dx.doi.org/10.1016/B978-0-12-375156-0.00032-1

TABLE 32.1 ICTV Classification of Human Picornaviruses

Genus	Species	Types (Many Are Known by Former Common Names)
Enterovirus	Human enteroviruses A, B, C, and D	Coxsackieviruses A1–24 (no A23), B1–6: Echoviruses 1–34 (no 10, 22, 23, or 28); Enteroviruses 68–71; polioviruses types 1–3
	Human rhinoviruses A, B, and C	More than 100 numbered human rhinoviruses (HRV 1–103)
Parechovirus	Human parechoviruses	Human parechoviruses 1–16 (includes former echovirus 22 and 23)
Hepatovirus	Hepatitis A virus	Hepatitis A virus 1
Kobuvirus	Aichi virus	

(Ig) superfamily, while other picornaviruses attach via other cell surface molecules including integrins, heparin sulfate, low-density lipoproteins, or extracellular matrix-binding proteins; similarly, different picornaviruses use different pathways to enter cells. In the case of poliovirus, attachment to the cell receptor leads to significant changes in the capsid; VP4 is released from the virion, and the N-terminal region of VP1 relocates from the interior to the virion surface. It is proposed that this hydrophobic N-terminal sequence is inserted into the plasma membrane, and the particle is internalized into endosomes. Subsequently, the inserted peptide facilitates the translocation of the viral genome across the endosomal membrane into the cytoplasm to initiate infection.

Following attachment, entry and viral genome uncoating, the VPg protein covalently bound to the 5′ end of the RNA is removed by cellular enzymes, and translation of the RNA begins. Initiation of translation does not follow the Kozak scanning model—first described by Marilyn Kozak, nucleotide sequences are scanned for consensus sequences known to represent ribosome binding sites—but relies on binding of ribosomes to an internal ribosomal entry site (IRES) within the 5′ UTR. This region of RNA can fold to form cloverleaf structures that bind host proteins, allowing the initiation of viral protein and RNA synthesis. Because the poliovirus genome does not contain internal translational stop codons, a single polyprotein is generated and this is subsequently cleaved at specific sites by virus-coded proteinases to generate 11 or 12 discrete proteins. The structural proteins that make up the virion capsid (VP4, VP2, VP3, and VP1 respectively) are coded by the 5′ end of the genome, while the various cleavage products of proteins from the 3′ end provide replicative functions, including protease and RNA-dependent RNA polymerase activities and the genome-associated protein VPg.

Virus replication takes place in a replication complex, consisting of RNA templates, the virus-coded RNA polymerase and several other viral and cellular proteins, all tightly associated with newly assembled smooth cytoplasmic membrane structures. Synthesis of the minus RNA strand is initiated at the 3′ end of the virion RNA, using the protein VPg as a primer. The completed minus strand is then used as the template for synthesis of the virion plus strand RNA. Most of the replicative intermediates (RIs) found within the replication complex consist of a full-length minus strand RNA from which several nascent plus strands are being transcribed simultaneously by the viral RNA polymerase. Newly synthesized plus strands and the structural capsid proteins then self-assemble into virions, undergoing a detailed stepwise process of morphogenesis that has been well documented (Fig. 32.3).

Picornaviruses have evolved various ways to rapidly and preferentially inhibit the translation, and the synthesis, of cellular mRNA. For example, the 2A protease of poliovirus cleaves eIF4G of the host cell's translation initiation complex, thereby affecting translation of 5′cap-bearing host mRNAs and allowing poliovirus mRNA (which lacks a 5′ cap) to be preferentially translated. Similarly, the poliovirus 3C^{pro} protease cleaves the Tbp subunit of the transcription factor TfIId thereby interfering with host cell RNA transcription by RNA polymerase II. This disruption of normal cell metabolism interferes with cellular antiviral responses, contributes to the cytotoxic effect of picornavirus replication, and ensures that cellular processes are re-directed to efficient production of new virions.

Poliovirus replication typically shows a short eclipse period of less than 3 hours, followed by the rapid exponential production of virus RNA, protein, and up to 10^5 newly assembled virions per cell by 6 hours following infection. However, the yield of infectious virus when assayed in cell culture is often 100-fold less, that is around 10^3 per cell, probably reflecting inefficiencies in both virus assembly and in the virus assay. By 6 hours, the cells are largely necrotic, and virions that have accumulated in large paracrystalline arrays within the cytoplasm are released as the cells begin to lyse (Fig. 32.4 and Table 32.2).

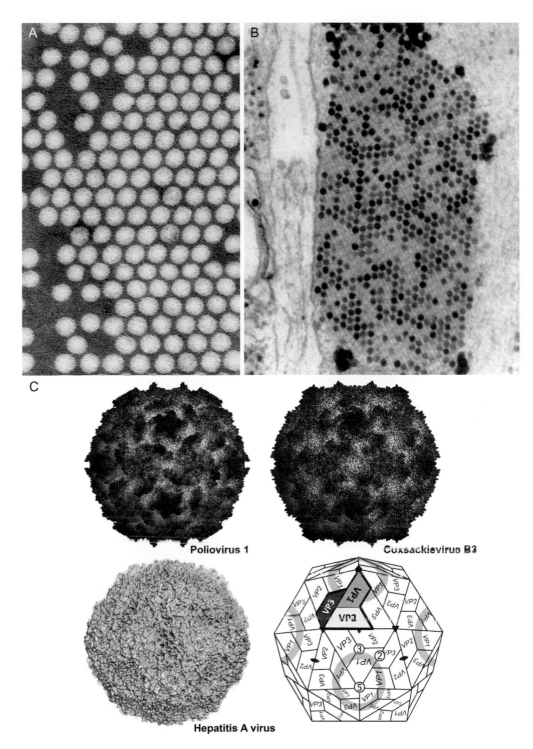

FIGURE 32.1 Morphology and structure of picornaviruses. (A) Negative contrast electron microscopy of poliovirus 1, in paracrystalline array—the virions are relatively smooth, presenting virtually no surface details. (B) Thin-section electron microscopy of Coxsackievirus A4 infection in the leg muscle of a mouse. (C) Models of the capsid surface of poliovirus 1, Coxsackievirus B3 and hepatitis A virus, constructed from X-ray crystallographic and cryoelectron microscopy data and analysis. The diagram of the icosahedral picornavirus virion shows the location of the viral proteins VP1 (blue), VP2 (yellow), and VP3 (purple) (capsid protein VP4 is located internally). The 5-fold, 3-fold, and 2-fold axes of symmetry are indicated in red. The canyon around the 5-fold axis of symmetry is indicated with a ring (gray). *(A) From Joseph Esposito (deceased), Centers for Disease Control and Prevention. For (C) from Eckard Wimmer: reproduced from Jiang, P., et al., 2014. Picornavirus morphogenesis. Microbiol. Mol. Biol. Rev. 78, 418–437, with permission.*

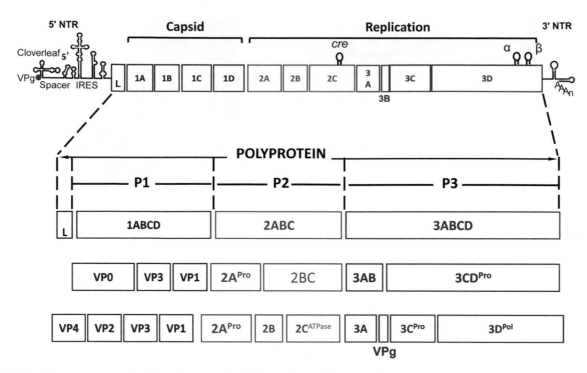

FIGURE 32.2   Genome structure and polyprotein processing of picornaviruses. The genome is a linear single-strand RNA ranging among the various viruses from 7.1 to 8.9 kb in length. The RNA has a covalently linked viral protein (VPg) at its 5′-end and poly(A) at its 3′-end. The single open reading frame (ORF) is flanked by a long 5′-non-translated region (NTR) and a short 3′-NTR. The long 5′-NTR contains an internal ribosome entry site (IRES) that directs polyprotein translation. Other essential RNA secondary structures have been identified in the open reading frame: polioviruses have a *cis* replication element (*cre*) and two RNA elements, α and β that play a role in replication. The single open reading frame is organized as 1ABCD-2ABC-3ABCD, with the numbers indicating the three different domains and each letter representing a protein. The P1 region encodes the capsid structural polypeptides. The P2 and P3 regions encode the non-structural proteins associated with replication. The polyproteins of many picornaviruses have an additional protein, the L protein, attached to the N terminus. Upon entry into the cell, the virion RNA acting as messenger is translated into a polyprotein that is rapidly cleaved into polypeptides P1, P2, and P3 by the viral proteases 2A and 3C. P1, P2, and P3 are subsequently cleaved by protease 3C. VP0 is cleaved into VP4 and VP2 by a third protease during capsid formation. *From Eckard Wimmer: reproduced from Jiang, P., et al., 2014. Picornavirus morphogenesis. Microbiol. Mol. Biol. Rev. 78, 418–437, with permission.*

POLIOMYELITIS

Poliomyelitis was once a greatly feared disease; its tragic legacy of paralysis and deformity a familiar sight in the 1950s. Today, by contrast, the disease is virtually unknown in nearly all parts of the world; circulation of wild-type poliovirus is limited to less than 10 countries worldwide, such has been the impact of the Salk and Sabin polio vaccines (see below). The foundation for these great developments was laid by John Enders, Thomas Weller, and Frederick Robbins in 1949, when they demonstrated the growth of poliovirus in cultures of non-neural cells. From this fundamental discovery flowed all subsequent studies on viral multiplication in cultured cell monolayers.

Pathogenesis and Immunity

Following ingestion, poliovirus multiplies first in the pharynx and small intestine. It is not clear whether the mucosa itself is involved, but lymphoid tissue (tonsils and Peyer's patches) certainly is. Spread to the draining lymph nodes leads to a viremia, enabling the virus to become disseminated throughout the body. It is only in the occasional case that the central nervous system becomes involved: virus is carried via the bloodstream to the anterior horn cells of the spinal cord, in which the virus replicates. The resulting lesions are widely distributed throughout the spinal cord and parts of the brain, but variation in severity gives rise to a spectrum of clinical presentations, with spinal cord involvement the most common form and bulbar involvement less so. Acquired immunity is permanent but specific to the poliovirus serotype. Circulating antibody if present helps restrict the spread of infection from the gut to the central nervous system, and the presence of neutralizing antibody against polioviruses is considered a reliable laboratory correlate of protection against poliomyelitis.

Pregnancy increases the incidence of paralysis, tonsillectomy increases the risk of bulbar paralysis, and inflammatory injections such as DPT vaccine increase the risk of paralysis in the injected limb, after the usual

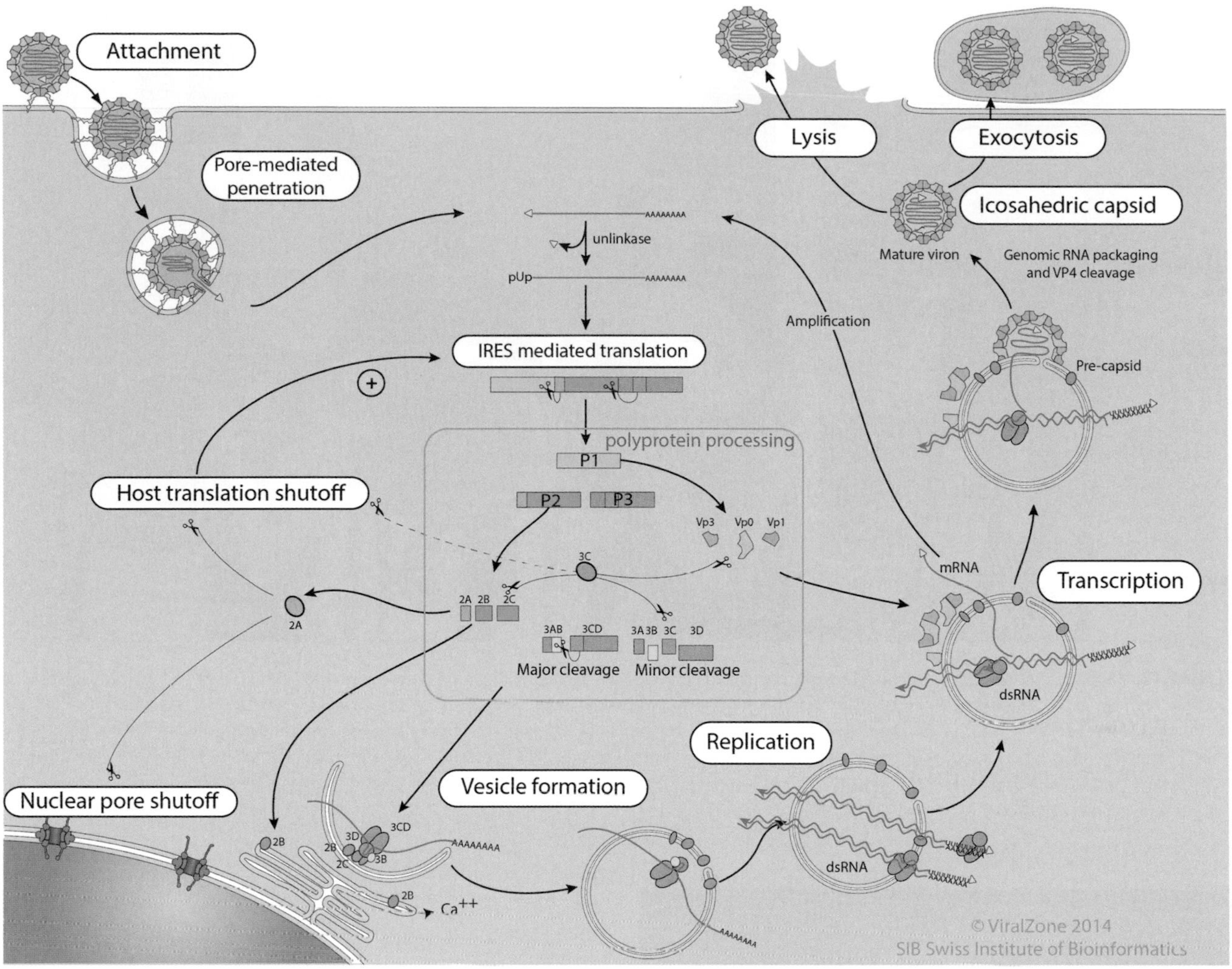

FIGURE 32.3 Replication of picornaviruses as exemplified by poliovirus. The virion first binds to a cellular receptor and enters the host cell in an endosome. The viral genome is released into the cytoplasm from early endosomes. The viral VPg protein is cleaved off the genomic RNA by a host protein called "unlinkase," and the RNA associates with ribosomes. Translation is initiated at an internal site near the 5′-end of the viral mRNA (IRES), and a polyprotein precursor is synthesized. The polyprotein is cleaved during and after its synthesis to yield the individual viral proteins. The proteins that participate in viral RNA synthesis are transported to membrane vesicles; RNA synthesis occurs on the surfaces of these vesicles. The (+) strand RNA is transported to these membrane vesicles, where it is copied into a double-stranded RNA (replicative intermediate). Newly synthesized (−) strands serve as templates for the synthesis of (+) strand genomic RNAs—with VPg acting as a replication signal. Structural proteins, resulting from cleavage of the polyprotein precursor, associate with (+) strand RNA molecules that retain VPg to form progeny virions. Construction of a virion requires 60 copies each of VP1, VP2, VP3, and VP4 proteins—that is, 60 polyproteins must be translated to produce each virion. Virions are released from the cell upon lysis. *From ViralZone, Swiss Institute of Bioinformatics, with permission.*

incubation period ("provocation"). A similar effect can be seen when intramuscular injections are given deliberately to treat a child who is incubating poliomyelitis ("aggravation"), a practice in countries where injections are regarded as the best kind of therapy for all manner of illnesses.

Clinical Features of Poliomyelitis

It is important to realize that paralysis is a relatively infrequent complication of an otherwise trivial infection. Of those few infections that become clinically manifest, most take the form of a minor illness ("abortive poliomyelitis"), characterized by fever, malaise, and a sore throat, with or without headache: vomiting may indicate some degree of aseptic meningitis. However, in about 1% of cases, muscle pain and stiffness herald the rapid development of flaccid paralysis. The incubation period of paralytic poliomyelitis averages one to two weeks, with outer limits of three days to one month. Legs are more often affected than arms. Bulbar poliomyelitis affecting the cranial nerves may accompany

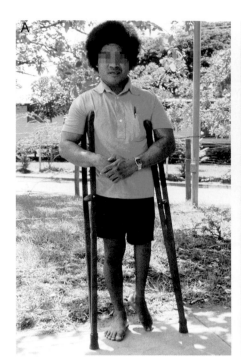

FIGURE 32.4(A,B) (A) Young man in Port Moresby, Papua and New Guinea, following the polio outbreak of 1962. The failure of limb movement leads to disuse atrophy, and a shortened, withered leg or a withered arm is frequently seen in the survivors as a result. (B) Future US president FD Roosevelt was diagnosed with polio in 1921, at the age of 39, which eventually left him partially paralyzed from the waist down and only able to walk with difficulty using a leg brace and a cane. *(A) Reproduced from Cooke, R.A., Infectious Diseases, McGraw Hill, with permission. (B), Roosevelt, Stalin, and Churchill on portico of Russian Embassy in Teheran, during conference—Nov. 28–Dec. 1, 1943. U.S. Library of Congress, public domain.*

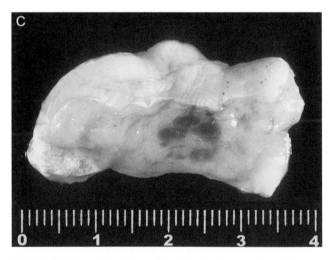

FIGURE 32.4(C) Transverse section of grossly atrophic calf muscle following poliomyelitis. Most of the muscle tissue has been replaced by fat. *Reproduced from Cooke, R.A., Infectious Diseases, McGraw Hill, with permission.*

extremity involvement or can occur as the sole paralysis. In bulbar paralysis, death may result from respiratory or cardiac failure. Otherwise, some degree of recovery of motor function may occur over the next few months, but paralysis remaining at the end of that time is permanent.

In some cases further muscle atrophy may be observed many years after apparent recovery ("late post-polio muscle atrophy" or "post-polio syndrome") (Fig. 32.4).

Clinical poliomyelitis can be caused by any of the three serotypes 1, 2, and 3 of wild-type poliovirus, and also rarely by attenuated vaccine-derived polioviruses that have acquired one or more back-mutations from the vaccine virus genetic sequence toward the virulent wild-type sequence.

Laboratory Diagnosis

Virus is readily isolated from feces for up to six weeks after the onset of illness, and sometimes from respiratory secretions during the first week, but rarely from CSF. Any type of human or simian cell culture is adequate for isolation, the virus growing so rapidly that cell destruction is usually complete within a few days. Early changes include cell retraction, increased refractivity, cytoplasmic granularity, and nuclear pyknosis. The serotype of the isolate can be identified by neutralization tests.

Distinguishing wild-type poliovirus isolates from vaccine-derived viruses poses an important responsibility for diagnostic laboratories during vaccination campaigns. All poliovirus isolates should be referred to, and characterized by laboratories of the Global Polio Laboratory Network.

TABLE 32.2 Properties of Picornaviruses

Virion	Non-enveloped, 30 nm diameter, icosahedral capsid, 60 protomers
Genera infecting humans	*Enterovirus* and *Hepatovirus* acid stable (pH > 3): *Rhinovirus* acid labile (pH < 5)
Genome	Linear, plus sense ssRNA genome, 7–8 kb, protein primer VPg at 5″ end, polyadenylated at 3″ end, infectious
Replication strategy	Replication takes place in the cytoplasm within a replication complex associated with smooth membranes. Virion RNA acts as mRNA and is translated into a polyprotein; this cleaves itself progressively to yield non-structural and structural proteins. Minus-strand RNA is synthesized using VPg as primer; replicative intermediates consist of full-length minus strand from which several nascent plus strands are being transcribed. Self-assembly of newly made plus strands and capsid proteins follows ordered steps of morphogenesis

The original protocol involved screening for vaccine-derived viruses using a combination of molecular and antigenic methods; this has now been largely replaced by reverse transcription–polymerase chain reaction (RT-PCR) targeted to nucleotide substitutions that occur early in reversion of vaccine-derived polioviruses. Candidate vaccine-derived viruses are sequenced in the VP1 region for routine analysis; the complete genome is sequenced if epidemiological investigation is initiated.

Epidemiology

As polioviruses are tropic for the gastrointestinal tract virus may be excreted for up to six weeks in feces, thus polioviruses spread mainly via the fecal–oral route. Direct fecal contamination of hands, thence food or eating utensils, is probably responsible for most case-to-case spread, especially in conditions of poor hygiene and sanitation. Uncommonly, explosive epidemics have resulted from contamination of water supplies by sewage. Respiratory spread from the pharynx may also occur. The disease is endemic in the tropics throughout the year. In temperate countries before the introduction of vaccination it classically occurred in sharp summer and autumn epidemics. Poliovirus infection is highly contagious and can have a high attack rate after introduction into a susceptible community, but the chain of infection is rarely clinically obvious as the vast majority of infections are asymptomatic.

Within the past 150 years, clinical poliomyelitis first appeared in epidemic form, progressed to become endemic on a global scale, but since has retreated to near-elimination. Two major developments have been largely responsible. The first of these was the introduction of modern standards of hygiene and sanitation to the more advanced countries of the world, which initially led to an increase in the incidence of paralytic poliomyelitis in older children and adults. This paradox is discussed in Chapter 13: Epidemiology of Viral Infections.

The second major influence on the epidemiology of poliomyelitis has been the development and widespread use of highly effective vaccines. In the United States, between 1955 when the inactivated poliovaccine was introduced, and 1973 when wild-type poliovirus no longer circulated in the community, the number of paralytic cases fell from >10,000 to around 10 annually. From 1973 onwards, the handful of cases that continued to be seen were either imported or due to vaccine-derived virus strains. Similar success has been achieved in many other countries. Buoyed by these successes, the World Health Assembly in 1988 set a goal for global eradication of wild polioviruses. Wild poliovirus was eliminated from the Americas in 1991, and the last case of paralysis due to type 2 wild poliovirus worldwide, was reported from Uttar Pradesh in 1999, reinforcing optimism that this goal could be rapidly realized. However, since the year 2000, as many as 2000 cases have been reported each year, and wild viruses have continued to be endemic in Nigeria, Pakistan, and Afghanistan. The problems, and current attempts to overcome them, are discussed below.

Poliovirus Vaccines—Two Alternative Strategies

In the late 1940s, the US National Foundation for Infantile Paralysis organized a nationwide appeal, "The March of Dimes." The response was overwhelming and the Foundation set about sponsoring a massive research drive on several fronts, in an attempt to turn Enders' then recent discovery to advantage by developing a poliomyelitis vaccine. Jonas Salk was commissioned to work toward an inactivated vaccine, and at the same time Albert Sabin worked towards a live attenuated vaccine. The formalin-inactivated (Salk) poliovaccine (IPV) was the first to be licensed and was enthusiastically embraced in North America, Europe, and Australia during the mid-1950s. Sweden, the country in which paralytic poliomyelitis first became apparent in epidemic form and which for many years continued to have the highest rate of poliomyelitis in the world, eliminated the disease by the use of Salk vaccine. Meanwhile, though, the live attenuated oral poliovaccine (OPV) of Sabin was also introduced and became the mainstay of the highly successful national vaccination programs in many countries over successive decades. The several advantages of such living vaccines over inactivated products are set out in detail in Chapter 11: Vaccines and Vaccination, and the global use of both types of vaccine has provided a unique opportunity to directly compare both strategies in real-life settings. In recent years many countries have re-instituted IPV for reasons discussed below. The near abolition of poliomyelitis in much of the Western World since the introduction of poliovaccine represents one of the great achievements of medical science.

OPV is a trivalent vaccine, consisting of attenuated strains of all three serotypes, derived empirically by serial passage in primary cultures of monkey kidney cells. Because of the diminishing availability of monkeys and the difficulty of ensuring that laboratory-bred animals are free of simian viruses, manufacturers now use human diploid fibroblasts as a substrate for the production of poliovaccine. The vaccine strains of types 1, 2, and 3 acquired 57, 2, and 10 mutations respectively, during serial passage in cultured cells, leading to attenuation which was confirmed by the absence of neurovirulence for monkeys. In all three types, attenuation has been shown to be related to mutations in the IRES region of the 5′ UTR, with other mutations in the capsid region helping to stabilize each phenotype. WHO provides written manufacturing and testing standards for OPVs, that include monitoring by neurovirulence testing in monkeys or in transgenic mice that express the human poliovirus receptor. Also recommended is the testing by RT-PCR and restriction enzyme cleavage of the products, to quantify in a particular virus preparation the proportion of single base mutations associated with attenuation. The initial course consists of three successive doses; although one successful "take" of the vaccine viruses is sufficient, concurrent infection with another enterovirus may interfere with the replication of the vaccine viruses, as commonly occurs in the developing countries of the tropics. Earlier concern that breast-feeding may represent a contraindication has not been substantiated—though colostrum contains moderate titers of maternal IgA, milk itself does not contain enough antibody to neutralize the vaccine virus.

One advantage of OPV over IPV follows from the fact that it multiplies in the alimentary tract. This subclinical infection elicits prolonged synthesis of local IgA antibodies, as well as the circulating IgG antibodies that protect the individual against paralytic poliomyelitis by intercepting wild polioviruses during the viremic phase. This mucosal antibody prevents primary implantation of wild virus in the gut and hence diminishes the circulation of virulent viruses in the community.

Fecal excretion of OPV viruses can continue for several weeks after vaccination, bringing the theoretical benefit that the vaccine may spread to contacts and extend the network of individuals protected from a single dose. However, an undesired consequence of vaccine spread to contacts is that it provides greater opportunities for selection of mutants displaying varying degrees of reversion toward virulence (see Chapter 7: Pathogenesis of Virus Infections). Within two to three days of administration, the OPV type 3 strain recovered from the feces of the vaccine recipient generally displays a single nucleotide change at one particular position in the 5′ non-coding region, representing a partial reversion to wild-type. Although these partial revertants can be shown to have partially regained neurovirulence for monkeys, only rarely do these revertant viruses produce paralysis in the vaccinated or among contacts. With the passage of days, the virus excreted by vaccinees accumulates further revertant nucleotide changes especially within the open reading frame for the capsid proteins, and acquires a greater neurovirulence. Very rarely (once per million vaccinees) the vaccinee, or an unvaccinated family contact, develops clinical poliomyelitis. Because vaccine-associated paralytic polio is 10,000 times more common in those who are immunocompromised in some way, such children should only be given IPV. Furthermore, because there are still significant numbers of parents who have never received poliovaccine, there is a strong argument for immunizing unvaccinated family contacts prior to, or simultaneously with, their infants. Nevertheless, it must be stressed that OPV has proven to be an outstandingly successful and safe vaccine.

Inactivated poliovaccine (IPV) became "the forgotten vaccine" in many countries following the development of OPV. The large amounts of virus required for IPV can be produced in non-malignant, aneuploid monkey kidney cell lines such as Vero, approved for this purpose. Poliovirus grown in such cells is purified by ultrafiltration, diafiltration, ion exchange chromatography, affinity chromatography on Sepharose-immobilized antibodies, and/or zonal ultracentrifugation. The importance of removing aggregates of virions before inactivation with formaldehyde was demonstrated many years ago following a disaster in 1955 in which clumped virus escaped inactivation and subsequently paralyzed numerous children in the United States. Two or three doses of IPV are sufficient to confer protection.

The particular advantages of IPV have led to its much greater use in recent years. Seroconversion rates to OPV can be less than optimal in developing countries for a variety of reasons (see below). However, of increasing concern is the fact that, in countries where wild-type poliovirus and its disease have been eradicated, even a very low rate of vaccine-associated paralytic poliomyelitis becomes unacceptable. For example, in the United States IPV was introduced universally from the year 2000 and no cases of vaccine-associated paralytic poliomyelitis have been reported since. Secondly, there have been a number of outbreaks since the year 2000 of poliomyelitis associated with vaccine-derived polioviruses thought to have originated from OPV usage. There is an increasing consensus that in countries where wild-type virus is no longer circulating, there must be a transition from OPV to IPV to eliminate vaccine-associated paralytic poliomyelitis and the circulation of vaccine-derived polioviruses.

Progress in Poliomyelitis Eradication

Wild polioviruses were eliminated from the United States and Canada during the 1970s, and in 1985 the Pan American Health Organization resolved to eradicate poliomyelitis from the western hemisphere by 1990. Regular vaccination with OPV under the Extended Programme of Immunisation (EPI) plan was supplemented by annual "national vaccination days," when all children under 14 years of age were given OPV, and by "mopping-up" vaccination in localities where cases of poliomyelitis had occurred. Both of these operations required much additional vaccine, the cost of which was largely covered by a special effort by Rotary International. The progress of eradication was monitored by considerably strengthening surveillance of flaccid paralysis, and the careful laboratory investigation of every such case.

Notwithstanding, wild polioviruses have continued to circulate in Nigeria, Pakistan, and Afghanistan, and there has been periodic fresh seeding of polioviruses from these sources to neighboring countries. The recent re-emergence of poliomyelitis caused by naturally circulating wild virus in Syria, Cameroon, Somalia, and China highlights the vulnerability to reappearance of this disease, particularly following disruption of society and community services. Two major factors are thought to play a role in this stalemate. First, vaccination programs have not always been pursued with sufficient rigor. Underutilization of vaccine has been well documented in Nigeria and shown to correlate with the attack rate of virologically confirmed poliomyelitis. Many children never receive a full course of OPV. Poverty, social dislocation, and religious objections may all have played a role, together with lack of a health service infrastructure adequate to ensure coverage of the whole population, so that OPV is not delivered satisfactorily, particularly in the more inaccessible rural villages. Second, seroconversion rates with OPV are sometimes less than optimal, because of loss of vaccine viability or interference by co-infecting gastrointestinal viruses present at the time of vaccination. Maintenance of refrigeration during storage and transport in hot tropical climates (the "cold chain") is essential to retain the viability of the vaccine virus, and this is not always achieved. However, it has been demonstrated in South and Central America that poliomyelitis can indeed be eradicated in developing countries by comprehensive coverage with OPV; in those regions no cases of poliomyelitis due to wild polioviruses have been seen since August 1991.

In the more developed nations poliomyelitis has been so effectively conquered by OPV that it is looked upon as a disease of the past. Nevertheless, importation of virus remains an ever present threat, especially to pockets of unimmunized people, such as immigrants and their preschool children, who are often concentrated in overcrowded urban areas. Thus it remains imperative that very high levels of immunization coverage are maintained to prevent such epidemics which could in turn be followed by re-establishment of endemicity.

ENTEROVIRUSES A, B, C, D, AND PARECHOVIRUSES

Pathogenesis and Immunity

Enteroviruses A, B, C, and D typically are ingested and grow to high titer in both the throat and the intestinal tract, but are shed in the feces for much longer compared to respiratory secretions. Dissemination via the bloodstream is doubtless the route of spread to the extensive range of susceptible target organs, giving rise to a pathology and clinical symptoms related to the particular target site. Little is known as to the factors that determine the tropism of different enteroviruses, for example, the predilection of coxsackie B viruses for muscle, or of enterovirus 70 for the conjunctiva. The incubation period is one to two weeks in the case of systemic enterovirus infections, but may be as short as a day or two for conjunctival or respiratory disease.

Immunity is type-specific and long-lasting. Antibody appears to be relatively more important in protection and recovery from picornavirus infections than for most other virus families. Perinatal infections tend to produce either no disease or mild respiratory or gastrointestinal symptoms in neonates with maternal antibody, but overwhelming acute disseminated disease may occasionally occur in those without protective antibodies. Furthermore, children with congenital B cell deficiencies may develop chronic disseminated enterovirus infections.

Parechoviruses present an interesting story of virus discovery. During the search for polioviruses in the 1950s, two new virus isolates were classified as enteroviruses and designated echoviruses 22 and 23. Despite showing differences in cytopathology from many other enteroviruses, it was not until nearly 50 years later that molecular sequencing revealed such marked genetic differences from enteroviruses that these viruses were placed in a new genus, *Parechovirus*, and designated human parechoviruses 1 and 2 (HPeV1 and HPeV2). Parechoviruses are not as readily isolated in cell culture as many enteroviruses, but over the past 10 years a further 14 types of human parechovirus have been identified. Protein sequences of human parechoviruses show less than 30% identity with the corresponding proteins of other picornaviruses, and the mature capsid comprises three, not four, protein cleavage products. However they share the same physical and biological characteristics. Human parechovirus infection is very widespread, particularly in children under two years of age.

Clinical Syndromes

Most enterovirus infections are subclinical, particularly in young children. Nevertheless, infection can induce a wide range of clinical syndromes involving many of the body systems (Table 32.3). However, it should be noted that each syndrome can be caused by several viruses, and that each virus can cause several syndromes even during the same epidemic. Rashes, upper respiratory tract infections, and "undifferentiated summer febrile illnesses" are common. Moreover, enteroviruses are the most frequent cause of meningitis, albeit a relatively mild form. In general, coxsackieviruses tend to be more pathogenic than echoviruses, causing a number of diseases rarely seen with echoviruses, for example, carditis, pleurodynia, herpangina, hand-foot-and-mouth disease, and occasionally paralysis.

Diseases Caused by Enteroviruses A, B, C, D, and Parechoviruses

Neurological Disease. Meningitis is more commonly caused by enteroviruses and parechoviruses than by all species of bacteria combined. The induced "aseptic" meningitis is fortunately not nearly as serious as many of bacterial etiology (see Chapter 36: Flaviviruses). Numerous enteroviruses have been implicated from time to time. The most commonly involved are echoviruses 3, 4, 6, 7, 9, 11, 16, 30, coxsackieviruses A7, A9, B1-6, enterovirus 71, HPeV3, and (in countries where they are still present) polioviruses 1 to 3.

Paralysis may be caused by polioviruses within endemic areas, but has also very rarely been associated with coxsackievirus A7. Some outbreaks of enterovirus 71 infection have been marked by meningitis, some encephalitis, and many cases of paralysis with a significant number of deaths. During the 1969 to 1971 enterovirus 70 pandemic, acute hemorrhagic conjunctivitis was complicated in a small minority of cases by radiculomyelitis, which manifested itself as an acute flaccid lower motor neuron paralysis, resembling poliomyelitis but often reversible. Virological investigation of all cases of acute flaccid paralysis is used to detect any residual circulation of poliovirus in a community. Parechoviruses have also been associated with encephalitis

TABLE 32.3 Diseases Caused by Enteroviruses

Syndrome	Viruses
Neurological	
Meningitis	Many enteroviruses, parechoviruses
Paralysis	Polioviruses 1, 2, 3; Enteroviruses 70, 71; Coxsackievirus A7
Chronic meningoencephalitis/dermatomyositis	Echoviruses, others
Cardiac and Muscular	
Myocarditis	Coxksackievirus B; some Coxsackievirus A and echoviruses
Pleurodynia	Coxsackievirus B
Skin and Mucosae	
Herpangina	Coxsackievirus A
Hand-foot-and-mouth disease	Coxsackieviruses A9, A16; Enterovirus 71; others
Maculopapular exanthem enteroviruses	Echoviruses 9, 16; many other
Respiratory	
Colds	Coxsackieviruses A21, A24; Echovirus 11, 20; Coxsackievirus B; others
Ocular	
Acute hemorrhagic conjunctivitis	Enterovirus 70; Coxsackievirus A24
Neonatal	
Carditis, encephalitis, hepatitis	Coxsackievirus B; Echovirus 11 and others

in the neonatal period, a more severe disease than is usually seen with enterovirus infection.

Chronic meningoencephalitis with juvenile dermatomyositis is a fatal condition seen in children with the B cell deficiency, X-linked agammaglobulinemia, or with severe combined immunodeficiency. Clinical features are varied and include headaches, seizures, ataxia, motor deficits, sensory disturbances, disturbed state of consciousness, and personality changes. In many cases an echovirus or other enterovirus can be recovered in high titer from CSF for months or years. Chronic hepatitis may also be present. More than half of these patients have an associated condition known as juvenile dermatomyositis.

Cardiac and Muscular Disease. Enteroviruses are the major cause of carditis in the newborn (see below), and are also an important cause of *carditis* at all ages, particularly in adolescents and physically active young adult males. The coxsackie B viruses and, to a lesser extent other enteroviruses such as coxsackieviruses A4 and A16, and echoviruses 9 and 22 have been directly incriminated by demonstration of virus, viral antigen, and/or viral RNA in the myocardium itself or in pericardial fluid. The disease may present predominantly as myocarditis, pericarditis, or cardiomyopathy characterized by a greatly dilated heart. Death is uncommon but recrudescences occur in about 20% of cases and permanent myocardial damage, as evidenced by persistent ECG abnormalities, cardiomegaly or congestive heart failure, may occur.

Pleurodynia, also known as Bornholm disease, is basically a myositis, characterized by paroxysms of stabbing pain in the muscles of the chest and abdomen. Seen mainly in older children and young adults, sometimes in epidemic form, it is caused principally by coxsackie B viruses. Despite the severity of the pain, a complete recovery is normal.

Exanthems and Enanthems. Rashes are a common manifestation of enteroviral infections. These are generally transient, always inconsequential, and significant only as an index of an epidemic which may have more serious consequences for other patients. Differential diagnosis from other rashes, viral or otherwise, is often difficult. For example, the fine maculopapular ("rubelliform") rash of the very common echovirus 9 may easily be mistaken for rubella. The "roseoliform" rash of echovirus 16 ("Boston exanthem") appears only as the fever is declining. Many other enteroviruses can also produce rubella-like or measles-like rashes with fever and sometimes pharyngitis, especially in children.

Vesicular ("herpetiform") rashes are less common but very striking. The quaint name of "*hand-foot-and-mouth disease*" is given to a syndrome characterized by ulcerating vesicles at these three anatomical sites. Coxsackieviruses A9 and A16 and the coxsackievirus A-like enterovirus 71 are most frequently responsible. Outbreaks of hand-foot-and-mouth disease are not unusual in childcare centers, causing consternation until the benign course of this infection is recognized. It is important to note that foot-and-mouth disease of cattle caused by an unrelated picornavirus is *not* transmissible to humans.

Herpangina is a severe febrile pharyngitis characterized by vesicles (vesicular pharyngitis) or nodules (lymphonodular pharyngitis), principally on the soft palate. Despite the confusing name, it is unrelated to herpesvirus infections but is one of the common presentations of coxsackie A virus infections of children.

Respiratory Disease. Enteroviruses are not major causes of respiratory disease but can produce a range of febrile colds and sore throats during summer epidemics. These come under the umbrella label of "summer grippe" or "undifferentiated febrile illness." Coxsackieviruses A21 and A24, and echoviruses 11 and 20 are prevalent respiratory pathogens; coxsackie B viruses and certain echoviruses are less common. Many countries since 2005 have reported increasing numbers of cases of severe respiratory disease caused by enterovirus D68, particularly in children and sometimes requiring intensive care and intubation. Acute flaccid paralysis and cranial nerve dysfunction in association with EV-D68 are also being reported.

Ocular Disease. Acute hemorrhagic conjunctivitis burst upon the world as a new disease in 1969, caused by the newly recognized enterovirus 70. Extremely contagious, with an incubation period of only 24 hours, the disease began in West Africa and swept across Asia to Japan, infecting some 50 million people within two years. As its name implies, this conjunctivitis was often accompanied by subconjunctival hemorrhages and sometimes a transient involvement of the cornea (keratitis). Mainly adults were affected. Resolution usually occurred within a week or two, uncomplicated by anything other than a secondary bacterial infection. However, a small minority of cases, particularly in India, were complicated by neurological sequelae, notably *radiculomyelitis*, as discussed above. In 1981 another major epidemic of acute hemorrhagic conjuctivitis arose in Brazil and swept north through Central America and the Caribbean to the southeastern United States and the Pacific Islands. The cause of the original pandemic was the previously unknown enterovirus 70, but recent epidemics have been caused by a variant of coxsackievirus A24, a picornavirus not nearly so often associated with conjunctival hemorrhage.

Neonatal Disease. Enteroviruses may infect newborn babies prenatally (transplacental), natally (fecal contamination of the birth canal) or, most commonly, postnatally (from the mother, or nosocomially in hospital nurseries). Coxsackie B viruses and certain echoviruses, notably type 11, can produce fulminant infections which are often fatal. Two major presentations may be distinguished, although there is a tendency for overlap. The first, caused by Coxsackie B viruses or echovirus 11, is the *encephalomyocarditis* syndrome. Classically, a neonate suddenly becomes dyspneic and cyanosed within the first week of life; examination reveals tachycardia and ECG abnormalities, often accompanied by other signs of overwhelming systemic infection, including manifestations of meningoencephalitis. The second presentation, sometimes called the *hemorrhage-hepatitis* syndrome, is associated mainly with echovirus 11 and less commonly with other echoviruses. The baby is lethargic and feeds poorly, then develops jaundice, followed by profuse hemorrhages, hepatic and renal failure leading to death within days.

Other Possible Associations

Juvenile-onset insulin-dependent diabetes mellitus (IDDM) has been speculatively linked with coxsackie B viruses for decades. For example, studies have shown IgM antibodies to coxsackieviruses in many cases of IDDM, and coxsackievirus B4 recovered from a fatal case reproduced a comparable diabetic condition when inoculated into mice. Coxsackie B viruses can infect isolated human pancreas, and enteroviral fragments have been reported in the beta cells of individuals with recently diagnosed IDDM. However, it remains to be demonstrated whether the association with IDDM is one of cause or effect. The apparent link between IDDM and HLA types Dr3 and Dr4 suggests the involvement of immunological mechanisms.

Chronic (postviral) fatigue syndrome, otherwise known as epidemic neuromyasthenia or myalgic encephalomyelitis, is a debilitating group of disorders characterized by persistent or fluctuating severe fatigue, unrelieved by rest and of greater than 6 months' duration. The onset frequently follows a flu-like illness, and a number of common infections including Epstein-Barr virus, coxsackieviruses, and Q fever have been implicated as either true etiological agents or triggers to the development of immune dysfunction. Recent research has led to clinical algorithms to be used to ensure consistent criteria in diagnosis, and different patterns of cytokine dysfunction have been identified in patients, providing hope for increased understanding of the pathogenic mechanisms involved.

Laboratory Diagnosis

Most but not all enteroviruses are readily isolated in cell culture, although molecular testing is now standard for the diagnosis of acute enterovirus infection. Primers spanning the conserved common 5' non-coding region allow amplification of a full range of Coxsackie A- and B-viruses, echoviruses, polioviruses, and enteroviruses 68 to 71 by RT-PCR. Commercial kits are available that can be used to test CSF, feces, bronchial and tracheal washings, serum or plasma, and autopsy samples. RT-PCR gives much faster results than virus culture, and for some types of sample has been found to be more sensitive. It does not yield virus for further identification, but reference laboratories may then identify the virus type by RNA sequencing. By modifying the design of primers, RT-PCR tests can be made selective for particular virus species or types.

Before the advent of RT-PCR, enterovirus infection was usually diagnosed by virus isolation from feces, or sometimes from the throat early in illness, or where appropriate from the eye, vesicle fluid, urine, CSF, blood, or organs such as heart or brain at autopsy. Isolation from "sterile" sites is much more indicative of a cause-and-effect relationship with the clinical illness than isolation from feces. Classically, primary cultures of monkey kidney were the standard substrate, but the diminishing availability of these cells has led to these being replaced in most laboratories with continuous cell lines. Polioviruses grow well in virtually any simian or human cell type. Human diploid embryonic lung fibroblasts (HDF or HEL) support the growth of most echoviruses but only a few coxsackieviruses. The monkey kidney cell line, BGM, is particularly susceptible to the coxsackie B viruses. A human rhabdomyosarcoma line, RD, supports the growth of some coxsackie A viruses previously regarded as not possible to grow in cell culture, in addition to many of the echoviruses. Enterovirus 70 is fastidious but can be isolated with difficulty in HDF cells. Cytopathic effects resemble those of poliovirus but develop more slowly, commencing with foci of rounded refractile cells which then lyse and detach from the substrate. Blind passage may be necessary to reveal the presence of virus, particularly with coxsackie A viruses. Provisional allocation to the *Picornaviridae* family is generally made on the basis of the characteristic CPE.

The use of infant mouse inoculation is now mainly of historic interest, although a few coxsackie A viruses (e.g., types 1, 19, 22) can only be isolated in this way. Newborn mice (less than 24 hours of age) are inoculated either intraperitoneally or intracerebrally, then observed for illness, killed, or subjected to histology.

Similarly, serology is of little value. The microneutralization test using known enterovirus serotypes has been used in the past, but it is labor-intensive, poorly standardized, relatively insensitive, and too type-specific. The finding of a single high antibody titer is of little practical clinical value, and paired specimens demonstrating a rising titer are not often available. In coxsackievirus carditis or enterovirus 70 hemorrhagic conjunctivitis, virus-specific IgM antibodies can be identified by IgM capture EIA any time up to two months after the appearance of symptoms, but cross-reactions with other enteroviruses limit the usefulness of this approach.

Assays for direct antigen detection have been developed for particular "clean" samples, for example, immunofluorescence to identify enterovirus 70 in conjunctival scrapings, or other enteroviruses in leukocytes from CSF. Type-specific monoclonal antibodies are useful only when the number of possible serotypes is strictly limited, such as identification of enterovirus 70 or coxsackievirus A24 in acute hemorrhagic conjunctivitis.

The typing of enterovirus isolates is frequently indicated, both for monitoring and analysis of disease outbreaks and for the identification of circulating poliovirus strains. Neutralization of virus replication was widely used to type the isolate, and this laborious procedure could be abbreviated by the use of an "intersecting" series of "polyvalent" serum pools. Each pool contains equine antibodies against, say, 10 of the 68 human enterovirus serotypes; the pools are mixed to ensure that antibodies to any given serotype are present in several and absent from several others, hence the isolate can be positively identified by scrutinizing the pattern of pools showing neutralization of infectivity. Aggregated virions may escape neutralization and hence appear impossible to type; this problem can be overcome by dispersing clumps with chloroform, or by using a plaque reduction assay. More recently, various schemes have been described using RT-PCR with appropriate primers, particularly focusing on the VP1 region, followed by the sequencing of PCR products.

Epidemiology

Enteroviruses are transmitted mainly by close contact, spreading rapidly and efficiently within families via the fecal–oral route. Echoviruses appear in the alimentary tract of most infants shortly after birth. In developing countries with low standards of hygiene and sanitation enteroviruses can be recovered from the feces of the majority of young children at any time—80% in one study in the Pakistani city of Karachi. In contrast, the New York "Virus Watch" program revealed an enterovirus carriage rate of only 2.4%; these differences are also reflected in the recovery of enteroviruses from sewage.

Droplet spread also occurs, the more so with the coxsackieviruses, and may be more relevant to the transmission of upper respiratory infections for which enteroviruses are often responsible. Acute hemorrhagic conjunctivitis is highly contagious, spreading by contact, with an incubation period of only 24 hours. Over half a million cases occurred in Bombay alone during the 1971 pandemic.

Over the years the commonest enteroviruses worldwide have included echoviruses 4, 6, 9, 11, and 30, coxsackieviruses A9 and A16, and coxsackieviruses B2–B5. Outbreaks of these serotypes, and others, often reach epidemic proportions, peaking in late summer/early autumn in countries with a temperate climate. A different type tends to become prevalent the following year as population immunity develops. Nevertheless, distinct serotypes do co-circulate to some degree, mainly among non-immune infants.

Young children constitute the principal target and reservoir of enteroviruses. It is significant that the prevalence of enteroviruses tends to increase at the time children return to school after holidays. Having acquired the virus from other youngsters they then take it home to the parents and siblings. In young children most infections are inapparent, or cause mild undifferentiated fevers, rashes, or upper respiratory tract infections. Disproportionately, the severe diseases—carditis, meningitis, encephalitis—are seen in older children and adults.

HUMAN RHINOVIRUSES A, B, AND C

The search for the common cold virus identified not one, but over 100 separate rhinovirus serotypes in the two species human rhinovirus A and B, and a further 60 or more distinct sequences within human rhinovirus C. Moreover, it is evident that rhinoviruses cause only about 50% of all colds. A vaccine for the common cold remains a distant prospect.

Pathogenesis and Immunity

Consistent with a predilection for replication at 33°C, rhinoviruses usually remain localized to the upper respiratory tract. Inflammation, edema, and copious exudation begin after a very short incubation period (2 to 3 days) and last for just a few days. Endogenous interferons may play a role in terminating the illness and in conferring transient resistance against infection with other viruses. Acquired immunity is type-specific and correlates more with the level of locally synthesized IgA antibodies that decline in titer within months of infection, rather than with IgG antibodies in serum which may persist for a few years. The typical progression of pathogenetic changes and accompanying clinical events is outlined in more detail in Chapter 8: Patterns of Infection.

Clinical Features

Everyone is all too familiar with the symptomatology of the common cold—profuse watery nasal discharge (rhinorrhea) and congestion, sneezing, often a headache, mildly sore throat, and/or cough. There is little or no fever. Resolution generally occurs within a week but sinusitis or otitis media may supervene, particularly if secondary bacterial infection occurs. Rhinoviruses have also emerged as the most frequently identified viruses causing disease exacerbations in asthma and in chronic obstructive pulmonary disease (COPD). Very occasionally rhinovirus pneumonia occurs, particularly among infants or young children.

Laboratory Diagnosis

Laboratory diagnosis of the common cold is usually not necessary, except in evaluating infection in severely immunocompromised patients or in the course of epidemiological research. Rhinoviruses are fastidious, growing only slowly, at 33°C, the temperature of the nose, in cell lines derived from embryonic human fibroblasts such as WI-38, MRC-5, or fetal tonsil, without conspicuous CPE.

Sensitivity can be enhanced by use of roller cultures and blind passage. Rhinovirus C species have not been grown in immortalized cell lines but only in organ or primary epithelial cell cultures.

RT-PCR has become the standard approach for diagnosis using primers for several short sequences in the 5′ noncoding region of the viral RNA that are almost completely conserved across more than 100 rhinovirus serotypes. By combining this with a semi-nested PCR and comparing the sequence of the PCR products with reference sequences, the particular rhinovirus serotype of the isolate can be determined. Care is needed in linking a positive PCR result to a current symptomatic infection, as shedding detectable by PCR has been reported for five to six weeks after a symptomatic infection.

Alternatively, a method of direct antigen detection in nasal washings by EIA has been described.

Epidemiology

Rhinovirus colds occur throughout the year, with peaks in the autumn and spring. Three or four rhinovirus serotypes circulate among the community simultaneously, one sometimes predominating; then a new set moves in after a year or so as herd immunity develops in the population. Pre-school children, who tend to be susceptible to almost all serotypes, commonly introduce the virus to the rest of the family after a two to five day interval, with a secondary attack rate of around 50%. Virus is shed copiously in watery nasal secretions for two to three days or more. Sneezing and coughing generate large- and small-particle aerosols, while "dribbling" and contamination of hands and handkerchiefs or paper tissues lead to contact spread via fomites to hands, nose, or eye.

The fact that many colds occur during winter and the change of season has encouraged the popular myth that one is most vulnerable if exposed to cold wet weather, but this is too simplistic a view. The influence of seasonality on transmissibility is discussed in Chapter 14: Control, Prevention, and Eradication.

Control

Clearly, the diversity of circulating serotypes makes the prospect of developing a useful vaccine a daunting challenge. Efforts continue to identify epitopes, particularly in the viral protein VP1, that may induce cross-reacting immunity against commonly circulating serotypes, but progress remains frustratingly slow.

Antiviral treatment has been steadily explored for decades. The prospect of considerable profits has encouraged many pharmaceutical companies to persist in the pursuit of that elusive goal, a cure for the common cold. However, problems in reaching this goal include: (1) viral replication reaches its peak within one to two days of onset of symptoms, and by the time any antiviral therapy is started it may be too late, and (2) because most common colds are not life threatening but inconvenient, an agent that is likely to be widely used in healthy people must have a very high safety profile. Several viral proteins, such as the polymerase, helicase, protease, and VPg, represent potential targets of attack, but much effort has also been concentrated on designing agents that block attachment or uncoating by binding directly to the virion itself. Some 91 of the 102 currently known rhinovirus serotypes use the widely distributed ICAM-1 as a receptor, and cryoelectron microscopy of complexes of rhinovirus 16 with the N-terminal domain of ICAM-1 have confirmed the precise location of the ligand, previously defined by X-ray diffraction on HRV-14 to lie within a "canyon" of the capsid protein VP1. In Chapter 12: Antiviral Chemotherapy, it is described how compounds bind to a hydrophobic pocket (beta barrel) immediately beneath the floor of that canyon, leading to blocking adsorption and particularly the uncoating of virus particles. It is postulated that these compounds stabilize the virion against a conformational change that occurs at pH < 6 and is required for intracellular uncoating. One such drug, "Pleconaril," was shown in human trials in 2001 to reduce the duration of symptoms by one day if oral administration was commenced within 24 hours of symptom onset; however, licensure was not granted due to concerns about safety and limited efficacy. Other agents that have been investigated include intranasal interferon, virus receptor blockers, antireceptor antibodies, soluble ICAM-1 (tremacamra), pirodavir, and 3C protease inhibitors, several of which have been shown to prevent colds if given prophylactically or to give a slight reduction in the duration of symptoms or viral shedding.

HEPATITIS A (HAV)

Careful human studies from the 1940s onwards identified and distinguished two major forms of hepatitis—an enterically transmitted disease known first as infectious hepatitis and later as hepatitis A, and the quite distinct serum hepatitis (hepatitis B). The causal agent of hepatitis A was unequivocally identified in 1973 by demonstration of a 27 nm icosahedral virus in patients' feces using immuno-electron microscopy. Biophysical and biochemical studies later established hepatitis A virus as a member of the family *Picornaviridae*.

Properties of Hepatovirus

For a time hepatitis A virus was classified as enterovirus type 72 but it was eventually accorded the status of a separate genus, *Hepatovirus*, on the basis of a number of differences from the enteroviruses, including stability of the virion at 60°C, lack of reactivity with an enterovirus group-specific monoclonal antibody, low percentage nucleotide homology with the genome of enteroviruses in spite of having the same gene order, and certain differences in the replication cycle (see below).

Antigenically, hepatitis A virus is highly conserved, there being only a single serotype in spite of the fact that four genotypes differing by around 20% in nucleotide sequence have been described; most human strains belong to genotypes I or III. An additional three genotypes have been isolated from monkeys. The majority of simian hepatitis A virus strains are thought to be host-restricted and differ from human hepatitis A virus at immunodominant antigenic sites, but at least one, the PA21 strain from *Aotus* monkeys, is very closely related antigenically to human hepatitis A virus genotype III.

Viral Replication

The replication of wild strains of hepatitis A virus in cultured primate cells is slow and the yield of virus is poor. This property has hindered the production of viral antigen for use as a vaccine, and has made the use of virus isolation impractical for laboratory diagnosis. Uncoating of the virion is inefficient. The 5′ non-translated region (NTR) contains an internal ribosomal entry site facilitating cap-independent initiation of translation. The processing of the P1 polyprotein and assembly of the cleavage products into virions follow a different pathway from that of enteroviruses. Very little complementary (minus sense) RNA is detectable in infected cells; most of the newly synthesized plus-stranded RNA becomes rapidly encapsidated into new virions. Cellular protein synthesis is not inhibited and there is little or no CPE; the infection of cell lines is non-cytocidal and persistent, with a restricted yield. After serial passage, rapidly replicating cytopathic variants emerge; they contain numerous mutations, including a 14 base reduplication in the 5′ NTR. Hepatitis A virus is resistant to several antiviral agents that inhibit the replication of enteroviruses.

Pathogenesis, Immunity, and Clinical Features

It is widely assumed that hepatitis A virus, known to enter the body by ingestion, multiplies first in the intestinal epithelium before spreading via the bloodstream to infect parenchymal cells in the liver, but this has yet to be formally proven. Virus is detected (up to 10^8 virions per gram) in the feces, and at a much lower titer in blood, saliva, and throat, during the week or two prior to the appearance of the cardinal sign, dark urine. Virus levels decline rapidly soon after the peak of serum transaminase levels. Hence the patient's feces are most likely to transmit infection both during the week before and the week after the onset of jaundice. The incubation period of hepatitis A is about four weeks (range two to six weeks).

The clinical features of hepatitis A closely resemble those already described for hepatitis B (see Chapter 22: Hepadnaviruses and Hepatitis Delta). The onset tends to be more abrupt and fever is more common, but the constitutional symptoms of malaise, anorexia, nausea, and lethargy which comprise the prodromal (preicteric) stage tend to be somewhat less debilitating and less prolonged. Hepatomegaly may produce pain in the right upper abdominal quadrant, followed by bilirubinuria, then pale feces and jaundice. Most infections worldwide occur in children, but these are generally subclinical, that is, asymptomatic, or anicteric, that is, symptomatic but without jaundice. The severity increases with age, about two-thirds of all infections in adults being icteric. The case–fatality rate is 0.5% (0.1% of all infections), and is the result of fulminant hepatitis leading to liver failure. The risk of fulminant hepatitis increases markedly in individuals over the age of 50. The illness of hepatitis A usually lasts about four weeks, but a minority relapse, typically after a premature bout of drinking or heavy exercise, and symptoms may continue for up to six months. Also, prolonged excretion of virus in feces has been observed and low levels of circulating virus may be detected in the serum by RT-PCR for several months. This finding is consistent with patterns of transmission of infection, including the occurrence of case clusters, sometimes associated with sharing of intravenous needles. However, in striking contrast to hepatitis B, all non-lethal infections resolve, with complete regeneration of damaged liver parenchyma, with no long-term sequelae or the development of a chronic carrier state.

The liver pathology in acute hepatitis A resembles that for hepatitis B. It is not striking until after viral replication has peaked and the immune response is underway, suggesting that hepatocellular injury is immunopathologically mediated. NK cells are mobilized and activated, as are CD8+ cytotoxic T lymphocytes, known to secrete interferon-γ that up-regulates expression of class I MHC protein on hepatocytes (a cell type that normally displays little of this antigen). The serum antibody response to hepatitis A virus is lifelong, declining significantly only in old age. As there is only one known serotype of hepatitis A virus, infection leads to lifelong immunity and second attacks of the disease are unknown.

Laboratory Diagnosis

Markedly elevated serum alanine and aspartate aminotransferase (ALT and AST) levels help to distinguish viral from non-viral causes of hepatitis, but do not discriminate between viral hepatitis A, B, C, D, and E. A single serological marker, IgM anti-HAV antibodies, is diagnostic for acute hepatitis A infection, and detection of this marker by EIA is

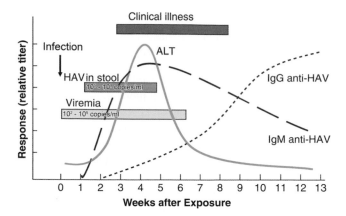

FIGURE 32.5 Typical sequence of events in acute hepatitis A virus infection. Biochemical markers (e.g., serum ALT levels, green line) and clinical symptoms follow the detection of hepatitis A virus RNA by real-time PCR in the stool (orange box) and serum (aqua box), and with the immune response as measured by anti-HAV IgM and IgG antibodies in sera detected by enzyme immunoassay. Three to four weeks after exposure to HAV, there is a rise in ALT and the onset of symptoms. Hepatitis A virus RNA can be detected in the blood and in the stools prior to the onset of clinical symptoms, and can remain detectable for varying times during convalescence. The IgM anti-HAV antibody titer peaks during the symptomatic phase and thereafter declines to baseline values within 3 to 6 months of symptomatic disease. IgG anti-HAV are detected for many years. ALT: Alanine aminotransferase. *Adapted from Vanessa Salete de Paula, 2012. Future Virology, 7(5), 461–472.*

the usual method of choice for diagnosis. IgM antibodies can be measured from the time symptoms and signs appear until about three to six months later (Fig. 32.5).

RT-PCR testing, based on the conserved 5′ non-coding region of the genome, can be used to detect virus RNA. It is usually used to test serum or plasma as fecal samples may contain interfering factors. It is positive earlier than IgM antibodies and may persist for months after recovery. RT-PCR is also used for the detection of trace amounts of hepatitis A virus in contaminated food or water. Virus isolation can be attempted early in infection if this is desired for research purposes; virus may be recovered from the feces in primary or continuous lines of primate cells, derived from monkey kidney or from human fibroblasts or hepatoma.

Epidemiology

Hepatitis A is an infection of humans and there is no known animal reservoir. Similar to poliovirus, the virus is spread via the fecal–oral route. As might be expected, therefore, the disease is hyperendemic in developing countries of Asia and Africa where overcrowding, inadequate sanitation, and poor hygiene are major problems. Where living conditions are poor, infection, usually subclinical, is acquired in early childhood, so that virtually all adults have protective antibodies. Clinical cases are infrequent, and most of those that do occur are seen in children or young adults, and in non-immune visitors from the more developed countries. Direct person-to-person contact spread is most important, but contaminated food and water are also major vehicles of spread. Major common-source outbreaks may occur, particularly when wells or other communal water supplies become polluted with sewage.

In highly developed countries, by contrast, the epidemiological picture is one of decreasing endemicity: the incidence of clinical disease has been gradually declining for decades, and only a minority of the population, notably the elderly, have antibodies. Since primary infections tend to be more severe with increasing age, the clinical disease is seen mainly in younger or middle-aged adults. Infection tends to be more prevalent in areas without mains sewerage and in lower socioeconomic groups, e.g., among American Indians and Hispanics in the Southwestern states of the United States, as well as those with high-risk behavior patterns, notably intravenous drug users and male homosexuals. In some Western nations most sporadic cases are seen in travelers returning from hyperendemic countries.

A third or intermediate pattern is seen in countries with more recently improved standards of living, for example much of South America, the Middle East, and Asia. In these areas virus still circulates widely, but less efficient transmission means that infection occurs later in life and clinical hepatitis among older children and adults is more common.

Outbreaks, which can continue in a community for months or even years, frequently originate in communal living establishments with marginal standards of hygiene or conditions that favor fecal–oral spread, such as children's day-care centers, homes for the mentally retarded, mental hospitals, prisons, and army camps. Infected handlers of food, especially uncooked or inadequately heated food such as salads, sandwiches, and berries represent a particular danger in fast-food outlets, restaurants, etc. Uncooked food grown using human feces as fertilizer constitutes another source. Hepatitis A virus can survive for months in water, hence sewage contamination of water supplies, swimming areas, or farms growing molluscs such as oysters and clams, can also lead to explosive outbreaks. Special problems arise in times of war or natural disaster.

Control

As hepatitis A is transmitted almost exclusively via the fecal–oral route, control rests on heightened standards of public and personal hygiene. Reticulated drinking water supplies and efficient modern methods of collection, treatment, and disposal of sewage (see Chapter 14: Control, Prevention, and Eradication) should be the objective of every local government. Where this is impracticable, as in remote rural areas, particular attention needs to be given to the siting, construction, maintenance, and operation of communal drinking water supplies such as wells; for example, these should not be downhill from pit latrines because of the risk of seepage, particularly after heavy rains. Public bathing and the cultivation of shellfish for human consumption should not be permitted near sewerage outlets.

Those employed in the dispensing of food should be required to observe high standards of hygiene, especially hand-washing after defecation. As far as children often contracting the infection at school, attention should be given to proper instruction of children in this matter. Especially difficult problems occur in day-care centers and homes for mentally retarded children. Routine precautions include separate diaper-changing areas, and frequent chemical disinfection of fecally contaminated surfaces and hands.

Passive immunization against hepatitis A is a well-established procedure to provide rapid protection that has been in use for many years. Normal pooled human serum globulin has up to this time contained sufficient anti-HAV antibodies to confer protection, and it was standard practice to immunize visitors or troops venturing into endemic areas using normal immunoglobulin prior to arrival and every four to six months thereafter. However this situation is now better dealt with using active immunization unless immediate protection is required. Normal immune globulin (0.02 ml/kg) is also effective in protecting family and institutional contacts following outbreaks in creches, schools, and other institutions, provided it is given within two weeks of exposure.

Both inactivated and live attenuated hepatitis A virus vaccines are in routine use for active immunization. The most widely used are formalin-inactivated preparations of virions grown by adaptation to the MRC-5 strain of human fibroblasts, and then adsorbed to alum as an adjuvant. Notably, the hepatitis A virus vaccine was the first human vaccine product to receive a licence before the completion of efficacy studies due to the overwhelming evidence as to the protective efficacy of anti-HAV antibodies. The vaccine is highly immunogenic and two doses injected 6 to 12 months apart regularly elicit an excellent immune response in 99% of recipients that is long lasting, possibly lifelong. The vaccine has an excellent safety profile based on experience of several hundred million doses. However, the resulting titers are usually below the detection limits of the routinely available commercial anti-HAV antibody tests, and hence routine serological testing to assess immunity after vaccination is of little value. Because the yield of virus from cultured cells is low, the vaccine is expensive and its use has generally been restricted to specialized target groups. These groups include: travelers or long-term visitors to countries in which hepatitis A virus is endemic; those whose occupation may put them at risk (military personnel, sewage workers, primate handlers, workers in preschool day-care centers, certain staff, and long-term residents of hospitals and institutions for the intellectually disabled, workers engaged in food manufacturing and catering, sex workers); those whose lifestyle puts them at risk (sexually active homosexual men, intravenous drug users), and those with chronic liver disease or chronic infection with hepatitis B or hepatitis C viruses; children in indigenous communities where hepatitis A virus is hyperendemic. Pre-vaccination testing is recommended in individuals from risk groups to exclude those already immune. More recently, universal vaccination of children under the age of one has been adopted in some countries, particularly those with intermediate endemicity. However, in many hyperendemic countries, clinical disease due to hepatitis A does not rank highly among other health priorities. It has been considered that, if post-vaccination immunity does not protect for life, but simply delays natural infection from childhood, when it is usually subclinical, to adulthood when it usually causes disease, then the use of vaccine in childhood could theoretically lead to an increase in adult disease; however, no evidence of this has been noted to date.

Several live vaccines based on attenuated strains are used in China, where more than 10 million doses are used annually in national immunization programs. These vaccines, given parenterally rather than administered orally, have a good record of efficacy and safety. However in general it has proved difficult to find the level of adequate attenuation without losing immunogenicity, perhaps because hepatitis A virus replicates mainly in hepatocytes.

There has been little progress in developing recombinant hepatitis A virus vaccines. However a number of combination vaccine formulations are available, for example combined hepatitis A/hepatitis B vaccines, or hepatitis A virus and typhoid Vi capsular polysaccharide vaccines, for use in appropriate settings.

FURTHER READING

Esposito, S., et al., 2015. Enterovirus D68 infection. Viruses. 7, 6043–6050.

Jacobs, S.E., Lamson, D.M., St. George, K., Walsh, T.J., 2013. Human rhinoviruses. Clin. Microbiol. Rev. 26 (1), 135–162.

Morbidity and Mortality Weekly Report (2011). Update on vaccine-derived polioviruses. 60(25): 846–850.

Nathanson, N., Kew, O.M., 2010. From emergence to eradication: the epidemiology of poliomyelitis deconstructed. Am. J. Epidemiol. 172, 1213–1229.

Chapter 33

Caliciviruses

It was not until 1972 that the first of the many types of viruses involved in gastroenteritis was identified, by immunoelectron microscopy (IEM; Fig. 33.1). It had been recognized for many years, however, that gastroenteritis could be transmitted by ingestion of bacteria-free filtrates of feces from patients with diarrhea. Norwalk virus, named after the town in Ohio that first recorded the virus, was the prototype of a succession of "small round-structured viruses" that were identified in the stools of young children in the United Kingdom, in Sapporo in Japan, and other centers. The taxonomic status of these agents remained unclear until 1991 when the sequencing of isolate genomes demonstrated all belonged to the family *Caliciviridae*. Caliciviruses had been well-known for years to veterinary virologists interested in major animal pathogens such as vesicular exanthema of swine virus and feline calicivirus. From the mid-1970s, other small round virions with a somewhat different, star-shaped outline had also been identified from gastroenteritis outbreaks in animals or humans; the identity of these distinct agents was confirmed in 1995 when these "astroviruses" were accorded the status of a distinct family, *Astroviridae* (see Chapter 34: Astroviruses).

The identification and characterization of many new viral agents causing gastroenteritis make clear that a significant proportion of enteric illnesses previously of unknown etiology are due to virus infection. Human caliciviruses are now the leading cause of gastroenteritis outbreaks from contaminated food in developed countries, and, along with rotaviruses and astroviruses, a major cause of severe childhood diarrhea worldwide.

CLASSIFICATION

Although human caliciviruses along with most other viruses causing gastroenteritis were initially discovered by electron microscopy, all defied cultivation. This meant initial attempts at classifying caliciviruses were based on distinct differences in appearance by electron microscopy, some of which differences are now thought to be due to bound coproantibody or partial proteolytic degradation. Initially, human viruses associated with disease outbreaks were also named by the location where each was first identified, for example

Norwalk, Manchester, Southampton, Lordsdale, Hawaii. Comparison of sequences has now led to identification of five genera within the family *Caliciviridae*, two of which (*Norovirus* and *Sapovirus*), contain human pathogens. The genus *Sapovirus* was named after the type species isolated from Sapporo, Japan, in 1977. The remaining three genera include important pathogens of animals; *Vesivirus* (vesicular exanthema of swine), *Lagovirus* (rabbit hemorrhagic disease) (RHD), and *Nebovirus* (viruses of calves) (Fig. 33.2).

Calicivirus genome sequences show considerable variation. Noroviruses are divided into six genogroups (GI to GVI), with viruses known to infect humans falling in genogroups I, II, and IV. Other noroviruses principally infect domestic animals and rodents, there being serological evidence that some animal strains may be able to infect humans. Within genogroups, various genotypes are distinguished based on a minimum sequence identity of 85% in the VP1 capsid protein. Genotypes are designated by Arabic numbers, such that a genogroup II, genotype 4 virus is designated GII.4. New variants can emerge as a result of genetic drift, and also due to recombination between genotypes, with the extent of mutation possibly correlating with population infection rates. Similarly, human sapoviruses fall into genogroups GI, GII, GIV, and GV. There are epidemiologic differences between the sapoviruses, which tend to be associated with gastroenteritis in infants, and noroviruses, which cause infections mainly in adults and older children.

VIRION PROPERTIES

The family *Caliciviridae* derives its name from the cup-shaped surface depressions that give the virion its unique appearance. The virion is 30 to 38 nm in diameter, slightly larger than the picornaviruses. The icosahedral capsid is constructed from 90 capsomers, each being a dimer of the major structural protein VP1. VP1 is composed of S (shell) and P (protrusion) domains, the latter forming dimeric arches and containing a P2 subdomain with sites involved in antigenic variation and receptor binding. Although relatively resistant to inactivation by heat (60°C), stomach acid (pH 3), or minimal levels of chlorination of drinking water, the virions of several caliciviruses are particularly

Fenner and White's Medical Virology. DOI: http://dx.doi.org/10.1016/B978-0-12-375156-0.00033-3

vulnerable to proteolysis and tend to lose capsid definition on purification or storage.

The plus sense linear ssRNA genome of 7.5 to 7.7 kb, similar to the picornaviruses, has a VPg protein covalently attached to the 5′ terminus and a polyadenylated 3′ terminus. The norovirus genome contains three ORFs. ORF1 encodes a non-structural polyprotein that is cleaved by the virus-coded protease to generate an ATPase/helicase, VPg, a protease and an RNA-dependent RNA polymerase. ORF2 encodes VP1, the major capsid protein, and ORF3 encodes VP2, the minor structural protein (Table 33.1). The sapovirus ORF1 encodes both the non-structural polyprotein and the major capsid protein within the same reading frame, and the minor structural protein VP2 is encoded by ORF2 (Fig. 33.3).

Virus Replication

Many caliciviruses show a narrow host range and are thus difficult to propagate in cell culture, and hence information about virus replication is sketchy for many caliciviruses. Recent progress has been assisted by the separate development of a human norovirus replicon system, the discovery of cultivable murine noroviruses, and finally the use of transient transfection assays.

Carbohydrate histo-blood group antigens present on epithelial cell surfaces act as norovirus receptors. Histo-blood group antigen expression is affected by an individual's ABO, Secretor and Lewis blood group antigen status, and there is an association between the pattern of histo-blood group

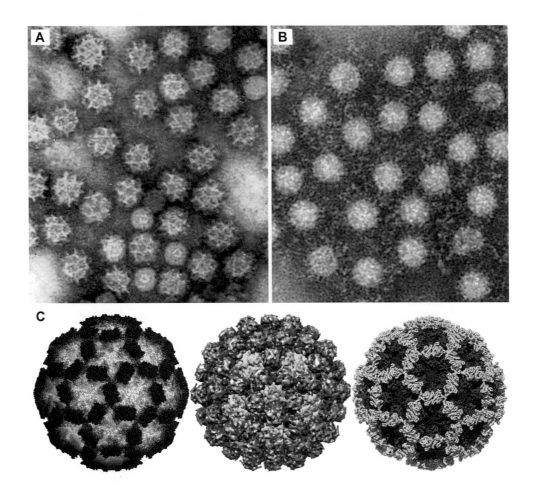

FIGURE 33.1 Calicivirus morphology and structure. Virions are non-enveloped, about 40 nm in diameter, with a capsid composed of 90 dimers of the major structural protein VP1. These form cup-shaped depressions or calices at each of the three- and five-fold axes of symmetry of the icosahedral capsid, giving a "Star of David" appearance. However, in some caliciviruses these depressions appear well defined, while in others they are hardly visible. (A) Negative contrast electron microscopy of Norwalk virus (norovirus)—virions free of antibody with characteristic surface pattern, said to resemble calices—hence the family name *Caliciviridae*. (B) Immunoelectron microscopy (IEM) of a norovirus with the surface pattern of virions obscured by bound antibody (homologous convalescent IgG). (C) Various models of norovirus—the different levels of resolution obtained and especially the various programs used to analyze X-ray crystallographic and cryoelectron microscopy images produce the many different virion graphics seen—in some instances it is difficult to discern that the same virus is illustrated. This is most notable with the small icosahedral RNA viruses, such as the caliciviruses. *(C) From Jean-Yves Sgro and B. V. Venkatar Prasad, with permission.*

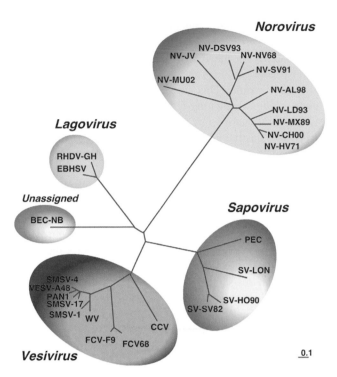

Norovirus

NV-DSV93
NV-JV   NV-NV68
            NV-SV91
NV-MU02
            NV-AL98
            NV-LD93
            NV-MX89
            NV-CH00
            NV-HV71

Lagovirus

RHDV-GH
EBHSV

Unassigned

BEC-NB

Sapovirus

PEC

SV-LON

SV-HO90
SV-SV82

SMSV-4
VESV-A48
PAN1
SMSV-17
SMSV-1   WV
                    CCV
FCV-F9   FCV68

Vesivirus   0.1

FIGURE 33.2 Phylogenetic relationships within the family *Caliciviridae*. Full-length capsid amino acid sequences were used for the analysis and included representative strains from each genus in the family. NV, Norwalk virus; PEC, porcine enteric calicivirus; SV, Sapporo virus; CCV, cetacean calicivirus; FCV, feline calicivirus; SMSV, San Miguel sea lion virus; PAN, primate calicivirus; VESV, vesicular exanthema of swine virus; WV, walrus calicivirus; BEC, bovine enteric calicivirus (now in the genus *Nebovirus*); RHDV, rabbit hemorrhagic disease virus; EBHSV, European brown hare syndrome virus. *Adapted from MacLachlan, N.J., Dubrovi, E.J., 2011. Veterinary Virology, fourth ed. Academic Press, London.*

antigen expression and susceptibility to particular strains of norovirus. Recent data indicate that noroviruses can bind to a wider range of blood group antigens than previously thought. Clathrin-dependent endocytosis then follows attachment.

Virus replication is confined to the cytoplasm. The full-length genomic RNA is translated to yield a protein precursor corresponding to ORF1, and this is subsequently cleaved to generate individual non-structural proteins. For

TABLE 33.1 Properties of *Caliciviridae*

Five genera: Two of these (*Norovirus* and *Sapovirus*) contain human pathogens
Non-enveloped spherical virion, 30 to 38 nm diameter, vulnerable to proteolysis
Icosahedral capsid with cup-shaped depressions; 90 capsomers, 1 major capsid protein VP1
Linear plus sense ssRNA genome, 7.5 to 7.7 kb, polyadenylated at 3′ terminus, protein VPg covalently linked to 5′ terminus, infectious
Three (norovirus) or two (sapovirus) open reading frames: 5′, non-structural proteins; middle, major capsid protein VP1; 3′, minor capsid protein VP2
Non-structural proteins are NTPase, cysteine protease, and RNA-dependent RNA polymerase, which have a sequence and motif similarity to poliovirus 2C, 3C, and 3D domains, respectively
Cytoplasmic replication; genomic RNA ORF1 is translated into polyprotein, cleaved into non-structural proteins; subgenomic mRNA encodes capsid proteins

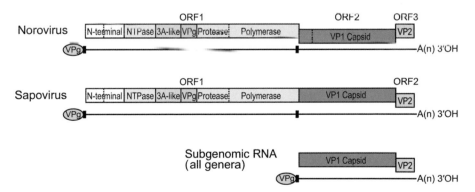

FIGURE 33.3 Genome organizations of viruses of the family *Caliciviridae*, genera *Norovirus* and *Sapovirus*. Sapoviruses contain a large ORF1 in which the non-structural polyprotein gene is continuous and in frame with the capsid protein (CP) coding sequence. Noroviruses encode the major structural CP in a separate reading frame (ORF2). The RNA helicase (HEL), protease (PRO), and polymerase (POL) regions of the genome are indicated. One RNA molecule corresponds in size to the full-length genome and the other, a subgenomic-sized RNA, is co-terminal with the 3′-end of the genome. The sgRNA is the template for translation of the major viral CP and the 3′-terminal ORF product that has been identified as a minor structural protein in some viruses. CP, capsid protein; Hel, helicase; Pol, polymerase; Pro, protease; sgRNA, subgenomic RNA; VPg, genome-linked protein. *Reproduced from King, A.M.Q., et al., 2012. Virus Taxonomy. In: Ninth Report of the International Committee on Taxonomy of Viruses, Academic Press, p. 808, with permission.*

most caliciviruses a separate subgenomic mRNA encodes the capsid protein VP1 (and with some caliciviruses the minor capsid component VP2). Because subgenomic minus sense RNA has also been reported, it is possible that the subgenomic mRNA as well as genomic RNA may be capable of replication via a complementary minus strand. Both genomic and subgenomic RNAs have the protein VPg covalently bound to each of the 5′ ends, which binds initiation factors involved in viral RNA translation and also acts as a peptide primer for replication. Genome replication then takes place in replication complexes associated with virus-induced membranous vesicles. Virions assemble in the cytoplasm, sometimes forming para-crystalline arrays associated with the cytoskeleton, and are released by cell lysis. Many details of calicivirus replication remain to be clarified, and further work could lead to the development of antivirals acting against unique targets such as VPg or the RNA-dependent RNA polymerase.

CALICIVIRUS (NOROVIRUS AND SAPOVIRUS) GASTROENTERITIS

Clinical Features

The disease caused by norovirus has been the most comprehensively studied, among volunteers as well as in natural outbreaks. Following a very short incubation period (generally 24 to 48 hours), there is an acute onset of nausea, vomiting, diarrhea, and abdominal cramps, with vomiting being most common in children and diarrhea among adults; headache, myalgia, and low fever are variable features. The severity varies widely from person to person and the duration ranges from 12 to 60 hours, but may be longer in infants less than one year of age. Most of the severe morbidity or mortality is seen at the extremes of age or in patients with co-morbidities, but recovery is complete with occasional complications of dehydration or aspiration of vomitus.

Pathogenesis, Pathology, and Immunity

Norovirus gastroenteritis is characterized by the sloughing of the tips of the villi in the jejunum, although these rapidly regenerate. There is infiltration of mononuclear cells and polymorphs into the mucosa, transient malabsorption, and delayed gastric emptying. The mechanism of diarrhea is not fully established, but suggestions include impaired absorption by the damaged mucosa, increased secretion by crypt cells, and reduced secretion of digestive enzymes leading to osmotic diarrhea.

Studies of immunity to noroviruses have produced conflicting results. Early volunteer studies showed short-term homologous immunity may develop following initial infection, but other studies found that susceptibility to illness was positively correlated with pre-existing antibodies. Indeed, when volunteers were challenged with norovirus twice, at intervals of two to three years, those who had contracted gastroenteritis on the first occasion and had developed an antibody response were the very ones who became sick on a second occasion. This suggested that (1) prior infection confers no significant long-term immunity, and (2) genetic or physiological factors may predispose certain individuals but not others to gastroenteritis following exposure to virus. It has since been shown that Norwalk virus-like particles (VLPs) bind strongly to the H type 1 and Lewis b carbohydrate blood group antigens, which function as virus receptors. Both antigens are expressed in some individuals, those possessing a functional α(1,2)-fucosyltransferase-2 (FUT2). Such individuals, known as secretors, are susceptible to norovirus, while persons who are homozygous recessives for the FUT2 gene and do not express H type 1 oligosaccharides (non-secretors), are resistant.

In contrast, antibodies to sapoviruses are generally acquired early in life, and in several studies the presence of antibodies was clearly correlated with resistance to illness (at least caused by genetically homologous sapoviruses), but not necessarily with resistance to reinfection. Susceptibility to sapovirus infection is not associated with particular histo-blood group antigen phenotypes.

Laboratory Diagnosis

Unfortunately, human caliciviruses are extremely fastidious and yet to be grown reproducibly in cultured cells. Early diagnostic approaches using electron microscopy or antigen detection using EIA have now largely been replaced by molecular assays including reverse transcriptase polymerase chain reaction (RT-PCR) where available. Many real-time RT-PCR assays have been developed, and multiplex assays for detecting noroviruses, sapoviruses, rotaviruses, astroviruses, enteric adenoviruses, and many bacteria are also commercially available. Sequencing of PCR products can then be undertaken for the purpose of strain identification, confirmation of the common source origin of case clusters, and for monitoring of the epidemiology and emergence of new strains. Isothermal amplification methods are also under development, including RT-loop-mediated isothermal amplification (RT-LAMP). In a possible outbreak situation, ready access to rapid diagnostic facilities is important for both assessment and ongoing management.

Epidemiology, Prevention, Control, and Treatment

Noroviruses are a common cause of epidemic gastroenteritis in both adult and pediatric age groups, and more sensitive molecular testing has shown noroviruses are also commonly involved in sporadic cases. Unlike bacterial gastroenteritis or food poisoning, norovirus infections occur year-round, with predominance in the winter months. About half of all adults have serological evidence of past infection. In developing countries these viruses are encountered earlier in life, with the pattern of endemic exposure resembling that of sapoviruses (see below). Common-source outbreaks of norovirus infection frequently occur via fecal contamination of food in restaurants, schools, camps, cruise ships, geriatric nursing homes, etc. Food contamination may occur either at the source of production or during food preparation. Several major outbreaks have been traced to consumption of uncooked or partially cooked oysters, cockles, clams, or other shellfish that have been harvested from sewage-polluted estuaries. Food-borne outbreaks have also been traced to the handling of foods that do not require cooking, for example salads and sandwiches, by infected food handlers.

Water-borne outbreaks have been attributed to the discharge of sewage into drinking water supplies, or into pools and lakes in which people swim. Secondary spread occurs from person to person within households or institutions, with transmission of up to two days after the resolution of symptoms. Secondary spread arising from contamination of the immediate environment sometimes occurs, as feces and vomitus can contain large amounts of virus and noroviruses are relatively resistant to environmental exposure: a small inoculum (less than 100 viral particles) is sufficient to initiate an infection. Aerosolized vomitus also can be a route of transmission. Nosocomial infections in hospitals and aged care facilities are a continuing problem, due to spread by many of the above mechanisms (Fig. 33.4). Water-borne outbreaks particularly tend to involve genotype I (GI) viral strains, while the predominating viral strains worldwide in food-borne and person-to-person transmission at present belong to genogroup II, genotype 4 (GII.4). GII.4 strains bind to a larger range of histo-blood group antigen genotypes, and it is proposed that this broader host range may be a factor in the epidemiological success of the GII.4 genotype.

Sapovirus infections are endemic worldwide, commoner in the winter, and cause disease more frequently in younger children than in older age groups; 90% of school age children have sapovirus antibodies. Similarly to, but less frequently than noroviruses, outbreaks have been reported in day-care centers, orphanages, maternity hospitals, schools, nursing homes, weddings, cruise ships, etc., with high attack rates. In surveys of acute sporadic gastroenteritis, however sapoviruses generally contribute significantly fewer cases than noroviruses.

The control of calicivirus gastroenteritis relies on preventing the contamination of food and water, application of strict hygiene by food handlers, and reducing secondary transmission by person-to-person contact. Applying such measures needs long-term national commitment and resources (see Chapter 14: Control, Prevention, and Eradication), which is out of the reach of many developing countries. Common-source calicivirus outbreaks can be reduced by enforcement of legislation relating to the preparation and supply of food for public consumption. This includes training food workers and managers in proper food safety practices, for example correct hand washing and not touching ready-to-eat food with bare hands; and requiring food workers with acute gastroenteritis to stay at home for two to three days after recovery from norovirus illness. Careful investigation of Norwalk virus outbreaks in the United States identified an ill food-handler as the probable source in 70% of instances. Of course viruses, unlike bacteria, cannot multiply in food or water. Because caliciviruses are inactivated at 60°C, well below normal cooking temperature, uncooked foods such as salads present the greatest risk. Shellfish are able

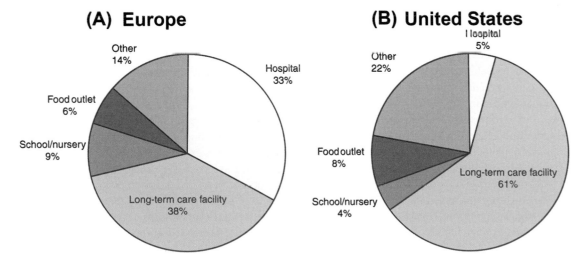

FIGURE 33.4 Setting of reported outbreaks of norovirus acute gastroenteritis in (A) Europe, 2004, and (B) the United States, 2006–08. In both settings, the majority of outbreaks occur in healthcare settings; in Europe many more outbreaks are reported from acute care hospitals. *Reproduced from Lopman, Ben A., Vinjé, J., 2014. Glass, Roger I. In: Viral Infections of Humans: Epidemiology and Control. Kaslow RA, Stanberry LR, LeDuc JW, editors. Springer, with permission.*

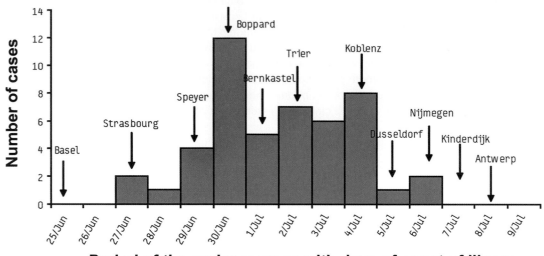

FIGURE 33.5   The typical problem in investigating cruise ship norovirus outbreaks: In 2006, the National Institute for Public Health and the Environment (RIVM) in the Netherlands was notified of an outbreak of gastroenteritis that had occurred during successive voyages of a river-cruise ship sailing through several European countries. The outbreak was one of a large cluster of cruise-ship-related outbreaks reported in Europe at that time, all of which had ascribed norovirus as the causative agent. A retrospective cohort study was performed among 137 people on board. Stool, food, water, and surface samples were collected: norovirus GGII.4-2006b was found to be responsible. *Adapted from Verhoef, L., et al., April–June 2008. Multiple exposures during a norovirus outbreak on a river-cruise sailing through Europe, 2006. Eurosurveillance 13, 4–6.*

to concentrate viruses from fecally contaminated water and are a notorious source of caliciviruses along with hepatitis A virus if served raw or inadequately cooked. Depuration, the prolonged flushing with clean water, is only partially effective in reducing the risk associated with live shellfish. Caliciviruses in drinking water can be inactivated using standard chlorination procedures. The US Environmental Protection Agency guidelines for municipal drinking water recommend a maximal residual chlorine concentration of 4 mg/L. Regular hand washing should also be insisted upon in settings that carry potential risk, for example on board cruise ships and in aged care facilities.

Management of an outbreak situation needs very prompt recognition and rapid diagnosis (Fig. 33.5). Patients with vomiting or diarrhea should be isolated (single room, gowns, and gloves), staff training should be immediately reinforced, and restrictions on ward admissions and movement of patients and visitors should be immediately considered and instituted where appropriate. Rapid surveillance of contacts by PCR testing can be considered. Environmental disinfection, particularly of high-touch surfaces, floors, and carpets, is recommended particularly following known high-level contamination, for example by vomitus, due to known extension of outbreaks by this route. Chlorine solutions at 1000 ppm are the preferred disinfectants. Similar principles

apply in managing outbreaks on cruise ships, and a ship experiencing an outbreak may be quarantined in port with restricted passenger movement until the outbreak has subsided.

It is clear that an effective norovirus vaccine would bring substantial public health and economic benefits. Candidate vaccines are based on virus-like particles (VLPs) assembled from recombinant capsid proteins that are morphologically and antigenically identical to native virus. Although the protective role of specific immunity following recovery is still not fully clear, at least 12 human phase I vaccine trials have been carried out using plant- or insect-cell-derived VLPs combined with various adjuvants. Systemic and mucosal antibody responses have been reported following both oral and intranasal administration. In several studies, a significant proportion of the vaccinees were shown to be protected against infection or against symptomatic disease when challenged later with live homologous or heterologous norovirus. Further work is focussed on the optimal strain composition of such vaccines. However, much more needs to be known about the long-term protection given by natural and vaccine-induced immune responses, and which are the most appropriate populations and age groups for possible vaccination. Furthermore, combination vaccines, including for example rotavirus and/or hepatitis E virus antigens, are under development and may ultimately be preferable for widespread immunization.

FURTHER READING

Glass, R.I., Parashar, U.D., Estes, M.K., 2009. Norovirus gastroenteritis. N. Engl. J. Med. 361, 1776–1785.

Oka, T., Wang, Q., Katayama, K., Saif, L.J., 2015. Comprehensive review of human sapoviruses. Clin. Microb. Rev. 28, 32–53.

Ramani, S., Atmar, R.L., Estes, M.K., 2014. Epidemiology of human noroviruses and updates on vaccine development. Curr. Opin. Gastroenterol. 30, 25–33.

Robilotti, E., Deresinski, S., Pinsky, B.A., 2015. Norovirus. Clin. Microb. Rev. 28, 134–164.

Tam, C.C., O'Brien, S.J., Tompkins, D.S., Bolton, F.J., Berry, L., Dodds, J., et al., 2012. Changes in causes of acute gastroenteritis in the United Kingdom over 15 years: Microbiologic findings from 2 prospective population-based studies of infectious intestinal disease. Clin. Infect. Dis. 54, 1275–1286.

Chapter 34

Astroviruses

The family *Astroviridae* was constituted in 1995 as representing a fourth family of non-enveloped viruses with a single-stranded, positive-sense RNA genome, along with *Picornaviridae*, *Caliciviridae*, and *Hepeviridae*. Astroviruses were first described in 1975 when small particles with star-like morphology were observed by electron microscopy in the feces of children with diarrhea. Human astroviruses are ubiquitous in young children and are one of the more important causes of pediatric gastroenteritis after rotaviruses and caliciviruses. Very recently, a number of new highly divergent human astroviruses have been described although these remain to be fully characterized.

PROPERTIES OF ASTROVIRUSES

Classification

Host-specific astroviruses were described from a wide variety of mammals and birds, mostly associated with gastroenteritis in younger individuals, soon after being first recognized in stool samples from young children. Two genera are now recognized with members grouped according to the hosts from which the virus was first isolated. Mamastroviruses infect mammals and comprise two large distinct genogroups, containing species that are defined by the human or animal host that they infect; eight different serotypes of human astroviruses have been recognized. Avastroviruses have been isolated from chickens, ducks, turkeys, pigeons, and guinea fowl. Phylogenies based on genetic relationships are presently being incorporated into the older host species-based classification, and a large number of newer isolates including many from bats have yet to be classified. In addition to the "classic" human astroviruses, since 2008 a number of highly divergent novel human astroviruses have been isolated from pediatric stool specimens in patients from Australia, the United States, Nigeria, Pakistan, and Nepal. Thus the number of astroviruses recognized to infect humans has nearly doubled in recent years, and the properties and significance of these newer viruses remain to be evaluated.

Virion Properties

The 28- to 30-nm spherical virion, though at first glance resembling a picornavirus, is in fact highly characteristic when viewed from a certain angle; a minority of the negatively stained particles reveal a five- or six-pointed star, without any stained central hollow. The capsid shell is formed by the VP34 protein, and proteins VP27/29 and VP 25/26 form dimeric spikes; all three proteins are derived by cleavage from the VP90 precursor (see below). The virion is resistant to pH 3 and to 60°C for 5 minutes (Fig. 34.1).

The human astrovirus genome is comprised of a single linear 7kb ssRNA molecule of positive sense, with a poly(A) tract at the 3′ terminus and VPg, a protein covalently bound to the 5′ end. The genome contains three open reading frames (ORFs): ORF1a and ORF1b at the 5′ end, which codes for non-structural proteins, and ORF 2 at the 3′ end, which codes for the structural proteins. Unlike the caliciviruses, astroviruses do not code for an RNA helicase, and a ribosomal frameshift between ORF1a and ORF1b allows for separate translation of the protease and RNA polymerase domains. The conserved N-terminal domain of the ORF2 protein contains one or more common epitopes that are recognized by monoclonal antibodies, while at least eight serotypes of human astroviruses can be distinguished with other monoclonal antibodies or by neutralization, mostly mapping to the hypervariable C-terminal region of ORF2 (Fig. 34.2; Table 34.1).

Viral Replication

Astroviruses were first isolated, in the presence of trypsin, in cultures of primary human embryonic kidney (HEK) cells and have been further adapted to growth in the LLC-MK2 continuous monkey kidney cell line. A wide variety of other cell lines can be infected with varying levels of efficiency, including Caco-2 and HEK293 cells. Following entry by clathrin-mediated endocytosis, the incoming virion RNA serves as a message for the non-structural proteins.

Fenner and White's Medical Virology. DOI: http://dx.doi.org/10.1016/B978-0-12-375156-0.00034-5

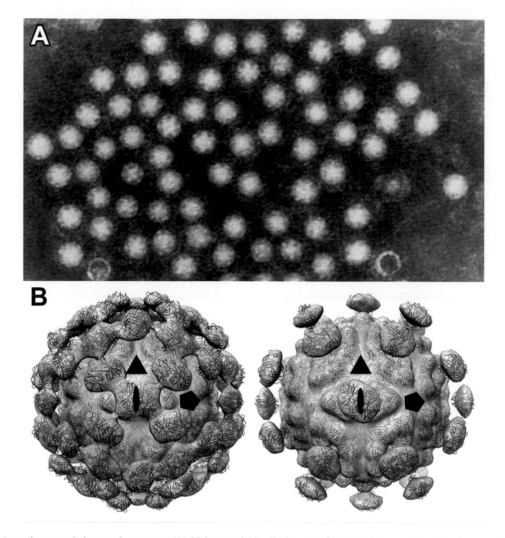

FIGURE 34.1   Astrovirus morphology and structure. (A) Virions variably display a surface morphology resembling five- or six-pointed stars, a manifestation of how the 180 copies of the capsid protein (CP) are arranged. (B) The virion undergoes dramatic structural change as it matures into the infectious form. (Left) Immature smooth-surfaced virion, exhibiting 90 dimeric spikes. The oval, triangle and pentagon symbols indicate the axes of 2-fold, 3-fold, and 5-fold icosahedral symmetry respectively. (Right) Mature infectious virion with 30 large projections at the icosahedral 2-fold axes. The immature particle has a remarkable resemblance to the hepatitis E virus particle. *Reproduced from York, R.L., et al., 2016. Structural, mechanistic, and antigenic characterization of the human astrovirus capsid. J. Virol. 90, 2254–2263, with permission.*

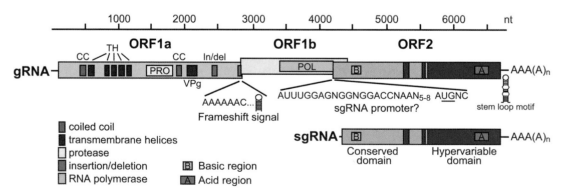

FIGURE 34.2   Genome organization and replication strategy of human astrovirus 1 (HAstV-1). The genome is arranged in three open reading frames (ORFs): ORF1a and ORF1b at the 5′-end coding for the non-structural proteins, and ORF2 at the 3′-end coding for the structural proteins. The virion RNA is infectious and serves as messenger RNA for the non-structural polyproteins. A frame-shift located between ORF1a and ORF1b is employed to translate the RNA-dependent RNA polymerase. A polyadenylated sgRNA corresponding to ORF2 is detected in the cytoplasm of infected cells. A full-length capsid polyprotein is translated from the sgRNA and forms particles in the cytoplasm of infected cells. In extracellular particles exposed to proteolysis this is cleaved into the three major capsid proteins resulting in an enhanced infectivity. Pol, polymerase; Pro, protease; panhandle, frameshift structure. *Reproduced from King, A.M.Q., et al., 2012. Virus Taxonomy, Ninth Report of the International Committee on Taxonomy of Viruses. Academic Press, with permission.*

TABLE 34.1 Properties of Astroviruses

The family comprises two genera, *Mamastrovirus* and *Avastrovirus*; *Mamastroviruses* include six species (*Human, Bovine, Feline, Porcine, Mink,* and *Ovine astroviruses*); *Human astrovirus* contains eight serotypes. Further novel human types have been recently described

Non-enveloped icosahedral virion, 28 to 41 nm, ~10% of virions resemble 5- or 6-pointed stars; 3 capsid proteins VP34, VP27/29, VP 25/26 in mature virion

Virion resistant to heat and acidity

Linear plus sense ssRNA genome, 6.4 to 7.7 kb, 3' polyadenylated, VPg at 5' end. Contains 3 open reading frames

Cytoplasmic replication; ORF1a and ORF1b at the 5' end of the genome encode the viral protease and RNA-dependent RNA polymerase respectively; a polyprotein nsp1ab is produced by ribosomal frameshift and then cleaved by a cellular protease

Sub-genomic mRNA (ORF2) encodes an 87- to 90-kDa polyprotein, which assembles into immature virions and is then successively cleaved both intracellularly and extracellularly into capsid proteins

This activity requires the presence of the 5' VPg. ORF1a gives rise to a protein product designated nsp1a, while a single long polyprotein nsp1ab is produced by ribosomal frameshift between ORF1a and ORF1b and is subsequently cleaved by the viral serine protease and cellular proteases to produce nsp1b, the RNA-dependent RNA polymerase and a number of other cleavage products.

In addition to the 7-kb genome, a 3'-coterminal 2.8-kb polyadenylated sub-genomic mRNA corresponding to ORF2 is synthesized. The VP90 polyprotein is synthesized from this sub-genomic RNA and its cleavage products make up the capsid proteins. The processing pathway of a human astrovirus serotype 8 has been described in detail. The primary polyprotein product is assembled into immature particles and then cleaved at the carboxyterminal region by intracellular caspases to a 70-kDa protein. The immature virions are then released from the cell, and further cleavages by extracellular trypsin yield proteins of 34, 27/29, and 25/26 kDa, leading to full activation of viral infectivity.

Less detail is known about replication of the viral genome, but it is presumed to occur using a full-length negative sense template. Assembly of immature virions is likely to take place in association with intracellular membranes. Clusters of virions can be seen accumulating in the cytoplasm in paracrystalline arrays. Finally, VP90 particles are processed to VP70 particles by cellular caspases and released into the medium before any further cleavages are initiated. Proteolytic cleavage of the non-infectious form exposes conserved epitopes involved in attachment to target cells and in eliciting the host immune response. This kind of post-assembly maturation is found with a number of viruses and has importance in the design of new antiviral drugs and vaccines.

CLINICAL FEATURES

Astrovirus gastroenteritis, seen principally in young children, is usually clinically mild. A mean incubation period of four to five days is followed by a watery diarrhea lasting from one to four days or more, abdominal discomfort, vomiting, fever, and anorexia. Vomiting is less common than in rotavirus or calicivirus infection, and the disease is usually milder. Disease is uncommon in healthy adults, but the elderly and immunosuppressed individuals are at risk. The full disease potential of the novel non-classic astroviruses remains to be clarified. Human astroviruses replicate primarily in the epithelial cells of the small intestine (Fig. 34.3), but systemic spread has been described in immunocompromised children. Very high levels of virus (up to 10^{13} genome copies/g) can be present in stools, and virus excretion can be detected for up to two weeks after symptoms have cleared.

Immunity probably lasts for many years, and most healthy young adults have antibodies to the prevalent classic human astrovirus serotypes and are thus resistant to further infection. Anamnestic responses develop following subsequent exposure to a heterologous serotype. However, the occurrence of symptomatic infections in the elderly suggests that immunity is not lifelong.

LABORATORY DIAGNOSIS

Although astroviruses are routinely found by electron microscopy in higher numbers in feces (~10^8 per gram) than are caliciviruses (<10^7 per gram), immune electron microscopy is still not optimally sensitive or convenient. Virus isolation using the human colon carcinoma cell line Caco-2 in the presence of trypsin has been used but is no longer practical for routine diagnosis. However, culture combined with reverse transcriptase polymerase chain reaction (RT-PCR) has been described, with a likely role for testing environmental samples or for investigating novel human astroviruses.

Antigen detection using commercially available EIA has been used widely for detecting classic human astroviruses; it requires approximately 10^5–10^6 particles/mL for a positive reaction, similar to electron microscopy, but is both less time consuming and is suitable for large batch testing. However, RT-PCR assays are now replacing many of the older approaches, and the greater sensitivity of molecular methods combined with the opportunity to simultaneously test for a number of gastroenteritis agents using multiplex assays, makes this an attractive choice.

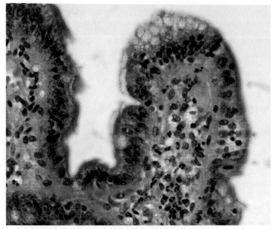

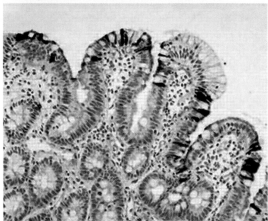

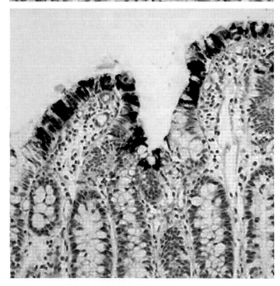

FIGURE 34.3 Jejunal biopsy from a bone marrow transplant recipient with astrovirus infection, demonstrating villus blunting and inflammatory infiltrate in the lamina propria. (Top) Epithelial damage and rarefaction necrosis, most pronounced at the villus tips. H&E. (Middle and bottom at higher magnification) Immunohistochemistry shows large amounts of viral antigen in the cytoplasm of epithelial cells, especially at the villus tips. *Reproduced from Sebire, N.J., et al., 2004. Pathology of astrovirus associated diarrhoea in a paediatric bone marrow transplant recipient. J. Clin. Pathol. 57(9):1001–1003, with permission.*

EPIDEMIOLOGY

Astroviruses are transmitted by the fecal–oral route; contaminated food and water are major sources, and person-to-person spread is also likely to be common. Astroviruses remain viable in the environment for long periods, and nosocomial transmission via fomites, including utensils for food preparation, is important in the case of institutional spread. Infections are endemic throughout the world causing a relatively mild form of gastroenteritis, mainly in young children under five years of age. In common with rotaviruses, infections occur year-round, but with a peak in winter in temperate regions and during the rainy season in the tropics. A majority of children have developed protective antibodies before entering school. Many infections may be sub-clinical, but in many studies 2 to 9% of infantile acute non-bacterial diarrhea is attributable to astroviruses. Epidemics also occur in the community, in day care centers and kindergartens, and nosocomially in pediatric wards, etc. Most children and adults are immune, but outbreaks have been recorded among immunosuppressed and geriatric patients in hospitals and nursing homes.

Serotype 1 is the most prevalent strain, representing around 50% of isolates in many countries, but other serotypes predominate in some studies.

PREVENTION

A vaccine is not yet available, although systems for production of virus-like particles (VLPs) have been described. Thus, measures to reduce fecal–oral transmission particularly between infants and young children are the mainstay of prevention and control. These include hand washing, care of nappies and soiled material, and the use of disinfectants to reduce spread by fomites. This is particularly important in institutions where outbreaks may occur, such as day care centers, institutions for the elderly, and hospitals with children and immunocompromised patients. In one study, the use of soapy water and ethanol wipes reduced the rate of diarrhea in a unit for immunocompromised children. Such measures of course apply equally to control of rotavirus and calicivirus transmission. Public health measures such as virus detection and inactivation in water and food also play a role.

FURTHER READING

Bosch, A., Pinto, R.M., Guix, S., 2014. Human astroviruses. Clin. Microb. Rev. 27, 1048–1074.

Chapter 35

Togaviruses

Epidemic equine encephalitis was first recorded in 1831 when an outbreak in Massachusetts was described. Since then, encephalitis in humans and horses caused by eastern equine encephalitis virus has been described all along the Atlantic seaboard of North America. A similar disease caused by western equine encephalitis virus occurs mainly in North American states/provinces west of the Mississippi River. This latter virus was first isolated from a horse in the San Joaquin Valley of California in 1931 by Karl Friedrich Meyer and his colleagues, who early on appreciated the role of mosquitoes in the viral transmission cycle and the risk to humans as well—in 1938 both viruses were isolated from human cases of encephalitis occurring in the same regions as equine cases. In 1936 an epidemic of encephalitis in both horses and humans occurred in Venezuela; the virus isolated was not neutralized by antibodies to the two known viruses and was thus named Venezuelan equine encephalitis virus. In succeeding years many other arboviruses were characterized; those similar to eastern, western, and Venezuelan equine encephalitis viruses were brought together in the family *Togaviridae*, genus *Alphavirus*.

Most togaviruses exist in geographically restricted habitats where defined species of mosquitoes and vertebrate hosts influence virus survival, geographic extension, overwintering, and amplification. Vertebrate reservoir hosts include wild birds and mammals, domestic animals and humans are usually not involved in primary transmission cycles in nature, although they may play roles in geographical extension and amplification events that lead to epidemics. For example, during epidemics of Venezuelan equine encephalitis a mosquito–horse–mosquito transmission cycle is responsible for explosive spread and spill-over into considerable numbers of humans.

Chikungunya (the name derived from the Makonde word meaning "that which bends up" in reference to the stooped posture resulting from the arthritis/arthralgia) was first recognized in 1952–53 in what is now Tanzania; the first urban outbreaks were recorded in Thailand and India in the 1960s–70s. A major emergence began in coastal Kenya in 2004 and moved to islands in the Indian Ocean where more than half a million cases occurred in the following two years. The virus then spread to India and throughout Southeast Asia and in 2013 to the Caribbean Islands and from there to North, Central, and South America. This dramatic geographic spread, with an estimated 7 million cases, has put a substantial economic drain on public health resources of affected countries.

In 1967 the similarities between rubella virus and the then known togaviruses prompted the question whether rubella virus should be considered an arbovirus, clearly indicative of faults in virus classification at that time; rubella virus is now classified as a non-arthropod-borne member within the family *Togaviridae*, genus *Rubivirus*.

PROPERTIES OF TOGAVIRUSES

Classification

Two genera have been recognized in the family *Togaviridae* based on virion properties and serologic relationships, the genera *Alphavirus* and *Rubivirus*. The latter contains a single member, rubella virus, and this infects only humans. There are at least 29 species within the genus *Alphavirus*; 10 can cause epidemics in humans and several others produce occasional disease (Table 35.1). Individual alphaviruses were first distinguished serologically, most importantly by neutralization tests, and now genetically, through genome sequencing.

Virion Properties

Alphavirus virions are spherical, enveloped particles 70 nm in diameter. The virion envelope contains heterodimers of the two envelope proteins E1 and E2 that form a regular icosahedral surface lattice; this envelope surrounds an icosahedral nucleocapsid that is 40 nm in diameter (Fig. 35.1). Enveloped virions and nucleocapsids have icosahedral ($T=4$) symmetry.

Rubivirus virions are slightly smaller, just over 60 nm diameter, and have T= 3 symmetry.

The RNA genome is a single molecule of linear, positive-sense, single-stranded RNA, 9.7–11.8 kb in size, with a 5′-methylated cap and a 3′-terminal poly-A tail (Fig. 35.2). The 5′ two-thirds of the genome codes for four non-structural proteins designated nsP1–4, that code for

Fenner and White's Medical Virology. DOI: http://dx.doi.org/10.1016/B978-0-12-375156-0.00035-7

TABLE 35.1 Diseases Caused by Alphaviruses

Virus	Geographic Distribution	Vertebrate Reservoir	Vector Mosquitoes	Disease	Epidemiologic Features
Eastern equine encephalitis	North and South America, Caribbean	Wild birds	*Culiseta melanura, Coquillettidea perturbans, Aedes* spp.	Encephalitis	Periodic small equine epizootics and human epidemics
Western equine encephalitis	North and South America—west of Mississippi	Wild birds	*Culex tarsalis*	Encephalitis	Periodic equine epizootics and human epidemics
Venezuelan equine encephalitis types IAB, IC	Venezuela, Colombia, Peru, Ecuador	Horses (epizootic)	*Aedes, Mansonia, Culex* spp. (epizootic subtypes)	Encephalitis	Periodic equine epizootics and human epidemics
VEE types II, ID, IE, IF	Enzootic subtypes in Florida, Central, and South America	Rodents (enzootic)	*Culex (Melanoconion)* (enzootic subtypes)		Subclinical in horses and humans
Chikungunya	Africa, south Asia, Philippines, Indian Ocean islands, Caribbean, North and South America	Monkeys, humans	*Aedes aegypti, Aedes furcifer-taylori, A. albopictus*	Fever, arthralgia, myalgia, rash	Urban epidemics like dengue, enzootic cycle-like yellow fever
O'nyong-nyong	East Africa	Unknown	*Anopheles funestus, Anopheles gambiae*	Fever, arthralgia, myalgia, rash	Large outbreak in 1959–62, again in Uganda in 1996–97
Mayaro	Tropical South America	Monkeys	*Haemagogus* sp.	Fever, arthralgia, myalgia, rash	Sporadic cases, limited outbreaks associated with deforestation
Ross River	Australia	Marsupials, rodents	*Culex annulirostris, Aedes vigilax*	Arthralgia, myalgia, rash	Annual outbreaks in humans and horses
	South Pacific islands	Humans	*Aedes polynesiensis*		Occasional introductions and epidemics
Barmah Forest	Australia	Marsupials, rodents	*Culex annulirostris, Ochlerotatus* spp.	Arthralgia, myalgia, rash	Annual outbreaks in humans
Sindbis	Africa, Asia, Europe, Australia	Wild birds	*Culex* sp., *Culiseta* sp., *Aedes* sp.	Fever, arthralgia, myalgia, rash	Sporadic cases, sometimes epidemics

Source: Data modified from Calisher, C.H., Monath, T.P., 1988. *Togaviridae* and *Flaviviridae*: the alphaviruses and flaviviruses. In: Lennette, E.H., et al. (Eds.), Laboratory Diagnosis of Infectious Diseases. Principles and Practice. Vol. II. Viral, Rickettsial and Chlamydial Diseases. Springer-Verlag, New York, NY, p. 414.

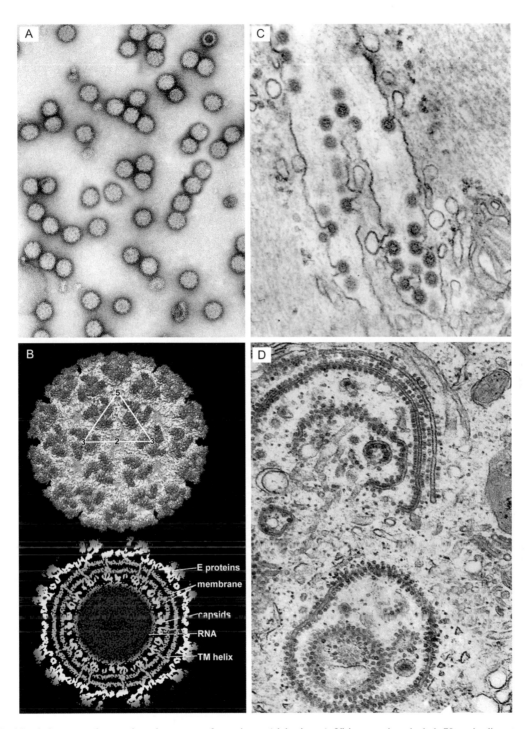

FIGURE 35.1 Morphology, morphogenesis and structure of togaviruses (alphaviruses). Virions are icosahedral, 70 nm in diameter, with a surface projection layer, an envelope and a 40 nm core surrounding the RNA genome. (A) Negative contrast electron microscopy of Sindbis virus, showing minimal evidence of envelope projections. (B) Model of chikungunya virus, showing (top) the surface structure, consisting of heterodimers of the E2 and E1 proteins that form a regular icosahedral surface lattice; and (bottom) a cross-section showing all the layers, as analyzed by image reconstruction from cryoelectron micrographs. (C) Ross River virus (strain NB5092) budding from the plasma membrane of a striated muscle cell in the leg of a mouse five days post-infection. (D) Eastern equine encephalitis virus infection, mouse brain at 48 hours post-infection—often an excess of nucleocapsids is produced on intracytoplasmic membranes. *(A) From Carl von Bonsdorff. (B) from Michael Rossmann: reproduced from Sun, S., et al., 2013. Structural analyses at pseudo atomic resolution of Chikungunya virus and antibodies show mechanisms of neutralization. Elife 2, e00435, with permission.*

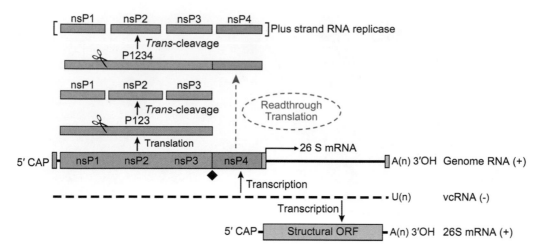

FIGURE 35.2 Togavirus genomes and coding strategies. The positive sense RNA genomes of alphaviruses and rubella virus (11.7 and 9.8 kb in length, respectively), are shown in the center, underneath which is shown the negative-sense complementary RNA which acts as the template for transcription of sub-genomic RNA. The first two-thirds is translated to produce the polyproteins P123 and P1234 which are processed to produce the non-structural proteins, nsP1–4. The structural proteins [the capsid protein (C), the two surface envelope glycoproteins (E1 and E2) and two small peptides (E3 and 6K) which serve as leader peptides for E2 and E1 respectively], encoded in the 3′-terminal one-third of the genome, are produced from a sub-genomic (26S) mRNA that is copied from a full-length replicative intermediate (–) strand RNA (vcRNA(–)). When low levels of P123 are present, cleavage of P1234 and read-through translation generates the nsP4 replicase (polymerase) of the virus. As the level of the protease P123 rises in the infected cell there is a shift to production of primarily positive sense full genome length RNA. Eventually, only positive sense RNA, new RNA genomes as well as the sub-genomic mRNA encoding the structural proteins is formed. *Reproduced from King, A.M.Q., et al., 2012. Virus Taxonomy, Ninth Report of the International Committee on Taxonomy of Viruses. Academic Press, London, with permission.*

the enzymes involved in virus replication. The 3′-one-third is not translated from the RNA genome, but is expressed via a 26S sub-genomic mRNA molecule, which itself is transcribed from a full-length negative-sense RNA replicative intermediate. This 26S sub-genomic mRNA codes for five structural proteins including a nucleocapsid protein (C, Mr 30–33,000), two envelope glycoproteins (E1 and E2, Mr 45–58,000), and two smaller peptides (E3 and 6K) not present in all alphaviruses. Virions contain lipids derived from host cell membranes and carbohydrates as side-chains on the glycoproteins. The viruses are not stable in the environment and are readily inactivated by disinfectants.

Viral Replication

Alphaviruses replicate to a high titer and cause severe cytopathic changes in many kinds of cells, for example Vero (African green monkey kidney), BHK-21 (baby hamster kidney), and primary chick and duck embryo cells. Alphaviruses also grow in mosquito cells without cytopathic changes, such as C6/36 cells derived from *Aedes albopictus*. In mammalian and avian cells, infection causes a complete shut-down of host cell protein and nucleic acid synthesis. In mosquito cells there is no shut-down and cell division is unaffected by infection.

Viral attachment to the host cell first involves interaction between the viral E1 glycoprotein with phospholipid receptors on the cell surface; entry involves binding of viral E2 glycoprotein to cellular proteins. This is followed by receptor-mediated endocytosis. Entry of viral nucleocapsids into the cell cytoplasm occurs by fusion of the viral envelope with the endosomal membrane. This fusion is triggered by low pH which disrupts E1/E2 dimers, leading to exposure of a fusion peptide in E1 which is otherwise masked by E2.

Upon entry into the cytoplasm, the virion RNA characteristically directs two rounds of translation. First, the 5′-end of the genomic RNA, serving as mRNA, is translated to produce a polyprotein which is then cleaved into four non-structural proteins, two forming the viral RNA-dependent RNA polymerase. This enzyme directs the transcription of full-length negative-sense (complementary) RNA that serves as a template for further positive-sense RNA synthesis. Two positive-sense RNA species are synthesized, full-length genomic RNA for amplification of the infection in the cell and for inclusion in progeny virions, and a 26S sub-genomic mRNA. Translation from the sub-genomic RNA template results in the production of large amounts of a polyprotein which is cleaved to form the individual structural proteins. Nucleocapsids are assembled in the cytoplasm upon endoplasmic reticulum membranes, and move to the plasma membrane where nucleocapsids align under patches containing viral glycoprotein peplomers. Finally, virions are formed by budding of the nucleocapsids through the peplomer-studded plasma membrane patches (Table 35.2).

TABLE 35.2 Properties of *Togaviridae*

Two genera: Alphavirus and Rubivirus

Spherical virion, enveloped, with an icosahedral surface lattice composed of E1 and E2 heterodimers, diameter 70 nm (*Alphavirus*), >60 nm (*Rubivirus*)

Icosahedral capsid, diameter 40 nm

Linear, plus sense ssRNA genome, 11 to 12 kb (*Alphavirus*), 10 kb (*Rubivirus*); 5′ capped, 3′ polyadenylated, infectious

Genes for non-structural proteins located at 5′ end of genome, those for structural proteins at the 3′ end

Envelope glycoprotein heterodimer E1+ E2 contains virus-specific neutralizing epitopes and alphavirus serogroup and subgroup specificities; nucleocapsid protein, C, with broadly cross-reactive alphavirus group specificity

Full-length and sub-genomic 26S RNA transcripts; post-translational cleavage of polyproteins

Cytoplasmic replication; budding from plasma membrane (alphaviruses) or intracytoplasmic membranes (rubella virus) of vertebrate cells, or from intracytoplasmic membranes of invertebrate cells, in which alphaviruses are non-cytocidal

DISEASES CAUSED BY MEMBERS OF THE GENUS *ALPHAVIRUS*

The majority of alphavirus infections are asymptomatic, with seroconversion as the only evidence of infection. However, when clinical disease is seen, it frequently takes the form of one of three characteristic syndromes: fever with non-specific influenza-like symptoms, acute neurological disease (encephalitis and meningitis), or a fever/rash/arthritis syndrome.

ALPHAVIRUSES CAUSING ENCEPHALITIS—EASTERN, WESTERN, AND VENEZUELAN ENCEPHALITIS

Clinical Features

Eastern equine encephalitis virus infection in humans is characterized by a four to ten day incubation period, followed by fever and drowsiness and neck rigidity. The disease may progress to confusion, paralysis, convulsions, and coma. The overall fatality rate among clinical cases is about 50 to 75% and many survivors are left with permanent neurological sequelae such as mental retardation, epilepsy, paralysis, deafness, and blindness. Western equine encephalitis virus is usually less severe: a high proportion of infections are silent or cause fever and myalgia only, and the case-fatality rate is about 3 to 10%. However it does cause high death rates for horses.

Venezuelan equine encephalitis virus (epidemic types) induces a systemic febrile illness in humans, and about 1%

of those affected develop clinical encephalitis. There is also abortion and fetal death when pregnant women are infected. In the absence of adequate medical care in less-developed areas, case-fatality rates as high as 20 to 30% have been reported in young children with encephalitis.

Pathogenesis, Pathology, and Immunity

Following viral entry via the bite of an infected vector mosquito, viral replication occurs in the vascular endothelium near the entry site and/or in regional lymph nodes. The resulting primary viremia allows virus to invade more distant sites including lymph nodes, bone marrow, spleen, and liver, where further viral replication provides the high titer secondary viremia, leading to both the invasion of target organs and the further spread of infection following fresh mosquito bites. Target organs may include muscle (myositis), connective tissue and joints (arthritis), skin (rash), central nervous system (encephalitis), and the reticuloendothelial system (especially dendritic reticulum and lymphoid cells). In the central nervous system infection involves neurons, but also the choroid plexus, ependyma, and meninges. Venezuelan equine encephalitis virus also infects the upper respiratory tract, pancreas, and liver.

Encephalitis due to alphaviruses results from the hematogenous spread of virus and subsequent entry of the central nervous system by one of several possible routes: (1) passive transfer of virus through the endothelium of capillaries in the central nervous system, (2) viral replication in vascular endothelial cells and release of progeny into the parenchyma of the central nervous system, (3) viral invasion of the cerebrospinal fluid with infection of the choroid plexus and ependyma, or (4) carriage of virus in lymphocytes and monocytes which may migrate into the parenchyma of the central nervous system. An alternative possibility, supported by experimental data, is that virus may replicate extensively in the olfactory epithelium in the nares, leading to invasion of the brain parenchyma via axonal spread to the olfactory bulbs. Once in the parenchyma of the brain there are no anatomical or physiological impediments to the dissemination of virus throughout the central nervous system. Typical pathology includes widespread neuronal necrosis with neuronophagia, intense perivascular and interstitial mononuclear inflammatory infiltration, and interstitial edema. The immunity that follows clinical or subclinical infection with an alphavirus probably lasts for life. Partial protection may also be conferred against serologically related viruses.

Laboratory Diagnosis

Diagnosis of the equine encephalitides has evolved, just as it has for all viruses, with serological methods giving way to reverse-transcription polymerase chain reaction (RT-PCR) assays for detection of viral RNA where these assays are available. IgM capture enzyme immunoassay is routinely employed to detect IgM antibodies in a single serum sample, or for detection of seroconversion (a fourfold or greater rise in antibody titer between acute and convalescent serum) an IgG antibody assay is used. Because the background prevalence of human eastern and western equine encephalitis in endemic areas is very low, a single serum demonstrating high-titer antibody is highly suggestive of recent infection. Where necessary, confirmation of positive antibody results is done by neutralization in cell culture, using paired sera. In complementary fashion, detection of viral RNA from the same samples through RT-PCR assay is now used in many laboratories.

The suckling mouse is a particularly sensitive host for virus isolation from clinical specimens (blood, brain) and is still used in some laboratories in the tropics. However, for practical reasons, in most instances cell cultures are used: Vero (African green monkey kidney) or BHK-21 (baby hamster kidney) cells and C6/36 (*A. albopictus* mosquito) cells are the most popular. Isolates are then identified by enzyme immunoassay. Where more than one alphavirus may be circulating in an area, isolate identification is confirmed by neutralization using reference monoclonal antibodies. This is extremely important in one particular circumstance: Venezuelan equine encephalitis viruses occur as epidemic and endemic types, so the identification of isolates to this

level of specificity is of considerable epidemiological importance. The types may be differentiated by special hemagglutination-inhibition assays, by neutralization assays, and now most often by genome sequencing. Types IAB and IC are equine-virulent, produce high-titered viremia and severe clinical disease in horses, donkeys, and mules. by contrast, type ID, IE, and II–IV viruses are equine-avirulent.

Epidemiology

Eastern Equine Encephalitis

Eastern equine encephalitis virus is endemic in North America along the Atlantic and Gulf coasts. In the United States, as many as 20 human cases are reported annually (average 6 per year). The virus is also endemic in the Caribbean, Central America, and along the north-eastern coast of South America. Less commonly, the disease occurs inland east of the Mississippi River in the United States. In northern regions cases occur in late summer and up until the time of the first frost, while in southern regions cases can occur throughout the year. In the United States the virus is maintained in freshwater marshes by *Culiseta melanura*. This mosquito, which feeds almost exclusively upon birds, is also responsible for amplification of the virus during the spring and summer months (Fig. 35.3).

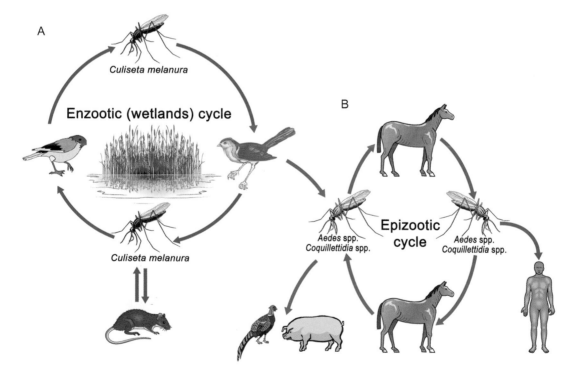

FIGURE 35.3 In North America, most human and equine Eastern equine encephalitis (EEE) virus infections occur near hardwood wetlands populated by the principal mosquito vector, *Culiseta melanura*, with passerine birds serving as asymptomatic amplification hosts (A). Such habitats occur along the Atlantic and Gulf coasts of the United States and some inland areas. In the southeast, where this mosquito is not abundant there are other vector mosquitoes. Bridge vectors (*Coquillettidia perturbans, Aedes vexans, Ae. Canadensis, Culex salinarius*) transmit the virus from reservoir bird species to humans, horses, pigs and pheasants (B). Although EEE virus infection is severe, often fatal, in a wide variety of domestic mammals and birds, none of these are believed to be efficient amplification hosts because viremia levels are generally low. Cases of EEE rarely occur far from enzootic foci. EEE outbreaks in horses and pheasants often precede human cases. *Adapted from the work of Scott C. Weaver, with permission.*

In areas near coastal salt-marshes *Aedes sollicitans* and *Aedes vexans* are responsible for transferring the virus from its endemic habitat species to nearby horses; these mosquitoes feed indiscriminately upon horses as well as birds and occasionally humans. In some other settings in eastern North America, the mosquito(es) responsible for outbreaks remains a mystery. *Coquillitidea perturbans* has been implicated in some settings by virus isolation and determination of a feeding preference for horses and birds: however, this mosquito does not exist at all locations where human disease is known to occur. The mystery is deepened by several lines of evidence opposing an old notion that virus is reintroduced from the tropics each year by migrating birds. Similarly, there is evidence against transovarial transmission and overwintering of virus in mosquitoes. Other notions, such as persistent infection of reptiles, amphibians, or birds appear unlikely.

Highlands J virus is a closely related, less-virulent alphavirus that is endemic on the east coast of the United States within the same areas as EEE virus.

Western Equine Encephalitis

Western equine encephalitis virus is widely distributed throughout the Americas, but especially in the western plains and valleys of the United States and Canada, where it is maintained year-round in an endemic cycle involving domestic and passerine birds together with *Culex tarsalis*, a mosquito particularly adapted to irrigated agricultural areas. Isolations have been made from *Culex stigmatosoma*, *Aedes melanimon*, and *Aedes dorsalis*, also competent vectors. *Culex tarsalis* mosquitoes may reach very high population densities when climatic conditions or irrigation practices are suitable. For example, in the 1930s equine epidemics of an enormous scale occurred in the western United States—between 1937 and 1939 more than 500,000 equine cases were reported along with many thousands of human infections. Since 1964 there have been less than 700 confirmed human cases.

Western equine encephalitis virus has been shown as a less virulent natural recombinant between eastern equine encephalitis and Sindbis viruses.

Venezuelan Equine Encephalitis

The endemic types of Venezuelan equine encephalitis virus that are mildly pathogenic, are perennially active in subtropical and tropical areas of the Americas: type II (Everglades virus) in Florida, types ID and IE in Central America, and types IF, III, IV, V, and VI in South America. Endemic types occur in silent, stable transmission cycles primarily involving *Culex* (*Melanoconion*) spp. mosquitoes and small mammals in tropical swamps in Central America and northern South America (Fig. 35.4).

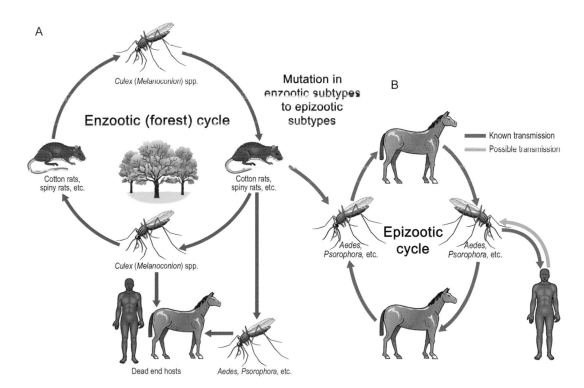

FIGURE 35.4 Enzootic and epizootic/epidemic transmission cycles of Venezuelan equine encephalitis (VEE) virus. (A) There is an enzootic transmission cycle involving virus subtypes ID–IF, II–VI in wet tropical forests and shaded swamp habitats in which there is transmission among rodents (genera *Sigmodon*, *Oryzomys*, *Zygodontomy*, *Heteromys*, *Peromyscus*, and *Proechimys*) and mosquitoes of the subgenus *Culex* (*Melanoconion*). Spillover to humans who live in proximity to infected equines results in epidemics responsible for many thousands of cases. (B) Epizootic/epidemic strains arise by mutation from time to time, causing severe clinical disease in horses and humans. *Adapted from Coffey, L.L., et al., 2013. Factors shaping the adaptive landscape for arboviruses: implications for the emergence of disease. Future Microbiol. 8, 155–176, with permission.*

Types IAB and IC are virulent—these viruses have been isolated only during epidemics. The epidemic types have been found primarily in Venezuela, Colombia, Peru, and Ecuador, causing major problems at approximately ten-year intervals. These viruses cause clinical disease, especially in children, with survivors sometimes suffering neurological sequelae. During epidemics horses are an important amplifying host and since nearly every horse in an epizootic area becomes infected, and there is always substantial associated human disease, the overall consequences may be devastating.

Although the interepidemic maintenance of the epidemic types of the virus had been a long-standing mystery, it is now understood that epidemic types emerge each time *de novo* from mutation of endemic types—the genetic difference between endemic and epidemic viruses is small. Phylogenetic and reverse genetic studies indicate that epidemic strains arise when enzootic forms adapt via positive selection to increase the viremia and virulence in equines, leading to highly efficient amplification in rural locations, and/or to mosquito vectors that undergo seasonal expansions. However, most of these epizootic/epidemic strains are short-lived, presumably due to a reliance on equids for amplification and the limited population turnover (and eventual high herd immunity level) of these relatively long-lived hosts (Fig. 35.5).

Independently, there is evidence that the use of formalin-inactivated vaccines containing residual infectious virus may have in the past been responsible for initiating some type IAB epidemics.

Prevention, Control, Treatment

In many areas, mosquito control programs operate when there is evidence of eastern or western equine encephalitis virus activity obtained either by surveillance, mosquito testing, or by use of sentinel animals. Aerial spraying with ultra-low-volume insecticides, such as malathion or synthetic pyrethrins, is employed.

In an unprecedented event in the history of this disease, an epidemic of Venezuelan equine encephalitis, which had started in northern South America in 1962 to 1964 and re-emerged again in 1969, moved north across Central America and Mexico in 1970 and reached Texas in 1971. There were many thousands of deaths of horses, mules and donkeys and a threat of spread further north into the equine industries of the southern United States. There were many thousands of human infections and many deaths, mostly among children in Central and South America. The epidemic was brought to a halt by an integrated international disease control program that included: (1) an active surveillance system (especially the rapid collection and testing of vector mosquito populations) to inform control strategies, (2) widespread use of the then experimental attenuated virus vaccine TC-83, (3) widespread use of ultra-low-volume aerial spraying of insecticides, and

(4) quarantine and prohibition of the movement of horses. With the end of the 1971 vector season, the virus disappeared. The control program was extremely expensive and demanding of human resources; however, much was learned that has been subsequently assimilated widely into the control of all mosquito-borne zoonoses.

The initial stages of a repeat of the above events occurred in 1995: a major epidemic occurred in Venezuela and Colombia, involving an unknown but very large number of horses, mules, and donkeys and an estimated 75,000 to 100,000 humans. This epidemic was remarkably similar in geographical localization and dynamics of spread to the earlier epidemic that occurred in the same regions of Venezuela and Colombia in the period 1962 to 1964. Viruses isolated during 1995 were antigenically and genetically similar to those obtained during 1962 to 1964. The lack of genetic change between the 1962 to 1964 and 1995 outbreaks is consistent with the slow rate of evolution of endemic alphaviruses. These molecular findings reaffirmed findings that epidemics emerge and re-emerge via a constant mutation of endemic virus types circulating continuously within endemic areas.

ALPHAVIRUS INFECTIONS MARKED BY FEVER, RASH, MYALGIA, ARTHRALGIA, AND ARTHRITIS

Fever, rash, myalgia, arthralgia, and arthritis characterize infections caused by chikungunya, o'nyong-nyong, Ross River, Mayaro, Igbo Ora, and Sindbis viruses (and Sindbis virus strains Ockelbo from Scandinavia and Babanki from Africa).

CHIKUNGUNYA

Chikungunya virus is the most important in terms of the scale of human illness, being widespread throughout Africa, India, and southeast Asia, including the Philippines, and more recently has spread throughout the tropical and subtropical Western Hemisphere. The explosive urban foci in this pandemic, observed particularly in Asia, have in many places infected the majority of the population within a few months.

Clinical Features

Clinically, chikungunya infection is marked by severe joint pain along with the acute febrile illness. Importantly, the illness induced by chikungunya virus cannot be distinguished clinically from that caused by dengue viruses as the early symptomology is almost identical. However, chikungunya virus disease has a shorter febrile phase but a much longer period of arthralgia that lasts well into convalescence, persisting even for years. Symptoms

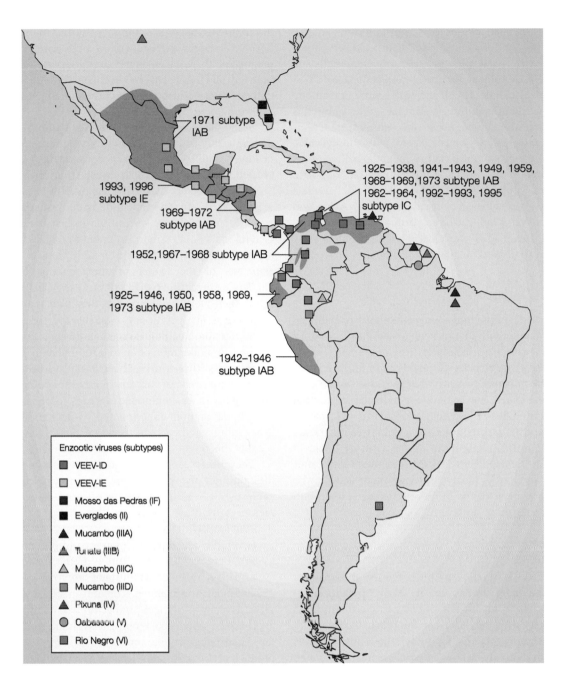

FIGURE 35.5   Locations of Venezuelan equine encephalitis (VEE) virus outbreaks in the Americas. Map showing the locations of all major outbreaks (purple) with the year of the outbreak and the VEE virus subtypes (color-coded) responsible for the outbreak. *Reproduced from Weaver, S.C., Barrett, A.D., 2004. Transmission cycles, host range, evolution and emergence of arboviral disease. Nat. Rev. Microbiol. 2, 789–801, with permission.*

develop after an incubation period of three to 12 days, with a headache, anorexia, and constipation accompanying a fever. A conjunctival suffusion or extreme hyperemia is a notable feature. Although the fever subsides after a few days, skin involvement is often evident in the form of a morbilliform rash that may become pruritic, sometimes with hyperpigmentation, blisters, and sloughing of the epidermis. This latter feature was particularly prominent during the Reunion outbreak, as was an increase in the proportion of patients with neurological complications and hepatitis. Also observed was a high incidence of neonatal infections, many of them serious. Chikungunya virus is often cited as a cause of hemorrhagic disease but in reality such a sequel is rare.

Pathogenesis, Pathology, and Immunity

Studies on the replication of chikungunya virus have focused on the effects of mutations and adaptation to new insect vectors and associated virulence. Acquisition of an A226V mutation in the viral E1 envelope protein has been shown as enhancing virus replication in *Aedes albopictus* cells and to lessen the requirement for membrane cholesterol. This mutation was also linked with increased virulence among patients in the Reunion outbreak. Parallel studies of Sindbis virus, another alphavirus, show that similar mutations decrease viral dependency on cholesterol within insect cells.

There is a paucity of information about the pathogenesis of chikungunya virus infection in humans. A long-held view was that alphaviruses from the Americas caused a predominantly neuroinvasive disease, whereas Asian and African counterparts induced largely arthralgia accompanied by a rash. However, it is now clear that chikungunya virus can also have some neurological involvement.

Laboratory Diagnosis

As already noted, chikungunya virus infection can readily be confused clinically with that of dengue and thus laboratory diagnosis is required to distinguish the two causes. IgM-capture and IgG EIA tests are the mainstay of initial testing, followed by confirmatory tests when positive. Some care needs to be exercised, however, owing to the cross-reaction with some antibodies to o'nyong-nyong virus. Ultimately neutralization tests are required for a definitive diagnosis. PCR is also useful, particularly as the virus is found at a high titer in the blood for at least four to six days after onset of symptoms. As always, however, care needs to be taken to ensure the selection of the appropriate primers and standardization against suitable reference reagents.

Epidemiology, Prevention, Control, Treatment

Much has been learned from the ongoing chikungunya pandemic: the explosiveness of its early spread caught international public health authorities off guard—in 2005 over 10,000 cases were recorded on the Indian Ocean island of Reunion, the numbers declining over the winter months only to resurge again the following year, when it is estimated that over 70,000 cases occurred. The infection disseminated quickly to Marotte, Mauritius, and The Seychelles. The case fatality rate of 1:1000 among the Reunion cases was unexpected.

Distinct differences exist between earlier outbreaks of chikungunya in Africa versus its spread to Asia and the Americas, reflecting genotypic differences between the two regions. In Africa the virus circulates through a sylvatic cycle with *Aedes furcifer-taylori* spp. mosquitoes being the principal vectors. It is thought that the virus is maintained in a wildlife animal reservoir, possibly a non-human primate. In contrast, chikungunya virus has adapted to an urban environment in Asia and the Americas, with *Aedes aegypti* and *A. albopictus* serving as the major vectors.

A. albopictus, the "Asian tiger mosquito," has progressively become a major determinant of both spread and human pathogenicity, being an aggressive feeder on humans during both day and night. This has occurred through adaptation of chikungunya virus to growth in the gut and salivary glands of *A. albopictus* (see below).

Some cases have been reported in Europe, notably in northern Italy and southern France where *A. albopictus* has become established and as a consequence opportunities created for human-to-human spread (Fig. 35.6).

In December 2013, local transmission in the Caribbean was reported—for the first time in the Americas—and by early 2015 more than 1.5 million human cases, in North, Central, and South American countries, had been reported to the Pan American Health Organization. Overall global estimates of the extent of the ongoing Chikungunya pandemic range from 6 to 8 million cases—in the absence of comprehensive surveillance data this is an approximation.

There is no specific antiviral therapy for chikungunya virus infection. Symptoms such as arthralgia can be lessened through administration of analgesics and anti-inflammatory compounds. Ribavirin has been found to be of little use. Passive transfer of immunoglobulin containing anti-chikungunya virus antibodies appears to prevent spread of virus to the brain in experimentally infected mice provided it is administered early in the acute phase, but it is hard to see that this might have any practical application.

The primary control method is avoidance of mosquito bite through the generic precautionary measures taken against all insect-borne infections. However, such measures are often impractical, especially in the control of *A. albopictus* with its increasing resistance to insecticides.

Given the developing importance of chikungunya virus as an emerging disease, effort is being made to develop an effective vaccine. Traditional approaches to vaccine development, for example formalin inactivated cell-grown virus, have not been encouraging, thus stimulating alternative approaches, such as developing a chimeric vaccine utilizing the Venezuelan equine encephalitis virus non-structural proteins and the structural proteins of chikungunya virus. The construction of virus-like particles is an alternative method of attempting to formulate a vaccine capable of safely inducing human neutralizing antibodies without side effects, particularly arthralgia.

OTHER ALPHAVIRUS INFECTIONS

O'nyong-nyong Virus, Mayaro, Ross River, and Sindbis Virus Infections

O'nyong-nyong virus is spread by *Anopheles funestus* and *Anopheles gambiae*, making it the only known human arboviral pathogen with an anopheline mosquito as a vector. Similarly, the few interepidemic isolates have also come from

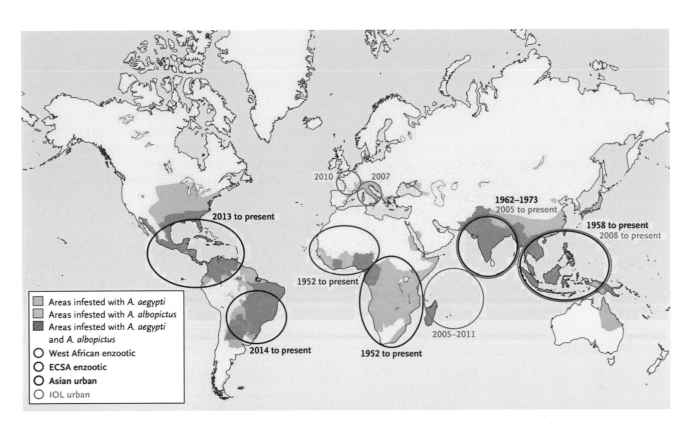

FIGURE 35.6 Origin, spread, and distribution of chikungunya virus and its vectors. The map shows the African origins of enzootic chikungunya virus strains and the patterns of emergence and spread of the Asian lineage and Indian Ocean lineage (IOL) of the virus during epidemics since the 1950s, based on phylogenetic studies. ECSA denotes the eastern, central, and southern African lineage. The distributions of the peridomestic vectors, *Aedes aegypti* and *A. albopictus*, are also shown. *Reproduced from Weaver, S.C., Lecuit M., 2015. Chikungunya virus and the global spread of a mosquito-borne disease. N. Engl. J. Med. 372, 1231–1239.*

anophelines, suggesting the virus is maintained enzootically by these mosquitoes as well. The vertebrate reservoir host of the virus has not been determined, but serological surveys suggest a role for domestic livestock and several species of rodents. The virus was responsible for an epidemic in East Africa extending from 1959 to 1962 affecting no less than 2 million people, and a second epidemic in Uganda in 1996 to 1997; as with the earlier epidemic, the virus moved rapidly through affected communities, impacting up to 70% of the population. Recently it has been shown that there is ongoing endemic human infection across coastal Kenya.

Mayaro virus occurs in forested regions of Central and South America and appears to be spread by *Haemagogus* mosquitoes, probably from a primate reservoir. Widespread human infection is demonstrable by seroconversion, and several epidemics have occurred in small towns in the Amazon basin.

Ross River virus produces a disease known as epidemic polyarthritis during late summer/early autumn in rural Australia. Several species of mosquitoes can act as vectors, and marsupials as reservoir hosts; domestic animals, rodents, and at times of epidemics, humans may act as amplifiers. In 1979 and 1980, following importation from

Australia, probably by air travel of persons incubating the infection, explosive epidemics occurred in Fiji and spread to other islands of the South Pacific, affecting up to half the indigenous population as well as tourists. Clinically, patients present with joint pain with or without fever and rash, and joint problems may persist for weeks or months.

Sindbis virus has long served as the prototypic alphavirus: laboratory strains become attenuated and so the virus has a reputation as a non-pathogen. However, wild-type Sindbis virus is another matter, with human cases accompanied by fever, rash and arthralgia recognized in many parts of Africa, northwestern Europe, India, and Malaysia. Epidemics have also been described in Egypt and South Africa. Various species of wild birds serve as reservoir hosts, with *Culex, Culiseta*, and *Aedes* mosquitoes as vectors.

Control of these mosquito-borne viruses rests largely upon avoidance of exposure, elimination of breeding sites, and the destruction of mosquitoes or their larvae. Control of alphavirus infections of humans is generally undertaken only in the face of threats of outbreaks, often after it is too late to have any practical effect. Mosquito larviciding programs may be carried out in short-term emergency situations, such as during an outbreak—aerial spraying is done with ultra-low-volume insecticides, such as malathion or synthetic pyrethrins.

DISEASE CAUSED BY THE RUBIVIRUS RUBELLA VIRUS

Rubella is a trivial exanthem of childhood. However, in 1941 an Australian ophthalmologist, Norman Gregg, noticed an unusual concentration of cases of congenital cataract among newborn babies in his practice—an epidemic of blindness. A diligent search of his records revealed that most of the mothers had contracted rubella during the first trimester of pregnancy. Further investigations revealed that these unfortunate children had also suffered a range of other congenital defects including deafness, mental retardation, and cardiac abnormalities (see Chapter 14: Control, Prevention, and Eradication).

Clinical Features

Rubella is such a mild disease that most adults are subsequently uncertain whether they have ever contracted it in childhood. The fine, pink, discrete macules of the erythematous rash appear first on the face then spread to the trunk and limbs and fade after 48 hours or less (Fig. 35.7). In nearly half of all infections there is no sign of a rash. Fever is usually inconspicuous, but a characteristic feature is that post-auricular, sub-occipital, and posterior cervical lymph nodes are enlarged and tender from very early in the illness. Mild polyarthritis, usually involving the hands, is a fairly frequent feature of the disease in adult females; it is usually fleeting but rarely may persist for up to several years.

Thrombocytopenic purpura and post-infectious encephalopathy are rare complications. Progressive rubella panencephalitis is an even rarer and inevitably fatal complication, developing insidiously in the second decade of life, usually in children with congenital rubella.

Congenital Rubella Syndrome

At least 20% of all infants infected *in utero* during the first trimester of pregnancy are born with severe congenital abnormalities, usually multiple, and most of the remainder have milder defects. The commonest congenital abnormalities are neurosensory deafness: this can be total or partial, due to cochlear degeneration, becoming progressively apparent in the early years after birth. Other common manifestations include blindness (total or partial, especially cataracts, but sometimes glaucoma, microphthalmia, or retinopathy), congenital heart disease (especially patent ductus arteriosus, sometimes accompanied by pulmonary artery stenosis or septal defects), and microcephaly with mental retardation. The congenital rubella syndrome may also include bone translucency and retardation of growth, hepatosplenomegaly, thrombocytopenic purpura, and cataracts (Fig. 35.8). Despite the diversity and severity of this pathology, the congenital rubella syndrome is sometimes missed at birth. About 10 to 20% of babies with congenital rubella syndrome die during the first year, and up to 80% develop some evidence of disease within the early years of life. Up to 20% of children with congenital rubella syndrome develop insulin-dependent diabetes mellitus as young adults.

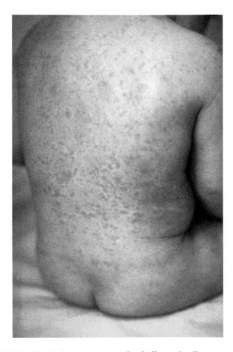

FIGURE 35.7 Typical appearance of rubella rash. *From* www.vaccine information.org, *courtesy of CDC.*

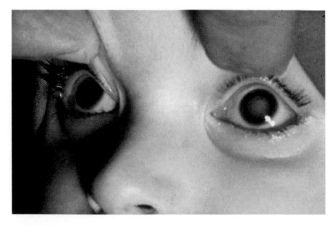

FIGURE 35.8 Cataract in a child with congenital rubella syndrome.

Pathogenesis, Pathology, and Immunity

The virus enters the body by inhalation to multiply asymptomatically in the upper respiratory tract and then spreads to regional lymph nodes via lymphatics and possibly by a transient viremia. After further replication for seven to nine days, virus then becomes disseminated to multiple sites within the body. The secondary viremia is greatest from days 10 to 17; viral shedding from the respiratory tract continues from 10 to 24 days post-infection. The rash typically develops 16 to 18 days after infection, coinciding with the appearance of antibody, and maximal respiratory transmission of infection typically occurs from five days before to six days after the appearance of a rash. Naturally acquired immunity to rubella lasts for many years; second infections occur occasionally but are usually subclinical and boost immunity further.

There is a high probability that the baby will suffer congenital abnormalities if rubella virus infects a woman during the first trimester of pregnancy. Severe damage (deafness, blindness, heart or brain defects) occurs in 15 to 30% of all infections during the first trimester, and in about 5% of those in the fourth month (usually deafness), but rarely thereafter. Minor abnormalities are even more frequent, and following spontaneous abortion or stillbirth, virus can be found in practically every organ.

What makes rubella virus teratogenic, when numerous non-teratogenic viruses are so much more pathogenic post-natally and so much more cytocidal for cultured cells? Paradoxically, this relative lack of pathogenicity may hold the clue to its teratogenicity. More cytocidal viruses may destroy cells and kill the fetus leading to spontaneous abortion (as does rubella occasionally). In congenital rubella, affected organs including the placenta are hypoplastic and show reduced cell numbers and necrotizing angiopathy in small blood vessels. Tissue damage is thought to be due to vascular injury and insufficiency, combined with a slowing of the rate of cell division, as has been demonstrated in cultured human fetal cells, leading to a decrease in overall cell numbers. Death of a small number of cells or slowing of the mitotic rate at critical stages in ontogeny might interfere with the development of those key organs which formed during the first trimester.

Neither the mother's nor the baby's immune response is able to clear the virus from the fetus. Although maternal IgG antibodies cross the placenta and the infected fetus manufactures its own IgM antibodies, cell-mediated immune responses are defective and remain so post-natally. Clones of infected cells may escape immune cytolysis even though maternal IgG might restrict systemic spread of virus. The rubella syndrome in the fetus is a true persistent infection, and throughout the pregnancy and for several months after birth the baby continues to shed virus in any or all of its secretions.

Laboratory Diagnosis

There are three common situations in which the clinician requires laboratory determination of rubella virus infection.

1. When a woman considering vaccination wishes to know whether she has ever had rubella and is now immune. As there is only one serotype of rubella virus, and specific IgG antibodies continue to be found in the blood for many years after clinical or subclinical infection, the detection of IgG antibodies is unequivocal evidence of past infection and subsequent immunity.

2. When an unimmunized woman develops a rash in the first trimester of pregnancy, or comes into contact with someone with rubella, and wishes to know whether she has active infection and whether she should have an abortion. Diagnosis of active infection requires the demonstration of either a rising titer of rubella antibodies in paired sera, or rubella IgM antibodies in a single specimen, or circulating virus RNA by RT-PCR. If paired sera are used, the first bleed must be taken during the first week after the onset of the rash, otherwise a fourfold rise in antibody titer may not be detected in the convalescent specimen taken 10 or more days later. On the other hand, if a pregnant woman is concerned about recent exposure to a known case of rubella, the first sample should be taken as soon as possible, but the second should be delayed for at least a month to allow for the two- to three-week incubation period of her putative infection. Rubella IgM antibody is generally detected by one week after the appearance of the rash and persists for at least one month but occasionally three months. IgM serology may be the only method of diagnosis in the case of the woman who first consults her doctor weeks after the rash has gone. RT-PCR may also be used if available, because of its sensitivity and the presence of viremia before any detectable IgM antibody response. Molecular typing can be used to track the origin of a new case.

   Importantly, reinfections in a patient who is already immune may also lead to a fourfold rise in maternal IgG antibody, but in such situations the risk of congenital disease in the infant is exceedingly low, and IgM antibody is rarely present in the mother.

3. When a baby is born with signs suggestive of the rubella syndrome, or its mother is believed retrospectively to have possibly contracted rubella during the first trimester. Diagnosis of congenital rubella infection in a newborn baby can be made by demonstrating rubella IgM antibodies in a single serum specimen from the baby. Many newborns will have rubella IgG antibodies, acquired transplacentally from the mother who may have been vaccinated or infected with the virus years earlier. As IgM antibodies do not cross the placenta, rubella IgM detected in umbilical cord blood must have been synthesized by

TABLE 36.1 Diseases Caused by Flaviviruses

Virus	Geographic Distribution	Vertebrate Reservoir	Principal Vectors	Human Disease	Epidemiologic Features
Dengue 1, 2, 3, 4	Tropics worldwide	Humans, monkeys	*Aedes aegypti*, other *Aedes* spp.	Fever, arthralgia, myalgia, rash	Urban epidemics
Zika virus	SE Asia, Africa, S and N America, Pacific Islands	Monkeys, ? humans	*Aedes aegypti*, other *Aedes* spp.	Fever, arthralgia, myalgia, rash. Infant microcephaly following infection in pregnancy	Sporadic human cases only until 2007. Since then, increasing dissemination in S America and SE Asia
West Nile	Africa, tropical Asia, Mediterranean, N and S America	Birds	*Culex* spp.	Fever, arthralgia, myalgia, rash	Endemic in tropics with sporadic cases, summer epidemics (Mediterranean, South Africa)
St. Louis encephalitis	Americas	Birds	*Culex tarsalis*, *Culex pipiens*, *Culex* spp. (tropics)	Encephalitis	Sporadic cases (tropical South America), periodic outbreaks (North America)
Japanese encephalitis	Asia	Swine, birds	*Culex tritaeniorhynchus*	Encephalitis	Endemic (Southeast Asia), summer epidemics (Northern Asia)
Murray Valley encephalitis; Kunjin	Australia, New Guinea	Birds	*Culex annulirostris*	Encephalitis	Periodic summer epidemics
Rocio	Southeastern Brazil	Birds	*Psorophora*, *Aedes* spp.	Encephalitis	Sporadic cases, occasional epidemics
Yellow fever	Tropical Africa and Americas	Humans, monkeys	*Aedes aegypti*, *Aedes* spp., *Haemagogus* sp.	Fever, hemorrhage, jaundice	Sporadic cases related to jungle exposure, periodic epidemics (Africa)
Kyasanur Forest disease	Southwest India	Rodents	*Haemaphysalis spinigera* ticks	Fever, hemorrhage, encephalitis	Sporadic cases, occasional outbreaks related to deforestation
Omsk hemorrhagic fever	Russia, Central Asia	Rodents	*Dermacentor pictus* ticks	Fever, hemorrhage	Sporadic cases, winter outbreaks associated with muskrat trapping
Alkhurma hemorrhagic fever virus	Egypt, Sudan, Saudi Arabia	Buffalo, camel, cattle	*Ornithodorus* spp. Ticks	Hemorrhagic fever	Sporadic cases in tourists, butchers
Tick-borne encephalitis	Russia, eastern Europe, Scandinavia	Rodents, birds, domestic animals	*Ixodes* ticks	Encephalitis	Sporadic cases with periodic high incidence, outbreaks from ingestion of raw milk
Louping ill	British Isles, Spain, Turkey	Sheep, birds	*Ixodes* ticks	Encephalitis	Sporadic cases, by direct contact with infected sheep or tick bite
Powassan	Canada, United States, Russia	Small mammals	*Ixodes* ticks	Encephalitis	Sporadic cases
Langat virus	Malaysia, Thailand, Siberia	Rodents	*Ixodes* ticks	No naturally occurring human disease	Possible vaccine candidate, possible treatment for malignancy

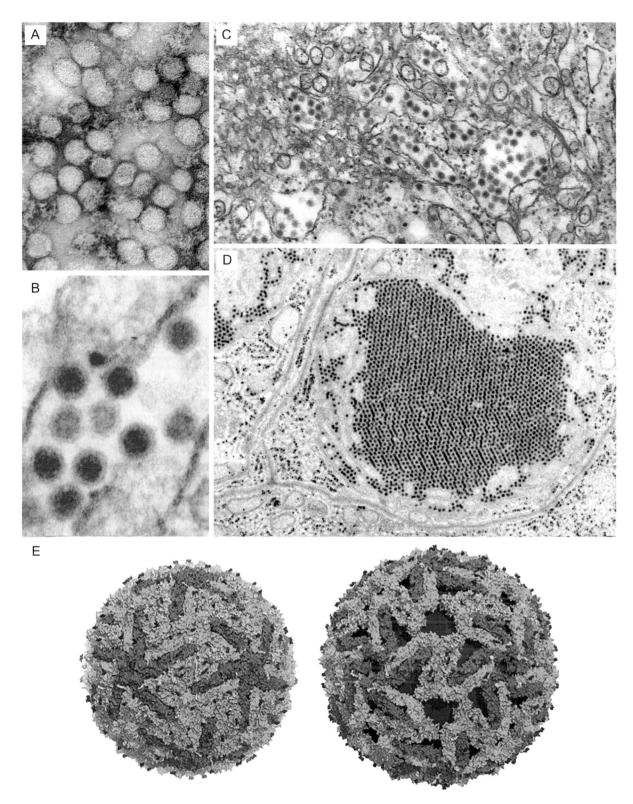

FIGURE 36.1 Flavivirus morphology and structure. (A) Negative contrast electron microscopy of yellow fever virus, showing the rather unstructured surface of 40 to 50 nm virions. (B) Thin-section electron microscopy of St. Louis encephalitis virus in BHK-21 cell culture, showing the electron dense core and thin envelope layer. (C) Thin-section electron microscopy of a mouse neuron infected with tick-borne encephalitis (Russian Spring-Summer encephalitis) virus, showing the typical extensive proliferation of endoplasmic reticulum membranes associated with virion accumulation. (D) Thin-section electron microscopy of the salivary gland of a *Culex pipiens pipiens* mosquito infected with St. Louis encephalitis virus at 21 days post-infection. There are massive numbers of virions, including many in paracrystalline array, in the lumen of the gland as a result of budding from the apical plasma membrane of infected cells. (E) Structure of dengue virus 1. (Left) When grown in mosquito cells at 28°C, mimicking infection in mosquitoes, mature dengue virus particles have icosahedral symmetry and a rather smooth, flat surface as a result of tight packing of E protein dimers. (Right) When exposed to 34°C or above, a temperature chosen to mimic human infection, virions irreversibly expand and the E protein dimers (olive and cyan) spread apart heterogeneously exposing deeper structures (purple) and ligands involved with virus binding to vertebrate cell receptors. *(E) Reproduced from Zhang X. et al., Dengue structure differs at the temperatures of its human and mosquito hosts. Proc Natl Acad Sci U S A. 2013; 110:6795–6799; and Fibriansah G. et al., Structural changes in dengue virus when exposed to a temperature of 37°C. J Virol. 2013;87:7585–7592, with permission.*

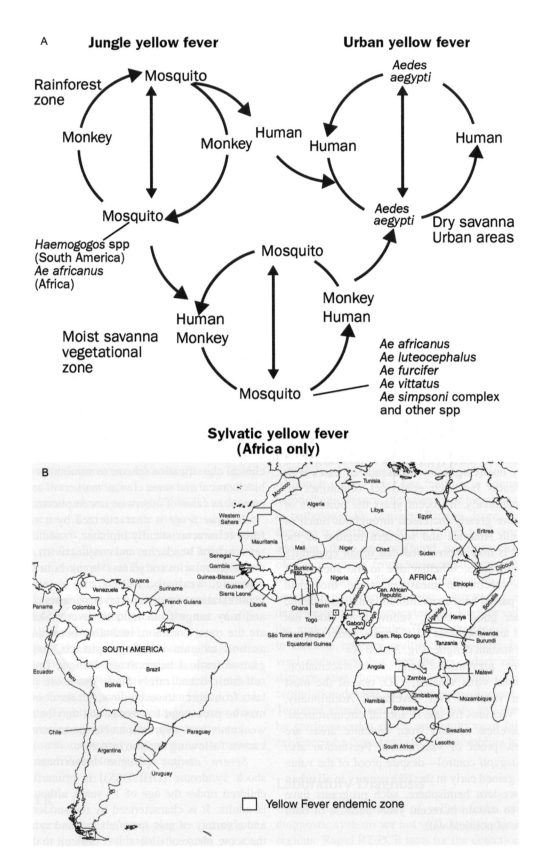

FIGURE 36.3   (A) Transmission cycles of yellow fever virus in different habitats, showing zoonotic and urban cycles. (B) Map showing regions endemic for jungle yellow fever. *(A) From Thomas P. Monath, with permission.*

circulatory failure usually occur. There is thrombocytopenia (platelet count <100,000/μL) and evidence of vascular leakage. Common hemorrhagic manifestations include skin petechiae, purpura, and ecchymoses. Epistaxis, bleeding gums, gastrointestinal hemorrhage, and hematuria occur less frequently. "The tourniquet test" is often used to indicate capillary fragility. Without early diagnosis and proper management, some patients experience hypovolemic shock from blood loss or more importantly from plasma leakage; this may be mild and transient or progress to profound shock with an undetected pulse and blood pressure. Coagulation system abnormalities suggest disseminated intravascular coagulopathy. This disease is generally observed during epidemics of dengue (often caused by dengue 2 virus) in a population with a previous history of infection with another dengue virus (e.g., dengue 1 virus) (see dengue pathogenesis below) (Table 36.3).

Clinicians and epidemiologists have classified severe dengue into four grades of illness:

Grade I is mild, the only hemorrhagic manifestations being scattered petechiae or a positive tourniquet test.
Grade II is more severe, with one or more overt hemorrhagic manifestations.
Grade III is characterized by mild shock with signs of circulatory failure; the patient may be lethargic or restless and have cold extremities, clammy skin, a rapid but weak pulse, narrowing of pulse pressure to 20 mmHg or less.
Grade IV is characterized by profound shock with undetectable pulse and blood pressure.

Dengue fever is typically a self-limiting disease with a mortality rate of less than 1%. When treated, severe dengue has a mortality rate of 2% to 5%, but when left untreated the mortality rate can be as high as 50%.

Pathogenesis, Pathology, and Immunity

Our understanding of the pathogenesis of uncomplicated dengue fever is based in large part upon extrapolations from clinical and pathological findings, with secondary contributions from experimental non-human primate experiments. Following virus entry during mosquito blood feeding, there is an initial infection of local fibroblasts and especially of dendritic cells, enhanced by active components of mosquito saliva. Virus is carried to local lymph nodes, spleen, and liver where cells of the reticuloendothelial system, including Kupffer cells, are infected—other cells, such as hepatocytes, are also infected. The viremia often rises to a very high titer within the first days of infection. Activation of a multi-compartment humoral and cellular adaptive immune response occurs quickly—many of the clinical signs and symptoms of infection are attributed to the exuberant cytokine/complement system/coagulation system responses. In most cases, the viremia is rapidly cleared and the illness, even the life-threatening illness, is resolved usually within a week to 10 days with lingering weakness for weeks following.

The pathogenesis of severe dengue (dengue hemorrhagic fever/dengue shock syndrome) is still a matter of controversy. Two theories, which are not mutually exclusive, are frequently cited to explain clinical and pathological observations: first, the theory of immune enhancement, whereupon patients experiencing a second infection with a heterologous dengue virus have a significantly higher risk of developing the vascular leak syndrome. These pre-existing antibodies can also come from transfer of maternal antibodies *in utero*. Pre-existing heterologous dengue antibodies recognize the infecting virus and form virus–antibody complexes, these are then opsonized and phagocytosed by macrophages via Fc receptors. However, because the antibodies are heterologous, the virus is not neutralized and is free to replicate once inside

---

TABLE 36.3 World Health Organization-Sponsored Revision (2009) of the Classification of Dengue Illness that More Accurately Reflects the Global Clinical Presentation Observed in Both Children and Adults

Probable Dengue	Warning Signs*	Severe

Live in/travel to dengue-endemic area
Fever and two of the following
  Nausea, vomiting
  Rash
  Aches and pains
  Tourniquet test positive
  Leucopenia
  Any warning sign
	1. Abdominal pain or tenderness
	2. Persistent vomiting
	3. Clinical fluid accumulation
	4. Mucosal bleed
	5. Lethargy, restlessness
	6. Liver enlargement >2 cm
	7. Laboratory increase in hematocrit concurrent with rapid decrease in platelet count
		1. Severe plasma leakage leading to shock (DHSS)
		  Fluid accumulation with respiratory distress
		2. Severe bleeding as evaluated by a clinician
		3. Severe organ involvement
		  Liver: AST or ALT >1000
		  CNS: Impaired consciousness
		  Heart and other organs

Laboratory-confirmed dengue	*Requires strict observation and medical intervention	

The classification includes two major categories, dengue with or without warning signs and severe dengue; the former includes dengue fever, and the latter includes dengue hemorrhagic fever/dengue shock syndrome (DHF/DSS) and other severe forms of infections previously also referred to as atypical dengue.

these macrophages. It can be demonstrated experimentally, that virus replicates to a high titer in macrophage cultures. This infection of macrophages initiates a cascade of events that produces cytokines and other vasoactive mediators, ultimately leading to increased vascular permeability, leakage, hypovolemia, shock, and in some cases death. The second hypothesis is that dengue viruses have greater or lesser capability to vary genetically and such genetic drift yields some variants with greater virulence and capacity to cause severe dengue disease. There is epidemiological and laboratory research evidence to support both of these theories. As more and more research results become available the balance between the two hypotheses moves back and forth: for example, some recent research suggests that only certain strains of the viruses may evoke severe disease by immune enhancement. Additionally, there still remains the fact that individuals may differ markedly in their innate physiological and/or immunological susceptibility to severe dengue—polymorphisms in particular genes have been linked with an increased risk of severe dengue complications.

Laboratory Diagnosis

Because of the importance of severe dengue and its rapid onset, dengue laboratory diagnostics are advanced in endemic regions. A definitive diagnosis of dengue infection depends nowadays mostly on detecting dengue virus-specific RNA sequences by PCR or similar assays. In some settings virus isolation in cell culture is attempted, using the C6/36 insect cell line derived from *Aedes albopictus*. Of the methods used for detecting specific antibodies in patient serum, the IgM capture ELISA assay is the method-of-choice. IgM antibodies against dengue viruses can be detected two to three days after infection and may persist for several months. Although tests designed for measuring anti-dengue virus IgG antibodies may be used, paired sera are required and care needs to be exercised to avoid the detection of cross-reactive antibodies elicited by other flavivirus infections.

Epidemiology, Prevention, Control, and Treatment

The original natural reservoir of the dengue viruses appears to have been African monkeys; reservoirs have also been established in monkeys in Southeast Asia. However, the important life cycle of the dengue viruses is mosquito–human–mosquito, involving the urban mosquito *Aedes aegypti*. *A. albopictus* and *A. scutellaris* also serve as efficient vectors. Although dengue fever has been known for over 200 years, prior to the 1950s outbreaks of dengue were rare and because of the slow transport of viremic persons between tropical countries, with epidemics in any particular locality occurring at intervals of decades. However, during and after the Second World War millions of people in all countries of the developing world moved to large cities, resulting in rapid and unplanned urbanization paralleled by an expansion of breeding places for *Aedes aegypti*. Furthermore, for the first time the near-simultaneous circulation of multiple viruses led to the appearance of an essentially new clinical disease, severe dengue. The first outbreak of severe dengue (dengue hemorrhagic fever and dengue shock syndrome), occurred in Manila in 1953 and 1954; by 1975 it was occurring at regular intervals in most countries of Southeast Asia and is today one of the leading causes of hospitalization and death among children in this region (Fig. 36.4).

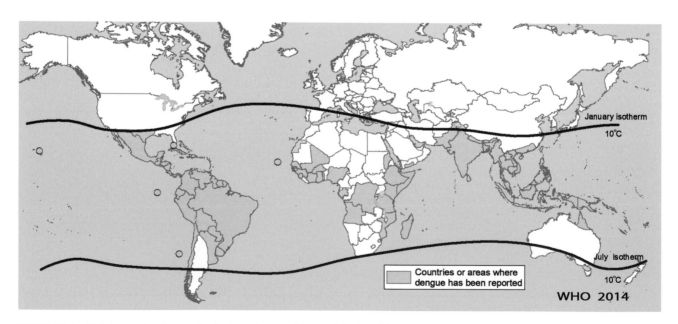

FIGURE 36.4   Global map showing the distribution of cases of dengue infection, 2013. The horizontal lines indicate the boundaries of the climatic region with the potential for year-round survival of *Aedes aegypti*. *Adapted from the World Health Organization, Geneva.*

In the Americas dengue had been known for over two centuries but did not become a major health problem until about 1975. Initially this was due to the same constraints that operated in Southeast Asia, notably the slow and infrequent movement of viremic persons. The explanation for the delay after the end of the Second World War, compared with the immediate post-war occurrence of the disease in South-East Asia, lies in the success of the campaign to eradicate *Aedes aegypti* from the Americas, implemented primarily to control urban yellow fever. Although good results were achieved in many countries, the aim of total eradication from the region failed and during the 1970s many cities in South American and Caribbean countries were re-invaded by *Aedes aegypti*. Coincidentally, the increased movement of people into slums of the urban fringes led to greatly increased breeding places for the mosquito. The development of air travel throughout the region led to increased movement of the dengue viruses, with the result that during the early 1980s there was a substantial increase in dengue virus transmission. Major epidemics involving many thousands of people occurred throughout the Caribbean (>1977) and South America (>early 1980s). This was exactly the situation that had led to the emergence of dengue hemorrhagic fever in Southeast Asia in the 1950s, with the same result. The first major outbreak of severe dengue in the Americas occurred in Cuba in 1981, and since then there have been epidemics throughout Central America and bordering countries with sporadic cases in many other countries. Thus dengue hemorrhagic fever is now well established as a major public health issue in countries of the Caribbean and in both South and Central America. Because of the increasing occurrence of *Aedes aegypti* and *Aedes albopictus*, it is also a potential threat to the southern states of the United States.

Aedes aegypti is the principal vector of the dengue viruses. Regular blood meals are required by the females for egg production, these being laid in small depressions or containers of still water close to human habitation. Once a female is infected, the virus is ingested and replicates first in the mid-gut, and thence in the salivary glands. The female remains infected for life. Increasingly, *Aedes albopictus*, the "Asian tiger" mosquito, has become an important vector in many regions, complementing the role of *Aedes aegypti*—in many places, *Aedes aegypti* presents an indoor risk while *Aedes albopictus* is most common around the outside of the house (Fig. 36.5). Because of the high level of viremia resulting from dengue virus infection of humans, the viruses are efficiently transmitted between mosquitoes and humans without the need for an amplification host.

There is evidence of a sylvatic (jungle) cycle for all dengue viruses, although this does not appear to be required for virus persistence in any given geographical region. The newest dengue virus, dengue 5 virus, is mostly known in its sylvatic environment.

At present, there is no vaccine generally available for dengue, although several candidates are at various stages of development. To be effective, a dengue vaccine will have to protect against all four major dengue viruses. For use in dengue-endemic countries, the vaccine will have to be safe for use in children 9 to 12 months of age; it will

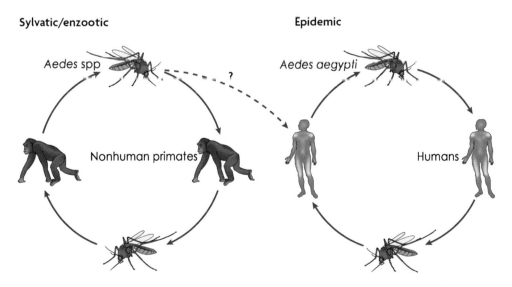

FIGURE 36.5 Transmission cycle of dengue viruses. The viruses are spread principally by *Aedes aegypti* mosquitoes, which breed in domestic and peridomestic water containers and live in or near human habitations. In addition, a sylvatic dengue transmission cycle, involving sylvatic *Aedes* spp. mosquitoes and non-human primates has been documented in western Africa and Southeast Asia. In contrast to the impact on yellow fever virus transmission, the contribution of the dengue sylvatic cycle is minimal, except perhaps for the initial transmission from a sylvatic habitat to various human habitats, rural or urban. *Reproduced from Whitehead, S.S., et al., Prospects for a dengue virus vaccine. Nat. Rev. Microbiol. 2007;5(7):518–528, with permission.*

have to be inexpensive, and provide long-lasting immunity. Virtually all modern molecular approaches are being used to develop dengue vaccines. These include: (1) tetrameric live-attenuated virus vaccines, by analogy with yellow fever and other viral vaccines, thought by many to provide the most complete and lasting immunity, (2) chimeric virus vaccines using various heterologous viruses, such as yellow fever virus, as backbone, (3) subunit vaccines, using various recombinant DNA technologies to express viral envelope antigens, and (4) DNA vaccine (and combination prime-boost vaccine). Without an animal model mimicking human disease, a reliable, ethical human challenge model, and comprehensive correlates of protection, advancing candidate vaccines through regulatory hurdles is proving difficult; despite this many authorities are optimistic.

Mosquito control is the principal tool currently available, but the common methods now used (spraying of insecticide aerosols from airplanes or trucks) fail to kill all *Aedes aegypti*, as the insects often rest indoors in places not reached by the aerosol. Furthermore, over the last several decades the extensive use of pesticides has led to pesticide resistance. Control of breeding is theoretically feasible, and governments in some countries legislate to make it the responsibility of citizens to remove abandoned containers, tires, etc. and to drain stagnant pools of water in the vicinity of dwellings. These measures are increasingly difficult in the slums of the urban fringes in expanding cities, but such a rigorous approach is effective (Fig. 36.6).

Personal protection measures should be encouraged where possible. These may include use of insect repellents, covering exposed skin (e.g., long-sleeved shirts), use of mosquito nets, and avoiding areas of high mosquito density.

ZIKA VIRUS

Zika virus was discovered in 1947 in the Zika Forest of Uganda during an intensive search for yellow fever virus. Virus was isolated from both a sentinel rhesus monkey and a pool of *Aedes africanus* mosquitoes; together with later virus isolation from humans in Tanzania and serological evidence of infections in forest-dwelling people, it was concluded that the virus existed in a sylvatic life cycle restricted to a belt across Africa into equatorial Asia. This seemed to be the case until 2007 when Zika virus began to follow the expansion of chikungunya virus around the world. The virus spread across the Indian Ocean to South East Asia and Polynesia, with a large epidemic reported on Yap Island in the Federated States of Micronesia in 2014: nearly 75% of the population was infected. From there the virus continued to spread across Pacific island nations—then, by 2015 Zika virus was being reported in 28 countries and territories in the western hemisphere. Modeling, using dengue dynamics, predicts many millions of cases in the Americas before this epidemic drifts into endemicity.

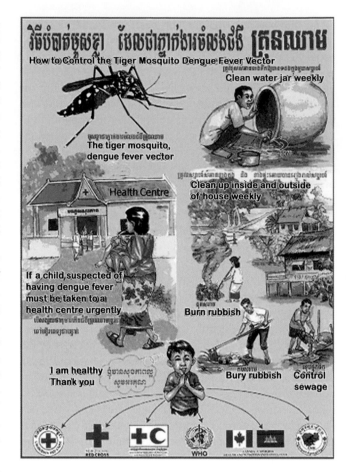

FIGURE 36.6 Public health poster about mosquito control, Cambodia.

Clinical studies in South America suggest that one in five adult cases are symptomatic. Clinical diagnosis is difficult, however, as early clinical signs of Zika virus resemble dengue and chikungunya. Patients may present with fever, headache, myalgia, joint pain, rash, and conjunctivitis. Severe disease requiring admission to hospital is uncommon. The major concerns are thrombocytopenic purpura with leukopenia, meningoencephalitis, microcephaly, an unusual incidence of Guillain-Barré syndrome and microcephaly in newborns (see below).

Where serological studies have been undertaken, it is clear that Zika virus infection has been substantially under-reported. Phylogenetic data from outbreaks in Africa, Asia, Polynesia, and now the Americas indicate Zika virus is evolving: three main phylogeographic lineages have been identified, one representing the original equatorial African location, the second likely originating in Malaysia and responsible for the 2007 outbreak in Micronesia, and the third the more recent expansion into the Americas. However, the divergence in nucleotide sequence between these lineages is less than 12%; therefore, currently available Zika virus-specific primers used for routine PCR testing should continue to be useful for diagnosis and distinction of Zika virus infections in populations continuing to experience multiple flavivirus infections. RT-PCR testing is recommended for samples taken within 14 days of onset of symptoms; later samples are tested initially for IgM antibody. PCR-based diagnostics have also been developed for use in blood banking.

During the global spread of 2014 to 2016 humans became the primary amplification hosts, with the urban mosquito, *Aedes aegypti*, being the principal vector. The virus is also transmitted by *Aedes albopictus*, the "Asian tiger mosquito." As already noted, this species complements *A. aegypti* in many urban areas, favoring outdoor habitats and day and night feeding, while *A. aegypti* favors habitats within dwellings. Importantly, there is worry that adaptation of the virus to *A. albopictus* may lead to the emergence of virus strains with a heightened pathogenicity for humans, as has been the case for chikungunya virus where a single amino acid change in the E envelope glycoprotein gene has furthered virus transmission rates by *A. albopictus*.

Little is known regarding the pathogenesis of Zika virus infection. There must be a substantial viremia following the incubation period to favor the success of the human–mosquito–human transmission cycle that is now so prevalent. Whether or not a high viremia level occurs in asymptomatic human infections remains to be established. There is also evidence of sexual transmission. High and prolonged virus excretion in urine for at least ten days after onset presents a further risk for containment. The virus grows well in cultured skin fibroblasts and Vero cells.

Of great concern has been the finding of microcephaly in newborns in Brazil and other countries following maternal Zika virus infection during pregnancy. By early 2016 Brazilian authorities reported more than 4000 cases with more and more cases anticipated as new surveillance systems are introduced in at-risk populations across the Americas. These findings have led to a review of case findings in earlier outbreaks in Asia and the Pacific islands—microcephaly was seen in these outbreaks but not fully appreciated until the very large Brazilian epidemic. There is also concern that Zika virus infection may be associated with the development of Guillain-Barré syndrome (Chapter 39: Viral Syndromes).

At present, control relies on the elimination of the urban vector *A. aegypti*. Spread of Zika virus into other *Aedes* species, notably *A. albopictus*, has already occurred and there is a serious risk of the virus spreading into multiple species of vectors in the same manner that West Nile virus disseminated across the United States. There is also a risk that the virus will adapt to new vectors and become more pathogenic for humans. There is neither a vaccine nor chemotherapy available, but there are ongoing emergency vaccine development programs ongoing in several countries.

JAPANESE ENCEPHALITIS

Japanese encephalitis virus was first isolated in 1935 from a fatal case of encephalitis although records show probable epidemics in Japan from the 1870s onwards. It is widely distributed throughout Asia, the Indian subcontinent, and the western Pacific. Japanese encephalitis virus gives its name to a serological complex containing three other significant human viruses, St. Louis encephalitis virus, Murray Valley encephalitis virus, and West Nile virus.

Clinical Features

Following exposure to mosquito bite there is a five- to seven-day incubation period and then a non-specific viral prodrome. Early clinical signs include lethargy, fever, headache, abdominal pain, nausea, and vomiting. The infection may then progress to nuchal rigidity, photophobia, altered consciousness, hyperexcitability, muscle rigidity, cranial nerve palsies, tremors, paresis, incoordination, convulsions, stupor, coma, and death. Fatality rates vary between 10% and 40% and are often reflective of medical care resources and capabilities. A high proportion, up to 40% to 70%, of survivors suffer persistent physical, behavioral, and/or psychological sequelae. The common focal flaccid paralysis is not dissimilar to that induced by polioviruses. The encephalitis is managed with supportive care as there are no specific antivirals demonstrating clear benefit. Lifelong neurological sequelae present a major economic and resource burden in poorer endemic regions.

Pathogenesis, Pathology, and Immunity

The pathogenesis of Japanese encephalitis involves the interplay between direct viral damage and the pathological effects of the host immune response. Infection is believed to initiate with the replication of virus in the skin at the site of being bitten by the mosquito and from there travel to regional lymph nodes to produce a primary viremia, which leads to infection of the reticuloendothelial system. A secondary viremia follows, which may result in systemic infection involving other organs and the central nervous system. The viremia is transient, lasting for only a few days. The capability to mount an early immune response plays a critical role in containing viral replication and preventing viral invasion of the spinal cord and brain.

Laboratory Diagnosis

The standard diagnostic procedure is testing of the patient's serum and cerebrospinal fluid for specific antibody by IgM antibody capture ELISA. Cross-reactions with other flaviviruses may be a problem particularly in patients with multiple previous flavivirus infections. As in other cases of arbovirus encephalitis the period of viremia is short, usually ending before the patient seeks medical care, therefore RT-PCR testing for viral RNA or virus isolation (using newborn mice or Vero or C6/36 cells) from blood or cerebrospinal fluid is usually not successful.

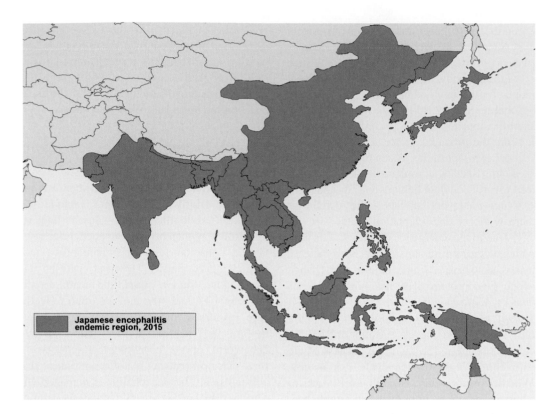

FIGURE 36.7 Japanese encephalitis reported cases, global distribution, 2014. *From the US Centers for Disease Control and Prevention.*

Differential diagnosis of Japanese encephalitis should include other causes of encephalitis, for example, other flaviviruses, Nipah virus and enterovirus encephalitis, herpes simplex encephalitis, as well as cerebral malaria, bacterial meningitis, and tuberculosis.

Epidemiology, Prevention, Control, and Treatment

Japanese encephalitis virus is the most important cause of viral encephalitis in China, Korea, Thailand, Indonesia, and other countries of Southeast Asia and in the past 20 years it has extended its range westward into India, Pakistan, Nepal, Myanmar, and Sri Lanka and eastward into the Pacific islands of Saipan and the northern Marianas, as well as the northern tip of Queensland, Australia. Some 50% of the world's population resides within endemic areas. It is estimated that there are more than 50,000 cases and 15,000 deaths annually (Fig. 36.7).

Japanese encephalitis virus is transmitted principally by the mosquito *Culex tritaeniorhynchus*, which breeds in fresh water and irrigated rice fields: this zoophilic (preferring animals to humans) mosquito feeds upon swine, horses, and humans, the latter being essentially a "dead-end" host (Fig. 36.8).

Human-to-human transmission is rare as the viremia in humans is of low titer and short duration. The increased virus activity in some areas has been ascribed to the recent upsurge in pork production—pigs are the most abundant species of domestic animals in many parts of Asia; swine have a short life span and continuously provide generations of susceptible hosts. The mosquito–swine–mosquito transmission cycle serves as an efficient mode of virus amplification. In tropical areas, outbreaks occur at the end of the wet season with sporadic cases appearing throughout the year. In temperate zones outbreaks tend to occur in late summer and early autumn.

The reservoir hosts are wading birds, such as egrets and herons. It is not known how Japanese encephalitis virus is maintained in the more temperate zones of China, Korea, and Japan. Overwintering in mosquitoes is a possibility or the virus may be reintroduced each summer by migratory birds from equatorial regions. Distinct genotypes of isolates from northern latitudes suggest overwintering to be the more likely.

At least five genotypes are recognized (Table 36.4). Genotype 3 is the most widely distributed, and the only genotype found in the Indian subcontinent. The evolutionary progression and geographical movement of different variants are complex, and homologous recombination between genotypes has been reported, resulting in many

FIGURE 36.8 Paddy fields. In the Japanese encephalitis endemic/enzootic regions of Southeast Asia, control has proven near impossible except through human (and in a few places swine) vaccination. *From Rice Wisdom, Save our Rice campaign, Penang, Malaysia.*

TABLE 36.4 Based on Comprehensive Genomic Sequencing, Five Genotypes of Japanese Encephalitis Virus Have Been Recognized, Each with a Geographically Distinct Distribution and Epidemic Potential

Genotype	Geographic Region	Epidemic/Endemic Pattern
Genotype 1 (G1)	Northern Thailand, Cambodia, Republic of Korea	Was thought to occur principally in temperate, epidemic areas
Genotype 2 (G2)	Southern Thailand, Malaysia, Papua New Guinea, Northern Australia	Was thought to occur principally in tropical, endemic regions [one strain in Republic of Korea 1951]
Genotype 3 (G3)	Japan, Republic of Korea, China, Taiwan, Philippines, India, Sri Lanka	Was thought to occur principally in temperate, epidemic areas [includes prototype Nakayama strain]
Genotype 4 (G4)	Only isolated in Indonesia between 1980 and 1981	Occurred principally in tropical, endemic region
Genotype 5 (G5)	China, Republic of Korea	[One strain from Malaysia 1952] Now reemerging in temperate, epidemic areas

Source: From Mackenzie, J.S. et al., Nat Med 2004;10 (12 Suppl): S98–109.

new, unpredictable chances for outbreaks in new settings in Asia. Genotype 2 occurs in Papua New Guinea and may form an evolutionary bridge with Murray Valley encephalitis virus.

Control has involved several integrated approaches, focusing on reducing mosquito populations wherever practical, the placing setting of pig farms and the use of human vaccines. Draining of rice paddies at the time when *Culex tritaeniorhynchus* normally breeds reduces the risk of transmission. The removal of the principal amplifier host, pigs, from areas of human habitation significantly reduces the risk of contact with humans. The combined use of environmental controls and the use of vaccination has stabilized, if not reduced, the incidence of Japanese encephalitis virus infection in many of the more economically advanced regions of Asia.

An inactivated vaccine has been available for some decades. Although relatively safe and effective, immunity is short, thus requiring annual booster doses. This vaccine has been expensive. A chimeric live vaccine using a yellow fever virus 17D backbone has been licensed recently in some countries. The incidence of Japanese encephalitis in China has been progressively reduced since the introduction of the SA 14-14-2 attenuated-live virus vaccine into the universal childhood immunization program—this inexpensive, highly efficacious vaccine is now being made available in other countries where the virus is endemic.

Vaccination of pigs would do much to reduce the risk of transmission to humans but the high turnover among pig populations means that vaccination of litters is often considered neither economically practical nor effective, especially if piglets acquire maternal antibodies.

WEST NILE VIRUS INFECTION

West Nile virus is a member of the Japanese encephalitis complex. First isolated in 1937 from a febrile patient in Uganda, few outbreaks were recognized until the mid-1990s when significant morbidity was recorded in Europe among both humans and horses. The virus suddenly appeared in North America for the first time in 1999 where it has since become established throughout the United States, Canada, the Caribbean, and in parts of South America, becoming a significant public health risk, especially to the immunosuppressed and the elderly. Presently the virus is also widespread throughout Africa, southern and Eastern Europe, Asia, and Oceania. In Australia the virus is identified as Kunjin virus, a minor variant of West Nile virus, that can cause mild non-specific symptoms but rarely causes encephalitis (Fig. 36.9).

Clinical Features

Most human infections are asymptomatic; clinical illness develops in only about 20% of infections. Less than 1% of infections progress to neurological disease. The incubation period varies from 2 to 14 days. The commonest presentation is a self-limited febrile illness known as West Nile fever, in which the rapid onset of fever is accompanied by myalgia, headache, anorexia, and back pain: symptoms may last for several weeks. Less often the gastrointestinal tract may be involved, leading to diarrhea and vomiting. About half of the symptomatic patients develop a macropapular rash. Infrequently, more severe infections may be accompanied by pancreatitis, hepatitis, and myocarditis. Careful follow-up study has revealed that many patients suffer continuing clinical signs from 2 to 4 weeks or longer.

Neuroinvasion is seen most frequently as meningitis, encephalitis, or focal flaccid paralysis. Movement and muscle tone are affected, with myoclonus and Parkinsonian features. There is some risk of confusion with Guillain-Barré syndrome. Viral meningitis/encephalitis occurs in as many as 40% of patients who are immunosuppressed.

Pathogenesis, Pathology, and Immunity

Following virus entry in the mosquito bite site and primary replication in regional lymph nodes, it is thought that the virus then disseminates systemically via the blood to the reticuloendothelial system. The resulting viremia is too low, however, to represent a significant risk that further insect bites might lead to human-to-human transmission.

A secondary viremia is most likely required for neuroinvasion to occur. The result is classical viral encephalomyelitis. It is not known why the elderly are at greater risk of neurological complications (in the United States in recent years the median age of patients who died was 78 years). It is most likely that the delay or decline in the host immune/inflammatory response associated with aging is responsible. It is known that West Nile virus directly infects and destroys, through necrosis and apoptosis, both neurons and glial cells, and the resulting classical mononuclear and T cell inflammatory infiltration (perineuronal and perivascular) is also known to be damaging.

Laboratory Diagnosis

As is the case with most arboviruses, specific serology is the most common approach to confirm cases—the preferred method is IgM capture ELISA. RT-PCR is an alternative, although the usual low-titered viremia means that its usefulness is limited to the most acute phase blood specimens and to cerebrospinal fluid. The presence of virus-neutralizing antibodies is the most specific evidence of West Nile virus infection, and is the method of choice for identifying individual viruses within the Japanese encephalitis complex of flaviviruses. These tests are time-consuming, however, and require specialist skills and reagents.

Epidemiology, Prevention, Control, and Treatment

Birds and mosquitoes maintain West Nile virus: more than 300 different bird species are susceptible, with infection yielding high, sustained viremia levels to infect blood-feeding mosquitoes and continue the virus transmission cycle. This is the case whether or not the infection causes clinical disease and death in the species of bird. In particular crows and related species in the family *Corvidae* have been conspicuous hosts of the virus because of their high mortality rate. Indeed, abnormal numbers of dead birds in urban areas was one of the first signs of the impending West Nile virus outbreak in New York in 1999.

In contrast to birds, the viremia in humans is of short duration and of relatively low titer, although there is a risk of virus entering the blood supply and thus screening of all

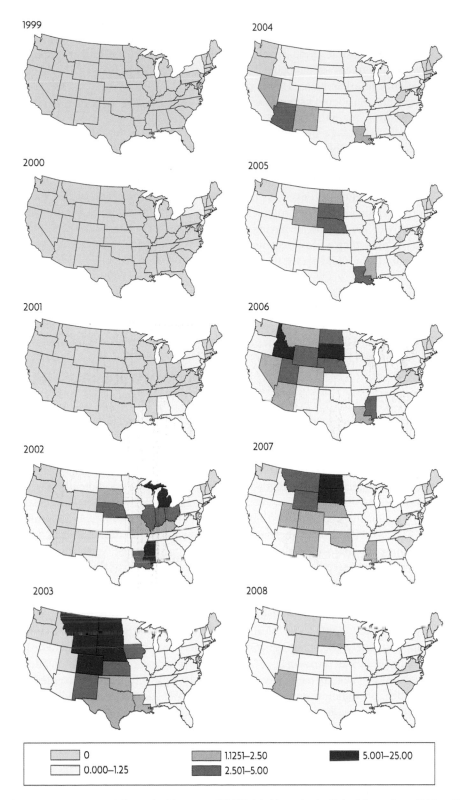

FIGURE 36.9 West Nile virus (WNV) annual disease incidence per 100,000 inhabitants in the United States from 1999 to 2008. Between 1999 and 2015, 42,462 cases were reported to the U.S. Centers for Disease Control and Prevention (CDC). Of these, 20,870 cases were neuroinvasive (mortality rate 9%) and 26,372 cases were non-neuroinvasive (mortality rate <1%). The rapid spread of the virus to every state in the United States within 4 years showed that ecological and environmental conditions had been permissive and all that was needed was introduction of the virus. A factor favouring this major expansion in range was the susceptibility of many different mosquito species—at least 60 species may serve as reservoir or bridging vectors in the Western Hemisphere. *From the US Centers for Disease Control and Prevention.*

blood donations has been done throughout the United States and Canada for the past decade.

The bird–mosquito–bird transmission cycle involves *Culex* spp. mosquitoes, with *C. pipiens* often being the major vector given its abundance in urban areas. However, over 40 species of mosquitoes are considered as competent vectors. Vector species differ from region to region, and this may explain any variation between the extent of human infection relative to virus activity in the local mosquito and bird populations. Both humans and horses are incidental hosts: equines do not represent a significant reservoir host.

Among the risk factors in humans, age is the most significant for the development of neurological complications, with an adult over the age of 80 having a greater than 30-fold chance of developing a life-threatening neurological condition compared to a child under the age of 10.

Several West Nile virus vaccines are in late-stage development and in clinical trials: a chimeric virus vaccine using yellow fever virus as backbone; several subunit vaccines, using various recombinant DNA technologies to express viral envelope antigens; and a DNA vaccine.

ST. LOUIS ENCEPHALITIS

St. Louis encephalitis has been reported throughout the majority of the United States, southern Canada, Mexico, and in limited areas of Central and South America. It is diffusely endemic over large rural areas with a low seropositive rate in humans, but with occasional epidemics of hundreds to more than a thousand human cases.

St. Louis encephalitis infection may be classified into three clinical syndromes: (1) febrile illness with headache and myalgia, (2) aseptic meningitis, and (3) encephalitis. In patients who develop severe disease, following an acute febrile prodrome, there may be an acute or subacute appearance of meningeal and other neurological signs and symptoms, including nuchal rigidity, disorientation, unsteady gait, apathy, somnolence, coma, and death. The case fatality rate is approximately 8%, ranging from approximately 2% in the young to 22% in the elderly.

The virus is transmitted among nesting and juvenile passerine birds by a number of culicine mosquitoes, including *Culex pipiens* and *Culex quinquefasciatus* in the eastern and urban areas of the western United States, *Culex tarsalis* in rural areas of the western United States, and *Culex nigripalpus* in Florida and southern areas of the United States. Depending on the climate of the region, maintenance of the virus can occur by year-round horizontal transmission from bird to mosquito, overwinter survival of infected mosquitoes, or venereal or transovarian infection.

The virus remains endemic in some regions, producing epidemics when temperature, moisture, and breeding conditions favor vector mosquito species. Breeding sites,

such as automobile tire dumps and tree-holes formed by tree cutting, influence vector population density but are not amenable to vector elimination programs because of cost and resource demands.

USUTU VIRUS INFECTION

Usutu virus is an African mosquito-borne flavivirus belonging to the Japanese encephalitis virus complex. It was discovered in South Africa in 1959 and among birds and *Culex* spp. mosquitoes in a number of African countries. In Europe, the presence of the virus was first reported in Austria in 2001 with a significant die-off of Eurasian blackbirds. In subsequent years, the virus has been found in Italy, Germany, Spain, Hungary, Switzerland, Poland, England, Czech Republic, Greece, and Belgium, where it has caused unusual mortality in birds. The first human cases of meningoencephalitis were seen in immunocompromised patients in Italy in 2009. The virus is considered to present a potential threat of further inroads in Europe and of introduction into Asian countries.

MURRAY VALLEY (AUSTRALIAN) ENCEPHALITIS

Murray Valley encephalitis virus, another member of the Japanese encephalitis complex, is endemic in Papua New Guinea, Irian Jaya, other areas of the eastern Indonesian archipelago, and in northern Australia, where outbreaks of encephalitis occur sporadically. The virus was first isolated from brain tissue in 1951 but it is clear that earlier epidemics were of the same etiology. In Australia, epidemics involving humans occur in the southeast of the country (Murray Valley) only in those very occasional summers with heavy rainfall and extensive flooding. These conditions encourage explosive increases in numbers of water birds, which are the principal vertebrate reservoirs of the virus, and its mosquito vectors, most notably in Australia *Culex annulirostris*. The encephalitis in humans is similar to that caused by Japanese encephalitis virus; like Japanese encephalitis, many infections are asymptomatic, but among patients who develop neurological disease, the case-fatality rate is about 24%, and >50% of survivors experience neurological sequelae.

ROCIO VIRUS ENCEPHALITIS

Rocio virus, another member of the Japanese encephalitis virus complex, was isolated from brain tissue in the course of investigating an explosive encephalitis epidemic in São Paulo State, Brazil, in 1975. The epidemic spread to more than 20 municipalities over two years with approximately 1000 cases, 100 deaths, and serious neurological sequelae (especially motor function and equilibrium deficits) in 20%

of survivors. Mortality in children under 1 year of age was 30% and 10% in persons over 60 years of age. Clinical manifestations were similar to those described for St. Louis encephalitis and Japanese encephalitis: fever, headache, vomiting, weakness, anorexia, abdominal distention, nausea, and oropharyngeal and conjunctival hyperemia, motor impairment, convulsions, coma, and death. The transmission cycle of the virus remains unknown, but is thought to involve wild birds and forest mosquitoes. Most cases have been in younger males (from 15 to 30 years of age) who have been involved in forestry and agricultural activities. It has been considered that this pathogenic virus presents important questions of "emergence" in a region undergoing rapid development from primal forest.

DISEASES CAUSED BY TICK-BORNE MEMBERS OF THE GENUS *FLAVIVIRUS*

Tick-Borne Encephalitis

Tick-borne encephalitis virus is a member of a genetically and antigenically linked complex of some nine mammalian tick-borne viruses. Many of these viruses are found across Eurasia; Kyasanur Forest disease virus occurs in India and recently has extended its range into the Middle East; Powassan virus occurs in North America (see Table 36.1). The diseases caused by these viruses have many local names: tick-borne encephalitis is the current name for the disease also widely known as Russian spring–summer encephalitis and Far Eastern encephalitis. Tick-borne encephalitis virus is an example of an emerging pathogen as it has spread into new geographical areas—over the past decades there has been a 400% increase in tick-borne encephalitis cases in Europe and the virus has spread to regions that were previously unaffected.

Clinical Features

There is a set of tick-borne encephalitis virus pathotypes that occupy separate niches across Eurasia: the European subtype, Siberian subtype, and Far Eastern subtype. The least virulent of these occurs in the west, in Europe, and the most virulent in the east, in Eastern Siberia, and Japan. Patients infected with the European subtype usually have the least severe disease (headache, myalgia, fatigue, nausea, malaise), with approximately 65% recovering after the first of the biphasic febrile episodes and 35% progressing to neurological involvement (meningitis, encephalitis) and neuropsychiatric sequelae. Most patients survive; the case-fatality rate is 1% to 2%. In contrast, the Siberian and Far East subtypes induce a monophasic disease, with rapid onset of high fever, headache, vomiting, myalgia, and indications of neurological infection including paresis, motor neuron paralysis, seizures, and coma. There is full-blown acute inflammatory encephalomyelitis; the case-fatality rate for the Siberian subtype is 6% to 8% whereas the Far Eastern subtype has a much greater and wider range of 20% to 60%. Survivors frequently develop permanent neurological sequelae, including paresis and atrophy of the neck muscles and paresis in the lower extremities, resulting in a polio-like syndrome.

Pathogenesis, Pathology, and Immunity

Tick-borne encephalitis virus is transmitted to humans through the bite of an infected tick. Virus is deposited in the skin where it infects dendritic cells or resident macrophages and is transported to draining lymph nodes. The virus is believed to replicate in lymphoreticular cells and yield a viremia characterized by a high virus titer, from where it rapidly penetrates the blood–brain barrier at specific sites in the brain. Infection produces lifelong protective immunity among survivors. Humoral and cellular immune protective responses are elicited, but there is some evidence that an exuberant CD8 T cell response may drive cytokine-mediated pathology.

Laboratory Diagnosis

IgM capture ELISA is the diagnostic method of choice using either blood or cerebrospinal fluid as substrate. Although viral RNA can be detected by PCR, the viremia inevitably declines to low or negligible levels by the time neurological symptoms develop. As with all suspected human flavivirus infections, information regarding the circulation of other flaviviruses at the time and place of exposure should inform the interpretation of serological results.

Epidemiology, Prevention, Control, and Treatment

Tick-borne encephalitis and closely related viruses are maintained in nature through a complex cycle involving *Ixodid* ticks and feral mammals, most notably rodents and deer. The different pathotypes (European, Siberian, and Far Eastern subtypes), important in considering the clinical features of human infection, also are important in considering differences in virus ecology, natural history, and epidemiology.

The epidemiology of the tick-borne flaviviruses is more complex than is the case for mosquito-borne flaviviruses, since ticks serve both as vectors and also as reservoir hosts. In contrast to mosquitoes, ticks live for several years, often longer than the generational time of a reservoir animal host. Ticks are often active from spring until the autumn in temperate climates. Ticks develop successively through stages (larva to nymph to adult) and a blood meal is generally required at each stage. Tick-borne flaviviruses are passed from one developmental stage to another (*trans-stadial transmission*) as well as from one generation of tick to the next (*transovarial transmission*). Some species of tick spend their whole lives on one vertebrate host, whereas others fall off, molt, then find a different host after each meal. The larvae and nymphs generally feed on birds or small mammals such as rodents, whereas adult ticks prefer larger animals.

Rodents in particular act as amplifying hosts for the virus but can also in some instances be considered long-term reservoirs, especially in expanding the geographical range of the virus. In some regions up to 15% of trapped rodents show evidence of infection.

Human infections follow exposure to adult ticks that became infected during the current or previous season, and human outbreaks have also occurred following ingestion of either unpasteurized sheep or goat's milk or cheese in endemic areas (Fig. 36.10).

Clinical management of tick-borne encephalitis cases is largely supportive.

Inactivated virus vaccines for humans are available for use against tick-borne encephalitis in Europe and eastern Asia, but are expensive. Vaccination is recommended for persons living in areas of endemicity, especially those likely to have high exposure to ticks due to their occupation or travel. The use of a vaccine in Austria since 1986 has substantially reduced the number of clinical cases.

Other Tick-Borne Flavivirus Infections of Humans

Louping ill virus is a tick-borne flavivirus found in the British Isles. It is maintained in an *Ixodes ricinus* tick–grouse cycle, from which virus is transmissible to sheep, in which it causes a rapidly lethal disease. Rarely, louping ill

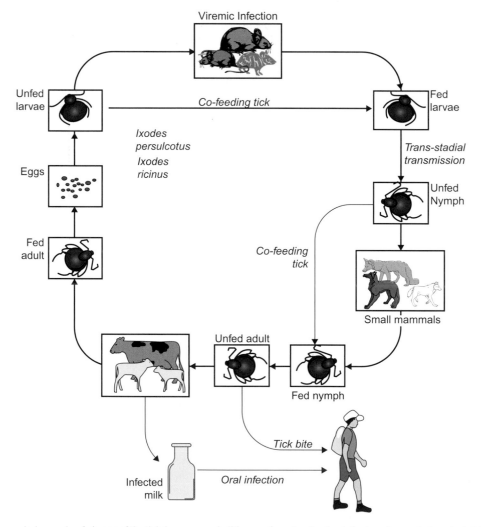

FIGURE 36.10 Transmission cycle of viruses of the tick-borne encephalitis complex, showing hosts for larval, nymphal, and adult ticks. Virus is passed to succeeding tick stages during molting, and is passed transovarially to progeny. Both male and female ticks are involved in transmission. *Reproduced from MacLachlan, N.J., Dubovi, E.J., 2011. Veterinary Virology, third ed., Elsevier, with permission.*

virus is transmitted to humans by ticks or occupationally by contact with infected sheep tissues, producing a relatively mild meningoencephalitis.

Powassan virus was named after a Canadian town in which the virus was isolated from a pediatric case of encephalitis. It is the only member of the tick-borne encephalitis complex found in North America. It is now found mainly in north central and north-eastern regions of the United States, and has also been found in the Primorsky region of Russia. This virus is principally transmitted by *Ixodes* spp. ticks, and includes two distinct lineages: one cycles through *Ix. cookie* ticks and woodchucks (*Marmota momax*) or skunks (*Mephitis mephitis*); the second circulates through the deer tick *Ix. scapularis* and the white-footed mouse (*Peromyscus leucopus*). A small number of human encephalitis cases over the past decade have been reported, due to either lineage.

Omsk hemorrhagic fever virus is a cause of hemorrhagic fever and encephalitis in western Siberia. The transmission cycle of the virus is between *Dermacentor* spp. ticks and water voles and common voles, but humans are usually infected by direct contact with the blood and tissues of muskrats (*Ondatra zibethica*) during trapping and skinning. Muskrats were imported from America in the early 1900s for their fur. Typical clinical signs of the disease in humans include petechial hemorrhages of the soft palate, nose, gums, uterus, lungs, and urinary and gastrointestinal tracts, but clinical symptoms also include diffuse encephalitis which disappears during the recovery period. The case-fatality rate is less than 3% and survivors have an uneventful recovery with no long-term sequelae. There is a relatively

high seroprevalence in the endemic region of central Russia suggesting more subclinical infection or mild disease than is appreciated.

Kyasanur Forest disease virus is transmitted primarily by *Haemaphysalis* spp. ticks in the tropical deciduous forests of the Karnataka State in South India, where episodically it is an important human pathogen (Fig. 36.11). A variant, Nanjianyin virus, has been found in Yunnan Province, China. This suggests that the virus(es) may have a wider geographical distribution than presently known. Persons with occupational or recreational exposure in forested endemic areas are at particularly high risk of exposure. Human disease has in several instances been accompanied by outbreaks of fatal hemorrhagic disease in forest monkeys. After an incubation period of 3 to 8 days, the clinical signs of infection are fever, headache, severe muscle pain, cough, dehydration, and gastrointestinal distress. After 1 to 2 weeks, some patients recover while most experience severe hemorrhagic fever with 2% to 10% mortality. A formalin-inactivated vaccine produced in chick embryo fibroblasts is licensed and available in India.

Alkhurma hemorrhagic fever virus (a distinct variant of Kyasanur Forest disease virus) has been isolated repeatedly since 1995 from the blood of patients with severe hemorrhagic fever and/or encephalitis in Saudi Arabia. The case fatality-rate has been above 30%. Subsequent cases have been documented among tourists to Egypt, extending the geographical range of the virus and suggesting that the geographical distribution of the virus is wide and infections are under-reported. The vectors are *Ornithodorus* spp. ticks of camels, sheep, and goats.

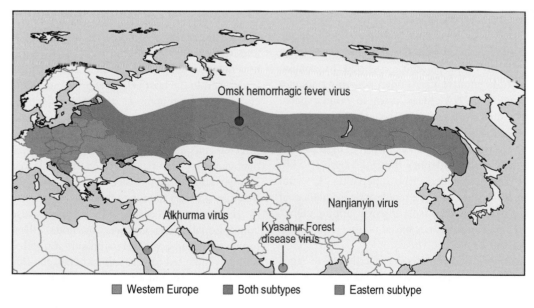

FIGURE 36.11 Geographic distribution of tick-borne encephalitis and other tick-borne flavivirus pathogens. *Reproduced from Mandl, C., Holbrook, M., Tropical Infectious Diseases, third ed., Saunders, with permission.*

DISEASES CAUSED BY MEMBERS OF THE GENUS *HEPACIVIRUS*

Hepatitis C

With the introduction of sensitive assays for screening blood for hepatitis B virus in the late 1970s, it was anticipated that post-transfusion hepatitis would be virtually eliminated, but this was not to be. There remained a substantial residue of cases that were called "non-A, non-B" hepatitis (NANBH). Nevertheless, in 1989, a team of molecular biologists in the United States succeeded in an ambitious assignment that seemed to many to be unachievable. The ingenious protocol they devised serves as a prototype applicable to the discovery of other currently unknown infectious agents that cannot be cultivated and is worth summarizing here.

Daniel Bradley and his colleagues had previously demonstrated that hepatitis could be transmitted to chimpanzees by inoculation of factor VIII known to have been contaminated with an agent that had caused non-A, non-B hepatitis among hemophiliacs. Filtration studies had indicated the size of the causative agent to be of the order of 40 to 50 nm, while a buoyant density of $1.1\,g/cm^3$ and sensitivity to chloroform suggested it to be an enveloped virus. The reasonable assumption was made that plasma from chimpanzees that had developed chronic hepatitis following inoculation with factor VIII would constitute a good source of the putative virus, and that virions could be concentrated from the plasma by ultracentrifugation. Total DNA and RNA were extracted from the pellet, denatured into single strands, and reverse-transcribed using random primers to initiate the transcription of any single-stranded nucleic acid. The resulting complementary DNA was cloned into the bacteriophage λgt11 expression vector, which was then used to infect bacteria. Bacterial colonies were then screened for production of any antigenic polypeptide sequence recognizable in an immunoassay by serum taken from chronically infected patients, assumed to contain antibody against the putative virus. After screening about a million such random cDNA clones, a single clone capable of binding antibodies from several infected individuals was found. This DNA was then used as a hybridization probe to derive a larger overlapping clone from the cDNA library, which in turn was used to identify the full-length (9.5 kb) positive-sense single-stranded RNA hepatitis C viral genome. Eventually the complete nucleotide sequence of the genome was determined by isolating overlapping cDNA clones using hybridization probes based on the sequence of previous clones.

Sequence analysis and transfection studies have since allowed designation of hepatitis C virus as a new genus within the family *Flaviviridae*, comprising at present one species that includes at least seven genotypes. The genome organization, and pattern of proteolytic cleavages of the primary polyprotein translation product, are similar but not identical to other flaviviruses (Fig. 36.2).

Clinical Features

The incubation period of hepatitis C, though ranging up to several months, averages 6 to 8 weeks. About 75% of infections are subclinical. Clinical infections are generally less severe than those caused by hepatitis B virus—there is a shorter pre-icteric period, less severe illness, absent or less marked jaundice, and levels of serum alanine aminotransferase (ALT) that are often lower, but may fluctuate widely. The case-fatality rate from fulminant hepatitis is 1% or less. However, hepatitis C virus infection leads much more frequently to chronic liver disease than hepatitis B virus. Long-term studies have shown that about 60% to 80% of all patients infected with hepatitis C virus remain continuously or erratically viremic, usually for decades unless treated. Among these chronically infected individuals, those with normal ALT levels (30% to 40%) have a good long-term prognosis, while those with persistently raised ALT levels have a 10% to 20% rate of progression to cirrhosis and a 1% to 5% rate of developing hepatocellular carcinoma over the ensuing 20 to 40 years. Indeed, end-stage liver disease due to chronic hepatitis C has become the commonest indication for liver transplantation in some Western countries, outstripping alcoholism and hepatitis B. In some countries hepatitis C has also now become the commonest identified risk factor for hepatocellular carcinoma, with more than 90% of hepatitis B virus-negative cases being hepatitis C virus antibody positive (Fig. 36.12).

Pathogenesis, Pathology, and Immunity

The pathogenesis of hepatitis C infection presents several remarkable puzzles, because its replication cycle does not include a DNA intermediate step with the capacity to integrate into the host cell. It is not clear whether long-term chronic infection involves persistent infection of individual cells, with or without virus production—unusual for an RNA virus—or whether it involves continuing turnover of infected cells and *de novo* infection of new cells. Similarly, oncogenesis without any defined viral DNA oncogene must involve additional indirect mechanisms (see Chapter 9: Mechanisms of Viral Oncogenesis).

Hepatitis C cannot be routinely isolated using conventional cell culture methods. However, virus replication can be studied using sub-genomic or full-length replicons adapted for growth in hepatoma-derived (Huh-7) cells by induced mutations in viral non-structural proteins, or using a unique genotype 2a isolate (JFH1) originally obtained from a case of fulminant hepatitis in Japan. An alternative model system is the use of mice in which the

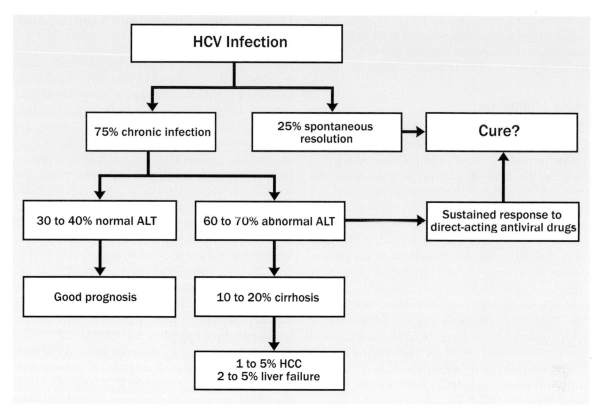

FIGURE 36.12 Natural history of infection with HCV without treatment. In the past, treatment with interferon and ribavirin led to a sustained response in 30 to 60% of individuals, depending on genotype and other variables. However, due to the spectacular success of direct-acting antiviral drugs, most patients can now expect elimination of circulating virus following a course of oral treatment that is well tolerated and does not include interferon. *Adapted from Sasadeusz, J. et al., 2008, HIV, Viral Hepatitis and STIs, A guide for primary care. ASHM, Sydney, pp. 19 (Chapter 1).*

liver has been populated with human hepatocytes that support the complete hepatitis C virus replication cycle.

Hepatitis C virus can gain access to the hepatocyte using two mechanisms, cell-free spread and cell-to-cell spread. Both CD81 and the scavenger receptor class B type 1 (SRB-1) proteins are required for entry of cell-free virus, although the requirement for the latter may differ between genotypes. The tight junction proteins claudin-1 and occludin also play a role in entry to polarized hepatocytes. The major target cell is the hepatocyte but there is also evidence suggesting viral replication in leukocytes and perhaps other cells, although this is controversial. Hybridization studies *in situ* and sophisticated immunofluorescence techniques have suggested 5% to 50% of all hepatocytes may be infected. Acute injury to hepatocytes is likely to be largely due to cytotoxic T cell mechanisms, and a complex inflammatory process leads to hepatocyte damage and fibrosis during chronic infection; steatosis is often also present, possibly through multiple interactions between the hepatitis C virus core protein, lipid droplets, and mitochondrial membranes.

Hepatitis C virus infection initially activates cellular signaling pathways that induce synthesis of type I interferon, but these pathways are later suppressed as viral proteins accumulate. There is evidence that viral proteins interact with cellular pathogen recognition pathways and interferon signaling pathways, to reduce the expression of type I interferon during infection and also to render the infected cell relatively resistant to the action of interferon. The CD4+ and CD8+ T cell-specific immune response is slow to develop, but then tends to be stronger and broader in those infections that resolve, compared with those that persist, suggesting that the cellular immune response may be at least partly responsible for virus clearance.

Antibodies have been shown to neutralize an infecting inoculum in chimpanzees. However, a protective role for neutralizing antibodies in natural infection is not clear. Antibodies continue to be made and most of the virus is bound in virus–antibody complexes but the infection is not eliminated as a consequence. Indeed, chimpanzees recovered from hepatitis C virus infection can be re-infected with the homologous (or heterologous) strain of virus. Highly exposed humans, for example, intravenous drug users, also frequently experience multiple episodes of acute hepatitis C, often due to exogenous reinfections with the same or another strain as well as reactivation of the original infection.

Finally, it should be noted that hepatitis C virus, similar to HIV, exhibits a high rate of genetic mutation during persistent infection by virtue of the error-prone NS5B

RNA-dependent RNA polymerase. The circulating virus population exists as a quasi-species of related sequences, and the emergence of escape mutants may also contribute to the ability of the virus to evade immune elimination.

Laboratory Diagnosis

Persons being tested for hepatitis C virus infection should have access to pre-test counseling about the implications of a negative or positive result, and post-test counseling (particularly in the case of positive results) about the significance of the finding, their prognosis, and the implications for spread to others.

In 1990, the first specific diagnostic test for hepatitis C virus antibody was licensed for the screening of blood donors in the United States. A recombinant fusion protein corresponding to a large portion of the non-structural protein NS4 was expressed from viral cDNA and used as an antigen in an enzyme immunoassay to detect viral antibodies. This first-generation enzyme immunoassay was quickly replaced by a more sensitive and specific second-generation version based on a recombinant yeast chimeric protein comprising the three most conserved hepatitis C virus proteins C, NS3, and NS4. Since then, a third-generation test which includes reconfigured core and NS3 antigens and a newly incorporated NS5 antigen, has become the dominant screening test; this has an estimated 98% sensitivity, but in acute infection there is a window period of 6 to 12 weeks before the test becomes positive. More recently, a number of point-of-care rapid immunoassays have been evaluated; these can be of value in resource-poor settings, but their performances vary widely. Because of the personal impact of a positive result and the low but measurable rate of false-positive reactions, all initially positive anti-HCV EIAs require supplemental testing, either by recombinant immunoblot assay (RIBA) or by confirming infection using a hepatitis C virus RNA test.

A confirmed positive result for antibody does not distinguish between acute, chronic, and past infection. Current infection is demonstrated by a positive RT-PCR test for HCV RNA; the quantitative version of this test is now highly sensitive, and can be used both to determine the presence or absence of infection and to quantitate the level of virus load in order to assist in patient management. Hepatitis C virus PCR tests may become positive around 2 weeks after infection, but assays for IgM antibody have not proved to be particularly useful.

Further investigation includes assessment of ongoing liver damage by tests of liver function, for example, transaminases; assessing the extent of cumulative liver damage by liver biopsy or non-invasive measurement of fibrosis: and testing for the hepatitis C virus genotype by line probe assay or PCR.

Epidemiology, Prevention, Control, and Treatment

In most Western countries the prevalence of antibodies to hepatitis C virus is around 1% among blood donors, but the prevalence in the community as a whole is estimated around 2.2% to 3.0%. Globally, it is likely that some 180 million individuals are infected with hepatitis C virus, with many being unaware of their infection until advanced liver disease and cirrhosis develop many years after initial exposure. While blood transfusion and factor VIII were major sources of infection in the past, the risk progressively declined in most countries as a result of more rigorous selection and testing of donors. Today in many countries the most clearly identifiable infected cohort is among intravenous drug users, in whom in some reports the rate of infection reaches 60% after 3 years of injecting. Intravenous drug users can account for up to 90% of new hepatitis C virus infections in some Western countries, with other routes of infection making up the remainder. Similar to hepatitis B virus, hepatitis C virus may be shed in genital secretions and saliva, but at levels that are several logs lower than in the blood. Although heterosexual promiscuity has been correlated with hepatitis C virus seropositivity, sexual transmission is thought to account for a very small fraction of new cases. Furthermore, documented transmission between discordant (i.e., one infected and one not) sexual partners occurs infrequently and significantly less than with hepatitis B. The same applies to perinatal transmission; it is relatively uncommon except when the mother is co-infected with human immunodeficiency virus. The risk from breast-feeding is negligible, and non-sexual spread within families is very rare.

Much higher rates of chronic infection are seen in certain countries, for example, Egypt, parts of the Middle East, and Vietnam. In these settings, nosocomial transmission due to re-use and inadequate sterilization of needles and syringes, particularly outside the formal healthcare system, is thought to have played a major role in transmission.

Hepatitis C virus is divided into 11 genotypes each with a unique global distribution. Multiple subtypes are found within each genotype. Although patients are normally only infected with a single genotype, each patient's sample consists of a mixture of closely related viruses referred to as quasispecies. Such quasispecies constantly change in composition during the course of a patient's infection, which compounds the challenge of developing satisfactory vaccines.

Genotype 1 represents the most common type of hepatitis C virus in the United States, and is the most difficult to treat. Patients with genotypes 2 and 3 are almost three times more likely to respond successfully to drug therapy, making genotype determination an important factor in clinical management. Most of the variability is associated with the E1/E2 genes, which differ in sequence by over 50% between some genotypes. The association of genotype 3 among drug

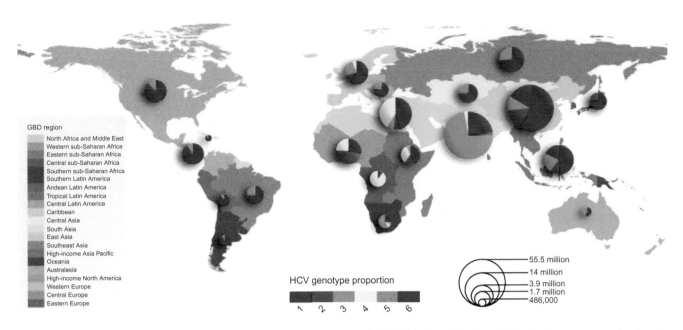

FIGURE 36.13 Relative prevalence of the different genotypes found in chronic HCV infection. The sizes of the pie charts are proportional to the estimated numbers of cases, as estimated from serology. *Reproduced from Messina et al., 2015. Hepatology 61, 77–87, with permission.*

users in Europe is distinct from the genotypes 1b and 2 found in other cohorts, and mapping genotype movement has been used to illuminate transmission networks (Fig. 36.13).

The immediate aim of hepatitis C virus treatment is the permanent loss of circulating HCV RNA, known as a sustained virological response (SVR). This is a predictor of normalization of ALT levels, arrest in the progression of liver disease, and even some reversal of fibrosis. From the 1990s, hepatitis C was treated with interferon-α and for the past 10 to 12 years standard treatment consisted of dual therapy with weekly subcutaneous injections of pegylated interferon (PEG-IFN) combined with oral ribavirin twice daily for 24 to 48 weeks. This reduced ALT levels to normal in about half of chronic hepatitis C patients, but all except about 20% to 25% of these showed a relapse after withdrawal of the drug. Response rates were higher among younger patients, among females, and in those with genotype 2 or 3 infections compared to genotype 1. Response rates are also significant among those with a lower viral load, and in those without cirrhosis. However, the incidence of disabling side effects from these two drugs (see Chapter 12: Antiviral Chemotherapy) frequently led to patients interrupting their treatment.

Antiviral treatment of hepatitis C has undergone an extraordinary revolution in the past few years. With the increase in basic knowledge about hepatitis C virus nonstructural proteins as potential targets, inhibitors of viral enzyme functions are rapidly being developed and introduced into the clinics. These drugs are principally targeted at the NS3-4A protease, the NS5A protein, and the RNA-dependent polymerase, NS5B. The protease inhibitors telaprevir and boceprevir, were licenced for use in the United States in 2011, but have now been largely replaced by second-generation protease inhibitors (simeprevir, asunaprevir, paritaprevir) with improved safety and dosage profiles. Inhibitors of NS5A (ledipasvir, daclatasvir) are potent and effective across all genotypes, but have a low barrier to resistance. NS5B polymerase inhibitors include both nucleoside (sofosbuvir) and non-nucleoside (dasabuvir) polymerase inhibitors. More than 30 other direct-acting antivirals (DAAs) are at different stages of development and clinical testing.

A number of clinical trials of these newer drugs have shown remarkable efficacy against genotypes 1, 2, and 3, when used in combination with either PEG-IFN or ribavirin. More recently, at least six different interferon-free combinations of direct-acting antivirals, with or without ribavirin, have been shown to be highly effective against infection with different genotypes. Due to the unpleasant side effects of interferon, this has been a long desired goal. Virtually 100% sustained virological responses are now achievable, although the high cost of these drugs creates a barrier to their wide-scale use in many countries unless funding subsidies are available. Many further drugs and combinations of the above are now being explored to define optimal treatment protocols and limit the emergence of resistance. This field is developing rapidly, and the reader should consult current guidelines issued by WHO (www.who.int/hepatitis/publications/hepatitis-c-guidelines-2016/en/) or by the American Association for the Study of Liver Diseases (AASLD) (http://www.hcvguidelines.org). The prospect that nearly all hepatitis C infections may now be curable by a simple oral drug regimen has transformed the hepatitis C landscape.

Prevention of hepatitis C depends on interrupting transmission. This includes measures to reduce spread between intravenous drug users, for example, by needle exchange programs, safe injecting rooms, education, and general community measures to address drug abuse. Policies to avoid accidental parenteral exposure to blood in the wards and operating theaters, must be universally understood and applied; the screening of blood donations and blood products should be universally available. Other settings of potential transmission, for example, tattoo parlors and prisons, need to be assessed and policies agreed. Finally, re-use of inadequately sterilized needles and syringes must be addressed by health authorities in those countries where this is practiced.

Prospects for a vaccine against hepatitis C virus are more distant. Chimpanzees vaccinated with different candidate vaccines have shown surprisingly poor immunity to challenge with homologous virus, and intravenous drug users who have cleared any detectable virus spontaneously or following treatment are found to be re-infected regularly. Despite many studies using envelope proteins in a variety of configurations with due attention given to the antigenic variation evident in natural strains, there is little optimism of hepatitis C vaccination becoming a reality in the next few years. However, the development of an effective vaccine is the "Holy Grail" that could ultimately control this infection worldwide.

OTHER HEPACIVIRUSES

In 1967, the serum of a surgeon (G.B.) acutely ill with hepatitis was inoculated into non-human primates. Nearly 30 years later, two flaviviruses were identified at the 11th passage of this inoculum having been passaged through tamarins. However, although evidence of extensive infection with either GBV-A or GBV-B was reported in a variety of New World monkeys, there is no evidence linking either virus with hepatitis in humans. A similar virus has been detected in fruit-eating bats trapped in Bangladesh (GBV-D). The suggestion is that these viruses represent a cluster of non-human flaviviruses, distantly related to hepatitis C virus, that have not crossed the species barrier into humans.

In 1995 a new related agent, hepatitis G virus (HGV, or GBV-C) was described in human sera. Although possessing the properties of a flavivirus, GBV-C is only distantly related to hepatitis C virus and does not cause hepatitis in humans. There is evidence of widespread infection among human populations, varying between 5% and 15%. It can be transmitted by blood and blood products, but normally is spread through sexual and mother-to-child contact. Infection is often associated with a high viremia that persists in up to 25% of cases. Despite the name, there is no evidence that this virus infects hepatocytes: some studies point to the virus persisting in lymphoid tissue. However, some but not all studies have found that individuals with persistent GBV-C and HIV-1 co-infection showed lower HIV-1 viral loads, higher CD4 T cell counts and longer survival, than those with HIV-1 infection alone; a possible mechanism is the finding that the GBV-C E2 envelope protein can interfere with HIV-1 entry, assembly, and release.

Classification of these agents is currently under discussion. Hepatitis G virus is the type species of the newly proposed genus *Pegivirus*, while GBV-B has been assigned to the genus *Hepacivirus*. Other candidate pegiviruses remain to be classified.

FURTHER READING

Guzman, M.G., Harris, E., 2015. Dengue. Lancet 385, 453–465.

Mackenzie, J.S., Gubler, D.J., Petersen, L.R., 2004. Emerging flaviviruses: the spread and resurgence of Japanese encephalitis, West Nile and dengue viruses. Nat. Med. 10 (12 Suppl), S98–109.

McMinn, P.C., 1997. The molecular basis of virulence of the encephalitogenic flaviviruses. J. Gen. Virol. 78, 2711–2722.

Monath, T.P., Chambers, T.J., 2003. Advances in virus research Volumes 59, 60, 61. The Flaviviruses: Structure, Replication, and Evolution. The Flaviviruses: Pathogenesis and Immunity. The Flaviviruses: Detection, Diagnosis and Vaccine Development. Elsevier, London.

Reed, W., 1902. Recent researches concerning the aetiology, propagation, and prevention of yellow fever, by the United States Army Commission. J. Hyg. 2, 101.

Samarasekera, U.,, Triunfal, M., 2016. Concern over Zika virus grips the world. Lancet 387, 521–524.

Webster, D.P., Klenerman, P., Dusheiko, G.M., 2015. Hepatitis C. Lancet 385, 1124–1135.

Chapter 37

Hepeviruses

In the immediate aftermath of the identification of hepatitis A and hepatitis B, cases of acute hepatitis continued to be recognized due to neither agent. Most cases of parenterally transmitted "non-A, non-B" hepatitis were eventually associated with hepatitis C virus, but outbreaks and sporadic cases of enterically transmitted non-A, non-B hepatitis (labeled at the time as "ET-NANBH") still occurred. The story of the original discovery of hepatitis E virus (HEV) is that in 1983 an intrepid Russian virologist, Mikhail Balayan, was investigating an outbreak of hepatitis in Tashkent, Uzbekistan; he needed to bring samples back to his Moscow laboratory but lacked refrigeration, so he mixed yogurt and a pooled filtrate of patient stools, drank it, went back to Moscow, and waited. When he became seriously ill a few weeks later, he started collecting and analyzing his own stool samples. Balayan found in his stools a new virus that produced biochemical evidence of hepatitis in non-human primates and could be seen by immunoelectron microscopy using convalescent human sera. Subsequently, in 1990, Gregory Reyes and his colleagues cloned and sequenced the genome of the virus. The virus was named hepatitis E virus and subsequently found to be the cause of major epidemics of severe hepatitis with high mortality in pregnant women. Later, it was discovered to be a zoonosis with reservoirs in swine and other animals.

Others recovered similar viruses from ET-NANBH outbreaks in India and many other countries of Asia as well as North Africa and Mexico. Daniel Bradley and colleagues then demonstrated that several of these strains reacted by immune electron microscopy with acute-phase sera from cases from many parts of the developing world.

Hepatitis E virus causes an acute and usually self-limiting infection, and is particularly prevalent in developing countries where it can be a major clinical and public health burden. It is notable for its high mortality in pregnant women. In contrast, in developed countries it is transmitted as a zoonosis from domestic animals, and can cause chronic infection with rapidly progressing cirrhosis in immunosuppressed patients. Hitherto it has been an under-diagnosed disease, whereas it may in fact now be the commonest form of viral hepatitis worldwide.

CLASSIFICATION AND PROPERTIES OF HEPATITIS E VIRUS

The family *Hepeviridae* contains viruses of humans, pigs, some rodents, monkeys and chickens. Based on natural hosts and sequence relationships, hepatitis E virus (HEV) has recently been classified within a new genus *Orthohepevirus* containing four species. *Orthohepevirus* species *A* includes viruses of humans, pigs, wild boar, deer, and rabbit. *Orthohepevirus B* viruses infect avians, *Orthohepevirus C* contains viruses infecting rats and ferrets, and *Orthohepevirus D* infects bats. There are four major genotypes of human HEV within the species *Orthohepevirus A*; HEV-1 and HEV-2 are restricted to humans and are found in developing parts of the world, while HEV-3 and HEV-4 are zoonoses and are seen in developed countries. A virus of cut-throat trout, with a different genomic organization and only about 13 to 27% amino acid similarity to mammalian and avian strains, has been assigned to a second genus *Piscihepevirus* (Fig. 37.1).

The spherical non-enveloped virions of HEV are 34 nm in diameter and resemble caliciviruses. The virions are extremely labile, tending to lose the outer layer even during storage either for a few days at 4°C, or following freeze–thawing. Infectivity is also lost during concentration by ultracentrifugation, or during purification by cesium chloride equilibrium gradient centrifugation owing to the high salt concentration. Since virions are present in low amounts in clinical samples and cell culture material, there have been few morphological studies of native particles. However, expression of the capsid protein in the baculovirus system produces both virion-sized, and slightly smaller, icosahedral virus-like particles (VLPs), which have been used for detailed structural studies (Fig. 37.2).

The genome is a single 7.5 kb molecule of single-stranded RNA of positive sense, polyadenylated at its 3′ end and having a 5′ methylated cap (a VPg protein primer similar to that of picornaviruses is absent). There are three separate but overlapping ORFs, with ORF2 the gene for the structural protein being located at the 3′ end of the genome and the short ORF3 partly overlapping the 5′ end of ORF

Fenner and White's Medical Virology. DOI: http://dx.doi.org/10.1016/B978-0-12-375156-0.00037-0

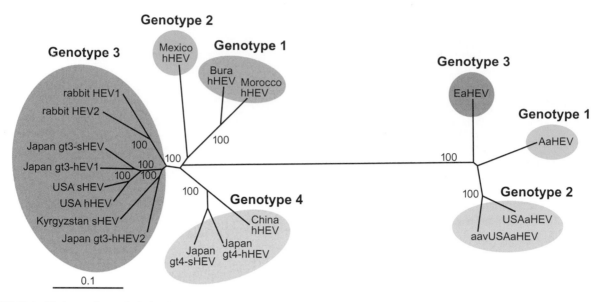

FIGURE 37.1 Phylogenetic tree depicting the relationship between strains of mammalian hepatitis E virus. This tree is based on the full-length genome sequences of representative hepatitis E virus strains including the four major genotypes of mammalian hepatitis E virus, a newly identified rabbit hepatitis E virus and three genotypes of avian hepatitis E virus (aavUSAaHEV, USAaHEV, AaHEV). *Reproduced from King, A.M.Q., et al., 2012. Virus taxonomy. In: Ninth Report of the International Committee on Taxonomy of Viruses. Academic Press, pp. 1025, with permission.*

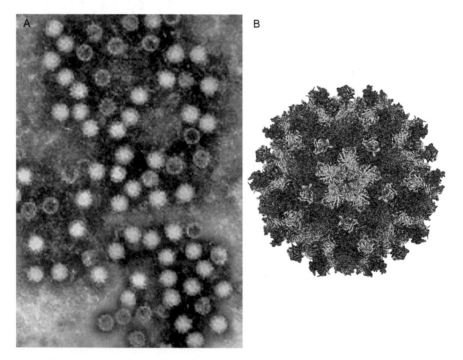

FIGURE 37.2 Morphology and structure of hepatitis E virus. (A) Negative contrast electron microscopy of hepatitis E virions, showing surface features. (B) Model of hepatitis E virus constructed from cryoelectron microscopy image analysis. Virions are icosahedral, non-enveloped (or pseudo-enveloped), spherical $T = 3$ particles with a diameter of 34 nm. The capsid is formed by homodimers of a single capsid protein (CP) in a complex intercalated arrangement. Here three colors are used to show the arrangement of these CP dimers on the surface of a $T = 3$ virion centered on its 5-fold axis of icosahedral symmetry. Because HEV is difficult to grow to titers needed for structural analysis, artificial virus-like particles (VLPs) have been commonly studied—however, VLPs assemble in a $T = 1$ structure and in some instances this has confused descriptions of this increasingly important pathogen. *(B) Reproduced from Guu, T.S., et al., 2009. Structure of the hepatitis E virus-like particle suggests mechanisms for virus assembly and receptor binding. Proc. Natl. Acad. Sci. U.S.A. 106, 12992–12997, with permission.*

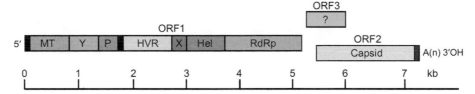

FIGURE 37.3 Organization of hepatitis E virus genome, comprising a short 5′ non-coding region (NCR), three ORF, and a 3′NCR. ORF2 and ORF3 overlap each other but neither overlaps ORF1. ORF1 encodes non-structural proteins, ORF2 encodes capsid protein, and ORF3 encodes a small phosphoprotein with a multi-functional C-terminal region. MT, methyl transferase; Y, "Y" domain; P, a papain-like cysteine protease; HVR, a hypervariable region that is dispensable for virion infectivity; Hel, helicase; RdRp, RNA-dependent RNA polymerase. *Reproduced from King, A.M.Q., et al., 2012. Virus taxonomy. In: Ninth Report of the International Committee on Taxonomy of Viruses. Academic Press, pp. 1023, with permission.*

2. In these respects HEV is quite unlike the picornavirus causing hepatitis, hepatitis A virus. Seven domains have been identified in the non-structural polyprotein ORF1: (1) methyltransferase, (2) "Y," a domain of unknown function, (3) a papain-like cysteine protease, (4) a hypervariable region, (5) "X," a domain of unknown function, (6) RNA helicase, and (7) RNA-dependent RNA polymerase. In terms of sequence homology within these domains as well as colinearity of genome organization (except for the position of the protease domain) the HEV genome closely resembles that of rubella virus, an enveloped virus currently classified in its own genus of the family *Togaviridae*. It is conceivable that HEV could have evolved from rubella virus (or its progenitor), either by deletion of the envelope glycoprotein genes or by recombination with a calicivirus-like genome (Fig. 37.3).

VIRAL REPLICATION

The viral replication cycle has not been characterized in detail because efficient cell culture systems have been developed only in recent years. Attachment appears to require heparin sulfate proteoglycans and as yet unidentified receptors. Following internalization and uncoating, viral RNA is translated to yield the ORF1 polyprotein. The RNA genome is transcribed to yield negative-sense RNA templates, which are then copied into both full-length and sub-genomic positive-sense RNA. ORF2 proteins and the RNA genome then assemble into new virions within the endoplasmic reticulum. Membrane-associated progeny HEV particles are finally released together with internal vesicles via the cellular exosomal pathway. The ORF3 protein modulates cellular signaling pathways, and may assist virus release. A proposed overview of HEV replication is shown in Fig. 37.4.

CLINICAL FEATURES

The clinical features of hepatitis E are similar to those of acute viral hepatitis caused by other hepatotropic viruses,

especially hepatitis A (see Chapter 32: Picornaviruses). The incubation period ranges from 15 to 60 days, with a mean of 40 days. Hepatitis E clinically manifests with icterus, malaise, anorexia, fever, hepatomegaly, and pruritus. In developed countries, clinical presentation may be indistinguishable from drug-induced liver injury, which also occurs in the elderly; thus an unknown number of cases of HEV infection may be misdiagnosed. HEV-infected persons exhibit a wide clinical spectrum, ranging from asymptomatic infection to fulminant hepatitis. The case–fatality rate is 0.5 to 3%. Patients with pre-existing liver disease are at increased risk of acute or subacute liver failure, and have a poor prognosis. However, the most striking feature of hepatitis E in developing countries is its extraordinarily high case–fatality rate of 20 to 25% among pregnant women in the final trimester (except in Egypt, for unknown reasons).

Laboratory findings of elevated serum bilirubin levels and markedly elevated levels of liver enzymes accompany these clinical signs, and small increases in alkaline phosphatase activity are seen. Studies of non-human primates have demonstrated that the clinical presentation, immunological response, severity of disease, and biochemical markers of liver damage, all increase with increasing amounts of inoculated virus (Fig. 37.5).

During acute HEV infection, enzyme immunoassays show that anti-HEV immunoglobulin M (IgM) can be detected days before the onset of symptoms and disappears over a four- to six-month period. Anti-HEV IgG appears soon after the IgM response and may persist at least 12 years after infection.

HEV in both developing and developed countries is associated with extrahepatic complications including a number of neurological conditions, glomerulonephritis, pancreatitis, thrombocytopenia, and aplastic anemia. HEV genotype 3 infections in immunosuppressed patients can lead to chronic infection. In one study of HEV-infected solid organ transplant recipients, 66% had infection persisting beyond 6 months; of these, a majority remained asymptomatic but 9.4% developed cirrhosis.

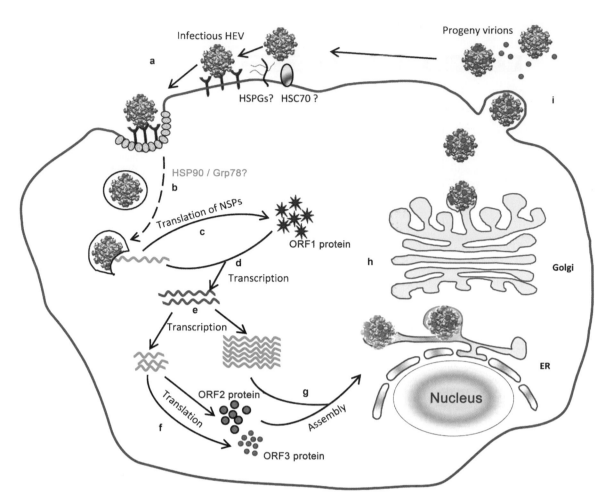

FIGURE 37.4 Proposed replication cycle of HEV. Step a: HEV attaches to the cell surface via HSPGs, HSC70, or other putative attachment receptor(s) and then enters the cell via an unknown specific cellular receptor. Step b: The HEV virion penetrates the membrane and enters the cells. HSP90 and Grp78 may be involved in this transport. The virion is then uncoated and releases the positive-sense genomic RNA into the cytoplasm of the cell. Step c: The positive-sense genomic viral RNA serves as the template to translate the ORF1 non-structural polyprotein in the cytoplasm. Step d: The viral RdRp synthesizes an intermediate, replicative negative-sense RNA from the positive-sense genomic RNA that (step e) serves as the template for the production of positive-sense, progeny viral genomes. Step f: The ORF2 and ORF3 proteins are translated from a subgenomic, positive-stranded RNA, and (step g) the ORF2 capsid protein packages the viral RNA genome and assembles new virions. Step h: The nascent virions are transported to the cell membrane. The ORF3 protein facilitates the trafficking of the virion, and (step i) the nascent virions are released from the infected cells. *Reproduced from Cao, D., Meng, X.-J., 2012. Emerg. Microbes Infect. 1, e17; doi:10.1038/emi.2012.7, with permission.*

PATHOLOGY, PATHOGENESIS, AND IMMUNITY

Most of our limited knowledge of the pathogenesis of HEV comes from experimental studies in monkeys or chimpanzees, animal species that are readily infected and in which HEV can be serially passaged. Macaques and tamarins begin to excrete virus into bile and feces about one month post-infection. Viral antigens can be demonstrated in the cytoplasm of a large proportion of hepatocytes by immunofluorescence without evidence of cytolytic damage, and it is thought that liver damage is eventually brought about by the cellular immune response against infected cells. Serum levels of liver enzymes such as alanine aminotransferase (ALT) then begin to rise, reaching a peak at about ten weeks. The infection then resolves; there is no evidence of chronicity. Antibodies rise slowly, do not reach high titers, and appear to fall to a low but persistent level fairly rapidly. The main site of replication is believed to be the liver, although in pigs experimentally infected with swine HEV, viral RNA is found in many tissues, particularly the small intestine, lymph nodes, colon, and liver.

The most conspicuous pathological feature of hepatitis E infection, in contrast to hepatitis A, is cholestasis. Histologically, the liver often shows intracanalicular stasis of bile and rosette formation of hepatocytes and pseudoglandular structures resembling embryonal bile ducts. It is not yet known whether HEV replicates initially

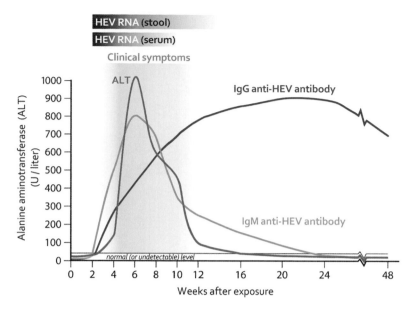

FIGURE 37.5 Course of acute hepatitis E virus (HEV) infection. Acute hepatitis E presents with fever, anorexia, vomiting, and jaundice several weeks after initial exposure. The onset of clinical symptoms coincides with a sharp rise in serum alanine transaminase (ALT) levels. Symptoms may persist for a few weeks to a month or more. ALT levels return to normal during convalescence. HEV RNA may be detected in both serum and stool early in the course of infection, but viremia may be difficult to detect by the time cases come to clinical attention. Anti-HEV IgM antibody titers increase rapidly and then wane over the weeks following infection, while anti-HEV IgG antibody titers continue to rise more gradually during the convalescent period, and anti-HEV IgG antibodies may be detected for months to years. *Reproduced from Krain, L.J., et al., 2014. Host immune status and response to hepatitis E virus infection. Clin. Microbiol. Rev. 27, 139–165, with permission.*

in the intestine, nor whether the observed sensitivity of the virion to proteolysis reflects a requirement for proteolytic disruption of the capsid by trypsin or other enzymes in the intestine prior to infection of the mucosa. An important question for future research is why the disease is so severe in pregnant women (see below).

LABORATORY DIAGNOSIS

Investigation of the patient with acute hepatitis will naturally begin with testing for hepatitis A (by IgM serology) and hepatitis B (HBsAg; anti-HBc IgM). However, HEV should not be ignored and specific tests should be used wherever patient history or local epidemiology indicates. EIAs and rapid immunochromatographic tests for HEV IgM and IgG antibodies, based on ORF2/ORF3 peptides or recombinant antigens, are available and react with all four genotypes. HEV IgM is usually greatest by the time of clinical presentation and thus is a reliable diagnostic marker. The presence of IgG antibodies is a good marker of past infection, and reverse transcriptase polymerase chain reaction (RT-PCR) testing has an important diagnostic role in immunocompromised patients with impaired immune responses, for monitoring chronic infections, and for genotype identification.

Immunoelectron microscopy, looking for aggregated calicivirus-like particles in an acute-phase fecal specimen using polyclonal sera, was the standard approach at first to diagnosis but has now been replaced by simpler and more sensitive new alternatives.

Growth of HEV in cell culture has been achieved recently, using both PLC/PRF/5 cells and A549 cells, but this approach is of limited diagnostic use. Non-human primates have been used as experimental models for studying all four genotypes, and pigs and rabbits have been used for infectivity studies for HEV-3 and HEV-4.

EPIDEMIOLOGY, PREVENTION, CONTROL, AND THERAPEUTICS

Human infection with hepatitis E virus has two quite distinct epidemiological patterns. In less-developed parts of the world, HEV-1 and HEV-2 are transmitted by the fecal–oral route, especially by contaminated water. Subclinical infection may be the rule in children, with icteric infection being mainly confined to young adults. Hepatitis E virus is now recognized to be the most important cause of epidemic hepatitis in Asia. In Central Asia hepatitis E, like hepatitis A, tends to peak in autumn, while in Southeast Asia it occurs particularly during the rainy season or following extensive flooding. The most notorious common-source epidemic resulted from fecal contamination of a drinking-water supply in New Delhi in 1955, causing 29,000 identified cases of icteric hepatitis. Ascribed to hepatitis A virus (HAV) at the time, the outbreak was demonstrated to have

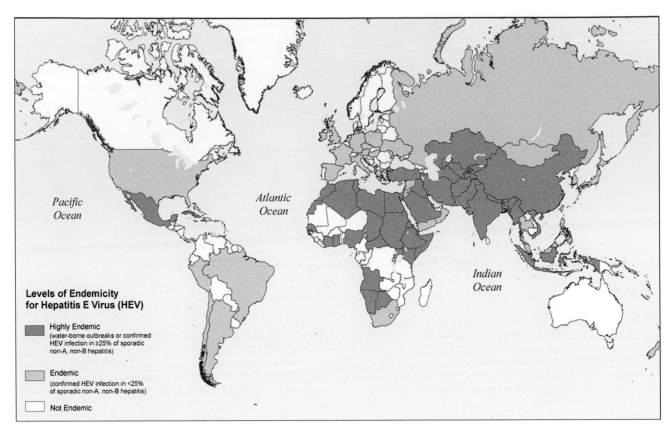

FIGURE 37.6 Global levels of endemicity of hepatitis E virus 2014. Human infection is common in developing countries with inadequate water supply and environmental sanitation. Large hepatitis E epidemics have been reported in Asia, the Middle East, Africa, and Central America. People living in refugee camps or overcrowded temporary housing after natural disasters can be particularly at risk. *From CDC, with permission.*

been caused by HEV through a modern diagnostic follow-up to a retrospective investigation of frozen stored paired sera for the presence of both HAV and HBV antibodies. Many similar water-borne outbreaks have been recorded subsequently, especially on the Indian subcontinent but also in Central Asia, China, Indonesia, North Africa, and Mexico. Particularly devastating outbreaks have occurred among Ethiopian refugees encamped in Somalia and South Sudan during the prolonged wars in the Horn of Africa (Fig. 37.6).

Clinical attack rates during epidemics range from 1% to 10%, with most of the cases occurring in young adults and most of the mortality among pregnant women. However, the secondary attack rate among household contacts of icteric patients is low (2.4% in one outbreak in Nepal, compared with 10 to 20% for hepatitis A at the same locality). This may be due to relatively low numbers of infectious virions shed in feces and/or their lability. Surprisingly, in many developing countries anti-HEV IgG seroprevalence is relatively low in children aged less than 10 years, rising to a peak by the age of 30.

In contrast, in developed countries HEV-3 and HEV-4 infection is a zoonosis with domestic pigs as the main

reservoir. An important route of transmission to humans is thought to be through eating inadequately cooked pork or game meat. In some countries, infection in pigs has a high prevalence but is largely asymptomatic. HEV-3 has been reported from many countries worldwide, while HEV-4 is found largely but not solely in China and Japan. Human cases of HEV-3 and HEV-4 infection occur sporadically or in small outbreaks, with a predominance of infections in males aged over 50 for reasons that are not understood. It is also not yet clear as to the full extent of asymptomatic infection, nor the reasons why, in some studies, human cases are more common among communities living near the sea.

Reliable reticulated and chlorinated water supplies and improved standards of sanitation (see Chapter 14: Control, Prevention, and Eradication) throughout the developing world must underpin any long-term program to control hepatitis E, or indeed any other enterically transmitted infectious disease. During an epidemic the usual rules apply: boil the water, and for food, "cook it or peel it." Avoiding the consumption of poorly cooked meat, especially pork, may reduce infection with HEV-3. Passive immunization is not effective.

Two recombinant HEV vaccine candidates, each based on a recombinant ORF2 protein from the HEV-1 genotype, have been clinically evaluated and found to be well tolerated, highly immunogenic, and to protect against natural exposure. One vaccine candidate, given as three doses, showed 95% efficacy in a Nepalese military population, while the other vaccine showed 100% efficacy after three doses in a Chinese population. The latter is licensed by China's Ministry of Science and Technology, and is being produced and marketed but has not as yet been approved by regulatory agencies of many other countries.

FURTHER READING

Cao, D., Meng, X.-J., 2012. Molecular biology and replication of hepatitis E virus. Emerg. Microbes Infect. 1, e17.Available from: http://dx.doi.org/10.1038/emi.2012.7.

Kamar, N., Dalton, H.R., Abravanel, F., Izopet, J., 2014. Hepatitis E virus infection. Clin. Microb. Rev. 27, 116–138.

Kmush, B.L., Nelson, K.E., Labrique, A.B., 2014. Risk factors for hepatitis E virus infection and disease. Expert Rev. Anti. Infect Ther. 1–13.

Teshale, E.H., Hu, D.J., Holmberg, S.D., 2010. The two faces of hepatitis E virus. Clin. Infect Dis. 51 (3), 328–334.

Chapter 38

Prions

Prions, also known as the agents of transmissible spongiform encephalopathies (TSE), cause at least four neurodegenerative diseases of humans: kuru, Creutzfeldt-Jakob disease (including variant Creutzfeldt-Jakob disease), Gerstmann-Sträussler-Scheinker syndrome, and fatal familial insomnia. Prions also cause several diseases of animals: scrapie of sheep and goats, bovine spongiform encephalopathy, feline transmissible encephalopathy, transmissible mink encephalopathy, and chronic wasting disease of deer and elk. In each of these diseases the common lesion is a spongiform degeneration of the gray matter of the brain with astroglial hypertrophy and proliferation. These uniformly fatal diseases are caused by prions, that is "infectious proteins, lacking a nucleic acid genome." The name prion is an acronym from *pr*oteinaceous *i*nfectious particle.

The etiology of this group of diseases posed an extraordinary biological enigma for many years, because it was understood that: (1) disease can be transmitted by tissues and tissue extracts in a similar way to conventional infections, (2) disease can also arise spontaneously, or as an inherited disorder, and (3) infectious inocula had been shown to lack nucleic acid, the physical basis for encoding all biological information in all living organisms and viruses. Several earlier hypotheses regarding the nature of the infectious moiety, the prion, were based in the disbelief that there could be an infectious protein—most of these promised that a nucleic acid-based moiety, some kind of cryptic virus, would eventually be found. Stanley Prusiner was awarded the 1997 Nobel Prize in Physiology or Medicine for his discovery of the nature of prions and the exceptional mechanisms of prion pathogenesis.

The prototype of the prion diseases, scrapie, was first described after the importation of sheep from Spain into England in the 15th century. The name reflects the characteristic scratching behavior of diseased animals. Scrapie is endemic in sheep in all countries except Australia and New Zealand.

In 1963 William Hadlow, working at the Rocky Mountain Laboratory in Montana, the United States, proposed that the human disease kuru was histopathologically and pathogenetically similar to scrapie in sheep and that it might be transmissible. Kuru, a fatal neurological disease, occurred only in the Fore tribe in the New Guinea highlands where ritualistic cannibalism was practiced on deceased relatives. Hadlow's idea led to the discovery by Carleton Gajdusek that kuru could be transmitted to chimpanzees, causing a disease indistinguishable from the human counterpart—for this discovery Gajdusek was awarded the Nobel Prize in Physiology or Medicine in 1976. The importance of this discovery became clear when it was shown that more widespread human diseases, such as Creutzfeldt-Jakob disease, and other animal diseases, such as bovine spongiform encephalopathy and chronic wasting disease of deer and elk, are also transmissible.

PROPERTIES OF PRIONS

Classification

Given the unique properties of prions, these agents have not been classified into families, genera, or species in the same way as viruses. Nevertheless, the International Committee on Taxonomy of Viruses (ICTV) manages the nomenclature. At the heart of this is a standard terminology for the pathological variants of the prion protein PrP (Table 38.1).

Prions are identified by host species and disease association (Table 38.2). Further distinctions are made according to molecular and biological properties. The primary amino acid sequence of prions mainly reflects the host from which they were isolated, but also registers mutations which define inherited variants, for example as seen in familial Creutzfeldt-Jakob disease in humans. The full amino acid sequence of all important prion variants has been determined.

Certain biological properties are used to distinguish strains of prions, particularly scrapie strains. Following intracerebral injection of prion-containing material into multiple strains of inbred mice, the following parameters are recorded: (1) incubation period and mortality pattern, (2) distribution and extent of spongiform lesions and prion protein (PrP) plaques in brains (assayed by immunohistochemistry using labeled anti-PrP antibodies), and in some cases (3) the titer of infectivity in brains. Prion strains "breed true," giving reproducible results in this kind of biological assay system. For example, prions from

Fenner and White's Medical Virology. DOI: http://dx.doi.org/10.1016/B978-0-12-375156-0.00038-2

TABLE 38.1 Prion Terminology

Term	Description
Prion	A proteinaceous infectious particle that lacks nucleic acid. Prions are composed largely, if not entirely, of PrP^{Sc} molecules
PrP^{Sc}	Abnormal, pathogenic isoform of the prion protein that causes sickness. This protein is the only identifiable macromolecule in purified preparations of prions
PrP^{C}	Cellular (normal) isoform of the prion protein
PrP$^{27-30}$	Digestion of PrP^{Sc} with proteinase K generates PrP$^{27-30}$ by hydrolysis of the N-terminal sequence
PrP^{res}	Alternative designation for PrP^{Sc}, that has been proposed to generalize the term for all types of prion diseases, not only scrapie
PrP^{sen}	Normal isoform PrP^{C} that is sensitive to hydrolysis by proteinase
PRNP	Human PrP gene located on chromosome 20
Prnp	Mouse PrP gene located on syntenic chromosome 2. *Prnp* controls the length of the prion incubation time and is congruent with the incubation time genes *Sinc* and *Prn-i*
PrP amyloid	Fibril of PrP fragments derived from PrP^{Sc} by proteolysis. Plaques containing PrP amyloid are found in the brains of some mammals with prion disease
Prion rod	An amyloid polymer composed of PrP$^{27-30}$ molecules. Created by detergent extraction and limited proteolysis of PrP^{Sc}
Protein X	A hypothetical macromolecule that is thought to act like a molecular chaperone in facilitating the conversion of PrP^{C} into PrP^{Sc}

TABLE 38.2 The Prion Diseases of Humans and Animals

Disease	Host	Source of Infection
Creutzfeldt-Jakob disease (CJD)	Humans	Iatrogenic—human prion contamination of dura mater grafts, therapeutic hormones, etc., all derived from cadavers
		Familial—germline mutation in *PRNP* gene
		Sporadic—unknown cause, perhaps somatic mutation in *PRNP* gene or spontaneous conversion of PrP^{C} into PrP^{Sc}
Variant Creutzfeldt-Jakob disease (vCJD)	Humans	Transmission of bovine spongiform encephalopathy prion to humans, epidemiologically proven due to eating beef products
Kuru	Humans	Ritual cannibalism in Fore people
Gerstmann-Sträussler-Scheinker syndrome	Humans	Familial—germline mutation in *PRNP* gene
Fatal familial insomnia	Humans	Familial—germline mutation in *PRNP* gene
Scrapie	Sheep, goats	Not certain, possibly scrapie prion contained in feed, but more likely by direct contact and contamination of pastures by placentas and fetal tissues
Bovine spongiform encephalopathy (BSE)	Cattle	Bovine spongiform encephalopathy prion contamination of meat-and-bone meal; some vertical transmission from cow to calf
Transmissible mink encephalopathy	Mink	Scrapie prion contamination of sheep carcasses and offal fed to mink
Chronic wasting disease	Mule deer, elk	Unknown in feral animals, contamination of paddocks in captive populations
Feline spongiform encephalopathy	Cats, felids in zoos	Bovine spongiform encephalopathy prion contamination of commercial cat food
Exotic ungulate spongiform encephalopathy	Greater kudu, nyala, oryx, and others in zoos	Bovine spongiform encephalopathy prion contamination of meat-and-bone meal

cattle, nyala, kudu, and domestic cats behave the same when subjected to this strain characterization protocol, indicating that all have been derived from the same source, namely cattle. Further, mice inoculated in the same way with material from cattle with bovine spongiform encephalopathy and humans with vCJD have behaved in the same way, yet differently from mice inoculated with material from sporadic human cases of Creutzfeldt-Jakob disease, including cases in farmers who had worked with cattle with bovine spongiform encephalopathy.

Similar results have been recorded by biochemical analysis of prions recovered from various sources: for example, when brain specimens are treated with proteinase K and the protease-resistant residues subjected to immunoblot analysis, four different staining patterns are found. Three patterns represent genetic, sporadic, and iatrogenic Creutzfeldt-Jakob disease in humans; the fourth represents all cases of variant Creutzfeldt-Jakob disease, bovine spongiform encephalopathy in cattle, and the similar feline and exotic ungulate diseases.

Prions from animals and humans can also be transmitted to various other animals (hamsters, rats, ferrets, mink, sheep, goats, pigs, cattle, monkeys, and chimpanzees), and again a spectrum of strain variation is seen. Some donor–recipient pairs lead to short incubation disease, others to longer incubation times, and yet in others disease is not apparent even after very long periods of observation or successive blind passage.

Prion Properties

Prions are normal cellular proteins that have undergone a certain conformational change as a result of post translational processing of a normal cellular protein and thereby have become pathogenic. The normal human prion protein, PrP^C, is composed of 208 amino acids. It is encoded in the genome of most mammals and expressed in many tissues, especially in neurons and lymphoreticular cells.

Experimental data implicate PrP^C in anti-apoptotic activities, cell signaling and adhesion, synaptic function, and protection against oxidative stress, although the exact physiological mechanisms of action for normal and altered forms remain elusive. Further, PrP^C has been found to bind copper but knockout mice lacking the gene for the protein appear normal. The amino acid sequence of PrP^C and the abnormal isoform of the protein, PrP^{Sc} in a given host are identical. Only the *conformation* is changed in the PrP^{Sc} isoform, from a structure made up predominantly of α-helices to one made up predominantly of β-sheets (Fig. 38.1). Monoclonal antibodies are available that can discriminate between PrP^C and PrP^{Sc}; others are available that can specifically precipitate bovine, murine, and human PrP^{Sc}, but not PrP^C, confirming the presence of an epitope common to prions from different species linked to disease

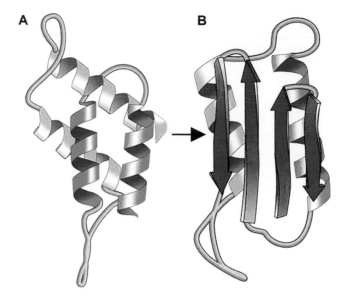

FIGURE 38.1 (A) The structure of normal prion protein (PrP^C, the normal isoform of PrP protein) with prominent α-helices, and (B) PrP^{sc} (the abnormal isoform) with prominent β-sheets. The transition from the normal to the abnormally folded isoform may be triggered by (1) contact with pre-existing PrP^{Sc}, for example as a transmissible condition (Creutzfeldt-Jakob disease and kuru); it may develop (2) in individuals carrying an autosomal dominant mutation in the *prnp* gene (familial spongiform encephalopathies); or it may (3) rarely arise spontaneously (sporadic CJD).

but different from the normal PrP^C protein isoform. PrP^C is encoded by a single gene on chromosome 20 of the human genome.

In humans, sporadic Creutzfeldt-Jakob disease (sCJD) is the most frequent form of prion disease and has demonstrated a wide phenotypic (e.g., clinical presentation, neuropathological lesion pattern), and wide PrP molecular variation. In contrast, variant Creutzfeldt-Jakob disease (vCJD) is a highly stereotyped disease, that, until now, only occurred in patients who are methionine homozygous at codon 129 of the *PRNP* gene (and codon 219 among Asian patients) (Table 38.3). This is the case in patients from the United Kingdom, France, and all other involved countries. Recent research has provided consistent evidence of strain diversity in sporadic CJD and, unexpectedly, in variant CJD—the methionine homozygosity effect at codon 129 is found across the strain diversity. Concerns have been raised about the possibility that the occurrence of vCJD in people who are valine homozygous or methionine-valine heterozygous at codon 129 might simply involve a much longer incubation period—although with each passing year this seems less likely.

When a given animal prion is passaged in mice or hamsters, the amino acid sequence of the recipient's PrP^{Sc} is that of the PrP^C of the recipient, not of the donor. In a particular host, there may be many different mutations in the PrP gene, each resulting in a slightly different PrP^{Sc}

TABLE 38.3 The Influence of Methionine/Valine Polymorphism at Codon 129 on the Various Forms of CJD and Kuru

Codon 129 Status of Various Cohorts

Cohort	Frequency of Codon 129 Polymorphism		
	MM (%)	MV (%)	VV (%)
Sporadic CJD	69	14	17
Variant CJD	100	0	0
Historical, Fore people with kuru	53	27	20
Historical, Fore survivors (males 20 years or younger)	0%	65	35
Contemporary, Fore survivors (women older than 50 years)	..	77	..
UK, control population	37	51	12
Japan, control population	92	8	0
Papua New Guinea, control population	32	43	24
Global, Caucasian population	52	36	12

Source: From Collins, S.J., Lawson, V.A., Masters, C.L., 2004. Transmissible spongiform encephalopathies. Lancet 363, 51–61.

conformation, and each in turn giving rise to a different lesion pattern, a variable incubation period, and a spectrum of mortality characteristics—this is the basis for the different prion strains. For example, prions from vCJD in humans have biological characteristics distinct from those in other types of Creutzfeldt-Jakob disease, but are similar to prions isolated from cattle, mice, cats, and macaques infected during the bovine spongiform encephalopathy epidemic in the United Kingdom from whence vCJD in humans evolved.

PrP^{Sc} protein is highly resistant to many environmental insults, chemicals, and physical conditions that would destroy any virus or microorganism; its high resistance to UV- and γ-irradiation indicates a very small radiation target size (Table 38.4). This resistance to degradation has major implications for the reuse of surgical instruments, especially those for tonsillectomy and appendectomy as evidence of abnormal PrP has been found in lymphoid tissue. There has been much debate in the United Kingdom and elsewhere as to the precautions required for surgical interventions among those considered at high risk of developing vCJD.

Prions can reach very high titers in the brains of infected hosts—laboratory strains passaged in hamsters can have titers of up to 10^{11} ID$_{50}$/gram of brain. Ultrafiltration has determined the diameter of prions to be about 30 nm. PrP^{res} is very resistant to endogenous proteases, which is the key to its accumulation into protein aggregates called PrP amyloid or scrapie-associated fibrils (SAF), which form neuronal plaques. Electron microscopy has shown that these fibrils, that is polymers of PrP^{res}, are formed as helically wound filamentous rods 4 to 6 nm in diameter. Prions do not evoke either an inflammatory or immune response in the host.

TABLE 38.4 Effects of Physical and Chemical Treatments on Prion Infectivity[a]

Treatment	Reduction of Infectivity
1 M NaOH	$>10^{6-8}$
Phenol extraction	$>10^6$
0.5% sodium hypochlorite	10^4
Histopathologic processing	$10^{2.6}$
3% formaldehyde	10^2
1% β-propiolactone	10^1
Ether extraction	10^2
Autoclave 132°C for 90 min	$>10^{7.4}$
Autoclave 132°C for 60 min	$10^{6.5}$
Autoclave 121°C for 90 min	$10^{5.6}$
Boiling 100°C for 60 min	$10^{3.4}$
Heating 80°C for 60 min	10^1

[a]Compilation of various experiments done with scrapie prion, infectivity assayed in mice.

Prion Replication

It is the presence of the horizontally or in some instances vertically transmitted PrP^{Sc} that catalyzes the conversion of normally encoded PrP^{C} molecules into more PrP^{Sc} molecules. While PrP^{Sc} acts as the template, the "seed-crystal," for the abnormal folding and polymerization of PrP^{C}—via an intermediate heterodimer with normal cellular PrP^{C}—there is evidence that another molecule, a hypothetical chaperone called protein X, is needed for prion replication when transmission occurs between distant host species. In any case, the process cascades exponentially, with newly formed PrP^{Sc} in turn serving as a catalyst for the conversion of more and more PrP^C molecules as they are produced in target cells such as neurons (Fig. 38.2). Eventually, so much PrP^{Sc} builds up that it polymerizes, forming fibrillar masses which become visible as plaques and cause neuronal degeneration and neurological dysfunction via mechanisms that are as yet poorly understood. In like manner, different isoforms of PrP^{Sc} breed true and are perpetuated even in mixed infections. In remarkable synthetic reactions, PrP^{res} can now be amplified and the reaction studied *in vitro* by subjecting normal brain homogenates or PrP^c, together with a PrP^{res} seed, to cycles of sonication and incubation.

Our understanding of the prion replicative process has also been extended by elegant studies using knockout and transgenic mice, that is mice lacking the *PRNP* gene or mice containing only the *PRNP* gene of another species. For example, mice lacking the *PRNP* gene do not develop disease when inoculated with the scrapie prion, whereas mice expressing reduced levels of the protein exhibit very long incubation periods. Furthermore, when normal brain explants are grafted into the brains of such knockout mice the animals develop lesions only in the normal graft tissue. Even more remarkable, transgenic mice carrying mutated *PRNP* genes that mimic those in human familial spongiform encephalopathies, show the neuronal degeneration typical of these diseases even without inoculation of exogenous prions. Finally, transgenic/knockout mice carrying the human *PRNP* gene, but not the mouse *PRNP* gene, develop neurological lesions when inoculated with the bovine spongiform encephalopathy prion, starting at about 500 days. This finding has been considered key evidence in associating vCJD with the bovine spongiform encephalopathy prion.

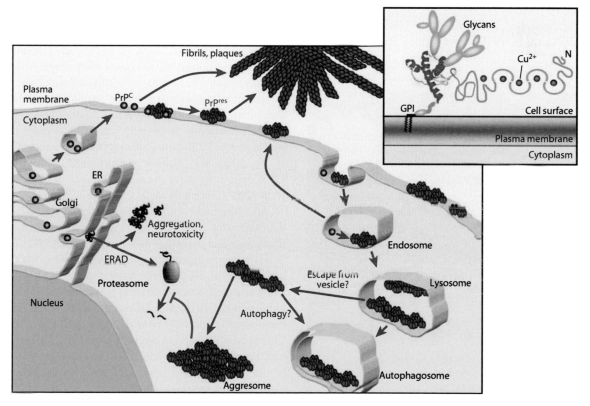

FIGURE 38.2 Model of biogenesis and accumulation of PrP^{res} in cells. PrP^c is a GPI-anchored plasma membrane glycoprotein (see inset), and is first synthesized in the endoplasmic reticulum (ER), then processed in the Golgi apparatus and transported to the cell surface (main diagram, left). When the abnormal protein PrP^{res}, together with co-factors, directly contacts PrP^c on the cell surface and/or in endosomes, it acts as a "seed crystal" to induce a cascade of transformation of PrP^c to PrP^{res}. PrP^{res} self-catalyzes rapid aggregation of itself, leading to a build-up of masses of fibrils which eventually harm neurons. Once PrP^{res} is made, it can accumulate on the cell surface, in intracellular vesicles (e.g., lysosomes) and aggresomes, or in extracellular deposits. Under conditions of mild proteasome inhibition, cytotoxic cytoplasmic PrP aggregates (e.g., aggresomes) can be found. Scrapie infection alone can inhibit proteasomes, apparently because of the presence of cytoplasmic PrP oligomers. Different prion strains may consist of the same protein misfolded in different ways; thus, each prion strain, each a differently folded version of the abnormal protein PrP^{res}, may transmit its characteristics separately to the normal prion protein, PrP^c, by forcing it to fold in its own unique way. ER, endoplasmic reticulum; ERAD, ER-associated degradation; GPI, glycophosphatidylinositol. *Reproduced from Caughey, B. et al., 2009, Annu. Rev. Biochem. 78, 177–204, with permission.*

HUMAN PRION DISEASES

Creutzfeldt-Jakob Disease

The majority of prion disease cases in humans are sporadic CJD, which has a worldwide incidence of 1 to 2 cases per million population per year. Sporadic CJD is not associated with any recognized transmission source nor with any typical *PRNP* gene mutation. It is not known whether it represents spontaneous misfolding of PrP^c, exposure to some unrecognized source of infection, or some other phenomenon. Around 5–10% of CJD cases are associated with a family history or evidence of a *PRNP* gene mutation, indicating a genetic basis. The typical age of onset is between 50 and 70 years, and primary clinical signs are of progressive dementia, sometimes associated with cerebellar signs. Death usually occurs within 1 to 12 months (Fig. 38.3).

Bovine Spongiform Encephalopathy (BSE) and Variant Creutzfeldt-Jakob Disease (vCJD)

Bovine spongiform encephalopathy (BSE) was first detected in cattle in 1986 in the United Kingdom; it was

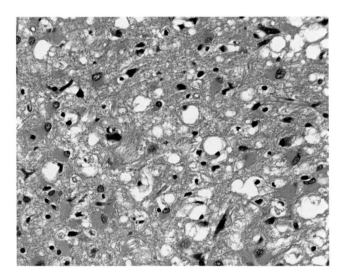

FIGURE 38.3 H&E stained section of brain showing typical changes of sporadic CJD. Note (1) most consistently, spongiform vacuolation throughout the cerebral gray matter, (2) reactive proliferation of astrocytes and microglia, (3) neuronal loss, (4) lack of inflammatory response, and (5) with disease caused by some prion genotypes, the formation and deposition of plaques within the brain. Plaque deposits of prion protein are seen in only 10% of sCJD cases. The spongiform change is characterized by large numbers of vacuoles varying from approximately 2 to 200 μm in diameter. The majority of vacuoles occur within neuronal processes (mainly neurites) and cell bodies; fewer are found in glia. In addition to the cerebral cortex, spongiform change is frequently observed in the basal ganglia and thalamus. Cerebellar involvement is present in most cases although less prominent and more variable than other regions of the brain. *From Sherif Zaki, CDC, with permission.*

called "mad cow disease" in the popular press. Officially recognized cases, confirmed by histopathology, amounted to about 280,000, but epidemiological modeling has indicated that upwards of two million cattle were infected. Epidemiological observations suggest that the cattle disease originated in the early 1980s when scrapie prions underwent a "species jump" and became established in cattle—rendered meat-and-bone meal produced from sheep carcasses and offal and fed to cattle is considered the probable source. A less likely alternative is that the cattle disease originated with a spontaneous mutation in a single bovine animal. In any case, as more and more diseased cattle were slaughtered and rendered to produce meat-and-bone meal a massive multiple-point-source epidemic in cattle followed. Export of meat-and-bone meal from the United Kingdom introduced the disease into Northern Ireland, the Republic of Ireland, Switzerland, France, Germany, the Netherlands, Belgium, and several other European countries, and the same source introduced the disease into zoo animals and cats in the United Kingdom. Control measures involved the rigorous exclusion of all meat, offal, and other cattle products from any cattle feed material, isolation and euthanasia of confirmed or probable cases, and restrictions on movement of animals. At the peak of the UK outbreak in 1992 approximately 37,000 cases were diagnosed in cattle in a single year; the number of confirmed cases was dwarfed by the actual number of infected animals that entered into the human food chain because of slaughter before the onset of clinical disease. Eventually effective control measures succeeded and full control was achieved by about 2008.

In 1996 the British government announced that 10 people may have become infected with the bovine spongiform encephalopathy prion through exposure to beef: "…although there is no direct evidence of a link, on current data and in the absence of any credible alternative, the most likely explanation is that these cases are linked to exposure to bovine spongiform encephalopathy… This is a cause for great concern." By 2016 the number of human cases of what is now called variant Creutzfeldt-Jakob disease (vCJD) had reached 178 (total confirmed and probable cases), but there were fewer and fewer cases in each year after 2000. Throughout this period pathological and molecular studies had strengthened the causative association between the bovine prion and the human disease. At the heart of this association were research breakthroughs on the nature of prions and associated pathogenicity.

vCJD is significantly different from classical CJD in both clinical presentation and pathology. The mean age of onset is 28 years, psychiatric and behavioral symptoms occur earlier and neurological symptoms later than in classical CJD, and there are different electroencephalographic changes from those seen in classical CJD. In contrast to the highly variable pathological picture seen in classical CJD, vCJD shows highly stereotyped pathology characterized by

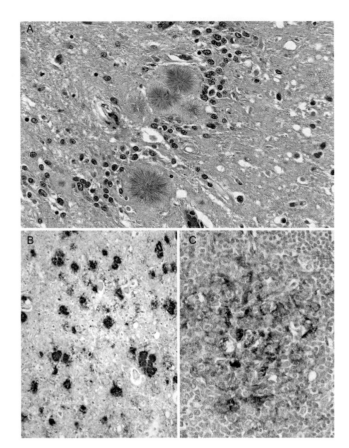

FIGURE 38.4 (A–C) Histopathology of variant Creutzfeldt-Jakob disease (vCJD), the form of disease seen in individuals exposed to the bovine spongiform encephalopathy (BSE) prion. In contrast to the heterogeneity in the neuropathological features of sporadic CJD, variant CJD is characterized by a highly stereotyped pathology. (A) The most striking feature is the deposition of "florid plaques." These are described as having a dense eosinophilic core surrounded by a pale region of radiating fibrils, which are encircled by a halo of microvacuolar spongiform change. H&E. (B) Variant CJD shows a distinct pattern of PrP deposition by immunohistochemistry. In addition to staining of the florid plaques, a large number of smaller plaque-like deposits are observed in intercellular spaces—here shown in the cerebral cortex. (C) Variant CJD is unique in that PrP^Sc is readily detected in non-neural tissues, especially in lymphoid follicles and other lymphoreticular tissues—shown here is an IHC-positive focus in a diagnostic tonsil biopsy. *(A) From Sherif Zaki, CDC, with permission.*

florid plaques surrounded by a spongiform halo. Protease-resistant prion protein is more readily detected in brain and lymphoid tissue including the tonsils (Fig. 38.4).

Between 1986 and 2016 there have been 178 cases of vCJD in the United Kingdom, 27 in France, 5 in Spain, 4 in Ireland, 4 in the United States, 3 in the Netherlands, 2 in Portugal, 2 in Italy, 2 in Canada and one each from Japan, Saudi Arabia, and Taiwan. Two of the four US cases, two of the four cases from Ireland, one of the two cases from Canada, and the single case from Japan, were likely exposed to the BSE agent while residing in the United Kingdom. In

the United Kingdom the number of cases of vCJD peaked in the year 2000: cases are now rare in the United Kingdom with only very occasional cases being reported (Fig. 38.5).

Kuru

Kuru occurred almost exclusively among the Fore ethnic group living in the Eastern Highlands of New Guinea, largely among women and children exposed by the oral route to the brains and viscera of deceased relatives. At its peak, the epidemic of kuru accounted for around 25% of deaths among females of these populations. Kuru progressively disappeared since the cessation in the 1960s of the cannibalistic practices by all indigenous ethnic groups of New Guinea, with the age of onset steadily increasing as each year passed. Strong genetic selection pressure on the *PRNP* gene was seen, especially for heterozygosity in codons 127 and 129, a known resistance characteristic; elderly survivors of the kuru epidemic are almost universally heterozygotic, unlike younger unexposed individuals.

Gerstmann-Sträussler-Scheinker Syndrome and Fatal Familial Insomnia

Gerstmann-Sträussler-Scheinker syndrome and fatal familial insomnia are very rare familial diseases caused by autosomal dominant mutations in the *PRNP* gene. In the former there is a single point mutation at codon 102 leading to a single amino acid substitution in the PrP^C protein. The animals develop typical spongiform encephalopathy lesions and disease when this single point mutation is introduced into the normal *PRNP* gene of mice.

Iatrogenic Prion Diseases

Iatrogenic CJD was first described in 1954 in a patient who received a corneal transplant. A number of cases have since been recorded of iatrogenic transmission of prion diseases, through the use of neurological instruments, transplantation of dura mater from cadavers, and contaminated pituitary-derived human growth hormone (hGH) and gonadotrophin hormone, both also from cadavers. These latter hormones are now produced by recombinant technologies, thus eliminating the risk of disease transmission. Before this change of source, it was estimated that around 1% of recipients developed CJD.

Iatrogenic infection is also thought possible via the use of the blood supply. A few well-documented cases of transfusion-associated vCJD have occurred in the United Kingdom leading to concern as to a possible risk among multiply transfused patients, especially those with hematological disorders and myeloid leukemias.

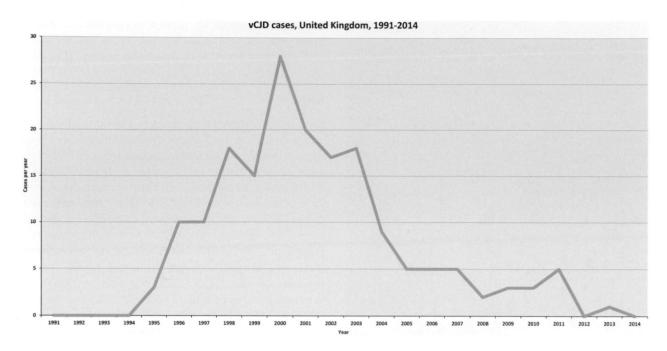

FIGURE 38.5 Incidence of variant Creutzfeld-Jacob disease (vCJD) in the United Kingdom, 1991–2014. The bovine spongiform encephalopathy (BSE) epidemic in cattle peaked in 1992, 4 years after the introduction of a ban on the inclusion of ruminant (cattle, sheep) materials in cattle feed. The associated human disease, vCJD, was not defined until 1996, 7 years after this ban was introduced. vCJD incidence peaked in 2000, when there were 27 diagnoses and 28 deaths, and then declined to an extent that many spokesmen for the livestock industry claimed it to be over by 2014. However, statistical modeling has produced evidence that the epidemic curve is skewed, with an extended tail. All cases of vCJD in the United Kingdom have been in people with a particular genotype (methionine homozygosity at codon 129 of the prion protein gene PRNP)—it is possible that there will be future peaks in people with other genotypes. To date, however, there is no evidence of a second wave. *Redrawn from UK surveillance data.*

DIAGNOSIS

A diagnosis of CJD is confirmed by demonstration in the brain tissue of typical neuropathological features, the presence of protease-resistant PrP on Western blot, or PrP^{Sc} by immunohistochemistry, or the detection of scrapie-associated fibrils by electron microscopy. Suggestive evidence that can be obtained without requiring brain tissue includes typical EEG and/or MRI changes, and a finding of 14-3-3 protein in cerebrospinal fluid. Serological tests, for example, conformation-dependent immunoassays, have been developed in an attempt to allow ante-mortem screening for infectivity, and testing tonsillar biopsy tissue for PrP^{Sc} by immunohistochemistry is useful in vCJD. More definitive prion strain characterization is performed using bioassays, usually in mice, which are not practical for routine use.

TREATMENT AND PREVENTION

No specific treatments are known to be of benefit against prion diseases once clinical disease has begun, although some experimental drugs have been identified that delay the onset of disease in animal models if given early enough.

Prevention of transmission revolves around avoidance of known transmission scenarios, for example, ensuring adequate sterilization of neurosurgical and EEG equipment, exclusion as corneal and dura mater donors of individuals with a history compatible with a prion disease. The lack of simple reliable tests to identify an infective source led to difficulties in devising blood transfusion screening policies to prevent possible spread of vCJD at the peak of the BSE epidemic in cattle. This was because the exact route and frequency of transmission from cattle to humans, and the extent of sub-clinical or latent infection in humans, could not be reliably determined. The decline to nearly zero in the number of new vCJD cases in the United Kingdom has lowered the priority for developing a test for blood donors.

OTHER CHRONIC NEURODEGENERATIVE DISEASES

Misfolding and aggregation of endogenous proteins in the central nervous system is a feature of several other neurodegenerative diseases. The molecular process is referred to as "nucleation-dependent aggregation." The proteins involved include amyloid-β (Aβ) and tau in the case of Alzheimer's disease, and α-synuclein in the case of Parkinson's disease. Intracerebral inoculation of aggregates of the relevant proteins has been shown to stimulate further aggregation and neuronal damage in animal models. However, no evidence for transmission of severe progressive disease in animal models has been found to date; in addition, epidemiological data do not suggest transmissibility of these diseases between humans. However, many fundamental biological puzzles remain unsolved, and only a deeper understanding may bring rational means of prevention and therapy.

FURTHER READING

Aguzzi, A., Calella, A.M., 2008. Prions: Protein aggregation and infectious diseases. Physiol. Rev. 89, 1105–1152.

Annual Report 2013. Creutzfeldt-Jakob disease surveillance in the UK. www.cjd.ed.ac.uk

Caughey, B., Baron, G.S., Chesebro, B., Jeffrey, M., 2009. Getting a grip on prions: Oligomers, amyloids and pathological membrane interactions. Annu. Rev. Biochem. 78, 177–204.

Erdtmann, R., Sivitz, L. (Eds.), 2003. Advancing Prion Science: Guidance for the National Prion Research Program. Institute of Medicine (US) Committee on Transmissible Spongiform Encephalopathies: Assessment of Relevant Science, National Academies Press (US), Washington (DC).

Imran, M., Mahmood, S., 2011. An overview of human prion diseases. Virol. J. 8, 559–567.

Prusiner, S.B., 1982. Novel proteinaceous particles cause scrapie. Science 216, 136–144.

Rutala, W.A., Weber, D.J., 2001. Creutzfeldt-Jakob disease: Recommendations for disinfection and sterilization. Clin. Infect. Dis. 32 (9), 1348–1356. Epub 2001 Apr 10.

Wadsworth, J.D.F., Joiner, S., Linehan, J.M., Asante, E.A., Brandner, S., Collinge, J., 2008. The origin of the prion agent of kuru: Molecular and biological strain typing. Philos. Trans. R. Soc., ser. B 363, 3747–3753. Available from: http://dx.doi.org/10.1098/rstb.2008.0069.

Chapter 39

Viral Syndromes

In the preceding 23 chapters, the role of various members of each family of viruses in the spectrum of human disease has been described. In this final chapter, the same scene is examined from the opposite aspect, namely, a focus in turn on each of the major clinical syndromes. Of course, this is not a textbook of medicine nor even of infectious diseases, but of virology. Thus space cannot be devoted to the detailed clinical descriptions of viral diseases and certainly not to any differential diagnosis from diseases caused by non-viral infectious agents, nor to the clinical management of infected patients. Appropriate reference works on infectious diseases have been listed under Further Reading. What follows is intended to provide only a bird's-eye view of the commoner syndromes so that the contribution of particular viruses to each one may be assessed. As the clinical features as well as the pathogenesis and epidemiology of all the major human viral infections were dealt with in the previous 23 chapters, this final chapter should be regarded as little more than an appendix which brings all these virus–disease associations together in a number of summary tables for ready reference.

THE HUMAN VIROME

First, however, the question is considered as to whether our bodies carry "normal viral flora," analogous to the bacterial microbiome. Particularly since the advent of molecular diagnostic and investigative techniques, and the use of next-generation sequencing (NGS), it has become more apparent that there are a number of different viruses, belonging to many different families, that can be regularly isolated from healthy individuals and that either are not known to cause disease or do so only under particular circumstances. These may represent lifelong persistent infections, or their presence may be due to an ultimately self-limiting, asymptomatic acute infection with prolonged excretion. Some are bacteriophages associated with gut bacteria. Some may even be animal viruses whose presence is due to the consumption of contaminated food. The composition of this virome is likely to vary between individuals, and over time in the same individual, and will be affected by the environmental exposure to exogenous viruses, and the age and immunocompetence of the host. As with the bacterial microbiome, some of these normally harmless commensal agents may be pathogenic under particular circumstances, for example, immunosuppression. Also similarly to the bacterial microbiome, discovery of one of these agents during disease or in a normally sterile site raises the question about its potential to occasionally cause disease. For example, a new cyclovirus has recently been reported in the cerebrospinal fluid of patients with acute CNS infection, and further work is needed to define its pathogenic role.

This area becomes even more complex when we consider co-infections. For example, co-infection with the non-pathogenic human flavivirus GBV-C (Chapter 36: Flaviviruses) has been reported to slow the progression of HIV infection. Progression of SIV infection in non-human primates is accompanied by a major expansion in the animals' enteric virome, which may in itself be involved in disease mechanisms. More extensive testing of children with respiratory illnesses, particularly using polymerase chain reaction (PCR), has revealed a significant percentage infected with two or more different viruses concurrently, and the roles of co-infection and interaction between multiple viruses are difficult to clarify. This subject becomes more complex again in immunosuppressed patients. On the other hand, co-infections involving HIV, HCV, and HBV are not unusual, and the implications for disease progression and antiviral treatment have been well studied.

Table 39.1 shows those viruses regularly found in each of the three main compartments in healthy individuals—skin, gut, and systemically. As with the persistent infections with pathogenic viruses that have been discussed in this book, persistence of these agents implies that mechanisms for evading elimination by the immune system must also be operating.

Several of the viruses listed have not been encountered elsewhere in this book. The member viruses of the family *Anelloviridae* are 30 nm in diameter with circular, negative-sense ssDNA genomes around 2 to 4 kb in size. They were discovered during studies of transfusion-transmitted hepatitis, but different genera are now known to infect more

Fenner and White's Medical Virology. DOI: http://dx.doi.org/10.1016/B978-0-12-375156-0.00039-4

TABLE 39.1 The Human Virome—Viruses Commonly Isolated from Healthy Individuals

Site	Virus	Possible Diseases
Skin	Betapapillomaviruses (types 5, 9, 49)	Skin cancer in patients with epidermodysplasia verruciformis
	Gammapapillomaviruses (types 4, 48 50, 60, 88)	None known
	Polyomaviruses (Merkel cell polyoma virus, HPyV6, HPyV7, HPyV9)	Merkel cell carcinoma; none known for other human polyomaviruses
	Circoviruses (human gyroviruses)	None known
Gut	Anelloviruses (Torque tenoviruses (TT), TT mini viruses and TT midi viruses)	None known
	Picobirnaviruses	None known (? diarrhea)
	Human enteroviruses and parechoviruses	Various syndromes (see Chapter 32: Picornaviruses)
	Human bocavirus, adenovirus groups C and F, Aichi virus, astrovirus, rotavirus	Gastroenteritis during acute infection
	Circoviruses (human gyroviruses)	None known
Systemic	Anelloviruses (Torque tenoviruses (TT) and TT-like mini viruses)	None known
	Herpesviruses (HSV and VZV in neurons; CMV, EBV, HHV-6, and HHV7 in circulating lymphocytes)	Various syndromes in acute infection and reactivation (see Chapter 17: Herpesviruses)

than 90% of humans and a number of animal species and to be present in plasma, saliva, and feces. They have not been reliably associated with any disease.

Viruses of the family *Picobirnaviridae* are non-enveloped 33 to 37 nm, containing a genome of two linear dsRNA segments. Different strains are widely distributed in humans, a range of mammals, and in birds and reptiles; they are excreted in the feces and not known to cause any disease.

Viruses of the family *Circoviridae* are tiny, 12 to 26 nm in diameter, containing a genome of circular, negative-sense ssDNA 1.7 to 2.3 kb in size. Chicken anemia virus (CAV) was the first member of this family discovered, in 1979, followed by the quaintly named beak and feather disease (BFDV). In 2011 the first human circovirus was detected by NGS in human skin; this is closely related to CAV within the genus *Gyrovirus*, and three further related human gyroviruses have since been described in feces of humans and/or chickens. No disease associations are known, nor is it completely clear whether these viruses are originally of human or chicken origin. This is another example of the new challenges in deciphering the epidemiological behavior and pathogenic roles of the plethora of new viral agents being identified in human samples.

VIRAL DISEASES OF THE RESPIRATORY TRACT

Respiratory infections are the most common afflictions of humans, and most are caused by viruses. Children contract on average about half a dozen respiratory illnesses each year, and adults about two or three. Admittedly these are mainly trivial colds and sore throats, but they account for millions of lost working/schooling hours and a significant proportion of all visits to family physicians. More serious lower respiratory tract infections tend to occur at the extremes of life, and in those with pre-existing pulmonary conditions. The most important human respiratory viruses are influenza and respiratory syncytial viruses (RSV), the former killing mainly the aged and the latter the very young. Of the estimated five million deaths from respiratory infections in children annually worldwide, at least one million are viral in origin.

Altogether, there are more than 200 human respiratory viruses, falling mainly within six families: orthomyxoviruses, paramyxoviruses, picornaviruses, coronaviruses, adenoviruses, and herpesviruses. Here we shall confine discussion to those that enter the body via the respiratory route and cause disease confined largely to the respiratory tract. Other viruses transmitted by the respiratory route are disseminated via the bloodstream and produce a more generalized disease, as is the case with most of the human childhood exanthems such as measles, rubella, and varicella. Other viruses, entering by non-respiratory routes, can reach the lungs via systemic spread, and pneumonia may represent the final lethal event, as in overwhelming infections with herpesviruses or adenoviruses in immunocompromised neonates or AIDS patients.

Systemic viral infections such as measles generate a strong memory response and prolonged production of IgG antibodies, which protect against reinfections for life.

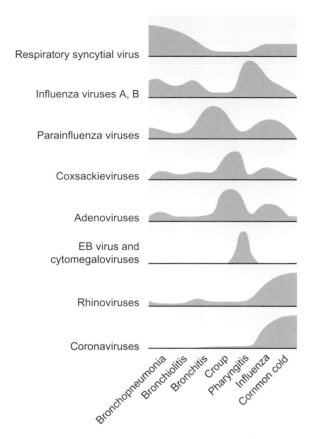

FIGURE 39.1 Frequency with which particular viruses produce disease at various levels of the respiratory tract. *Modified from Dr. D.A.J. Tyrrell.*

perhaps a mild sore throat or cough, but little or no fever. Rhinoviruses are the major cause, several serotypes being prevalent year-round and accounting for about half of all colds. Coronaviruses are responsible for about another 15%, mainly those occurring in the winter months. Certain enteroviruses, particularly coxsackieviruses A21 and A24 and echoviruses 11 and 20, cause febrile colds and sore throats, especially in the summer. In children, respiratory syncytial virus (RSV), human metapneumovirus, parainfluenza viruses, and the low-numbered adenoviruses are between them responsible for up to half of all upper respiratory tract infections (URTI). Specific viral diagnosis is not usually required for immediate patient management, but with increasing availability of rapid point-of-care multiplex PCR testing for the most common agents, laboratory diagnosis is being done more frequently to provide immediate information about ongoing epidemic situations, for example, caused by emergent influenza virus strains. In individuals with recurring or prolonged symptoms restricted to the upper respiratory tract, a possible allergic basis should not be overlooked.

Otitis media or sinusitis sometimes complicates URTI. Bacterial superinfection is generally the ongoing driver of the process, but a prior viral URTI may be the initiating mechanism and viruses have also been recovered from effusions. Respiratory infections with RSV, influenza, parainfluenza, adenovirus, or measles viruses predispose to otitis media. Indeed, repeated viral infections can precipitate recurrent middle ear infections, leading to progressive hearing loss.

In contrast, viruses that cause infection localized to the respiratory tract with little or no viremia, such as RSV or rhinoviruses, induce only a relatively transient mucosal IgA antibody response and a less robust systemic antibody response; hence, reinfections with the same or a somewhat different strain can recur repeatedly throughout life. In addition, different strains of viruses like influenza viruses arising by antigenic drift may cause sequential episodes of the same disease in a single patient.

Some viruses have a predilection for one particular level of the respiratory tract, but most are capable of causing disease at any level, and the same virus can produce different syndromes in different individuals, even within one family (Fig. 39.1). Conversely, similar clinical syndromes in different patients may be due to infection with unrelated viruses. Nevertheless, for ease of description we will designate six basic diseases of increasing severity as we descend the respiratory tract: rhinitis, pharyngitis, croup, bronchitis, bronchiolitis, and pneumonia (Table 39.2; see also Fig. 7.2).

Rhinitis (Common Cold)

The classic common cold (coryza) is marked by copious watery nasal discharge and obstruction, sneezing, and

Pharyngitis

Most pharyngitis is of viral etiology. Upper respiratory infections with any of the viruses just described can present as a sore throat, with or without cough, malaise, fever, and/or cervical lymphadenopathy. Influenza, parainfluenza, and rhinoviruses are common causes throughout life, but other viruses are prominent in particular age groups: RSV, human metapneumovirus and adenoviruses in young children, herpesviruses in adolescents and young adults. Adenoviruses, though not major pathogens overall, are estimated to be responsible for about 5% of all respiratory illnesses in young children. Pharyngoconjunctival fever is just one particular presentation, as described in Chapter 18: Adenoviruses—others include "acute respiratory disease" (ARD) as seen in US military camps. In patients without accompanying nasal symptoms, additional causes should be considered including group A and non-group A streptococci, mycoplasmas, and chlamydias. Primary infection with herpes simplex virus (HSV), if delayed until adolescence, presents as a pharyngitis and/or tonsillitis rather than as the gingivostomatitis seen principally in younger children; the characteristic vesicles, rupturing to form ulcers, can be confused only with herpangina, a common type of

TABLE 39.2 Respiratory Viral Diseases

Disease	Virus		
	Common >15%	**Less Common 5–15%**	**Some Cases**
Rhinitis (common cold)	Rhinoviruses	RSV, hMPV, coronaviruses	Parainfluenza, hMPV, influenza viruses
		Coxsackie viruses A21, A24; echo 11, 20	Adenoviruses
Pharyngitis		Parainfluenza viruses 1–3	RSV
		Influenza viruses	Cytomegalovirus
		Rhinoviruses	Coronaviruses
		Adenoviruses 1–7	Herpes simplex viruses
		Epstein-Barr virus	
		Coxsackie A viruses	
Laryngotracheobronchitis (croup)	Parainfluenza viruses 1, 3	RSV, influenza viruses, coronaviruses, adenoviruses	Rhinoviruses
Bronchitis	Influenza viruses	Parainfluenza virus 3	Parainfluenza virus 1, 2
			Rhinoviruses
		RSV	Adenoviruses
Bronchiolitis	RSV	Parinfluenza virus 3	Influenza A viruses
		Adenoviruses	Rhinoviruses
		hMPV	Enteroviruses
			Parainfluenza viruses 1, 2
Pneumonia (children <5)	RSV	Influenza viruses	Parainfluenza viruses 1, 2
	Parainfluenza virus 3	Adenoviruses	Rhinoviruses
		CMV	Varicella-zoster virus
			Enterovirus D68
Pneumonia (adults)	Influenza viruses	Adenoviruses	RSV,
			Measles virus
			Coronaviruses
			Varicella-zoster virus
Pneumonia (immune-compromised)	CMV	Adenoviruses	Influenza viruses
		RSV	Parainfluenza virus 3
		Herpes simplex viruses	

vesicular pharyngitis caused by coxsackie A viruses (see Chapter 17: Herpesviruses and Chapter 32: Picornaviruses). Infectious mononucleosis (glandular fever) is usually seen in adolescents and young adults, and is marked by a very severe pharyngitis, often with a diphtheria-like membranous exudate, together with cervical lymphadenopathy and fever (Chapter 17: Herpesviruses). This syndrome is generally caused by Epstein-Barr virus (EBV), but occasionally by cytomegalovirus (CMV), especially if lacking the sore throat, swollen glands, and heterophil antibody.

Laryngotracheobronchitis (Croup)

Croup is one of the serious manifestations of parainfluenza and influenza virus infections, predominantly in children less than 3 years old. Typically following symptoms of rhinorrhea and sore throat, the child develops fever, a "barking" or "metallic" cough, inspiratory stridor, and respiratory distress, sometimes progressing to complete laryngeal obstruction and cyanosis. Parainfluenza viruses are responsible for about 75% of all cases, type 1 being commoner than others. Influenza viruses and RSV are important causes during winter epidemics, and rhinoviruses, adenoviruses, and *Mycoplasma pneumonia* are less commonly responsible.

Tracheitis and Tracheobronchitis

Influenza, parainfluenza viruses, and RSV are the main viral causes of acute bronchitis. There is also evidence that chronic bronchitis, which is particularly common in smokers, may be exacerbated by acute episodes of infection with influenza viruses, rhinoviruses, or coronaviruses.

Bronchiolitis

Respiratory syncytial virus is the most important respiratory pathogen during the first year or two of life, being responsible for nearly all cases of bronchiolitis in infants during winter epidemics and about three-quarters of hospitalized cases overall. Human metapneumovirus, parainfluenza viruses (especially type 3) and influenza viruses are the other major causes of this syndrome. The disease is often preceded by symptoms of rhinitis but can then develop with remarkable speed. Breathing becomes rapid and labored with marked expiratory wheezing, accompanied by a persistent cough, cyanosis, a variable amount of atelectasis, and hyperinflated lung fields visible by X-ray (fig. 26.4). The infant may die peracutely, and hence RSV is one of the causes of unexplained "cot deaths," otherwise known as the sudden infant death syndrome (SIDS). Children suffering an episode of severe bronchiolitis have been noted to have higher rates of subsequent wheezing episodes or asthma. Whether the initial illness leads to a subsequent predisposition to wheezing, or whether children with such a predisposition suffer more severe initial RSV infections, is not clear.

Viral Pneumonia

The impact of viral pneumonia is greatly dependent on the age and immunocompetence of the patient. Whereas viruses are relatively uncommon causes of pneumonia in immunocompetent adults, accounting for approximately 8% of cases, they are very important in young children. RSV, human metapneumovirus and parainfluenza virus (mainly type 3) are between them responsible for 25% of all pneumonitis in infants in the first year of life. Influenza also causes a considerable number of deaths during epidemic years. Infections with adenoviruses 3 and 7 are less common but can be severe, and long-term sequelae such as obliterative bronchiolitis or bronchiectasis may permanently impair lung function. Up to 20% of pneumonitis in infants has been ascribed to perinatal infection with cytomegalovirus (see Chapter 17: Herpesviruses). CMV may also cause potentially lethal pneumonia in immunocompromised patients, as may measles, varicella, and adenoviruses. Moreover, viral pneumonia occasionally develops in adults with varicella and in military recruits during outbreaks of adenovirus 4 or 7 disease. In contrast measles has been often complicated by bacterial pneumonia, especially in malnourished children in Africa and South America. In the elderly, particularly in those with underlying pulmonary or cardiac conditions, influenza is a major cause of death, either via influenza pneumonitis or, more commonly, via secondary bacterial pneumonia attributable to *Staphylococcus aureus, Streptococcus pneumoniae*, or *Haemophilus influenzae*.

Viral pneumonitis often develops insidiously following URTI, and the clinical picture may be atypical. The patient is generally febrile, with a cough and a degree of dyspnea, and auscultation may reveal some wheezing or moist rales. Sputum may be scanty, and laboratory diagnosis of the responsible agent may be confused by the frequent asymptomatic shedding of some viruses, for example, herpesviruses or adenoviruses. Unlike typical bacterial lobar pneumonia with its uniform consolidation, or bronchopneumonia with its streaky consolidation, viral pneumonitis is usually confined to diffuse interstitial lesions. The radiologic findings are not striking; they often show little more than an increase in hilar shadows or, at most, scattered areas of consolidation.

VIRAL GASTROENTERITIS

By no means do all viruses found in feces cause gastroenteritis. Some do, but others cause "silent" infections of the gastrointestinal tract; some may then move on, usually via the bloodstream, to target other organs elsewhere in the body. For this reason, it has not been easy to define those viruses actually causing gastroenteritis, especially as enteritis is so common and not always easy to distinguish from minor changes in bowel habits arising from time to time due to dietary, psychological, or other reasons. Moreover, co-infections involving more than one agent occur frequently and it can be difficult to identify the causative pathogen, or to prove in mixed infections a cooperative role involving more than one agent.

Four groups of viruses have been proven to cause symptomatic gastroenteritis; rotaviruses, human caliciviruses, enteric adenoviruses, and astroviruses. Assiduous searches have revealed a fascinating range of other viruses in feces, some of which are not human infections, for example, bacteriophages infecting enteric bacteria and plant or animal viruses from ingested food. Miscellaneous "small round viruses" and "parvovirus-like" agents have been carefully described but a clear etiologic association with disease has yet to be demonstrated. The same applies to the enteric coronaviruses, picobirnaviruses, enteroviruses, the Aichi agent, and enteric toroviruses, which have been isolated or visualized by electron microscopy in feces

TABLE 39.3 Virus Infections with a Proven Association with Gastroenteritis

Causative Agent	Patient Age Groupings	Selected Symptoms		Incubation Period	Duration of Illness	Characteristics
		Vomiting	Fever			
Group A rotaviruses	Infants and toddlers	Common	Common	1 to 3 days	5 to 7 days	Commonest cause of severe childhood diarrhea; outbreaks in high-risk groups of adults. Now vaccine-preventable
Group B rotaviruses	Children and adults	Variable	Rare	56 hours (average)	3 to 7 days	Associated with cholera-like disease in adults in China
Group C rotaviruses	Infants, children, and adults	Unknown	Unknown	24 to 48 hours	3 to 7 days	Sporadic cases; rare outbreaks in children
Adenoviruses (enteric)	Young children	Common	Common	7 to 8 days	8 to 12 days	Endemic in children
Caliciviruses (Sapovirus)	Infants, young children, and adults	Common in infants; variable in adults	Occasional	1–3 days	1 to 3 days	Endemic in children
Caliciviruses (Norovirus)	All ages	Common	Rare or mild	18 to 48 hours	12 to 48 hours	Commonest cause of outbreaks in adults
Astroviruses	Young children and elderly people	Occasional	Occasional	1 to 4 days	2 to 3 days; occasionally 1 to 4 days	All children infected in first 3 years of life; us. mild disease; outbreaks in day-care centers, nursing homes

from patients or even outbreaks of human gastroenteritis (especially in psycho-geriatric patients, AIDS patients, or immunocompromised children), but have not been proven to cause the disease from which they were recovered. It is probable that some human enteric viruses are harmless passengers in most people most of the time but are capable rarely of causing diarrhea under certain circumstances in certain individuals, especially those immunocompromised. Table 39.3 lists only those enteric viruses that unequivocally cause gastroenteritis in humans.

Gastroenteritis vies with upper respiratory infection as the commonest of all infectious diseases and is the greatest cause of death. It has been estimated that 5 to 10 million children die each year in developing countries from diarrheal diseases, rotavirus infections in malnourished infants being a major contributor. Rotaviruses cause severe diarrhea in young children, which may last up to a week and lead to dehydration requiring fluid and electrolyte replacement. Most infections are sporadic, but nosocomial outbreaks occur frequently in hospital nurseries. Nearly all rotavirus infections are caused by group A serotypes and occur mainly in infants under the age of two years; group B rotaviruses have been associated with some very large waterborne outbreaks in China, affecting adults as well as children; group C rotaviruses cause only occasional zoonotic infections.

Similar to other groups of viruses causing gastroenteritis, enteric adenoviruses were first visualized in feces by electron microscopy and identified on the basis of their characteristic morphology; because of the copious numbers of virions excreted, enteric adenoviruses were demonstrated by direct immunoassay to be distinct from other known members of that family. Later, suitable techniques were developed for growing these "fastidious" or "enteric" adenoviruses belonging to species F (types 40 and 41). These particular serotypes of human adenoviruses (but not the numerous other types that replicate in the respiratory and/or gastrointestinal tract but cause no disease in the gastrointestinal tract) have turned out to be common causes of gastroenteritis, especially in young children, for example, in outbreaks in day-care centers.

Among the *Caliciviridae*, sapoviruses are also common, especially in young children. In contrast, noroviruses tend to infect older children and adults, and are now recognized as the most common cause of outbreaks of non-bacterial gastroenteritis in many developed countries. The illness consists of an explosive episode of nausea, vomiting, diarrhea, and abdominal cramps, sometimes accompanied by headache, myalgia, and/or low-grade fever. Their high transmission rates have led to their causing significant outbreaks on holiday cruise ships and in institutions for the elderly.

TABLE 39.4 Distinguishing Characteristics of Endemic and Epidemic Gastrointestinal Virus Infections

	Childhood Diarrhea (endemic)	Outbreaks (epidemic)
Viruses	Group A rotaviruses, adenoviruses, astroviruses, noroviruses, and sapoviruses	Human noroviruses, group B and C rotaviruses
Age	<5 year	All ages
Mode of transmission	Fecal–oral (hand-to-mouth, environmental contamination)	Common source (contaminated food, shellfish, water). Environmental contamination, person-to-person
Prevention and control	Improved living conditions (sewerage, running water, handwashing facilities, education) Vaccine against rotavirus available	Public health regulations for food and water quality; case and contact tracing, quarantine; environmental disinfection

Astrovirus infections display many of the epidemiologic and clinical features of rotavirus infections but are not as virulent. They appear to be endemic worldwide, with occasional epidemics, causing a relatively mild form of enteritis with watery diarrhea, mainly in young children; outbreaks have also occurred among immunosuppressed and institutionalized geriatric patients.

There are some distinctions between those infections that occur more typically by person-to-person spread, and those that occur in common source outbreaks (Table 39.4).

VIRAL DISEASES OF THE CENTRAL NERVOUS SYSTEM

Most meningitis and almost all encephalitis, where a cause is found, is of viral etiology (Table 39.5). Infections of the CNS arise, in the main, as a rare complication of a primary infection established elsewhere in the body which fortuitously spreads to the brain, usually via the bloodstream. Sometimes they may involve reactivation of a latent herpesvirus or papovavirus infection, particularly following immunosuppression. Overwhelming disseminated infections acquired perinatally may also involve the brain.

Certain viruses have a predilection for particular parts of the CNS, and the clinical signs of the resulting disease often reflect this. For example, most enteroviruses do not go beyond the meninges, but polioviruses invade the anterior horn of the spinal cord and the motor cortex of the cerebrum, herpes simplex virus commonly involves the temporal lobes, and so on. Some viruses induce neuronal necrosis/apoptosis directly, and there is abundant evidence of inflammation in the brain; others do their damage in more subtle ways, leading to demyelination of nerves, sometimes involving immunopathologic processes.

One must distinguish between neurovirulence, that is, the ability to cause neurologic disease, and neuro-invasiveness, that is, the ability to enter the nervous system. Mumps virus, for example, displays high neuroinvasiveness, in that evidence of very mild meningitis accompanied by changes in the cerebrospinal fluid (CSF) are detectable in about half of all infections, but low neurovirulence, in that it rarely causes much damage. In contrast, herpes simplex virus (HSV) displays low neuroinvasiveness, in that it rarely invades the CNS, but high neurovirulence, in that when it does it often causes devastating damage. Thus, *neurotropism*, the ability to infect neurons, is the product of neuroinvasiveness and neurovirulence. Moreover, not all neurotropic viruses are neuronotropic, that is, able to infect neurons, as is the case with rabies virus, polioviruses, togaviruses, flaviviruses, and bunyaviruses; some viruses, such as JC polyomavirus, preferentially replicate in non-neuronal cells like oligodendrocytes, causing demyelination. Destruction of neurons has the most serious consequences, as lost neurons are not replaced.

The blood–brain barrier, which tends to exclude viruses from the CNS, also limits access of lymphoid cells, antibodies, complement, etc.; only when inflammation disrupts the blood–brain barrier does the immune response come into play. Thus the barriers that inhibit virus invasion also deter virus clearance, accounting for the high frequency with which persistent virus infections involve the CNS.

The many and varied neurological syndromes caused by viruses include meningitis, encephalitis, paralysis, myelitis, polyneuritis, and several unusual demyelinating and degenerative syndromes. Using PCR-based diagnostic tests it is now possible to identify a cause in at least half the cases of meningitis, but a viral etiology is still not confirmed in a majority of patients with encephalitis.

Meningitis

Viral meningitis is much commoner than bacterial meningitis but is much less severe. Only meningeal cells and ependymal cells are involved, and recovery is almost always complete. The patient presents with headache, fever, and neck stiffness, with or without vomiting and/or photophobia. Lumbar puncture reveals a clear CSF, perhaps under slightly increased pressure, with near normal protein and glucose concentrations, and

TABLE 39.5 Viral Diseases of the Central Nervous System

Disease	Viruses[a]
Meningitis	**Enteroviruses**
	Mumps virus (in countries that do not immunize) West Nile virus
	Herpes simplex virus type 2; other herpesviruses rarely Lymphocytic choriomeningitis virus
Paralysis	**Enteroviruses 70, 71**; Coxsackie virus A7
	West Nile virus
	Polioviruses (in those countries where still circulating)
Encephalitis	**Herpes simplex viruses**
	Mumps virus (in countries that do not immunize)
	Arboviruses (togaviruses, flaviviruses, bunyaviruses; see Tables 29.1, 35.1, 36.1)
	Arenaviruses, rabies virus, enteroviruses, adenoviruses, influenza viruses, other herpesviruses
Post-infectious encephalomyelitis	Measles virus, varicella-zoster virus, rubella virus, mumps virus, influenza virus, (vaccinia virus), others
Guillain-Barré syndrome	Epstein-Barr virus, cytomegalovirus, HIV, influenza viruses
Reye's syndrome	Influenza viruses, varicella-zoster virus
Subacute sclerosing panencephalitis	Measles virus
Progressive multifocal leukoencephalopathy	JC polyomavirus
AIDS encephalopathy (AIDS dementia complex)	**HIV**
Tropical spastic paraparesis	HTLV-I
Subacute spongiform encephalopathy	**Prions**

[a]The commonest causal agents are in bold type.

only a moderate pleocytosis; the white cell count may range from normal ($<10/mm^3$) to over $1000/mm^3$, but is usually $30–300/mm^3$, with lymphocytes predominating after the first day or so. This is generally referred to as "aseptic" meningitis.

By far the most important etiologic agents are the numerous enteroviruses, including all the Coxsackie B viruses, Coxsackie A7 and A9, polioviruses (in countries where they still circulate), and many echoviruses which are listed in Chapter 32: Picornaviruses. Mumps virus remains an important cause in those countries that do not immunize against mumps. The herpesviruses, HSV, EBV, and CMV, are rare sporadic causes, as are certain arboviruses in endemic regions. Lymphocytic choriomeningitis virus can be acquired from laboratory or pet mice or hamsters.

Meningitis may be the only clinical evidence of infection with these viruses. For example, only half of all cases of mumps meningitis follow typical parotitis. Enteroviral meningitis often occurs during a summer/autumn epidemic in which other patients experience rashes, myositis, or other common manifestations of infection with the prevalent agent, but meningitis is often the sole presentation.

Specific viral diagnosis can usually be made using PCR testing of cerebrospinal fluid. The most crucial diagnostic imperative is to exclude the less-common, life-threatening, and treatable condition of bacterial meningitis.

Encephalitis

Encephalitis is one of the most serious of all viral diseases. The illness often begins like meningitis with fever, headache, vomiting, and neck rigidity, but alteration in the state of consciousness indicates that the brain parenchyma itself is involved. Initially lethargic, the patient becomes confused then stuporose. Ataxia, seizures, and paralysis may develop before the victim lapses into a coma and dies. Survivors may often be left with a pathetic legacy of permanent sequelae, including mental retardation, epilepsy, paralysis, deafness, or blindness. Globally, rabies and Japanese B encephalitis constitute the majority of cases of viral encephalitis, while in developed countries where mumps and poliovirus vaccination is practiced, herpes simplex virus is the most commonly identified cause of severe sporadic encephalitis.

Still, despite the improved diagnostic results of PCR testing, it is a challenge to virologists that a viral etiology is never found in a majority of cases.

Encephalitogenic mosquito-borne or tick-borne togaviruses, flaviviruses, and bunyaviruses, endemic to particular regions of the world, cause epidemics of encephalitis from time to time when the appropriate combination of ecologic circumstances develops. The ecology of each of these arboviruses and features of the disease(s) they cause are described in detail in Chapter 29: Bunyaviruses, Chapter 35: Togaviruses, and Chapter 36: Flaviviruses (see also Tables 29.1, 35.1, and 36.1). Encephalitis is also an irregular feature in certain hemorrhagic fevers (see Table 39.6). Rabies causes an invariably lethal and distinctive form of encephalitis, described fully in Chapter 27: Rhabdoviruses. In temperate regions of the world where mumps vaccination is not used, mumps is the commonest cause of encephalitis, but it is generally a relatively mild meningoencephalitis with only rare sequelae, mainly unilateral deafness.

Herpes simplex encephalitis is a very unpleasant disease indeed, infecting both neurons and glia to produce a focal encephalitis generally localized to the temporal lobes in immune adults, but diffuse necrotizing encephalitis in the newborn. Untreated, HSV encephalitis carries a 70% case–fatality rate (see Chapter 17: Herpesviruses), but this has significantly improved with routine early treatment with acyclovir. In neonates or immunocompromised patients, HSV and the other herpesviruses, and occasionally enteroviruses or adenoviruses, are also capable of causing encephalitis, generally as part of a widely disseminated and often fatal infection. *Chronic meningoencephalitis* is a fatal condition seen in children with the B cell deficiency, X-linked agammaglobulinemia, or severe combined immunodeficiency. Enteroviruses are the usual causal agent (see Chapter 32: Picornaviruses). The majority of these children also have a condition known as *juvenile dermatomyositis*.

Paralysis

Enterovirus 71 and Coxsackievirus A7 are rare sporadic causes of a paralytic disease essentially indistinguishable from poliomyelitis. The radiculomyelitis associated with enterovirus 70 infection is generally reversible. Rarely, clusters of cases of acute flaccid paralysis with anterior myelitis still occur in regions where polioviruses have been eliminated, for example, in California in 2012–13; thorough investigation is necessary to exclude polioviruses, and other enteroviruses are often implicated.

In countries from which polioviruses have not yet been effectively eliminated by vaccination, these viruses remain a cause of both aseptic meningitis and paralytic poliomyelitis. Very rarely indeed, the oral polio vaccine itself can cause paralysis, mainly in immunocompromised

individuals. Systems are in place in many parts of the world to detect, investigate, and monitor cases of acute flaccid paralysis (AFP). This is very important to obtain early warning of unsuspected circulation of polioviruses, but also to improve our knowledge about other viruses that might cause this syndrome.

Post-infectious Encephalomyelitis

Post-infectious encephalomyelitis is a severe demyelinating condition of the brain and spinal cord which occurs as an occasional complication presenting one to two weeks after any of the common childhood exanthemata (measles, varicella, rubella), influenza or mumps or other infections. Prior to the eradication of smallpox, it also occurred as a complication of vaccination with live vaccinia virus in approximately 10 to 300 cases per million recipients. Measles infection is followed by this complication in approximately 1 in 1000 cases, compared with 1 in 1 million recipients of live measles vaccine. The pathology of post-infectious encephalomyelitis resembles that of experimental allergic encephalomyelitis, giving rise to the hypothesis that this is an autoimmune disease in which virus infection provokes an immunologic attack on myelin. Certainly there is little virus demonstrable in the brain by the time post-infectious encephalomyelitis develops, and the major histologic finding is perivenous inflammation and demyelination. The clinical severity can vary greatly, spontaneous recovery is the rule but permanent neurological deficits may occur in up to 40% of cases.

Guillain-Barré Syndrome

Guillain-Barré syndrome (GBS) is an acute inflammatory demyelinating polyradiculoneuropathy which follows exposure to any one of several viruses. Epstein-Barr virus (which has also been associated with transverse myelitis and Bell's palsy) is most commonly implicated, the Guillain-Barré syndrome appearing one to four weeks after infectious mononucleosis. Partial or total paralysis develops, usually in more than one limb. Complete recovery occurs within weeks in most cases, but 15% retain residual neurologic disability. The syndrome is also seen with cytomegalovirus, influenza, and early in some HIV infections.

An outbreak of Guillain-Barré syndrome in the United States in 1976 was traced to the introduction of a formalin-inactivated vaccine against the so-called swine strain of influenza. The vaccine was withdrawn, and the syndrome has not been associated with any subsequent vaccine. A review by the Institute of Medicine in 2003 concluded that recipients of this vaccine were at an increased risk of GBS of 1 in 100,000, but the exact reason remains something of a mystery. It does prove, however, that live virus is not

that it is quite common for more than one, and sometimes all six, of the herpesviruses (HSV-1 and HSV-2, VZV, CMV, EBV, and HHV-6) to be reactivated in bone-marrow transplant recipients. In recipients of allogeneic marrow transplantation, mortality due to CMV pneumonia was 10% to 30% until the prophylactic, or early pre-emptive, use of ganciclovir lowered this to 2% to 5%. Similarly, HIV/AIDS patients characteristically suffer successive reactivation, sometimes in a roughly predictable order as their CD4+ T cell count drops, of any or all of the herpesviruses, plus polyomavirus, papillomaviruses, adenoviruses, and hepatitis B virus (see Chapter 23: Retroviruses). The pathogenesis, clinical manifestations, and management of immunocompromised patients by antiviral chemotherapy, active or passive immunization, and appropriate virologic and immunologic screening, discussed for each of the herpesviruses individually in Chapter 17: Herpesviruses; chronic infection with parvovirus B19 is discussed in Chapter 21: Parvoviruses, reactivation of human polyomaviruses is addressed in Chapter 20: Polyomaviruses, and that of adenoviruses in Chapter 18: Adenoviruses.

DISEASES OF UNKNOWN ETIOLOGY

Viruses are frequently suspected, and sometimes blamed, for causing a range of diseases whose causation is uncertain. Many examples could be highlighted, and some of the more prominent ones are discussed in the relevant sections of this chapter. Other conditions not already mentioned include Kawasaki disease, systemic lupus erythematosus, pityriasis rosea, inflammatory bowel disease, obesity, brain tumors, and Bell's palsy; in these examples a number of different viruses have been investigated to varying degrees for a possible causative role, without conclusive proof emerging. The careful reader should by now appreciate the rigorous evidence-based approach needed to confirm a true causative role for a particular virus, and the wealth of sophisticated molecular, epidemiological, and clinical tools available for such work. Herein lies one of the many ongoing challenges for virologists of the future.

FURTHER READING

Bennett, J.E., Dolin, R., Blaser, M.J., 2014. Mandell, Douglas, and Bennett's Principles and Practice of Infectious Diseases: 2 Vol., 8th Edn. Saunders/Elsevier, Philadelphia, ISBN-13: 978-1455748013.

Kimberlin, D.W., Long, S.S., Brady, M.T., Jackson, M.A., 2015. Red Book 2015: Report of the Committee on Infectious Diseases, 30th Edn. American Academy of Pediatrics, Elk Grove Village, Illinois, ISBN-13: 978-1581109269.

Lecuit, M., Eloit, M., 2013. The human virome: new tools and concepts. Trends Microbiol 21, 510–515s.

Lipkin, W.I., Briese, T., Hornig, M., 2011. Borna disease virus—fact and fantasy. Virus Res 162, 162–172.

National Academy of Medicine/Institute of Medicine, 2015. Beyond Myalgic Encephalomyelitis/Chronic Fatigue Syndrome: Redefining an Illness. National Academies Press, Washington, DC. Available from: http://dx.doi.org/10.17226/19012.

Relman, D.A., 2002. New technologies, human–microbe interactions, and the search for previously unrecognized pathogens. J Infect Dis. 186, S254–S258.

Richman, D.D., Whitley, R.J., Hayden, F.G., 2009. Clinical Virology. ASM Press, Washington.

Tan, le. V., van Doorn, H.R., Nghia, H.D.T., et al., 2013. Identification of a new cyclovirus in cerebrospinal fluid of patients with acute central nervous system infections. mBio 4 (1–10), e00231–13.

Virgin, H.W., 2014. The virome in mammalian physiology and disease. Cell 27 157 (1), 142–150.

Virgin, H.W., Wherry, E.J., Ahmed, R., 2009. Redefining chronic viral infection. Cell 10 138 (1), 30–50.

Wylie, K.M., Mihindukulasuriya, K.A., Sodergren, E., Weinstock, G.M., Storch, G.A., 2012. Sequence analysis of the human virome in febrile and afebrile children. PLoS One 7, e27735.

Wylie, K.M., Weinstock, G.M., Storch, G.A., 2013. Virome genomics: a tool for defining the human virome. Curr Opin Microbiol 16, 479–484.

Zuckerman, A.J., Banatvala, J.E., Griffiths, P., Schoub, B., Mortimer, P., 2009. Principles and Practice of Clinical Virology, 6th Edn. John Wiley & Sons Ltd, Chichester, UK.

Index

Note: Page numbers followed by "*b*," "*f*," and "*t*" refer to boxes, figures, and tables, respectively.